FROM

Phyllis M. Ellis.

3/9/01.

oad

CW01659419

Norfolk Holiday 2001.

A Flora of Norfolk

Gillian Beckett
Alec Bull
Robin Stevenson

First published in Great Britain in 1999
by **Gillian Beckett**

ISBN 0 9534999 0 1 (hardback)

Design and Production: **Paul Westley**, Norwich, Norfolk
Maps: **Bob Ellis,** Norwich, Norfolk
Photography: see listing in Photographic Acknowledgements
Colour Scanning: **BD&H**, Norwich, Norfolk
Main Cover Scan: **PDQ (Norwich)**, Norwich, Norfolk
Imagesetting: **Laser Reprographics**, Norwich, Norfolk
Dust Jacket Printing: **Rapide**, Watton, Norfolk
Bookmark Printing: **Rapide**, Watton, Norfolk
Printed and Bound: **Jarrold Book Printing**, Thetford, Norfolk

Front cover images: Main photograph courtesy of The Norfolk Wildlife Trust
Heath Spotted Orchid *Dactylorhiza maculata*
Inset photograph courtesy of Bob Ellis - Bee Orchid *Ophrys apifera*

Contents 3

Sponsors of the Flora **4**

Preface
Richard Mabey **5**

The Flora: Aims, Methods and Acknowledgements
Gillian Beckett and Alec Bull **6**

The Setting for the Flora
Gillian Beckett **10**

The Soils of Norfolk: Their Nature and Distribution
W. M. Corbett **12**

Man, the Landscape and Plants... an Historical Perspective
Gillian Beckett **16**

Habitats in Norfolk
Peter Lambley **25**

Nature Conservation in Norfolk
Mark Cocker **31**

The Norfolk Botanists
Alec Bull **35**

Nature Reserves in the County **41**

Higher Plants Formerly Recorded in the County **43**

Bryophytes Formerly Recorded in the County **44**

Photographic Acknowledgements **44**

Bibliography **45**

Abbreviations used in the text **47**

Map of Places Mentioned in the Text **48**

Key to Maps and Species Entries **50**

(Higher) Plants and Ferns - maps and text
Gillian Beckett and Alec Bull **51**

(Bryophyte) Liverworts and Mosses - maps and text
Robin Stevenson **266**

Index **288**

SPONSORS OF THE FLORA

H.M. The Queen

Mr J. Austen
Sir Jeremy Bagge
Mr & Mrs P. Banham
Mr & Mrs K. Beckett
Ms C. Bennett
Mr H.C. Brun
Mr & Mrs A.L. Bull
Mr R.E. Buscall
Mr R. Chancellor
Mr T. Cook
Mr R.S. Cross
Mrs R. Don
Mrs U. Doyle
Dr E. Duffey
Mrs J.E. Gaffney
Mr J. Goldsmith
Mrs R. Hadman
Lady Rose Hare
Mrs C. Harmer
Mr & Mrs M. Helliwell
Miss G.B. Hilbert
Mr S.R. Martin
Mr & Mrs D. Mathias
Mr E. Pank
Mr & Mrs R. Payne
Mr J. Piper
Mr J.F. Pope
Pratt Family Partners
Mr E. Rogers
Mrs M. Salter
Mrs F. Schumann
Mr J. Snape
Mr H. Van Cutsem
Mr & Mrs J. Welsh

Botanical Society of the British Isles
Broads Authority
Cadbury (Sarnia) Trust
Courtyard Trust (Lord Melchett)
Heacham & West Norfolk Naturalists' Society
National Rivers Authority
Norfolk County Council
Norfolk & Norwich Naturalists' Society
Wild Flower Society

Norfolk has a reputation as Britain's premier wildlife county, an accolade it won because of its remarkable range of habitats, which range from sandy grass heaths to prehistoric limewood. It's also a county where natural and human life have been intricately entwined for millennia. Staying on the north coast in the 1960s I saw locals still gathering wild plants for the table, enacting one of the most ancient aspects of this relationship. It was an experience which awakened my own love of plants, previously in a very poor second to birds.

Over the years that followed I have been drawn to many of the county's botanical Meccas, whose history and ecology are so enthrallingly spelt out in this FLORA.
I have ambled in spring through the bird cherries and mauve-tinged wood anemones of Wayland Wood, gazed across the great sweeps of sea-lavender of Stiffkey marshes, nosed a boat amongst the loosestrifes and flowering rushes of the Broads. Yet Norfolk is a county of richly unpredictable nooks and niches as well as great set-pieces.
You may see grape hyacinth on sandy roadsides in the Brecks and a single plant of sea-kale on Blakeney Point.

This new Norfolk FLORA wonderfully encompasses this huge diversity. The first comprehensive account since 1968, it brings together descriptions of the county's habitats with a species by species survey of all its plants. A sophisticated innovation is to show the distribution of many of the species printed over maps with informative backgrounds including soils and rivers. The result is what all local Floras aspire to - a document which is a historic and geographical record, as well as a botanical one. It charts the persistent but fragile continuity of nature, and the perennial fascinations of naturalists, and in so doing it will be an inspiration to all who love Norfolk and seek out its special plants.

Richard Mabey

The Flora: Aims, Method and Acknowledgements

Gillian Beckett & Alec Bull

This project started informally after the end of the B.S.B.I.'s Monitoring Scheme in 1987 when comparisons of data collected made it obvious how much the county's flora had changed since Petch and Swann's *Flora of Norfolk* of 1968. It is the first time the tool of mapping has been used for a Norfolk Flora and it has meant that all parts of the county have been visited, rather than just those of botanical interest as was usual in the old-style floras. The 2 x 2km square (tetrad) has been chosen as the most convenient unit and the wide coverage of the county has resulted in an abundance of new records. Although the project did not begin until 1987, it was decided to back date it to 1985 so that records made between 1985 and 1987 could be incorporated.

The area covered by this survey includes all the land lying within both present and past political boundaries of Norfolk as well as within the botanical boundaries of the Watsonian VC27, East Norfolk and VC28, West Norfolk. Because of this it incorporates small areas of Cambridgeshire, Lincolnshire and Suffolk. This gives a total of 1441 tetrads. To make this number more easily manageable, the county was divided into two using the 100km line which neatly bisects the county and approximates to the botanical division of the two vice-counties. One of us was responsible for recording each half, each taking our respective vice-county (VC27, ALB and VC28, GB) and gathering our own group of recorders, though in practice there was often considerable overlap. Many of these recorders remained with us throughout the project. For the first few years, each recorder agreed to cover a number of tetrads, visiting each on at least three different occasions. Because of the diversity of the area covered, no attempt was made to limit the amount of time spent on any tetrad as has been suggested in recent statistical surveys. Here we felt it would have resulted in more complex areas being under-recorded, while spending a prescribed amount of time in almost entirely arable tetrads with limited access would have been very soul-destroying for the recorders as well as a waste of precious time.

Field meetings were held throughout the project targeting either very rich areas or ones for which access was limited to short periods, while in the later stages of the survey most meetings were taken up with visiting less well recorded areas. Bob Ellis designed a computer programme TETRAN to help us discover which were under-recorded rather than just poor in species and this helped us to direct recorders to the areas where they were most needed. Consistency of recording proved difficult to obtain in the treatment of casuals, garden escapes and trees, some recorders including only those plants which they felt were truly established, others recording everything found outside gardens. As this did not affect the native and truly established weeds, no effort was made to correct this. For the distribution map below, however, records for casuals and garden escapes have been omitted to give a fair idea of the spread of native plants across the county.

Record sheets were collected in annually and were fed into computers using Alan Morton's MAPRECS. The coloured base maps used throughout the book were largely derived from the Soil Survey's Norfolk map with soils of like type grouped together for convenience. Roads and towns have also been used as base maps where appropriate. All maps were produced using Alan Morton's mapping programme DMAP. The maps include only those records made since 1985 and cover all native and some introduced species except for those which are virtually ubiquitous or very rare. The text, however, covers all plants which are known to have been recorded within the county at any time. Details of place, recorder and date are given only for records made since the production of Petch & Swann's *Flora of Norfolk* in 1968. Nomenclature follows Stace 1997 unless later information is available.

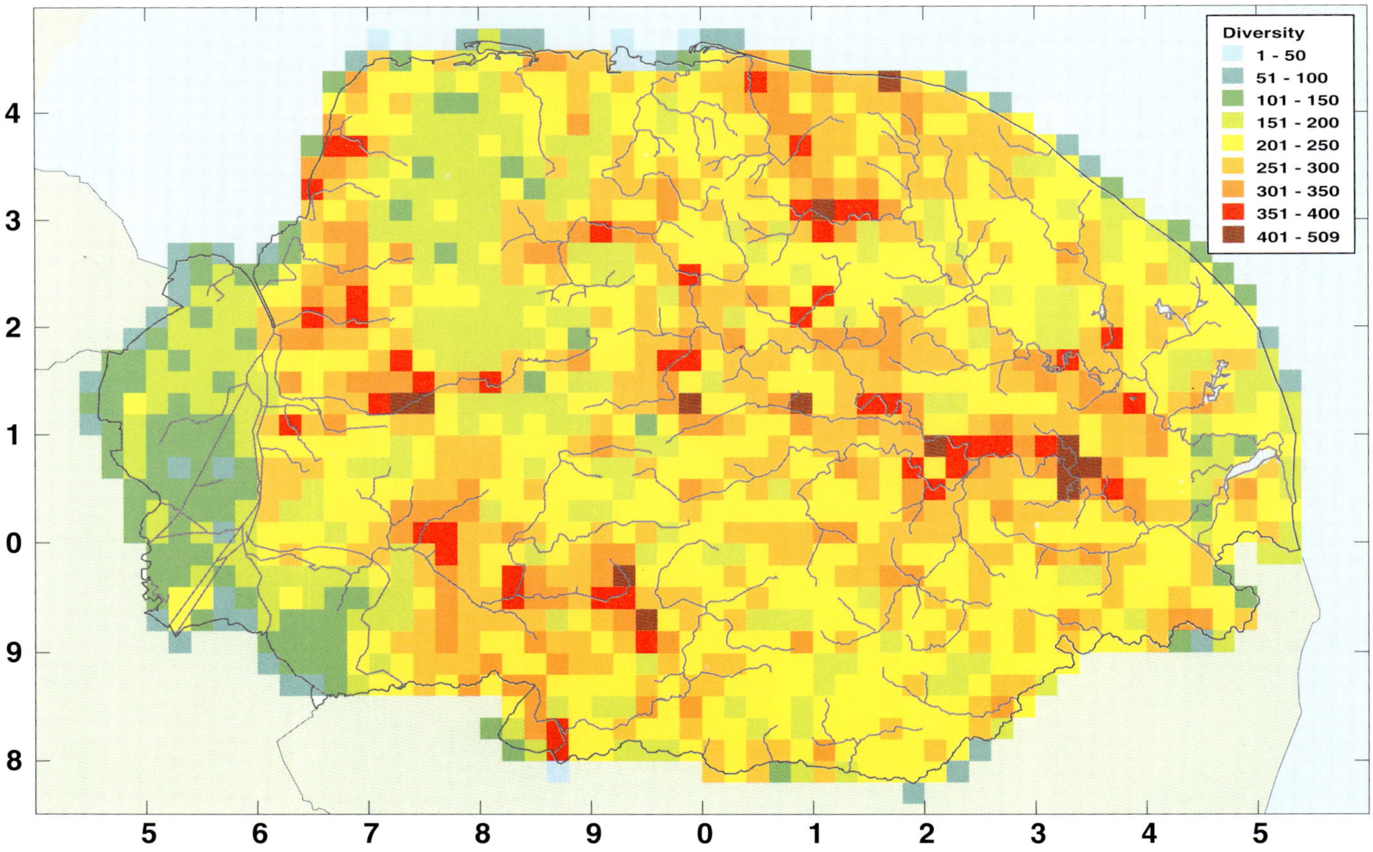

Total Numbers of Native and Well-established Plants in the County

Higher Plants

It seems invidious to pick out any people in a book which has been the work of so many, but the following must be mentioned, for without them the whole project would surely have foundered long ago. First, John Snape who took on the position of treasurer, without at first being quite aware of what he was letting himself in for. He didn't flinch at difficulties, however, even when we increased the costs by 50% by using colour throughout. Without his handling of our finances, VAT etc., as well as his confidence in all of us we would certainly have had great problems. Alan Morton not only provided the excellent programme, DMAP, but also gave us a great deal of help, especially in the early days of the survey when we were relative beginners at computing. Robert (Bob) Ellis joined us first as a recorder, but his expertise in computing found us using him more and more during the project. He spent hours of his own time devising programmes, producing lists and maps and finally took on the creation of the maps which are used in the book, then marrying the coloured backgrounds with the dot maps produced by DMAP. He also did a great deal of recording in the field. We would never have had the courage to venture into colour without Bob's help. Richard Hobbs was responsible for suggesting John as our treasurer and since then has attended all our committee meetings and used his contacts and knowledge to ease our path in many ways. Finally Paul Westley whom we employed as a Graphic Designer to help with the production of the book and who put up with our erratic timing without complaint and put far more into the final work than his contract with us asked for. He should take all credit for the appearance of the book.

Along the way we have had advice and help from David McCosh who read and commented on the *Hieracium* text, Andrew Dudman who did the same for *Taraxacum* and Ron Payne and Arthur Copping who checked the section on grasses and helped with identification. We would also like to thank Dr. J. Akeroyd, J. Bevan, E.J.Clement, Dr T.A.Cope, A.C. Jermy, J.M.Mullin, A.Newton, B.Ryves and all those who helped with identification; the Soil Survey for permission to base our soil maps on their information, also Bill Corbett for advice with our soil maps; Mike Hall, John Goldsmith and Peter Lambley for help with photocopying and Peter also for general advice and information, also all those who wrote Introductory chapters for this Flora. To these names must be added, anonymously as there have been so many, all those landowners who allowed us access to their land, both as individuals or as a party.

Of recorders, a separate mention must be made of Shane Plant who made available to us his on-going survey on a 1km. basis covering almost every tetrad in East Norfolk and a good number in the West. This resulted in the region of 20 000 records being added, also R.M.Payne whose *Flora of King's Lynn* was published in 1995 and *A Flora of Walls in West Norfolk* in 1998. We had access to all the records he made for these as well as helpful information from both surveys. The Environment Agency (then the N.R.A.) also allowed us to use the records collected for their River Corridor survey which gave us many records from areas which would have been difficult to access and other survey information was given us by the Norfolk Wildlife Trust; English Nature especially C.Doark's survey of Broadland dykes; the R.S.P.B., the Wildfowl and Wetlands Trust and Anglian Water. Finally, though really the most important of all must be the recorders who probably tramped far more miles along dull roads than they spent in more rewarding areas, but they did it with no more than an occasional good humoured grumble.

First, those who made a major input to recording.

Beckett, K.A. & Mrs G.
Brewster, Mrs M.A.
Bull, A.L.
Cooper, Mrs M.
Copping, A.
Crewe, I.M.
Daniels, E.T.
Ellis, R.W.
Farrow, F.J.L.
Ferroussat, Mrs K.
Gaffney, Mrs J.E.
Green, P. & I.
Hall, L.
Harrison, K.K.
Hawksford, Mr J.
Haycock, Miss B.M.
Helliwell, Mrs M.
Howlett, D.
Humphris, Mrs J.
Keene, M.
Lambley, P.W.
Leech, Dr A.R.
Lester, D.
Martin Dr S.R.
Mathias, D. & Mrs B.
Parker, R.
Payne, R.M.
Plant, S.
Richmond, D.I. & Mrs R.M.
Rivett, K. & Mrs M.
Rogers, E.V.
Schumann, Mrs F.
Williamson, Mrs H.

The following were also contributors

Abdullah, Mrs M.
Abel, Dr
Akeroyd, Dr J.
Allen, B.
Andrews, Mrs A.
Andrews, Mrs M.W.
Andrews, R.
Anglian Water
Armitage, P.
Ashwell, Mrs O.
Austen, J.
Austin, Mrs M.
Banham, P.R. & Mrs E.
Barrett, R.
Barsted, C.
Bateman, Mrs E.
Beaumont, Dr E.
Benn, M.
Boyd, W.A.
Brandon-Jones, Ms E.
Brendell, Ms T.
Bristow, C.E.
Broom, Ms A.
B.S.B.I. party
Buckingham, D.
Buckle, P.
Bull, Ms C.
Bull, K.H.
Bull, Mrs R.M.
Buttle, C. & Mrs P.

Cadbury, J.
Carmichael, N. & Mrs
Chancellor, R.J.
Cheyne, Miss D.
Child, Mrs J.
Cobb, P.R.
Cocker, M.
Coombe, Dr D.E.
Corne, Mrs G.
Coupland, G.
Coupland, R.
Coward, J. & Mrs
Crompton, Mrs G.
Cross, E.
Cross, R.
Dack, C.
Darbyshire, Mrs
Davis, J.
Dawson, Mrs J.
Dennis, G.
Doarks, C.
Dolman, P.
Doncaster, T.
Doyle, U.
Driscoll, R.
Duffey, Dr E.
Dunsford, D.
Durrant, K.C.
Eaglen, J.
Ebrehart, T.
Edler, S.
Ellis, Mrs J.P.
English Nature
Evans, Mrs E.
Evans, Mr R.E. & Mrs L.
Everett, S.
Fellowes, Dr M.
Fisher, P.
Fisher, S.
Fitzgerald, Lady R.
Florance, G.
Flowers, D.
Forster, C.H.
Garrod, G.
George, Dr M.W.
Gibbons, N. & Mrs M.
Goldsmith J. & Mrs S.
Gough, Ms S.
Green D.
Griffin, Mrs B.
Gurney, M.
Hadman, Mrs R.
Hailey Ms P.
Haines Mrs C.
Hall M. & Mrs S.
Hancy, Mrs B.J.
Hanson, C.G.
Harding, M.
Hare, A.
Hare, Lady R.
Harrap, S.
Harris, Ms J.
Harrold, R.
Harwood, T.
Heath, J.
Henson S.
Hirons, G.
Hobbs, Ray
Hobbs, R.J.T.
Hoblyn, R.
Holve, H.
Hooton, G.
Humphrey, A.W.
Hyde. Mrs E.
Jeffreys, Mrs M.
Jermy, A.C.
Johnson, I.
Kelly, Ms J.
Kennison, G. & Mrs W.
Kimber, R.
Lake, Mrs S.
Lambdon, P.
Land, R.
Lark, A.
Last, W.
Laurence, K.
Lawson, P.G.
Laxton, Ms L.
Leaney, Dr R.M.
Leonard, D. & Mrs Y.
Levien, Mrs J.
Lewis, A.
Ley, Ms C.
Lilley, G.
Lughadha, Miss E. ni
Madders, Miss R.
Madgewick, Ms J.
Maidstone, R.
Martin, B.
Masson, J.
Maskall, R.
Matthews, Mrs J.
Maxey, D.M.
Maybury, G.
McCallom, J.
McCosh, D.J.
McSweeney, P.
Merson, Mrs M.
Miller, J.
Mitchell, W.
Moore, Ms L.
Morgan, Ms V.
Morris, G.
Mott, J.
Mountford J.O.
Mountney Mrs D.
Mower, D. & Mrs J.
Muddeman J.
Murrell, Ms J.D.
Mussell, Mrs D.
Negal, P.
Newton, A.
Nichols, B.
Nobbs, D.
Nobes, G.
Norfolk Wildlife Trust
NRA (Now Environment Agency)
Nuns of All Hallows
Ochyra, R.
O'Donnell, S.
Oswald, P.
Owen, N.
Oxenford, J.
Parmenter, Ms J.
Partridge, J.
Peck, G.
Penny, C.
Petch, Dr C.P. (to 1986)
Preston, C.D.
Primavesi, A.
Regan, C.
Reynolds, F.P.J.
Rich, Dr T.C.G.
Rihan, J.R.
Rix, Mrs S.
Robson, C.
Rollins, P. & Mrs P.
Rose, Dr F.
Ross, Mrs J.
Rothera, Dr S.
Rumsey, F.J.
Russell, H.
Salter, Mrs M.
Sampson, Mrs D.
Saul, K.
Saunders, M.
Schneidau, Ms L.
Secker, J.
Sell, P.D.
Showler, D.
Simper, I. & Mrs F.
Simpson, G.
Smart, S.
Smith, Ms J.
Smith, P.A.
Smith, W.
Snape, J.
Southwood, R.
Spicer, Mrs S.
Squirrel, Ms J.
Starling, Ms A.
Stevenson, C.R.
Stewart, N.F.
Strauss, D.
Street, L.
Taylor, B.
Taylor, Mrs G.
Taylor, Mrs S.J.
Telfer, M.
Thompson J.A. (E.H.)
Tilbrook, Mrs R.
Tofts, R.
Tollhurst, Ms S.
Trist, P.J.O.
Walker, P.M.
Watkinson, Dr A.
Watts, Dr G.D.
Wells, T.C.E.
Westley, P.
White, I.
W.F.S. Party
Wildfowl & Wetland Trust
Williamson, J.
Wilson, P.
Woodell, S.R.J.
Woodman, J.
Wood, Mrs M.
Woods, D.C.
Woods, R.
Yaxley, R.

Bryophytes

The first systematic account of Norfolk's bryophytes took the form of a list of names, located within four subdivisions of the county, produced by Mundford (in White 1863). This was followed by Burrell (in Nicholson 1914), who produced a much more detailed account, including localities and estimates of abundance.

E.L.Swann (in Petch & Swann 1968), summarised the state of knowledge thirty years ago. He included many records by his contemporaries, notably Francis Rose. He also commented briefly on doubtful and missing records. Many additional records were presented in the *Supplement to the Flora* (Swann 1975). Finally he gave a summative account of the bryoflora as he knew it in the *Journal of Bryology* in 1982. Since then two check lists have been published; one for the mosses (Stevenson 1992), and one for the liverworts (Stevenson 1993). These were used to introduce nomenclatural changes, as well as to clarify some of the minor sources of confusion in the records, mainly relating to the records for which no voucher specimens were available. Stevenson (1994) gave an account of the changes in the Norfolk bryoflora over the last century.

Any flora is, at best, a snapshot of what existed in an area during the period of its compilation. This account of the Norfolk bryophytes will give a very blurred picture. Recording has been going on since 1980, so the time scale is rather longer than that for the recording of the higher plants. The resolution of the mapping, (see below), is not as detailed as that for the vascular plants, but none the less, when taken in conjunction with earlier accounts, it does provide a picture against which changes can be measured. In a number of cases plants mentioned by Swann have not been refound. Where this is the case, the fact that there are no recent records is noted and this is followed in square parentheses by such comments as were made by Swann (1968:1982). Where there are grounds for believing that a species has been lost, this has been stated.

It has not been practicable to quote all old records, however, in the majority of instances the date of the last record has been given, e.g. E20 (1967). It has not been possible to do this consistently and reference to the earlier literature is advisable if such details are required. Individual records have not been attributed to specific recorders, except for new county records established since Swann (1982), and for any notable rediscoveries of older records .

The scale of mapping has been the 10km square. In a county as large as Norfolk and given the paucity of recorders, no other option was possible. Some areas have, inevitably, been covered more thoroughly than others and much of the county is undoubtedly underworked. None the less, the degree of coverage has been sufficient to enable at least some patterns of decline and spread to be recognised.

Norfolk lies within 70 10km squares, although only about 41 of these are whole squares, the rest containing varying portions of sea or of adjacent counties. For the sake of simplicity, however, the full 70 squares have been considered in arriving at the following measure of geographical abundance.

<3 squares [< 4%] .... = very rare
4-6 squares [5-8%] .... = rare
7-12 squares [9-17%] .... = occasional
13-24 squares [18-34%] .... = frequent
25-48 squares [35-68%] .... = widespread
>48 squares [more than 68%] ... = very widespread.

Maps have not been produced for the very common species occurring in more than 68% of squares, nor for those rare and very rare species which occur in fewer than 8%. For the latter, the squares have been listed. Squares in which they have been recorded in the past are shown in square parentheses, thus enabling readers to identify possible shrinkage in frequency (or fertile areas for further field work). The accounts of those species that occur in more than 48 squares are prefixed by the words 'Very widespread'.

The bryophytes, and in particular the mosses, have been the focus of much taxonomic revision lately, and consequently there have been a lot of name changes. The nomenclature and the order and presentation of species used in this account follows that to be adopted in the forthcoming second edition of Smith (in press). In the case of the hepatics, (liverworts and hornworts) the nomenclature and order follows that to be used by Paton (in press).

Acknowledgements & List of Recorders for Bryophytes

The late Richard Libbey encouraged me to start collecting systematic records, and offered invaluable help in the field, and wise advice and comments in general. Similarly Alex Smith, and Harold Whitehouse, have always given freely of their time and advice on technical problems, as have many BBS referees. Peter Lambley has helped in locating landowners, and in suggesting places to record.

In the field, Richard Fisk and David Strauss, both based in Suffolk, and Alec Bull have contributed massively to the recording process - particularly in the east of the county. Gary Kennison, of the Broads Authority, supplied invaluable records from Broadland, as did Jo Parmenter. Latterly John Mott and Bob Ellis have made significant contributions. Harold Whitehouse, Chris Preston, Mark Hill, and other members of the 'Cambridge Group', have visited Norfolk on many occasions to help with recording; for their help & companionship I am grateful. As I am to all those recorders whose names are listed below.

The text has been proof read by Alex Smith and David Strauss, without whose care & attention to detail this flora would read less well. Needless to say, all dubious opinions, factual errors, inconsistencies, and infelicities are mine alone.

I must also extend my thanks to those numerous landowners, or their agents, who have allowed me access to their land. To the others I can only plead: Forgive me my trespasses...

Lastly, I have to thank my wife Wendy, who endures bryology with fortitude and good humour.

Robin Stevenson

Appleyard J.
Boyce D.
Brewster M.A.
Bull A.L.
Corley M.F.V.
Ellis R.
Finch R.A.
Fisk R.J.
Hancy B.
Hill M.O.
Hodgetts N.G.
Holyoak D.L.
Kennison G.C.B.
Lambley P.W.
Libbey R.P.
Lowe C.J.
Mott J.
Ochyra R & H.
Parmenter J.M.
Preston C.D.
Rose F.
Southwood R.
Smith A.C.
Stanley P.
Stern R.C.
Stevenson C.R.
Strauss D.F.
Swann E.L.
Timms M.
Whitehouse H.L.K.
Wright P.A.

The Setting for the Flora

Gillian Beckett

Norfolk is a county of remarkably diverse habitats, lacking only mountains. Geologically it is young with its oldest outcrop the Kimmeridge Clay of Jurassic age in the west of the county, though even this is largely hidden by the later silts and peats of the Fenland. Because of the simplicity of its solid geology, Norfolk can be divided into two main divisions, those which lie below or above the chalk deposits.

Hunstanton Cliffs, Greensand and Chalk

Greensand

The first strata of ecological importance are those of the sands and clays largely of the Lower Cretaceous period though the Sandringham Sands overlap with the upper layers of the Jurassic. These Sandringham Sands contain beds which are very fine grained and prized for glass making. Above them lies the iron-rich strata which continues the Lower Greensand, here hard enough to produce a local building stone, the carstone. These deposits are both acid and generally free draining and are associated with a heathland flora, now much overtaken by trees, both from commercial plantations and by self-seeding. In a few valleys where the iron has leached downwards to form an ironpan or where the water table is high, bogs have developed. Immediately above the carstone is a band of gault clay, so thin at its northern extremity to be virtually untraceable, but widening southwards to form a clay valley which becomes recognisable by Sandringham and, like the Lower Greensand, trends westwards and is eventually buried under the Fenland near Stoke Ferry.

Fenland; Peat and Silt Fens

All the deposits mentioned so far are present under the much later layers of the Fenland. This area divides into two, the Silt Fens and the Peat Fens. The more northerly parts of the region are made up largely of silt deposited by the great rivers which flow through it to the Wash. Immediately behind the silt barrier is a string of villages between King's Lynn and Wisbech, settled in Saxon times while to the south of these lies Marshland, also chiefly silt but sometimes interleaved with the residue of saltmarsh, representing the times in the past when it became inundated by the sea. To the south of both are the Peat Fens. These have been reclaimed more recently and consist of intensively farmed land, divided with dykes. The only settlements here are on former river banks or as scattered farms positioned where there is some firm ground. Botanically it is the Peat Fens which are now very poor in species, but the flood lands which make up the Ouse Washes give an idea of the riches this area once held.

The Chalk and the Deposits Above

The scarp edge of the chalk was rounded off during the Anglian glaciation and its subsequent gentle relief has caused few problems for the plough, so only a few areas of true chalk grassland remain.

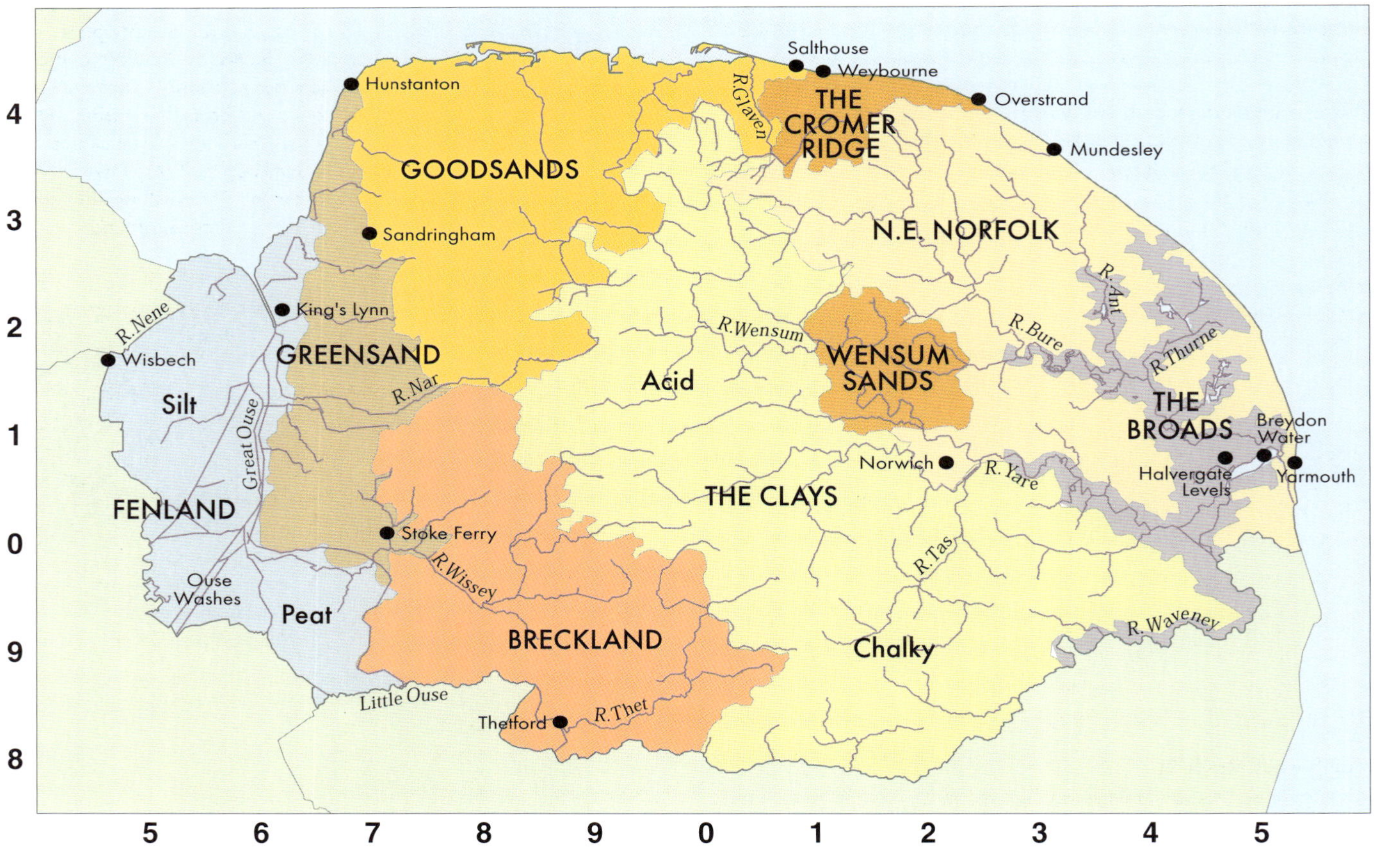

The covering of glacial till which deepens eastwards has, however, been eroded away on steep slopes leaving the influence of the chalk below traceable in the flora of some roadside banks and the rare pieces of uncultivated land. Because of the glacial deposits, the rest of West Norfolk can be categorised by the nature of this till rather than the solid geology beneath with the chalk dipping eastwards until the top of the deposit lies at sea level on the beach at West Runton. Farther east it disappears under the later crags.

Breckland

In south-west Norfolk, adjoining the Fenland, is Breckland. Here the chalk is covered by varying depths of sand giving very sharp drainage. The climate is the closest in England to being continental with higher summer temperatures, low rainfall and a tendency to frequent night frosts. Most conspicuous are the large areas of recent afforestation but until recently agriculture has been extensive rather than intensive. Only in the last twenty years has supermarket-led agriculture brought vegetable growing and subsequent irrigation to the region. The typical plants of the grass-heath of the traditional Breckland now survive only on reserves, the wider ride verges in the forest and in the large military training area where agriculture is represented by sheep grazing.

The Goodsands

To the north of Breckland lies north-west Norfolk, an area called 'Goodsands' by Arthur Young in the early years of the 19th century. Perhaps it is not the ideal name, but it is one which has not been bettered. Although with many characteristics of the Breck, this area has a marly layer between sand and chalk and digging into this layer and spreading it over the land at the time of enclosure gave the soils both body and nutrients. This method is now replaced with artificial fertilisers, but the clay layer still prevents the soils from drying as rapidly in drought conditions. Being at a greater elevation and closer to the coast, rainfall too is also notably higher than in Breckland or Fenland. This is now prime barley and sugar beet country, but has been saved from becoming an arable prairie by game shooting interests which have maintained hedges and copses for pheasants and areas of grassland for partridges. There are also several large parks with lakes, grassland and woods.

The Cromer Ridge and the Wensum Sands

A second region of acid sands lies in east Norfolk running from the coast between Salthouse and Overstrand southwards to just beyond Norwich. Its heathland flora and its subsequent history of land use is very similar to that of the greensand, but here the soils are of glacial origin.

Central Norfolk, the Clays

In the centre of the county, the dominant soils are clays, tending to acidic in the north but becoming more limy as they merge with the chalky boulder-clays of south-east Norfolk. Much of this clay lies on the county's watershed and the often level ground can still become waterlogged if drainage is neglected. In the past it tended to be avoided for agriculture for as long as possible, especially in its least fertile parts and it is here that the greatest concentration of ancient woodland is to be found, sometimes adjacent to the heaths of the previous region. The chalky boulder clay region is now wheat country and it has probably suffered more than any other part of the county from hedgerow removal and the ploughing of its former wide verges. This is much to be regretted as it has a most distinctive flora which now only survives, and then often perilously, by green lanes and on the few remaining areas of common land.

North-east Norfolk

This is another area of fertile soils which is highly agricultural though some of its many streams still have grazing land beside them. Its proximity to the north-east coast gives it a strong maritime air, (in both senses), but this is another area which has lost a great number of its hedgerows and verges and it can no longer be said to have any distinctive botanical character away from the coast.

Broadland

This must be the most familiar part of Norfolk, even to those who don't even know the county. The botanical interest lies chiefly in the areas of marsh and fen which lie on the level ground close to the rivers. Several of these areas are still cut traditionally for reed and others used for grazing. Some, however, have been drained and are now arable, bringing increasing problems with nutrient-rich water reaching the dykes, broads and rivers. Many of the broads, largely man made, have become polluted, particularly with phosphorus from sewage works and nitrogen from arable land. Cleaning them up and returning them to their original state is proving a major task, but one that is being tackled energetically. Included in Broadland are the Halvergate Levels, a large area of grazing marsh lying between Breydon Water and the main rivers. The efforts to keep this area as grassland have reached the national press and with the problems which have beset cattle farmers in the last years of the 20th century, it seems that for some of it at least, purchase by conservation bodies may be the only certain solution.

Rivers and Sea Coast

Having water on all sides - the only way to get into Norfolk without crossing water is in the hundred yards or so which separate the headwaters of the Little Ouse and Waveney - the county has many rivers. Only the Great Ouse is a large river with its source outside the county, all the others are local rivers and many still have small areas of fen and mire along their courses, others, especially in Fenland are largely artificial in their courses and tend to run in straight sided channels which have little in the way of waterside vegetation. Rivers run through most of the regions described and are in many ways quite separate from them. The coast too is a region of its own, though a very narrow one. Beside the southern end of the Wash there has been a great deal of reclamation which has quickly been turned into arable land and has as yet a remarkably poor flora, but once the shingle bank is reached at Snettisham the flora improves dramatically. The north coast has nationally important saltmarsh as well as calcareous dunes, shingle banks and brackish marshes until the cliffs at Weybourne are reached. The cliffed coast continues around to Mundesley where acidic dunes begin. These have quite a different character to the north coast dunes. Saltmarsh occurs again around Breydon Water while the last stretch of coast around Yarmouth is very largely built up.

Shingle Bank - Cley

The Soils of Norfolk: Their Nature and Distribution

W. M. Corbett

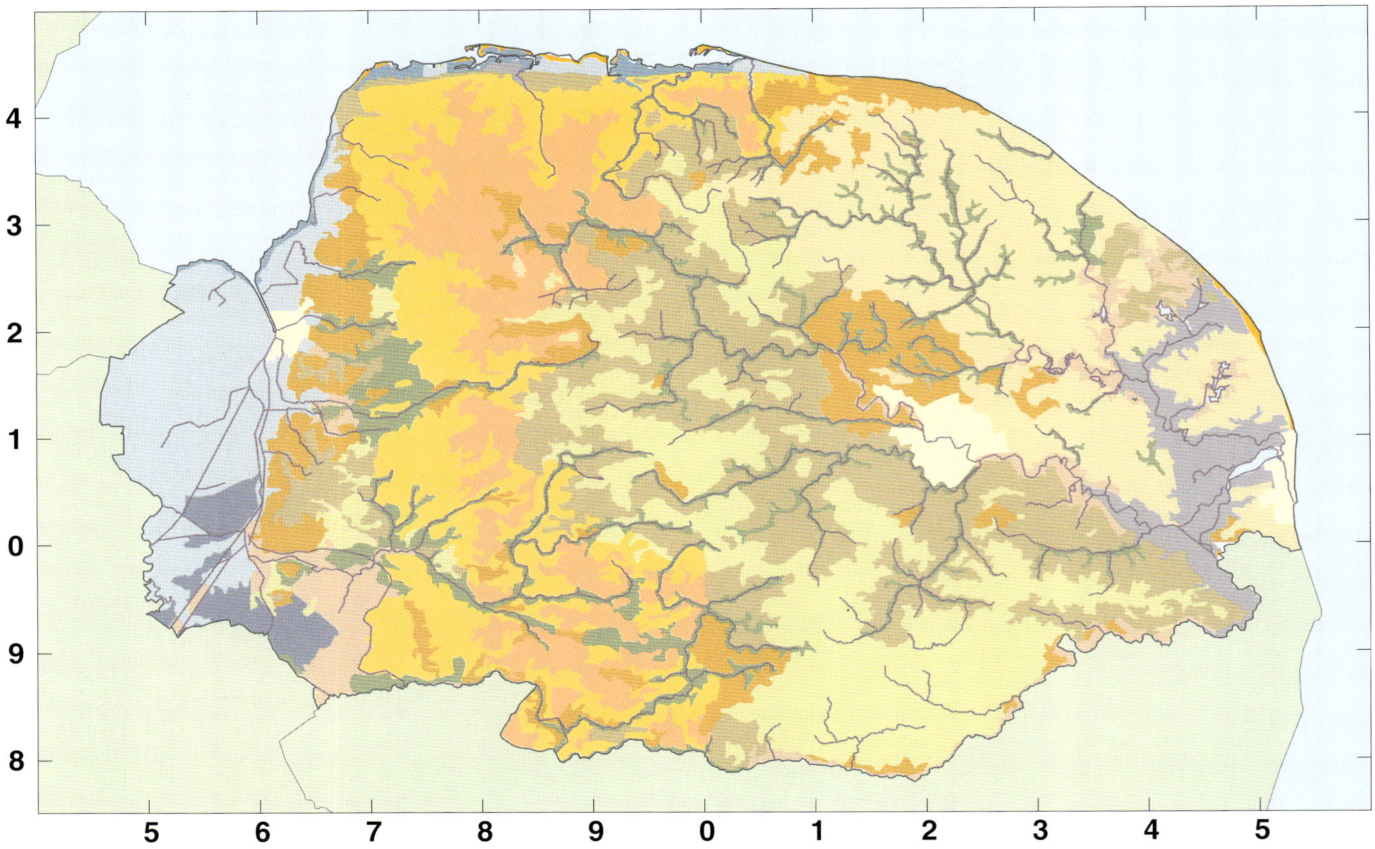

Figure 1 - Digitized Soil Map of Norfolk

1 Clayey marine alluvium, locally silty and calcareous. Some fen peat.

2 Fen peat. locally very acid and locally thin over sands and chalk.

3 Marine alluvium, clayey or silty some calcareous.

4 Silty or clayey lake marl and marine alluvium, the former extremely calcareous.

5 Wet sandy or silty soils with, in parts, a thin peaty surface and some peat.

6 Seasonally waterlogged soils of chalky-boulder clay/Lowestoft Till with a loamy or clayey surface. Some groundwater sites on Jurassic or Cretaceous rock have a thin peaty cover.

7 Well drained, loamy over clay soils on chalky drift with some seasonal waterlogging.

8 Deep, well-drained sandy and loamy soils, the former distributed in small scale hexagonal patterns.

9 Deep, mainly well-drained, coarse loamy soils with a loessal component.

10 Shallow, calcareous, well-drained, sandy and loamy soils in association (in a stripe pattern) with deeper sandy soils.

11 Deep, well-drained sandy soils, some very acid (podsolized).

12 Dune sand and marine shingle.

13 Unripened marine alluvium (mainly saltmarsh).

14 Urban areas.

1. Soils and Geology

The soils of Norfolk and their distribution can be outlined in terms of their parent materials, the rocks or sediments from which they have been formed. A primary division is between the upland surfaces, mainly of Pleiostocene age, and the post-glacial Holocene (Flandrian) surfaces of Fenland and the coastal and estuarine marshes.

Central & East Norfolk

The Pleistocene deposits are either basal tills laid down underneath an ice sheet, or glaciofluvial drift derived from its outwash. These deposits were laid down in the Anglian stage of the Pleistocene, approximately 400,000 years ago (Rowe & Atkinson pers.comm.).

There are two major till facies: (1) the chalky till (Chalky Boulder Clay/ Lowestoft Till) with chalk stones in a matrix mainly of Mesozoic clay and chalk flour, (Perrin et al. 1979); and (2) the less calcareous North Sea Drifts with a matrix derived, at least in part, from the sandy deltaic deposits then flooring the North Sea (Corbett 1996). The Chalky Boulder Clay/ Lowestoft Till forms a broad, flat central plateau which rises to about 60m O.D., and creates a north-south watershed, Fig.1, Unit 6. The North Sea Drift forms most of the lowland in the north-east of Norfolk, Fig.1, Unit 9. The northern parts of these lowlands are capped by glaciofluvial drift, sands and gravels that form the surface over most of the feature known as the Cromer Ridge.

The coastal platform to the north of the Cromer Ridge is cut into a range of materials including tills, North Sea Drift, Marly Drift (see below) and Glaciofluvial outwash. From the western part of the Cromer Ridge, a gently graded outwash plain, on the southern part the Wensum sands, Fig.1, Unit 11, extends as far south as Norwich, separating the chalky till plateau to the west from the North Sea Drift lowland to the east, (Cox & Nickless 1972). Further late glaciofluvial sands and gravels form terraces flanking the Holocene floodplain in most of the major valleys. On slopes above these terraces, certainly on the eastern side of the central plateau, preglacial sands and gravels outcrop from below the chalky till.

The Chalky Boulder Clay/Lowestoft Till with its predominantly clay matrix is almost impermeable, but on the northwest side of the plateau, centred on the headwaters of the Wensum, is a much more permeable till with a higher content of chalk, particularly flour or silt-sized particles in the matrix, marly drift (Perrin et al.1979).

West Norfolk

The above generalised stratigraphic pattern applies to the uplands of central and east Norfolk but not to West Norfolk. On the northwest coast from Wells to Hunstanton, loamy, decalcified till belonging to the Hunstanton till of Devensian age (10/60,000 years B.P.) forms a low coastal platform (Straw 1960). To the south, between the western edge of the central plateau and Fenland, Fig.1,Units 10-11 the West Norfolk lowland, there is an escarpment of Cretaceous rocks: Chalk, Gault Clay and Lower Cretaceous Sandstones. This sandstone includes both iron-rich cemented Carstone and the almost pure quartz sands of the Leziate Beds. These Cretaceous beds dip gently to the east, but further south, in South Norfolk where their strike changes from NNW/SSE to NE/SW their dip is almost horizontal and the chalk, capped on the high ground by thin chalky drift, forms, at 30-40m O.D., a low, dissected plateau, Breckland (Corbett 1973). On the Breckland plateau, the chalky drift is capped by up to metre of aeolian sand, a coversand probably blown in by saltation from the floor of the North Sea (Corbett 1996). Outside Breckland to the east and north, on the central chalk till plateau, similar material is incorporated in the surface of the chalky tills. In northeast Norfolk on the North Sea Drift, the equivalent surface drift has a more silty, loessal character, (Catt et al.1972, Perrin et al.1974, Corbett, 1977).

Fenland and Broadland

The Holocene surfaces of Fenland and the coastal and estuarine marshes have marked spatial variation in stratigraphy and lithology (Gallois 1979, Coles & Funnell 1981, Waller 1994). This stratification was caused by fluctuation in sea level over the last ten thousand years resulting in the deposition of fresh water peats and marine muds, silts and sands, (Coles & Funnell 1981).

In Fenland, the more southerly parts of the Romano-British silts and clays in the north overlie, at depth, thin layers of peat. These thicken southward and eventually outcrop in southwest Norfolk. Over a large part of the outcrop however, pump drainage, cultivation, weathering and wind erosion have destroyed the surface peat so that the underlying fen clay now forms the surface. (Seale 1974). Likewise, on the east coast estuary marshes where the complete stratigraphic sequence is a five layered sandwich with peat at the top, the bottom and in the middle separating beds of marine clays and silts (Coles & Funnell 1981), the surface peat on pump drained sites under grass or arable has oxidised and the present land surface is the upper layer of marine sediment.

In the downstream part of estuaries, this marine sediment spans or almost spans the wide floodplains, but upstream, as it nears its landward limit, it tapers in towards the present river channel (Corbett & Tatler 1970, Hodge et al., 1984). In section, near the limit, the marine sediment feathers out over the underlying middle peat. This peripheral or feathering-out zone marks the limit of pump drainage and on the undrained sites upstream, where the surface is still the upper peat, the vegetation includes alder carr, *Phragmites*, sedge and grass communities.

Marine sediments contain sulphides and near the periphery of the drained zone, where the sediments are thin and decalcified after drainage and oxidation, pH values well below 3.5 may occur, (Dent 1986). Such values are particularly common on the flood plain of the River Ant. Where such pH values occur beyond the spread of marine alluvium, as in the upper Waveney valley, the probable source of sulphides is the Mesozoic clay in the matrix of chalky till (Catt pers.comm. in Corbett 1979).

2. Soils and Geomorphology.

Descriptions so far have been of the variation in the soil parent material of Norfolk which is related to the geology, a scale of spatial variation measured in kilometres, On the upland, further divisions within these major divisions, can be described and defined in terms of the processes which have affected surfaces since their deposition (their geomorphology).

The relevant processes operated at least during the Devensian age, when the maximum ice advance just reached the North Norfolk coast and Norfolk experienced periglacial or tundra conditions, a frozen soil in winter and surface melt in summer. The nature and effect of these processes is closely related to topographic position: the effect is spatial variation in soil type on two scales (1) on a scale of tens to hundreds of metres and (2) a scale measured in metres.

Solifluction operated at the larger scale causing summer downslope movement of melted surface layers over a still frozen subsoil (Corbett 1973). This occurred even on gentle slopes of less than 2 degrees and it has since been supplemented by present day erosion. Cryoturbation operated at the smaller scale, this is the upward rupturing or mushrooming of sub-surface layers into the weathered surface horizons following freezing and expansion. The centres of upthrust have a spatial interval of 5-7m and the effect of this process is most prominently expressed at the surface by soil and vegetation where the surface and sub-surface layers have a contrasting lithology. The Breckland landscape with its stripes and polygons is a good example of both periglacial processes, solifluction and cryoturbation, and their interaction, (Corbett 1973).

Breckland

The small scale cryoturbation effect on the Breckland plateau has the form of isolated polygons, their centres of chalk drift upthrust 5-7m apart, being separated by troughs, more than one metre deep of decalcified sand. The equivalent features on the slopes, where there has also been solifluction, have a stripe form, sub-surface ridges of chalky drift running down the slope being separated by roughly parallel troughs, again more than one metre deep of decalcified sand. Up until the arrival of myxamatosis in the 1950s, Breckland heaths were heavily grazed by rabbits and, under this grazing pressure acidophylous plant communities picked out accurately in plan the distribution of the underlying periglacial features, Fig.2.

Since the disappearance of the rabbits these vegetation patterns are not so distinct, having often been swamped by plant communities with less specific site requirements.

Stone Striping - near Grimes Graves

On the Breckland plateau, the aeolian or wind-blown coversand is up to one metre thick, but less than 30cm thick on the peripheral slopes. Below the coversand, the cryoturbation features in the chalky drift, polygons on the plateau and stripes on the slopes Fig.2. in vertical section at depth, are structurally similar and comparison of the depths at which the structure of similarity occurs indicates that slopes have been truncated to a depth of 2-3 m (Corbett 1973). This weathered or partially weathered soliflucted matieral has accumulated below the slopes on the floors of dry valleys. The full spatial range of the variation from plateau to valley floor caused by solifluction may extend over several hundred metres.

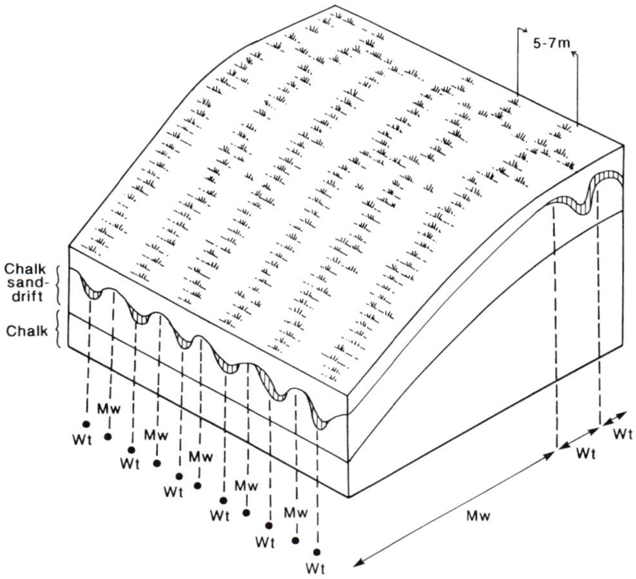

Figure 2 - Breckland Stripes and Polygons
Mw Shallow Sandy Calcareous Soils
Wt Deep Sandy Acid Soils

Central & South Norfolk

On the chalky till plateau of central Norfolk, as in Breckland, periglacial features are represented by variation in the depth of coversand. However, in the surface layers of the chalky till plateau, there has been much incorporation and intimate mixing of the coversand with the underlying clayey till. In general the thicker or deeper the depth of incorporation of coversand, the coarser its texture. In the centre of the plateau, the layer of incorporated coversand can be up to be 1m thick and its texture, coarse loam, Fig. 3A. In contrast, in the very gently sloping land near the periphery of the plateau, coversand is represented only by a sand fraction in the plough layer, a fraction which is absent in the immediately subsurface clay. Inspite of this sand, the plough layer is still clayey, Fig 3b. Between these two extremes most of the plateau surface has coversand incorporation to a depth of 40-80cm and the texture is fine loam (Corbett 1979).

The underlying chalky till with its clay matrix, is impermeable and in the winter months, when rainfall exceeds evapotranspiration, a perched water table develops, (Corbett 1979). However, on almost all sites, land drainage for arable farming holds this seasonal water table at depths below 50cm. At the edge of the plateau where the chalky till is less than 3m thick, and, throughout its depth, partially weathered or fissured, this seasonal water table is only transient. On such peripheral sites the surface horizons have loamy texture and they represent the solifluction product from the chalk till plateau above. These surface loams or head deposits extend downslope from the weathered chalk till fringe and cap the upper parts of the preglacial sand outcrops on the sides of valleys and re-entrants. All of this spatial variation in depth and texture from the centre of the plateau to valley sides, distances of several hundred metres, is interpreted as solifluction effect (Corbett 1979).

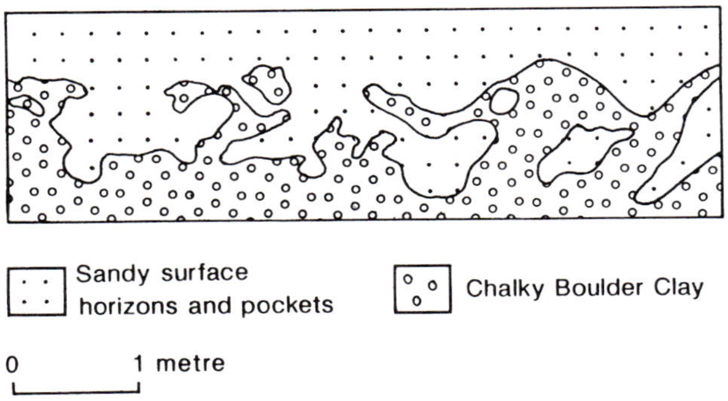

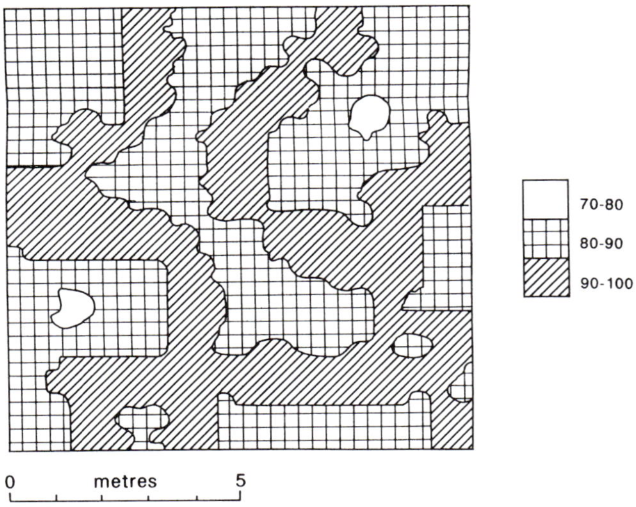

Figure 3a - Coversand Patterns on Chalky Boulder Clay/Lowestoft Till. Central Plateau Site, Section and Map

Superimposed on the above general trend is variation in depth of coversand incorporation on a horizontal scale of a few metres. On central plateau sites with the deepest incorporation of surface coversand, even deeper troughs, filled with relatively pure coversand, interlock and isolate crests in the till surface, a coarse equivalent of the Breckland polygons, Fig 3a. Near the plateau periphery, with coversand represented only by the presence of a sand fraction in the plough layer, there are occasional shallow holes in the till surface immediately below the plough layer and these holes are filled with relatively pure coversand, Fig.3b. Between these two locations, in the centre and periphery of the plateau, over most of the plateau surface, the variation in depth again takes the form of more sandy troughs, but these do not interlock and, as far as the water table is concerned, each sandy trough forms a small, independent catchment a few metres long, (Corbett 1979).

These local patterns on the chalky till plateau are not as symmetrical as those in Breckland and between the three forms of pattern on the ground there is no sharp boundary, one type merging into the other, parts of a continuum. The subsurface features have little or no surface expression in land use, site conditions being dominated by the hydrological condition which is common to the whole plateau. There is however, some tendency for the occasional patches of moist oak woodland to be concentrated on central plateau sites, the most difficult sites for the disposal of surface drainage water.

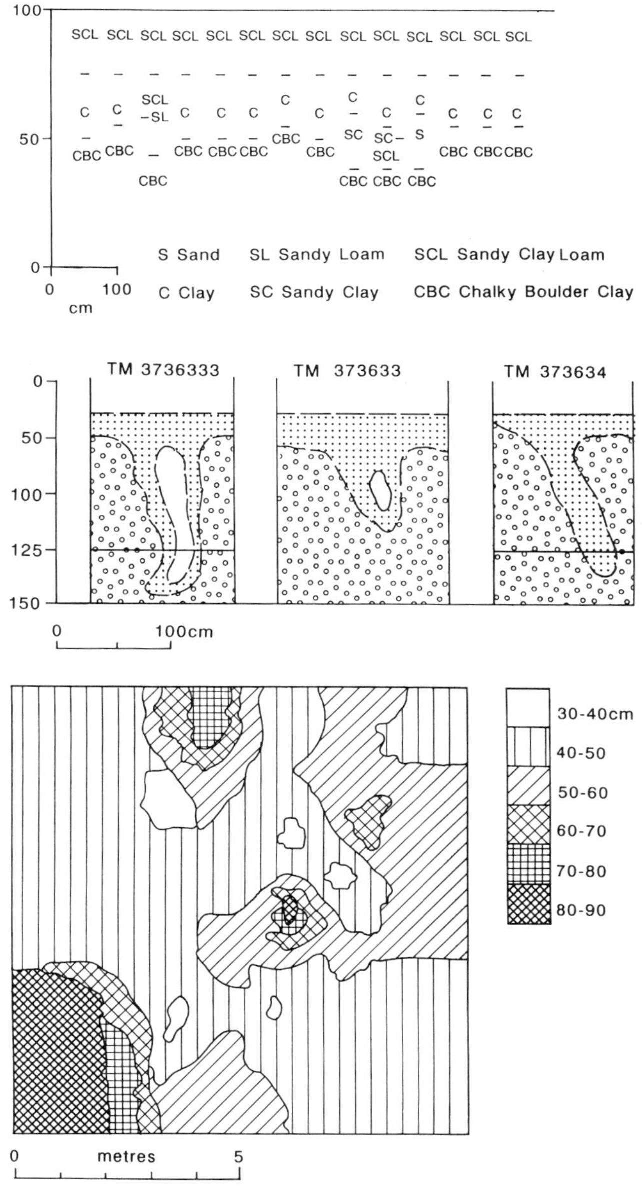

Figure 3b - Coversand Patterns on Chalky Boulder Clay/Lowestoft Till. Peripheral Plateau Site, Sections and Map

The Cromer Ridge

A third major landscape, affected markedly by periglacial processes, is the glaciofluvial sands and gravels capped by silty coverloam in the north of north-east Norfolk, the Cromer Ridge. Here solifluction and, more recently, erosion have stripped the loessal coverloam from most slopes and this material has accumulated in valley floors and re-entrants to depths locally exceeding two metres. At such depths the coverloam is very silty but near the surface, because of recent erosion, it becomes markedly more sandy. Above the slopes on the gently rounded interfluves, the coverloam is 40-80cm thick. The variation in depth occurs within a few metres and in plan, as shown on aerial photographs, it takes the form of a complex and intricate pattern of striations, Fig 4. (Corbett & Tatler 1974). These are interpreted as the combined effect of cryoturbation and solifluction, individual striations being local flow channels.

The reason why the striations are delineated so accurately on aerial photographs is the marked contrast in water holding capacity between the underlying sands and gravels and the silty coverloam, the latter holding more than double the amount of plant available water. This site factor is reflected strikingly in plant growth, particularly in dry years when aerial photographs clearly delineate by darker tones the more vigorous plant growth in valleys and re-entrants and on the interfluve striations, Fig 4.

Figure 4 - Coverloam (loessal) Pattern in northeast Norfolk

From these three examples of major soil landscapes in Norfolk it is clear that variations within them, caused by periglacial processes, is common to the whole of upland Norfolk. The surface expression of these variations, particularly with regard to plant growth, is most obvious where there is a marked contrast, chemically as in Breckland or physically as in north-east Norfolk, between the aeolian surface layers and the underlying drifts or tills.

Fenland

Soil patterns on equivalent scales to those described above for Pleistocene surfaces also occur on the Holocene, post-glacial surfaces of Fenland and marsh, but on these latter sites this process is purely depositional. During deposition, drainage channels, now fossil features termed rodhams (roddons), had a sorting effect on the lithology of the alluvium. The land surface expression of rodhams is now low, meandering ridges and while the crests of these ridges, which enclose the original channel, tend to be silty and calcareous, towards the back swamp position they are less calcareous and more clayey, (Seale 1975).

The ridges, 10-100m wide, now stand a few metres above the land surface. On fenland peat they are particularly conspicuous and their distribution as seen on aerial photographs, is a very fine branching network which coarsens towards the larger channels and rivers. On minerogenic marshes, tonal contrast on aerial photographs is slight but the roddams can still be recognised in the field by their relief.

Man, the Landscape and Plants... an Historical Perspective

Gillian Beckett

Until the advent of self-powered machinery, man farmed within the limits imposed by the county's soils; he did not attempt to grow water-demanding crops on the dry soils of the Breckland, nor monocultures of wheat on the ill-drained clays of central and south-east Norfolk. Of course, he knew he could not succeed. Only within the last sixty years have the combination of the application of business driven economics and necessary technical expertise made it possible. Before that, for some ten thousand years, man moulded the face of the county using only the power of his own muscles and that of amenable animals. He used its varied soils to his advantage by developing the forms of land- use which would enable him to get the best from the conditions he encountered. We have inherited the countryside he created and the patterns of his work are still visible not only in what we see in the landscape, but also in the surviving flora.

The last ice advance only just reached the north-west coast of the land we now know as Norfolk, coming to a halt against the earlier morainic ridge which now runs from Stiffkey to Mundesley, but the rest of the area was subject to tundra conditions with resulting frost heaving and blocked drainage. This left the contours much sharper than they are now, the valley slopes steeper and streams and rivers more clearly defined. Ground surfaces too will have been much more irregular. Outflow from glaciers brought mixed materials along the rivers depositing first of all gravels, then sands and silt as the water courses made their way to the far off sea. At the height of the ice advance, sea levels were low and the Norfolk coast will have stood somewhere near the present Dogger Bank with a continuous land bridge from the east coast to the continent. The land was, however, approaching its present outline by the time man started actively cultivating the soil some eight thousand years ago. By this time too the land surface had already been thoroughly colonised by plants and most of it will have been tree covered. This cover, however, was far from uniform and like the ground flora, depended largely upon the moisture content and acidity or alkalinity of the soils. This is the state which we call the natural vegetation. Since then the history of this vegetation has been one of man's efforts to use it or change it to his own advantage which he has done with increasing success.

The Wildwood

The make up of the wildwood before man's intervention has been worked out through the use of pollen deposits, helped by the survival in the fenland of the remains of some of this woodland still lying beneath the covering of peat. The tree trunks which come to the surface as the base of the peat is reached are locally known as bog oaks, but show a wide range of tree species. Certainly Lime, Oak, Wych Elm, Ash, Maple and Hazel were all important at this period together with some Hornbeam, Yew and Pine. The extent, timing and results of man's clearance of this tree cover was largely dictated by the fertility of the soil. Grazing within the woods was valuable from an early date for cattle are naturally woodland creatures, but the effect of such grazing in the long term was that the trees failed to regenerate leading eventually to rough grassland. So land was cleared both deliberately for farming and by the action of stock. As the remaining amount of woodland became smaller and woods had become islands in a sea of agriculture and settlement, so management became essential if sufficient woodland products were to be available. Over the centuries the method of management which we call coppice with standards was gradually developed. This involved the retention of sufficient timber trees as standards to satisfy needs, chiefly for building, and coppicing the rest, this involved cutting the trees to ground level in a regular cycle to provide the small wood which was used for tools, fencing, firing etc. The woods were divided into a number of compartments, each was cut to the ground in turn and once the re-growth was out of the reach of animals, cattle were allowed in, the rest of the wood being fenced with hurdles. The need to balance the number of standard trees which cast shade and partly suppressed growth of the coppice stools below with the need for small wood is seen in a proposed agreement of 1574. This was between Sir Nathaniel Bacon who required large timbers for his new house at Stiffkey, and the Dean and Chapter of Norwich Cathedral who owned two woods at Hindolveston. Bacon was offered a 21 year lease for the right to cut oaks, paying £100 down and an annual rent of £5 but had to preserve the 'springs' which were the young coppice shoots and noted that the right of grazing (the liberty of the feed) was already in someone else's hands, so the cathedral was getting three separate incomes from the wood.

Old Coppice Stools - Castle Rising

Although throughout history settlements have encroached into these woods and then later been abandoned, a core always remained as woodland and these areas have never been felled or cultivated. The evidence for this can be seen in the uneven floor and varied soil types which reflect the chaotic soils of the surface of glacial deposits and in their floras which typically include Early Purple Orchid, *Orchis mascula*, Bluebells, *Hyacinthoides non-scripta*, Nettle-leaved Bellflower, *Campanula trachelium*, Giant Bellflower, *Campanula latifolium*, Herb Paris, *Paris quadrifolia*, Hairy Wood-rush, *Luzula pilosa*, Wood Sedge, *Carex sylvatica*, Greater Butterfly Orchid, *Platanthera chlorantha* and Wood Melick, *Melica uniflora*.

The Clays

Much of the county's woodland had been cleared by the beginning of the Iron Age and by then the make up of the remaining woodlands was almost as much to do with man's selective felling as with the natural forest components. Tree species which had failed to respond to coppicing were lost, notably pine. Climate too took its toll and the number of lime trees was much reduced as they regenerated less well in a cooling climate. Oaks, having the most valuable timber, were favoured for standards and those which responded well to coppicing, notably Ash, Maple, Hazel and Lime where it was present were the main constituents of the shrub layer. Clearance was slowest on the poorly drained soils and a comparison of the maps of surviving woodland noted in the Domesday Book of 1086 with that surviving today shows that the pattern of distribution is remarkably similar. On ill-drained soils not only was the initial labour far heavier but the resulting soils very difficult to work with existing technology. On the other hand, trees grew well there and woodland products were of correspondingly high value so the owners of this sort of land included the woodland as part of the cash value of their estate. These poorly drained areas were largely on clay soils, and by the early Middle Ages, the economy found to be most suited was one with small areas of arable on the better drained slopes and cattle grazing combined with woodland products as the cash crops. Some old woods still have tan-pits in them where the bark of the felled oaks was steeped to provide the material for working the hides of the cattle. As cattle needed more continuous care than sheep, driving to and from their pastures and regular milking, a family could manage a small herd, so small farms became typical of these areas with plenty of grazing land which was often held in common.

Chalky Boulder Clay

The part of Norfolk where this sort of economy was best developed was on the fertile chalky boulder-clay. Here the level areas of heavy clays long remained as woodland while grazing animals moved between the woods and pastures. Much has been written on the movement of settlements in the Middle Ages especially from the 13th century onwards when many settlements, probably because of population pressure, moved to new sites around the commons and greens. Many of these greens funnelled in to the areas of settlement, often with the roads flanked by broad bands of grazing to facilitate the movement of animals from one area to another and eventually to market. These greens, commons and roadsides developed a rich grassland flora, typically with Dyers' Greenweed, *Genista tinctoria*, Sulphur Clover, *Trifolium ochroleucon*, Stone Parsley, *Sison amomum* and Green-winged Orchid, *Orchis morio*. This grassland type must have covered large areas in this part of the county with outliers where similar soils occurred, and both commons and greens survived long enough to be recorded by Faden on his map of 1797. Typical was the complex south of Wymondham which included commons at Besthorpe, Bunwell, Wattlefield and Forncett all of which were interconnected and such long, narrow grasslands as that at Tibenham Long Row. Enclosure began seriously not long after Faden's map was made and all these had gone by the time Bryant made his Norfolk map in 1826. Corn prices went up twenty-fold between 1795 and 1820 because of the Napoleonic Wars, but wages remained almost the same. With such cheap labour it became possible to construct complicated drainage systems across the area, first with bush drains and later with earthenware ones, so commons and greens were taken in hand by their manorial owners and the wide verges were hedged and incorporated into the fields. The rich assemblage of plants remained for a while along the reduced road verges, but most have now been lost through the pressures of modern traffic and farming and these communities can now be seen only on a few surviving commons as at New Buckenham and along the margins of some green lanes and the wide verges of roads as at Wacton. These merit high conservation status. When the old meadows of this area were enclosed, a few Lammas meadows remained, there were areas divided into strips for hay cutting, then open to grazing from Lammas, that is August 1st. Shelfanger meadow is one of the few survivors and retains its rich flora. On these fertile soils, many of the woodlands have been lost too, but happily a few fine examples survive, notably Tindalls Wood and Ashwellthorpe Wood (NWT) where the flora notably includes Small-leaved Lime, *Tilia cordata*, Wild Service-tree, *Sorbus torminalis* and Wood Spurge, *Euphorbia amygdaloides*.

Acid Clay

Where the soils were more acid and so less fertile, the pressures to reclaim land for arable were less strong and relatively more woodland has survived in one form or another. On the glacial ridge between Holt and Cromer the dominant oak of some of the woods is the Sessile Oak *Quercus petraea*, but these woods tend to be on lighter soils and the largest number of old woods left today is to be found in the central parts of the county. The area between Fakenham and Holt, referred to on Faden's map as Stock Heath, was a great mixture of coppiced woods, wood-pasture, where the trees were pollarded to keep the young growth above the reach of grazing animals, and heathland. The 'Stock' refers not to the cattle but was an early name for the tree stumps which were conspicuous from felled trees. This suggests that this area moved from woodland to heath from Saxon times onward though that does not mean that parts of it hadn't been cleared and abandoned before. Thursford Wood with its pollards is a remnant of this former wood pasture. One of the largest of the woods was Horningtoft Great Wood which was shown by Faden as being a mile across, and which by tracing the plants on a deep bank and ditch, was clearly once even larger. In 1826 it was all still there, but by the end of the 19th century was reduced to three small, separate woods one of which has since been felled. Testerton and Colkirk woods nearby were also felled at this time, but enough woods do remain for us to appreciate their wide range of plants. These woods are very varied in their flora and as well as the typical woodland species often support Lily-of-the-valley, *Convallaria majalis*, Greater Woodrush, *Luzula sylvatica*, Wood Sorrel, *Oxalis acetosella* and until recently Crested Cow-wheat, *Melampyrum cristatum* while the most acid also have Climbing Corydalis *Ceratocapnos claviculata*. The traditional use of these woodlands for coppice more or less died out in the 18th and 19th centuries, but many survived, some because of their value as game cover, others because the effort to clear them was not considered worth while and a few of the smaller ones because they still provided free materials and firewood for the farm. Sadly in the last 50 years of the 20th century many were coniferised, though a more recent trend has been to reverse this and the original ground flora has often been found to have survived surprisingly well. Foxley Wood (NWT) and Honeypot Wood,Wendling (NWT) can be visited. A similar group of woods also occurs along the line of the gault clay in West Norfolk from Wolferton southwards and of these Reffley Wood (WT) is open to the public.

Valley Grassland

Away from the clay areas the best grazing has always been found in the river valleys where grassland developed early at the expense of carr and woodlands of elm, alder, willow and poplar. These grasslands were often too valuable to be used for general grazing and most were cut for hay each summer and only grazed once the crop had been taken. The resulting hay was essential for the overwintering cattle and sheep and where meadows were rushy and rough, they were cut similarly but for bedding material for the animals.

The valley of the Little Ouse and Waveney is particularly wide and shallow with the land on either side of the river waterlogged and peaty. Here the resulting fens provided not only grazing but reed and willow and peat for fuel. Many of these fens have been drained and ploughed but the best of them, that at Redgrave and South Lopham (SNT), which straddles the county border, is managed for its wildlife interest. It has suffered seriously in recent years from water abstraction and although recovering, it has not yet regained its former richness. Because the county's rivers in geological times flowed right across the county from west to east, many of the present river valleys are almost linked across the watershed, giving a once near continuous stretch of grazing land which in turn often joined up with the commons and greens. Typical are the valley of the Nar where the grazing land of Bradenham Common and the Grazing Grounds at its headwaters once adjoined Shipdham Common at the head of the Blackwater and a source of the Yare. Many of these valley meadows survived into this century, only to be drained and ploughed in the 1950s and 1960s when financial incentives were too extravagant to be missed. At the same time the water authorities had a policy of deepening and straightening even the smallest streams which effectively dried and ruined botanically many of the best of the sites which had survived. The fens adjacent to the Gaywood River, Sugar, Derby and Leziate for example which feature so frequently in Petch & Swann 1968, still survive, but lost most of their rarities after the deepening and virtual canalisation of their river. The richest of the river valley grasslands were those where the river water was chalky, so some of the finest developed in the valleys of the Nar, Wissey and Thet. These valleys run down to the Fenland and once had a similar flora to that area.

Pingoes

Most remarkable are the series of commons which lie at the headwaters of these valleys and which have always proved difficult to reclaim because of their strange configuration owing to that curious glacial phenomenon, the pingo. Pingoes were formed when temperatures fell low enough for the chalk and covering soils to freeze to a depth of several feet, creating permafrost conditions. Springs, however, welled up from deep below ground and the continuous upward movement of their water did not freeze until it reached the surface layer. There it formed a lens of ice just beneath the ground. This grew steadily because of pressure from below, forming a dome of ice thinly covered with soil. These features are still to be seen in the Arctic where they are of great size and the name pingo comes from the language of the Inuit people of Canada. Each summer as the surface layers melted with the sun, the chalky soils slumped down the side of the dome, only to re-freeze the next winter. In this way the dome became surrounded by a circle of earth. When temperatures warmed, the ice finally melted leaving the springs emerging at the base of saucer-shaped depressions surrounded by raised chalky rings. These circular banks with good drainage were soon colonised by chalk loving species while the water in the depressions held a wide range of aquatic species notably the Fen Pondweed, *Potamogeton coloratus* and least Bur-reed, *Sparganium minimum* as well as the Fen Orchid, *Liparis loeselii.* These areas were used for regular grazing and retained much of their richness until the late 1960s when increased water abstraction and overdeepened streams began to dry out the pingoes. The cessation of grazing too, notably by rabbits after myxomatosis, allowed scrub and trees to take over. Today, many of these areas are being managed and still support a wide range of plants, the best of the accessible areas being Thompson Common (NWT) and Foulden Common (privately owned but open).

Pingoes - East Walton

Sandy Soils

On the lighter, drier soils of the county the tree cover will always have been thinnest. The soils being easier to work than the clays meant that even if temporary clearances of shifting cultivation were abandoned, the return of trees could either be kept at bay by grazing stock or easily re-cleared. Not all the woodland was cleared, however, some had to be retained as every property needed productive woodland to grow timber for its important buildings as well as smaller wood for every day use. After the Norman conquest and during the Middle Ages, individual land holdings grew in size and each estate was able to convert more land to agriculture, keeping only those woods which produced the best timber. These were often on properties which lay on heavier land. The others were cleared and their sites put down to grass or included in arable fields until the very existence of most was lost, or remained as no more than a curious field name such as Hartswood Field at Great Massingham. Interestingly, what is probably the original outline of that wood is still traceable by the Dog's Mercury, *Mercurialis perennis* in adjacent hedges though the wood was felled before 1600. A search of maps will reveal many examples of Wood Farm, Wood Lane and similar names now lying within arable fields.

One far reaching result of the early clearance of these soils was that before long the best top soils, exposed to weathering, were washed downslope into the valleys. Because primitive, shallow cultivation was all that could be managed at the time, minerals from lower levels were not brought to the surface and the shallow, surface soils became leached of their minerals to such an extent that they became acid and agriculturally very unrewarding except as rough grazing. These leached soils soon supported extensive heather heaths. Kirby Trimmer writing in the introduction to his Flora of Norfolk in 1866 comments that 'In the reign of Charles the Second, a heath extending with little interruption from Dunston south of Norwich, northward to Hevingham and from there westwards to Lynn and then computed to be a hundred miles in circumference is now reduced to comparatively small, detached portions of that description of wild'. Much of this heath was once held in common between local parishes and was grazed with the Norfolk sheep which Marshall in 1818 described as being able to 'thrive upon heath and barren sheep walks where nine-tenths of the breeds in the kingdom would starve'. Although still sometimes thought of, even today as wasteland, these heaths were used for grazing all sorts of stock including the all important plough oxen as well as sheep, donkeys and the geese for which the county later became famous. Geese were grazed by many of the poor who could not afford more valuable stock and it was no accident that the goose was chosen for the anti-enclosure poem:

The law will punish man or woman
Who steals the goose from off the common
But lets the greater villain loose
Who steals the common from the goose.

Heather was used for thatching, bees which fed on its nectar produced first class honey, the damper areas gave rushes for rush lights and even the cutting of Bracken was regulated because of its value for animal bedding and for its ash which produced an alkali used amongst other things in soap and glass making. A few trees remained and these were usually pollarded, keeping the young growth well away from grazing animals whether cattle or deer. The huge pollard oaks seen by so many in the grounds of Sandringham House are growing on what in medieval times was part of Rising Chase, a hunting property of the Dukes of Norfolk, valued in the days when the castle at Castle Rising was still in use. Later growing upon open heath and sheep walk they remained a valuable resource for centuries. Boundary banks, like the one which runs across the edge of Roydon Common and separates the heathland belonging to the parishes of Roydon and Grimston, give an indication of the way the owners of the lands of adjacent villages viewed possible encroachments on their territory.

In Norfolk there are four main areas of such light, sandy soils, those of north-west Norfolk, referred to as the Goodsands by Arthur Young in 1795; the Breckland; the sands of the greensand which outcrops from Hunstanton south to Shouldham and the rather similar soils of glacial origin from Cromer southwards by Holt to Poringland south of Norwich. Of these areas of light soils, the most acid were those of the Greensand and of the East Norfolk glacial soils where the main vegetation was made up of both Common Heather (Ling in Norfolk), *Calluna vulgaris* and Bell Heather, *Erica cinerea* as well as both Common Gorse, *Ulex europaeus* and Western Gorse, *Ulex gallii*. The Goodsands and the Breckland, which were less acid, had Ling and Gorse growing together with various heath grasses making up what is usually called grass heath.

Calcareous Sands - The Goodsands and Breckland

In the Goodsands and Breckland, both of which were large enough to support communities without any land on better soils, the pattern of local farming was adapted to allow for local needs of firing and wooden products. Small wood for tools and everyday use were obtained from scrub or pollarded trees left on the edge of the grazing land (the field name Frith may indicate this use), or from small areas of coppice grown on the crofts adjacent to the village housing. Gorse too was widely used for firing, especially for bread ovens, a use which continued in Syderstone at least until the first half of the 20th century. Apart from the areas used for arable, the rest was largely given over to sheep or rabbit grazing. Much of the open grazing in this area was enclosed for private use as early as the 1500s as great landowners built up very large flocks of sheep. Hedges were only to be found immediately around the villages or sometimes between the great areas of sheep grazing known as foldcourses. On the slopes where the soil was shallower a chalk flora developed, but this was usually where the best arable was to be found so soon only the baulks between the strip fields or man-made features such as the many tumuli (prehistoric burial mounds), the dividing banks (known as dykes) and fortifications of earlier ages were ornamented with Rockrose, *Helianthemum chamaecistus*, Salad Burnet, *Sanguisorba minor*, Marjoram, *Origanum vulgare*, Thyme, *Thymus pulegoides* and other valued herbs.

Chalk Grassland - Grimes Graves

Botanically, probably the most important feature of the agriculture of these areas was the temporary ploughing or breaking of portions of the heath. These areas, called Brecks, were cultivated for a few years until their fertility was lost, then allowed to return to the general grazing. This meant that there were always areas of open ground, rather akin to that of the post glacial before the trees returned, and some of the plants such as Field Southernwood *Artemisia campestris*, now confined to the Breckland, may have survived in the area for 12,000 years. Others may have been introduced by very early farmers as weeds in their seed corn, but all owe their survival to this farming practice which, although now associated with the Breckland was once in use as far north as the coast. Spanish Catchfly, *Silene otites* for example, survived at Gayton until 1961 safe from the plough in a chalk pit until that was abandoned and became overgrown with trees and scrub, while the Proliferous Pink, *Petrorhagia prolifer* was found in the Bircham area in the 1840s by K.Trimmer.

Throughout the area land was used for both sheep and corn growing until the 18th and 19th centuries when the population began to rise. Then enclosure began in earnest for arable cultivation and by 1820 much of the former heathland had been divided into fields. The fields were ploughed and the practice of marling using the chalk dug from just below the ground often in the fields themselves, gave the soil fertility and the heathy vegetation quickly disappeared. Within a few years the heathland flora was gone and only names survive. Ling Farm and Ling Common are reminders of a wide open landscape which gave poor returns by our standards, but supported wildlife as dramatic as the flocks of bustard last seen in the 1830s. Until the 1939-1945 war there were a few areas of this landscape left including not only Syderstone Common which survives, but areas such as Docking and Massingham Commons which were ploughed in the 1940s. The flora of chalky areas has proved a little more tenacious and still remains on the verges of roadsides where they are survivors from the pre-enclosure landscape, as well as on the few places which were too steep to plough as at Ringstead Downs and Cockthorpe Common or on prehistoric earthworks such as Warham Camp, an Iron Age fort and Grimes Graves, the site of prehistoric flint mines.

Breckland after 1795

A similar pattern of events happened in the Breckland, but much of that area lacks the marly layer beneath the sand and consequently dries up even more rapidly. During the Middle Ages the farming pattern was similar to that of the Goodsands, but the keeping of rabbits in warrens gradually became more important. At first they were managed within the general farming system, but the success of the animals, kept for their meat and fur, gradually came to dominate the land use. Enclosure was again widespread between 1795 and 1815 and the new fields were often planted with pine hedges. Arable crops now had to compete against the rabbits as well as the poor soils and once prices fell after the end of the Napoleonic war, farming became ever more difficult. Many of the large estates found salvation in the Victorian gentleman's fondness for shooting and the rabbits, along with pheasants and particularly partridges were managed for their sporting potential. The pine hedges were no longer cut and remain as the now familiar pine rows (locally deal rows), their trunks still curiously contorted where for a few years they had been kept trimmed, and many shelter belts were put in as game cover. The collapse of farming fom the 1860s onwards (caused by the import of cheap corn, wool and meat largely from the new colonies of Canada and Australia) meant that the land would no longer yield a profit, even rabbit skins were imported and by the start of the 20th century much of the land was abandoned. This failed landscape was described by W.G.Clarke in his evocative book of 1926, In Breckland Wilds and was still dominated by the rabbit. The plants of shifting cultivation found plenty of scope in the sandy warrens and so, unlike those of the Goodsands, were able to hold their own.

At the end of the 1914-1918 war the sight of a large area of apparently waste land attracted government attention and it was decided to use it to build up the nation's stock of timber which had fallen so disastrously low. The Forestry Commission was set up in 1921 and began planting over the abandoned arable and heaths as near-bankrupt landowners greeted the possibility of selling their useless land with delight. Much of the land was planted by the 1930s. Land which was not afforested was brought into agricultural use during and after the 1939-1945 war. When Garboldisham Heath was visited by W.G.Clarke in 1911, he wrote of the spiked speedwell, *Veronica spicata*, 'We estimated that in an area about 100 by 50 yards there were at least 1000 specimens'. This heath was ploughed in the 1950s and the species now has only one natural site in the county. North of Thetford the Stanford Training Area, an area set aside for military training use in 1941, still gives some idea of the state of the 19th century unimproved Breckland, but access is severely restricted, even to those with permits. Today, with financial help and water taken from the diminishing water table, Breckland fields can grow any crop which the climate will allow, and there is no room in the fields for the plants of shifting cultivation and open warrens. Not all has however been lost. Happily a few naturalists were aware of the rapidity with which the old Breckland was disappearing and the newly formed Norfolk Naturalists' Trust (now Wildlife Trust) was able to buy both Weeting Heath and East Wretham Heath in the 1930s while Barnham Cross Common at Thetford also survives. In the present day forest too, some grass rides, especially on chalk soils, are remnants of the old grass heaths and this is where the Purple- stemmed Catstail Grass, *Phleum phleoides* and some of the best colonies of Spanish Catchfly, *Silene otites* can be found, while the Smooth Rupturewort, *Herniaria glabra*, survives on the surface of the rides themselves. Since the late 1980s when the early plantings reached maturity, large areas have been clear-felled and in a very short time many of the typical plants of grass heaths have appeared again, growing from dormant seed which, as as is typical of the seeds of plants of dry ground, is very long-lived. Colonies of the Sand Pansy, *Viola tricolor* ssp. *curtisii* and of the Smooth Cats-ear *Hypochaeris glabra* can already be found where pines have been growing for 60 years. Hopefully their seed will remain in the ground until the next forest cycle is complete.

Forest Ride - Cranwich

Acid Sands - The Cromer Ridge and the Greensand

The two remaining areas of light sandy soils are quite different in origin but have a similar flora and history. These are the greensand of West Norfolk, widening southwards from Hunstanton to Shouldham and the glacial sands of East Norfolk which stretch from the ridge behind Cromer southwards beyond the River Yare and Norwich to Poringland. Both of these lack the chalky base of the Goodsands and Breckland. Descriptions of the East Norfolk heathlands have come down to us from the writings of that propagandist of enclosure, Arthur Young. Writing in 1805 he describes Felthorpe's 1500 acres of common heath as 'a place which harboured poachers' and complains that 'the farmers could not get their work done' presumably because local inhabitants were out on the heath gathering fuel, looking after their animals and tending their bees, all tasks which might benefit them but brought no direct gain to the landowners! Not surprisingly landlords could see higher profits after enclosure and these heaths received similar treatment to all the others. Because of the poor soil, however, many of these areas were afforested by their owners during the 19th century while Mousehold Heath, the common heath of the City of Norwich whose remaining 180 acres are within the city ring road, once extended north-eastwards for seven miles, its extremity traceable today by names such as Mousehold Farm and Mousehold Plantation at Woodbastwick. Poringland Heath too is largely under houses or the plough. Salthouse and Kelling Heaths near the coast with fine Bell Heather, *Erica cinerea*, Western Gorse, *Ulex gallii* and Burnet Rose *Rosa pimpinellifolia* still remain as true heath together as does Marsham Heath and smaller areas elsewhere. The extensive warrens and heaths at Dersingham, Sandringham and Wolferton with their distinctive heathland flora escaped early enclosures, but again much of the land was planted during the 19th century for amenity and as game cover, notably with rhododendrons. As late as 1921 Dutt, writing of the view from the edge of the woods at Sandringham, could say 'One looks across wide stretches of heathery moorland to the green pastures which border the sea'. This area, like Grimston Warren has been planted with commercial forestry since 1950 and neither view nor heath survive. Other heaths, as at North Wootton, described by Dutt as 'a sandy heath with patches of heather' have been allowed to become engulfed with birch and bracken, and this has been the fate of most of the smaller pieces of remnant heath.

Although the loss of such fine areas of heather heath is sad, their great botanical richness lies in the poorly drained areas which have developed in the valleys where the drainage is impeded and often where iron from the surface sands had leached downwards to form an impermeable iron pan. A few of these bogs survive and are quite extensive, notably on the greensand at Dersingham, Wolferton and Roydon where relics of a late-glacial flora remain with acres of Cotton-grass, *Eriophorum angustifolium* and *E. vaginatum*, Yellow Asphodel, *Narthecium ossifragum* and Cranberry *Vaccinium oxycoccus*, a flora more associated with upland Britain. At Holt Lowes, Buxton Heath (at Hevingham) and Beeston too are to be found three species of Sundew, *Drosera*, Marsh Gentian, *Gentiana pneumonanthe*, Butterwort, *Pinguicula* and other typical bog plants. How many other species of wet heaths grew when areas like this were continuous along the valleys rather than small, relict populations we shall never know, but earlier recorders noted three species of Clubmoss, *Lycopodium selago*, *L. clavatum* and *L. inundatum*, all of which have been lost as permanent members of our flora.

Cotton Grass on bog - Dersingham NNR

Sand pits have long been a feature of many of these heaths. Small pits used locally are to be found in most of the parishes in these areas. Many have been used as rubbish dumps and deliberately levelled or become scrub-filled but a few survive and in them plants needing thin, dry soil where there is no serious competition, have

Marshland, Reeds and River - How Hill

been able to find a home. On the greensand east of King's Lynn, however, large operations on an industrial scale are to be found, providing fine sand for much of the country's glass production. The outworked areas at Bawsey are now a large recreation area and although much of it is now up to 10 metres below the old ground level, many elements of the original flora of Bawsey Heath are still to be found, notably red sheets of Mossy Stonecrop, *Crassula tillaea* and colonies of the scarce Smooth Catsear, *Hypochaeris glabra* as well as many small clovers and a sometimes colourful display of plants of short grassland. It is an excellent example of the tenacity of these plants with their generally long-lasting seeds.

Fenland and Broadland

The status and use of land in the Fenland has always depended largely upon changing sea levels. Many earlier clearance phases were later engulfed by flooding, sometimes from land waters, others by the sea, waterlogging the area and allowing the formation of peat. The river water is very limy and areas which were flooded produced excellent quality reeds and sedge while those which were wet only in winter gave excellent crops of grass during the spring and summer as the regular cutting for hay or grazing encouraged the growth of grasses. In a few places where the fen peat continued growing above the water table, it became acid and raised bogs developed. Wet lands which were not regularly managed grew trees which were tolerant of water-logging and sometimes fen carr developed. One of the most fertile areas of the Fenland built up along the margins of the Wash as a result of the amount of silt which came down the rivers from the Chilterns and the limestone country of central England when they were first opened up for arable cultivation. This silt was deposited where the fresh waters of the rivers met the sea and built up a barrier of fertile silts between the peat lands and the sea. A look at the churches of this area of the fens, the silt fens, shows very graphically the value of the land in the Middle Ages, but not so much for cultivation as for grazing. In Broadland the rivers flowing on the glacial valley floors reached what was at first a wide estuary, but silt soon built up in a similar way near the coast, creating what is now the wide area of marshland around Breydon Water while the waterlogged upper valleys gradually filled with peat as the estuary became more restricted. Here too reed and sedge were important with pasture and meadowland wherever conditions made this possible.

One of the great differences between the two areas is, however, not a natural but a historical one and that is in the exploitation of peat. The Fenland area was large, stretching south into Cambridgeshire and northwards into Lincolnshire with enormous deposits of peat up to 10m deep in some areas. The population was, however, very small with no major city nearby, so although peat was dug, used and sold throughout the area until the early years of the 20th century, it did no more than scratch at the surface of the land. In Broadland on the other hand, the areas of peat were small, of parish rather than regional size, and in medieval times the area was one of the most densely settled parts of England with Norwich the country's second city on its main waterway. Here peat was dug in quantity. The records of Norwich Cathedral Priory as late as 1330 show that 400,000 turves were supplied to them in that one year alone. This represents 100,000 cubic feet (8500 cu.m.) of peat. It is not surprising that the diggings became so deep that in a time of flood tides, the rivers spilled over into the deep diggings, broke the baulks alongside the rivers and formed what we now know as the Broads.

The areas of open water interspersed with higher land made the Broadland area very difficult to reclaim for cultivation except on a local scale and unlike the Fenland, the area belonged to a large number of individual owners. The vegetation of the marshland and pastures in the two areas remained, however, very similar until the middle of the 18th and 19th centuries. As early as the seventeenth century the great expanses of the Fenland with its basically very fertile soils had attracted the attention of men who could see great profits from its development and great schemes of drainage had begun. At that time the intention was merely to control the flood waters sufficiently to prevent them from drowning the land in the summer months and the whole area remained famous for its grazing as did the Broadland marshes, notably the Halvergate Levels between the Rivers Bure and Waveney. This grazing land was vital to the supply of meat to the growing London market and it is recorded that in 1720, 40 000 head of Scots cattle were driven into the county for fattening on these marshes. In Fenland, however, drainage plans became ever more grandiose and although often suffering from setbacks, by the middle of the nineteenth century most of the fens and marshes had dried out, the rivers been canalised and the special plants of the area were reduced to smaller and smaller areas. Early records seldom mention plants but a list of birds taken in the fens in 1300 include bitterns, storks and cranes, not the records of naturalists but food for one of the monasteries of the area. One of the first to note down the plants seen in the Norfolk fens was W.Skrimshire who collected plants in the Wisbech area at the turn of the 18th-19th century. His list of plants, especially from Bardolph Fen, is one we now associate chiefly with the Broads with Marsh Orchids, *Dactylorhiza fuchsii* and *D. incarnata*, Saw Sedge, *Cladium mariscus*; Marsh Pea, *Lathyrus palustris*, Greater Spearwort, *Ranunculus lingua* and Water Soldier, *Stratiotes aloides* while Walsoken fen, nearer Wisbech, had Greater Water Parsnip, *Sium latifolium*; Marshwort, *Apium inundatum* and Milk Parsley, *Peucedanum palustre*. At Bardolph Fen too an area of bog plainly survived as he recorded Sundew, *Drosera rotundifolia*; Lesser Butterfly Orchid, *Platanthera bifolia* and Bladderwort, *Utricularia vulgaris*. Of all his records of species found in wet places about the only one of interest which still survives is the Fringed Waterlily, *Nymphoides peltata* which he noted between Outwell and Nordelph in 1818 and which is still there. Little else of this original flora now remains and the flora of the parishes of the peat fens is the poorest in the county. In Broadland when windpumps came into use in the 19th century, the water levels of the grazing marshes were controlled, but not at that time for arable. Much of this marshland remained until some was put under the plough in the 1970s and 1980s with the help of misplaced grants. Happily that trend has recently stopped and in Broadland many of the plants which also once grew in the Fenland have survived, particularly in the many fine reserves, so it is in places like Upton Fen (NWT), Hickling (NWT), How Hill (Broads Authority) and Strumpshaw (RSPB) that this flora can still be seen.

The Coast

In the past both Fenland and Broadland merged gradually into the coastal marshes and saltings as in fact did the land around all but the cliffed shores, with the sea penetrating far inland along the estuaries and river valleys. This changed gradually over the centuries. Every time a river was embanked, however far inland, it was prevented from spreading its silt over adjacent marshes and pastures, so the amount it carried in its flow to deposit nearer the sea increased. As the rivers continued to silt up, those who owned rights over the salt marshes built new banks to reclaim more land, at first for freshwater marsh and then for grazing. This reclamation was not always straight forward and many floods were recorded, but these set-backs were usually only temporary. By the time the first useful maps of the county appeared, the land and the salt marshes were almost entirely separated by a man- made earthen sea wall. The habitat which lay in the intermediate zone has been almost completely lost and survives only in a few areas alongside tidal rivers where its special plants are still to be seen. Marsh mallow, *Althaea officinalis* appears to belong here as do Slender Hare's-ear, *Bupleurum tenuissimum*, Round-fruited Rush, *Juncus compressus* and Saltmarsh Rush, *Juncus gerardii*. Just inside the sea walls where the waters can still be a little brackish, excellent grazing marshes have developed and these have a flora which includes such species as Grass-leaved Vetchling, *Lathyrus nissolia*, Strawberry Clover, *Trifolium fragiferum*, and Hairy Buttercup, *Ranunculus sardous*. The last fifty years of the 20th century have seen many of these marshes ploughed and almost all the recent intakes from the Wash have moved from saltmarsh to arable in one step, but fine examples are still to be seen at Snettisham and along the north coast as well as on the Halvergate Levels.

Salt Marsh - Burnham Overy

Because Norfolk has such a low lying coast, the relative positions of sea and land even on the open shore have always fluctuated, partly because of a natural cycle of silting and erosion, but also because of drainage, reclamation and sea defence works. The outer barrier is still largely natural made up of shingle banks with their typical flora of Horned Poppy, *Glaucium flavum*, Sea Campion, *Silene uniflora* and occasionally Sea Kale, *Crambe maritima* and of sand dunes ranging from mature dune systems to the moving sand of the outer beaches. The north coast sands are largely of shell sand and consequently limy so the dune flora includes Autumn Gentian, *Gentianella amarella*, Pyramidal Orchid, *Anacamptis pyramidalis* and in wetter parts Marsh Helleborine, *Epipactis palustris* while those of the east coast are acidic and the dunes support Heather, *Calluna vulgaris*, Cross-leaved Heath, *Erica tetralix* and Royal Fern, *Osmunda regalis*. On the north coast a wide band of saltmarsh lies between land and dunes and this has prevented the destruction by development of the botanically interesting juxtaposition of marsh and dune which has happened almost everywhere else around the south-east coast of England. The wide band of Shrubby Seablite, *Suaeda vera* with Sea Heath, *Frankenia laevis* and the Tiny matted Sea lavender, *Limonium bellidifolium* make a community now unique to this coast. Considerable areas of saltmarsh also occur along the southern edge of the Wash and around Breydon Water. All these marshes were once extensively grazed by the small and nimble Norfolk sheep and were managed as part of the manorial grazing. Flocks were driven on to the Burnham marshes until the 1930s. The 17th and 18th centuries saw sea banks being built at Cley and Burnham, first across the mouths of the estuaries to keep the salt water from the land behind and add to the permanent grazing land, later and into the 19th century to reclaim the land between the mainland and the dunes. A number of these reclamations were ploughed up from the 1960s, but most have returned to grassland and those within the Holkham National Nature Reserve are now managed for their wildlife, chiefly for birds, but their special flora is returning.

Introduced Plants

As well as altering the natural vegetation of the county out of all recognition, man has also been responsible for the introduction, both deliberately and accidentally, of a surprising number of species. Knowing which species are truly native and which introductions made in prehistoric times can only be surmised, often by looking at their native habitats on the continent and seeing whether similar sites are or have been present in Britain. An increasing amount of botanical information is becoming available from archeological digs as well as from late glacial sites, but evidence that a plant has been found in a Roman site but not in earlier ones cannot be taken as absolute proof that it was not present before that.

Deliberate Introductions

Very early introductions of plants from mainland Europe will have been of two sorts, those which were brought by early settlers and traders because they were valued as a food or because they supplied a useful raw material and those which arrived accidentally, for example weeds with the seed corn. The ancestors of many of our fruit and vegetables came from western Asia, from the stretch of land which lies between the Caspian and Black Seas, but by the time they reached Britain they were of selected strains and hybrids already far from their native parents in appearance as well as distance. The most common of these to escape into our hedgerows is the domestic apple and it is also considered likely that some willow species, notably the osiers came from the same area and were introduced across Europe to Britain for their basket-making value (Meikle 1984). Pollen from many cornfield weeds has been found in late glacial sites where presumably they throve on open ground, but others came with early farmers and it is probably these, notably Corn Cockle, *Agrostemma githago* and Cornflower, *Centaurea cyanus* which have become virtually extinct as their artificial habitat has been lost because of herbicides and modern farming methods. Other cornfield weeds may have originally moved into the fields from adjoining habitats as perhaps has happened with Corn Caraway, *Petroselinum segetum* and Mousetail, *Myosurus minimus*, both plants recorded in the county in the 19th century as weeds of arable land, but now found in well-grazed grassland. Probably the most critical factor in the disappearance of some arable weeds has been the practice of buying in seed corn. Until such corn, expertly sieved and cleaned, was easily available, farmers gathered their own seed and then sowed it, often together with the weeds which had been harvested at the same time. In this way, local variations in the weed flora became perpetuated and this may still be the explanation of some of the odd distributions to be found today.

With the Romans came the Sweet Chestnut, *Castanea sativa*, probably a disappointment as they rarely fruit as successfully as they do in the warmer summers of Italy, also Alexanders, *Smyrnium olusatrum* which was eaten as a vegetable; Fennel, *Foeniculum vulgare* for use with the fish dishes of which they were apparently very fond; Deadly Nightshade, *Atropa belladonna*, used as a cosmetic; Opium Poppy, *Papaver somniferum* as a pain killer and probably also Ground Elder, *Aegopodium podagraria* for which gardeners will always mutter imprecations. Other plants will have spread widely with the newly made Roman roads and their wide verges giving weed species new opportunities. Increasing numbers of sheep and then rabbits through the Middle Ages may have caused an explosion in the population of species with seeds capable of being carried on their coats. Corn buttercup, *Ranunculus arvensis* and Shepherd's Needle, *Scandix pecten-veneris* may have been moved about the fallow fields in this way. The latter species has recently shown signs of making a come back from near extinction and this may be because of the increasing number of sheep being moved around the county.

As the world was opened up by explorers, so new plants were discovered and brought back to Britain, but most of these were destined for gardens and only a few managed not only to escape from their confines, but to find themselves a niche in the countryside. Feverfew, *Tanacetum parthenium*, Dame's Violet, *Hesperis matronalis* and Honesty *Lunaria annua* are all plants which are widespread outside gardens but only rarely manage to establish themselves, relying upon a constant supply of seeds to maintain their hold. Others such as Slender Speedwell, *Veronica filiformis*, in spite of not setting seeds in this country, has now moved from the rock garden to well-mown grassland, but happily not yet to really wild places while Ivy-leaved Toadflax, *Cymbalaria muralis* and Wallflower, *Cheiranthus cheiri* although abundant throughout the county are almost always in man-made habitats. Plants which are both useful and ornamental have perhaps had more success and Soapwort, *Saponaria officinalis* is not infrequent by roadsides as are a number of mints, notably the Hybrid Apple Mint, *Mentha x villosa*. Most conspicuous of all these introductions, however, are the Giant Hogweed, *Heracleum mantegazzianum* which maintains itself by seed and Japanese and Giant Knotweeds, *Fallopia japonica* and *F. sachalinensis* which have spread far since their introduction a hundred years ago and are particularly harmful to the native vegetation in wet habitats.

Wet areas have always been most susceptible to introduced species. Attempts to 'beautify' ponds often bring in plants such as Yellow Flag, *Iris pseudacorus* and White Waterlilies, *Nymphaea alba* but also unwisely Reed-mace (Bulrush), *Typha latifolia* which will soon choke a small pond and the variegated form of Reed Canary Grass sometimes called Gardeners' Garters, *Phalaris arundinacea* 'Picta'. Monkeyflower, *Mimulus guttatus* from North America and the Balsams. Indian Balsam, *Impatiens glandulifera* from the Himalaya and Orange Balsam, *Impatiens capensis* from North America are now abundant waterside plants having found their way into our river systems. More dangerous, however, are the true water plants which have been grown in garden ponds or aquaria and then been thrown out into ponds and streams. These are plants which are happy in the nitrogen-rich waters which are found so widely today and they spread rapidly at the expense of the native species. One of the first was Canadian Waterweed, *Elodea canadensis*, noted in Britain as an escape in 1842 and more recently its sister species *Elodea nuttallii* which was first noted in the country in 1966 and in Norfolk in 1978. Both are now abundant. The smallest of the Duck-weeds, *Lemna minuta* is another very recent arrival which is spreading very fast, especially in the Broads while the New Zealand Pigmyweed, *Crassula helmsii* did not arrive until 1987 and is only just beginning to spread. It is already proving a problem elsewhere.

Accidental Introductions, the Weeds

Of the herbaceous plants, it is the true weeds which have been the most successful and many of these have been helped on their way, first in the 19th century by the expansion of railways which opened cuttings and tracks to passing seeds and then helped them on their way with the draught of passing trains. The most famous example of plants which have mastered this habitat must be Oxford Ragwort *Senecio squalidus* which escaped from the Oxford Botanic Garden in 1794 and spread at first slowly, then rapidly along the new tracks soon appearing in all the towns served by the rail network. It is still scarce in some rural areas. Less conspicuous but equally successful has been the American Willowherb, *Epilobium ciliatum* first seen in Britain in 1891 and in Norfolk in 1952, and now the commonest of the small willowherbs right across the county. The 20th century has seen the creation of many airfields with the opening up of gravel pits for concrete runways, and the enormous expansion of roads. New bypasses and trunk roadways have been built (motorways have not reached Norfolk but the effect of those in adjacent counties has been felt) and there has also been a vast increase in the speed of the vehicles which travel along them, so plants able to take advantage of these factors have also moved with much greater speed. Prickly Lettuce, *Lactuca serriola* and Canadian Fleabane, *Conyza canadensis* have extended their ranges enormously in the last twenty years, but the most conspicuous plants to have spread along our roads are those which are tolerant of the high mineral content of the strips alongside the tarmac. The spread of reflexed Salt-marsh Grass, *Puccinellia distans,* Early Scurvy-grass, *Cochlearia danica* and such coastal species as the two Oraches, *Atriplex littoralis* and *A. prostrata* have been observed in many counties and Norfolk is no exception. The grass was first noted away from the coast in 1978 and was joined by early scurvy-grass in the 1980s. These have remained chiefly on trunk roads, but the oraches are now to be found fringing many other roads, usually together with couch grass, *Elytrigia repens.* It has been pointed out (Coombe 1994) that these plants have also been recorded in the past around manure heaps and it may be that they are tolerant of soils with a high mineral content rather than just salt. As the spread of these and other coastal plants across Norfolk has almost coincided with this survey, it has been possible to plot their advance and it is certain that they have moved in from outside the county, along the A17 from Lincolnshire, the A10 and A45 from Cambridgeshire and the A11 and A12 from Suffolk, not from the Norfolk coast.

Introduced shrubs and trees have been widely planted, mostly from the 18th century onwards, though the Sycamore, *Acer pseudoplatanus* was here earlier as Gerard in 1597 described it as a scarce introduction. With the planting of parks and plantations, for ornament, timber and game cover, a much greater range of trees and shrubs was added to the small woody flora of the county. The status of some of these plants is not always obvious and many self-sow. Box, *Buxus sempervirens* and Yew; *Taxus baccata* were probably once native, but have been so much planted that we no longer know if there are any survivors of the original stock. There is no doubt, however, about Rhododendron, *R. ponticum* and Oregon Grape, *Mahonia aquifolium* both of which can be found maintaining themselves in abundance where conditions are favourable. Note has been made under individual species entries of many other woody introductions.

The status of grasses has always presented a problem when they occur as weeds in arable land. Most arrive accidentally but others are sometimes sown in strips on field edges for game feeding as are the Millets, *Panicum* spp. These, along with odd plants of Cockspur Grasses, *Echinochloa crus-gallii* and *E. colona* and Bristle-grasses, *Setaria* spp. are sometimes found on field edges and by coverts, but these rarely survive long. Bulbous Canary-grass, *Phalaris aquatica,* a perennial which probably derived from this source has been more successful and is being found increasingly as a weed of corn fields. It was first recorded in this habitat in the county in1991. Hungarian brome, *Bromus inermis,* another relative newcomer was first noted in 1958 and is proving almost ineradicable on a few roadside verges while *Ceratochloa carinata,* an escape from Kew in 1919, first recorded here in 1986 is showing signs of spreading rapidly. *Bromus secalinus,* once known as a pest of wheat fields, appears to be making a come-back in East Norfolk.

One very much man-made habitat, of great interest until the early 1970s, was the refuse tip. In the days when rubbish was simply spread and flattened, seeds from a wide variety of sources from bird cages to garden rubbish germinated. Each autumn the resulting plants were eagerly sought out by botanists who gathered and identified them, hence the range of unusual plants listed for Harford Tip, the main refuse site for Norwich. With modern hygiene, all such rubbish is now bagged in plastic and deeply buried so the autumn search for exotics is no more.

It is the unpredictability of introduced species which makes the recent practice of sowing 'native seed' mixes on roadside verges and in fields all the more dubious. Certainly they can produce colourful displays, at least for a few years, but among the seeds there are far too often both species and strains of native species which in fact come from mainland Europe. The problem is largely one of seed collection. Gathering large quantities of native seed from wild sites is never easy, so most come from fields where plants have been treated as crops, but to sow them in the first place seed has to be bought in and it is the original source this seed which can be doubtful. Examples which have been noted in the county in recent years are a larger-flowered European variety of Bird's-foot Trefoil, *Lotus corniculatus* var. *sativus* found on some new road verges and cultivated strains of Red Fescue *Festuca rubra* or another Fescue, *F. brevipila* which comes from Eastern Europe. These have the potential of being not only the weeds of the future, but may hybridise with and eventually dominate our own native stock.

The Future

Over the centuries there have been enormous losses from our native flora and the worry that these losses will increase has a scientific base in the principle of island biogeography. This has arisen from the study of native communities on islands and the realisation that the smaller the island, the more likely extinctions become. It is a sad fact that our wild places, be they nature reserves or not, are increasingly becoming islands themselves, lying in a sea of cultivation and urbanisation and unless the land between is managed more sensitively a great deal more will be lost. Perhaps we should count ourselves lucky that so much has survived, but there is no doubt that we all have a duty to the authors of the next Norfolk Flora as well as to all the inhabitants of this county that when they start recording they do not have many more losses to report.

Habitats in Norfolk

Peter Lambley

This section provides a brief account of the main present-day plant communities of Norfolk using the National Vegetation Classification (NVC) as set out in British Plant Communities (Rodwell 1991 *et seq.*) where possible. It is confined to semi-natural vegetation (that composed largely of native species) though modified from their original state by man's activities and to the small number of types (mostly coastal) which can still be regarded as natural.

WOODLANDS AND SCRUB

Norfolk has a low proportion of woodland cover compared with many English counties and a large percentage of this is in the form of the recently planted conifer plantations in Breckland. Never-the-less there is a surprising diversity of woodland in the county although ancient woodland, which has not been replanted, occupies only a small area. Rackham (1986), states that of 64 types of ancient woodland so far recognised 34 are known in Norfolk. This diversity is demonstrated at Great and Little Woods, Swanton Novers where soils as different as boulder clay and sands and gravels occur in close proximity to one another and support a range of woodland types including ash-maple-hazel, limewood, plateau and valley alderwood (in bird cherry variants) oakwood and several intermediates.

The main areas of ancient woodland are in south-east Norfolk with a band extending north through the middle of the county toward the north coast. In addition to the ancient woodlands, the cessation of traditional practices has resulted in the development of extensive areas of secondary woodland on heathland and in Broadland since the beginning of the century. Hornbeam woods are largely confined to south-east Norfolk with sessile oak woods on the Cromer-Holt ridge. Broadland has extensive areas of wet woodlands or carrs, whilst other examples occur elsewhere. Plantations, especially those which are broadleaved, on ancient woodland sites may still retain many of the plants associated with these woods. For example poplar plantations at Honingham support a good population of *Paris quadrifolia*.

Woodland Community - Blickling

Oak woodland

This woodland (W10), is developed particularly on sands and gravels in north of the county. Bracken and brambles are the main associates with either *Quercus petraea* or *Q. robur* as the dominant tree. These woodlands are sometimes in the form of oak-pasture woodlands with large pollards as in the Hull woods near Glandford, whilst in places such as the Pond Hills near Edgefield associated species include *Sorbus aucuparia* and *Ilex aquifolium*. The ground flora is often dominated by bluebells, bracken and in places *Convallaria*. In Swanton Novers Great Wood the *Maianthemum bifolium* is associated with this community. *Quercus petraea* occurs in coppiced stands reminiscent of western oak woods with little ground cover except the moss *Leucobryum glaucum* at Swanton Novers Great Wood and Edgefield Little Wood

Beech woodland

Fragments of beech woodland (W14) occur on the gravels of the Cromer-Holt ridge at Felbrigg and near Northrepps They are probably remnants of old wood pasture and are possibly the furthest north locality for native beech in Britain. (Rackham, 1986). The ground flora is largely absent except for mosses, though with brambles in places.

Mixed broadleaved woodland

This community (W8) encompasses most of the woods which are developed on the boulder clays including ash-oak- maple woods particularly in the middle and south-east of the county. Within this overall grouping a number of other distinct types can be recognised most notably limewoods and oak-hornbeam woods. Lime woods with *Tilia cordata* are generally developed on the better soils and include Hockering Wood, the third largest lime wood in the country. These woods generally have few shrubs as hazel in particular does not flourish under the deep shade of lime. Hornbeam woodlands are largely restricted to south-east Norfolk and probably reach their natural limit in the British Isles just south of Norwich.

The ground flora of these ancient woods is often very rich with species such as *Mercurialis perennis, Primula vulgaris, Hyacinthoides non-scripta, Anemone nemorosa, Allium ursinum* in the wetter parts. Rarer plants such as *Gagea lutea, Melampyrum cristatum, Paris quadrifolia, Platanthera chlorantha* and *Campanula trachelium* are associated with this community. The shrub layer includes stands of *Prunus padus* particularly in wetter more calcareous flushes and rarely *Viburnum lantana* and *Sorbus torminalis.*

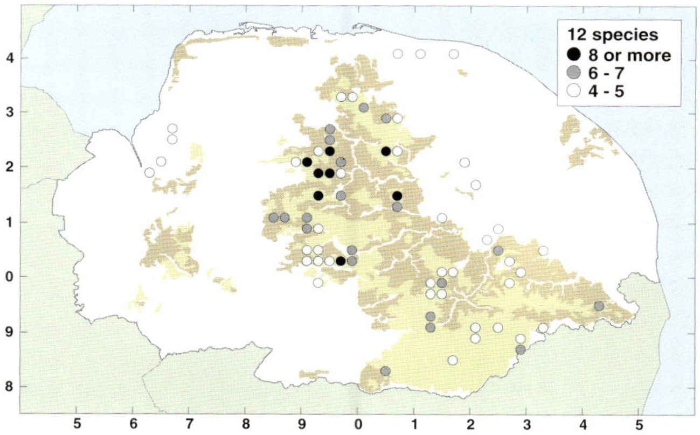

Map showing sites with 4 or more of the 12 species typical of the Deciduous Woodland community (W8 Fraxinus/Acer Woodland) listed below. Rodwell 1991

Anemone nemorosa, Campanula latifolia, Campanula trachelium, Carex sylvatica, Convallaria majalis, Lamiastrum galeobdolon, Luzula sylvatica, Melica uniflora, Milium effusum, Orchis mascula, Platanthera chlorantha, Ranunculus auricomus

Oak-Birch woodland

These are largely secondary and are generally developed on heathland which is no longer managed. The ground flora is generally dominated by *Pteridium* and *Rubus*.

Alder woodlands

Plateau alder woodland is a rare stand type found in some Norfolk woodlands such as Swanton Novers. Elsewhere alder forms extensive stands of largely recent origin particularly in the flood plains of Broadland. These belong to the *Alnus glutinosa-Carex paniculata* (W5) community of which the *Lysimachia vulgaris* sub-community (W5 b) is largely confined to Broadland in Britain. Woodlands dominated by *Alnus glutinosa* with *Urtica dioica* abundant in the herb layer (W6) are widespread, whilst the *Alnus glutinosa-Fraxinus excelsior-Lysimachia nemorum* (W7) community in Norfolk is confined to a few limited stands in Broadland.

Willow carr

Most of the fen carrs with sallow come under the *Salix cinerea-Betula pubescens-Phragmites australis* (W2) community. The distinct sub-community (W2b) is the main habitat for *Dryopteris cristata* and is characterised by the development of cushions of *Sphagnum fimbriatum, S. palustre, S. recurvum* and other *Sphagnum sp., Drosera rotundifolia, Pyrola rotundifolia* and a range of other ferns including *Osmunda regalis, Thelypteris palustris, Dryopteris dilatata* and *D. carthusiana* are frequent associates.

Scrub

A number of scrub communities are widespread including *Crataegus,* this like the oak-birch woodland is generally developed on unmanaged commons and other grasslands as at Whitwell Common and Leziate Fen. Unmanaged Broadland fens and those on greensand often develop a dense, impenetrable *Myrica* scrub. This community is described by Wheeler (1978).

GRASSLAND

Unimproved or semi-improved grasslands are rare in Norfolk and largely confined to the following areas: the Stanford Practical Training Area, the main river valleys, north and west Norfolk, grazing marshes on the north Norfolk coast and in Broadland with a few scattered sites in mid and south Norfolk.

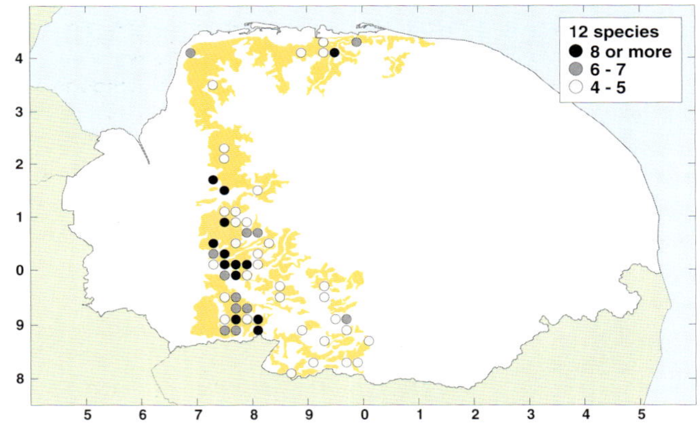

Map showing sites with 4 or more of the 12 species typical of the Chalk Grassland community (CG2 Festuca/Avenula Grassland) listed below. Rodwell 1992.

Asperula cynanchica, Astragalus danicus, Avenula pratense, Campanula glomerata, Carex ericetorum, Cirsium acaule, Filipendula vulgare, Gentianella amarella, Helianthemum nummularium, Hippocrepis comosa, Sanguisorba minor, Scabiosa columbaria

Chalk grassland

Despite the fact that chalk underlies about 60% of the county there is very little typical chalk grassland. This is because it is covered generally by drift deposits of varying depths and slopes are generally gentle allowing easy cultivation. The main sites which survive are in north-west Norfolk and Breckland where natural features such as steep slopes at Ringstead and Cockthorpe Common or patterned ground in Breckland support chalk grassland. Alternatively man-made features, both ancient like the Bitcham Ditch near Marham and Warham Camp, or more recent such as the old railway line at Narborough and quarries at Alderford and Wells also support calcareous grasslands. Elsewhere a calcareous variant of mesotrophic grassland occurs on. chalky boulder clay particularly in south-east Norfolk. For a fuller description see Moore (1997).

The main chalk grassland communities are a *Festuca ovina - Avenula pratensis* (CG2) community with species such as *Scabiosa columbaria, Briza media, Carex flacca, Anacamptis pyramidalis, Asperula cynanchica, Hippocrepis comosa* and *Linum catharticum* which occurs at Warham Camp, East Walton Common and other sites. In Breckland the *Festuca ovina-Hieracium pilosella-Thymus pulegioides* (CG7) community is characteristic and is better developed here than elsewhere in the country as on parts of the Stanford Training Area and Weeting Heath. The *Helictotrichon pubescens* (CG6) community is rare though present at a number of sites including the Stanford Training Area.

Mesotrophic grassland

Old unimproved meadows are rare in Norfolk and have usually survived through accidents of ownership. They tend to occur on the heavier clays of south-east Norfolk as at Shelfhanger (a lammas meadow), New Buckenham Common and some of the better roadside verges and churchyards. Elsewhere a few scattered meadows survive as eg the Brinks at Northwold and Rosie Curston's meadow at Mattishall. The *Cynosurus cristatus-Centaurea nigra* (MG5) community is a major component of these meadows and is usually species rich with species such as *Centaurea nigra, Cynosurus cristatus, Ophioglossum vulgatum, Silaum silaus, Trisetum flavescens, Leucanthemum vulgare* and *Primula veris. Orchis morio* is a feature of this community. The *Cynosurus cristatus-Centaurea nigra, Galium verum* (MG5b) sub-community occurs in places on chalky loams on boulder clay and explains the presence of species like *Anacamptis pyramidalis* and *Blackstonia* away from the main chalk areas. The sandy, slightly calcareous soils of the Breckland support a second type of mesotrophic grassland, the *Festuca ovina-Galium verum* (SD8) fixed dune grassland which is well developed in the Stanford Training Area. It is characterised by abundant *Bromus hordeaceus.* Some Breckland specialities such as *Artemisia campestris* appear to be associated with these communities, as they are in dune areas in southern Scandinavia. It is of interest that the normally coastal *Lathyrus nissolia* occurs in this community at the Stanford Training area.

Acid grassland

Acid grassland is mostly present in the Breckland where variants of the *Festuca ovina-Agrostis capillaris-Rumex acetosella* (U1) are quite widespread, especially in the Stanford Training Area. Some of these including the *Coelocaulon aculeatum-Cladonia sub-community* (U1a) and the *Erodium cicutarium-Teesdalia nudicaulis* (U1c) sub-community are largely or entirely confined to the Breckland in Britain. These grasslands grade into the CG7 chalk grasslands and are sometimes difficult to distinguish from them.

HEATHLAND

Heathland in this context is defined as vegetation dominated by *Calluna, Erica* species or *Ulex* species. In Norfolk it is developed on the sandier parts of the greensand series, and on glacial gravels and sands especially on the Cromer- Holt ridge, and between Holt and Norwich. Areas of *Calluna* heath also occur amongst the grass heaths of the Breckland. Elsewhere *Calluna* heath may once have been widespread on the sides of the Broadland valleys but only fragments remain as at Barton Broad.

Heathland - Marsham

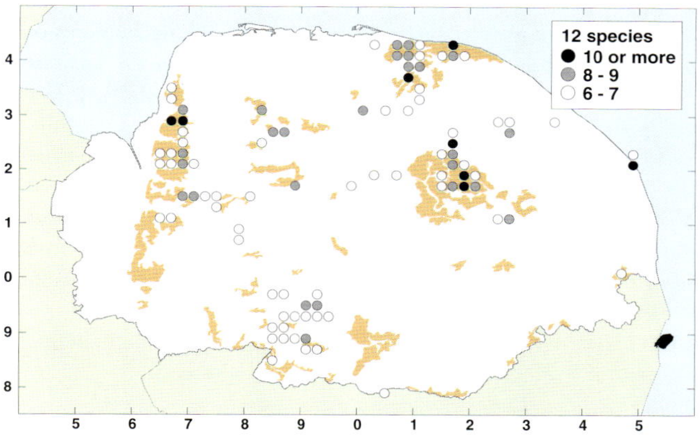

Map showing sites with 6 or more of the 12 species typical of the Heathland community (H8 Calluna/Ulex gallii Heath) listed below. Rodwell 1991.

Calluna vulgaris, Carex binervis, Carex pilulifera, Deschampsia flexuosa, Erica cinerea, Erica tetralix, Galium saxatile, Molinia caerulea, Nardus stricta, Potentilla erecta, Teucrium scorodonia, Ulex gallii

The north-east Norfolk heaths are an outlier of a community largely dominated by *Calluna vulgaris* and *Ulex gallii* (H8), which is far more widespread in western Britain. This provides a wonderful display of purples and yellow in late summer, the main flowering time of *Ulex gallii*. On the greensand ridge and in parts of Breckland the dry heath is dominated by the *Calluna vulgaris-Festuca ovina* (H1) community, this is probably a reflection of the dry climate in Breckland but it is less easy to explain the difference between the greensand heaths and those of north-east Norfolk. A possible explanation is that some of the north-eastern heaths eg Buxton have a deposit of clays below the sands and gravels and this may result in a less well drained more humid heath. In many places lack of management in recent years has led to a succession to birch and oak woodland. These drier heaths often grade down into wet heath which is characterised by the M16 *Erica tetralix-Sphagnum compactum* community. This is usually developed along seepage lines and is the habitat for *Gentiana pneumonanthe, Drosera rotundifolia* and formerly *Lycopodiella inundata*, now probably extinct. In places the wet heath grades into valley mires as at Buxton Heath, Holt Lowes, Sheringham and Beeston Common, Roydon and Dersingham.

WETLANDS

MIRES AND SWAMPS

Although rainfall totals are low in the county it has a wealth and diversity of wetlands which are probably unmatched by any other county in lowland Britain. This is a consequence of topography and geology with chalk overlain by a complex series of both acid and calcareous drift deposits. There are thus many NVC communities present. They broadly group into flood plain wetlands which are primarily irrigated by a river and valley mires which occupy slopes as well as the valley floor and are maintained primarily by springs and seepages. The main floodplain wetlands are concentrated in Broadland. Valley mires are scattered through the rest of the county, but especially near the headwaters of the main rivers and on the greensand. An unusual topographical variant of these mires are those developed in the ground ice depressions (pingos) in or near the Breckland as at Thompson and East Walton and Adcocks Commons.

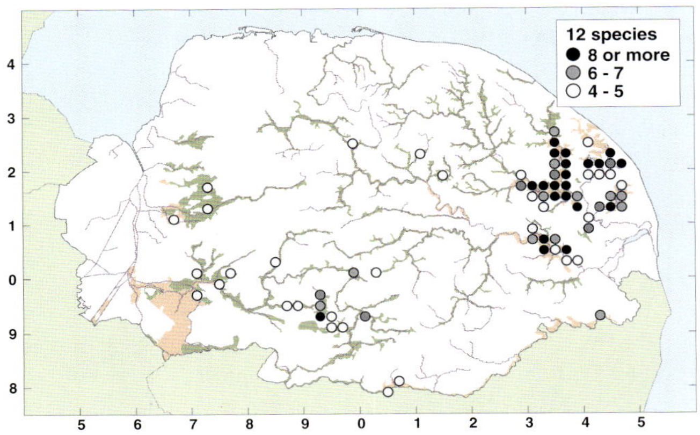

Map showing sites with 4 or more of the 12 species typical of the Swamp Communitiy (S24 Phragmites/Peucedanum Tall-Herb Fen) listed below. Rodwell 1995.

Calamagrostis canescens, Carex appropinquata, Carex elata, Carex pseudocyperus, Cladium mariscus, Cicuta virosa, Peucedanum palustre, Ranunculus lingua, Scutellaria galericulata, Sium latifolium, Thelypteris palustris, Typha angustifolia

Broadland Floodplain Fens and swamps

The herbaceous fens of Broadland form the largest expanse of species- rich fen in lowland Britain. and have been described by Wheeler (1978) and recently surveyed by Parmenter (1995). Particularly noteworthy is the *Phragmites australis-Peucedanum palustre* (S24) community and sub-communities which are almost wholly confined to the Broads. This covers over 550 hectares, 20% of the total fen area. A number of rare species are associated with this community including *Peucedanum palustre, Cicuta virosa, Sium latifolium* and *Ranunculus lingua*. The *Phragmites australis-Eupatorium cannabinum* (S25) community is widely distributed in the Broads. It is commonly found on freshwater ronds which are influenced by winter flooding. In the Yare valley the *Glyceria maxima* swamp community (S5) is present, though apparently less extensive than formerly. The *Cladium mariscus* (S2) community and sub-communities are restricted in distribution mainly to the Ant and Thurne valleys. There are also small areas of communities predominantly found in the north and west of Britain - the *Carex rostrata-Potentilla palustris* (S27) and the *Carex rostrata-Sphagnum squarrosum* (M5) community. The *Carex rostrata-Calliergon cuspidatum* mire (M9) is important for a number of rare species including *Liparis loeselii*. Its dominant species is usually *Phragmites australis* or *Cladium mariscus*, but *Carex diandra* and *C. lasiocarpa* are also constant and in the case of *C. lasiocarpa* sometimes dominant. Associated species include *Juncus subnodulosus, Peucedanum palustre, Carex elata, Mentha aquatica, Epipactis palustris* and *Hydrocotyle vulgaris*.

In addition fen meadow communities occur which include small areas of the *Schoenus nigricans-Juncus subnodulosus* (M13), *Juncus subnodulosus-Cirsium palustre* (M22) covering 110 hectares and *Molinia caerulea-Cirsium dissectum* (M24) covering 34 hectares. These communities are also found on the peaty soils of the drained marshes.

Valley Mires

The valley mires in Norfolk are only matched in their number and interest by those of Anglesey and the Lleyn peninsula in North Wales. They can be sub-divided broadly into three groups: calcareous valley mires, mixed valley mires and acid valley mires though it is probably best to consider them as a continuum from acid to calcareous depending on the nature and source of the ground water feeding the site. They generally have a range of vegetation types with some zonation from the wettest to the driest areas. This is reflected in the vegetation which often has areas of tall herb or swamp vegetation, mire and fen meadow. On acid sites also wet heath grading to a dry heath. The lack of management on many sites this century has resulted in the development of *Salix* carr often along the central stream. They have been surveyed in detail by Smart (1992-3).

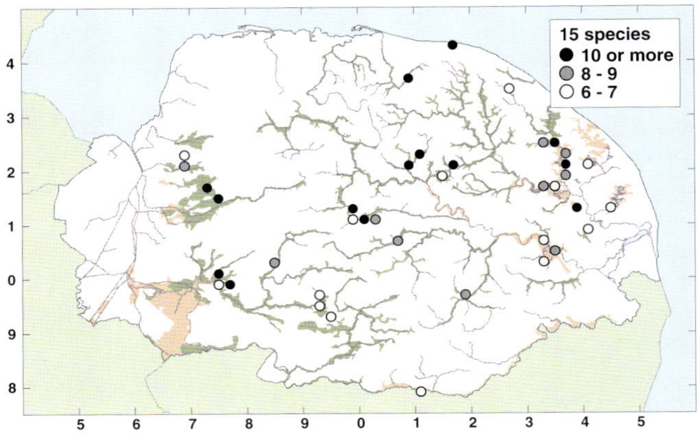

Map showing sites with 6 or more of 15 species typical of the Valley Mire communities (M9 & M13 Carex rostrata/Calliergon cuspidatum and Schoenus nigricans Mires) listed below. Rodwell 1991.

Carex diandra, Carex dioica, Carex rostrata, Carex viridula ssp. **brachyrhyncha, Dactylorhiza traunsteineri, Epipactis palustris, Galium palustre, Gymnadenia conopsea, Hypericum elodes, Juncus subnodulosus, Menyanthes trifoliata, Parnassia palustris, Pinguicula vulgaris, Potentilla palustris, Schoenus nigricans**

The complex nature of the drift geology over the chalk means that water derived from several sources can feed a site. This gives rise to mixed mires which exhibit a perplexing range of calciphilous and acidiphilous communities adjacent to one another with some of the resultant communities not fitting readily into the NVC as at Holt Lowes and Sheringham and Beeston Commons. The communities include mire, fen meadow and tall herbaceous fen including reed swamp. The most extensive of the mire communities is probably the *Schoenus nigricans-Juncus subnodulosus* (M13) community including its variants M13b and c which is of major importance for its floristic richness with species such as *Epipactis palustris, Parnassia palustris, Briza media, Carex dioica, Pinguicula vulgaris, Carex viridula* sub.sp. *brachyrrhyncha, Gymnadenia conopsea* ssp. *densiflora* and *Dactylorhiza traunsteinerii*. This is well developed for example at Buxton Heath and Badley Moor. Other mire communities include the *Carex rostrata-Calliergon cuspidatum* mire, *Carex diandra-Calliergon giganteum* sub-community (M9b) which is confined to wet ground with species such as *Menyanthes trifoliata, Potentilla palustris, Galium palustre* and *Carex diandra*. Some communities are rare in Norfolk for example the M29 which supports *Hypericum elodes* when it occurs as at Swannington Upgate Common.

The *Juncus subnodulosus-Cirsium palustre* (M22) fen meadow community is characteristic of the drier parts of calcareous flushes. The *Molinia caerulea-Cirsium dissectum* fen meadow (M24) and its variants b and c is developed in slightly more acidic situations with species like *Sphagnum* and *Drosera rotundifolia* present. At Southrepps Common *Gymnadenia conopsea* ssp. *densiflora*, *Triglochin palustris* and *Parnassia* occur in this community. Tall herb fen communities often associated with a lack of grazing include the *Filipendula ulmaria-Angelica sylvestris* mire and the *Urtica dioica-Vicia cracca* sub-community (M27).

Acid mires

The valley mires developed on the greensand are more acid than elsewhere, but even here as at Roydon Common patches of chalky boulder clay influence the vegetation. The more acid mires have a series of communities which are well developed at Roydon Common and Dersingham with the *Sphagnum auriculatum* (M1) and the *Sphagnum cuspidatum/recurvum* (M2) bog pool communities present in the wettest areas. with species such as *Eriophorum angustifolium* and *Drosera rotundifolia* present. The *Schoenus nigricans-Narthecium ossifragrum* (M14) mire is a rare community which occurs where there are areas of flushed calcareous groundwater. Species associated with it include a rich bryophyte layer with *Aneura pinguis, Sphagnum subnitens,* and *Campylium stellatum.* Other species include *Carex lepidocarpa, Potamogeton polygonifolius, Menyanthes trifoliata* and *Drosera anglica.* The *Narthecium ossifragrum-Sphagnum papillosum* (M21) mire community is well developed at these two sites with *Vaccinium oxycoccus, Eriophorum angustifolium, Drosera rotundifolia* and *Sphagnum recurvum.* Two rarities are associated with this community: *Hammarbya paludosa* and *Rhynchospora alba.*

Pingo systems

These are developed at a number of sites in or around Breckland as at Thompson Common, East Walton, Adcocks Common and Foulden Common. They are characterised by mosaics of tall herb fen, swamp fen, fen meadow, calcareous mire developed in the pingos and seepage zones. There are often stands of *Cladium* swamp (S2) and *Phragmites* swamp (S4). The calcareous flushes have *Schoenus nigricans-Juncus subnodulosus* (M13) communities and the fen meadow communities: *Juncus subnodulosus-Cirsium palustre* (M22) and the *Molinia caerulea-Cirsium dissectum* fen meadow (M24). Smart has observed that there is a strong similarity between the M22 community as present at East Walton and the dune slack community SD17.

OPEN WATERS

Open water communities occur in the Broads, (though affected by pollution) and dyke systems of the fens and grazing marshes. The rivers which range from calcareous fast flowing chalk streams to those which are slow flowing in the Fens are also surprisingly diverse.

Dyke systems in the Broadland fens and some of the grazing marshes support *Spirodela polyrhiza-Hydrocharis morsus-ranae* (A3), *Hydrocharis morsus-ranae-Stratiotes aloides* (A4) and the *Potamogeton natans* (A9) communities. Rare species associated with these communities include *Potamogeton acutifolius* and *Stratiotes aloides.* The dyke systems were recently re-surveyed by Harris et. al. (1997). Brackish dykes in coastal grazing marshes have a community with Ruppia species.

The best surviving aquatic communities in the Broads occur in the Thurne catchment especially Hickling Broad where a charophyte dominated community has re-established with particularly *Chara intermedia* in abundance together with *Najas marina.* Other species characteristic of the less eutrophicated Broads include *Myriophyllum spicatum, Potamogeton pectinatus, Ceratophyllum demersum* and *Hippuris vulgaris.* Less common species include *Nuphar lutea, Nymphaea alba, Potamogeton crispus, P. friesii, P. pusillus* and *Zannichellia palustris.* Elsewhere, Perch Lake at Westwick has a single species community of *Littorella.*

The Breckland meres provide an unusual habitat of fluctuating water bodies which can be dry for several years. They generally exhibit a clear zonation and are often dominated by pure stands of *Myosoton aquaticum* in dry times. The grasses *Alopecurus aequalis* and *A. geniculatus* often dominate large areas of the beds when dry as at Fowlmere. *Rumex maritimus* is a feature of the margins of the some of the meres along with the rare moss *Physcomitrium eurystomum.* The fast flowing middle reaches of the Nar, Tud and upper reaches of the Wensum support communities with *Ranunculus pencillatus* ssp. *pseudofluitans* (A17) and on the Nar also *Hippuris vulgaris.* On the middle reaches of the Wensum the faster flowing sections below mills are dominated by communities in which *Potamogeton perfoliatus* and *P. pectinatus* (A11) are important, the slower reaches above mills have *Sagittaria sagittifolia* and *Nuphar* (A8) dominated communities.

COASTAL

SAND DUNES

Norfolk has about 1100 hectares of dune system located in two main areas between Holme and Blakeney Point on the north coast and on the east coast from about Eccles to Great Yarmouth. The dunes range from being slightly calcareous at Holme to very acid at Winterton where there are extensive areas of dune heath. A feature of the dunes on the north coast is that they include low dunes. They have been surveyed in detail by Radley (1994).

Drift line and foredunes

Well developed drift line communities occur along many parts of the north Norfolk coast with *Honkenya peploides* and *Cakile maritima* forming a characteristic community (SD2) together with other annuals such as *Salsola kali* and *Atriplex laciniata.* The dynamic foredunes are dominated by *Elytrigia juncea* (SD4) and *Leymus arenarius* (SD5), the latter which is an open often quite floristically rich community with elements from the strandline like *Cakile maritima* and others such as *Eryngium maritimum.* A community dominated by *Ammophila arenaria* (SD6) is characteristic of the higher mobile dunes.

Semi-fixed dunes

The more fixed dunes are dominated by a number of communities including the *Ammophila arenaria-Festuca rubra* (SD7). This is well developed at the western end of the Holkham dunes where it occurs in rather species poor variant, the *Hypnum cupressiforme* sub-community which is rich in winter and spring annuals like *Myosotis ramosissima* and *Erophila verna*; and the *Elymus pycnanthus* sub-community which is dominated by this grass.

Dune grasslands

The *Festuca rubra-Galium verum* community (SD8) occurs in small areas of a number of dunes including the western part of the Holkham Dunes with characteristic species such as *Galium verum, Pilosella officinarum, Crepis capillaris* and *Polypodium vulgare.* More widespread are areas of short dry lichen-dominated *Carex arenaria-Coelocaulon aculeatum* (SD11) dune which is particularly characteristic of areas with sand over shingle as at Blakeney Point. The rare grass *Corynephorus canescens* is frequent in this latter community. At Hunstanton and Holme dunes support dense thickets of *Hippophae* scrub (SD18) whilst areas of *Ligustrum vulgare* scrub occur more generally. At Holkham a large proportion of the dunes were planted in the 18th and 19th centuries and now support naturally regenerating *Pinus nigra* ssp. *maritima* and *Quercus ilex* woodland in which *Monotropa hypopitys* and *Goodyera repens* occur.

Dune slacks

Slacks are relatively rare and limited in extent compared with dunes on the west coast but do support the *Prunella vulgaris* sub-community of the *Festuca rubra-Galium verum* dune grassland and also the *Salix repens-Holcus lanatus* dune slack community (SD16b). These hold a number of interesting species including *Salix repens* (rather rare), *Gnaphalium luteoalbum, Botrychium lunaria, Dryopteris cristata, Pyrola rotundifolia, Aceras anthropophorum, Dactylorhiza incarnata* var. *coccinea, Epipactis palustris* and the rare liverwort *Pelatophyllum ralfsii.*

Dune heath

Winterton supports the largest area of this habitat on the east coast. The community (H1d) is characterised by dominated by *Calluna vulgaris* together with *Festuca ovina* and *Carex arenaria* at times frequent or co-dominant. There are few vascular plant associates, but lichens (*Cladonia* species) and bryophytes can be frequent.

SHINGLE

Extensive areas of shingle occur between Weybourne and Blakeney Point with small but not insignificant areas at Scolt Head Island and Snettisham. It also often underlies dunes elsewhere on the north Norfolk coast and is exposed in lows and on the fringes of the dunes. Where it is disturbed as between Cley and Weybourne it is either unvegetated or supports a very open community with little other than *Glaucium flavum*. On the wider beach along the spit a very open pioneer community occurs (SH6a) with *Silene uniflora* and *Glaucium flavum* with few associates other than occasional *Rumex crispus* and *Senecio jacobaea*. Where there is more sand a community of *Honkenya peploides* and *Silene uniflora* is developed (SH26) with associated species such as *Sedum acre, Phleum arenarium, Ammophila arenaria* and *Glaucium flavum*. More stable areas with a higher sand content support an *Ammophila arenaria-Ceratodon purpureus* assemblage (SH52) with main associates including *Carex arenaria, Sedum acre* and *Phleum arenarium*. An important community (SM 25) dominated by *Suaeda vera* occurs along the saltmarsh interface and is largely restricted to Norfolk. Associated species include *Silene uniflora, Rumex crispus* and *Elymus pycnanthus*. On sheltered lateral spits a stable maritime grassland community (SH33) develops with *Plantago coronopus, Armeria maritima* and *Festuca rubra*. *Limonium vulgare, Limonium binervosum* and *Silene uniflora* are often associates. Lichens (*Cladonia* species) are often important. In very stable areas this develops into a lichen (*Cladonia*) dominated community (SH51) with *Festuca rubra* and *Cochlearia danica* together with associates including *Silene uniflora, Armeria maritima, Crepis capillaris* and *Sedum acre*. At Snettisham 12 communities have been identified including extensive areas of an *Arrhenatherum elatius-Silene uniflora* grassland (SH37 and a grassland (SH43) with *Dicranum scoparium, Festuca rubra* and *Plantago lanceolata*, this grades where protected from trampling into a particularly rich (over 20 species per quadrat) community with *Dicranum scoparium, Rumex acetosella* and *Aira praecox*. For further details see Sneddon & Randell, (1994).

SALTMARSH

Norfolk has about 2800 hectares of salt marsh the sixth largest overall total for this habitat in Britain (Burd, 1987). This is concentrated in three main areas - the Wash, the north Norfolk coast and a small area within Breydon Water. Those in north Norfolk in particular have significant areas of nationally scarce communities including the *Limonium-Armeria* which occupies 16%, 94 hectares of *Suaeda vera* (3.3%) and small but frequent stands of the rare *Frankenia laevis-Limonium bellidifolium* community.

Those on the Wash are mostly developed on muddy substrates and were formerly grazed, this management is now being revived. There are some pioneer communities mainly *Spartina* and *Suaeda-Salicornia* with *Aster*. However, they are predominately low-mid saltmarsh communities with *Puccinellia* the predominant community because of recent land reclamations

The north Norfolk marshes are on sandier substrates and whilst historically they were grazed this has not happened in recent times. The lowest end of the salt marsh is occasional stands of *Spartina* (SM6), further upshore communities dominated by the annuals *Salicornia* (SM8) and annual seablite *Suaeda maritima* (SM9) become frequent, with stands of *Aster tripolium* (SM11 and 12). The low-mid marsh is dominated by sea purslane *Atriplex portulacoides* (SM14) or with communities with salt marsh grass *Puccinellia* (SM10 and 23). This is followed by a zone dominated by a number of sub-communities of the saltmarsh grass *Puccinellia maritima* (SM13) including the *Limonium vulgare-Armeria maritima* and the *Plantago maritima-Armeria maritima*. In late summer these communities are coloured with the purple haze of flowering sea lavender. Small areas of upper saltmarsh are dominated by sea rush *Juncus maritimus* (SM18) as at East Hills. At Brancaster and Thornham, tidal reedbeds occur at the landward edge of the saltmarsh where freshwater springs emerge, these can be assigned to the *Atriplex prostrata* sub-community (S4d) of the *Phragmites* community, and have a variety of associates such as *Bolboschoenus maritimus, Aster tripolium* and *Glaux maritima*. The drift lines and transitions to dune and shingle are noteworthy in that they are restricted to north Norfolk These are the *Suaeda vera-Limonium binervosum* (SM21) salt marsh and the *Suaeda vera* strandline (SM25). The former is the distinct community in which both *Frankenia laevis* and *Limonium bellidifolium* occur and although very limited in total area it is quite widespread where shingle is exposed on the saltmarsh - sand dune transition. The most abundant communities are the *Atriplex portulacoides* and low marsh *Puccinellia* (24% and 20% respectively), occurring in significant quantities on all sites with the exception of the young marshes at Blakeney Point and Scolt Head Island.

CLIFFS

The coastal cliffs between Weybourne and Happisburgh are generally composed of Pleistocene sands and clays. Where little water drains from them they are relatively vertical though still prone to erosion, Where water issues from them, especially between Cromer and Mundesley, they are very unstable with frequent landslips creating a mosaic of habitats from bare mud which is colonised by ruderal species such as *Tussilago farfara*, through relatively stable areas of grassland which are often dominated by a community (SD7) in which *Festuca rubra* and the colourful *Anthyllis vulneraria* are conspicuous, to scrub with *Hippophae*, as near Cromer. The flushes often have *Equisetum palustre* and *Juncus* species with some orchids such as *Dactylorhiza incarnata* ssp. *coccinea*. In places perched *Phragmites* beds occur in areas of impeded drainage on the cliffs. The rare *Orobanche purpurea* appears to be largely confined to mesotrophic grassland on the cliff tops, never occuring on the slopes.

Meadow - Mattishall

Nature Conservation in Norfolk

Mark Cocker

The county of Norfolk has had a key role in the evolution of national conservation policies during the twentieth century. This essay presents an outline of the major developments in the last half century and pays particular attention to those occurring since publication of the Petch and Swann flora in 1968. For this purpose the county has been divided into four separate regions: the North Coast including the Wash, the Broads, Breckland and an area entitled Agricultural Norfolk.

Separate treatment of the first three has a certain logic, given their ecological unity, distinct histories of land-use and the often unique environmental problems confronting each of them. The fourth is a more arbitrary designation, referring essentially to the rest of the county - the geographical matrix in which the other three distinct regions are situated. However it does embrace a number of habitats, like valley mires and bogs, that are critical elements in Norfolk's overall biodiversity, but which frequently lie outside the strong legislative/administrative cordons protecting the other three areas. This wider countryside is very much the environmental frontline and has suffered the deepest losses largely because of agricultural intensification. But it has also been an important theatre for a number of creative initiatives and partnerships and for this reason alone it warrants independent treatment.

The North Norfolk Coast and the Wash

There is a certain chronological value in beginning an account of Norfolk conservation with its North Coast, since this is where local environmentalists took those first critical steps. The acquisition of two of the county's environmental jewels - Blakeney Point (1912) and Scolt Head (1923) by the National Trust - had a major impact not only locally but nationally. The creation of a reserve at Cley Marshes (1926), originally bought by members of the Norfolk and Norwich Naturalists' Society, heralded the foundation of the Norfolk Wildlife Trust (NWT) and helped trigger the development of a nationwide county trust network, currently under the aegis of The Wildlife Trusts.

After the Second World War further sections of the north coast were added to these original core reserves. The National Trust acquired saltings and marshes at Morston and Brancaster (1968), Stiffkey (1969) and Salthouse (partly with the NWT); the NWT itself obtained Holme Dunes in 1955; English Nature (EN) assumed management of Holkham National Nature Reserve in 1967 while the Royal Society for the Protection of Birds (RSPB) purchased a large block of saltmarsh and coastal lagoon at Titchwell in 1973.

Collectively these reserves represent much of the 40 km stretch of coastline between Hunstanton and Weybourne and are an outstanding mosaic of sand- and mudflats, shingle ridge, saltmarsh, sand dunes, reedbed, grazing marsh and lagoons. A total area of 7,700 hectares constitutes one of the most unspoilt coastal landscapes in Britain, whose unity and importance were recognised by its notification as a single Site of Special Scientific Interest (SSSI) in 1986. The area has been protected by further national and international designations, becoming an Area of Outstanding Natural Beauty (1968), a Heritage Coast, a UNESCO Biosphere Reserve, a Wetland of International Importance under the Ramsar Convention and a Special Protection Area, SPA, under the European Union's Directive on the Conservation of Wild Birds (1989).

Only a short strip of coastline around Hunstanton separates the North Coast from that outstanding coastal feature, the Wash.

Coastal Grazing Marsh - Burnham Norton

Britain's largest and most important estuary is shared between Norfolk and Lincolnshire and covers a total of 620 km. Much of it comprises zones that are either intertidal or permanently covered by water, but overall it does contain almost 10 percent of Britain's entire saltmarsh. Agricultural intensification has increasingly encroached on the Wash's southern fringe, especially to the north-west of Kings Lynn, where land reclamation for arable farming continued until the late 1960s with heavy use of fertilisers that has destroyed much of the rich flora of marshes and water courses. Botanically the most rewarding area is between Heacham and Wolferton and in 1972 the RSPB assumed management of a stretch at Snettisham, extending to 1,420 ha and embracing all of the Wash's major habitat types while most of the remaining area, although in private hands, is managed sympathetically.

One of the problems of the coast in recent years, especially the North Coast, has been its sheer popularity and the pressures of visitor numbers. The Norfolk Coast Project was established in 1991 to address problems of this nature and has developed a strategy document to help reduce the impacts of traffic and tourism.

Another threat to the Coast has come not from land-based activity but from the North Sea itself. This low-lying coastline has always been highly mobile and subject to alteration from large surge tides. Three major incursions have occurred in the period - in 1953, 1978 and 1996, and the ruin caused to large parts of Cley's visitor infrastructure by the 1996 flood is typical of the damage inflicted. Such storms have also caused the steady erosion of the shingle bank between Salthouse and Blakeney and that from Heacham to Snettisham and of the vulnerable sand-dune sections between Holme and Wells. Combined with the possible effect of sea-level rises as a consequence of global warming (a rise of 50 cm within 50 years is widely predicted), future storms could cause lasting damage to stretches of the coast, particularly to the grazing marsh and freshwater habitats. These losses would have major impact on all the coast's wildlife as it is no longer possible for the plants of salt marsh or grazing marshes to spread back inland again as would have happened in the past. Perhaps even more vital, however, is the stretch of coast from Happisburgh to Yarmouth North Denes, an area which includes Winterton National Nature Reserve. Here, coastal defences protect not just the local marshes, but the whole of the northern Broadland which could be inundated with salt water should they be breached. This would be disastrous to an area which is of national importance and not only for its wildlife.

The Broads

From the Victorian period onwards the Norfolk Broads have held a special place in the national consciousness as a recreational paradise and have vied with the North Coast as the county's most cherished wildlife region. The NWT reflected this attitude by acquiring seven Broadland reserves in the half century from 1928. These comprised a total of 1,150 ha of open water, flower-rich fen, reedbed and carr woodland and included the largest freshwater body in Norfolk, Hickling Broad. From 1970 to the nineties the RSPB followed this lead, with the acquisition by 1997 of 640 ha of the Halvergate Marshes at Berney, including parts of Breydon Water (with an ongoing policy of increasing this whenever possible), and also a complex of habitats in the Mid Yare totalling almost 800 ha. However, despite owning some of the most prized areas, conservationists were forced to realise that they had not secured the Broads from environmental deterioration.

In 1976 an alarming report by EN highlighted a number of major problems. Originally medieval peat-diggings, the Broads are subject to succession and many of the practices that had maintained the area's environmental diversity, like reed- and hay-cutting, had become uneconomical and had been discontinued. Neglect and over-extraction of water for domestic and agricultural use meant that many botanically important habitats were drying out, scrubbing up and reverting to woodland. Since the war the Broads water system had also become heavily contaminated by phosphates from domestic sewage and nitrates from agrochemicals. These nutrient-rich conditions led to unnaturally high levels of vegetative growth and a proliferation of phytoplankton, which in turn caused decreased light on the water bottom, a build up of sediments and reduced water flows. By the 1970s it was discovered that many broads supported few or no flowering plants at all.

Intense pressure from many quarters eventually led to the foundation of the Broads Authority (BA) in 1978, which acquired statutory powers a decade later. The Broads themselves were elevated to a status equivalent to a national park and the new authority, with an innovative emphasis upon partnership, set about tackling the problems with great dynamism.

Of the three principal habitats in the authority's 303 km^2 executive area - open water, fen and grazing marsh - it was the last which threw up the most pressing problems. Britain's entry into the Common Market in 1974 brought profitable opportunities to Norfolk farmers with guaranteed prices for cereals. New European milk quotas, by contrast, were forcing some dairy farmers into herd reductions. The local consequence of these international pressures was a switch from traditional grazing practices to arable production. Between the 1930s and 1984, 37 percent of the Broadland grazing marshes went under the plough, with attendant consequences of deeper drainage, lowered water-tables, increased use of agrochemicals and a deteriorating dyke flora. Eventually the Halvergate Marshes, the largest block of lowland grazing meadows in eastern England, became an emotive and symbolic battleground between the forces of conservation and modern agriculture.

In the end a remarkable compromise was its outcome. By 1985 local farmers were being paid subsidies not to convert to arable, but to continue traditional grazing practices. This pilot project, known as the Broads Grazing Marsh Conservation Scheme, eventually blossomed into the system of Environmentally Sensitive Areas (ESA). By 1998 the scheme is set to involve 918 Norfolk farmers and cover 18,000 ha, mainly of grazing marsh with some fen and carr woodland. As with Cley in 1926, local conservationists had established precedents at Halvergate which had profound and lasting consequences at a national level. Today there are 22 ESA schemes in Britain helping to safeguard 426,000 ha of wildlife-rich landscape, including Norfolk's own Breckland. Yet at Halvergate itself, there is little room for complacency. In the 1990s there have still been some conversions to arable owing to changes in the beef market caused by BSE.

The other major challenge for the BA has been reversing the deterioration in water quality. Working in close collaboration with a number of partners, including Anglian Water, EN, NWT and the EA, it has pursued a range of pioneering techniques. These have included the more efficient stripping of phosphates from water effluent at local sewage plants. With the flow of enriching pollutants controlled, it has been possible to clean out certain broads by suction dredging the suffocating beds of sediment. Another experimental technique has been bio-manipulation - the removal of fish and larger predators in order to encourage populations of zooplankton, such as water fleas, which at relatively low densities rid the system of large quantities of algae.

By 1997 fifteen restoration projects have been conducted. These ranged from mud-pumping smaller isolated broads like Cockshoot and Alderfen, to the much more ambitious restoration project at Barton, the second largest of all the broads. The successes to date have been exciting and impressive. They are powerful proof of the effectiveness of a collective approach and they offer the Norfolk Broads - the most important wetland in lowland Britain - a future which seemed impossible only 20 years before.

Breckland

As long ago as the late nineteenth century, W.G. Clarke drew attention to the unique ecological conditions and exceptional botanical importance of Breckland. Yet, it has taken almost a century for the region to receive the kind of urgent environmental protection which its wildlife riches undoubtedly deserve.

Part of the reason for this oversight may be Breckland's geographical position, straddling portions of west Norfolk and Suffolk and the eastern fringes of Cambridgeshire. The fact that no one county could claim it outright left the area, to some extent, administratively orphaned, while its light soils, good drainage and marginal agricultural value made it an early target for Forest Enterprise (FE) in the 1920s. This newly founded organisation, then the Forestry Commission, was given the task of making good a national shortage of timber following the First World War, and Breckland became one of its most important forests. Eventually 292 km^2 of the old Breck heathlands and abandoned arable were replaced by plantations of Corsican and Scots pines.

Although afforestation was an obvious cause of habitat loss, it has been intensive farming again which has totally destroyed plant habitats. The entire Breckland heath and grassland areas dropped from an estimated 29,000 ha at the turn of the century to just 7000 ha today. Much of this lies within the large section owned by the Ministry of Defence known as the Stanford Training Area. This embraces over 9000 ha and includes by far the largest Breckland SSSI (4597 ha). This area was taken over by the military in 1941, but in spite of some problems, wildlife and soldiers coexist and much Breck-type grassland remains in this area.

Since the Second World War, large areas of agricultural land have been intensively cultivated for arable crops and the recent trend of growing vegetable crops for supermarkets using heavy irrigation is as environmentally disastrous as it is economically successful. On those areas of heathland which survived, traditions involving sheep or commercially farmed rabbits were discontinued, but it was the advent of myxomatosis and the virtual disappearance of the rabbit population of the area which allowed rank grasses to thrive and together with scrub invasion brought about a process of rapid habitat deterioration. Reflecting these wider environmental changes has been the incremental loss of Breckland's botanical specialities like Spiked Speedwell and Spanish Catchfly. Clarke once found the first in stands numbering more than 1000 spikes, and described the second as so abundant as to appear a hay crop growing to a height of nearly 2 feet. By 1994, Norfolk had just a single area of heath with Spiked Speedwell, while the catchfly was known from just three sites, down from 30 in the 1950s.

The NWT was one of the first agencies to attempt to safeguard some of the better remaining stretches of habitat, purchasing East Wretham (147 ha) in 1938, Weeting Heath (138 ha) in 1942 and Thetford Heath (100ha) in 1948 (now in Suffolk). However coordinated measures commensurate with Breckland's national importance did not really get underway until much later, with the establishment of the Breckland Environmentally Sensitive Area in 1987. This embraces 94,000 ha in both Norfolk and Suffolk and is funded by the Ministry of Agriculture, Fisheries and Food (MAFF). The scheme provides a range of financial incentives for landowners to implement wildlife friendly measures, from the maintenance or re- creation of heather and grass heathland, to the creation of uncropped wildlife strips and headlands. Within the Norfolk part of the ESA there are 21 SSSIs and about 170 County Wildlife Sites and the ESA subsidies have been important in bringing these areas into more favourable conservation status.

In the last eight years two other key schemes have been developed which complement the ESA. The Breckland Countryside Project was established in 1992 and supported jointly by the local councils, the Countryside Commission and EN. The Breckland Wildlife Partnership was launched in the following year with initial funds from the European Union, involving a coalition of the RSPB and the Suffolk and Norfolk Wildlife Trusts. Often working in close collaboration with FE, the Partnership has been instrumental in developing a number of valuable initiatives. Recently 300 ha of FE lands have been earmarked for reversion to heathland, targeting sections in proximity to areas of importance for wildlife. Another valuable project has been the preparation of action plans for 20 rare Breckland plants, which have been incorporated into the FE conservation plan. FE has also drawn up a management plan for the rides and existing open spaces amongst their plantations.

The massive loss of Breckland habitats in the last eighty years leaves no room for complacency as the century draws to a close. The region is still assailed by a range of pressures. It is, for example, an important area for recreation and tourism and visitor numbers are high. At the same time, the full range of measures being pursued in the last decade suggests that the prospects for wildlife conservation and enhancement in Breckland have probably never been better.

Agricultural Norfolk

Fifty percent loss of ancient woodland and hedgerows, 75 percent loss of heathland, 40 percent loss of wet grassland and the virtual disappearance of old, species-rich meadows - the grim statistics of habitat destruction since the early part of the twentieth century have become a familiar litany to everyone concerned for Britain's countryside. Nowhere do they relate more closely in Norfolk than to its agricultural heartland, which is now sometimes written off as a beet-and-barley prairie almost devoid of wildlife interest.

Like many generalisations it contains a portion of truth. The increased demands for food during the Second World War, the arable intensification resulting from the Agriculture Act of 1947 and the cereal incentives provided by the European Union's Common Agricultural Policy have all had profound consequences for the wider Norfolk landscape. In addition to the piecemeal erosion of ordinary features that are the very fabric of a botanically rich countryside, like field headlands or borders, hedges, woods and permanent pastures, several important habitats of this county region have also been seriously reduced.

Many heaths and commons, valley mires, bogs and fens - in some ways the silent partners in Norfolk's rich wildlife heritage - have been destroyed during the course of the century. Typical of the losses was the destruction in the 1950s and 1960s of heaths in the Wensum Valley around Sparham, Lyng and Elsing, these commons succumbing either to the plough or to gravel-extracting operations. Similarly, the larger heathland area extending between Cawston and Marsham in the north, and from Horsford and Horsham St Faith in the south, has gradually been fragmented or has deteriorated because of afforestation, urban development, agricultural improvement or lack of management. Of those fens and wet heaths which have survived, many have lost much of their unusual flora because of the lowering of the water table through the deepening of water courses and over-extraction of the ground water and this remains the greatest threat to wildlife sites today, even those which are National Nature Reserves.

Local conservation bodies have pursued a number of measures to safeguard some of the most important sites. Between 1961 and 1980 the NWT acquired heathlands or valley mires at Scarning (4ha), Roydon Common (109ha), East Winch (32ha), Syderstone (24ha) and Booton Common (8ha), together with the very rich and varied site at Thompson which contains both chalk grassland, heath and fen.

One critical protective mechanism has been the system of SSSIs, which was initiated with the foundation of the Nature Conservancy in 1949 (now EN). Although this measure was further strengthened with the Wildlife and Countryside Act of 1981, and the SSSI network

Conservation Headland - Didlington

is seen as a cornerstone of all environmental policy in Britain, the system has not been fail-safe. In the first place, only the most important habitats were designated, and of these it was intended to cover only a representative sample. Equally, while it gave legal protection to sites, it made few or no provisions for their proper management, and neglect has often been as critical to the loss of seral habitats, like heath and fen, as wilful destruction. In the 1990s, however, positive management agreements between EN and owners of SSSIs are increasingly common.

Another project intended to correct some of the original deficiencies of the SSSI system and to protect areas essential to the overall biological richness of the county was the Norfolk Habitat Survey conducted in1985 and again in 1993-8. These surveys identified 1200 County Wildlife Sites (CWS), most of which are on private land and without public access. They represent most of Norfolk's remaining semi-natural habitats and amount to 2.5 percent of the county's total area. Protection of these sites has been assisted by policies in the County Structure Plan and other Local Plans which ensure that county and district councils have regard for their wildlife value and both EN and NWT should be consulted on any proposed developments relating to them though in reality some are still at risk from unsuitable development by their owners.

The reaction of land owners to species-rich habitats on their properties has varied widely, but an increasing number of such sites are being looked after and managed in a responsible manner. In this they have been assisted by another valuable mechanism for habitat enhancement in the wider Norfolk countryside, that is the Countryside Stewardship Scheme, administered by MAFF. This scheme was initially intended to provide financial subsidies to landowners to implement favourable management practices of semi-natural habitats such as old pasture and commons in South Norfolk, many of which are County Wildlife Sites. It has also provided funds for habitat restoration schemes and has now been widened to cover more arable habitats, like field margins and hedgerows, throughout the county.

In the late 1980s and 1990s a number of new initiatives evolved both nationally and locally were aimed at bringing important wildlife habitats, including under-managed SSSIs and County Wildlife Sites, back into a favourable conservation status. The North Norfolk Heaths Forum is a typical example. This regional coalition of EN, NWT, Norfolk County Council, the Countryside Commission and the Farming and Wildlife Advisory Group, has been successful both in focusing attention on Norfolk's heaths and improving management of important sites, like those at Buxton, Cawston and Marsham as well as NWT reserves at Roydon, East Winch and Syderstone.

Often overlooked but equally critical to the maintenance and enhancement of many important habitats in Norfolk is the practical work performed by the county's army of volunteers. Some of these operate under the direction of a nationally co-ordinated body like the British Trust for Conservation Volunteers (established in 1959), or are a regionally organised task force, like the Broads Conservation Volunteers. But others are independent initiatives run by committed individuals. Such projects operate very successfully on Beeston Common and Buxton Heath. Scrub-bashing is the classic staple of these unsung heroes, but without their perennial cycle of hands-on effort Norfolk wildlife undoubtedly would be much the poorer.

Acknowledgements

I am extremely grateful to John Davies, Chris Durdin, Pete Hayman, Richard Hobbs, Peter Lambley, Reg Land, John Morley, Lisa Schneidau and John Snape, who were all very helpful in providing me with information and/or commenting on drafts of the text.

The Norfolk Botanists

Alec Bull

Although a Botanist is described in the dictionary as a person who is skilled in the study of plants, botany itself comes from the Greek 'grass' or 'fodder'. The first 'botanists' would have been the 'wise men' or 'wise women' of the village or tribe who made a study of all the natural world around them, and whatever properties each plant might have that would be of benefit to their own community. To quote just one example, we know from accounts by the Roman invaders that tribes such as the Iceni decorated themselves with the blue dye from Woad, before going into battle. Woad produced an excellent long lasting dye, and it is perhaps no coincidence that experiments are currently going on with a view to producing the dye in sufficient quantities for use in inkjet printers.

Each succeeding invasion of Britain saw additions to our flora, though we may not now know to whom each introduction is attributable. The Romans brought Alexanders which they used as a vegetable, and Sweet Chestnut which was fuel, food and building material. No doubt the Saxons and Normans both added to the list of species found in these islands, and subsequent 'friendly' invasions added even more. When Edward III (1327-77) was on the throne and sheep were of prime importance, especially their wool, Flemish weavers came over in considerable numbers and set up their looms. They also brought with them lilac, carnations, Provence rose, asparagus, carrots, celery and cabbages. (Swann 1968)

James Edward Smith - Early Beginnings

Where shall we start in our quest for the first botanists as we know them? No further from home than Gentleman's Walk in Norwich in 1759. At No.37, opposite St Peter Mancroft, there stood a woollen drapers owned by William and James Smith. James lived over the shop and was married to Frances, daughter of the Rev. John Kinderley, perpetual curate of St.Helens in the city, rector of South Walsham and chaplain to the Countess of Leicester. To her was born, on the 29th of December in that year, her firstborn child who was christened James Edward. Though the eldest of seven, he was their only child for five years and so had all of his mother's attention, and it was from her that he obtained his love of flowers. 'I can just remember tugging ineffectually with all my infant strength at the tough stalks of the wild succory (chicory) which grew on the chalky hillsides about Norwich.' (Geldart 1914). Not strong enough to attend school, he was visited by the best tutors in the city and acquired a knowledge of French, Italian, Mathematics and rudimentary Latin. He was much interested in Natural History but had neither teacher nor books. 'I wandered long in the dark until some of the principal elementary works, the publications of Lee, Rose, Stillingfleet and a few others as came my way, and were devoured over and over again. At length, however, I found I wanted something more, and to apply to practice what had thus been acquired. I was then furnished with systematic books and introduced to Mr Rose, (Hugh Rose, apothecary in Norwich 1707-92) whose writings had long been my guide. I was shown the works of Linnaeus. Nor shall I ever forget the feelings of wonder excited by my finding his whole system of animals, vegetables and minerals contained in three 8vo volumes.'

So, equipped with Berkenhout's *Botanical Lexicon*, and doubtless carrying a bunch of plants for study, Smith regularly attended at the house of Hugh Rose at No.8, Tombland, long since demolished The same house where Rose had written his *Elements of Botany*, assisted by the Rev. Henry Bryant. This was a translation of *Philosophia Botanica and other Treatises of the Celebrated Linnaeus.* Rose had written this with the more learned of Norwich's botanical students in mind. The appendix described 'some plants lately found in Norfolk and Suffolk'. The Norfolk plants were 'a Geranium found near Spixworth church in 1771 by Mr William Humphrey; *Holosteum umbellatum* first noticed and examined by Mr John Pitchford in spring 1765; *Tillaea muscosa* found on Drayton Heath and several other places near Norwich, first examined and ascertained by the Rev.Mr Bryant in 1766; *Ophrys paludosa* found on Felthorpe Bogs by Mr Charles Bryant (brother of the above) and *Hydnum auriscalpium* (Earpick fungus) found last autumn near Norwich in a small pine plantation by Mr Rose himself'.

Rose became afflicted with incurable blindness in 1779, but his interest in botany remained until his death. Smith always showed the greatest consideration to his old teacher. When away from home, he sent him messages by way of mutual friends rather than writing what Rose could no longer read.

Smith senior had intended that his son should be engaged in the importation of raw silk, but he must have been remarkably forbearing and understanding of his son's desires, for, in a letter which survived for some considerable while, he stated that his son, being determined to study botany, must also study medicine, as only a medical course included botany as one of its subjects. In October 1781, James Edward set out for Edinburgh where his botanical lecturer was to be Dr.John Hope, the first teacher of the Linnaean system in Scotland. Already we see the 'Norwich School of Botanists' appearing on the scene. When Smith senior and his second son, John, arrived home after accompanying James to Edinburgh, he wrote to him "Mr Crowe and Mr Pitchford and many of your friends have called and enquired after you" and a few days later, "Mr and Mrs Rose are well, as also is Mr Humphrey."

The Founding of the Linnaean Society

Carl Linnaeus had died in January 1778. His only son, Charles, had been trained to succeed him in his scientific work, and should have inherited his collections, but his mother compelled him to purchase them at her own price. In May 1781, Charles was welcomed by surviving friends of his father in London, and especially by Sir Joseph Banks the botanist, who had travelled round the world with Captain Cook. In November 1783, Charles died unmarried and the Linnaean museum reverted to his mother and sisters. According to Swedish law, the state had the right to inherit part of the possessions of a University professor. As this would fall within the sphere of influence of the king, who was then out of the country cruising in his yacht, Linnaeus' widow wrote instantly to Sir Joseph Banks offering him the collections. Of all the people she knew, he was the one most likely to pay the high price she was asking. By this time, James Smith had finished his course in Edinburgh and was continuing his studies in London. He was a frequent visitor to the Banks house and was breakfasting there when the letter arrived asking £1,000 for the entire collections and library. Banks declined, but urged Smith to acquire the treasures. Father was even more indulgent this time and advanced the £1,000 necessary and also paid for transport of the collections which amounted to a further £88/5s/0d.

In April 1784, the collections now safely in London, Smith wrote to T.J.Woodward, Bryant (Charles?) and John Pitchford and sent messages to James Crowe and Hugh Rose telling them of his purchase. Pitchford replied at length for all of them concluding -"I shall now look up to you as a second Linnaeus - I think you highly deserving of being the possessor of such remains - the English Botanist will now have an opportunity of knowing what natives of his own country are in the Species Plantarum."

Smith hired apartments in Paradise Row, Chelsea, to house the Linnaean collection and the following year arranged them in order with the help of Banks and Daniel Solander who had been a botanist with Captain Cook in the Pacific. Smith then toured abroad for a while, and when he returned in 1787, he arranged a meeting for February 26th 1788 which was attended by seven people, and was followed on March 18th by a second meeting at which the Foundation Roll of the new Linnaean Society was announced. This consisted of 20 Fellows, 3 Honorary Members, 39 Foreign Members and 11 Associates. Among these were James Crowe, Robert Stone and T.J.Woodward, all Fellows; the Associates were John Pitchford, who became a Fellow in 1797, the Rev. Charles Sutton, Lilly Wigg and the Rev. William Kirby. Dr J.E.Smith was elected President, a post he held until his death in 1828.

While in London, Smith frequently visited William Hudson, apothecary, in Paxton Street, Haymarket. He had been much influenced by Benjamin Stillingfleet, a Norfolk man educated at Norwich Grammar School, whose *English Calendar of Flora* was published in 1761. Hudson had been induced to become the author of *Flora Anglica* (1767) "the first true attempt to set forth the British Flora according to the system and nomenclature of Linnaeus". The foreword was written by Stlllingfleet.

The Norwich School of Botanists

Smith wrote "In my young time, this circle (of Norwich botanists) was peculiarly enriched by the possession of Mr Rose, Mr Bryant and Mr Pitchford, and was often favoured with the society of the learned and amiable Stillingfleet, and the correspondence of Hudson, and they may altogether be considered as the founders of Linnaean Botany in England."

The Norwich botanists corresponded with the Rev. John Lightfoot (1735-88), author of 'Flora Scotica'. In Lightfoot's copy of Dillenius '*Ray*' at Oxford Botanic Gardens are many MS records of Norfolk plants found by Bryant and Crowe and four of Pitchford's. One of the latter was the first Norfolk record for *Liparis loeselii*, found on the bogs at Newton St.Faiths in 1767. After a deal of haggling, one would imagine, Pitchford exchanged his only specimen of Liparis with Mr Lightfoot" for above 60 of the rarest British plants."

John Pitchford was a surgeon and also a Roman Catholic, but botany was a great leveller at a time when non-Anglicans were still frowned upon. His two major finds have been mentioned, though he specialised in mints and sedges. A contemporary botanist in the far west of the county, and also a surgeon, was William Skrimshire (1766-1830) of Wisbech, and it may be to him whom T.J. Woodward is referring in a letter to Dr.Smith, "Mr Pitchford has been to Wisbech to sell his estate, (presumably his wife's as she was recently deceased) and has met with a brother surgeon who has found some good things in that country." Skrimshire's Herbarium is preserved in Wisbech Museum, though his Catalogue of Plants seems to have been lost in relatively recent times.

In 1796, at the age of 37, Smith married Pleasance Reeve of Lowestoft. Disliking London, she persuaded her husband in the autumn of that year to return to Norwich where they set up home at 29, Surrey Street. They brought the Linnaean collections with them, and it was here that Smith spent the fruitful last 30 years of his life. It was also to Norwich that botanical pilgrims had to journey if they were desirous of worshipping at the shrine of Linnaeus.

It was here too that most of Smith's writings were scripted, including the first four volumes of the *English Flora*. The mosses were covered in Volume V after his death by W.J.Hooker, his protege, another Norwich lad, and were followed by Volume V Part II, the fungi by M.J.Berkeley. Smith also wrote the text for 'English Botany' 1790-1814, invariably known as 'Sowerby's English Botany' from the illustrator.

After his return to Norwich, he continued his friendship with James Crowe of Tuckswood House, Old Lakenham, also a surgeon. Crowe owned other properties including some at Askwicken in the west of the county and a considerable proportion of the parish of Saham Toney near Watton. At Lakenham, he had a large garden running down to the river and, as he rode about Norfolk on his horse, he collected cuttings of any interesting looking willows and arranged

Liparis loeselii - Catfield

them in his garden, each one carefully labelled and bearing either a number or a letter. Number 12 was *Salix croweana* and the letter M was *S.smithiana*. *Salix forbyana* "the very cutting sent by the Rev.J.Forby was there", (Forby was Rector of Fincham). Smith spent long hours in Crowe's garden studying the willows and, while writing Volume III of Flora Britannica in 1800 and 1801 he wrote, "I am investigating the difficult genus Salix and have already written this twice over." In 1819 he enumerated 141 species of willow in Rees 'Cyclopaedia' based on his researches at Lakenham, but many of Smith's 'species' are now recognised as hybrids. e.g. *Salix x smithiana*. Crow died in 1807 and in 1819, Smith reported that his living collection of Salices was still carefully preserved, but by 1841, Grigor (author of *The Eastern Arboretum*) stated that only two of Crowe's specimens remained.

Thomas Jenkinson Woodward of Bungay, who also had a house in Ditchingham, was described by Smith as 'one of the best of English botanists'. Born at Huntingdon in 1745 and educated at Eton and Cambridge, he moved to Bungay in 1770 and later married a local girl. He was a lawyer by profession and a great friend of Robert Stone who also lived in Bungay at that time, but later moved to Gorleston where, at some date unspecified, he discovered *Erythraea pulchella*. In 1797, Stone was visited there by a European botanist, H.A.Noeden, who had 'a great botanical excursion with him, seeing many rare plants'. In about 1809, Stone moved to the family estate at Bedingham Hall, 4 miles north west of Bungay. He had a nearly complete herbarium of British plants and paid especial attention to fungi, discovering *Hydnum imbricatum* and *Lycoperdon coliforme.* Late in the year 1801 or early in 1802, the Woodwards moved to Walcot Hall just north east of Diss and much improved the estate planting many trees. He wrote nostalgically to Smith in 1811 that "I so seldom see Dawson Turner since my removal from Bungay."

The Forbys and Dawson Turner

The Forby family of Fincham has been mentioned. The family held the living of that parish from 1723 to 1825. The Rev. Joseph Forby was another willow man, specialising in the study of osiers for practical purposes. He was succeeded in the living on his death in 1799, by his nephew, the Rev. Robert Forby who had been educated at Lynn Grammar School and Caius College, Cambridge. Though Vicar of Horningtoft, he lived first at Barton Bendish and later at Wereham, taking in pupils at both places. Amongst these, between 1790 and 1793 was Dawson Turner who later wrote 'To Botany he was more particularly attached during the time which I spent with him, and it was no less my pleasure and my pride to accompany him on his botanical rambles, than it is at present to acknowledge that I am indebted to his precepts and example for any proficiency which I may have in this delightful pursuit.' Dawson Turner and L.W.Dillwyn published *The Botanist's Guide through England and Wales* in 1805 in two volumes. Many of the older records in Floras published during the 19th century were taken from this work.

In 1797, the Rev. William Kirby of Barham near Ipswich, accompanied by Thomas Marsham, first secretary of the Linnaean Society, visited his sister, wife of the Rev. Charles Sutton, rector of Holme-next-the Sea and Thornham, while they were living at 197 King Street, Norwich. Sutton was an expert on Orobanche and discovered *Orobanche elatior.* Kirby wrote in his diary - 'July 12th. We breakfasted and spent the morning with Dr.Smith and examined the Linnaean Apes (bees). The Smiths dine with us and in the evening we go to Thorpe by water. July 13th. Engage in a botanical expedition to Newton St.Faiths. Our party consisted of Dr.Smith, Mr Marsham, Mr Pitchford, Mr Foster, Mr Weston junior (Smith's brother in law), Mr Sutton and myself. The plants gathered were numerous, and many rare. *Polypodium oreopteris, P.filix-mas, Osmunda spicant, Brachythecium glaucum, Satyrium viridis, Orchis conopsea, Carex vesicaria, C.limosa, C.hirta, Linum radiola, Serapias palustris, Peplis portula, Comarum palustre, Menyanthes trifoliata, Gentiana pneumonanthe, Alisma ranunculoides, Tillaea muscosa* etc. We returned in the evening after being consulted by the villagers as learned doctors.'

W.J.Hooker - Another Norwich Boy

William Jackson Hooker was born July 6th 1785 in St.Saviours parish, Norwich, his father a clerk in a firm of wool staplers and a keen gardener. At that time, Norwich was known as the city of gardens. Hooker's first tentative steps along the Linnaean pathway were taken in the garden of the Mill House at Old Costessey, owned by his friend Mr Simon Wilkin. A proportion of the grounds was laid out as an experimental botanic garden with Hooker's help and advice, and together they 'superintended it with great interest'. By the age of 20, Hooker was well versed in flowering plants, mosses, hepatics, lichens and freshwater algae. In December 1805 he discovered the rare and ephemeral moss *Buxbaumia aphylla* in a wood at Sprowston, its only Norfolk record ever, and took it to Smith. Smith suggested that he send some to Dawson Turner, the latter inviting Hooker to visit him in Gorleston. This meeting resulted in Hooker meeting his future wife for the first time.

One only has to examine some of the illustrations executed by Hooker to realise that, through him, there is a direct link between the Norwich School of Botanists and the Norwich School of Painters. He applied their craftsmanship to the illustration of the plant world in general, and it can be no accident that his mother was the daughter of one James Vincent of Norwich, grandfather of George Vincent, reckoned by some to be one of the most talented of the Norwich School of Painters. Hooker, as everyone knows, was director of Kew Gardens from 1841 to 1865.

Referring to the Great Yarmouth district in 1834, James Paget tells us that 'probably no neighbourhood has been investigated so completely as this, which has had the good fortune to have been for nearly a century, the constant stage for the action of some enquiring mind. Long ago, Dr Sims, Dr Aikin and Joseph Sparshall were engaged in the observation of our plants by the feeble light which the science then, comparitively speaking, in its infancy afforded them'. Sparshall was a grocer, and in 1767 had a shop at a corner of one of the Rows facing the Market Place. John Aikin MD 1747-1822, was in Yarmouth for a year in 1784 when he took a London practice, but returned to Yarmouth in 1786 for six years before finally returning to London in 1792. He was the author of *The Calendar of Nature* and *The Woodland Companion.* James Paget himself and his brother Charles were both eminent botanists, and together published *A Sketch of the Natural History of the Yarmouth District* from which the above quote was taken. James Paget's herbarium is in Norwich Castle Museum. He is more widely known as Sir James Paget, knighted for his services to surgery, and it is to him that the Gt Yarmouth and District hospital owes its name.

In 1818, the quay at Yarmouth was said to be one of its glories. James Paget had been born in a house at the seaward end four years earlier, whilst at the town end, facing the quay, stood the Bank House, where Barclays Bank now is, the home of Dawson Turner. It was there in 1793 that Smith met Lilly Wigg for the first time. The latter, born at Smallburgh on Christmas Day 1749 to John Wigg and his wife Sarah nee Lilly, had been a shoemaker, later a schoolmaster, and at the time that Smith met him, a clerk in Dawson Turner's bank. On July 20th 1793, Smith wrote to T.J.Woodward 'We have been at Yarmouth and I have seen the puritanic brown locks of Mr Lilly Wigg which so much belie his name. I am, however, not the less satisfied with himself.' Woodward replied, 'I am glad you saw Turner at Yarmouth. He is a very clever young man and will make an excellent botanist. Wigg's locks certainly do not partake of the lilly, nor do they give him a very promising appearance, but the exterior is not always the interior, and he has a great knowledge and indefatigable industry under that uncouth appearance and harsh address.' Though Robert Forby had instructed Dawson Turner in botany, the latter mentions Lilly Wigg as 'my instructor and co-adjutor in marine algae'. Dawson Turner's great work *Fuci* issued between 1808-19 contains some plates by his wife, but the majority of the illustrations were by his son-in-law, W.J.Hooker.

Thus was the Norwich School of Botanists engaged for well over half a century in discovering the Flora of Norfolk. For instance, the Rev.Henry Bryant wrote to Smith from Heydon in 1782, 'Crowe and I have made good use of our time and have found a multitude of new things, many of which are not described.' Their efforts at recording are still to be found in the annals of the Linnaean Society, and in various Herbaria both locally and nationally. Smith was knighted in 1814 and died in 1828, but his influence on botany in Norfolk continued for long after his death. Dawson Turner followed him as President of the Linnaean Society from 1828-33. Smith's widow edited and published two volumes of his letters in 1832 entitled *The Memoirs and Correspondence of the late Sir James Smith.* Lady Smith had the more important of the 5000 letters he had preserved in his lifetime, bound and presented to the Linnaean Society in 19 volumes in 1857. She continued to live at 29, Surrey Street for some years before going back to Lowestoft to live with her brother.

H.D.Geldart and the Rev.Francis Galpin

Joseph Geldart succeeded Lady Smith at Surrey Street. To the many strange plants to be found growing in that garden is attributed the first interest in botany exhibited by that gentleman's son, Herbert Decimus Geldart, one of the outstanding botanists in Norwich and Norfolk in the latter half of the l9th century. Swann refers to his many records in Nicholson's *Flora of Norfolk* of 1914; his many papers in *Transactions of Norfolk & Norwich Naturalist's Society*; his articles on botany in Mason's *History of Norfolk*, 1883 and in Dutt's *Norfolk* in 1900 and the botanical section for the county in the *Victoria County History of Norfolk* in 1901.

In addition, the Rev.Francis William Galpin in the foreword to his *Flowering Plants and Birds of Harleston*, Norfolk published in 1888 says that the list of Norfolk plants, with additions published yearly, is already approximately complete, under the direction of a friend interested in the club, Mr H.D.Geldart of Norwich. The club referred to was the Harleston Botanical Club, founded in 1883 to record all the plants growing within a four mile radius of Harleston railway station. This resulted in 532 species being recorded during the first year, 1884. The following year, the circle was extended to a five miles radius, resulting in 80 new species being added, and in 1886 the circle was pushed out a further mile resulting in a further 40 species being found. In 1887, a general review of the whole six mile radius circle added only 24 more new records. Literature was searched which added over 100 further species within the survey area, though 18 of these were presumed extinct as they had not been seen for many years. The final total was 791 species. Though the records were not mapped as such, the whole survey only covered four years and would appear to have been the earliest attempt at systematic recording within the county. Given the circumference of the circle, the survey extended over the county boundary into the 'Saints' district of north Suffolk, but most of the recorders lived within Norfolk. In addition to casual records such as those supplied by Mrs Sancroft Holmes of Gawdy Hall, there were 15 members of the club, plus a further 15 who sent in records but were not club members. Bearing in mind the number who have helped with our county wide survey, it would appear that the pursuit of botany was a much more acceptable pastime in the 1880s for those with an interest in the countryside than it is today. Would this have been because they had no TV at the time, so they had to go out and search for things themselves?

Amongst the MS records perused was *A Catalogue of Plants found in the neighbourhood of St.Margarets, South Elmham, Suffolk, and Brockdish, Norfolk* by the Rev. E.A.Holmes who was rector of the former parish from 1833 to 1885; a *Catalogue of Plants found in Dickleburgh* by 'D.C.' 1860-70; *A Catalogue of Plants found in Shimpling* by Rev.H.W.Millard (no date) and a *Catalogue of Plants found in Hoxne, Suffolk, Billingford, Scole and Wacton'* by J.C.Collins of Diss. Among other former botanists in that area of south Norfolk, Galpin mentions T.J.Woodward and also Daniel Stock of Bungay who recorded *Helleborus foetidus* from Bath Hills, Ditchingham circa 1805 where it is still abundant. Galpin, however, says that Stock's records are deprived of much of their value by his vagueness, as many of his records are just annotated 'Bungay' where he lived. Though of short duration, Galpin's contribution was probably the most important study ever carried out on the south Norfolk claylands.

Kirby Trimmer - The First County Flora

Meanwhile in 1866, the Rev.Kirby Trimmer published the first Flora of the whole county, followed by a supplement in 1884. Trimmer was curate at Stanhoe, home of compiler of the west Norfolk half of the present Flora, for a period of 12 years in the 1830s and 1840s, from whence he moved to Crostwick in the east of the county as vicar, and finally to Norwich where he became perpetual curate of St.George's, Tombland. It was while he was in Norwich that he wrote his Flora with assistance from the many botanists scattered about the county, but still drawing heavily on his own catalogues of the plants found in the two parishes where he had served. He was another student of the genus Mentha and left his collection to the British Museum.

In 1910, W.G.Clarke and W.H.Burrell claimed the first Norfolk record for *Veronica spicata* when they found 'thousands of flowering spikes on Garboldisham Heath'. They were apparently not aware of the record in the Diary of J.D.Salmon (1802-59) which is in Norwich Castle Museum with his Herbarium. His Herbarium sheets for the species all came from between Barnham and Euston in Suffolk where the plant was then abundant, however his diary for July 30th 1837 records 'Brettenham. A fine day in ye afternoon took a walk towards Rushford found *Veronica spicata* in flower but sparingly against the Fir Plantation not far from the Oak Wood although I looked over a considerable portion of the heath I could not find a single specimen anywhere else. The space occupied does not exceed half a dozen square yards.'

Veronica spicata - Weeting

H.G.Glasspoole was born at and lived in Ormesby St.Michael. Like most of his contemporaries, he left a large herbarium which he bequeathed to H.D.Geldart. In 1869, he found the first and only British record of *Carex trinervis* which Swann mentions, adding that 'it is since considered to have been a hybrid', however, in the *BSBI Handbook on Sedges* (Jermy and Tutin 1968) A.C.Jermy, himself a Norfolk man, clarifies the issue. He writes 'Herbarium material was variable and the species was probably hybridising with *Carex nigra* as much material labelled *Carex trinervis* is closer to that species. Living material found its way to C.E.Salmon (author of the *Flora of*

Surrey and a visitor to Norfolk from time to time,) who sent some to Arthur W.Stelfox, then in Dublin, who has the plant still alive in his garden in County Down. That plant has the aristate female glumes and inrolled leaf of *C.trinervis*. A thorough search of Ormesby Common, an atypical habitat, has failed to reveal it in recent years." This tale resembles that of the Dickleburgh Oxlip (q.v.) and it would appear that Glasspoole really did find *Carex trinervis.*

A New Discovery by Applied Science

As the study of botany went on apace, the application of scientific principles started to bear fruit in a rather unexpected manner. Arthur Bennett (1843-1929) who was a frequent visitor to Norfolk from his home in Croydon, read a paper to the Norfolk and Norwich Natural History Society in 1909. 'In the winter of 1882, I had been comparing a map of the Dutch flora with our eastern counties and made a list of plants that might possibly occur in Norfolk, one of which was *Naias marina*. On July 20th 1883, I and my daughter had tea with Mr A.D.More in London, who then saw us onto the mail train for Yarmouth. We tried to persuade him to join us, but he replied that he would only come if we could guarantee that we should find a new British plant. Reaching the draining mill below Heigham bridge, I engaged the lad there to take us to Hickling. We passed through Kendal Dyke into the Sounds and dragged for plants finding many Chara species. Passing through Deep Dyke into Whitesley, we took the channel to Hickling near the keeper's house and, exactly at the spot marked BM4'6" on the 6" Ordnance Survey map, my daughter at the bow brought up a lot of aquatics with the drag. She passed it to me and I at once saw that we had Naias! Three days later, Mr H.Grove went over the same ground with the same lad and found it again. The lad had said to him 'Be you from Lunnon arter weeds? Ar! Yer too late!'

Bennett was also amongst the first to realise, publicly at least, that many less common species were heading for extinction, and he warned of this in another paper read to the Naturalists' Society in 1912 called *A Review of 28 species of East Anglian Plants made between 1899 and 1909*. In it he said, 'I am by no means satisfied with the result, as it shows how little we really know of the history of the majority of these species, many being very rare and enumerated in Flora after Flora without a scrap of additional information. I can only hope that what I have been able to place on record will stimulate East Anglian botanists to study the life histories of these species as year by year it becomes more apparent that some are gradually tending toward extinction, either from drainage, cultivation, enclosures or other causes, and among ferns, by the senseless gathering of huge quantities for sale. Mr Davey, in his *Flora of Cornwall*, 1909, relates that 'a collector boasted to me that he had recently despatched a truckload of roots of *Osmunda regalis* weighing over five tons from one of our local stations.' One wonders how much that fern in Norfolk may have suffered a similar fate in the same way that bluebells, snowdrops and some aquatics such as water soldier still do at the present time.

The Second County Flora

Another 'foreigner' who moved to Norfolk and made his botanical reputation here was William Nicholson (1858-1935). Born in Birkenhead, he joined the staff of Gurney's Bank (now Barclays) in Norwich and took up botany. He worked for thirty years on a new Flora for the county which was published in 1913. It must be said that Nicholson's *Flora of Norfolk* is a very important volume for those interested in historical records as, for a variety of reasons, many of these were not carried forward into Petch & Swann's 1968 Flora. For instance, though Nicholson inexplicably says 'not native' under his entry for *Antennaria dioica*, in our neighbouring county, Simpson's *Flora of Suffolk* in 1982 states 'native' unequivocally. Yet Nicholson gives 'Stratton Strawless, J.E.Smith; Westwick, D.Stock and K.Trimmer and Swaffham Heath, J.Pitchford', to which can be added Roughton Heath, mentioned in Pigott's *Flowering Plants and Ferns of Cromer and its Neighbourhood*. This was published in 1885, not as a standard Flora, but as a series of rambles through the district describing the plants to be found, but it is none the less interesting for that. Petch and Swann ignore the plant altogether. When asked on one occasion why this was, Swann replied that Petch thought that nobody would be interested in what was growing before Nicholson's time. Mention must be made of the work of W.G.Clarke and W.H.Burrell in the early years of the present century especially. Clarke was born in Yorksire in 1877 and died in 1925 whilst Norfolk born Burrell lived from 1865 to 1945. Clarke, apart from his special interest in Breckland, travelled all over Norfolk by bicycle with his friend Burrell, though one must assume that one or the other would have taken their bicycles on the train to each meeting as the former lived in Norwich and the latter in Thetford. Clarke is associated especially with Breckland, a name which he himself coined, and particularly with his magnum opus *In Breckland Wilds*, but his work was not confined to that district. In 1914, he read a paper to the Norfolk and Norwich Natural History Society in which he described how, following a suggestion made in 1908, he and Burrell had devoted part of their leisure over the intervening years towards a Vegetation Survey of Norfolk. The area studied comprised 54 square miles. Without listing all the parishes, it was roughly from Crostwick to Saxlingham Thorpe, to Ashwellthorpe to East Tuddenham and Weston Longville. Ninety years on, one or two passages help to make sense of some of today's plant distributions. 'Roadside timber is mainly oak, elm and ash. Quickset fences are mainly blackthorn and whitethorn, with holly, maple, elm and tea plant being used to repair weak spots. Woods mostly planted except for a few alder copses. Some woods very diverse, with birch, black and white poplar, horse and sweet chestnuts, sycamore, alder, robinia, beech, hornbeam, lime, maple, cherry, holly, spruce, elm, silver fir, larch, Austrian pine, Weymouth pine, yew and Douglas fir and scrub of whitethorn, blackthorn, hazel, rhododendron, box, privet, holly, Portugal laurel, common laurel, (cherry laurel) elder, snowberry and bramble.'

An Introduction to Mapping

Whilst a number of other undoubtedly excellent botanists deserve a mention, such as Fred Robinson from Watton who was a friend of G.C.Druce and made a number of important discoveries, whilst at the same time, contributing vastly to the Botanical Exchange Club, it is time to move on and consider when mapping first appeared on the scene. As far as Norfolk is concerned, this appears to have been early in 1914, when the Naturalist's Society were regaled with a lecture by Dr.Winifred Brenchley from Rothampstead Experimental Station on *Mapping as an Ecological Instrument*. She demonstrated how mapping techniques could be used to plot both plants and soils onto a single map. To do this, she used the work of Professors Oliver and Salisbury on Blakeney Point in l912. Incidentally, Professor Oliver was apparently responsible for the intoduction onto the Point of a number of plants which are still thriving there. His introductions of *Crambe maritima* and *Lathyrus japonicus* may have been lost since the 1996 surge, but included among the sandunes the straggly mass of semiprocumbent white poplar known as 'the wood', *Tamarix gallica* by the hut used as a laboratory by the two professors and plants of *Kniphofia uvaria* and *Yucca* (J.Read pers.comm.).

20th Century 'Greats'

Five botanists have dominated the late middle part of the 20th century, and one specialist bryologist who also knew his plants. These were Charles Plowright Petch ,1911-87; Eric Lister Swann, 1904-89; Richard Pearse Libbey, 1911-87; Edward A.Ellis, 1909-86 and Ernest Daniels, born in 1911 and still with us and currently celebrating 70 years as a member of the county Natural History Society. The Bryologist was G.H.Rocke, a fruit farmer from Lingwood who did much excellent work in the Broads, including the discovery of *Eriophorum gracile* in Norfolk which he found in 1955. Sadly this had disappeared with the deterioration and drying out of the site by 1959. In correspondence with Dr.Francis Rose, himself a frequent visitor to the county and provider of many records, Rocke attributed this to drainage operations taking place on the far side of the Halvergate levels at a time when draining and arablisation of the levels was an attractive proposition. He was probably right as, on the levels at that time, the lowering of the water level in the dykes in one area resulted in a corresponding drop throughout the system from Acle to Breydon Water. In another letter to Dr Rose dated 1960 he wrote 'Swann says Norfolk will not be worth botanising in in 50 years time and I am inclined to agree with him'. When asked in 1998 whether he supported this theory only 12 years short of Swann's deadline, Ernest Daniels gave his qualified approval, citing among a number of other points habitat deterioration; bad verge management and especially damage from winter salt and random planting of new verges and amenity areas with seed which is often from imported stock .

Following the wonderful age of discovery from approximately 1780 to 1830, botanical research in the county moved progressively forward through the l9th century through people such as H.D.Geldart, and especially through some clergymen who were only in the county for a few years, such as Galpin whose incumbency at Harleston only lasted for five years, or E.F.Linton who-held the living of Sprowston from 1879-89, during which time he rewrote the known boundaries of Hieracia and Rubus within the county and added immeasureably to the already copious literature on Salix. This age seemed to come to an end with the 1914-18 war after which botany seemed to fall into the doldrums somewhat. Was there less to discover? Or was it, as has been suggested, that more attention was now being paid to the budding science of ecology? Certainly, much attention has had to be devoted to a changing pattern in the countryside, more especially in the years since 1945, with the onset of hedgerow removal and the devastating effect on our flora from herbicides and other forms of pollution. Swann in 1960 was already commenting on what he saw as the future of botany in the county and, like many botanists, he has to be seen as being pessimistic, but not without cause.

Eric Swann was probably the best, and certainly the most meticulous botanist of his generation. He was elected a Fellow of the Linnaean Society in 1972 and was President of the Botanical Society of the British Isles from 1975-77. He refused promotion within Barclays Bank as he did not want to leave King's Lynn. 'They wanted me to move to Wisbech, but I told them I was a Botanist and that there were no plants worth looking at round Wisbech!' He kept a full and immaculate card index of everything, and a large Herbarium which is now in Norwich Castle Museum.

Edward A.Ellis, Ted to everyone who knew him, was keeper of natural history at the Castle, and botanical recorder for east Norfolk. It has to be acknowledged that he made his mark mainly in Mycology, but he knew his higher plants as well, though his 'card index' was kept within his remarkable memory, records which came to him second hand being recorded on odd scraps of paper or cigarette packets. This seemingly haphazard method of filing records did not endear him to Swann when the latter needed east Norfolk records for his Flora. As both could be very outspoken, even in public, a clash of swords could frequently be heard when the two of them met. This gave rise to a local meaning of the saying 'East is east and west is west and never the twain shall meet!'

Swann's co author for the 1968 Flora, Charles Petch, was a respected member of the medical profession with a hospital consultancy and a private practice in Harley Street. He lived in Surrey and was an active member of the Flora Committee for that county. He also retained the Manor House as the family home at Wolferton and spent a good proportion of his weekends there, though not all were devoted to botany as he also went beagling and sailing in season as well as playing the 'cello for relaxation. He was an excellent field botanist as well as a mycologist and the two authors regularly conferred when they found species of critical identification. He also wrote one of the introductory chapters. He had, however, little patience with the current fashion for subspecies and varieties and suggested to the present author on one occasion that there was only one bramble. Neither would he venture over the vice county boundary into east Norfolk if he could avoid doing so, possibly making Swann even more parochial than he otherwise would have been. He kept a firm hand on the planning for the 1968 Flora and it was he who insisted on the categories for Extinct and Casual plants being put within separate sections at the end of the book. This made that flora very annoying to use by anyone seeking an overview of any particular genus within the county as they found that they had to go to three different sections within the book to achieve that end.

Richard Libbey came to Norfolk from Yorkshire when he was appointed to the County War Agricultural Executive Committee at the outbreak of war in 1939. When the war was over, the family returned to Yorkshire for a while, but returned and settled in west Norfolk where Richard became a frequent companion of Eric Swann, and sometimes with the present author also during one of Swann's frequent 'B.E.'s (Botanical excursions.) He added many records to the supplement to Swann's Flora, and especially to the Bryological records chiefly for the west of the county. His bryophyte collection is in Norwich Castle Museum, though his 12,000 sheet herbarium of higher plants is in the Herbarium at the University of Leicester.

One other of Swann's contemporaries must be mentioned, though again his working life was spent outside his native county. Charles Hubbard was born on the Sandringham estate at Appleton where his father was head gardener. His working life was centred on Kew from whence he paid frequent visits to his native county at weekends and for holidays. To this end he maintained a holiday home at Gressenhall. He and Swann also had frequent 'B.E's and corresponded regularly between times. Following Swann's death, the resultant letters from his end of the correspondence were deposited at Kew. Hubbard's visits were responsible for the many records of grass hybrids and variations recorded for west Norfolk, and his recently revised Pelican handbook *Grasses* is still regarded as the best grass reference work by many botanists.

The Last Amateur Flora

All of the above, and many of our present team contributed to the BSBI Atlas project which was published in 1961, but Swann admitted that he hated dot maps, and refused to include any in his Flora seven years later. His final project, however, was a dot map flora of the Stanford Training Area for the MoD on the basis of the lkm square. This was done with the help of a number of our present recorders between 1979 and 1984. His final five years were dogged by failing health. The main link between Swann's Flora and the present one has to be Ernest Daniels (Bull 1998). During the first two years of the present effort he tramped the streets and byways of the city of Norwich, the riverside and the Sweetbriar Marshes to record the four main tetrads covering the southern half. He was almost 80 at the time and has added records every year since.

So we come to the end of another century, and back into a state of happy co-operation between both halves of the county. Most of us are amateurs, as were those botanists of past centuries. We can boast of one or two doctors among our ranks, but no active clergymen that can be recalled. Our everyday skills are different only in that they reflect the age in which we live, ranging from teachers to experts in computer technology. Some have said that ours will be the last amateur flora, and indeed, there are some floras already published that have been produced in the main by professional botanists, but whilst the spark of enquiry is still within us, there will doubtless be some who will start, as did the present author, in the vein of the quote reproduced below, by Professor F.O.Bower, writing in *Makers of British Botany*, 1913, 'It is the boys who are touched with the love of organic nature from their earliest years, who grub about hedgerows and roads, and by a sort of second sight, appear to know instinctively, as personal friends, the things of the open country who provide the material from which our little band of workers may be recruited.'

Are there any such boys, or girls, out there in the countryside who will carry on our investigations into the 21st century for the love of it? It should be a labour of love for each one of our recorders who will be listed hereafter, to try and make sure that there is at least one that they have encouraged in the right direction in their own district so that the authors of the next Flora will have a continuing source of inspiration.

Nature Reserves in the County

Alderfen Broad	TG 355 195 NWT *broad, fen & carr woodland*
Ant Broads & Marshes NNR	TG 356 225 NWT & Butterfly Conservation (includes Barton Broad & Catfield Fen) *broad, fen, grazing marsh, dykes*
Ashwellthorpe Lower Wood	TM 140 980 NWT *ancient woodland*
Barnham Cross Common	TL 865 815 LA *heathland, scrub*
Berney Marshes	TG 465 055 RSPB *grazing marsh*
Blakeney NNR	TG 007 443 NT (includes Morston & Stiffkey Marshes) *saltmarsh, grazing marsh*
Booton Common	TG 112 228 NWT *fen and scrub*
Brancaster Marshes	TF 772 450 NT *saltmarsh and dunes*
Brettenham Heath NNR	TG 337 166 EN *grass heath, scrub, woodland*
Bure Marshes NNR	TG337 166 EN *swamp alder carr, woodland, fen, open water*
Burgh Common	TG 448 131 NWT *fen, marsh, wet meadow, dykes, woodland*
Calthorpe Broad NNR	TG412 258 EN *carr woodland, grazing marsh, fen, open water*
Cley & Salthouse Marshes	TG 054 441 NWT *Visitor Centre* *shingle bank, grazing marsh*
Cockshoot Broad	TG 343 163 NWT *broad, dykes, carr woodland*
Dersingham NNR	TF665 283 EN *acid valley mire, heathland*
East Winch Common	TF 698 160 NWT *open heath, some bog, pools, woodland*
East Wretham Heath	TL 913 887 NWT *open grassland, meres*

Felbrigg Park
TG 204 403
NT
parkland and lakeside

Foxley Wood
TG 049 229
NWT
ancient woodland

Hickling Broad NNR
TG 428 222
NWT *Visitor Centre.*
broad, fen, grazing marsh, dykes

Hoe Rough
TF 978 168
NWT
unimproved grassland, scrub

Holkham NNR
TF 892 448
EN
saltmarsh, dunes, grazing marsh, woodland

Holme Dunes NNR
TF 697 438
NWT *Visitor Centre*
sand dunes, grassland, salt marsh

Holt Lowes
TG 082 376
NWT
heath, valley bog

Honeypot Wood, Wendling
TF 932 142
NWT
ancient woodland

How Hill
TG 370 191
BA - How Hill Trust
reeds, marsh, carr woodland

Litcham Common
TF 888 173
LA
heath, seasonally wet, woodland

Lolly Moor, Dereham
TF 992 104
NWT
grassland, scrub, woodland

Ludham & Potter Heigham
TG400 180
EN
dykes, grazing marsh

Marston Marshes, Norwich
TG215 055
LA
wet grassland

Martham Broad NNR
TG 466 203
NWT
broad, reed & sedge fen, carr woodland

Narborough Railway Line
TF 750 118
NWT
good chalk flora on disused railway bank

New Buckenham Common
TM 090 906
NWT
wet and dry grassland, scrub

Pensthorpe
TF 948 291
PRIVATE *Visitor Centre*
river, marsh, wet woodland

Ranworth Broad
TG 357 149
NWT *Visitor Centre*
fen, carr, woodland, open water

Reffley Wood
TF 656 223
WT
part ancient wood, part woodland on heath

Ringstead Downs
TF 705 400
NWT
chalk downland with typical species

Roydon Common NNR
TF 697 228
NWT
heathland, valley bog, carr woodland

Scolt Head
TF 465 810
EN
sand dunes, salt marsh, shingle, mud flats

Snettisham
TF 648 333
RSPB
shingle bank and grassland

Southrepps Common
TG 261 351
LA
calcareous fen

Sparham Pools
TG 073 178
NWT
gravel pits and remnant heathland

Strumpshaw Fen
TG 342 066
RSPB
fen, woodland, grazing marsh

Swanton Novers
TG 013 320
EN
ancient woodland (permit only)

Syderstone Common
TF 835 316
NWT
heathland with seasonal pools

Thompson Common
TL 943 967
NWT
grassland, fen, pingoes, woodland

Thursford Wood
TF 979 333
NWT
ancient wood-pasture, ponds

Titchwell Marsh
TF 749 436
RSPB *Visitor Centre*
sand dune, salt marsh and pools

Tyrell's Wood, Long Stratton
TM 205 896
WT
ancient deciduous woodland

Upton Fen
TG 379 137
NWT
fen, dykes, woodland and pools

Wayland Wood
TL 924 995
NWT
ancient woodland

Weeting Heath NNR
TL 756 881
NWT
grass heath (No access beyond hides without permit)

Ouse Washes (Welney)
TL 471 861
RSPB
grazing marsh, part in Norfolk

Wheatfen
TG 325 057
TED ELLIS TRUST
marsh, fen, wet woodland

Winterton Dunes NNR
TG 495 195
EN
sand dunes, heather, pools

Higher Plants Formerly Recorded in the County

This list does not include hybrids or casuals.

Plants last recorded before 1914

	Last known sighting
Andromeda polifolia	19th century
Antennaria dioica	19th century
Atriplex pedunculata	19th century
Carex trinervis	1869
Chenopodium chenopodioides	19th century
Chenopodium vulvaria	19th century
Cuscuta epilinum	19th century
Cuscuta europaea	19th century
Cuscuta trifolium	19th century
Epipactis purpurata	1910
Herminium monorchis	1914
Holosteum umbellatum	1887
Huperzia selago	1903
Hypochaeris maculata	1904
Melampyrum arvense	19th century
Orchis ustulata	19th century
Orobanche ramosa	19th century
Pulicaria vulgaris	pre 1918
Pulsatilla vulgaris	19th century
Ranunculus omiophyllus	19th century
Ranunculus parviflorus	1894
Roemeria hybrida	19th century
Scirpus sylvaticus	19th century
Senecio paludosus	1876
Tephroseris palustris	1898
Teucrium scordium	19th century
Trifolium squamosum	1830

Plants last recorded between 1915 and 1968

	Last known sighting
Adonis annua	1915
Alyssum alyssoides	1951
Arnoseris minima	1953
Carex limosa	1963
Chamaemelum nobile	1964
Chenopodium glaucum	1962
Chenopodium urbicum	1964
Cicendia filiformis	1928
Coeloglossum viride	1954
Eriophorum gracile	1959
Filago pyramidata	1958
Himantoglossum hircinum	1956
Iberis amara	1943
Linum perenne	1920
Mentha pulegium	1961
Mertensia maritima	1931
Ophrys insectifera	1930s
Potentilla neumanniana	1955
Primula elatior	1963
Scleranthus perennis	1961
Spartina maritima	1968
Thesium humifusum	1965
Valerianella rimosa	1941
Veronica verna	1946
Viola persicifolia	1936

Plants seen since 1968 but not recorded during this survey

	Last known sighting
Alchemilla glabra	1983
Barbarea stricta	1983
Bromus interruptus	1972
Deschampsia setacea	1974
Lycopodiella inundata	1971
Potamogeton praelongus	1976
Utricularia intermedia	1979
Vulpia unilateralis	1979

Plants believed lost in Petch & Swann 1968 since refound

Equisetum hyemale
Erophila praecox
Lycopodium clavatum
Muscari neglectum
Orobanche rapum-genistae
Viburnum lantana

Pulsatilla vulgaris

Bryophytes Formerly Recorded in the County

Plants last recorded before 1914

	Last known sighting
Amblyodon dealbatus	early 19th century
Antitrichia curtipendula	early 20th century
Bryum mamillatum	1902
Bryum torquescens	?
Bryum turbinatum	1849
Buxbaumia aphylla	1805
Ditrichum heteromallum	1909
Entosthodon fascicularis	early 20th century?
Entosthodon obtusus	1909
Eurynchium schleicheri	early 20th century
Hamatocaulis vernicosus	1910
Neckera pumila	early 20th century
Orthotrichum pumilum	1907
Sanonia uncinata	early 20th century
Tritomaria exsectiformis	1911
Tortella inclinata	1900
Tortula winsonii	1907
Ulota coarctata	1907

Plants last recorded between 1915 and 1968

	Last known sighting
Acaulon muticum	1967
Aphanorhegma patens	1955
Barbilophozia barbata	1967
Brachythecium glareosum	1957
Bryum intermedium	1959
Cladopodiella franciscii	1962
Cephalozia pleniceps	1967
Cephaloziaella elachista	1967
Cephaloziaella rubella	1952
Ceratodon conicus	1963
Cryptothallus mirabilis	1967
Dicranella rufescens	1956
Dicranum spurium	1961
Drepanoclados sendtneri	1956
Leptodontium flexifolium	1956
Nardia geoscyphus	1930
Preissia quadrata	1956
Pterygoneuron lamellatum	1967
Rhytidiadelphus loreus	1962
Riccia rhenana	1967
Scapania nemorea	1967
Scapania undulata	1966
Scleropodium tourettii	1956
Warnstorfia exannulata	1956

Plants seen since 1968 but not recorded during this survey

	Last known sighting
Calypogeia arguta	1977
Campylopus brevipilus	1968
Didymodon ferrugineus	1979
Lophozia incisa	1968
Lophozia longiflora	1968
Plagiothecium ruthei	1960s?
Ptilium crista-castrensis	1960s?
Racomitrium ericoides	1960s?
Racomitrium lanuginosum	1969
Riccia subbifurca	1979
Splachnum ampullaceum	1973
Tomenthypnum nitens	1970

Photographic Acknowledgements

Paul Banham
p .240

Gillian Beckett
pp. 10, 11, 14, 16, 18, 19, 20 both, 21, 22,, 31, 34, 54, 57, 58, 59, 60, 63, 64, 67, 72, 73, 74, 75, 81 both, 82 left, 85, 87, 88, 91, 92, 93, 97, 99, 102 both, 103, 106, 108, 110, 111 top, 126, 129, 131, 133, 134 bottom, 138, 139, 140, 145, 149, 150, 153, 155 bottom, 156, 157, 158, 159, 167 both, 172, 177, 181, 184, 185, 192, 196, 197, 199, 201, 203, 205, 208 left, 209, 214, 215, 216, 222, 226, 227, 228, 231, 232, 234, 235, 243, 246, 248, 257, 258, 261, 263 right.

Alec Bull
pp. 55, 90, 118, 120, 143, 155 top, 164, 171, 191, 213 top.

Ursula Doyle
pp. 134 top, 217, 256 left.

Bob Ellis
pp. 79, 82 right, 101, 109, 137, 147, 166, 168, 173, 178, 194, 195, 200, 206, 207, 211, 213 bottom, 220, 225, 236, 242, 247, 252, 253, 254, 263 left, 264, 265.

Francis Farrow
pp. 161, 163.

Peter Lambley
pp. 27, 30, 36.

Stephen Martin
pp. 25, 38, 107, 135, 142, 152, 176, 180, 187, 188, 256 right.

Norfolk Wildlife Trust
p. 43.

Eric Rogers
p. 68.

Robin Stevenson
pp. 105, 111 bottom, 268, 269, 270 both, 271, 272 both, 274, 280, 286, 287 both.

Stella Taylor
p. 208 right.

John Williamson
pp. 94, 189, 255.

Bibliography

Alison, H. & Morley, J. 1989 *Blakeney Point and Scolt Head Island* The National Trust.

Ando, H. 1992 Studies in the genus *Hypnum* Hedw. (VIII) *Hikobia* **11**: 111-123

Bartlett, D. 1997 *The Broads Plan 1997* The Broads Authority, Norwich.

Beckett, G. 1994 Plants of the Breckland, 1869-1994 *Trans. Nfk. and Norw. Nat. Soc.* **30**(1), 53-63.

Blom, H.H. 1996 A revision of the *Schistidium apocarpum* complex in Norway and Sweden. *Bryoph. Bibl.* 49. Berlin: J.Cramer

Brenchley, W.E. 1920 *Weeds of Farm Land* Longmans

Burd, F. 1987 *Salt marsh survey of Great Britain: Norfolk.* Nature Conservancy Council unpublished.

Catt, J.W., Corbett, W.M., Hodge, C.A.H., Madgett, P.A., Tatler, W. & Wier, A.H. 1972. Loess in the Soils of North Norfolk. *Journal of Soil Science,* **22**, 444-452.

Clapham, A.R., Tutin, T.G. & Moore, D.M. 1987 *Flora of the British Isles* 3rd Edition C.U.P.

Clarke, W.G. 1925 *In Breckland Wilds* Robert Scott, London.

Clement, E.J. & Foster, M.C. 1994 *Alien Plants of the British Isles* B.S.B.I.

Coles, B.P.F., Funnell, B.M. 1981 Holocene palaeoenvironments of Broadland, England. *Special Publications of the International Association of Sedimentologists,* **5**, 123-131.

Coombe, D.E. 1994 'Maritime' plants of roads in Cambridgeshire *Nature in Cambridge* No.36.

Corbett, W.M. 1973 Breckland Forest Soils. Special Survey No.7. Harpenden.

Corbett, W.M. 1977 Distribution of Loess in North East Norfolk. *Bulletin of the Geological Society of Norfolk,* **29**, 3-22.

Corbett, W.M. 1979 Soils in Norfolk IV. Soil Survey record No.60. Sheet TM28 (Harleston) Harpenden.

Corbett, W.M. 1987 Soils in Norfolk VII. Soil Survey record No.106. Sheet TF82 (Helhoughton) Harpenden.

Corbett, W.M. 1996 *Till Facies and Glaciation in Parts of East Anglia.* Ph.D, Thesis (unpublished)

Corbett, W.M. & Tatler W. 1970 Soils in Norfolk 1. Soil Survey Record No.1. Sheet TM49 (Beccles North), Harpenden.

Corbett, W.M. & Tatler W. 1974 Soils in Norfolk II. Soil Survey Record No.21. Sheets TG13-14 (Barningham/Sheringham), Harpenden.

Corley, M.F.V. & Hill, M.O. 1981 *Distribution of Bryophytes in the British Isles* Cardiff: B.B.S.

Cox, F. & Nickless E.F.P. 1972. Some Aspects of the Glacial History of Central Norfolk. *Bulletin of the Geological Survey of Great Britain,* **42**, 79-98.

Dent, D.L. 1986 Acid sulphate soils: a base for research and development. *International Institute for Land Reclamation and Improvement* Publication 39.

Dudman, A.A. & Richards, A.J. 1997 *Dandelions* B.S.B.I.

Edees, E.S. & Newton, A. 1988 *Brambles of the British Isles* Ray Society.

Ellis, E.A. 1965 *The Broads* Collins

Gallois, R.W. 1979 Geological Investigations for the Wash Water Storage Scheme. Report Institute Geological Sciences, No. 87/19.

Galpin, F.W. 1888 *The Flowering Plants, Ferns and Allies of Harleston* Bartlett and Company

Geldart, A.M. 1913-1914 Sir James Edward Smith and some of his Friends *Trans. Nfk. and Norw. Nat. Soc.* **5**(9)

George, M. 1992 *The Land Use, Ecology, and Conservation of Broadland.* Packard

Godwin, H. 1975 *History of the British Flora.* C.U.P.

Graham, G.G. & Primavesi, A.L. 1993 *Roses* B.S.B.I.

Grimmett, R. & Jones, T. 1989 *Important Bird Areas in Europe* International Council for Bird Preservation. International Waterfowl and Wetlands Research Bureau, Cambridge.

Harris, J. et al. 1997 *Grazing Marsh Dyke Survey 1997* English Nature, Broads Authority, Environment Agency. unpublished

Hill, M.O., Preston C.D. & Smith A.J.E. 1992 *Atlas of the Bryophytes of the British Isles* Vols. 1-3. Colchester: Harley Books.

Hodge, C.A.H., Burton, R.G.O., Corbett, W.M., Evans, R. & Seale, R.S. 1984 Soils and their uses in Eastern England. Bulletin No.13. Soil Survey of England and Wales, Harpenden.

Hubbard, C.E. 1984 *Grasses* 3rd Edition. Pelican Books, London.

Ingrouille, M. 1995 *Historical Ecology of the British Flora* Chapman & Hall

Jermy, A.C. and Tutin, T.G. 1968 *British Sedges* B.S.B.I.

Keble-Martin, W. 1965 *Concise British Flora in Colour* Ebury Press

Kent, D.H. 1992 *List of Vascular Plants of the British Isles* B.S.B.I.

Lambley, P. 1994 Heathlands in North and West Norfolk *Trans. Nfk. and Norw. Nat. Soc.* **30**(1) 53-63.

Lousley, J.L. 1981 *Docks & Knotweeds* B.S.B.I.

Mabey, R. 1996 *Flora Britannica* Sinclair-Stevenson

Marshall, W. 1818 *Review & Abstract of the County Reports to the Board of Agriculture...Norfolk*

Moore, L. 1997 Chalk grassland in Norfolk. *Trans. Nfk. and Norw. Nat. Soc.* **31**, 1-30.

Mundford, G.H. 1863 List of bryophytes in White's *Directory of Norfolk*

Nicholson, W.A. 1914 *A Flora of Norfolk* West, Newman & Co. London.

Norfolk Naturalists Trust 1988 *Guide to the Nature Reserves of Norfolk* Norwich.

Norfolk Wildlife Trust 1998 *Norfolk County Wildlife Sites Handbook.* English Nature and Norfolk County Council, Norwich.

Parmenter, J. 1995 *Report of the 1991-1994 Broadland Resource Survey.* E.N., B.A. Unpublished.

Paton, J.A. 1995 A new Combination for a Variety of *Leiocolea rutheana* J.Bryol.

Paton, J.A. (in press) *Liverwort Flora of the British Isles*

Payne, R.M. 1995 The Flora of King's Lynn *Trans. Nfk. and Norw. Nat. Soc.* **30** (3) 317-342

Payne, R.M. 1998 *A Flora of Walls in West Norfolk.* Privately printed.

Perrin, R.M.S., Davies, H. & Fysh, M.D. 1974 Distribution of Late Pleistocene Deposits in Eastern and Southern England. *Nature* **248**, 320-324, London.

Perrin, R.M.S., Rose, J. & Davies, H. 1979 The Distribution, variation and origin of pre-Devensian tills in Eastern England. *Philsophical Trans. Royal Soc. London.B.* **287** 535-570.

Perring, F.H. & Farrell, L. 1983 *British Red Data Book:* 1 Vascular Plants 2nd Ed. N.C.C.

Perring, F.H. & Walters, S.M. eds. 1962 *Atlas of the British Flora* Nelson

Petch, C.P. and Swann, E.L. 1968 *The Flora of Norfolk* Jarrold and Son, Norwich.

Petch, C.P. and Swann, E.L. 1962 West Norfolk Plants Today *Proc. B.S.B.I.* **4**, Part 4

Pigott, B.A.F. 1880s *Flowers and Ferns of Cromer and its Neighbourhood* Jarrold and Sons London.

Preston, C.D. 1995 *Pondweeds* B.S.B.I.

Rackham, O. 1986 The ancient woods of Norfolk. *Trans. Nfk. and Norw. Nat. Soc.* **27**, 166-177.

Radley, G. 1994. *Sand dune vegetation survey of Great Britain: a national inventory: part 1, England.* J.N.C.C. Peterborough

Rich, T.C.G. 1991 *Crucifers of Great Britain & Ireland* B.S.B.I.

Rich, T.C.G. & Jermy, C. 1998 *Plant Crib, 1998* B.S.B.I.

Rihan, J.R. & Gray, A.J. 1991 The Distribution of Hybrid Marram Grass, *X Calammophila baltica* (Flugge) Brand, in the British Isles *Watsonia* 18: 369-379

Rodwell, J.S. ed. 1991-1996. *British Plant Communities*, 1-4. C.U.P.

Rose, F. 1981 *The Wild Flower Key.* Warne

Rose, F. 1989 *Grasses, Sedges, Rushes & Ferns.* Viking

Ryves, T.B., Clement, E.J. and Foster M.C. 1996 *Alien Grasses of the British Isles* B.S.B.I.

Sales, F. 1993 Taxonomy and Nomenclature of Bromus sect. Genea *Edinburgh J.Bot.* **50**: 1-31

Salisbury, E.J. 1931-1932 East Anglian Flora *Trans. Nfk. and Norw. Nat. Soc.* **13**: Part III.

Seale, R.S. 1974 Geology of the Ely District. *Bull. Geological Soc. Norfolk* **25**, 21-36.

Seale, R.S. 1975 Soils of the Ely District, sheet 173. Memoirs of the Soil Survey of England and Wales, Harpenden.

Sell, Peter & Murrell, Gina. 1996. Flora of Great Britain and Ireland Vol.5. C.U.P.

Simpson, F.W. 1982 *Simpson's Flora of Suffolk* Suffolk Naturalists Society, Ipswich

Smart, S. 1992-3 *Norfolk Valley Fen Survey.* English Nature unpublished.

Smith, A.J.E. 1978 *The Moss Flora of Britain and Ireland* C.U.P.

Smith, A.J.E. 1990 *The Liverworts of Britain and Ireland* C.U.P.

Sneddon, P. & Randell, R.E. 1994. *Coastal vegetated shingle structures of Great Britain: Appendix 3. Shingle sites in England.* J.N.C.C Peterborough.

Stace, C.A. 1975 *Hybridisation and the Flora of the British Isles* Academic Press

Stace, C.A. 1997 *New Flora of the British Isles* 2nd Edition C.U.P.

Stevenson, C.R. 1992 A Check List of the Mosses of Norfolk. *Trans. Nfk. and Norw. Nat. Soc.* **29**(3) 189-195

Stevenson, C.R. 1993 A Check List of the Liverworts of Norfolk. *Trans. Nfk. and Norw. Nat. Soc.* **29**(5) 367-370

Stevenson, C.R. 1994 Changes in the Status of Norfolk Bryophytes. *Trans. Nfk. and Norw. Nat. Soc.* **30**(1) 32-43

Stewart, A, Pearman, D.A. & Preston, C.D. eds. 1991 *Scarce Plants in Britain* J.N.C.C. Peterborough

Straw, A. 1960 The Limit of the Last Glaciation in North Norfolk. *Proceedings of the Geological Association* **71**, 379-390

Swann, E.L. 1975 *Supplement to the Flora of Norfolk* F.Crowe & Sons, Norwich

Swann, E.L. 1982 Norfolk Bryophytes Today. *J.Bryol:* 12(1) 77-112

Tatler, W. & Corbett, W.M. 1977 Soils in Norfolk 111. Soil Survey Record No.41. Sheet TG31 (Horning), Harpenden.

Trimmer, K. 1866 *Flora of Norfolk* Hamilton Adams and Co.London.

Trimmer, K. 1885 *Supplement to the Flora of Norfolk*

Trist, P.J.O. 1979 *An Ecological Flora of the Breckland* Academic Press

Waller, M. 1994 The Fenland Project No.9: Flandrian Environmental Change in Fenland. East Anglian Archaeology Report No.70. Fenland Project Committee, Cambridge County Council.

Warburg, E.F. 1963 *Census Catalogue of British Mosses.* B.B.S.

Washbourne, R. ed 1976 *Nature in Norfolk: A Heritage in Trust* Jarrolds, Norwich.

Watson, W.C.R. 1956 *Rubi of Great Britain and Ireland* C.U.P.

Wheeler, B. 1978 The Wetland Plant Communities of the River Ant Valley, Norfolk. *Trans. Nfk. and Norw. Nat. Soc.* **24**, 153-187.

Williamson, T. 1997 *The Norfolk Broads, A Landscape History.* Manchester University Press.

Young, A. 1969 reprint *General View of the Agriculture of the County of Norfolk, 1804.* David & Charles

Abbreviations used in the text.

A.C.	Arthur Copping
A.C.S.	Alex Smith
A.L.B.	Alec Bull
C.E.H.	Charles Hubbard
C.J.L.	C.J. Lowe
C.P.P.	Charles Petch
C.R.S.	Robin Stevenson
D. & B.M.	David & Barbara Mathias
D.F.S.	David Strauss
D.L.H.	D.L. Holyoak
D.M.M.	Dorothy Maxey
E.L.S.	Eric Swann
E.T.D.	Ernest Daniels
E.V.R.	Eric Rogers
F.F.	Francis Farrow
F.S.	Frances Schumann
G.B.	Gillian Beckett
G.M	George Maybury
H.L.K.W.	Harold Whitehouse
H.W.	Heather Williamson
I. & F.S.	Ian & Frances Simper
I. & P.G	Ian & Paul Green
J.A.	J. Appleyard
J.E.G.	Jean Gaffney
J.H.	Joyce Humphris
J.M.	John Mott
J.M.P.	Jo Parmenter
J.W.	John Williamson
K.A.B.	Kenneth Beckett
K.K.H.	Kerry Harrison
M.A.B.	Ann Brewster
M.F.V.C.	M.F.V. Corley
M.K	Michael Keene
M.O.H.	Mark Hill
N.G.H.	Nick Hodgetts
P.A.W.	P. Wright
P.G.L.	Peter Lawson
P.R.B.	Paul Banham
P.R.C.	Paul Cobb
P.W.L.	Peter Lambley
R.C.S.	R.C. Stern
R.J.F.	Richard Fisk
R.M.P.	Ron Payne
R.P.L.	Richard Libbey
R.W.E.	Bob Ellis
S.F.	Stuart Fisher
S.P.	Shane Plant
S.W.	Stan Woodell
W.A.B.	Bill Boyd
W.M.	Bill Mitchell

A.W.	Anglian Water
B.A.	Broads Authority
B.B.S.	British Bryological Society
B.S.B.I.	Botanical Society of the British Isles
E.A.	Environment Agency (formerly National Rivers Authority)
E.N.	English Nature
E.S.A.	Environmentally Sensitive Area
F.E.	Forest Enterprise (formerly Forestry Commission)
J.N.C.C.	Joint Nature Conservation Council
M.A.F.F.	Ministry of Agriculture, Food and Fisheries
N.C.C.	Nature Conservancy Council, (also Norfolk County Council.)
N.W.T.	Norfolk Wildlife Trust (formerly Norfolk Naturalists' Trust.)
R.S.P.B.	Royal Society for the Protection of Birds
S.S.S.I.	Site of Special Scientific Interest
W.T.	Woodland Trust

E.	East Norfolk VC27
W.	West Norfolk VC28

W. & E.	Numbers in the text refer to 10km squares. Numbers at the end of non-mapped entries are the number of tetrads in which the plant has been recorded.

Map of Places Mentioned in the Text

Honeypot Wood - Green place names represent nature reserves

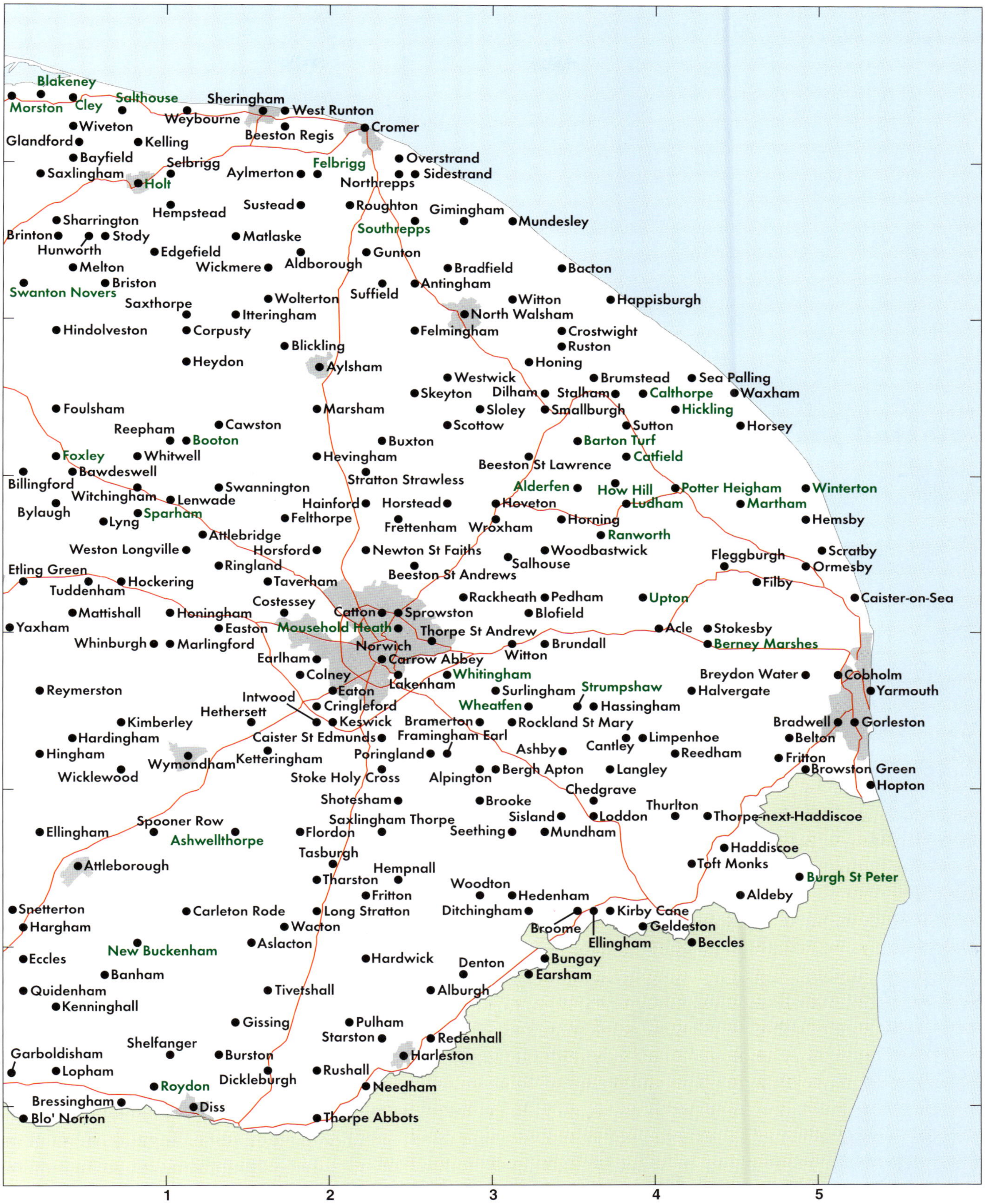

Honeypot Wood - Green place names represent nature reserves

Key to Maps and Species Entries

Map Backgrounds

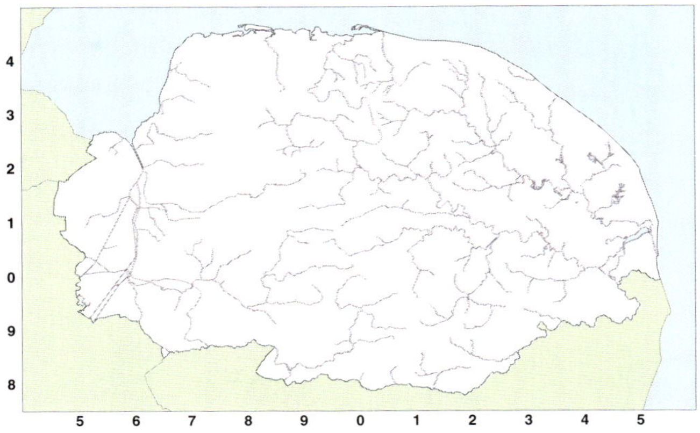

Rivers

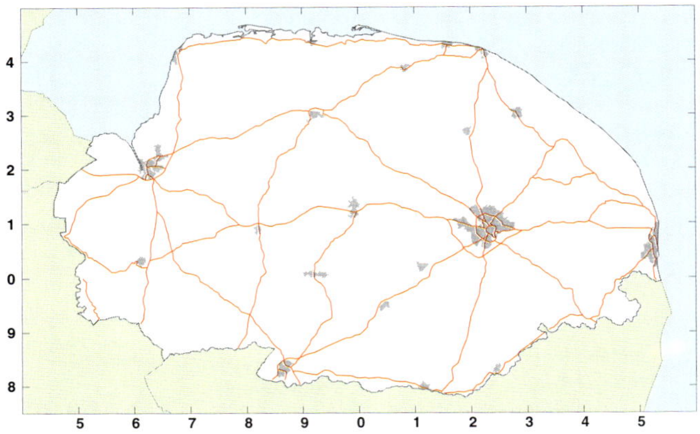

Roads with urban areas

Selected soils

Clay and peat

Earthy peat

Estuarine alluvial soils, mainly acidic

Estuarine alluvial soils, mainly calcareous

River valley alluvial soils

Wet clays, mainly calcareous

Clay loams

Brown sands and earths over chalk

Brown earths over drift

Calcareous sands

Acidic sands

Sand dune and shingle

Saltmarsh

For fuller details see map and key on page 12.

Mowing Marsh - How Hill

■ before an entry indicates that the plant has not been seen during this survey.

W. & E. numbers at the end of non-mapped entries are the number of tetrads in which the plant has been recorded.

W = West Norfolk, V.C. 28

E = East Norfolk, V.C. 27

LYCOPODIACEAE

■ **Huperzia selago** (L.) Bernh. ex Schrank & Mart. **Fir Clubmoss**

Syn. *Lycopodium selago* L.
Recorded by Trimmer, 1866, on heaths from Edgefield to Felthorpe E03, 13, 23. Last seen in the Holt area E03 in 1903. Now considered extinct not only in Norfolk, but in most of lowland Britain.

■ **Lycopodiella inundata** (L.) Holub. **Marsh Clubmoss**

Syn. *Lycopodium inundatum* L.
Given 10 localities by Trimmer in 1866, most of which have been afforested or become overgrown. Last records are from Holt E03 in 1959; North Wootton W62 from 1949 to 1962; Felmingham E22 and Hevingham E12 in 1971.

Lycopodium clavatum L. **Common Clubmoss**

Recorded by Trimmer, 1866, in 10 sites on heaths. Recent records are from Lenwade E11 in 1969 and North Walsham E22 in 1971. The only extant sites are at Beeston E14, where two plants were found in 1994 by D.Mower during scrub clearance. It was recorded at the same site at the beginning of the 20th century (comm. KC.Durrant) and may possibly have survived unnoticed as may the plant found at Leziate W61, in 1997 by SF, alternatively they may have been the result of the arrival of chance spores. 1W. 1E.

SELAGINELLACEAE

Selaginella kraussiana (Kunze) A.Braun **Krauss's Clubmoss**

Introduced to gardens from tropical and southern Africa and recorded as well established at Felbrigg, E14 by Dr E.Beaumont, also at Fincham, W70 by CRS, both in 1985. 1W. 1E.

EQUISETACEAE

Equisetum hyemale L. **Dutch Rush, Rough Horsetail**

Winter green perennial native in northern England and Scotland, but considered introduced in Norfolk. A few localities were given in earlier floras, all of which appear to refer to finds made in the 18th century. Found at Stibbard W92 in 1974, R.Harmer and at Shipdham W90 in 1988, DMM, both in wet ditches on heavy soils and associated with woodland plants, still present. W2. E0.

[**E. variegatum** Schleich. ex Web. & Mohr **Variegated Horsetail**

This northern species was reported from Croxton W88 in the 1890s, but the record must be considered an error.]

E. fluviatile L. **Water Horsetail, Toad Pipes** (old local name)

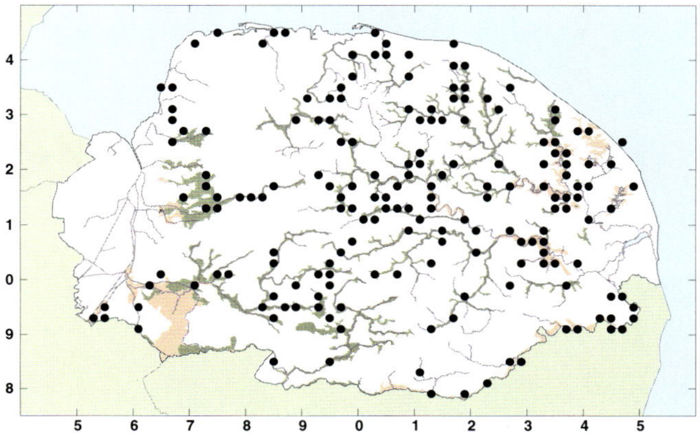

Deciduous perennial occurring in ditches, streams and occasionally in ponds, always in standing water.

E. arvense L. **Field Horsetail**

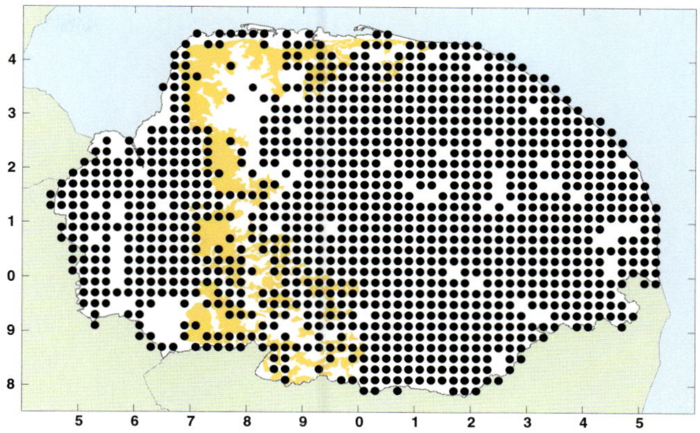

Deciduous perennial, absent from only the driest soils, a pernicious weed in cultivated land.

E. sylvaticum L. **Wood Horsetail**

Deciduous perennial, recorded from a number of sites in the 19th century including Holt Lowes E03 where it was listed in the Botanists' Guide of 1805. It is still to be found there in what is now its only known locality in the county. W0. E1.

E. palustre L. **Marsh Horsetail**

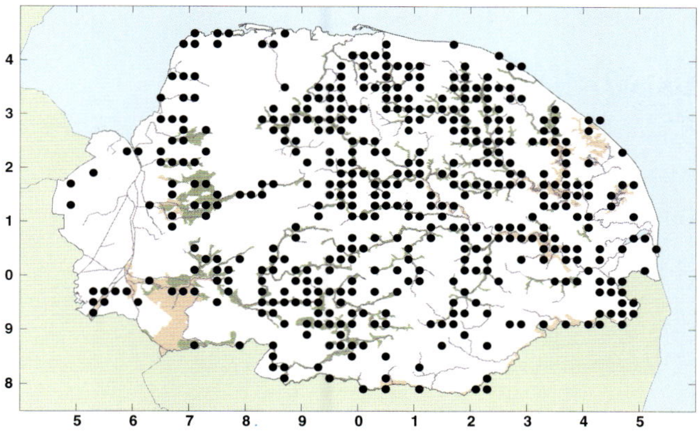

Deciduous perennial, frequent in marshy ground and ditches.

E. telmateia Ehrh. **Great Horsetail**

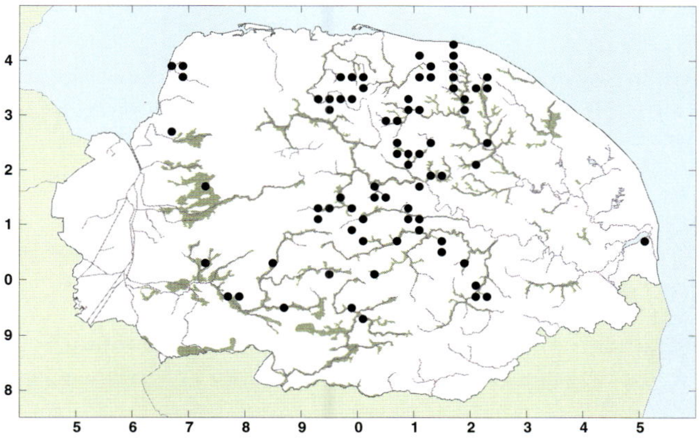

Deciduous perennial of deep, wet soils, frequently associated with spring lines. In the past the silica-rich stems were dried and powdered as an abrasive for fine polishing (EAE in litt.).

OPHIOGLOSSACEAE

Ophioglossum vulgatum L. **Adder's Tongue**

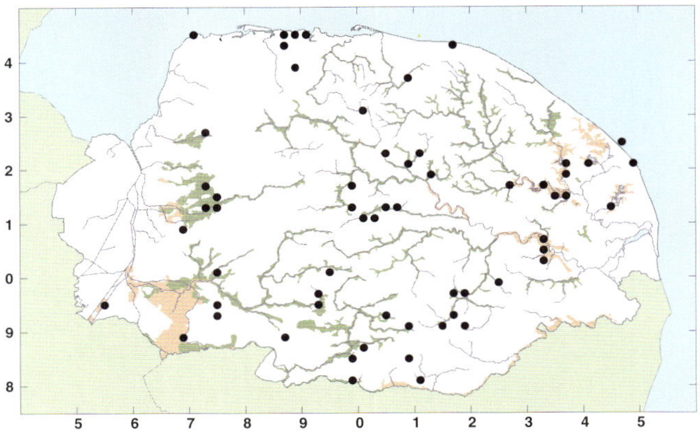

Herbaceous perennial, confined to undisturbed sites, but in a wide variety of habitats including moist woods, marshes, old pastures, chalk pits, sand dunes, fens and reed beds.

Botrychium lunaria (L.) Sw. **Moonwort**
Deciduous perennial of light, sandy soils with 15 sites in Nicholson's 1914 Flora, mostly on heaths and dry grassland. Found in two sites during the present survey, in dune slacks at Wells-next-the-Sea W94 in 1996 by P&E.Banham and on Weeting Heath W88 in 1989, P.Dolman. Of the old sites, Kelling E04 and Ringland E11 are in east Norfolk, all the others are in Breckland as are other records, Harling W98, 1969 and near Thetford W88, 1972. W2. E0.

OSMUNDACEAE

Osmunda regalis L. **Royal Fern**

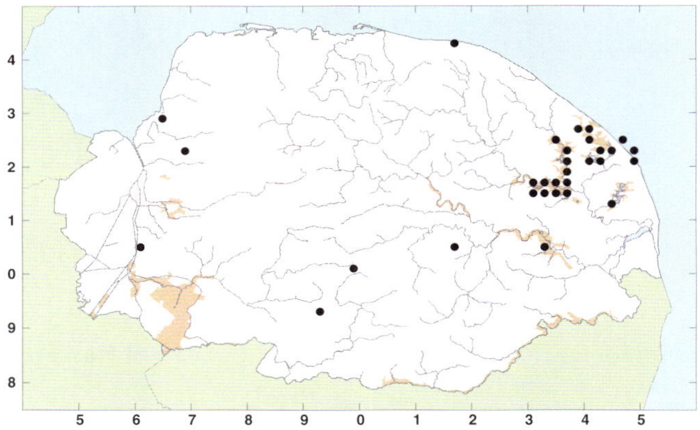

Deciduous perennial of deep, wet, acid soils, also on the acidic sand dunes of the east coast. Still frequent in Broadland but elsewhere often crowded out by scrub and trees and generally decreasing. The plant recorded from a roadside hedgebank at Wimbotsham W60 in 1995, RMP presumably came from a chance spore.

ADIANTACEAE

Adiantum capillus-veneris L. **Maidenhair Fern**
Introduced to gardens from warm temperate and tropical areas, not fully hardy. One plant seen in Elm Hill, Norwich, E20 in 1986 by KH.Bull. Reported as 'tidied away' the following spring. W0. E1.

PTERIDACEAE

Pteris cretica L. **Ribbon Fern**
Evergreen, sub-tropical fern recorded from King's Lynn W62 in 1993, RMP on a wall above basement steps. The plant was destroyed the following year. W1. E0.

MARSILIACEAE

Pilularia globulifera L. **Pillwort**
Formerly on heathy commons both north of Norwich and in the King's Lynn area, seen at Horsford E11 in 1951 and Derby Fen W72 in 1960. Usually found growing at the margins of pools. It was observed at Lound Waterworks, Hopton E50 by K.Rivett in 1991, carpeting the sandy foreshore of the reservoir, where it had been noted previously by EAE in 1964. It was also found forming a 2m wide fringe around a small pond in a horse pasture at Rollesby E41 in 1998, Flora Group. W0. E2.

POLYPODIACEAE

Polypodium sp. **Common Polypody**

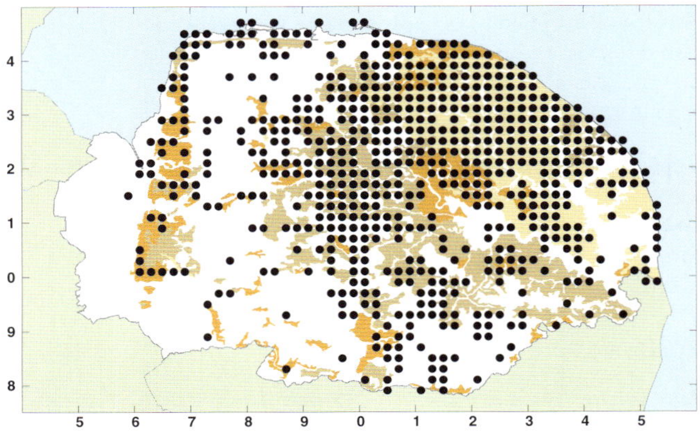

Early in the survey determination of the two species was often based on the shape of the fronds alone. The need to use spore characters means that these records cannot be allowed, so our understanding of the distribution of the two species is incomplete.

P. vulgare s.s. L.
Winter green perennial found on shady banks on moderately acid soils, sand dunes and occasionally on walls, particularly on the greensand. Apparently by far the less common species on walls. W105. E248.

P. interjectum Shivas
Winter green perennial, apparently the dominant species in Norfolk including most records from walls and those on hedgebanks in the north-east of the county, perhaps where they get sufficient minerals from winds off the sea. W122. E240.

P. x mantoniae Rothm. & U. Schneider **Hybrid Polypody**
First recorded at Felmingham E22 in 1972, it has been found during the survey at a number of sites and probably occurs wherever the two parent species grow together. It can form extensive colonies and may sometimes be more abundant than either parent. W6. E7.

DENNSTAEDTIACEAE

Pteridium aquilinum (L.) Kuhn **Bracken**

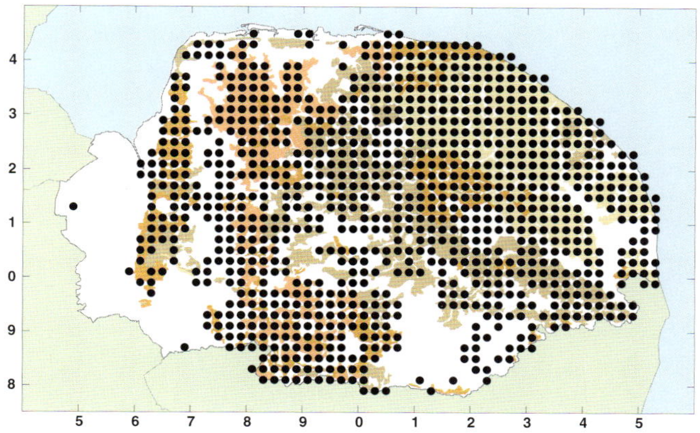

Deciduous and frost-tender perennial, absent only from waterlogged and very calcareous ground. It is most successful where there is a little cover from trees and scrub which protect the young fronds from the effects of spring frosts. The covering of heaths with trees and scrub, whether intentional or not, has been very much in its favour. In parts of the county it is still possible to trace former heathland by the bracken which remains in hedgerows and on roadside verges. It is also a feature of urban walls in King's Lynn, W61 and W62, RMP and Norwich E20, ETD.

THELYPTERIDACEAE

Thelypteris palustris Schott **Marsh Fern**
Syn. *T. thelypteroides* Michx.

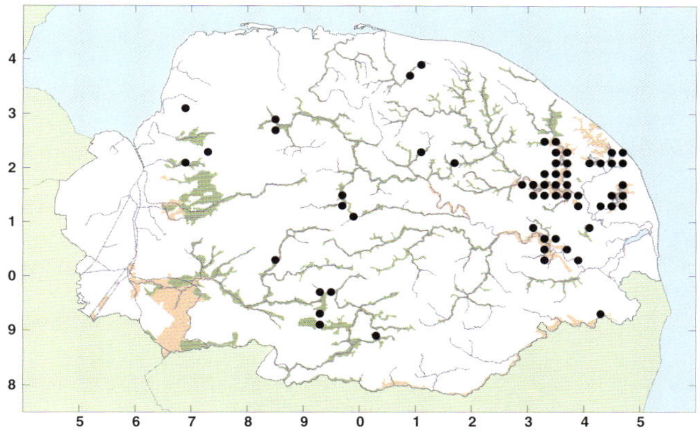

Frequent on peat in open marshes, carr and alderwoods where it can sometimes be the dominant ground cover, particularly so in carr which has developed over open water. A nationally scarce species.

Oreopteris limbosperma (Bellardi ex All.)Holub
Lemon-scented Fern, Mountain Fern
Syns. *Thelypteris limbosperma* (Bellardi ex All.) H.P.Fuchs, *Thelypteris oreopteris* (Ehrh.) Sloss.
Chiefly a northern and western species it was described as rare by Nicholson in 1914 but in 19 localities, most in east Norfolk where it has yet to be re-found. During this survey apparently confined to moist, acid woods at Shouldham and Wormegay, W61 and Runcton Holme W60, present in all sites in 1997, RMP. W4. E0.

ASPLENIACEAE

Phyllitis scolopendrium (L.) Newman **Hart's-tongue Fern**
Syn. *Asplenium scolopendrium* L.

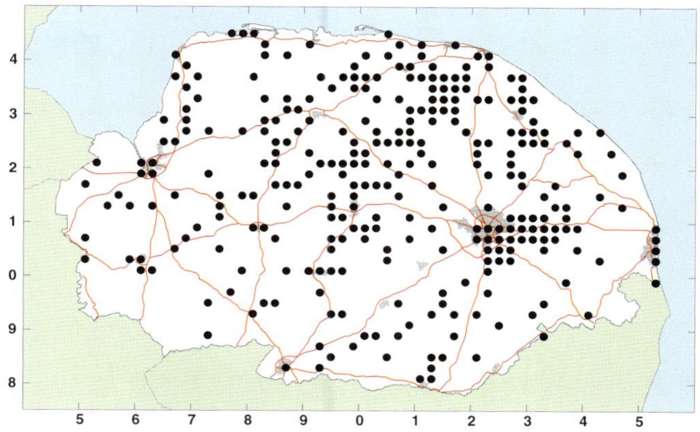

Winter green perennial, in Norfolk most frequently found on walls and once considered a characteristic plant of well shafts. It also occurs, though less frequently, on moist banks, ditch sides and in woods.

Asplenium adiantum-nigrum L. **Black Spleenwort**

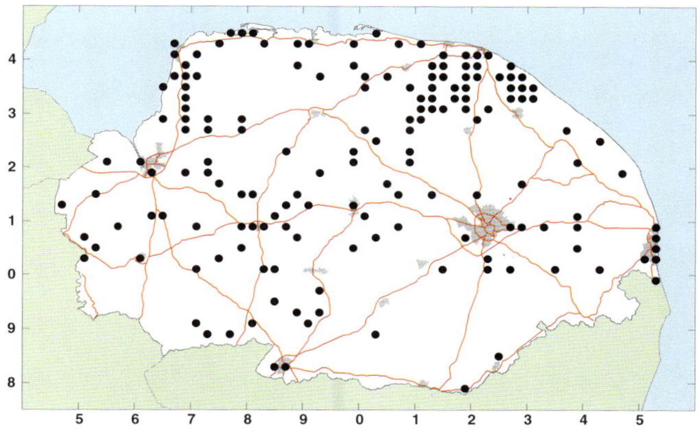

Winter green perennial, almost all records are from walls but also found very occasionally on shady, moist banks.

A. trichomanes L. **Maidenhair Spleenwort**

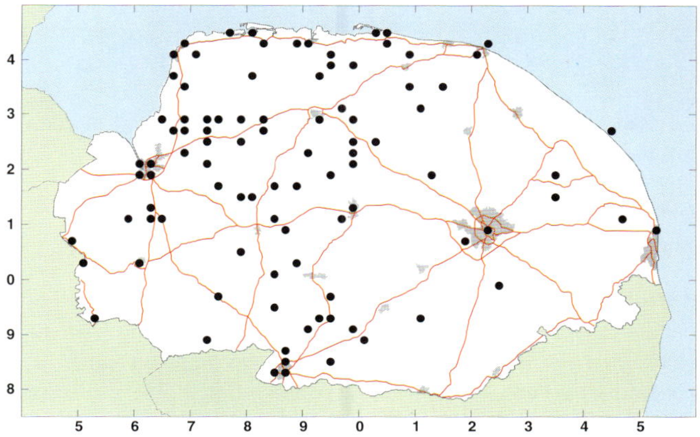

Winter green perennial almost confined to walls, still known from Stanhoe church W83 where it was recorded by Trimmer before 1842. All our plants are ssp. **quadrivalens**. D.E.Mey.

A. ruta-muraria L. **Wall Rue**

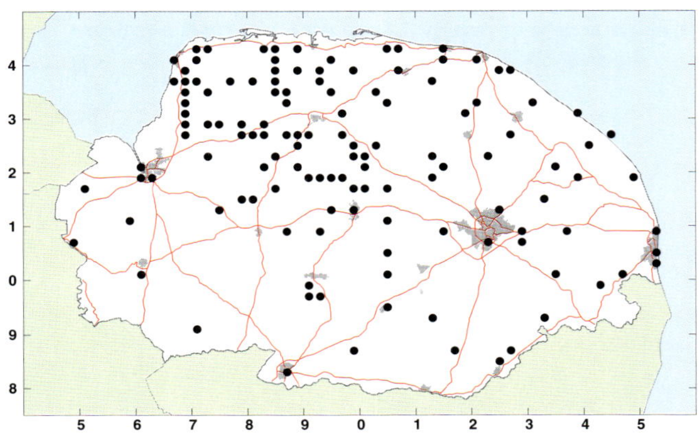

Winter green perennial. All records received are from walls.

Ceterach officinarum Willd. **Rusty-back**
Winter green perennial with a south-western distribution in Britain. Only 12 localities have ever been recorded from Norfolk. All have been on walls. At present it is known from Heacham W63; Sedgeford and Shernborne W73; Great Snoring W93; Croxton W88; Loddon E39; Norwich riverside E20 and Heydon E12 where it was recorded by Trimmer, 1866, 'above the reach of ladders', as is still the case. W5. E3.

Ceterach officinarum - Sedgeford

WOODSIACEAE

Matteuccia struthiopteris (L.) Tod. **Ostrich Fern**
Deciduous fern native to North America and introduced to gardens. Naturalised on Dersingham Fen W62, in damp areas amongst scattered birch trees where it was first observed in 1978, also beside the lake at Thickthorn Hall, Hethersett, E10. W1. E1.

Athyrium filix-femina (L.) Roth **Lady Fern**

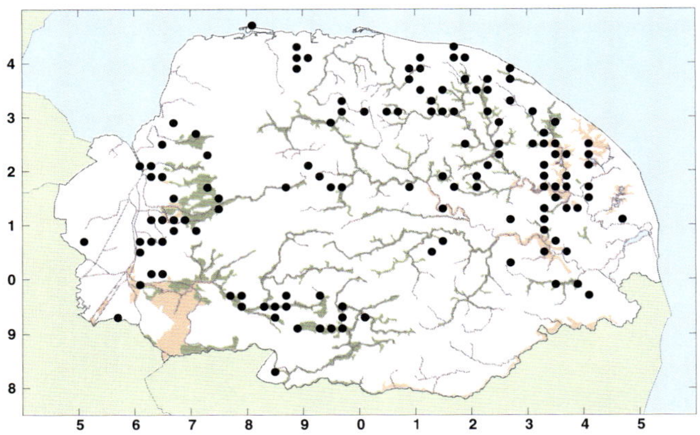

Deciduous fern of shady woods and moist sites on acid soils, also occasionally on old walls, often where water runs down from faulty guttering.

Gymnocarpium dryopteris (L.) Newman **Oak Fern**
Syn. *Thelypteris dryopteris* (L.) Sloss.
Deciduous fern found chiefly in northern England, not recorded for Norfolk in earlier floras. A casual from chance spores. Found on a disused railway bridge at Mundesley E33 in 1959 and still present in 1965, also in conifer forest near Brandon W88, in 1989, MK.W1. E0.

■ **G. robertianum** (Hoffm.) Newman **Limestone Fern**
Syn. *Phegopteris robertiana* (Hoffm.) Sloss.
Deciduous fern native to limestone areas. Known from the railway bridge near St Helen's Well, Santon W88 before 1909 and until 1929 when it was last seen. The old bridge was demolished and a new one built during 1939-1940.

■ **Cystopteris fragilis** (L.) Bernh. **Brittle Bladder Fern**
Deciduous fern recorded three times this century from walls, at Salhouse station E31 in 1958; the disused airfield at Hardwick E29 in 1962 and Sheringham Hall, E14 in 1974.

DRYOPTERIDACEAE

Polystichum setiferum (Forssk.) T. Moore ex Woyn.
Soft Shield Fern

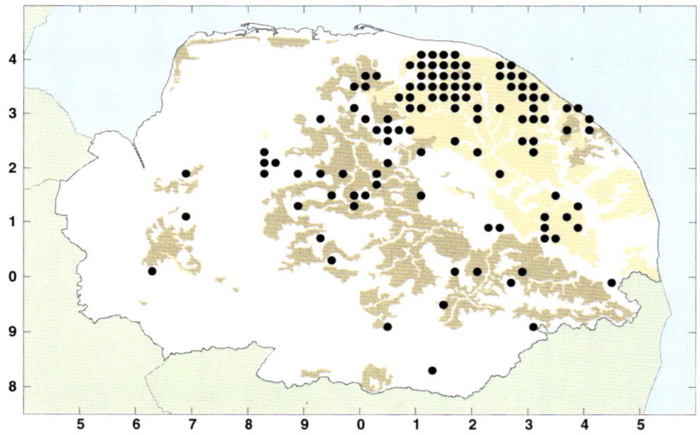

Winter green fern found in wet ditches, hedgerows and woods on moist, acid soils, very local in its occurrence.

P. aculeatum (L.) Roth **Hard Shield Fern**

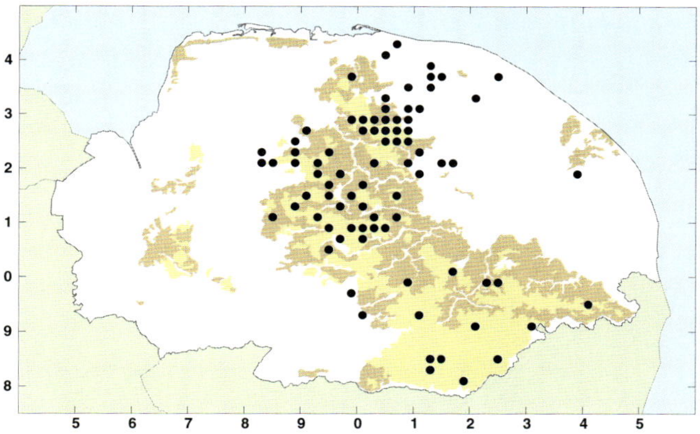

Winter green fern found in similar sites to *P. setiferum*, often growing with it, but generally less frequent except on clay soils in central Norfolk. Generally a plant of northern Britain. In spite of the close proximity of the parents, their hybrid, *P. x bicknellii* does not appear to have been recorded from Norfolk.

Dryopteris filix-mas (L.) Schott **Male Fern**

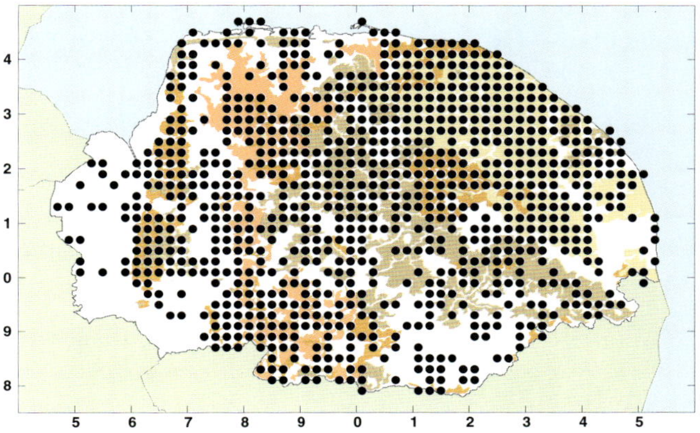

Deciduous fern occurring widely in both moist and dry woods, hedgerows, sand dunes and on walls, but absent from the peat fens. Generally more tolerant of open sites than *D. dilatata*.

D. affinis (Lowe) Fraser-Jenk. **Scaly Male Fern**

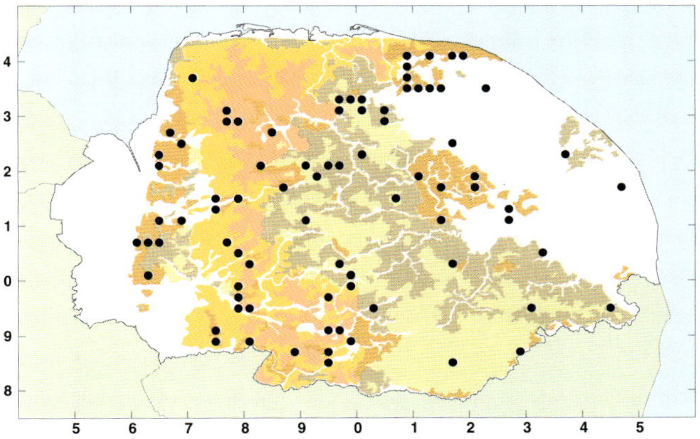

Mainly winter green perennial, found in wetter sites than *D. filix-mas*. Ssp. **affinis** (syn. *D. pseudomas* (Woll) Holub & Pouzar) appears to be confined to ancient woodland while ssp. **borreri** (Newman) Fraser-Jenk., syn. *D. borreri* (Newman) Newman ex Oberh. & Tavel occurs more widely in shady sites.

D. cristata (L.) A. Gray **Crested Buckler Fern**

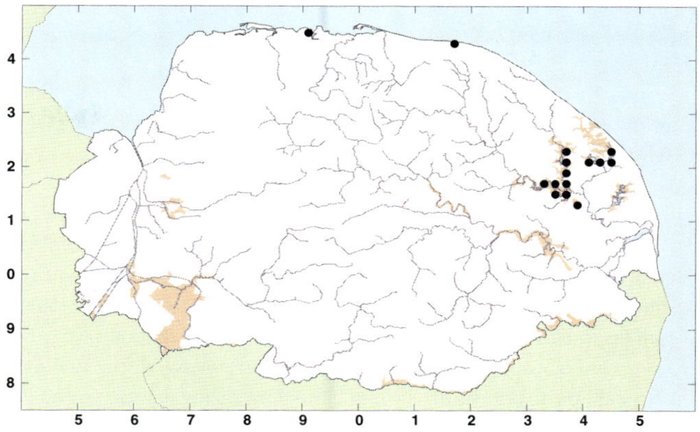

Once frequent in suitably wet sites throughout the county, but already in 1914 said by Nicholson to have become rare and quoting from a letter of 1868 "nearly exterminated by the fern dealers". Found in the 19th century at Edgefield and Holt E02; Dersingham W62; Bawsey W62; Leziate W62 and at Scoulton W90 until 1977, but now frequent only in Broadland where it thrives best in acid soils, particularly where sphagnum mosses are present. Also found at two sites on the north coast. Red Data Book Species.

Dryopteris cristata - Horsey

D. x uliginosa (A. Braun ex Döll) Kuntze ex Druce
(D. cristata x D. carthusiana)
Not infrequent where both parents are to be found, now only in the Broads. WO. E1.

D. x deweveri (J.T.Jansen) Wachter
(D. carthusiana x D. dilatata)
Rare, shown to RWE & ALB at Hickling E42 by AC.Jermy in 1997 and subsequently found at a number of other sites in east Norfolk. WO. E10.

D. carthusiana (Vill.) H.P. Fuchs **Narrow Buckler Fern**
Syns. *D. spinulosa* Kuntze, *D. lanceolata-cristata* (Hoffm.) Alston

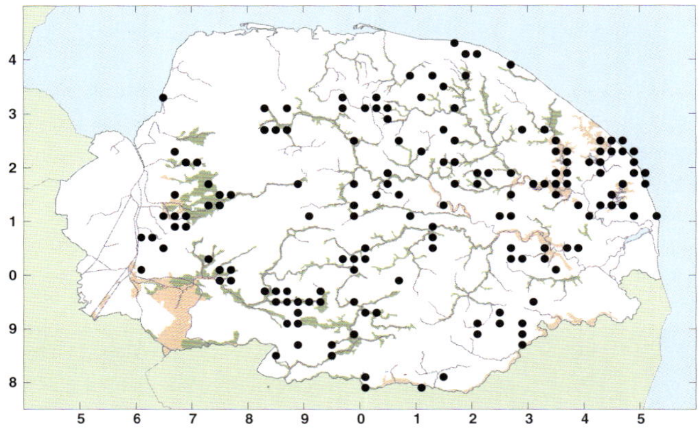

Deciduous perennial of damp, shady areas in marshes and woods where it can be abundant, but much less frequent than *D. dilatata.*

D. dilatata (Hoffm.) A. Gray **Broad Buckler Fern**
Syn. *D. austriaca* Woynar ex Schinz & Thell. non Jacq.

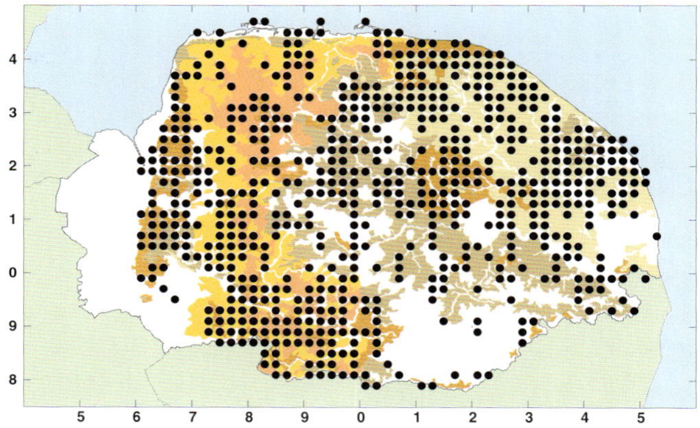

Deciduous perennial with a wide tolerance of moisture. Abundant on the dry soils of sand dunes at Holkham Meols W84; Wells-next-the-Sea W94 and Winterton E41 and, apart from bracken, the most abundant fern in the coniferous plantations of the Breckland, but also frequent in slightly acid reed beds as at Hickling and Horsey E42. Usually in at least partially shaded sites.

BLECHNACEAE

Blechnum spicant (L.) Roth **Hard Fern**

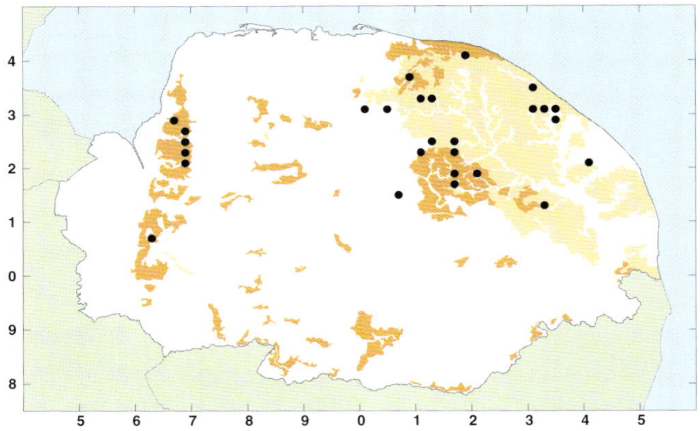

Almost confined to moist, acid soils, occurring on wet heaths and damp woods, especially along ditch-sides. Scarce in central and eastern Britain.

AZOLLACEAE

Azolla filiculoides Lam. **Water Fern**
Syn. *A. caroliniana* Willd.

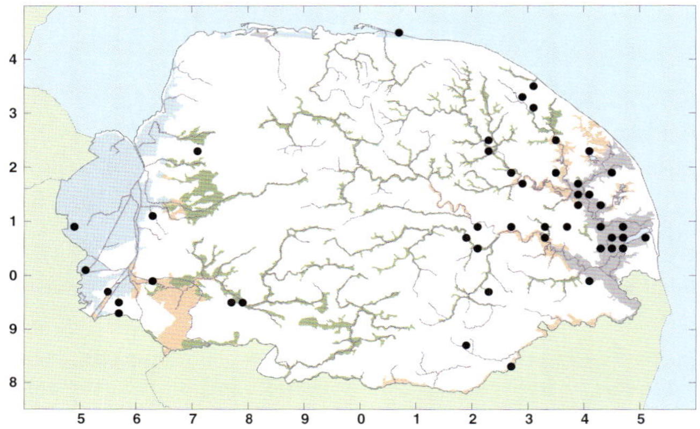

First recorded as introduced to a pond in Chapel Field Gardens, Norwich in 1903, also at Woodbastwick. Spread from there by the 1912 flood and since then frequent in the Bure and Wensum. Occasionally in isolated ponds, often a throw out from garden ponds and aquaria but possibly also distributed by ducks. Unlikely to survive in still water in very hard winters unless in very sheltered sites.

CONIFERAE

All conifer species now growing in Norfolk were originally planted, but many can be found in semi- natural situations such as parks and "improved" woodland or as ornamentals bordering more modern commercial forestry. Only those species observed to be regenerating naturally have been included in the main list, but the following can also be found growing in such sites:- *Abies procera* (syn. *A. nobilis*), *A. concolor, Picea sitchensis, Araucaria araucana, Cedrus deodara, C. libani, C. atlantica, Tsuga canadensis, Pinus radiata, P.contorta, P. pinea, P. wallichiana, Cupressus macrocarpa, Chamaecyparis nootkatensis, X Cupressocyparis leylandii, Sequoia sempervirens, Sequoiadendron giganteum, Cryptomeria japonica, Metasequoia glyptostroboides* and *Taxodium distichum.*

PINACEAE

All winter green conifers apart from Larix.

Abies grandis (Douglas ex D. Don) Lindl. **Grand Fir**
Introduced from western North America, this is the most commonly planted silver fir in Norfolk, producing thickets of seedlings, some several years old, in plantations on acid soil at Garboldisham W98 and Lynford W89. W15. E17.

A. cephalonica Loudon **Greek Fir**
Up to 10 old trees scattered around Stanford parish (Stanford Training Area) where seedlings have been found, though sheep grazing will probably prevent their survival. W1. E0.

Pseudotsuga menziesii (Mirb.) Franco **Douglas Fir**
Introduced from western North America and widely planted commercially. Successfully regenerating from seed in shady, woodland sites with acidic soils. W47. E63.

Tsuga heterophylla (Raf.) Sarg. **Western Hemlock**
Introduced from western North America and sometimes planted for commercial forestry. Regenerates freely in moist, acid, shady sites. W13. E14.

Picea abies (L.) H.Karst. **Norway Spruce**
Introduced from mainland Europe, this spruce, most familiar as the Christmas tree, is not generally a successsful tree in East Anglia and is only occasionally found regenerating. W56. E87.

Larix decidua Mill. **European Larch**
Deciduous conifer from Europe, formerly much planted for timber and sometimes regenerating. Now very largely replaced by the hybrid which follows. W64. E114.

L. x marschlinsii Coaz. **Hybrid Larch**
Syn. *L. x eurolepis* A.Henry nom. illeg. *(L. decidua x L. kaempferi)*
Deciduous conifer, now the chief planted larch. A moderately fertile hybrid, regenerating though not freely. W51. E44.

L. kaempferi (Lindl.) Carrière **Japanese Larch**
Deciduous conifer from Japan, occasionally planted and regenerating. W7. E7.

Pinus sylvestris L. **Scots Pine**
Widely planted both commercially and in parks for ornament for several centuries. In the Breckland area it was planted early in the 19th century for hedging and for a while kept regularly cut. These hedges were soon abandoned and the resulting rows of gnarled trees are characteristic of the area and known locally as deal rows. The tree regenerates freely from seed especially in open, heathy soils where it is often invasive. W314. E540.

Pinus sylvestris - Foulden

P. nigra Arnold **Corsican Pine**
Introduced from Europe, the straight trunked ssp. **laricio** Maire, (ssp. *corsica*) is now more widely planted commercially than other pine. Ssp. **nigra** (ssp. *austriaca*), the Austrian pine was planted from 1855 onwards on sand dunes at Holkham and Wells-next-the-Sea W84 and 94 along with some Corsican pine and there both regenerate well. In forest conditions seedlings are less common than those of *P. sylvestris*. W131. E140.

P. pinaster Aiton **Maritime Pine**
Introduced from the Mediterranean, planted from 1878 onwards on sand dunes at Wells-next-the-Sea and Holkham W84 and 94 along with *Pinus nigra* and locally regenerating freely. W2. E0.

P. strobus L. **Weymouth Pine**
Introduced from North America. Occasionally planted and regenerating in a trial plantation at Lynford, W89. W1. E2.

CUPRESSACEAE

All evergreen conifers.

Chamaecyparis lawsoniana (A. Murray bis) Parl. **Lawson's Cypress**
Introduced from western North America, planted in mixed and conifer woodland and occasionally in field hedges, regenerating freely in moist, acid, shady sites. W17. E77.

Thuja plicata Donn ex D. Don **Western Red Cedar**
Introduced from western North America, occasionally planted and regenerating freely, noted in gardens, cracks in paving and even on wall tops (RMP), but only likely to survive in moist, acid, shady sites. W34. E48.

Juniperus communis L. **Juniper**
Not considered native in its long standing site at Stanford W89 as the woodland in which it grows was planted after 1824, (JM.Schofield 1967). Also found established at Scolt Head W74 in 1997, SW, presumably from seed. W2. E0.

TAXACEAE

Taxus baccata L. **Yew**
Winter green conifer widely planted for many centuries in churchyards and parks where some very old specimens are to be seen. Self-sowing freely, the fruit are widely eaten by birds and seedlings are often found at some distance from the parent trees. They are particularly conspicuous on old walls, but here the seedlings are seldom able to reach maturity. More successful in woodland. In churchyards the Irish yew, *T.b.* 'Fastigiata' was commonly planted in the 19th century and a few very old trees of this cultivar can be found as in Stanhoe churchyard, W83, KAB. W165. E401.

LAURACEAE

Laurus nobilis L. **Bay**
Shrub or small tree introduced from the Mediterranean to gardens for its use as a herb. Old trees persist for many years, one at Dereham W91 has been surrounded by a tarmac carpark for 20 years, and another grows near the church at Thurlton E49, ALB. Seedlings sometimes occur. W0. E3.

ARISTOLOCHIACEAE

Aristolochia clematitis L. **Birthwort**

First recorded at Carrow Abbey E20 by the Rev Mr Salton in 1793 (Eng.Bot.) where it may be a relic of plants in the nuns' herbal garden. Still thriving in the same site. Also long known near Sturston Hall W89 in Stanford Training Area where it is unchecked and has become rampant after 45 years. It now occurs in two distinct colonies occupying 984 and 1258 square metres respectively, ALB 1997. W1. E1.

NYMPHAEACEAE

Nymphaea alba L. **White Waterlily**

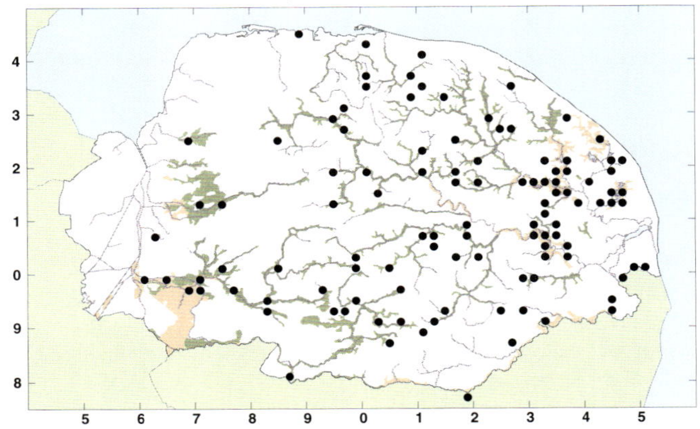

Widely distributed. The native stock is almost always white and typically smaller flowered. Formerly much commoner, especially in the Broads area, but recent Broadland recovery schemes have involved some introductions from native stock. Plants which have been introduced to gravel pits and village ponds are frequently coloured cultivars belonging to the hybrid groups *N. x laydeckeri* and *N. x marliacea*. The botanical variety *rubra* has not been observed.

Nuphar lutea (L.) Sm. **Yellow Water-lily, Brandy-bottle**

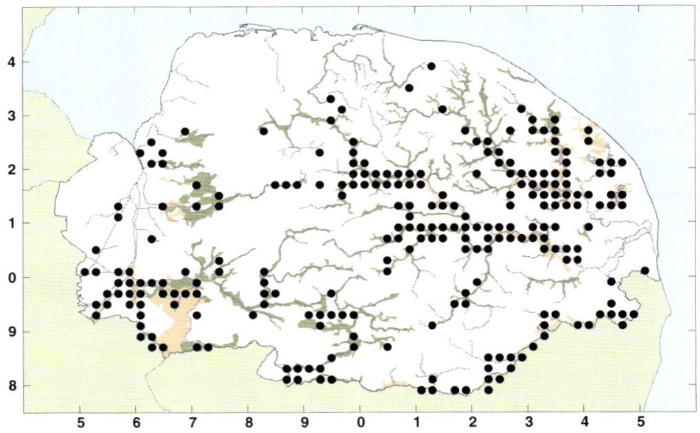

Generally found in gently moving water such as streams and dykes as well as ponds.

Nuphar advena (Aiton) W.T. Aiton **Spatterdock**

A large colony naturalised for some years near Shropham W99 was first noted by DMM in 1992 and still survives. W1. E0.

CERATOPHYLLACEAE

Ceratophyllum demersum L.
Rigid Hornwort, locally **Ferret's-tail**

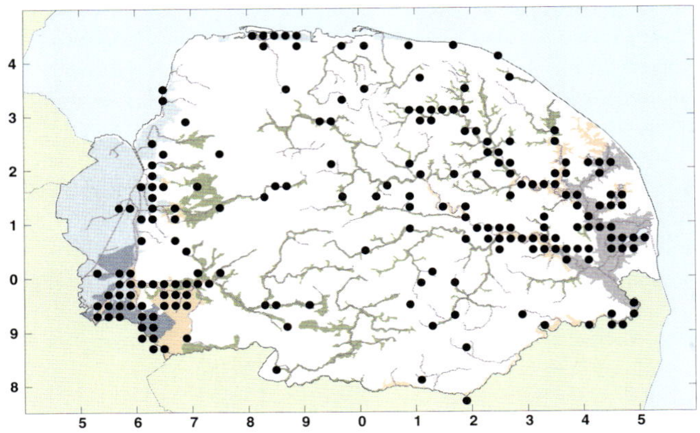

Winter green perennial of unpolluted ditches, ponds and streams.

C. submersum L. **Soft Hornwort**

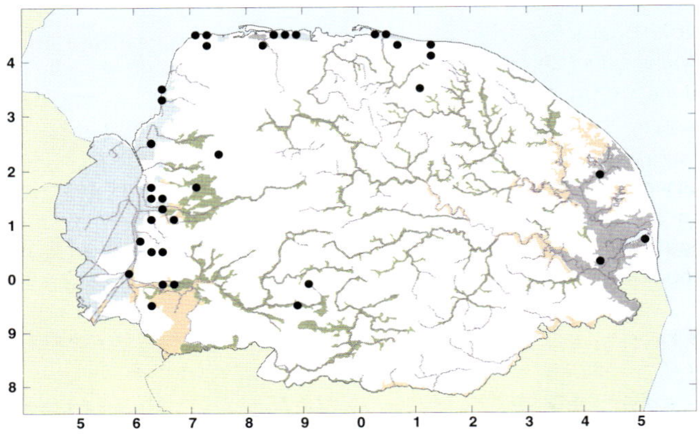

Winter green perennial not uncommon in unpolluted ditches and ponds especially near the sea.

Nuphar lutea - How Hill

RANUNCULACEAE

Caltha palustris L. **Marsh Marigold, Kingcup**

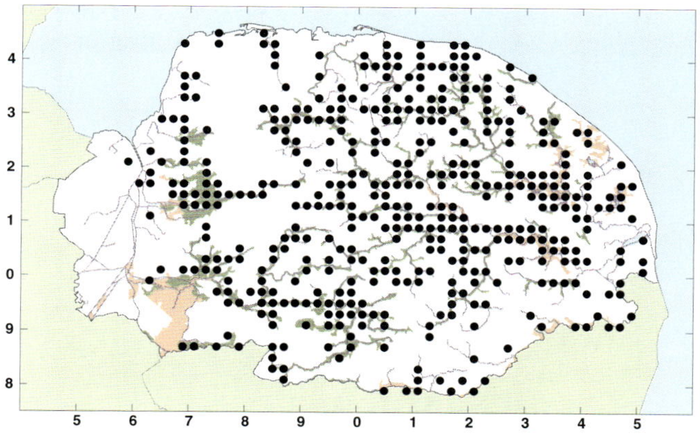

A deciduous perennial of marshes, pond margins, streamsides and wet woods, very variable in flower size. Double-flowered plants have occurred at Gayton W61 and an autumn flowering clone grew at Roydon W62 until destroyed by development.

Helleborus foetidus L. **Stinking Hellebore**
Winter green perennial, perhaps native to woods and scrub in parts of the county. In south-east Norfolk abundant all along the wooded scarp of the Bath Hills at Ditchingham and first recorded there by Daniel Stock *circa* 1800, also in old woodland at Hedenham E39. Otherwise usually in parks, shrubberies, churchyards and near gardens where it was probably originally introduced. Very persistent, it has been known in some sites as on a roadside bank at Shernborne W73, for more than 50 years. W8. E17.

■ **H. viridis** L. **Green Hellebore**
Partially winter green to deciduous perennial, probably formerly native in woods but now considered extinct in the county. Recorded first in 1837 at Ingoldisthorpe W63, also later from Bradenham W90 and Brinton E03 where it was last seen in 1953.

Eranthis hyemalis (L.) Salisb.
Winter Aconite, old local name, **New Year's Gift**
Summer dormant, tuberous perennial introduced to cultivation from southern Europe and now widely naturalised in copses, plantations and churchyards. W69. E123.

Nigella damascena L. **Love-in-a-mist**
Annual from southern Europe, occasionally occurring on rubbish tips and with garden waste, sometimes persisting for a number of years, ETD. The species has been recorded on roadside banks and verges in several sites during the current survey. W5. E7.

Aconitum napellus L. **Monk's-hood, Wolfbane**
Deciduous perennial, a certain native of southwest England and Wales it has long been known from north- east Norfolk and parts of Broadland where it is of uncertain origin. The plant grown in gardens and escaping is frequently the hybrid **A.x cammarum** L. which is now the commonest taxon in cultivation. W13. E16.

Consolida ajacis (L.) Schur **Larkspur**
Syn. *Delphinium ambiguum* L.
Annual from the Mediterranean, once a rare weed of cornfields, now usually a garden escape. Recorded once in beet at Harling W98 in 1986, CPP, and three times in corn, in North Creake TF83 in 1989, R.Woods; Hargham E08 in 1993, ALB and Garboldisham W98 in 1996, JH. All other records appear to refer to garden escapes. The members of this genus are much confused and because of re-classification it is difficult to separate old records of this, *C. regalis* Gray and the following species with any confidence. W30. E53.

C. orientalis (J.Gay) Schroedinger **Forking-larkspur**
Syn. *Delphinium orientale* J.Gay.
One recent confirmed record at King's Lynn Docks W62 in 1991, RMP. W1. E0.

Anemone nemorosa L. **Wood Anemone**

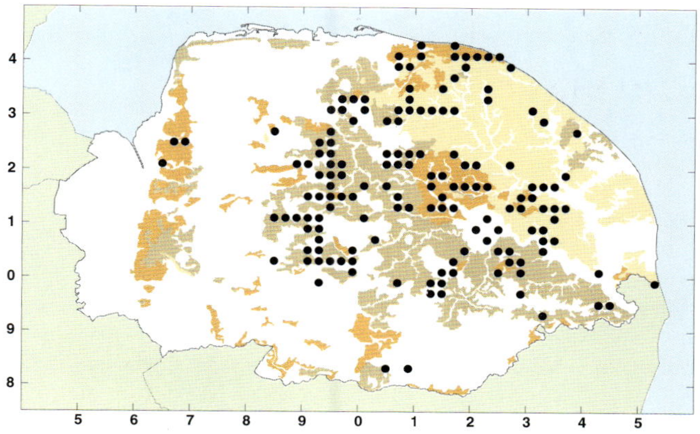

Summer dormant, rhizomatous perennial of ancient woodland, occasionally in long-established meadows. Has survived in woods replanted with conifers for more than 30 years.

Anemone nemorosa - Great Ryburgh

A. apennina L. **Apennine Anemone**
Summer dormant, rhizomatous perennial from southern Europe, occasionally naturalised in parks and churchyards, known from Stanhoe W83 since the 1880s. W1. E4.

A. ranunculoides L. **Yellow Wood Anemone**
Summer dormant perennial from mainland Europe, occasionally persisting in shrubberies and parks. Known from four sites before 1970 persisting at Ovington W90 where it was recorded in 1916, and at Reymerstone, E00 in 1998, J.Eaglen. W2. E1.

A. blanda Schott & Kotschy **Balkan Anemone**
Summer dormant perennial from southern Europe, occasionally surviving from garden throwouts as on a roadside verge far from houses at Sedgeford W73 in 1988 KAB, also in a churchyard at Framingham Earl E20, ALB. W1. E1.

■ **Pulsatilla vulgaris** Mill. **Pasque Flower**
Syn. *Anemone pulsatilla* L.
Deciduous perennial recorded from the adjacent parishes of Sporle and Lexham, W81 in the 18th and 19th centuries. Extinct.

Clematis vitalba L. **Travellers' Joy, Old Man's Beard**

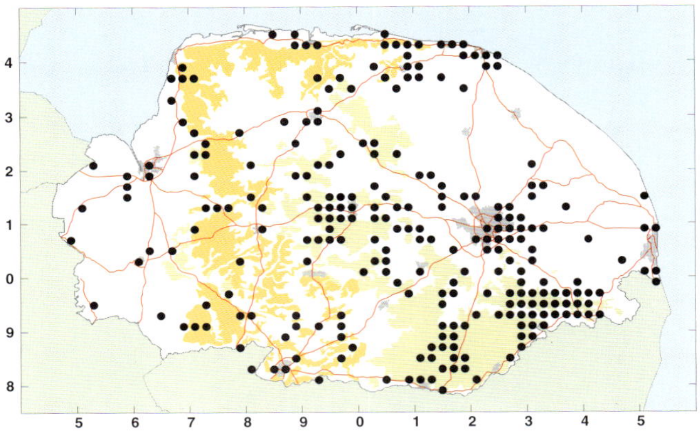

Deciduous, woody-stemmed climber of mainly calcareous soils, apparently increasing its range. A very old plant found in Tindall's Wood, Hedenham E39 had a trunk 42.5cm in circumference 1m. above the ground, in 1998, ALB. The wild species is sometimes used as a rootstock for garden clematis and some plants may come from this source especially in churchyards and hedgerows near gardens.

C. tangutica (Maxim.) Korsh. **Orange-peel Clematis**
Syn. *C. orientalis* ssp. *tangutica*
Deciduous, woody-stemmed climber occurring in Norwich E20, a flowering seedling in rough grassland was found seventy metres from its parent plant in 1989, ETD. W0. E1.

Ranunculus acris L. **Meadow Buttercup, Crowfoot**

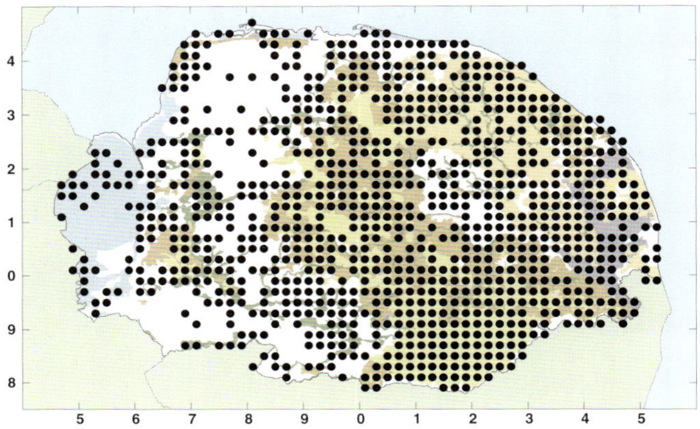

Winter dormant perennial of moist, established neutral to basic grassland.

Ranunculus arvensis - Brisley

R. repens L. **Creeping Buttercup**
Winter green perennial common in a wide variety of habitats both wet and dry and a rampant weed in some gardens. W696. E714.

R. bulbosus L. **Bulbous Buttercup**

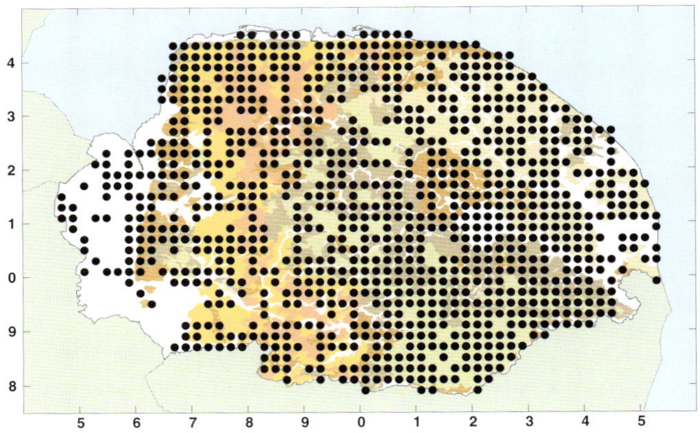

A late-summer dormant perennial of dry, established alkaline to neutral grassland. Away from the chalkier west, usually findable in spring in churchyards which may have the only remaining example of this type of grassland left in their parishes. Ssp. **dunensis** Druce has been recorded from Cley E04 in 1967, PD.Sell; Snettisham W63 and Santon W88, both in 1988, PD.Sell & JD.Murrell.

R. sardous Crantz **Hairy Buttercup**

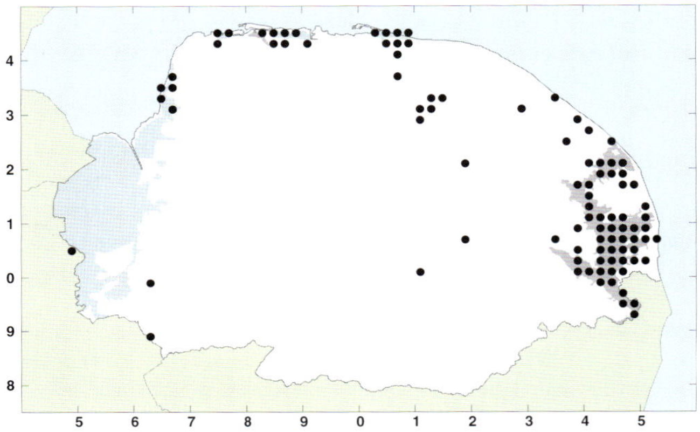

A frequent annual of coastal grasslands, the common buttercup of the grazing marshes from Acle and Beccles to Yarmouth. Occasionally found inland. A record from Outwell W50 in 1994, K&GB may be a relic of former maritime influence in the area.

■ **R. parviflorus** L. **Small-flowered Buttercup**
In the 19th century this annual was widely distributed, but with no records since 1894 it must now be considered extinct.

R. arvensis L.
Corn Buttercup, Corn Crowfoot, Bur Buttercup
Annual long decreasing. It was recorded as 'frequent in cultivated fields' by Galpin in 1888, 'rather rare' in 1914 but in 21 parishes, and 'rare' in 1968 with 11 sites. Since then it has been seen only six times, five of those sites being on heavy clay soil within 10 miles of Dereham, W91. W3. E3.

R. auricomus L. **Goldilocks**

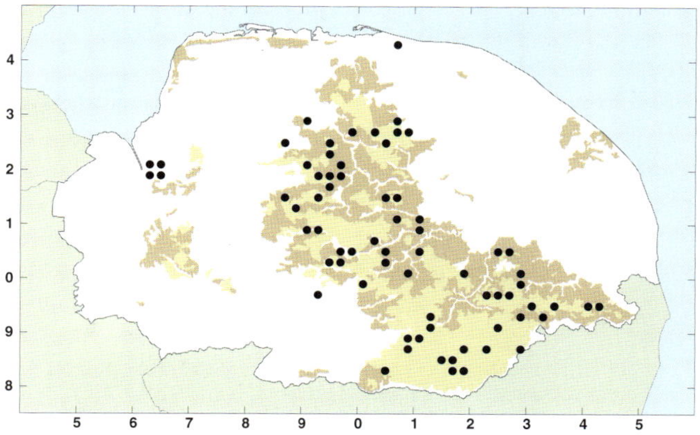

Summer dormant perennial of ancient woodlands, also in undisturbed grass verges on clay, probably as a relic of former adjacent woodland.

R. sceleratus L. **Celery-leaved Buttercup**

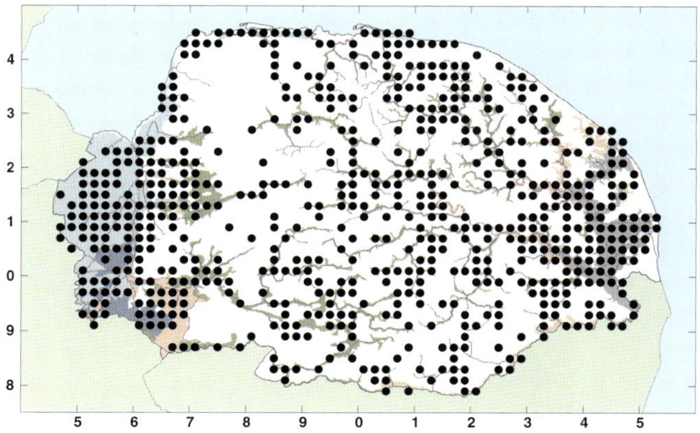

Autumn germinating annual, tolerant of some pollution and seasonal dryness, the only flowering plant in many fenland ditches.

R. lingua L. **Greater Spearwort**

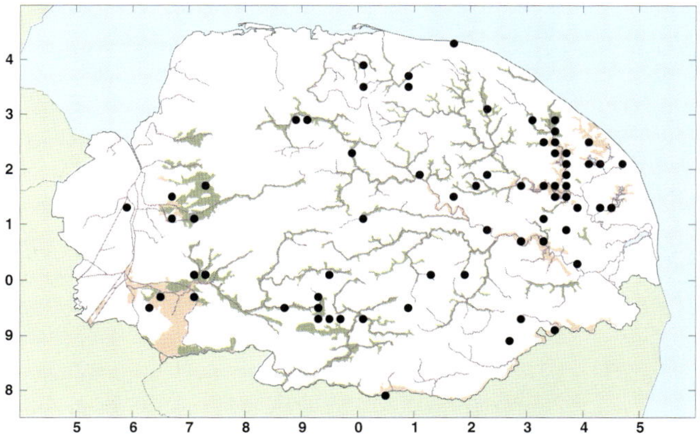

Herbaceous perennial thriving in peat and deep mud, once abundant in Fenland and the Broads but in west Norfolk disappearing from many of its old sites through increasing dryness.

R. flammula L. **Lesser Spearwort**

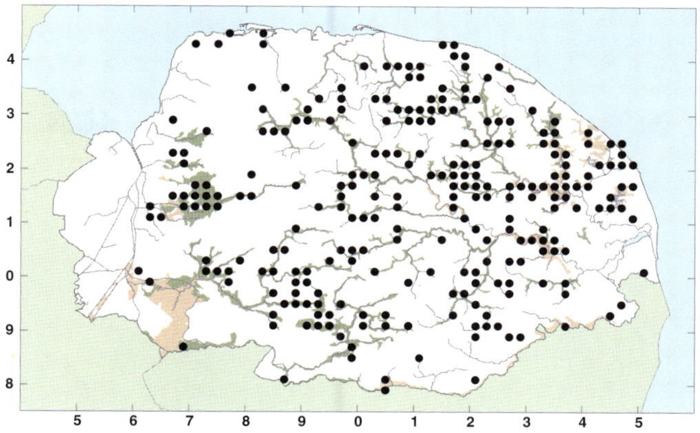

Winter green perennial found in wet areas on heaths, meadows, pastures and bogs. Our plants belong to ssp. **flammula**.

R. ficaria L. **Lesser Celandine, Pilewort**

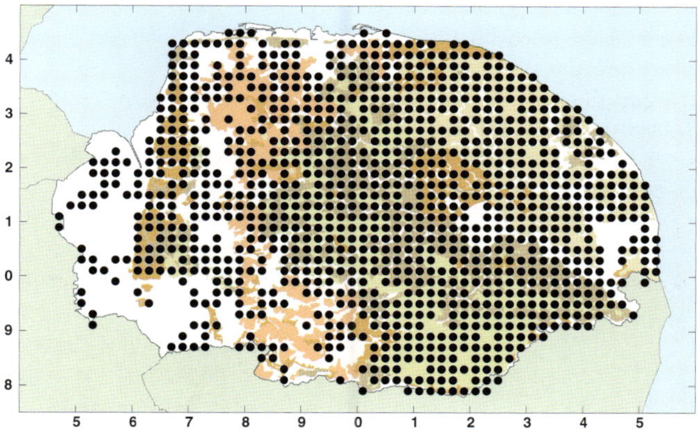

Summer dormant, tuberous rooted perennial, common on heavier soils with all the characteristics of a woodland species and perhaps a survivor from former woodland cover. Absent from areas that were once well-grazed sheep walk. Notably frequent in villages where it was once used herbally and perhaps sometimes introduced for this purpose. Frequently throws up 'sports' in the form of doubles, plants with white, pale yellow or orange flowers or strongly marked leaves. Many of these persist in gardens, but are generally short-lived in the wild. Ssp. **bulbilifer** Lambinon, usually occurs in deep shade and near villages, ssp. **ficaria** in more natural sites.

R. hederaceus L. **Ivy-leaved Crowfoot**

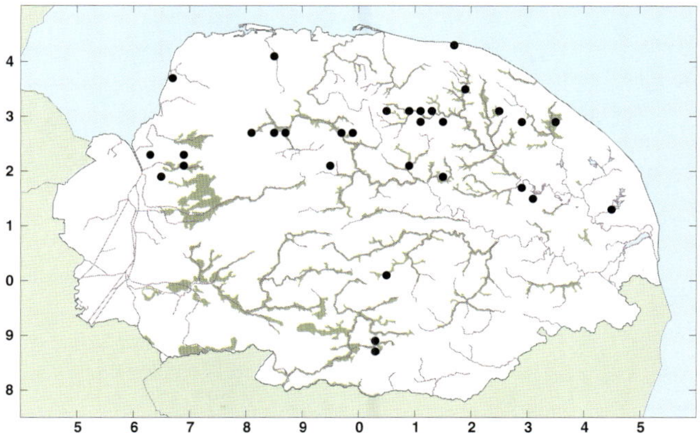

Perennial growing in shallow ponds and on surrounding mud.

■ **R. omiophyllus** Ten. **Round-leaved Crowfoot**
Syn. *R. lenormandii* F.W.Schultz
Perennial recorded at West Runton, E14 in the 19th century. Considered extinct.

R. baudotii Godr. **Brackish Water Crowfoot**

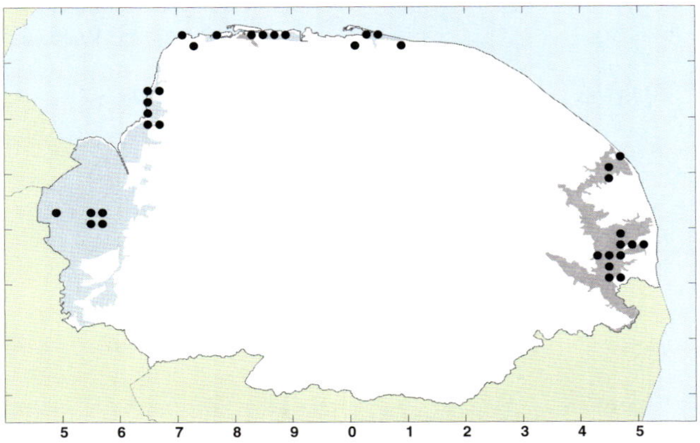

Perennial almost confined to brackish streams and ditches within the influence of the sea.

R. trichophyllus Chaix
Thread-leaved Water Crowfoot, Water Fennel

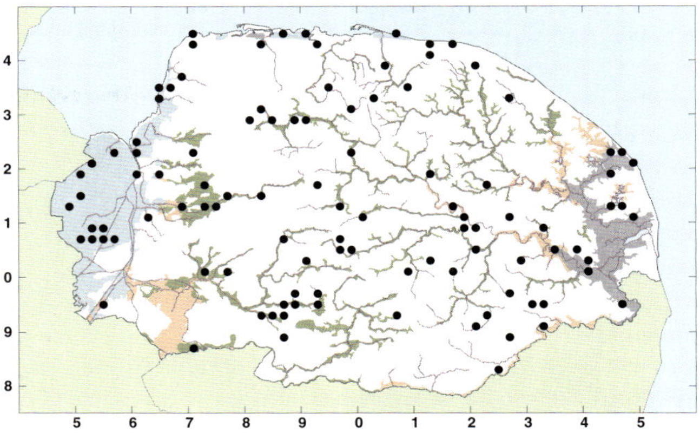

Perennial scattered in ponds, clay pits, woodland pools, meres, marshes, dune slacks and dykes.

R. aquatilis L. **Common Water Crowfoot**
Syn. *R. heterophyllus* Weber

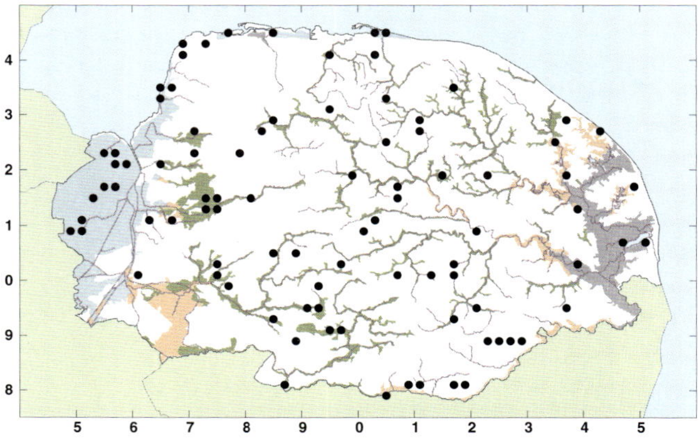

Perennial frequent in ponds and ditches, widely scattered over the county.

R. peltatus Schrank **Pond Water Crowfoot**

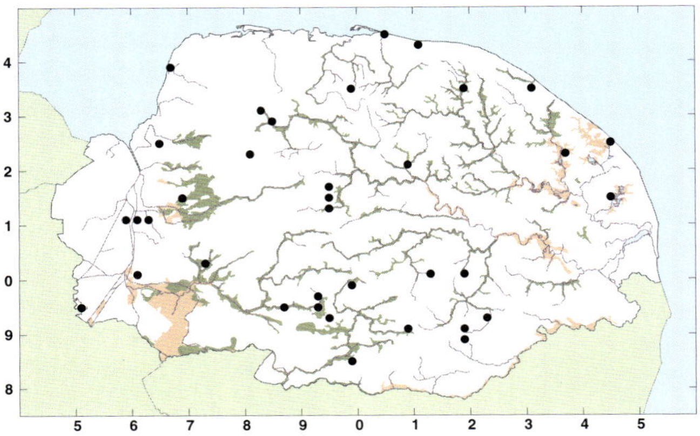

Perennial, occasional in ponds and dykes, strangely scarce in the Broads.

R. penicillatus (Dumort.) Bab. **Stream Water Crowfoot**

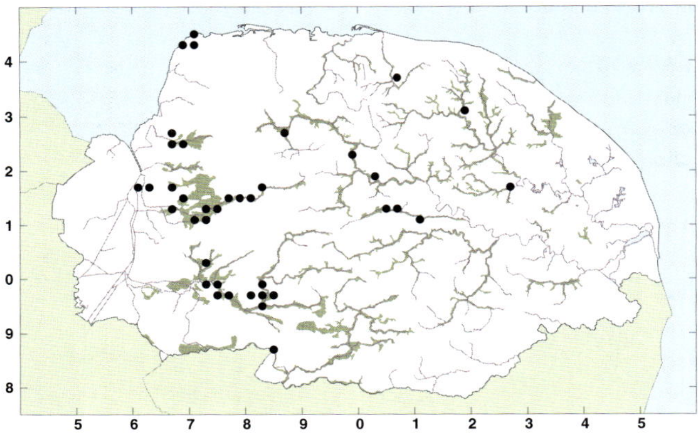

Perennial of chalk streams, our plant is ssp. **pseudofluitans**. The variety *pseudofluitans* is the most frequent taxon, but var. *vertumnus* was also seen at Castle Rising W62 in 1986.

R. fluitans Lam. **River Water Crowfoot**

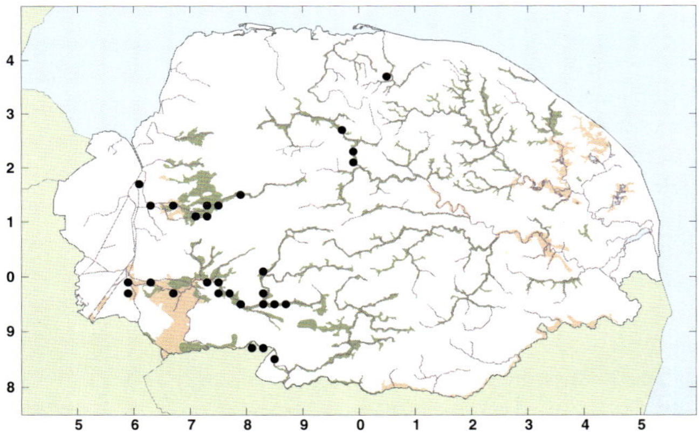

Perennial found in rivers of moderate flow with a chalky origin. Reported as 'proving a hindrance to navigation in the Breckland rivers' by H.Dixon Hewitt in 1948, but never as frequent today. In east Norfolk found in the R.Glaven E03 in 1996, ALB.

R. circinatus Sibth. **Fan-leaved Water Crowfoot**

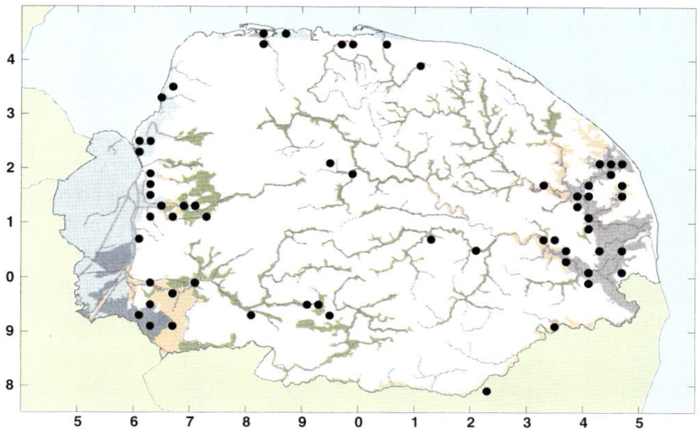

Perennial typical of dykes, disused gravel workings and the deeper, slow-flowing rivers of east Norfolk.

■ **Adonis aestivalis** L.
Annual, one record, King's Lynn Docks in 1888.

■ **A. annua** L. **Pheasant's-eye**
Annual known until about 1900 in cornfields, the last record being in barley from foreign seed near Norwich in 1915. Almost extinct nationally.

Myosurus minimus L. **Mousetail**

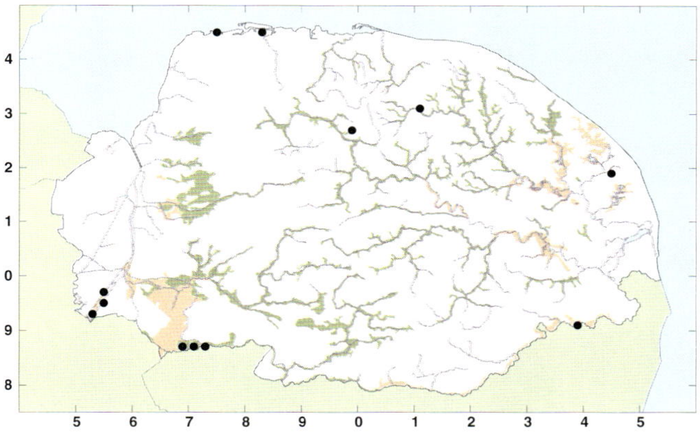

A scarce but scattered annual usually occurring in damp gateways into pastures where its hard fruits can be broken up by trampling animals. Once found in arable fields, perhaps relying on the plough horses to open the fruits, now rarely in this habitat.

Myosurus minimus - Burnham Norton

Aquilegia vulgaris L. **Columbine**
Deciduous perennial native to fens and marshes, once gathered in bunches in the Foulden area W79, but now scarce everywhere because of drainage and scrub encroachment. Also on boulder clay banks in mid-Norfolk where it has retreated since the regular use of herbicides. Most native stock is blue, or occasionally chocolate coloured. More recent garden escapes now confuse its natural distribution. W23. E61.

Thalictrum flavum L. **Greater Meadow-rue**

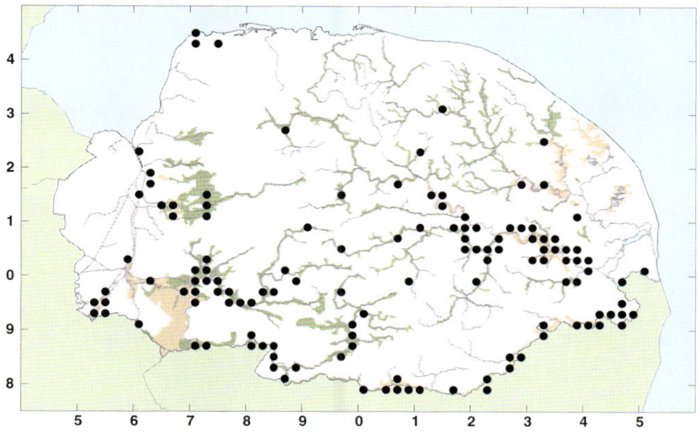

Deciduous perennial of streamsides, wet fens and meadows, it is often the first of the larger plants to disappear when the water table is lowered and this may account for its somewhat patchy distribution.

T. minus L. **Lesser Meadow-rue,** locally **Wild Maidenhair**

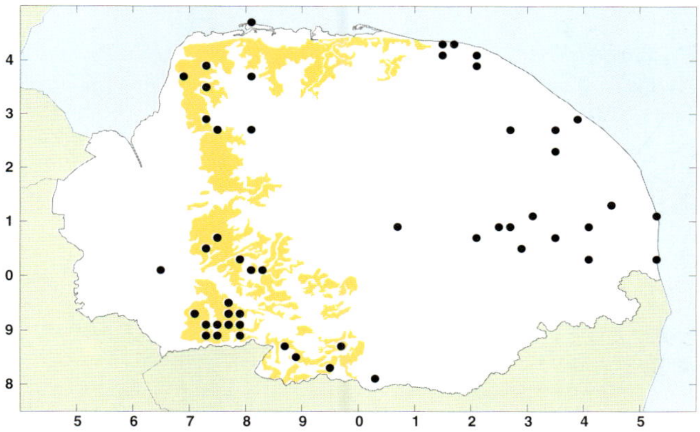

Deciduous perennial widely distributed in well-drained grassy sites on chalk but never common and lost from many sites where it has been crowded out by coarser vegetation. A complex species, but most of our plants appear to be ssp. **minus**. Ssp. **arenarium** (Butcher) Clapham, is still to be found at Scolt Head W74 where it has been known for 60 years and at Caister-on-Sea E51 where it was first noted in 1840 and still occurs on sand dunes near Caister golf clubhouse, 1995, L.Laxton.

BERBERIDACEAE

Berberis vulgaris L. — **Barberry,** locally **Pipperidge**

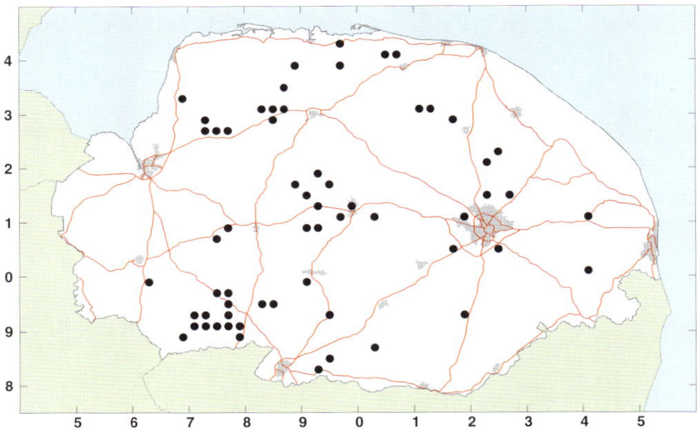

Deciduous shrub of hedgerows, not welcomed near wheat fields as it provides the alternative host for wheat rust.

B. thunbergii DC. — **Thunberg's Barberry**
Deciduous shrub from Japan, occasionally escaping from gardens. Found by Marriott's Way near Whitwell E02 in 1994, J.Humphris. W0. E1.

B. wilsoniae Hemsl. — **Mrs Wilson's Barberry**
Deciduous shrub from western China, occasionally bird sown. Noted in King's Lynn W61 in 1995, RMP; Weasenham Woods W81 in 1996, ALB; Swaffham, W80 in 1997, KAB and Roughton E23 in 1997, ALB. W3. E1.

B. julianae C.K. Schneid. — **Chinese Barberry**
Evergreen shrub from south-east Asia, found as a seedling on walls in King's Lynn, W61 in 1996, RMP. W1. E0.

B. buxifolia Lam. — **Box-leaved Barberry**
Evergreen shrub from South America, long established on East Walton Common, W71, but showing no sign of regeneration. W1. E0.

B. darwinii Hook. — **Darwin's Barberry**
Evergreen shrub from South America occasionally naturalised on walls and waste ground, as in Norwich, E20, ETD and King's Lynn W61and W62 in 1994 and 1995, RMP. W4. E2.

B. x lologensis Sandwith — **Hybrid Barberry**
(B. darwinii x B. linearifolia)
Partially fertile evergreen shrub, seen self-sown from plantings in a wood at Rougham W82 in 1995, KAB. W1. E0.

Mahonia aquifolium (Pursh) Nutt. — **Oregon Grape**
Evergreen suckering shrub introduced for ornament from western North America. Formerly planted for game cover and particularly frequent in the Breckland where it thrives in plantations; near Norwich and in the Waveney valley, reproducing both vegetatively and by seed. The fruits, which are eaten by birds, make good jam and jelly. W127. E128.

■ **M. x decumbens** Stace — **Hybrid Oregon Grape**
(M. aquifolium x M. repens)
In 1967, JE.Lousley stated that mahonias in Norfolk belonged to this hybrid but this has not been substantiated during the present survey and no specimens have been found.

PAPAVERACEAE

■ **Roemeria hybrida** (L.) DC
Four records from fields, all in the 19th century, none since, presumably only ever a casual.

Papaver pseudoorientale (Fedde) Medw. — **Oriental Poppy**
Syn. *P. orientale* auct. non L.
Herbaceous perennial from western Asia and an occasional garden throwout, perhaps not very long lasting. Confused with the very closely related *P. bracteatum* Lindl. Recorded in Norwich E20 in 1990, ETD; Beeston Regis Common E14 in 1991, FF; Hempton W92 in 1993, SF; Stody E03 in 1996 and Gimingham E23 in 1997, both by ALB and on a roadside verge at Gayton Thorpe W71 in 1995, K&GB. W4. E8.

P. atlanticum (Ball) Coss. — **Atlas Poppy**
Old records for this herbaceous perennial, introduced to cultivation from Morocco, were often erroneously named *P. lateritium* K.Koch. It occurs as an occasional escape or throwout from gardens. W6. E15.

P. somniferum L. — **Opium Poppy**
An annual from the eastern Mediterranean grown both as a field crop and in gardens. Widely cultivated in previous centuries, particularly in the Fenland as a cure for ague, but now no commoner in that area than elsewhere which suggests that all our plants are recent escapes from cultivation. W235. E351.

P. rhoeas L. — **Common Poppy**
Autumn and spring-germinating annual still found in cornfields but now most often in rape, also on waste ground especially on light soils. Pink and white flowered plants occur naturally though rarely. W674. E702.

P. dubium L. ssp. **dubium** — **Long-headed Poppy**

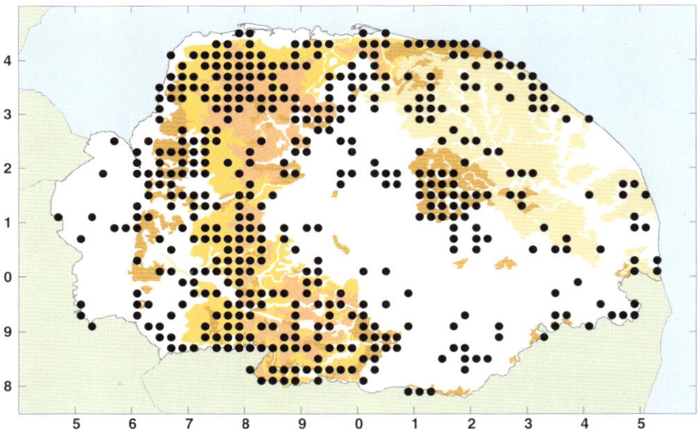

Autumn-germinating annual found in similar sites to *P. rhoeas* and often mixed with it, but less common.

Glaucium flavum - Cley

P. dubium ssp. **lecoqii** (Lamotte) Syme **Yellow-juiced Poppy**

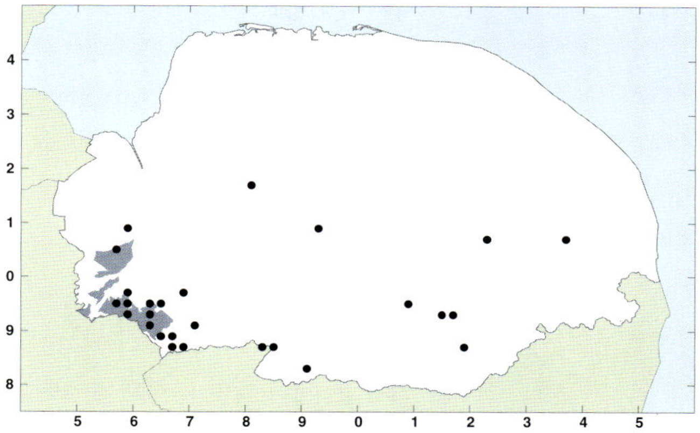

Annual, widely found in the Fenland where it can be frequent and sometimes replaces ssp. *dubium*.

P. hybridum L. **Rough Poppy**

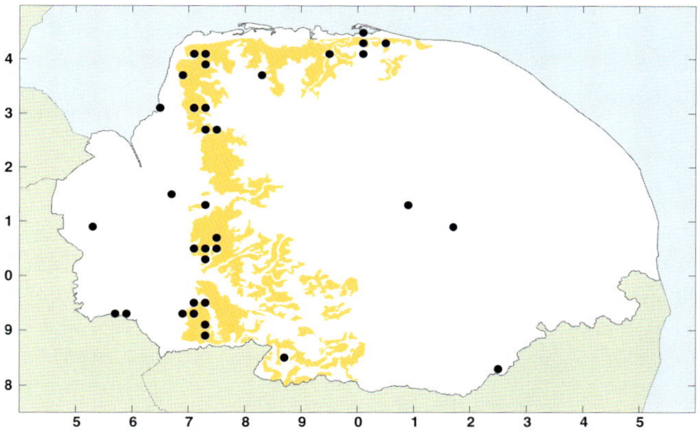

A scarce autumn-germinating annual, found chiefly but not always on chalky soils.

P. argemone L. **Prickly Poppy**

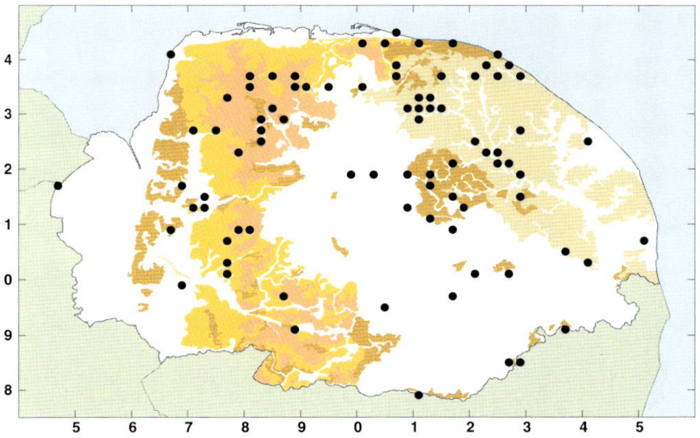

Scarce spring germinating annual found both as a weed of cultivation on light, sandy soils and on old walls where it can be very persistent.

Meconopsis cambrica (L.) Vig. **Welsh Poppy**
Herbaceous perennial, occasionally escaping from gardens, but rarely persisting. W9. E16.

■ **Glaucium corniculatum** (L.) Rudolph **Red Horned-Poppy**
Annual from the Mediterranean, of casual occurrence, found in 1775 near Norwich and again in 1915 and 1924, also in King's Lynn in 1885 and again in 1964. Not seen during this survey.

Glaucium flavum Crantz **Yellow Horned-Poppy**
Short-lived partially evergreen perennial of shingle banks by the sea. Records from Wells-next-the-Sea, Burnham and Brancaster in about 1796 and Holme in 1964 show this habitat was once more widespread. Now known from Wolferton to Heacham W63 on the Wash coast, at Scolt Head W84 and between Blakeney and Weybourne E04 and 14. W9. E8.

Chelidonium majus L. **Greater Celandine**

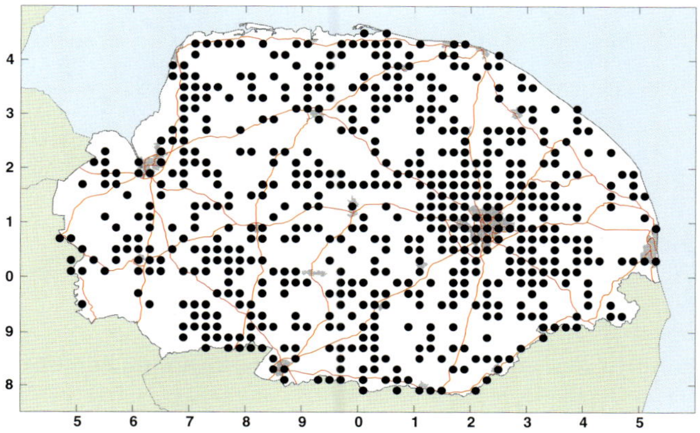

Short-lived winter green perennial native to Eastern Europe, formerly widely grown and usually still found near habitation. Known in cultivation since 1548.

Eschscholzia californica Cham. **Californian Poppy**
A western north American species behaving as an annual. Most records are from waste tips but also found on earthy cliffs at Sheringham E14 and Cromer E24 and on sunny banks in some villages. Surviving in pavement and road cracks amongst houses on the former RAF station at Sculthorpe W83 in 1995, before redevelopment. W5. E37.

FUMARIACEAE

Corydalis solida (L.) Clairv. **Bird-in-a-bush**
Syn. *C. bulbosa* (L.) DC. nom.illeg.
Deciduous perennial introduced to gardens from mainland Europe and occasionally naturalised as at Ovington W92 between 1973 and 1990; in parkland at Geldeston E39 and in the Rosary Cemetery, Norwich E20, ETD. W1. E4.

C. cava (L.) Schweigg.& Korte **Hollow-root**
Syn. *C. bulbosa* auct. non (L.) DC. nom. illeg.
Deciduous perennial introduced to gardens from mainland Europe. There has been much confusion between this species and the last. Naturalised at Ketteringham E10 1991, R.Tilbrook. W0. E1.

Pseudofumaria lutea (L.) Borkh. **Yellow Corydalis**
Syn. *Corydalis lutea* (L.) DC.
Partially winter green perennial from the southern Alps, well established on old walls, rarely occurring in other habitats. W115. E140.

P. alba (Mill.) Lidén **Pale Corydalis**
Syn. *Corydalis ochroleuca* Koch
Deciduous perennial introduced to gardens from south-east Europe. Naturalised at Holme Hale W80 in 1992, D&BM; on the churchyard wall at Cawston E12 and perhaps overlooked elsewhere. W1. E1.

Ceratocapnos claviculata (L.) Lidén **Climbing Fumitory**
Syn. *Corydalis claviculata* (L.) DC.

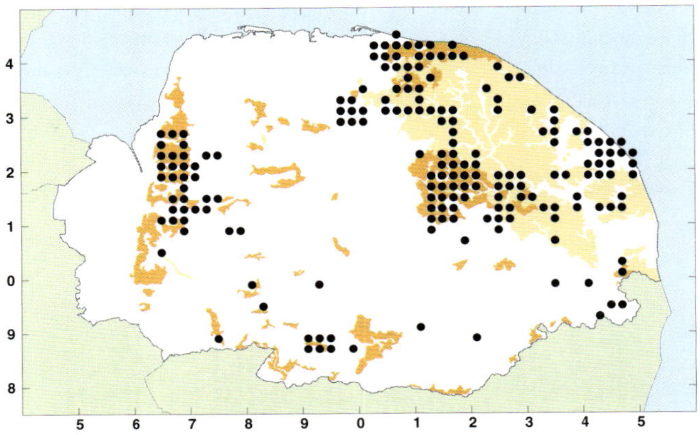

Over-wintering annual of shady places on acid soils and on embanked tracks and pathways in the Broads, curiously local in its distribution.

Fumaria capreolata L. **White Ramping Fumitory**
Annual described as 'rather common around Norwich' in the 1914 Flora. Found in a few scattered sites, but never more than a few plants in each. During this survey seen at Cley E04; Aldborough E13; Bradwell E50; Stanhoe W83; Southery W69; Rocklands W99, and Stalham E31. W3. E4.

■ **F. bastardii** Boreau **Tall Ramping Fumitory**
Annual with a largely western distribution in Britain. Recorded as a casual at Ranworth E31 in 1915. More recent records are believed errors for the following species.

F. muralis Sond. ex W.D.J.Koch **Common Ramping Fumitory**

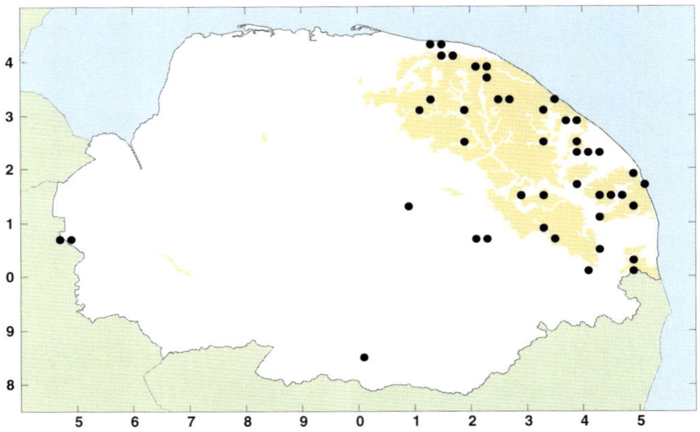

Annual very scarce in eastern England but scattered throughout east Norfolk. Our plants belong to ssp. **boraei** (Jord.) Pugsley. In the west, found only in the parish of Emneth W40 at separate sites in both Norfolk and Cambridge in 1994 and 1995, K&GB.

F. officinalis L. **Common Fumitory**
Annual, sometimes overwintering, frequent in cultivated and waste ground. Ssp. **wirtgenii** (W.D.J.Koch) Arcang. occurs chiefly on chalky soils, ssp. **officinalis** is widespread. W386. E477.

F. densiflora DC. **Dense-flowered Fumitory**
Syn. *F. micrantha* Lag.
Annual, native to south-eastern England, in Norfolk found on arable ground, notably in hot, dry summers. Seen in 1976 at Corpusty E13 and Boughton W70. During this survey at Garboldisham E08 and Antingham E23 in 1996 and Ditchingham E39 in 1997. W0. E6.

F. parviflora Lam. **Fine-leaved Fumitory**

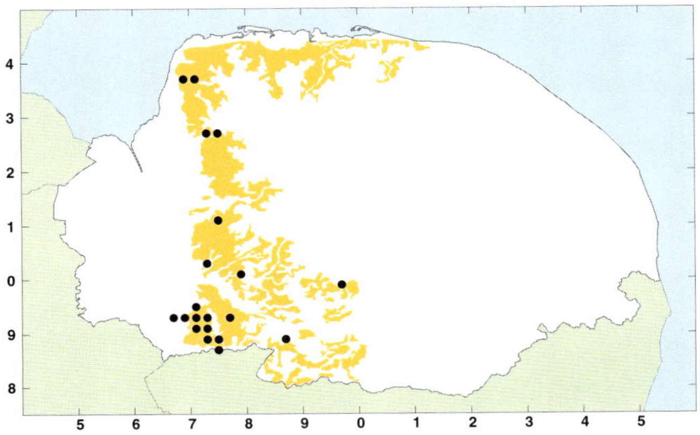

Annual of disturbed habitats on light, chalky soils, not uncommon in the Feltwell area W79. A nationally scarce species.

F. vaillantii Lois. **Few-flowered Fumitory**
Annual of similar disturbed habitats to *F. parviflora* though less confined to chalk and far more scarce, recorded only three times before 1960, the last at Narborough W71, 1958, but appeared in a garden at Fakenham W93 in 1997, MK. W1. E0.

PLATANACEAE

Platanus x hispanica Mill. ex Münchh. **London Plane**
Deciduous tree of uncertain origin, some authorities consider it may be a sport rather than a hybrid. It has been widely planted in urban areas and parks. Seedlings are seen occasionally as in Norwich E20, ETD. W3. E31.

ULMACEAE

Elms have always caused problems with their identification, but never more so than during the preparation of this Flora when the majority of elms of the county have been reduced to sucker shoots which are difficult, if not impossible, to identify. For the purposes of recording, we have used a simplified classification of three species, *U. glabra, U. procera* and *U. minor ssp minor* and their hybrids. Of these, *U. glabra* is quite distinct and *U. procera* almost completely wiped out by Dutch Elm Disease, *U. minor ssp minor* and the hybrids are treated below.

Ulmus glabra Huds. **Wych Elm**

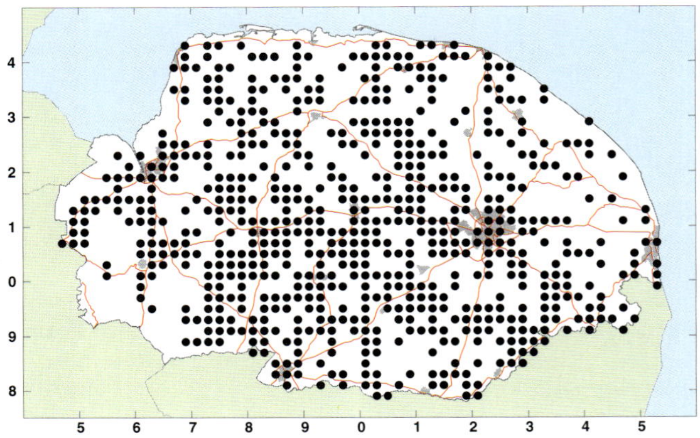

Deciduous tree widespread except in the peat fens. A few mature trees remain as at Holt E04, EVR, but there are many small trees, especially in isolated sites as well as hedgerow shrubs.

U. x vegeta (Loudon) Ley **Huntingdon Elm**
U. glabra x minor ssp. minor

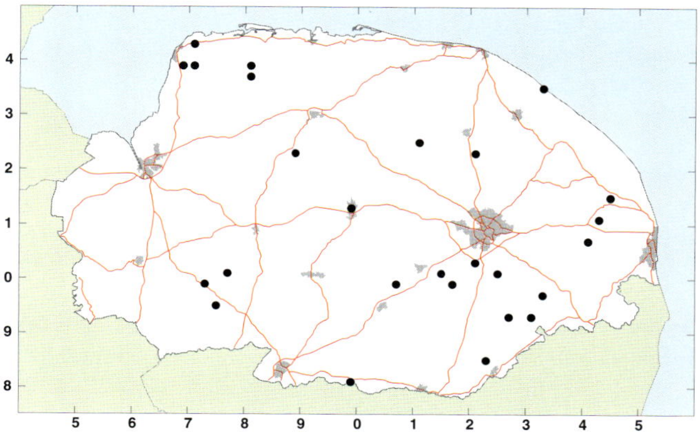

Deciduous tree. There are a number of cultivated clones of the basic hybrid between subspecies of *U. minor* and *U. glabra*. Many of the few surviving mature trees belong to the Huntingdon Elm clone as at Stanhoe W83, a 28m tree in 1996, K&GB; Mundham E39; Blo Norton E07 and Stokesby E40 the last three in 1997, ALB. Other records belong to the Dutch Elm, *U. x hollandica* with a similar but uncertain parentage.

U. procera Salisb. **English Elm**
Deciduous tree. In the early 1980s, EL.Swann collected many elm specimens and after discussion with R.Melville, came to the conclusion that there were very few trees of this species in the county and most were within the city of Norwich. Since then, these appear to have been totally wiped out by the last epidemic of Dutch Elm disease.

U. minor Mill. ssp **minor** **East Anglian Elm**
Syn. *U. carpinifolia* Gled
Deciduous tree surviving mostly at shrub size in hedges and copses though there are a few medium-sized trees. This group is difficult to identify as mature specimens are scarce, if indeed they survive, but is by far the commonest hedgerow elm in the county, especially on drier soils. W535. E644.

Ulmus minor before Dutch Elm disease - **Reepham**

CANNABACEAE

Cannabis sativa L. **Hemp**
Annual introduced from southern Asia, formerly widely grown in the county as a fibre crop and still grown under strict Home Office licence. Cultivars with a negligible narcotic content (less than 0.3%) are used for a variety of purposes ranging from high-quality paper to horse-bedding. Casuals may occur either from this source or, more usually, from imported birdseed. W7. E1.

Humulus lupulus L. **Hop**

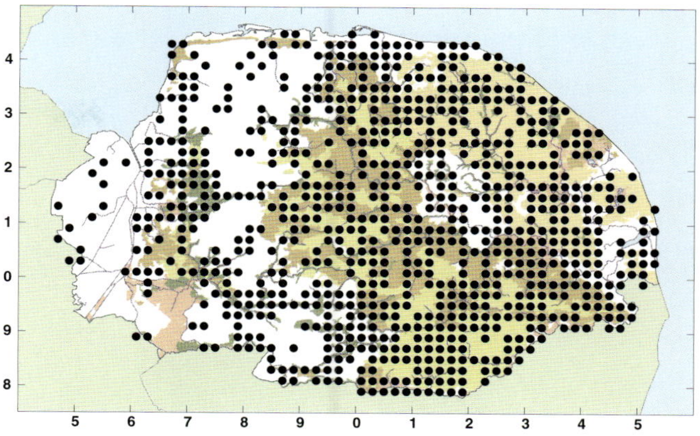

Herbaceous perennial twining climber, native in fen carr where it can festoon shrubs and trees. The cultivated form introduced in the early 16th century from Flanders and used in commercial beer production does not appear to have been grown on a large scale in Norfolk. It is possible, however, that some of our hedgerow colonies are relics of local cultivation.

URTICACEAE

Urtica dioica L. **Stinging Nettle**
Ubiquitous deciduous perennial. A stingless diploid form, variously given varietal, sub-species or species rank as var. *subinermis*, ssp. *galeopsifolia* and *U. galeopsifolia* respectively occurs in damp areas. Its correct identity has been confirmed genetically only at Coldham Hall Marsh, Surlingham E30. Plants with long, narrow, stingless leaves have also been found in other areas of fen and marsh in Broadland and in similar sites in west Norfolk, where they reached 2.5m at Hilgay W69 in 1996, HW& FS. These may be derived from hybrid plants, but genetically are probably tetraploids as is *U. dioica*. W708. E718.

U. urens L. **Small Nettle**

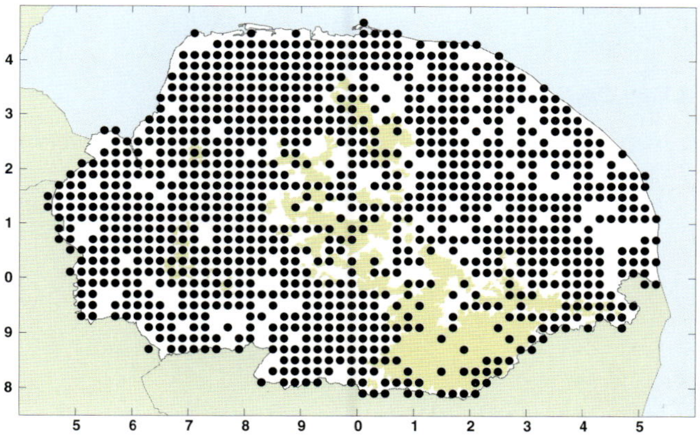

Annual of cultivated land on all but the heaviest soils, often overwintering.

■ **U. pilulifera** L. **Roman Nettle**
Annual from the Mediterranean, formerly found around Yarmouth where it was first recorded in 1745 and King's Lynn where it was last seen in 1900. Now considered extinct in Britain.

URTICACEAE

Parietaria judaica L. **Pellitory-of-the-wall**
Syn. *P. diffusa* Mert. & W.D.J.Koch

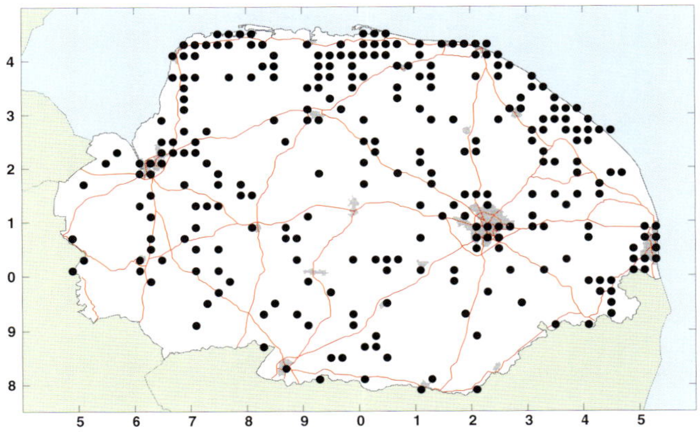

Winter green perennial frequent on old walls, particularly of churches and churchyards, but also occasionally found on dry, grassy banks as at East Rudham W82.

Soleirolia soleirolii (Req.) Dandy **Mind-your-own-business**
Syn. *Helxine soleirolii* Req.
Winter green perennial introduced to gardens from the Mediterranean islands and occasionally a persistent weed of paths and lawns in sheltered sites, not fully hardy in the open. W22. E34.

JUGLANDACEAE

Juglans regia L. **Walnut**
Deciduous tree native to south-east Europe and Asia. Single trees found far from habitation as in the Breckland forests and Fenland droves may be either relics of abandoned cultivation or perhaps grown from nuts dropped by rooks or jays. W16. E34.

Pterocarya fraxinifolia (Poir.) Spach **Caucasian Wingnut**
Deciduous tree introduced from the Caucasus. In Norwich E20 a planted tree has a surround of more than a hundred suckers, ETD. W0. E1.

MYRICACEAE

Myrica gale L. **Sweet Gale, Bog Myrtle**

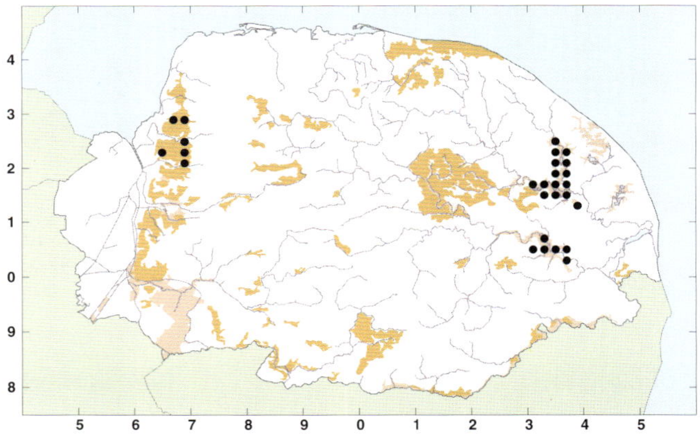

Deciduous shrub sometimes frequent on wet, acid soils, particularly on the greensand and in Broadland where it can form an impenetrable fringe around the edge of very wet areas dominated by saw sedge, *Cladium mariscus*.

FAGACEAE

Fagus sylvatica L. **Beech**
Deciduous tree, doubtfully native in Norfolk but frequently self-sowing. Often grown in the past in areas of wood pasture where it was often pollarded, notable trees of this origin can be seen at Blickling E12; Felbrigg E23 and Thursford W93. W338. E580.

Castanea sativa Mill. **Sweet Chestnut**
Deciduous tree from mainland Europe, introduced to southern Britain by the Romans who cultivated it for food and fuel. Often self-sowing on light, acid soils. W182. E357.

Quercus cerris L. **Turkey Oak**
Deciduous tree from the eastern Mediterranean, often self sowing. W73. E130.

Q x hispanica Lam. **Lucombe Oak**
(Q. cerris x Q. suber) Syn. *Q. x pseudosuber* Santi
A spontaneous and fertile hybrid tree, semi-deciduous but probably always planted. Fine examples are to be seen at Blickling E12 and at Roydon E08 near Diss. W0. E7.

Q. ilex L. **Evergreen Oak**
Evergreen tree from southern Europe, known to be at Holkham W84 since 1729. It will self sow freely though is not fully hardy inland in severe winters. Forms dense, self-sown thickets under pines on the sand dunes between Holkham and Wells-next-the-Sea W84 and 94. W29. E146.

Q. petraea (Mattuschka) Liebl. **Sessile Oak**

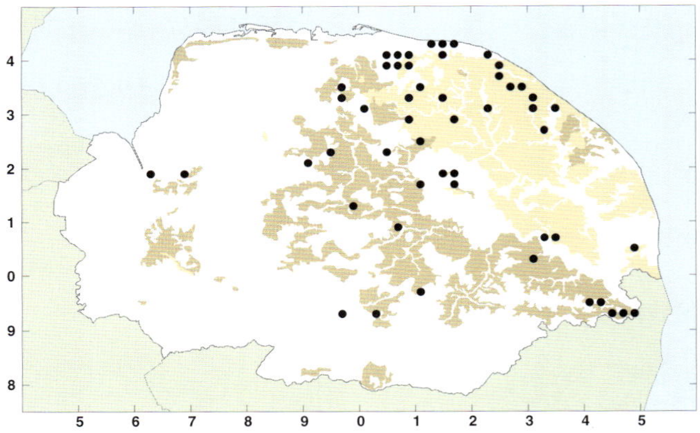

Deciduous tree, a native of the acid, gravelly soils of north and north-east Norfolk, notably at Holt Wood, E04 and Edgefield Little Wood E13, it was rarely coppiced, but stools survive at Swanton Novers E03 and Holt E04. On other soils it is almost certainly of planted origin.

Quercus petraea - Holt

Q. x rosacea Bechst. **Hybrid Oak**

(Q. petraea x Q. robur)

Deciduous tree likely to arise where both parents grow, today sometimes found where *Q. petraea* is no longer to be found. Horningtoft, hedgerow W92 in 1996, KAB and Bayfield E04 in 1995, EVR. W2. E3.

Q. robur L. **Pedunculate Oak**

Deciduous tree, common in hedgerows and old woods and tolerant of both wet and dry soils. It was regularly pollarded in areas of wood pasture and ancient trees of this origin often survive in parkland as at Sandringham W62. It has also frequently been planted, today, sadly often with saplings grown from imported seed. W608. E703.

Q. rubra L. **Red Oak**

Deciduous tree from North America, planted for ornament, especially by Breckland roads. Most trees are still too young to produce viable seed. W5. E20.

BETULACEAE

Betula pendula Roth **Silver or Warty Birch**

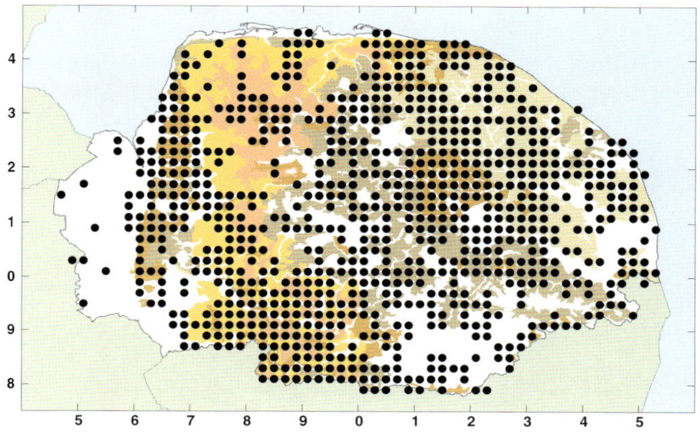

Deciduous tree of light, acid soils, rapidly invading former heathland as grazing pressure is removed. Trees on heavy soils are usually planted.

B. x aurata Borkh. **Hybrid Birch**

(B. pendula x B. pubescens)

Deciduous tree occurring with the parents and sometimes forming a continuous range between them. W1. E6.

B. pubescens Ehrh. **Downy Birch**

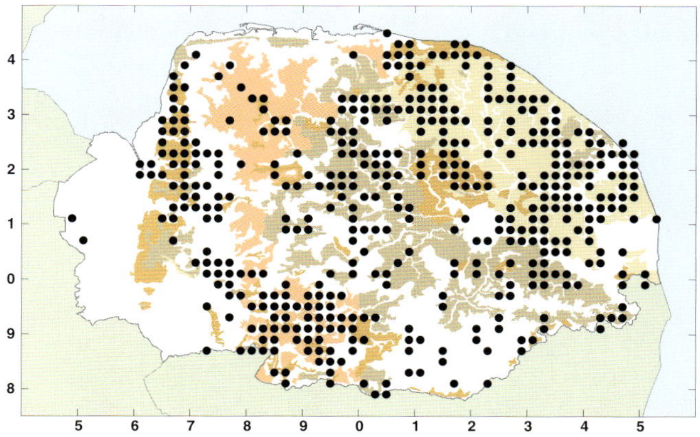

Deciduous tree of light acid soils, more tolerant of moisture than *B. pendula* and less common. Our trees are ssp. **pubescens**.

Alnus glutinosa (L.) Gaertn. **Alder**

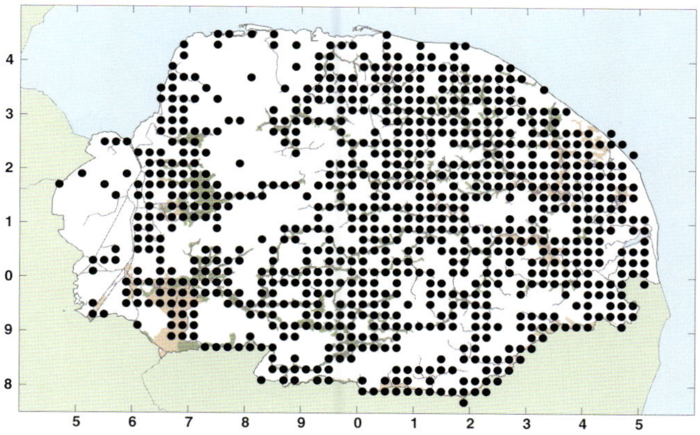

Deciduous tree of wet places, taking the place of birch as a first coloniser tree on fen and bog and soon forming almost impenetrable alder carr as has happened in wide areas of former peat cutting in Broadland. It also occurs in some ancient woods, often where there is a spring line. In these sites it may be native, but its wood has always been valued for its ability to persist for long periods when water-logged, so planting is likely to have occurred. It is also occasionally found as a result of recent tree-planting schemes.

A. incana (L.) Moench. **Grey Alder**

Deciduous tree from North America, occasionally seen mixed in shelter belts of *A. glutinosa* as at Walsoken W41 or as an 'amenity tree', as beside the Loddon bypass E39 where it is spreading rapidly by suckering. Seedlings have not yet been reported. W4. E24.

A. cordata Desf. **European Alder**

Deciduous tree from Italy and Corsica, planted for ornament and as shelter belts, particularly around orchards where they can be planted very thickly. Seedlings have not been found. W0. E14.

Carpinus betulus L. **Hornbeam**

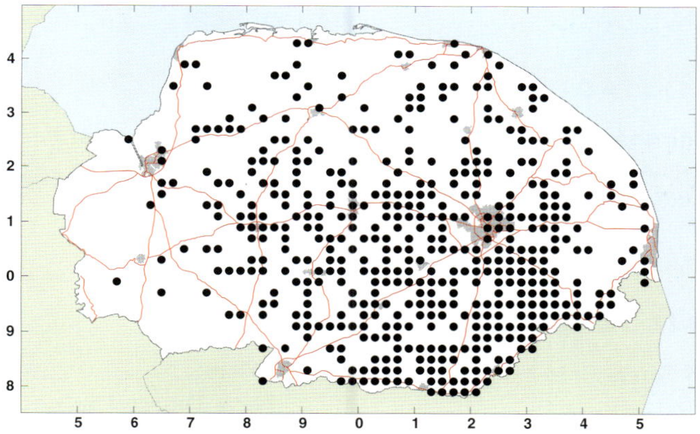

Deciduous tree native in old hedges and woods on the boulder clay of south-east Norfolk, also much planted in parkland and elsewhere.

Corylus avellana L. **Hazel**

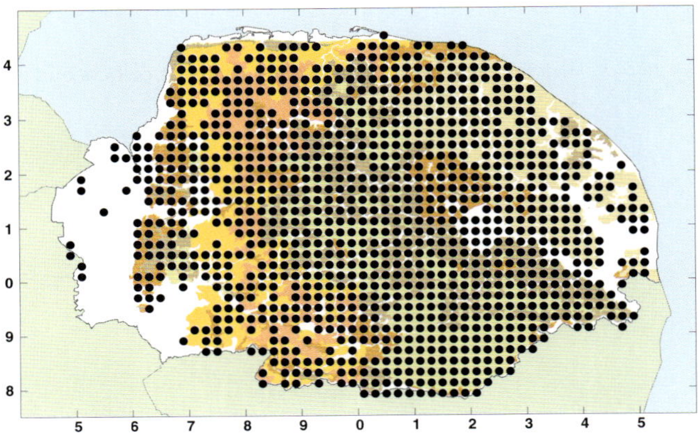

Deciduous tree, a native of hedges, copses and woods, long valued for its coppice poles as well as its nuts. At one time, rent for Wayland Wood W99 was paid in sacks of nuts. Grey squirrels make these quantities impossible to gather today. The tree is absent from extreme coastal areas.

C. maxima Mill. **Filbert**

Deciduous tree from southern Europe, occasionally found as a relic of cultivation as at Mouse Hall W89. A self-sown seedling was found at Watlington W61 in 1996, RMP. Where both species occur, bushes can be found with fruits which are intermediate between the parents and especially with the fused surrounding bracts open at the top and not contracted beyond the apex of the nut, ALB. W2. E0.

PHYTOLACCACEAE

Phytolacca acinosa Roxb. **Indian Poke, Pokeweed**

Herbaceous perennial introduced to gardens from Asia. Occasionally found in wild sites, presumably from bird-sown berries as in Swannington churchyard E11, R.Tilbrook and an arable field at Massingham W72 in 1989, E.Cross. W1. E1.

NYCTAGINACEAE

Mirabilis jalapa L. **Marvel of Peru**

Tender annual from tropical America found half-a-mile from the nearest house at Earsham E38 on waste ground in 1997, ALB. W0. E1.

AIZOACEAE

■ **Tetragonia tetragonioides** (Pall.) D. Kuntze **New Zealand Spinach**

Annual from the south Pacific and Japan, grown as a vegetable and seen on Harford tip Norwich E20 in 1972, PGL, the only record.

CHENOPODIACEAE

Chenopodium bonus-henricus L. **Good King Henry**

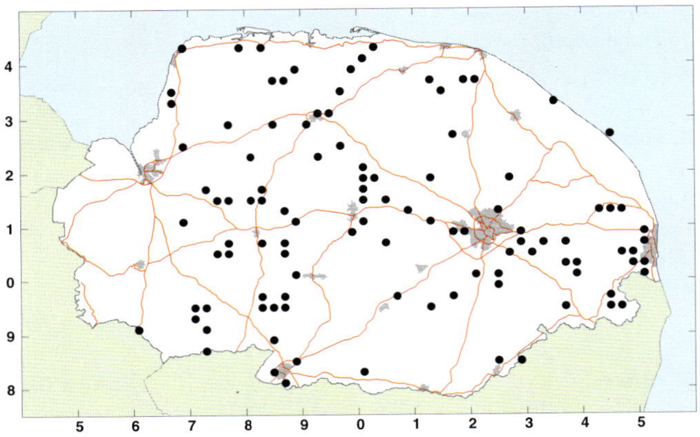

Herbaceous perennial, formerly grown as a pot herb. Curiously local in some areas including the old settled area of the Fenland, the Broads and parts of north-east Norfolk.

■ **C. glaucum** L. **Oak-leaved Goosefoot**

Annual, recorded twice, at North Walsham E23 in 1959 and Winterton Ness E51 in 1962.

C. rubrum L. **Red Goosefoot**

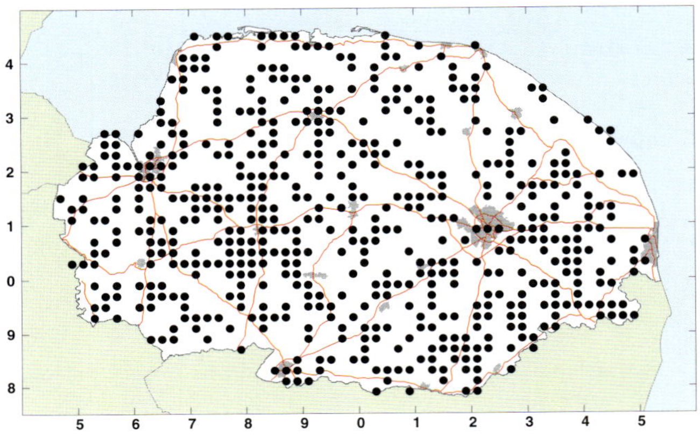

Annual, at its most lush on and around old manure heaps so not restricted by soil type. Also occurs in the beds of dried-up ponds when small plants can carpet the ground.

■ **C. chenopodioides** (L.) Aellen **Many-spiked Goosefoot**

Syn. *C. botryoides* Sm.

A former native, known in the 19th century from east Norfolk.

C. polyspermum L. **Many-seeded Goosefoot**

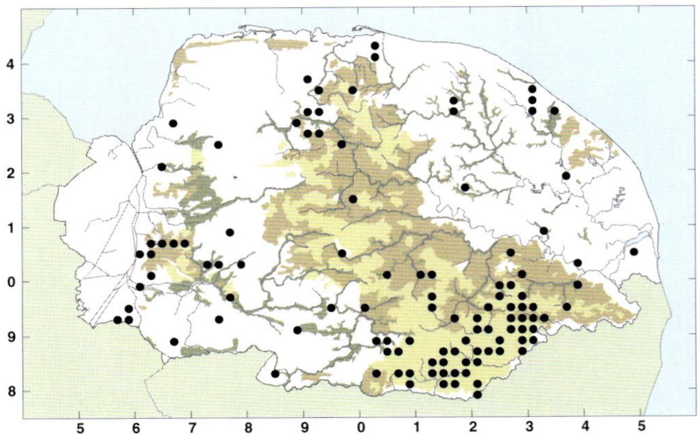

Annual, always in disturbed ground and common throughout the boulder clay areas of south Norfolk as a weed of arable crops, scarce elsewhere. A red coloured variant was found in the Fordham area W68 in 1997, RMP.

■ **C. vulvaria** L. **Stinking Goosefoot**

Annual, formerly native, recorded in the 19th century from King's Lynn and Yarmouth.

C. hybridum L. **Maple-leaved Goosefoot, Sowbane.**

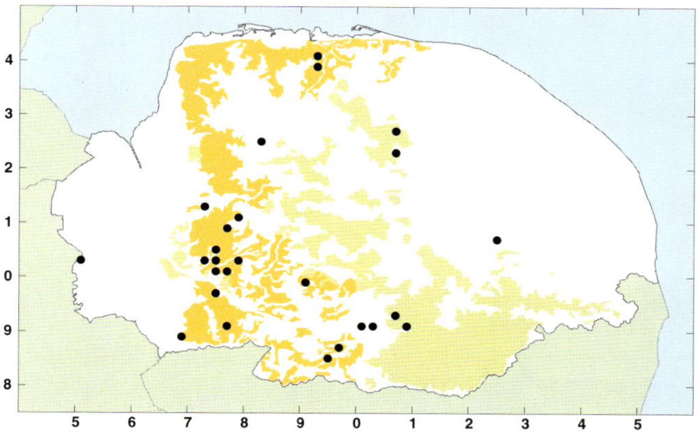

Scarce annual of cultivated ground, possibly increasing in frequency since 1968. Has appeared in some fields annually for many years.

■ **C. urbicum** L. **Upright Goosefoot**

Scarce annual with four records only between 1800 and 1964.

C. murale L. **Nettle-leaved Goosefoot**

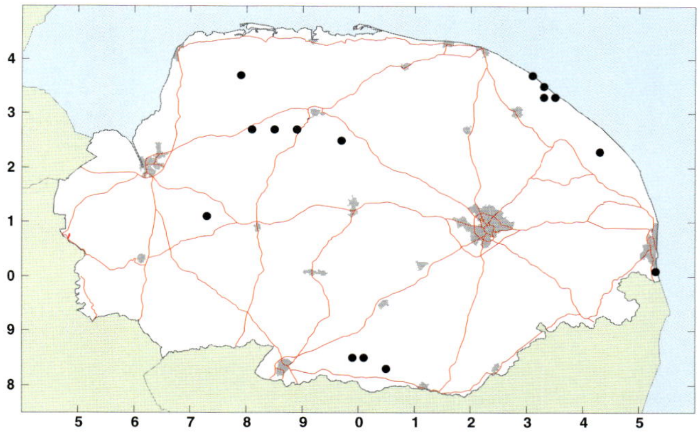

Scarce annual of cultivated ground, recorded in scattered sites through the county.

C. ficifolium Sm. **Fig-leaved Goosefoot**

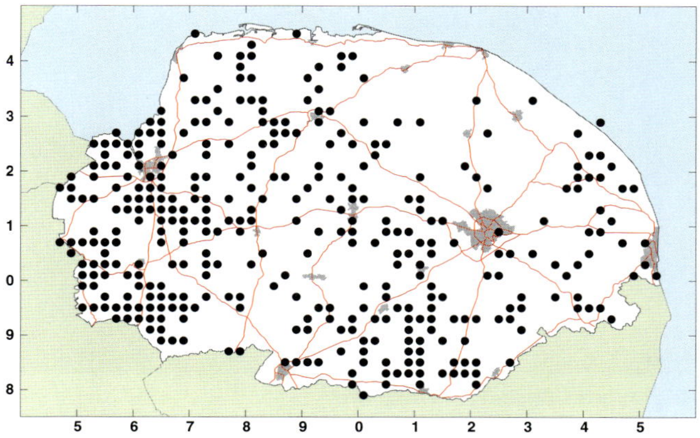

Annual of waste and cultivated land, particularly on heavy soils, often found together with fat hen, though never as abundant.

C. album L. **Fat Hen**

Annual of cultivated ground and waste places, occasionally overwintering. Known to have been used as a vegetable since the Iron Age and still occasionally even today, though regarded by most people as a pernicious weed. W689. E697.

■ **C. hircinum** Schrad.

Annual, a casual recorded at Whitlingham E10 in 1917 and Geldeston E39 in 1930.

■ **C. suecicum** Murr **Swedish Goosefoot**

Annual introduced from northern Europe, two records, the first from Walsoken Marsh in 1796 by W.Skrimshire (as *C. viride*), the second at West Runton E14 in 1983 by ETD, a colony of over 20 plants at the base of the cliff. It would be nice to think that the seeds had survived the North Sea crossing.

■ **C. probstii** Aellen **Probst's Goosefoot**

Annual from North America recorded from Harford tip, Norwich E20 in 1969 by GB and from Beetley tip W94 in 1974 by RPL & ELS.

■ **Bassia scoparia** (L.) Voss **Burning Bush, Summer Cypress**

Syn. *Kochia scoparia* (L.) Schrad.

Occasionally seen on rubbish tips but not recorded during this survey. This name also includes **Kochia densiflora** Turcz., recorded as a carrot field alien in the early 1970s.

Spinacea oleracea L. **Garden Spinach**

Annual introduced from Asia as a crop plant, found surviving on Harford tip, Norwich E20 in 1971, GD. Watts and at Heacham, W63 in 1988, PRC. Recently found in its non-succulent wild form growing from bird seed at Tuddenham E01 in 1995, ALB. W1. E1.

Atriplex hortensis L. **Garden Orache**

An ornamental annual, usually seen in its red form. Found away from habitation at Kettlestone W91 in 1992, MK and Watlington W61 in 1993, RMP. W6. E3.

A. prostrata Boucher ex DC.

Halberd or **Spear-leaved Orache.**

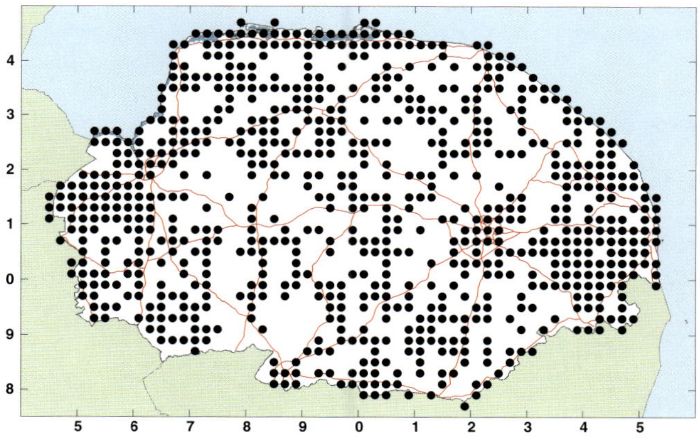

Annual of the high tide line on salt marshes and sea banks, also long found around manure heaps. More recently it has colonised mineral-rich roadside verges and is rapidly becoming abundant through the county.

A. x gustafssoniana Tascher. **Kattegat Orache**
(A. prostrata x A. longipes)
First found at Snettisham W63; Terrington W52; Burnham Overy W84 and Brancaster W74 in 1977, all by RPL, ELS and PM.Taschereau, also at Horning E31 in 'reed bed of fen carr', 1978 ELS & RPL. During this survey noted at Berney Arms E40 in 1997 by R.Ochyra and AC and subsequently found to be present on all the saltings between Berney Arms and Reedham, but not those downstream, ALB and RWE. ELS commented in 1982 that the frequency of this 'suggests that *A. longipes* has been with us for a long time'. W0. E4.

A. x hulmeana Tascher.
(A. prostrata x A. littoralis)
Hybrid recorded from Wolferton W62; The Point, King's Lynn W62; Heacham W63; Burnham Deepdale W84 all by PM.Taschereau and from Burnham Overy Staithe W84 in 1995, MK with the Flora Group. W1. E0.

A. glabriuscula Edmondston
Mealy Orache, Babington's Orache

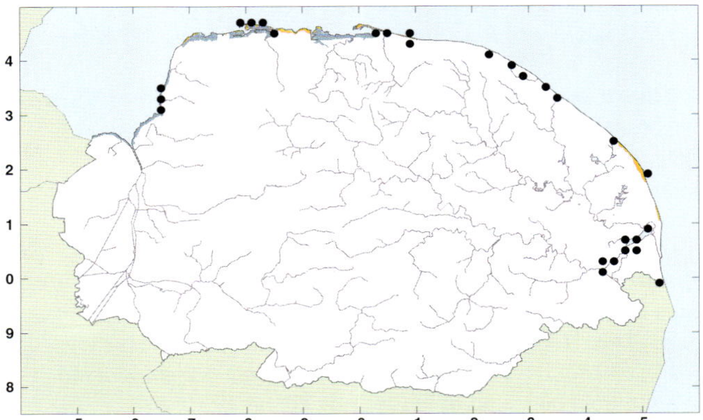

Annual of waste ground near the sea, usually found above the level of normal tides, often where the upper salt marsh adjoins dry land. Found growing beside the coast road at Salthouse E04 after the 1996 flood.

A. longipes Drejer **Long-stalked Orache**
Annual, its second British record was made in Norfolk at Brancaster W74 by PM.Taschereau with ELS & RPL in 1976, also found at Terrington Marsh W52 and Heacham W63 in 1977 and Snettisham W63, 1991. EL.Swann notes that it is found 'not near the sea, but occurs in brackish dykes and often associated with reeds'. A nationally scarce species. W6. E0.

A. littoralis L. **Grass-leaved Orache**

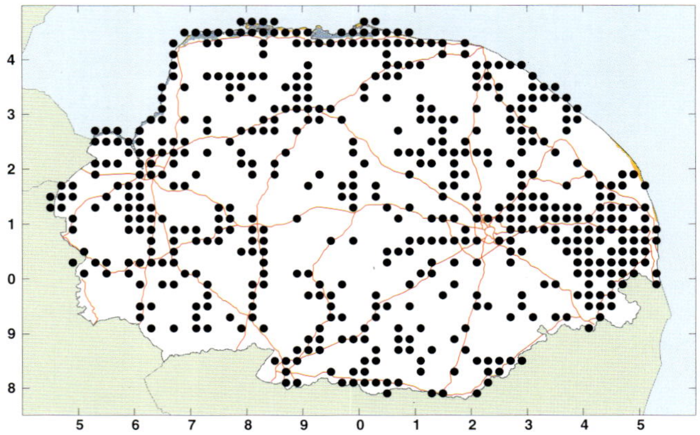

Annual of the high tide line on salt marshes and sea banks, now, like *A. prostrata*, spreading rapidly by roadsides inland and found on almost all classes of road which are salted in winter.

A. patula L. **Common Orache**
Annual of waste ground and arable fields, also by roadsides.
W514. E592.

A. laciniata L. **Frosted Orache**

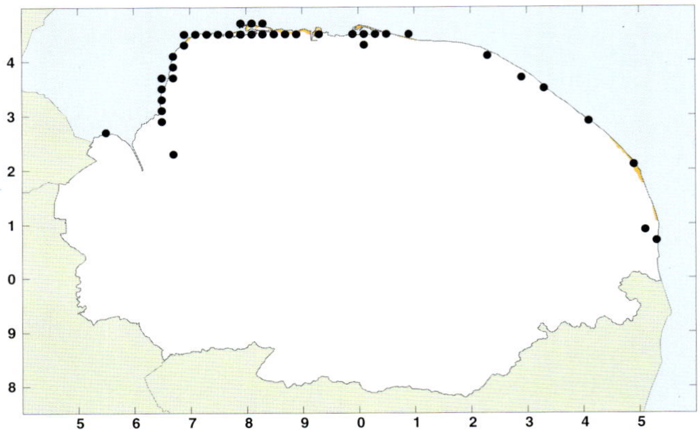

Annual of beaches, usually found near the tideline but rarely present in any quantity.

A. portulacoides L. **Sea Purslane**
Syn. *Halimione portulacoides* (L.) Aellen

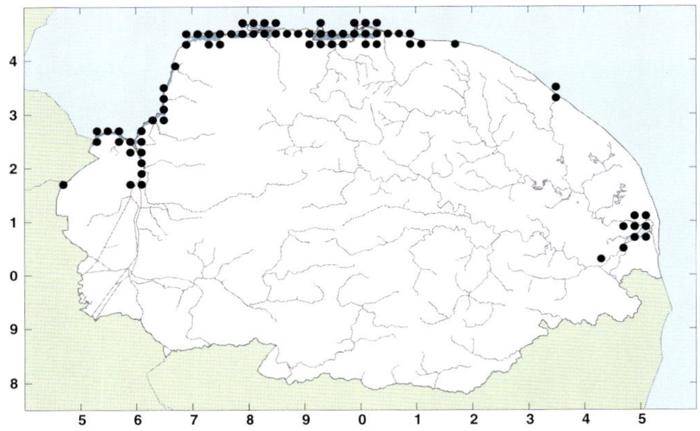

Woody evergreen perennial of salt marshes, fringing pools and creeks that are flooded at high tide.

Atriplex portulacoides - Holkham

■ **A. pedunculata** L. **Pedunculate Sea Purslane**
Syn. *Halimione pedunculata* (L.) Aellen
Annual of salt marshes and tidal estuaries. First discovered at Great Yarmouth about 1776, noted in The Monthly Magazine of April 1st 1824. Five records were made before 1900. Now known from only one site in the British Isles. Red Data Book Species.

Beta vulgaris L. ssp. **maritima** (L.) Arcang **Sea Beet**

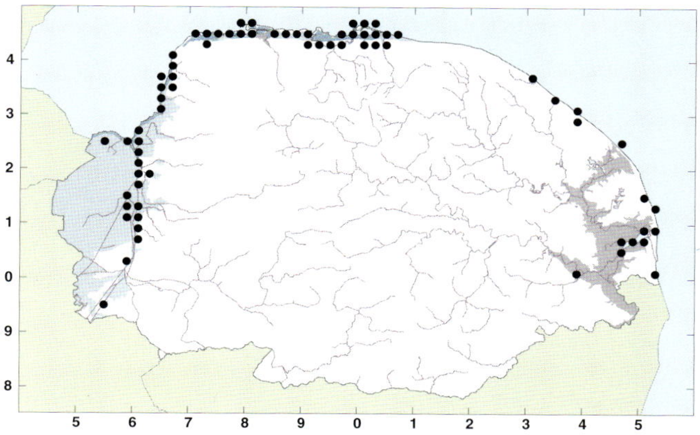

Annual species, occuring on salt marshes and tidal estuaries. From it ssp. **vulgaris**, root or sugar beet has been developed, this is an occasional casual of roadsides or volunteer in other crops.

Sarcocornia perennis (Mill.) A.J. Scott
Perennial Glasswort locally **Sheep's Samphire**
Syn. *Salicornia perennis* Mill.

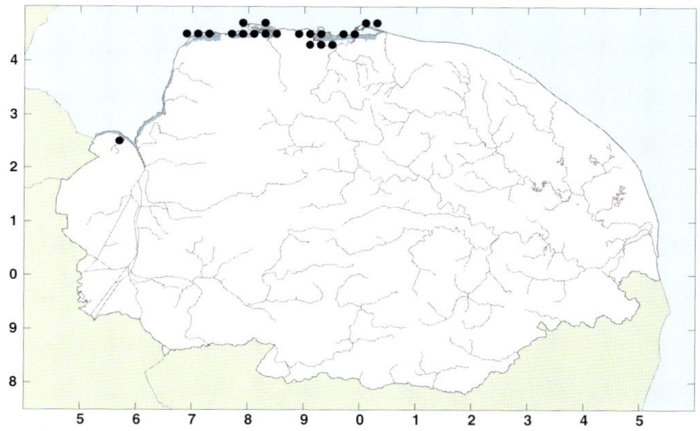

Perennial of drier areas in established salt marsh, the only marsh samphire not gathered for eating, hence its local name. A nationally scarce species.

Salicornia pusilla Woods **One-flowered Glasswort**
Syn. *S. disarticulata* Moss

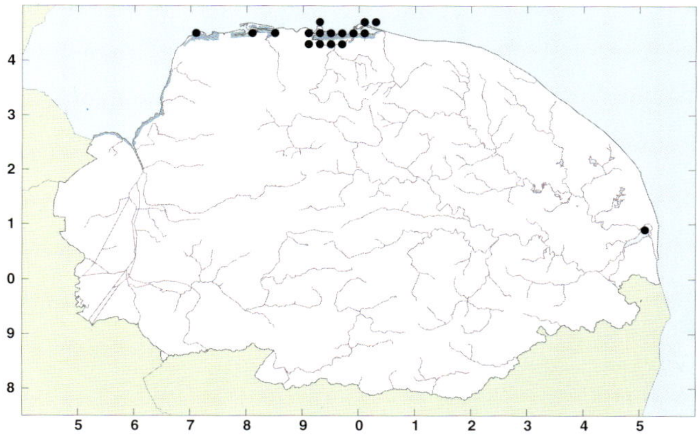

Annual, found on the drier edges of the salt marsh, particularly where it meets dry ground, occurring on the north coast marshes and on the edge of Breydon Water. A nationally scarce species.

S. ramosissima Woods **Purple Glasswort**

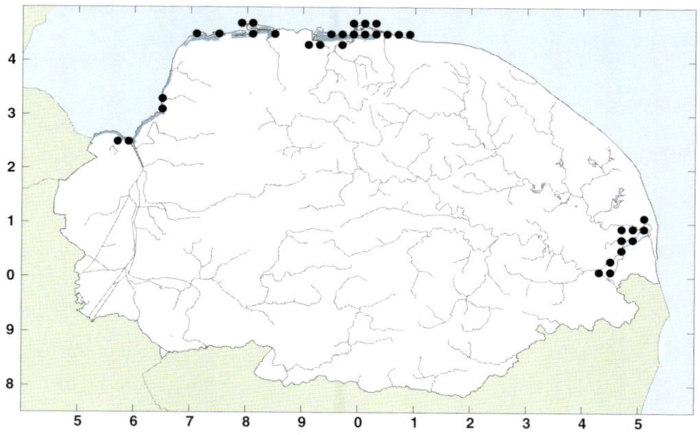

Annual, found on the Wash saltings as well as around the north coast and Breydon Water. Our most widespread species. In the Breydon area it occurs on riverside ronds as high upriver as saline influence extends.

S. europaea L. **Common Glasswort**
Syn. *S. stricta*

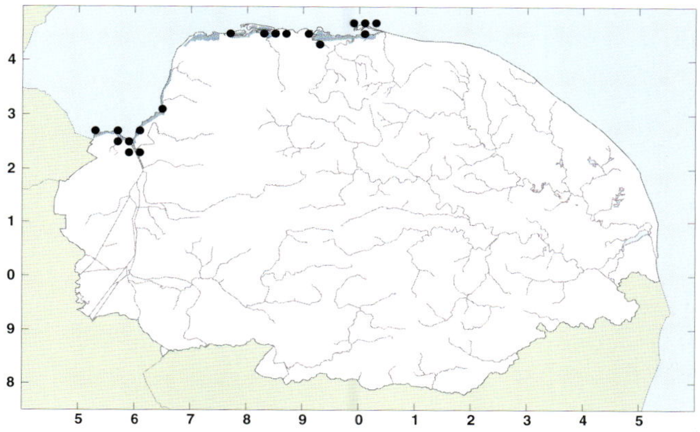

Annual, frequent throughout the main north coast areas of salt marsh and in the Wash, but not found elsewhere.

Salicornia europaea - Burnham Overy

S. obscura P. Ball & Tutin **Glaucous Glasswort**
Annual, found only on marshes at Blakeney Point W94 and E04 and Scolt Head W74 and 84. W3. E1.

S. nitens P. Ball & Tutin **Shining Glasswort**
Annual, found only once during this survey, at Blakeney Point E04.
W0. E1.

S. fragilis P. Ball & Tutin **Yellow Glasswort**
Annual, apparently restricted to a small area of the north coast marshes between Wells-next-the-Sea W94 and Blakeney, E04.
W2. E3.

S. dolichostachya Moss **Long-spiked Glasswort**

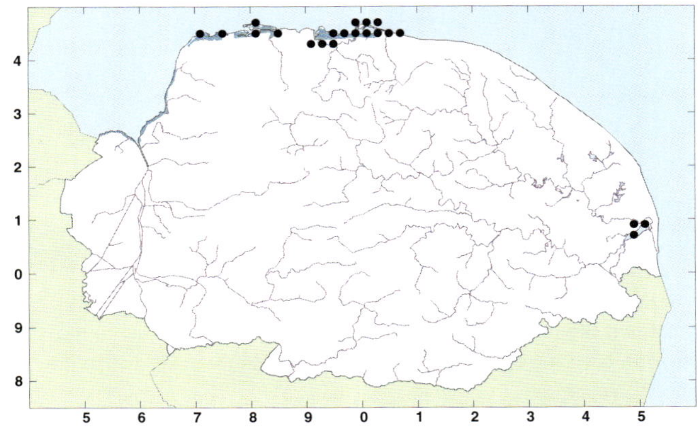

Annual, found where the mud is deep and regularly inundated.

Suaeda vera Forssk. ex J. Gmel. **Shrubby Seablite**
Syn. *S. fruticosa* sensu Coste non Forssk.

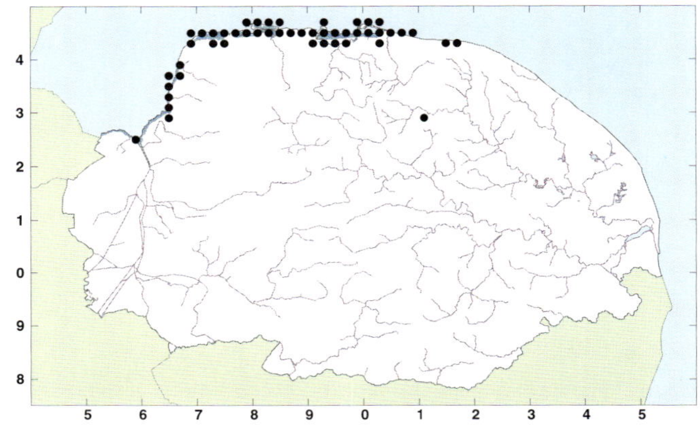

Evergreen shrub of the high tide line of sandy beaches, salt marshes and sea banks. A single bush was found beside a main road at Corpusty E12 in 1996, MAB. A nationally scarce species, uncommon outside East Anglia mainly because of the destruction of its habitat through coastal development.

Suaeda vera - Holkham

S. maritima (L.) Dumort. **Annual Seablite**

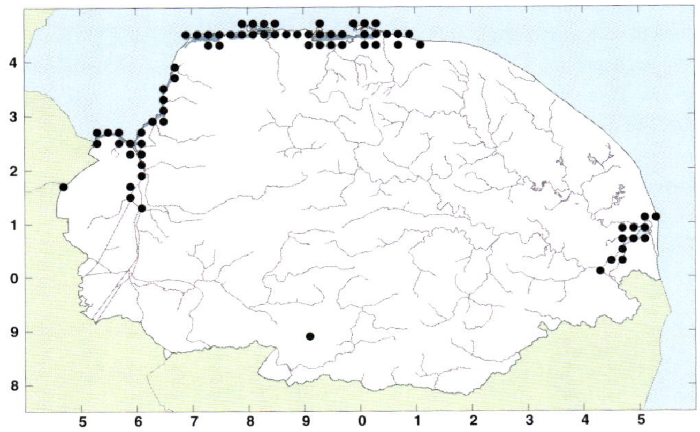

Annual of tidal salt marshes extending along rivers as far as the limit of saline influence and on sand and shingle to the limit of spring tides where it makes a distinctive purple splash of colour in autumn. Recorded once inland at East Wretham W98 in 1996, EVR.

Salsola kali L. **Saltwort**

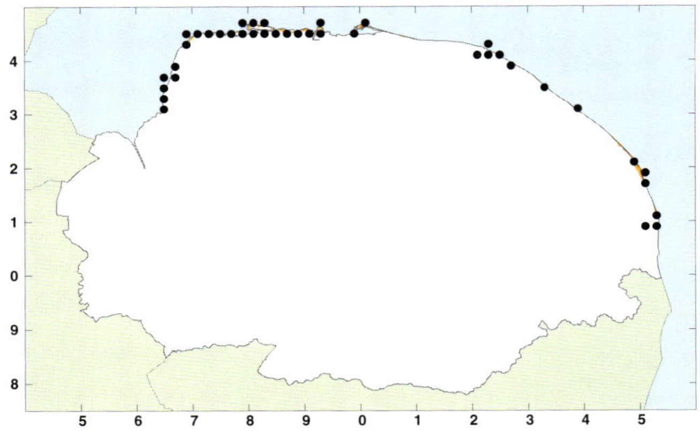

Annual of the tide line, especially on sand, this is ssp. **kali**. Ssp. **ruthenica** (Iljin) Soo, (syn. *S. pestifer* Nelson) occasionally occurs in arable, found in kale at Weeting W78 in 1976, in the same field in sugar beet in 1996, both JEG and at Hargham E09 in 1996, EVR.

AMARANTHACEAE

Amaranthus retroflexus L. **Common Amaranth**

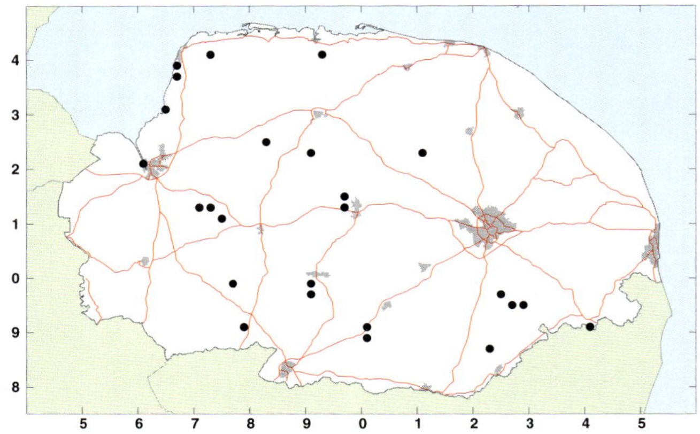

Scattered, chiefly as an arable weed, often associated with cattle, but also on waste ground.

A. hybridus L. **Green Amaranth**
Syn. *A. hypochondriacus* L.
Scattered as an occasional weed of arable, but more usually on waste ground. W12. E3.

■ **A. cruentus** L.
Syn. *A. hybridus var cruentus* Thell.
Annual from South America found once, at Houghton W72 before 1968.

■ **A. caudatus** L. **Love Lies Bleeding**
Annual from South America introduced to gardens and reported from Harford tip, Norwich E20 in 1960 and 1971.

■ **A. quitensis** Kunth. **Mucronate Amaranth**
Annual from South America recorded at Harford tip, Norwich E20 and Taverham E11 in 1959, also on the old tip at Lenwade E01 in 1978, ETD.

A. bouchonii Thell.
Bouchon's Amaranth locally **Factory Weed**

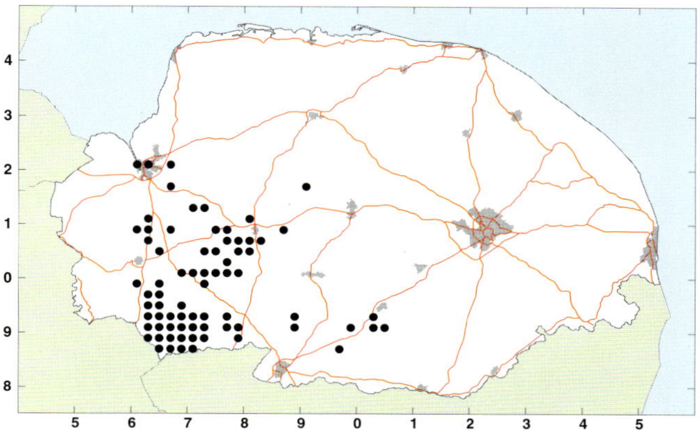

Annual of uncertain origin, well established as a field weed in lighter soils. It has persisted in the same field at Weeting since 1967, JEG. On Methwold and Feltwell Fens, it has been observed locally that it appears after waste material from the sugar beet factory at Wissington W69, has been spread over the fields. By this means, the seed, which has been carried to the factory with beet from fields where it is established, is being moved to new sites, hence it is known locally as Factory Weed (HW). Since 1995, with the closure of the beet factory at King's Lynn, a much larger area is being served by Wissington and it will be interesting to see if the plant takes advantage of this.

Amaranthus bouchonii - Cockley Cley

A. blitoides S.Watson **Prostrate Pigweed**
Spring germinating annual occurring in arable land, first seen at Appleton W72 in 1949. During this survey found at Warham W93 in 1993 by PR.Banham where Mr Green, who farms the field, reported having seen it in the same site since 1986, also at Stow Bedon W94 in 1997, EVR. W2. E0.

A. albus L. **White Pigweed**
Spring germinating annual, found in a carrot field at Garboldisham W98 in 1975 and a single plant beside roadside farm buildings at South Lopham E08 in 1994, ALB. W0. E1.

PORTULACACEAE

Claytonia perfoliata Donn ex Willd.
Spring Beauty, Miners' Lettuce, Buttonhole Flower
Syn. *Montia perfoliata* (Donn. ex Willd.) Howell

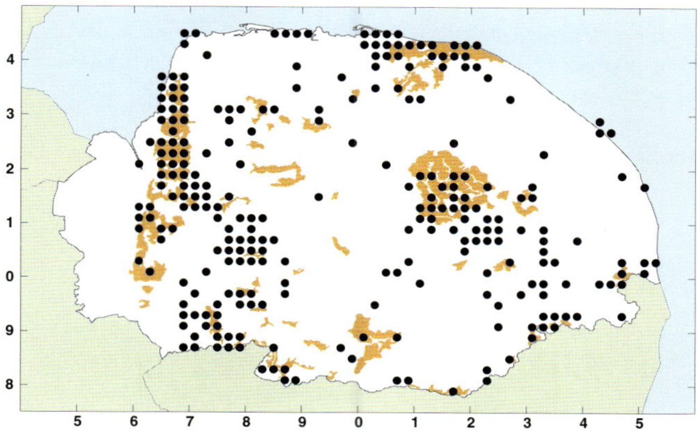

North American annual naturalised and sometimes a troublesome weed in light, sandy soils, particularly in shady sites. Typical of Breckland pine belts in spring. Increasingly grown as a winter salad crop.

C. sibirica L. **Pink Purslane**
Syn. *Montia sibirica* (L.) Howell
Annual from north-west America, sometimes cultivated and occasionally naturalised in shady places, sometimes well away from gardens. W10. E25.

Montia fontana L. **Blinks**

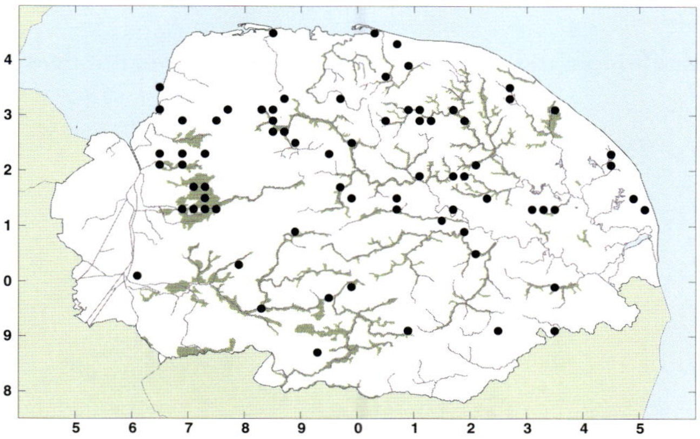

Annual or short-lived perennial of seasonally damp, usually acidic soils. Four subspecies are recognised of which ssp. **chondrosperma** (Fenzl.) Walters (ssp. *minor* W.R.Wayw. nom. illeg.) is found throughout the county and ssp. **amporitana** Sennen (ssp. *intermedia* (Beeby) Walters) has been found on acid ground at Roydon W62; East Winch W61 and Denver Common W51 in 1995 by J&HW.

CARYOPHYLLACEAE

Arenaria serpyllifolia L. ssp. **serpyllifolia**
Thyme-leaved Sandwort

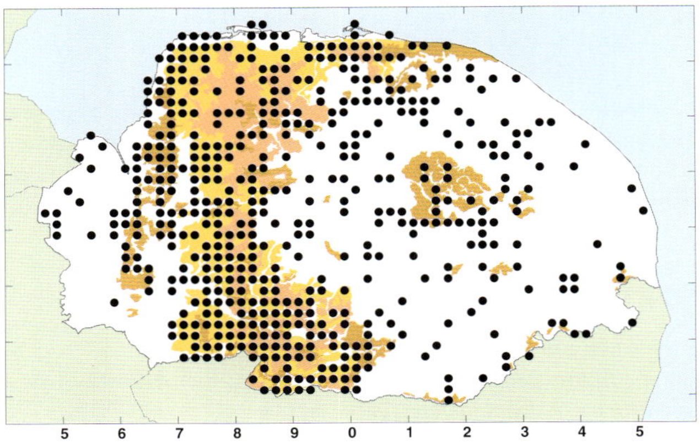

Annual or biennial, sometimes over-wintering. Frequent on dry soils, paths and walls. This appears to be the most frequent subspecies, especially on calcareous soils.

Arenaria serpyllifolia L. ssp. **leptoclados** (Rchb.) Nyman

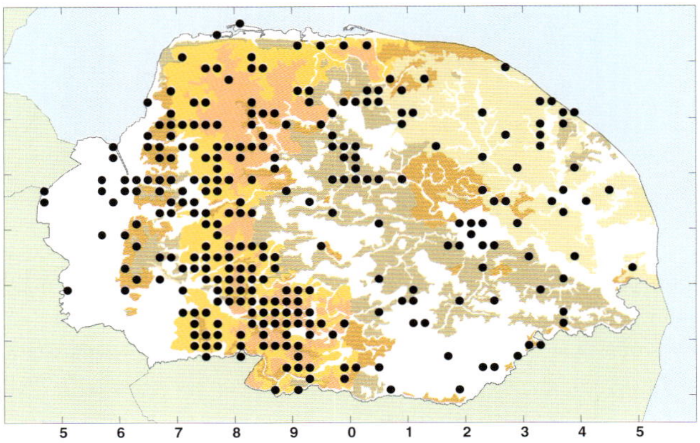

This subspecies is less common but still widespread, especially on lighter soils.

■ **A. balearica** L. **Mossy Sandwort**
Perennial introduced to rock gardens from the western Mediterranean islands and ocasionally established on walls as at Overstrand E24 in 1977, EAE and Sidestrand E23 in 1980, ETD.

Moehringia trinervia (L.) Clairv. **Three-nerved Sandwort**

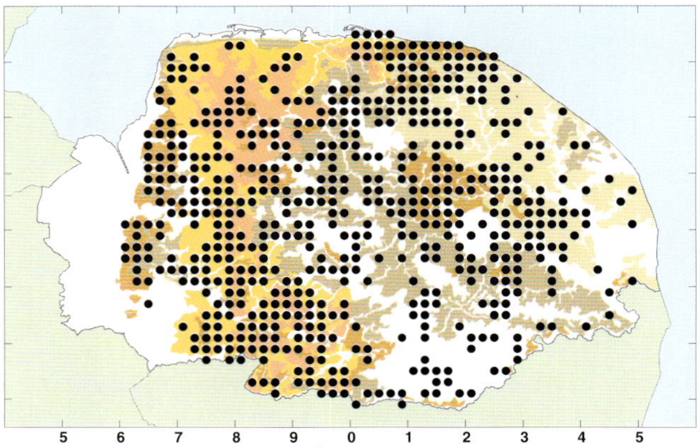

Short-lived perennial or over-wintering annual of shady sites. A regular colonist of newly planted woods, a shade rather than a woodland species. Least common on wet or heavy soils.

Honckenya peploides (L.) Ehrh. **Sea Sandwort**

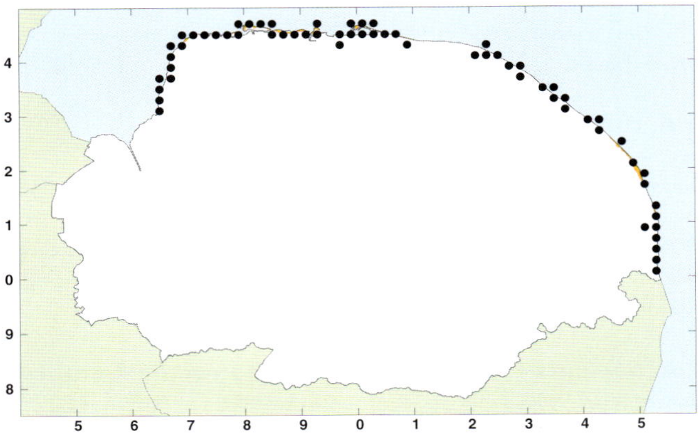

Deciduous perennial confined to sand dunes and shingle. Plants can be killed to ground level if a sufficiently severe spring frost occurs to affect their coastal habitat.

Minuartia hybrida (Vill.) Schischk. **Fine-leaved Sandwort**

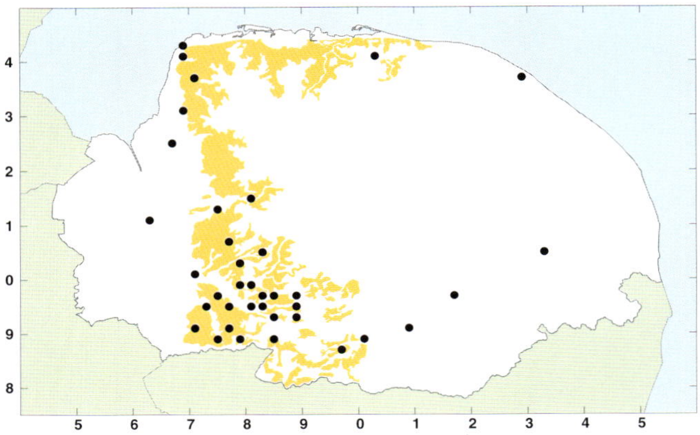

Annual, native to dry grass heaths, also seen on old walls but nowhere frequent. A plant which apparently needs open ground to become established. A nationally scarce species.

[**Stellaria nemorum** L. **Wood Stitchwort**
Recorded in the 1952 Atlas for Salthouse, E04 but not refound and considered an error.]

S. media (L.) Vill. **Chickweed**
Ubiquitous annual or biennial of disturbed ground, germinating whenever weather is favourable. W711. E697.

S. pallida (Dumort.) Crép. **Lesser Chickweed**

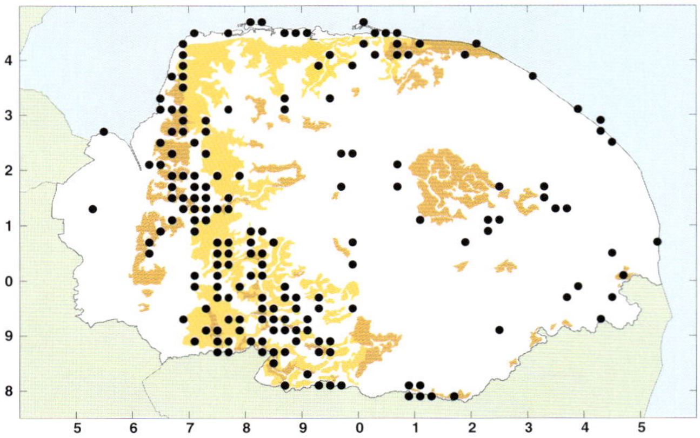

Annual often over-wintering. Occurring on sandy and shingly soils, particularly in woodland giving light shade as in Breckland shelter belts and forest rides. Also found on walls.

S. neglecta Weihe **Greater Chickweed**

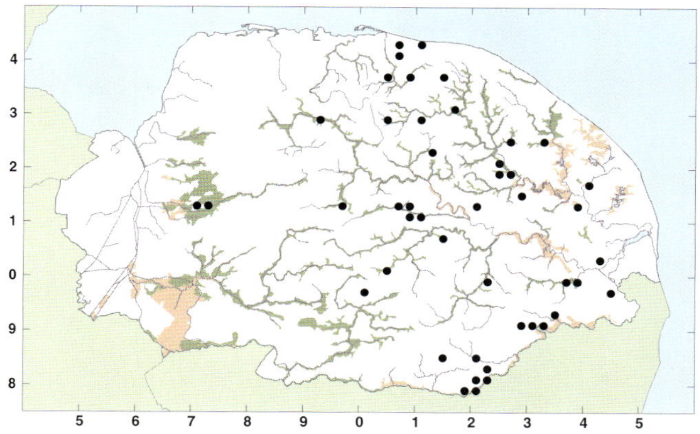

Annual or short-lived perennial of damp, shady places, usually on peaty alluvium in river valleys.

S. holostea L. **Greater Stitchwort**

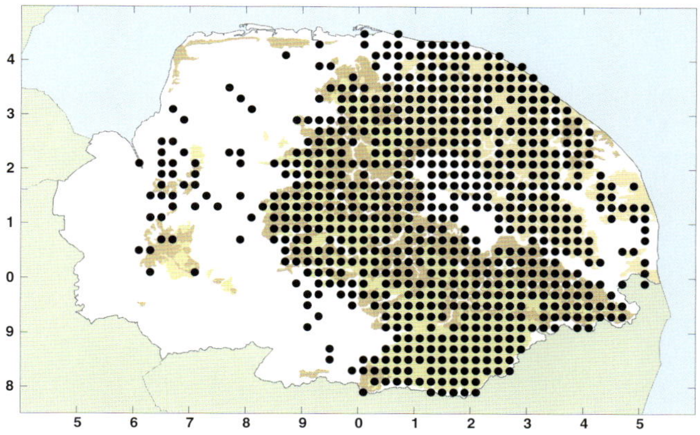

Scrambling winter green perennial of partially shady places especially hedgerows, mostly on heavier soils. Particularly associated with sites which are or have been ancient woodland so very scarce or absent from Breckland, Broadland and both former and existing heathland sites.

S. palustris Retz. **Marsh Stitchwort**

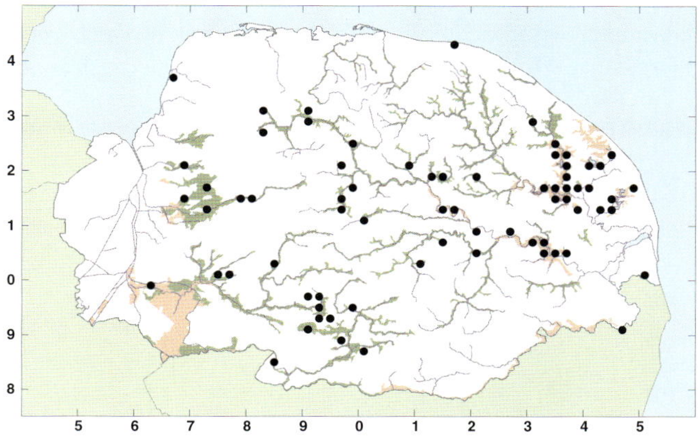

Winter green perennial of marsh and fen, rather scattered and never common.

S. graminea L. **Lesser Stitchwort**

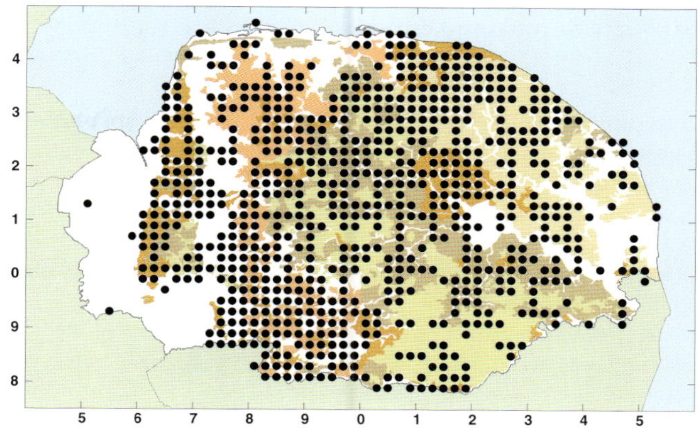

Winter green perennial of well-drained, grassy places. Winter leaves are shorter and broader than normal.

S. uliginosa Murray **Bog Stitchwort**
Syn. *S. alsine* Grimm nom. inval.

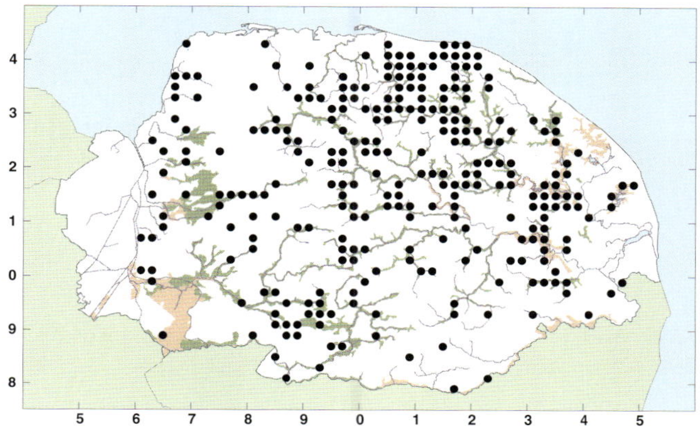

Winter green perennial of soils which are moist in winter. In Breckland, often found beside forest rides, even on dry, sandy soils, presumably because of the dampness retained by the compacted soils and run-off from them. Also similarly along stream margins where the soil is compacted by cattle trampling.

■ **Holosteum umbellatum** L. **Jagged Chickweed**
Annual, once found in dry places. First recorded in 1775 on the city walls in Norwich, but not seen since its sighting by HD.Geldart in 1887. Extinct in the British Isles.

Cerastium arvense L. **Field Mouse-ear**

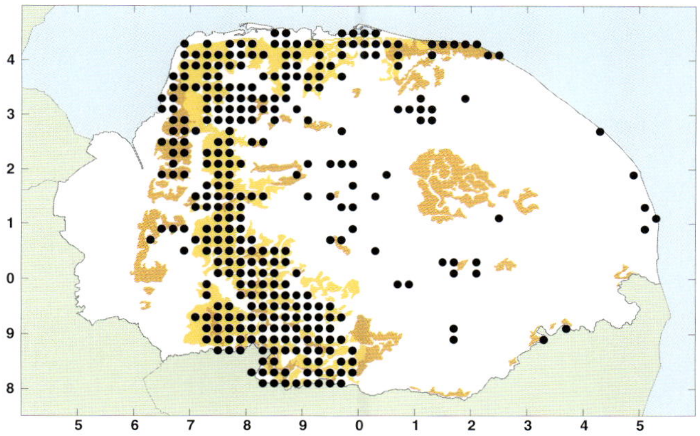

Winter green perennial, frequent in dry, neutral to alkaline grassland, much commoner in west Norfolk than in most of England.

C. arvense x C. tomentosum
Syn. *C. x mauereri* M.Schulze, nom. nud.
Observed growing with both parents on verges at Binham W93; Warham W94 and Raynham W82, KAB, 1995-1997. W3. E0.

C. tomentosum L. **Snow-in-summer**
Winter green perennial introduced from Italy and Sicily to gardens, now well established on roadside banks and found not infrequently on walls, also at Heacham, W63 on shingle. W92. E305

C. fontanum Baumg. **Common Mouse-ear**
Syn. *C. holosteoides* Fries.
Winter green perennial of grassland and a common weed of cultivated ground on all soils. Our plant is ssp. **vulgare** (Hartman) Greuter & Burdet (syn. ssp. *triviale* (Murb.) Jalas). W681. E694

C. glomeratum Thuill. **Sticky Mouse-ear**

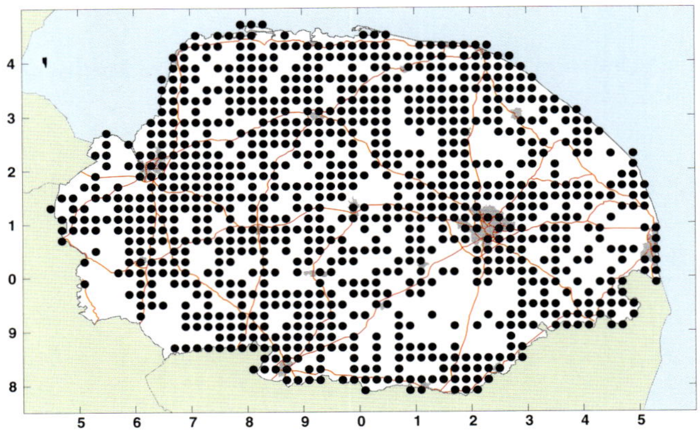

Mainly over-wintering annual of disturbed ground, the commonest *Cerastium* on walls (RMP). Appears to avoid wetter soils.

C. diffusum Pers. **Sea Mouse-ear**
Syns. *C. atrovirens* Bab., *C. tetrandrum* Curt.

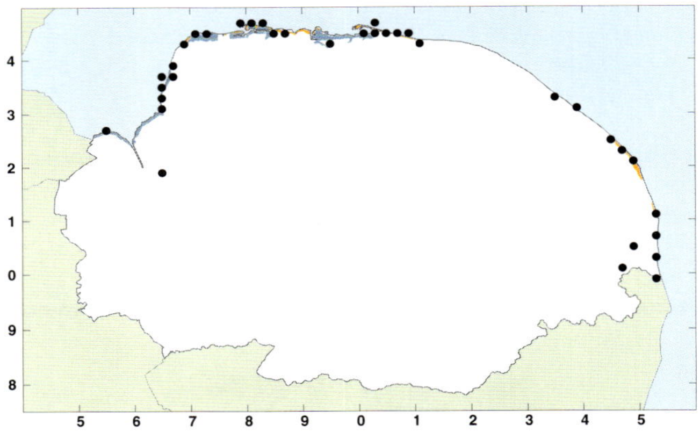

Annual of sand and shingle near the sea, never common.

C. semidecandrum L. **Little Mouse-ear**

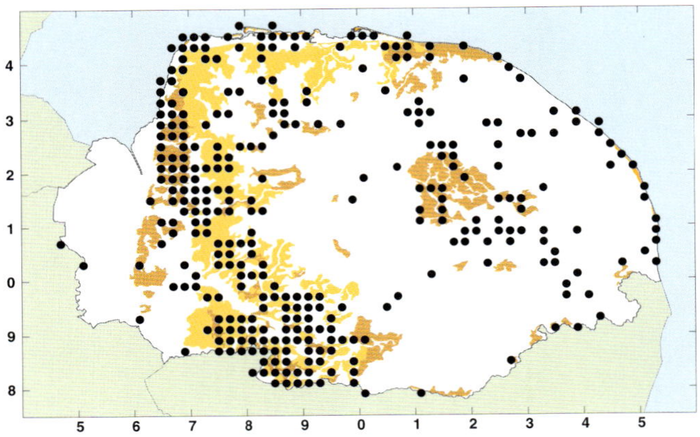

Over-wintering annual of dry, bare places both on coastal dunes and inland.

Myosoton aquaticum (L.) Moench **Water Chickweed**

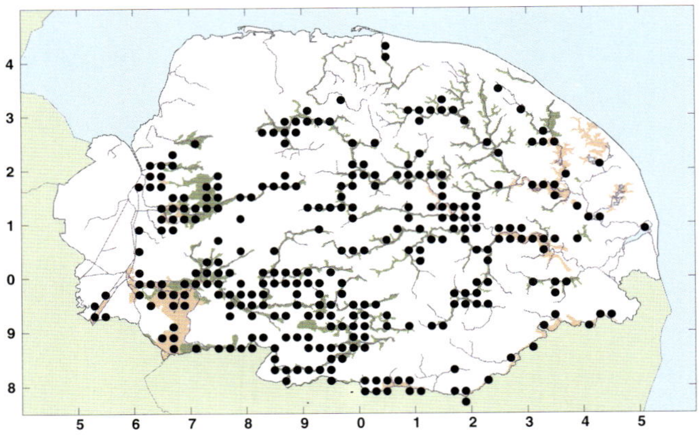

Winter green perennial of the middle and upper reaches of rivers and streams. More or less confined to peaty alluvium, but increasingly in drier habitats such as newly surfaced forest rides (JEG) and on the Devil's Dike at Beechamwell W70, HW.

Moenchia erecta (L.) P. Gaertn., Mey. & Scherb. **Upright Chickweed**
Scarce annual of consolidated, sandy soils, formerly recorded both from coastal and inland locations. During this survey seen only once, at Snettisham W63 in 1996, SF. W1. E0

Sagina nodosa (L.) Fenzl **Knotted Pearlwort**

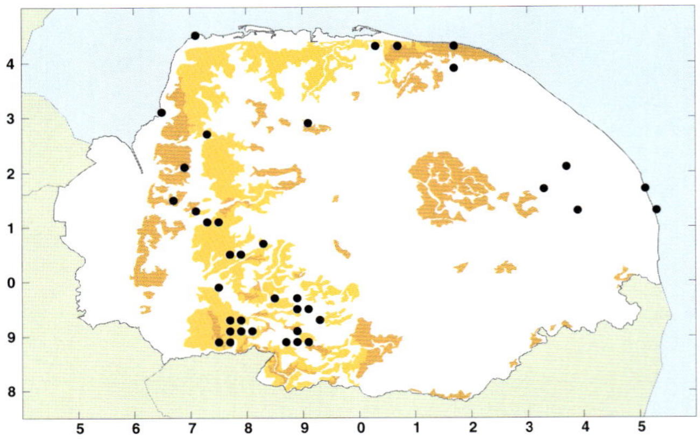

Scarce perennial of sandy soils, especially where there is some moisture present in winter. Most frequent in the Breckland where it is found on heaths and can be conspicuous around the meres in their first and second seasons after drying out. Also on moist, open forest rides. Infrequently on the coast as at Holme W74 in 1996, JEG.

■ **S. subulata** (Sw.) C. Presl **Heath Pearlwort**
There are two records of this species in the 1914 Flora, but EL.Swann, who examined the specimens in 1964, considered them misidentified so it cannot be claimed as a Norfolk native. The golden-leaved cultivar ***S. s.*** 'Aurea' is, however, widely cultivated and can produce normal green seedlings, these can escape and may be the source of a recent record at Overstrand E24.

S. procumbens L. **Mossy Pearlwort**

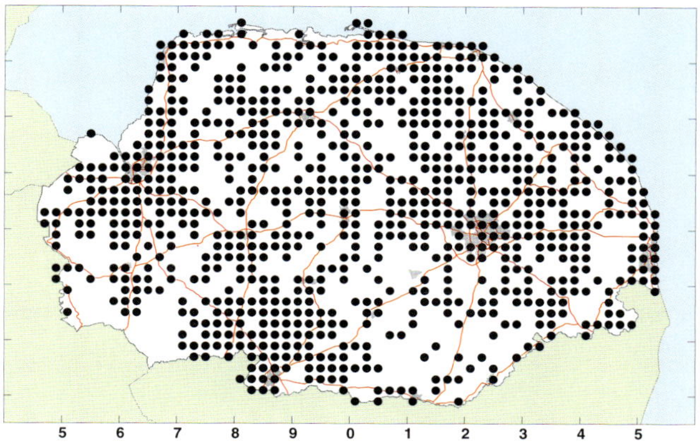

Winter green perennial common on paths, tracks, walls and short turf including lawns.

S. apetala Ard. **Annual Pearlwort**

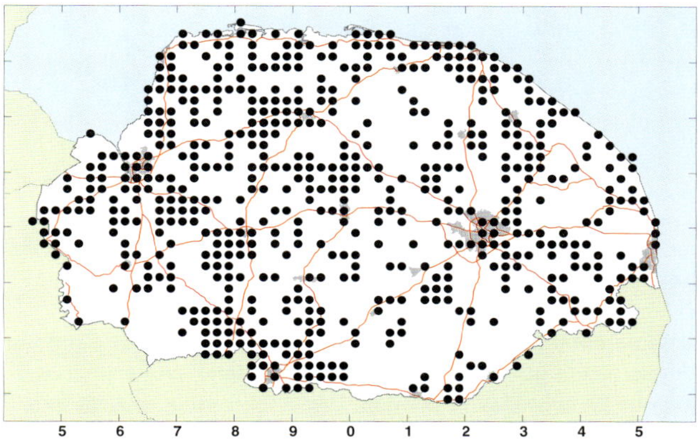

Overwintering annual, frequent on dry bare ground, often on walls and tracks within the confines of those villages and towns where not too much 'tidying up' is carried out. Of the three subspecies, ssp. **erecta** is by far the commonest, and ssp. **filicaulis** is rare, having been observed in three sites on dry, grass heaths in Breckland in TL78 and 88. Ssp. **apetala** (syn. *S. ciliata* Fr.), which was first recognised for Britain from Thetford in 1847, is scattered in a variety of habitats from tracks on heathland to the tarmac of a road, most of its sites having compacted soil. It is also found on walls.

S. maritima Don **Sea Pearlwort**
Annual, scattered along the coast on sand from Wolferton W62 to Burnham Overy W84 and on shingle at Cley E04 but nowhere common. Also found in 1992 on the bank of the new enclosure (bund) at Terrington W52, I&F.Simper. W9. E8.

■ **Scleranthus perennis** L. **Perennial Knawel**
Winter green perennial recorded on the coast at Snettisham beach W63 before the 1953 floods and at Cley E04 in 1954. Inland it occured at Barnham Cross Common W88 until 1943 and Weeting Heath W78 in 1961. Our plants are ssp. **prostratus** P.D.Sell. Not seen in the county since 1961, although a few recent introductions have taken place in Breckland. Red Data Book species

S. annuus L. **Annual Knawel**

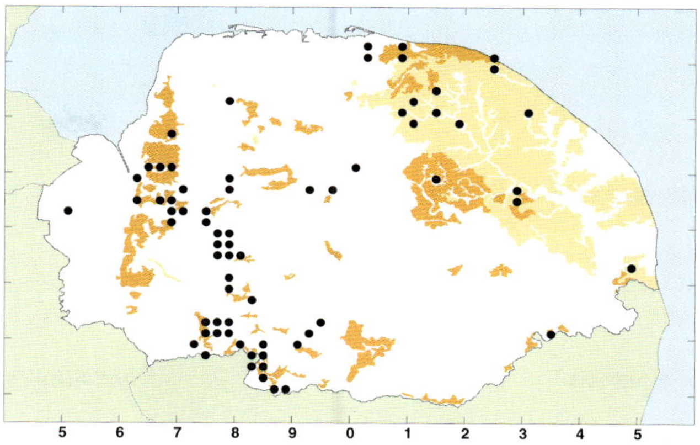

Annual of open sandy ground sometimes overwintering, described as 'widespread and frequent' in 1968, but now far less common and rare in the east of the county. Our plants all seem to be ssp. **annuus**. Ssp. **polycarpos** (L.) Bonnier & Layens, frequent in the Suffolk Breckland, has been found only once in Norfolk at Santon W88 in 1976 by ELS.

[Corrigiola littoralis L. **Strapwort**
There is a specimen of this in the herbarium at Liverpool marked 'Norfolk, 1887' with no further details. Its origin must be considered very dubious as this is not an eastern species.**]**

Herniaria glabra L. **Smooth Rupturewort**

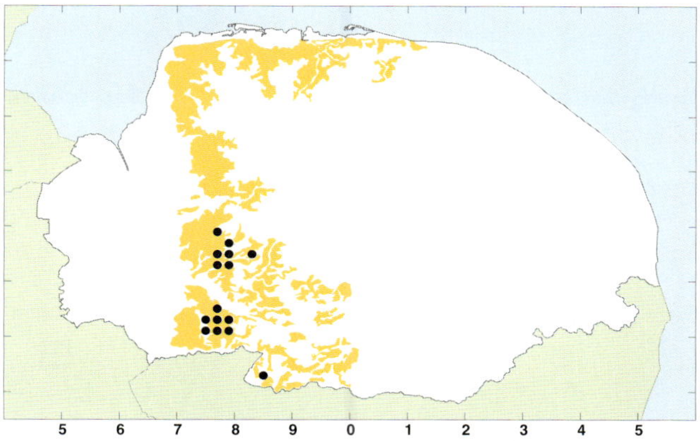

Short-lived winter green perennial of slightly calcareous, sandy ground, rare nationally and in Norfolk almost confined to compacted rides in Breckland forests, JEG. Plants grown in gardens self sow freely in paths as has been observed at Sisland E39 and Stanhoe W83. Red Data Book species. For the two east Norfolk records of this species in Petch & Swann, 1968, see over.

Herniaria glabra - Cranwich Heath

■ **H. hirsuta** L. **Hairy Rupturewort**

Short-lived winter green perennial recorded from Horstead E22 in 1917 and in gravel pits at Frettenham E21 in 1945. The identity of these plants has given rise to much discussion but the latter was determined as this species by NY.Sandwith in 1962 (pers.comm. EAE to ELS 1968).

Spergula rupicola Lebel ex Le Jolis **Rock Sea Spurrey**

A specimen collected at Hunstanton,W64, in 1908 by FR.Tennent is in the Herbarium at Cambridge, comm. G.Crompton. Also seen beside a mineral sheep lick in pasture at East Tuddenham E01 in 1996, forming a sizeable patch by autumn 1997, ALB. W0. E1.

S. arvensis L. **Corn Spurrey**

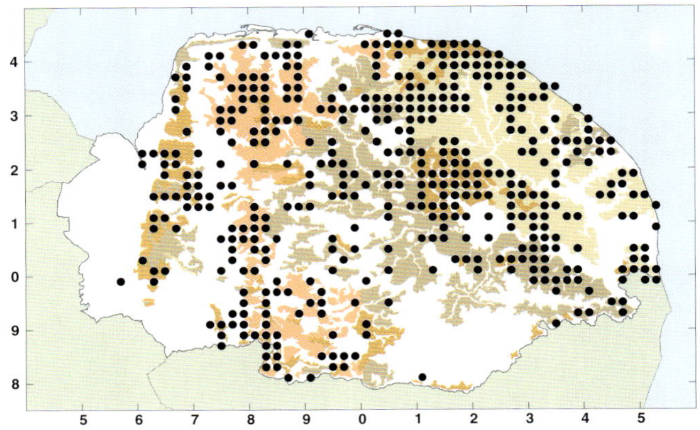

Annual of disturbed ground, sometimes overwintering. Considered an indicator of soil acidity on medium or light sandy soils.

Spergularia media (L.) C. Presl **Greater Sea Spurrey**

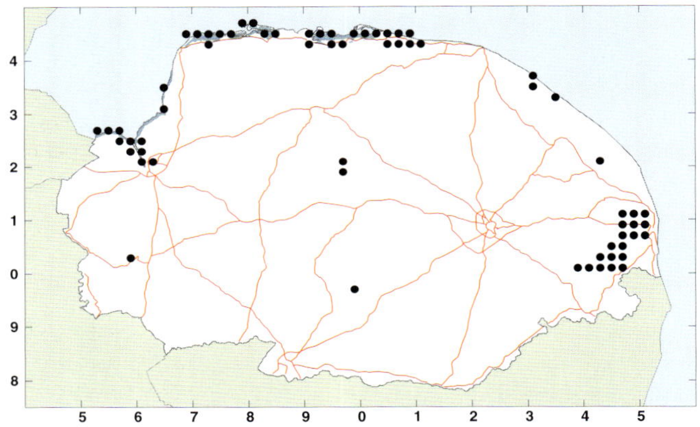

Perennial of salt marshes, spreading inland where, like *S. marina*, it is occasionally found in bare places on roadside verges. Both appear to have spread not from local coasts but from the south being first recorded along the A146 from Suffolk in the 1980s, ETD and along the A10 from Cambridgeshire in the early 1990s, DE.Coombe.

S. marina (L.) Griseb. **Lesser Sea Spurrey**

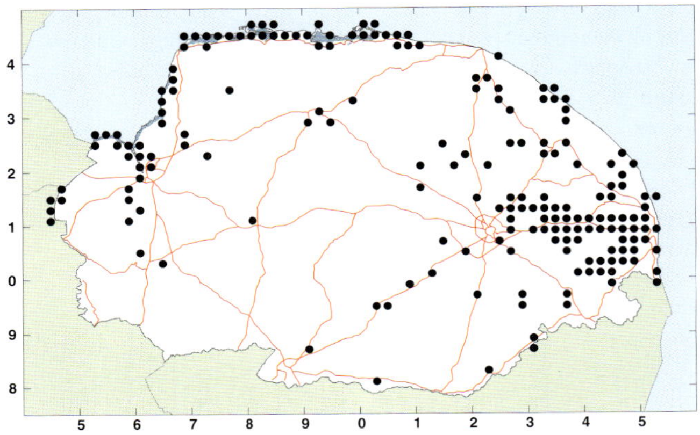

Perennial of salt marshes, generally commoner than *S. media*. See comments above under that species.

S. rubra (L.) J.S. Presl & C. Presl **Sand Spurrey**

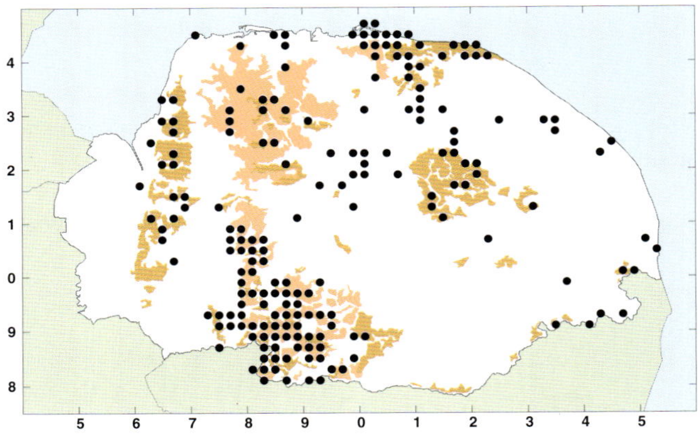

Overwintering annual or biennial of neutral to acid, sandy soils, frequent only on tracks on heaths and in the more acid areas of the Breckland forests.

Lychnis coronaria (L.) Murray **Rose Campion**

An occasional garden escape but very persistent once established and one that can survive mowing. More frequent in east Norfolk and perhaps becoming more common. W3. E64.

L. flos-cuculi L. **Ragged Robin**

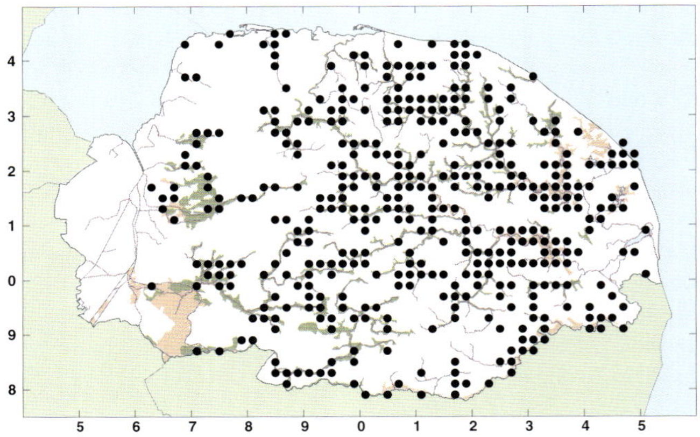

Partially winter green perennial of marshes, stream banks and woodland glades, shade tolerant. Quite absent from apparently suitable sites in the Fenland. White-flowered forms sometimes occur.

Agrostemma githago L. **Corn Cockle**
An annual from central Europe, once a frequent cornfield weed, now almost always the result of deliberate sowing, usually amongst so-called native plants as on the verges of the Norwich southern bypass. Very often these plants belong to the larger cultivated form 'Midas'. W3. E12.

Silene otites (L.) Wibel **Spanish Catchfly**
Winter green perennial typical of and always confined to the Breckland. Over 20 sites were formerly known, but it is now reduced to four, two at Thetford W88; Harling W98 and Cranwich W79 where it can still occasionally be found in abundance in thin grassland, especially after spring droughts have reduced the competing herbage. Although able to survive in denser grassland for a number of years, it needs open ground for the establishment of seedlings. A Red Data Book species. W5. E0.

Silene otites - Cranwich

S. vulgaris Garcke **Bladder Campion**

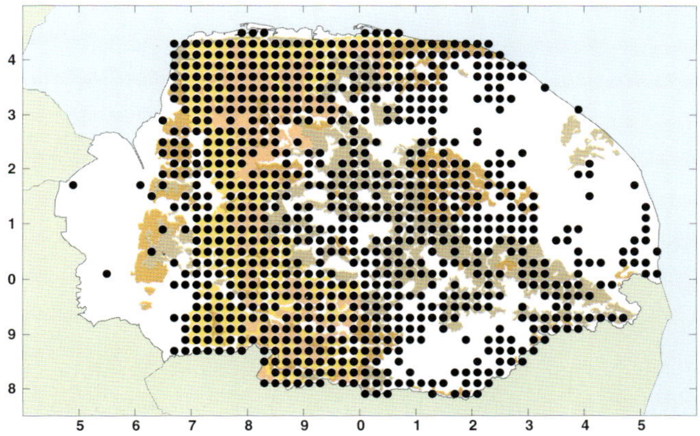

Partially winter green perennial of established grassland, scarce on heavy soils, largely absent from Broadland and replaced in coastal tetrads by *S. uniflora*.

S. uniflora Roth **Sea Campion**
Syn. *S.maritima* With.

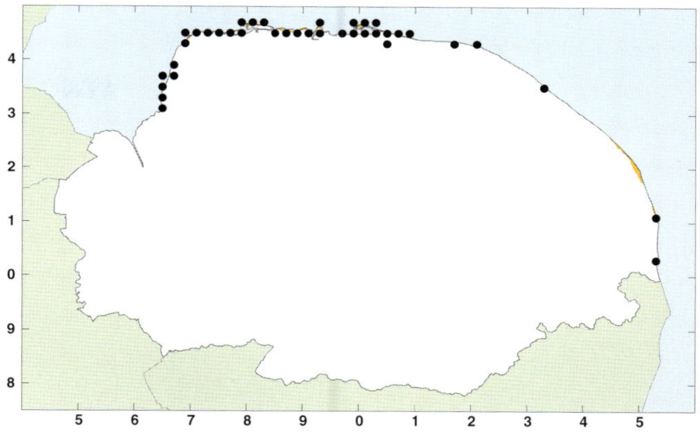

Partially winter green perennial of established dunes and shingle. A double-flowered form has been known for some years on the salt marshes at Morston E04, J.Oxenford.

S. noctiflora L. **Night-flowering Catchfly**

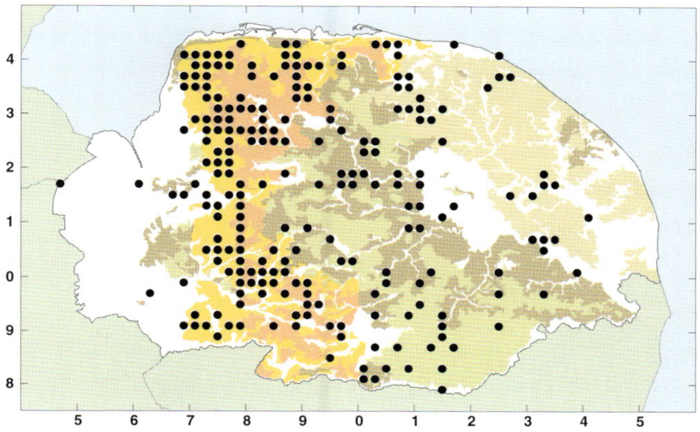

Spring germinating annual of light soils sometimes found in spring sown crops, particularly sugar beet but has become less frequent in the county since 1985. Scarce outside Norfolk.

S. latifolia Poiret **White Campion**
Syn. *S. alba* (Mill.) E.H.L.Krause nom.illeg.
Winter green perennial of roadsides and cultivated ground.
W646. E667.

S. x hampeana Meusel & K. Werner **Hybrid Campion**
(S. dioica x S. latifolia)
A frequent hybrid, sometimes found where *S. dioica* no longer occurs. W112. E179.

S. dioica (L.) Clairv. **Red Campion**
Syn. *Melandrium rubrum* (Weig.) Garcke.

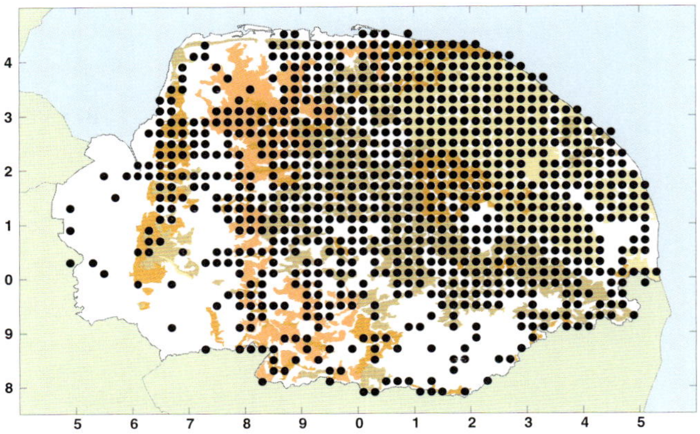

Winter green perennial of woods and hedgerows, many plants show signs of introgression with *S. latifolia* and only in old woodlands are the flowers of this species truly red. Absent from recently reclaimed land and old sheepwalk areas, but also surprisingly scarce on the boulder clay.

S. coeli-rosea (L.) Godr. **Rose of Heaven**
Annual introduced to gardens from southern Europe, observed growing near the slipway to the sea at West Runton E14 in 1993, CG.Hanson. W0. E1.

S. gallica L. **Small-flowered Catchfly**

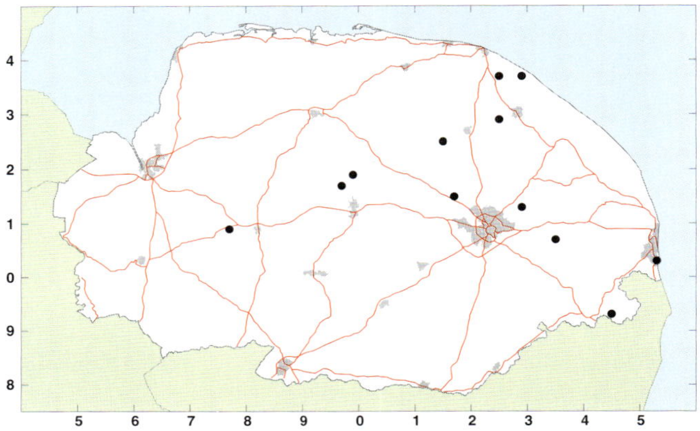

Scarce annual, sometimes overwintering. A particularly large colony was noted on the banks of a disused railway track at Gimingham E23 in 1995, ALB & WM. A nationally scarce species.

S. conica L. **Sand Catchfly**

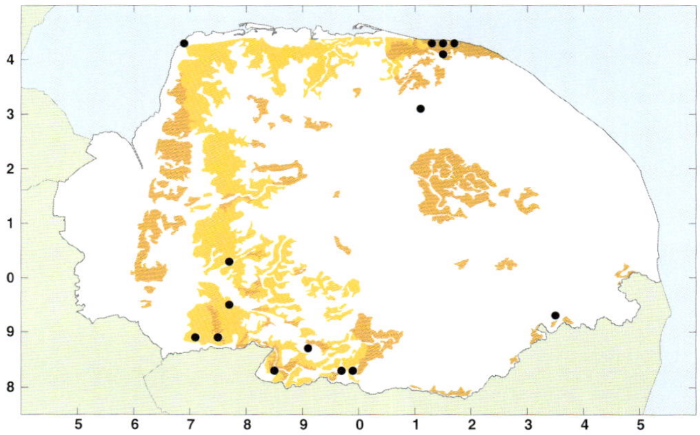

Spring germinating annual of light sandy soils now typical of the Breckland, though it has become much reduced even there in recent years. Needs open ground for satisfactory germination. Appeared in thousands in a bare conservation headland at Harling W98 in 1992, but has been reduced to less than twenty plants as the grass sward has closed in. Also on the coast and inland at Corpusty E13, MAB and Broome E39 R.& JP.Ellis. A nationally scarce species.

Cucubalus baccifer L. **Berry Catchfly**
Herbaceous perennial introduced from eastern Europe in the 19th century to Merton W99 (pers.comm. Lord Walsingham) and now well established at four sites in the area. Well able to hold its own in rank grass and open areas in carr. W4. E0.

Saponaria officinalis L. **Soapwort**

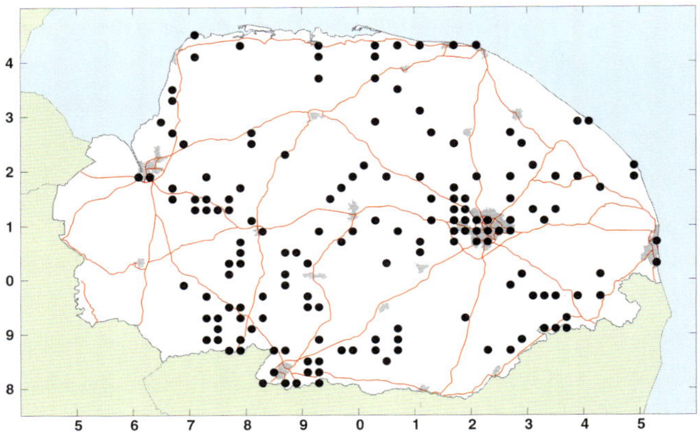

Herbaceous perennial introduced to gardens from mainland Europe and well naturalised by some roadsides. Both single and double forms occur in shades of pink to red. Plants with double pink flowers (*flore-pleno*) are the commonest. The rarer white-flowered forms have not been observed.

S. ocymoides L. **Alpine Soapwort**
Herbaceous perennial from the mountains of Europe, an occasional escape from gardens. One plant was found on a steep bank at the edge of a cemetery in King's Lynn W61 in 1992, RMP. W1. E0.

■ **Vaccaria hispanica** (Mill.) Rauschert **Cowherb**
Syn. *V.pyramidata* Medik.
Annual introduced from the Mediterranean, probably mainly with bird seed. Recorded eight times, the last at Brancaster W74 in 1978, R.Kimber.

Petrorhagia prolifera (L.) P. Ball & Heywood **Proliferous Pink**
Syn. *Kohlrauschia prolifera* (L.) Kunth
Autumn germinating annual, its Norfolk site in W79 is one of only two remaining in Britain. Recorded from the Bircham area, W73 by K.Trimmer, probably before 1840. Red Data Book Species. W1. E0.

Petrorhagia prolifera - Cranwich

■ **Dianthus plumarius** L. **Pink**
Winter green perennial introduced to gardens from southern Europe, occasionally escaping. Recorded from Upton E31 in 1974, ETD; Weybourne E14 in 1975, KH.Bull and Gt Yarmouth E50 in 1976, M.Benn.

D.deltoides L. **Maiden Pink**
Winter green perennial of dry grassland, much reduced in frequency through destruction of habitat and the rankness of much remaining grassland. Now confined to the Breckland at Ickburgh W89; Weeting W78 and Thetford W88, but formerly also at Cley E04. A garden escape at Heacham W63 noted amongst the beach huts in 1992, PRC. A nationally scarce species. W4. E0.

D. barbatus L. **Sweet William**
Short-lived perennial from southern Europe introduced into gardens and sometimes thrown out on roadsides, though rarely persisting long. W4. E6.

D. superbus L. **Superb Pink**
A winter green perennial from southern Europe found in 1992 at Grimes Graves W89, probably from introduced seed. W1. E0.

■ **D. armeria** L. **Dept ford Pink**
Annual or biennial, perhaps native to grassland in south Norfolk. Last seen at Pulham in 1964. Also an occasional garden escape.

POLYGONACEAE

■ **Persicaria campanulata** (Hook.f.). Ronse Decr. **Bell-flowered Knotweed**
Syn. *Polygonum campanulatum* Hook.f.
Herbaceous perennial native to the Himalaya, occasionally escaping from gardens, recorded from Ashby E41 in 1963.

P. wallichii Greuter & Burdet **Himalayan Knotweed**
Syn. *Polygonum polystachyum* Wall ex. Meissn.
Herbaceous perennial native to the Himalaya and noted as well established at Happisburgh E33 in 1991; King's Lynn W61 and 62 in 1992 and Kirby Cane E39 in 1995. W4. E2.

P. bistorta (L.) Samp. **Common Bistort**
Syn. *Polygonum bistorta* L.

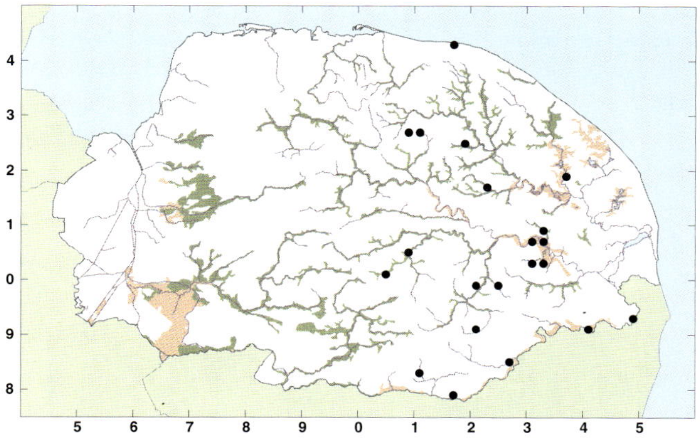

Herbaceous perennial still found in a few old herb-rich damp meadows as at Newton St. Faith's E21 and Shelfanger Town Meadows E18. Some other sites indicate possible introduction, perhaps for culinary purposes. A former site at Cranworth W90 was along a stream bank forming one side of the moat of long lost manor house. Generally rare in Britain apart from the north-west.

P. amplexicaulis (D.Don) Ronse Decr. **Red Bistort**
Syn. *Polygonum amplexicaule* D.Don
Herbaceous perennial native to the Himalaya, occasionally naturalised. Recorded for Norwich E20 in 1990, ETD and Cley E04 in 1991, MK. W0. E2.

P. amphibia (L.) Gray **Amphibious Bistort**
Syn. *Polygonum amphibium* L.

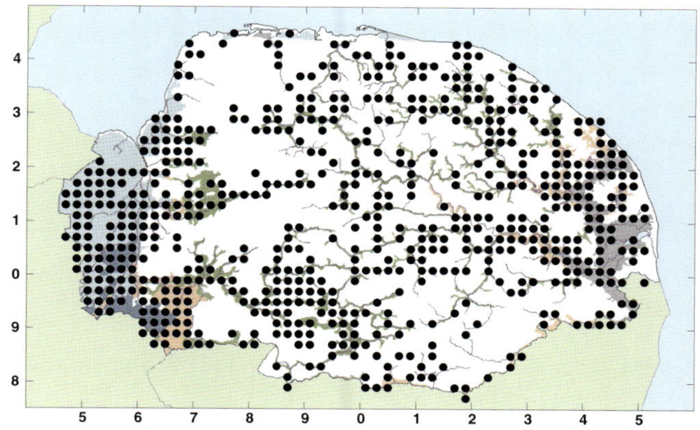

Herbaceous perennial of still and slow-moving water, but also frequent in non-aquatic habitats where it produces quite distinctive foliage and is sometimes confused with *P hydropiper.*

P. maculosa Gray **Redshank**
Syn. *Polygonum persicaria* L.
Spring germinating annual, a widespread weed of fields and waste places, commonest on moist soils. W588. E621.

P. lapathifolia (L.) Gray **Pale Persicaria**
Syns. *Polygonum lapathifolium* L., *P. nodosum* Pers.

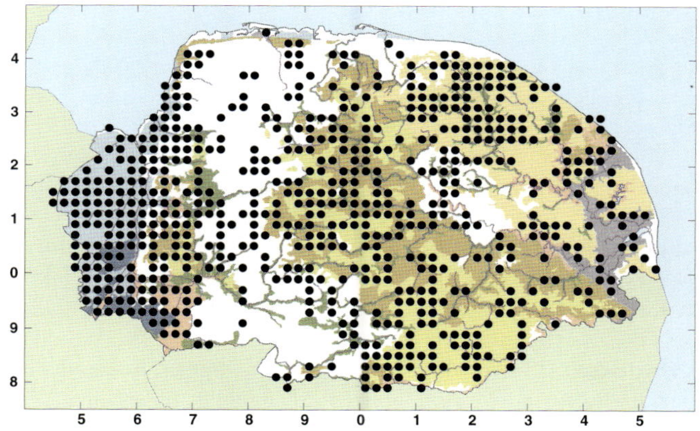

Spring germinating annual in similar habitats to *P. maculosa* but less frequent, also commoner on heavier soil.

■ **P. patulum** M.Bieb.
Casual recorded as a weed of carrot fields before 1950 and at Beetley gravel pit W91 in 1976, CE.Hubbard.

■ **P. pensylvanicum** (L.) M.Gómez
Casual found at Beetley pit W91 in 1975, RPL.

■ **P. senegalense** Meisn.
Casual from Africa recorded twice, both on rubbish tips, at Keswick E20 in 1963 and at Harford, Norwich E20 in 1971, RPL & ELS.

P. hydropiper (L.) Spach **Water Pepper**
Syn. *Polygonum hydropiper* L.

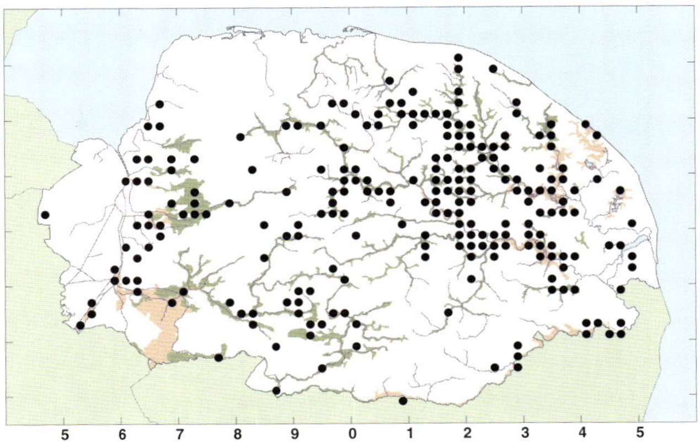

Spring germinating annual of damp places chiefly in woodland and woodland rides, much less common than *P. amphibium*.

P. mitis (Schrank) Opiz ex Assenoi **Tasteless Water-pepper**
Syn. *Polygonum laxiflorum* (Weihe) Opiz
Spring germinating annual in similar sites to *P. hydropiper* but now very scarce. Recorded from nine sites this century but only three since 1950, Cringleford, E10 in 1981 and Sweet Briar Marshes, Norwich E20 in 1990, both ETD and Hassingham E30 in 1992, J. Harris. W0. E3.

P. minor (Huds.) Opiz **Small Water-pepper**
Syn. *Polygonum minus* Huds.
Spring germinating annual, rare. Fulmodeston E03 in 1992, J.Akeroyd; Welney Washes W59 in 1993, S Everett and Hoveton Broad E31 in 1991, R.Southwood. W1. E2.

P. capitata (Buch.-Ham. ex D.Don) H.Gross
Syn. P*olygonum capitatum* Buch.-Ham. ex D.Don
Half-hardy perennial from the Himalayan region grown chiefly as a summer annual and used in window boxes and hanging baskets. An apparently increasing pavement weed observed in Diss E17 in 1994, AC; Kenninghall E08 and Fakenham W92, both in 1997, KAB and King's Lynn W62 in 1998, EVR. W2. E2.

Fagopyrum esculentum Moench **Buckwheat**
Spring germinating annual from Asia, occasionally as a casual from pheasant feed or garden bird seed. W10. E1.

F. tataricum (L.) Gaertn. **Green Buckwheat**
Annual introduced from Asia, occasionally found as an impurity in buckwheat sown for game-birds. Found once only, at Lenwade E11 in 1993, J.Humphris. W0. E1.

Polygonum oxyspermum C.A. Mey.& Bunge ex Ledeb.
Syn. *P. raii* Bab. **Ray's Knotgrass**
Spring germinating annual to short-lived perennial occasional just above the high tide line, chiefly on shingle beaches. Recorded twice only, at Blakeney E04 in 1959 and during this survey, at Snettisham W63 in 1991, MK. Our plant is ssp. **raii**. W1. E0.

P. arenastrum Boreau **Equal-leaved Knotgrass**

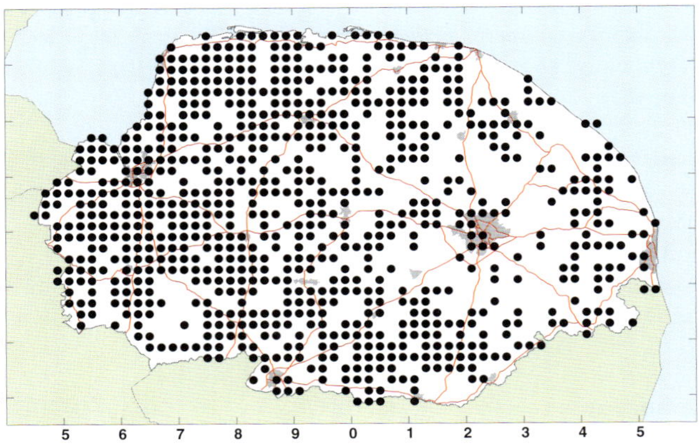

Spring germinating annual, a frequent weed of tracksides and waste ground on dry soils, commonest on compacted ground.

P. aviculare L. **Knotgrass**
Spring germinating annual, common as a weed of field edges, roadsides and waste ground, found in both prostrate and erect forms which have no apparent ecological preferences. W698. E704.

P. rurivagum Jordan ex Boreau **Cornfield Knotgrass**

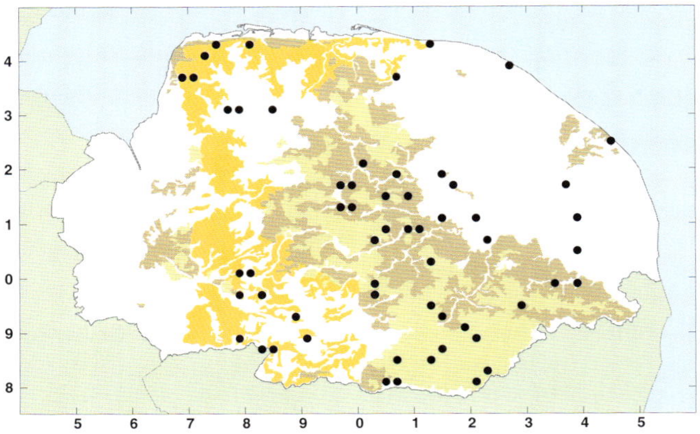

Spring germinating annual, an uncommon but perhaps still overlooked weed of calcareous land on both light and clay soil, not always easy to distinguish from *P. aviculare* with which it sometimes seems to intergrade.

Fallopia japonica (Houtt.) Ronse Decr. **Japanese Knotweed**
Syns. *Polygonum cuspidatum* Siebold & Zucc., *Reynoutria japonica* Houtt.

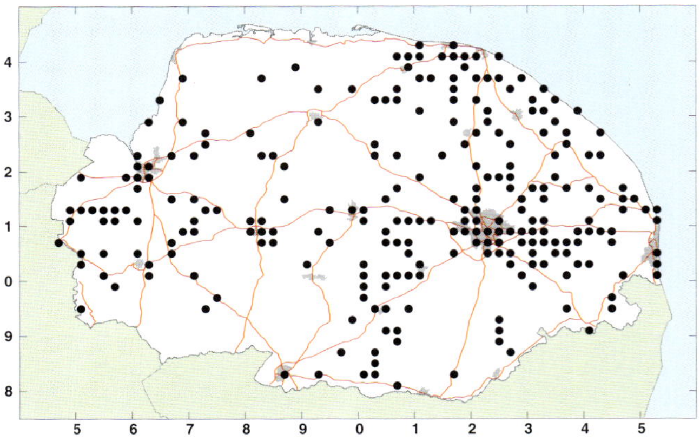

Rhizomatous, herbaceous perennial native to Japan and south-east Asia, originally introduced to gardens and long established in waste places. Very difficult to eradicate when well established.

F. sachalinensis (F.Schmidt ex Maxim.) Ronse Decr. **Giant Knotweed**

Syns. *Polygonum sachalinense* F.Schmidt ex Maxim, *Reynoutria sachalinensis* (F.Schmidt ex Maxim) Nakai.

Rhizomatous, herbaceous perennial native to eastern Asia, originally introduced to gardens and long established in waste places, especially in moist ground, much less frequent than *F. japonica*. A colony in wet woodland at Southrepps covers 50m by 20m with stems 3-4m high and 3cm diameter at the base, ALB. W7. E15.

F. baldschuanica (Regel) Holub **Russian vine**

Syns. *F. aubertii* (L.Henry) Holub., *Polygonum baldschuanicum* Regel., *Bilderdyckia aubertii* (L.Henry) Moldenke, *Polygonum aubertii* L.Henry.

Woody-stemmed twining climber native to eastern Asia, growing well in drier soils than the preceding species. Originally introduced to gardens and occasionally escaping. W12. E52.

F. convolvulus (L.) A. Löve **Black Bindweed**

Syn. *Polygonum convolvulus* L.

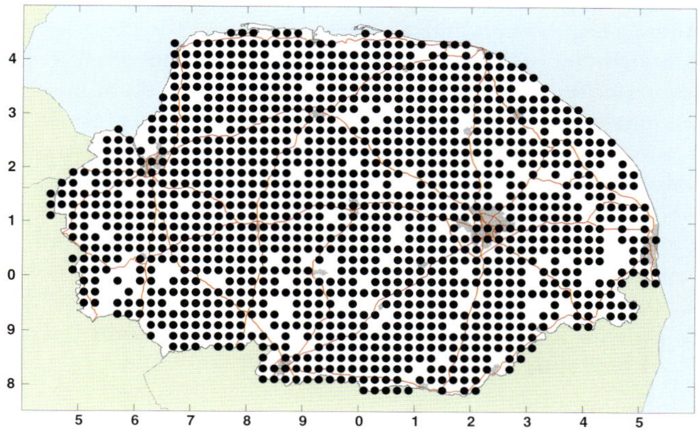

Annual of arable ground, abundant on all soils.

Rheum x hybridum Murray. **Rhubarb**

Syns. *R x cultorum* Thorsrud & Reis nom.nud, *R. rhabarbarum* auct.non L.

Herbaceous perennial, a long-lasting relic of cultivation, most frequent in the Fens. A large colony found in the middle of a marshy meadow at Itteringham E12 in 1995 is in a curious site. W6. E3.

Rumex acetosella L. **Sheep's Sorrel**

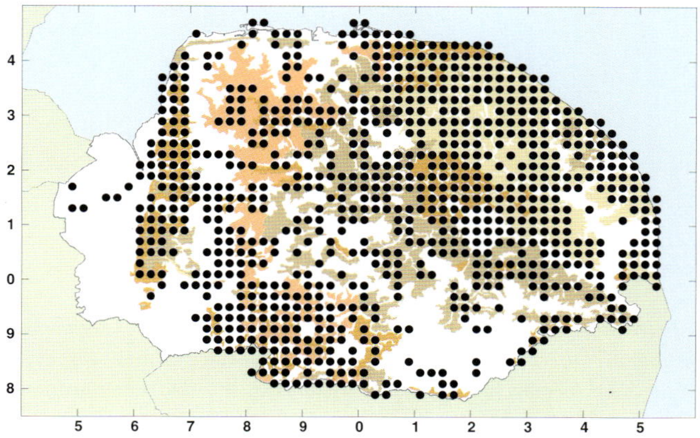

Winter green perennial of acid sandy soils. Where its growth is unrestricted, as at Brettenham NNR W98, individual plants will form a mosaic of differing shades, each a metre or more across. Ssp. **acetosella**, including the former ssp. *tenuifolia* with very narrow, rolled leaves, is abundant in very shallow, dry soils, especially in Breckland and on the greensand. Other material in the county is ssp. **pyrenaica** (Pourret) Akeroyd, (ALB).

R. acetosa L. **Common Sorrel**

Winter green perennial of established, neutral and alkaline grasslands. Slow to re-colonise after disturbance. Our plants are all ssp. **acetosa**. W592. E639.

R. hydrolapathum Huds. **Great Water Dock**

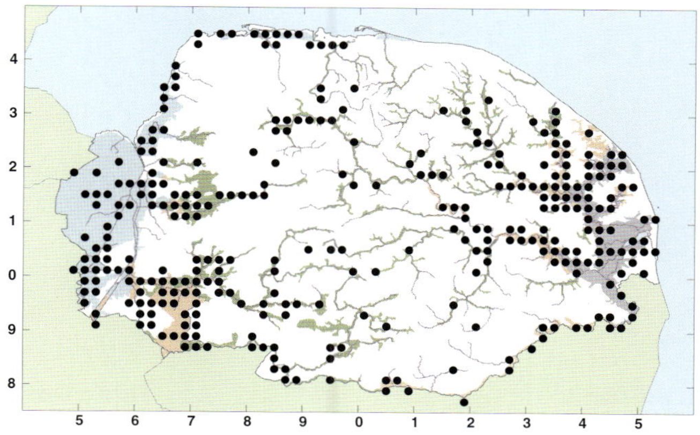

Herbaceous perennial of shallow water up to about 30cm in depth and the margins of ponds, streams, dykes and rivers.

R. crispus L. **Curled Dock**

Herbaceous perennial of roadsides and rough ground. Ssp. **littoreus** Hardy, **Shore Dock**, is a more condensed, succulent form growing on shingle and stony banks close to the sea, often where it is washed over by spring tides. Recorded from Wells-next-the-Sea W94, PB; Cley E04, ALB and Holme W64. W685. E701.

R. conglomeratus Murray **Clustered Dock**

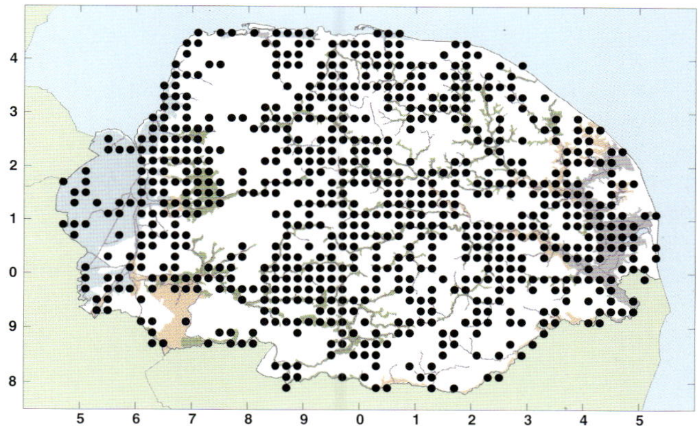

Herbaceous perennial of damp, grassy places.

Rumex acetosella - Brettenham

R. sanguineus L. **Wood Dock**

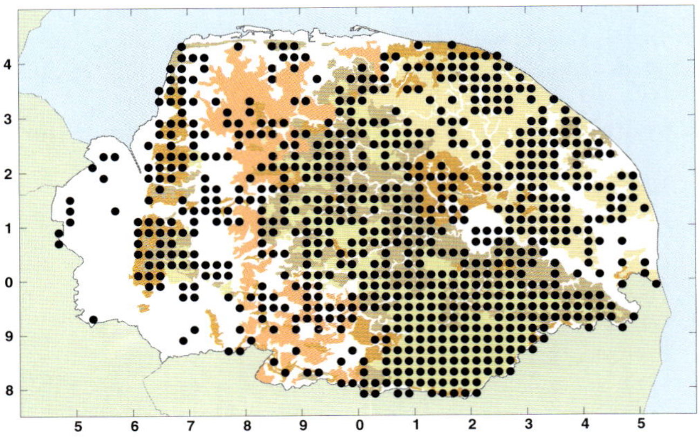

Semi-deciduous perennial of damp, shady places, especially on bare soils.

R. pulcher L. **Fiddle Dock**

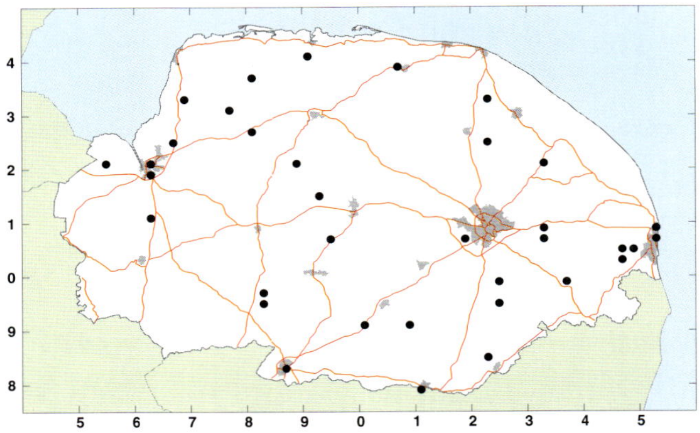

Scarce perennial of dry, grassy places, occurring as a single plant in many of its recorded sites.

R. obtusifolius L. **Broad-leaved Dock**
Semi-winter green perennial, ubiquitous. W697. E712.

R. palustris Sm. **Marsh Dock**

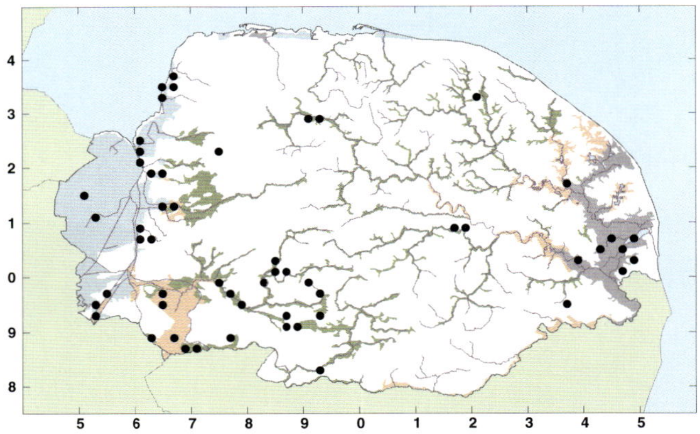

Spring germinating biennial or short-lived perennial growing on bare, wet mud, especially where ponds have dried out.

R. maritimus L. **Golden Dock**

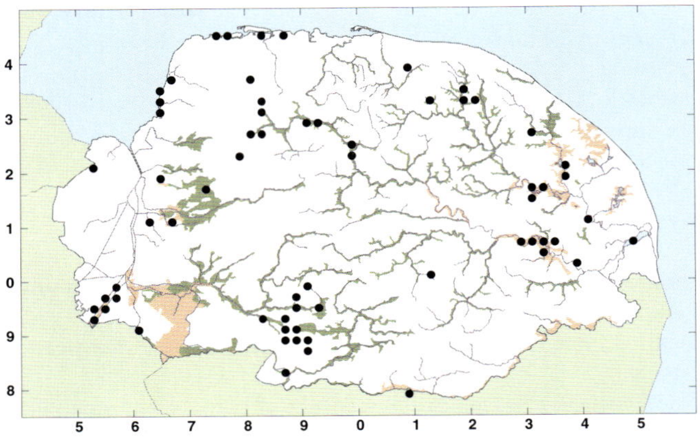

Spring germinating annual, sometimes overwintering, growing on bare, wet mud, much more frequent in dry summers when ponds dry out. Sometimes on soil dredged from dykes and ponds as at Rockland St Mary E30, 1991 and Stanhoe W83 in 1997.

The following *Rumex* hybrids have been recorded during this survey, **R. x schulzei** Hausskn. (**R. crispus x R. conglomeratus**); **R. x sagorskii** Hausskn. (**R.crispus x R. sanguineus**); **R. x pratensis** Mert. & W.J.D.Koch, (**R. crispus x R. obtusifolius**); **R. x wirtgenii** Beck (**R. conglomeratus x R. palustris**) and **R. x dufftii** Hausskn. (**R.obtusifolius x R. sanguineus**). Other hybrids recorded before 1975 by E.L.Swann are: **R. x lingulatus** Jungner (**R. hydrolapathum x R. obtusifolius**); **R. x muretii** Hausskn. (**R. conglomeratus x R. pulcher**); **R. x knafii** Celak (**R. conglomeratus x R. maritimus**); **R. x oluginensis** Borbàs (**R. pulcher x R. obtusifolius**) and **R. x callianthemus** Danser (**R. obtusifolius x R. maritimus**).

PLUMBAGINACEAE

Limonium vulgare Mill. **Common Sea Lavender**

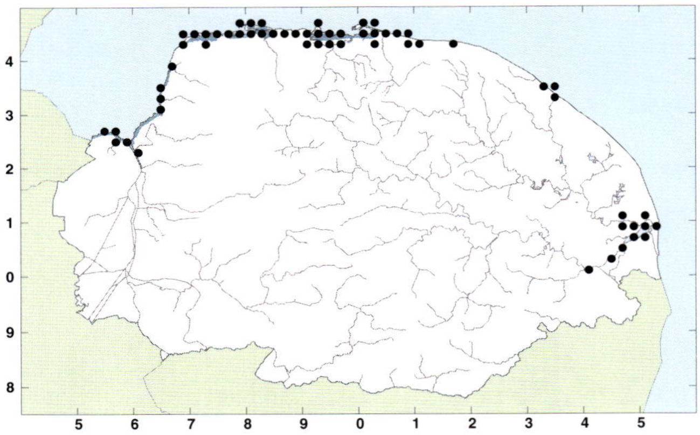

Perennial, frequent wherever there is salt marsh.

L. humile Mill. **Lax-flowered Sea Lavender**
Perennial, rare, all sites are on the north coast salt marshes in deep mud. Seen recently at Stiffkey W94 in 1990, MK; Blakeney harbour E04, in 1991, K.Ferrousat; Scolt Head W84 in 1993, SW and Wells-next- the-Sea W94 in 1994, J.Austen. A nationally scarce species.
W4. E3.

L. bellidifolium (Gouan) Dumort. **Matted Sea Lavender**

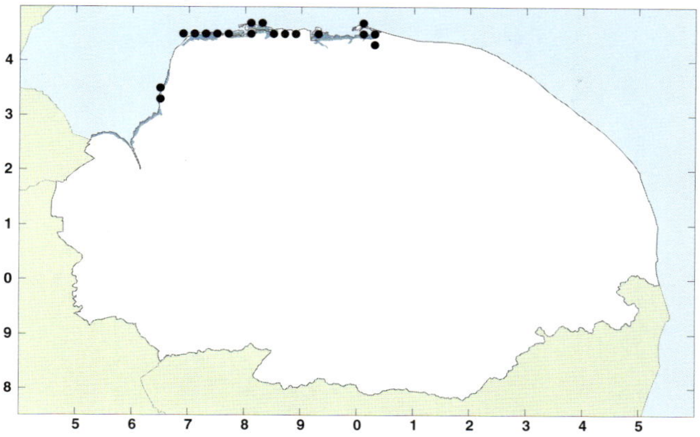

Low-growing, deciduous perennial confined to the north coast marshes from Holme to Blakeney, though it has in the past been found as far east as Cromer. It also has two sites in W63 on the Wash coast. It requires a firm substrate and usually grows where the saltmarsh lies over shingle. Where it occurs it can be quite plentiful and will quickly colonise any new suitable areas, these often replacing sites where it has become covered with sand from winter storms. Red Data Book Species.

Limonium bellidifolium - Brancaster

L. binervosum (G.E.Sm.) Salmon **Rock Sea Lavender**

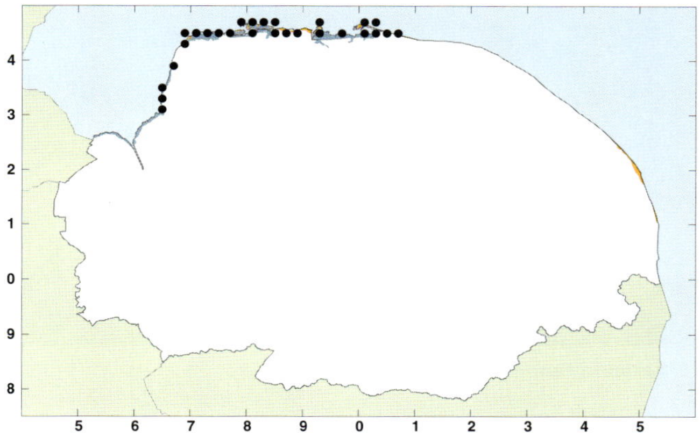

Winter green perennial, often found growing with *L. bellidifolium*. In the absence of rocks it grows where the sand overlies shingle and also on sandy edges of the marsh. It occurs along the coast from Wolferton W63 to Cley E04 and was also seen on the cliffs at Sheringham E04 in 1976, ETD. Our plants belong to ssp. **anglicum**.

Armeria maritima Willd. **Thrift**

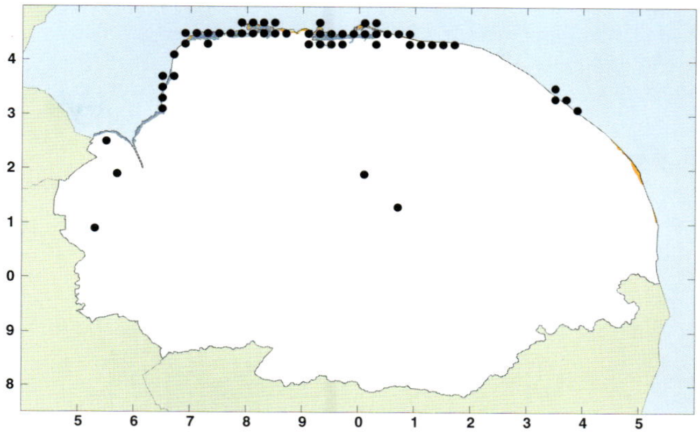

Winter green perennial found on the drier parts of salt marshes and on cliffs. Larger and more colourful forms occasionally found naturalised on walls and near habitation, even by the sea as on the shingle bank at Heacham W63, may possibly be natural but are more likely to derive from garden cultivars as does that in E01, DMM. A plant was found beside the Hockering bypass E01 in 1997, S.Rix.

ACTINIDIACEAE

Actinidia deliciosa (Chevalier) Liang & Ferguson
Syn. *A. chinensis* Planchon **Chinese Gooseberry, Kiwi Fruit**
Deciduous, climbing perennial introduced to gardens from China. A seedling was found in a garden where the plant has never been grown, at Stow Bardolph W60 in 1996, Lady Rose Hare, comm. RMP. W1. E0.

CLUSIACEAE

Hypericum calycinum L. **Rose of Sharon**
Winter green perennial from Bulgaria and Turkey, rarely naturalised but persisting once established as in the pinewoods at Holkham W84 where it has been known for more than 40 years. W9. E48.

H. androsaemum L. **Tutsan**

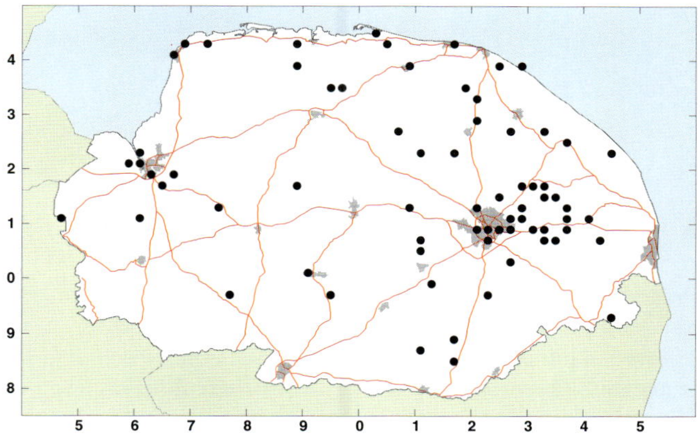

Perennial, partially evergreen in mild winters but usually deciduous in Norfolk. Its status is doubtful though it is probably an escape from gardens except perhaps in some woodland sites in the east. In the past it was grown for its herbal values, its name coming from *toute saine*, heal-all.

H. hircinum L. **Stinking Tutsan**
Deciduous perennial from the Mediterranean, long naturalised by the lake in Holkham Park W84. W1. E0.

H. x inodorum Mill.
(H. androsaemum x H. hircinum)
Perennial, an escape from gardens where it is frequently grown. Found well naturalised in a conifer plantation at Corpusty E13 in 1977, MAB and in Earlham cemetary E20. S.Henson 1998. W0. E1.

H. perforatum L.
Common St John's-wort, Perforate St John's-wort

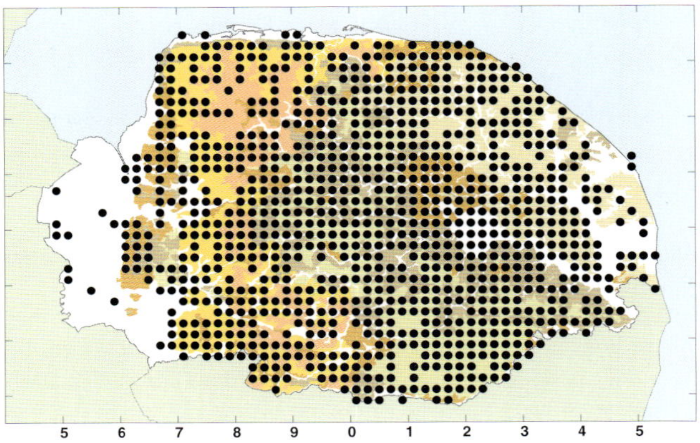

Herbaceous perennial, spreading by underground rhizomes to form large patches on dry grassland, especially on road verges. Not infrequent but seems to be scarce in Broadland.

H. x desetangsii Lamotte
(H. perforatum x H. maculatum)
Herbaceous perennial found in damp grassland at Lolly Moor W91 in 1993, RWE; Wellingham W82 in 1994, MK; Harling Road W98,1994, K&GB; Ditchingham E39 in 1995, ALB; Carbrooke W91 and Bradenham W90 in 1996, MK and Stoke Holy Cross E20 in 1997, R.Chancellor & RWE. W4. E2.

H. maculatum **Imperforate St John's-wort**

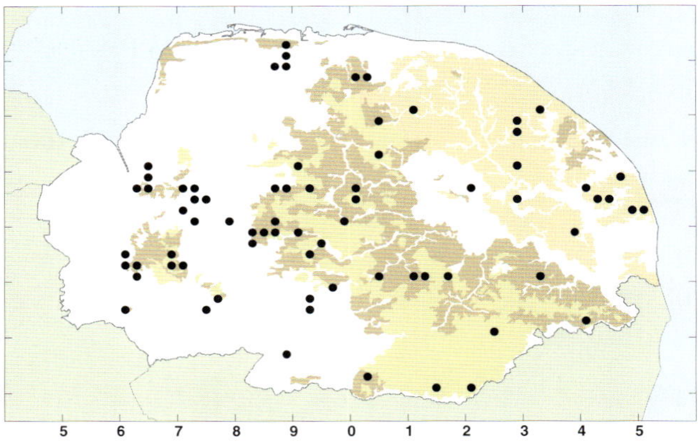

Herbaceous perennial of damp grassland, far less frequent than *H. perforatum*, perhaps overloooked but its leaves are distinctive, narrow and parallel-sided. In east Norfolk sometimes found on drier, sandy soils. Our plants belong to ssp. **obtusiusculum** Crantz.

H. tetrapterum Fr. **Square-stemmed St John's-wort**

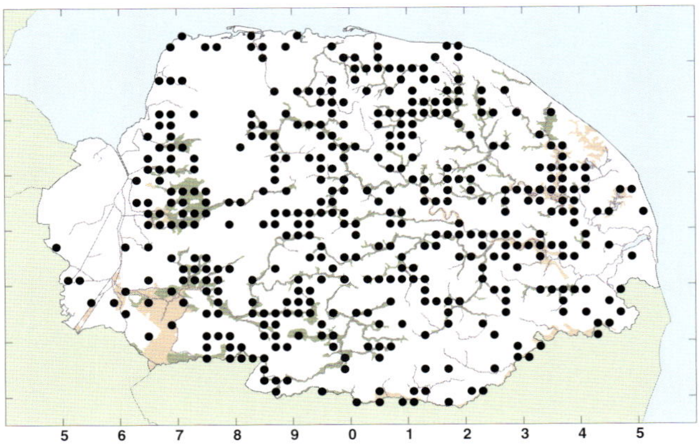

Herbaceous perennial of marsh and fen, also frequent along dyke banks with standing water.

H. humifusum L. **Trailing St John's-wort**

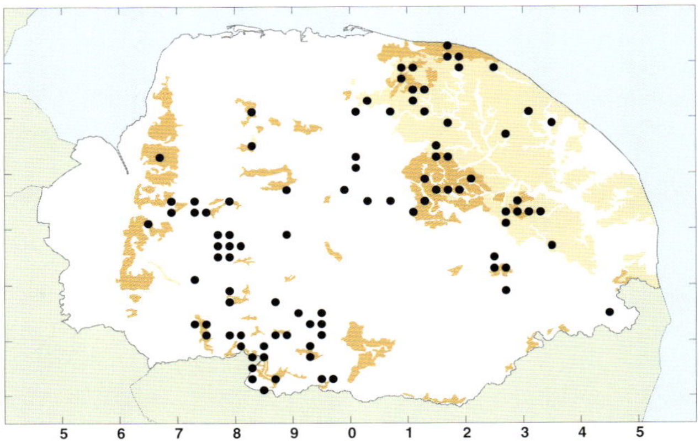

Winter green perennial confined to dry, acid soils.

Hypericum humifusum - Mundford

H. pulchrum L. **Slender St John's-wort**

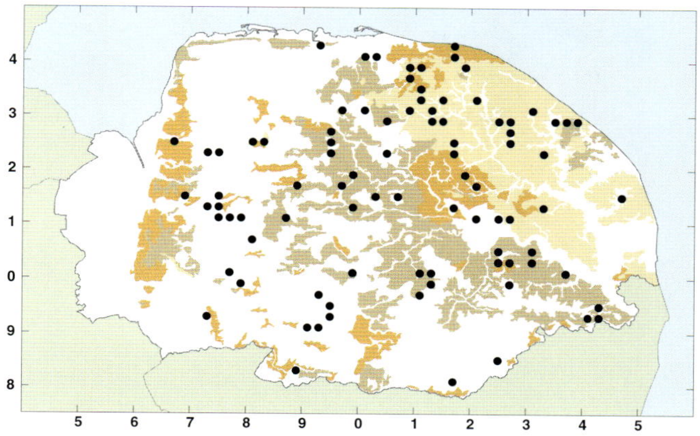

Herbaceous perennial, confined to acid grassland on heaths, woodland rides and hedgebanks, occasionally on clay in some east Norfolk sites.

H. hirsutum L. **Hairy St John's-wort**

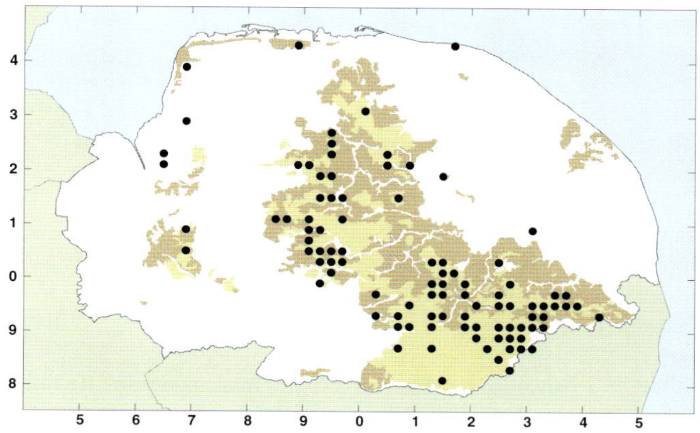

Herbaceous perennial chiefly in old woods but also scattered in damp, calcareous sites especially on the boulder clay.

H. montanum L. **Pale St John's-wort**
Rare herbaceous perennial of chalky soils. Found during this survey only at Holkham W84 in 1989, B.Taylor; Norwich E20 in 1990, ETD and in 1997 at Weybourne Springs E14 where it has been known for at least 80 years, FF. W1. E2.

H. elodes L. **Marsh St John's-wort**

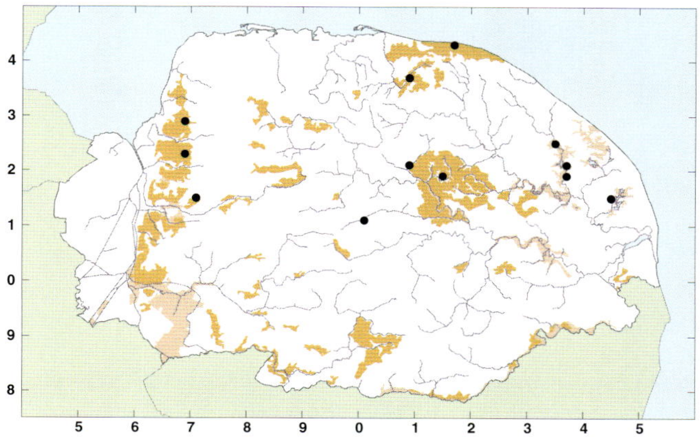

Herbaceous perennial of wet, acid soils, occcasional in bogs, also by ponds and dykes where it can form floating mats and look like an aquatic. A chiefly northern and western plant in Britain, rare in eastern England.

TILIACEAE

Tilia platyphyllos Scop. **Large-leaved Lime**
Deciduous tree, scarce and probably always planted. W54. E57.

T. x europaea L. **Hybrid Lime, Common Lime**
Syn. *T. x vulgaris* Hayne (*T. platyphyllos x T. cordata*)
A deciduous tree much planted, especially in avenues and parks where it can usually be distinguished by the often dense, twiggy outgrowths around its bole. It is partly fertile and seedlings can occasionally be found nearby. W127. E429.

T. cordata Mill. **Small-leaved Lime, Linden, Pry**

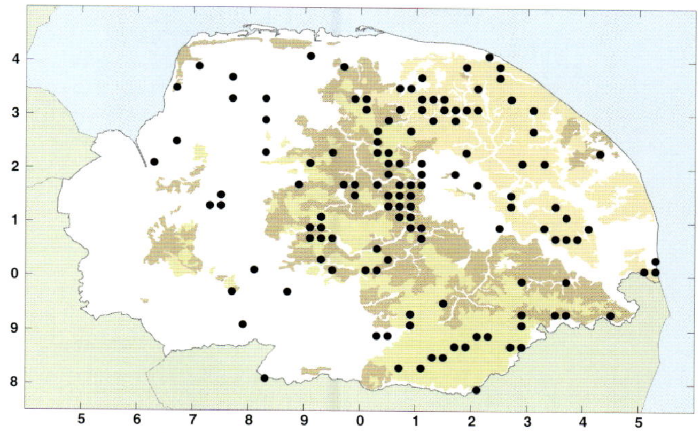

Deciduous tree, native in a few ancient woods and hedgerows. It is the dominant tree in a few woods for example at Hockering Wood E01 where it is also present in hedges and woodland within a two-mile radius, ALB. It is also an important tree at Swanton Novers Great Wood E03. Its natural distribution away from such woodland sites is much masked by later planting.

MALVACEAE

Malva moschata L. **Musk Mallow**

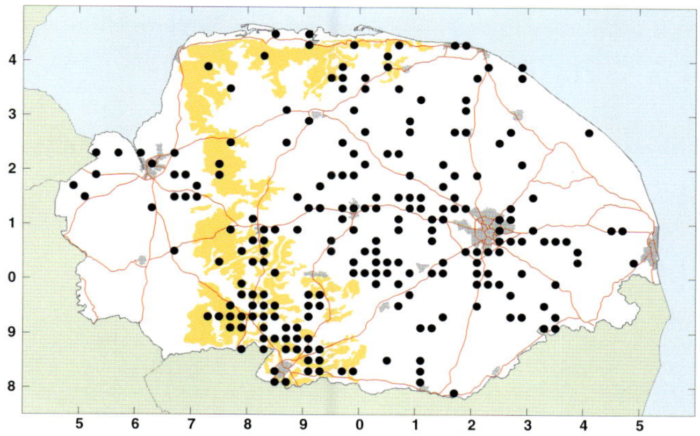

Herbaceous or semi-winter green perennial of dry grasslands. White flowered forms are not uncommon. It is often included in 'wild flower mixes' so its appearance on new roadside verges is likely to be from this source.

M. sylvestris L. **Common Mallow**
Winter green to semi-winter green perennial frequent and apparently increasing on roadsides perhaps because of a tolerance of mineral-rich conditions. The small, round fruits were once eaten by children and known locally as 'pick-cheeses'. W652. E688.

■ **M. parviflora** L. **Least Mallow**
Annual from southern Europe, a rare casual recorded twice, the last at Houghton W72 in 1950.

M. pusilla Sm. **Small Mallow**
Annual, introduced from Europe, rare. Found at Oxborough W70 in 1994, JEG. W1. E0.

M. neglecta Wallr. **Dwarf Mallow**

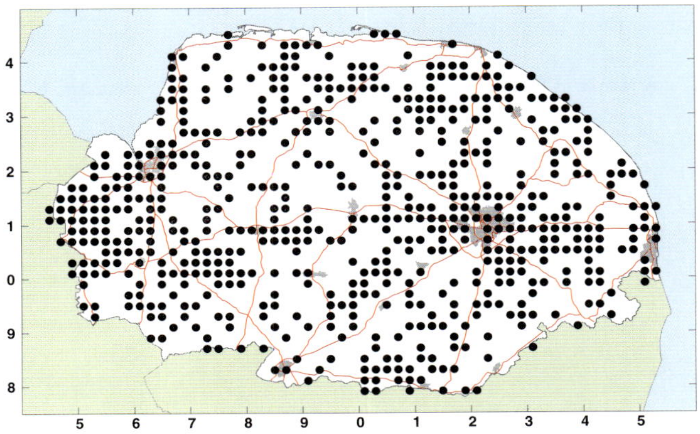

Annual, sometimes overwintering, a weed found particularly near habitation.

Lavatera arborea L. **Tree Mallow**
Woody based perennial native to western Britain. Introduced some years ago near the Beach Road at Salthouse E04 and rapidly spreading nearby in 1997, perhaps assisted by the sea flood of early 1996. Occasional elsewhere. W0. E17.

L. olbia L. **Rose Mallow**
Shrub or subshrub widely grown, originally from the western Mediterranean and Portugal, now represented in gardens by a series of cultivated varieties derived from hybridisation with the smaller, herbaceous *L. thuringiaca*. These are partially sterile and are not known to regenerate naturally, so plants in the wild, as at Bawsey Pits W62 and Cranwich Heath W79, are most likely to have been deliberately planted. *L.o.* 'Rosea' is the cultivar most often seen. W2. E0.

Althaea officinalis L. **Marsh Mallow**

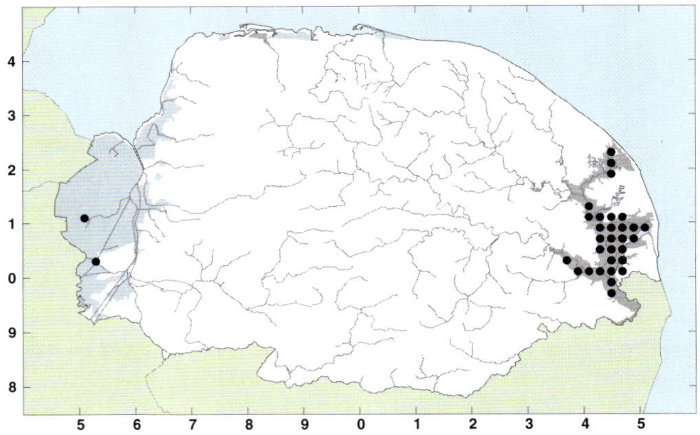

Herbaceous perennial confined to grasslands near but not by the sea. Once frequent in the northern fenland, where it was recorded from Tilney St Lawrence W51 in 1976, CPP; Outwell W50 in 1987, HW and Marshland St James, W51 in 1992, RMP, but in this area is now close to extinction. It continues to thrive, however, on the banks of the rivers and marsh dykes of the grazing levels adjacent to the Yare, Waveney and Bure, ALB. A nationally scarce species.

Althaea officinalis - Thurne Bank

■ **A. hirsuta** L. **Rough or Hispid Marsh Mallow**
Annual from mainland Europe, a rare casual of field edges recorded once, at West Newton W63 in 1949.

Alcea rosea L. **Hollyhock**
Syn. *Althaea rosea* (L.) Cav.
Herbaceous perennial introduced to gardens probably from western Asia and grown mainly as a biennial, sometimes established at the base of old walls and waste places, occasionally persisting for several years. W26. E60.

Abutilon theophrasti Medik. **Velvet Leaf**
Spring germinating annual from south-east Europe, occurring as a casual in fields on three occasions during this survey and once on waste ground. W2. E2.

Hibiscus trionum L. **Flower of an Hour, Bladder Ketmia**
Annual introduced to gardens from warm temperate and subtropical areas of the old world. Occasionally escaping on to waste ground as at Etling Green E01 in 1994 and Heacham W63 in 1993. W1. E1.

DROSERACEAE

Drosera rotundifolia L. **Round-leaved Sundew**

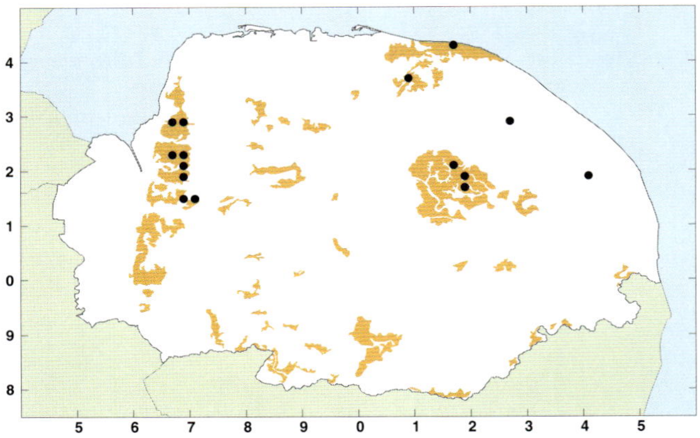

Not as frequent as formerly due to loss of habitat, but still in some abundance in bogs. Occasionally also found in waterlogged ground on acid soils.

D. anglica Huds. **Great Sundew**
Syn. *D. longifolia* L.
Occurs in mildly acid, wet areas as well as true bogs, but much reduced in frequency. Recorded from sixteen 10km. squares in 1968, but during this survey found only at Roydon W62; Scarning W91; Badley Moor, Dereham E01; Beeston Regis E14 and Holt Lowes E03. W3. E3.

D. intermedia Hayne **Oblong-leaved Sundew**
Largely confined to wet acid ground, so with a more limited range than the preceding species. Found at Roydon and Dersingham W62; East Winch W71; Buxton Heath E11 and Holt Lowes E03. W5. E2.

■ **D. x obovata** Mert. & W.D.J.Koch (*D. rotundifolia x D. anglica*) and **D. x beleziana** E.G.Camus (*D. rotundifolia x D. intermedia*) were both listed from Dersingham W63 in the early years of the 20th century, but there seems some confusion over the records so neither can be considered as certain.

CISTACEAE

Helianthemum nummularium (L.) Mill. **Rock Rose**
Syn. *H. chamaecistus* Mill.

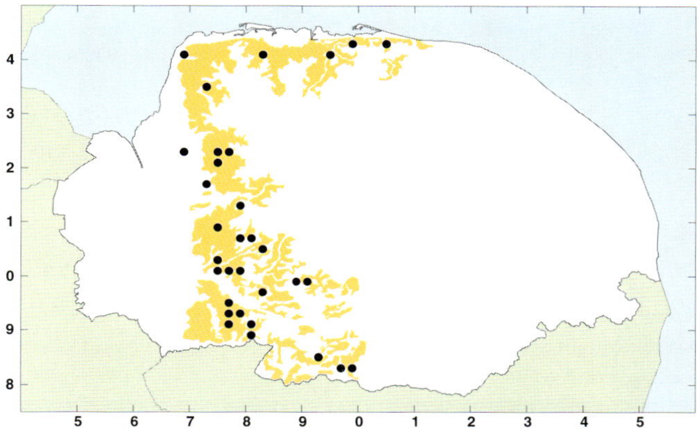

Evergreen mat-forming shrub confined to short turf on well-drained chalky soils, chiefly in west Norfolk. Its farthest eastward extent is at Cley E04, KKH.

Helianthemum nummularium - Grimes Graves

VIOLACEAE

Norfolk violets were well covered in a study published by Mrs E.S.Gregory in 1912 *(British Violets)*. Many references are made to this in the 1968 *Flora of Norfolk* but it has not been thought relevant to repeat the details again here.

Viola odorata L. **Sweet Violet**

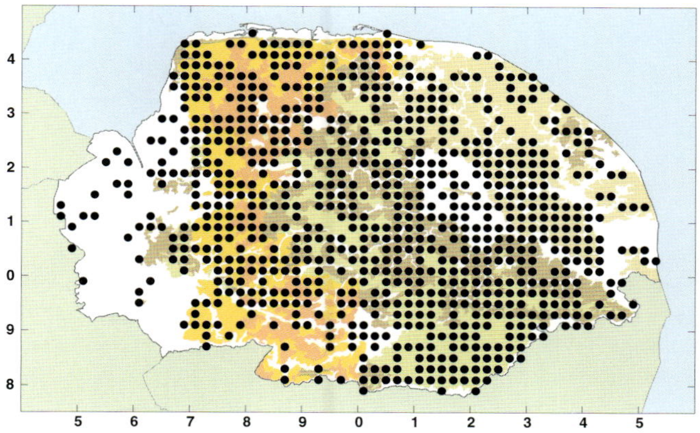

Herbaceous perennial most frequently seen on dry chalky soils where it occasionally forms large patches in woodland and on banks. Also associated with villages, churchyards etc. White-flowered forms with purple spurs and partly coloured lateral petals known as var. *dumetorum* are frequent in the wild. Cultivated forms can be found in shades of pink and apricot, a colony of the latter occurring in a shrubbery at Ketteringham E10 in 1991, R Tilbrook.

V. hirta L. **Hairy Violet**
Herbaceous perennial, recorded from Hunworth E03 in 1995, KKH; Beechamwell W70 in 1996, HW, a long established site; Ringstead W64 in 1997, TCE.Wells and South Lopham E07 and 08 in 1997, ALB. A nationally scarce species. W4. E4.

V. riviniana Rchb. **Common Dog Violet**

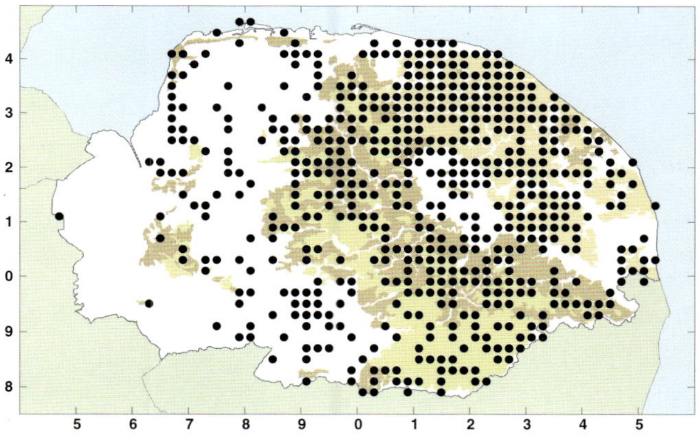

Winter green perennial of woodlands and shady sites, it also occurs rarely with pure white or pink flowers. Plants in dry, open grassland often belong to var. *minor* (Murb.ex Greg.) Valentine, a variety which was once a feature of some Breckland grass heaths but is now rare.

V. cornuta L. **Horned Pansy**
Perennial introduced to gardens from the Pyrenees and occasionally escaping. Recorded from Banham E08 in 1990, R.Parker. W0. E1.

V. reichenbachiana Jord. ex Boreau **Early Dog Violet**

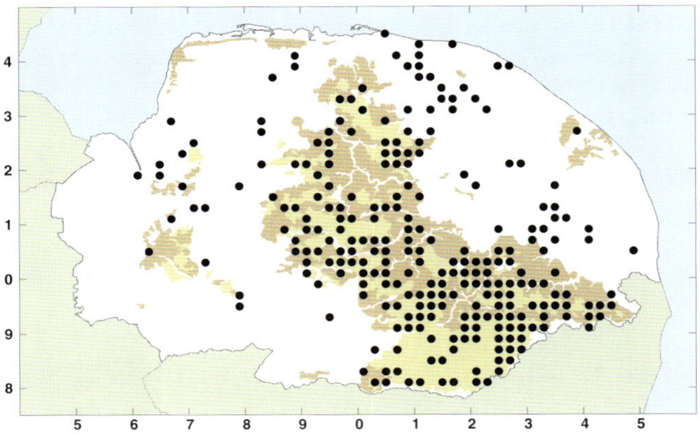

Winter green perennial, less frequent than *V. riviniana* and much more closely associated with woodland, also on shady hedgebanks on clay. The hybrid between these two species was recorded at Horningtoft, W92 in 1954, but could occur wherever the two species grow together.

V. canina L. **Heath Dog Violet**

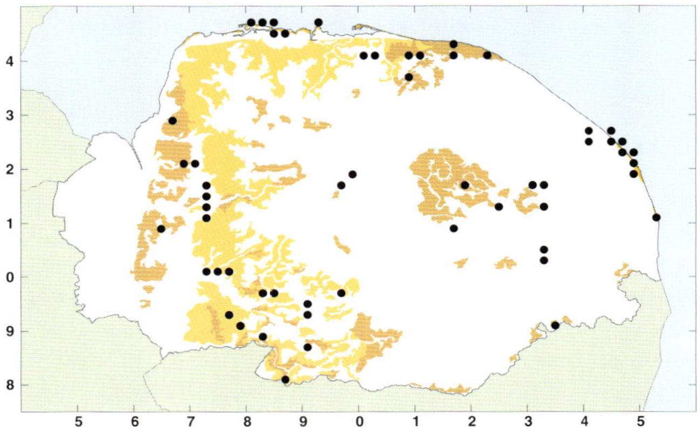

Perennial of grassy heaths and sand dunes described as 'frequent' in 1968, but hardly so now except on coastal dunes.

■ **V. persicifolia** Schreber. **Fen Violet**

Syn. *V. stagnalis* Kit.

Herbaceous perennial once occurring in a fen near West Dereham W60. Last recorded in 1936, the area being drained shortly afterwards and the plant must be considered extinct. A Red Data Book species.

V. palustris L. **Marsh Violet**

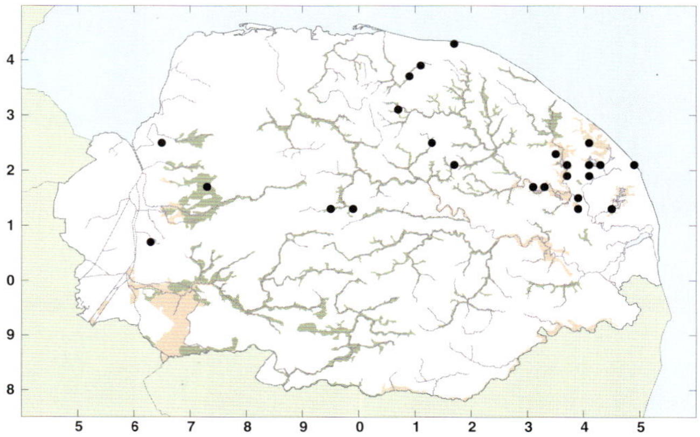

Herbaceous perennial, scarce in marshy, usually acid sites. Still holding its own in Broadland but struggling in west Norfolk.

V. tricolor L. ssp. **tricolor** **Wild Pansy**

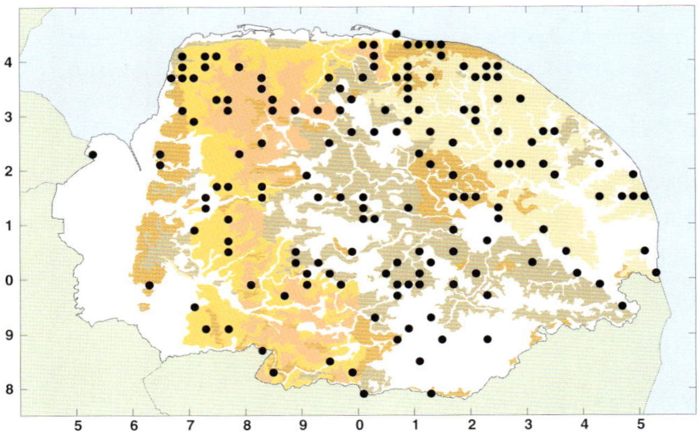

Annual, a weed of arable fields chiefly on light soils. It is not often found in its pure state as it hybridises freely with *V. arvensis*.

V. tricolor L. ssp. **curtisii** (E.Forster) Syme **Sand Pansy**

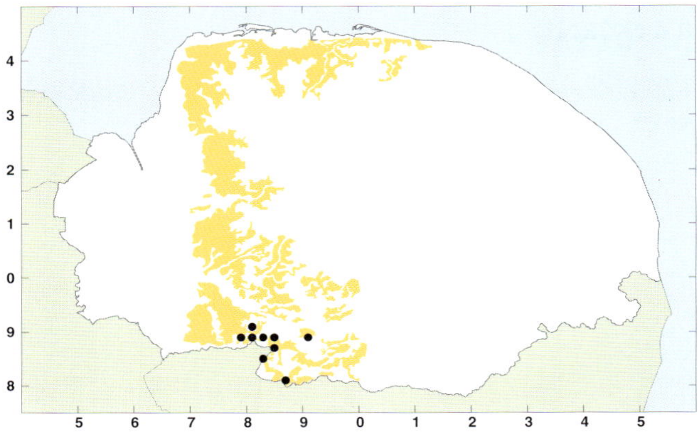

A short-lived perennial which is scattered around Britain on coastal dunes. In Norfolk, however, it is confined to dry, acid grasslands in the Breckland where it also occurs on forest rides and in clear-fells.

Viola tricolor ssp. **curtisii - Santon**

V. x contempta Jord.

(V. tricolor x V. arvensis)

Annual to short-lived perennial, frequent with both the parents and sometimes in the absence of one. W6. E20.

V. x wittrockiana Gams ex Kappert **Garden Pansy**
A garden hybrid between cultivated forms of *V. tricolor, V. lutea* and *V.cornuta.* Occasionally escaping and sometimes back-crossing to *V. tricolor.* The resulting plants have a wide range of colours and flower sizes, and can be very difficult to distinguish from true *V. tricolor.* W8. E43.

V. arvensis Murray **Field Pansy**
Annual of arable fields, particularly on light soils, remarkably resistant to present strains of herbicide. W555. E639.

TAMARICACEAE

Tamarix gallica L. **Tamarisk**
Syn. *T. anglica* Webb
Deciduous shrub from south-western Europe introduced to gardens as an ornamental wind-break, especially near the sea. Planted in 1912 on Blakeney Point E04 by Prof. FW.Oliver and still surviving. W1. E4.

FRANKENIACEAE

Frankenia laevis L. **Sea Heath**

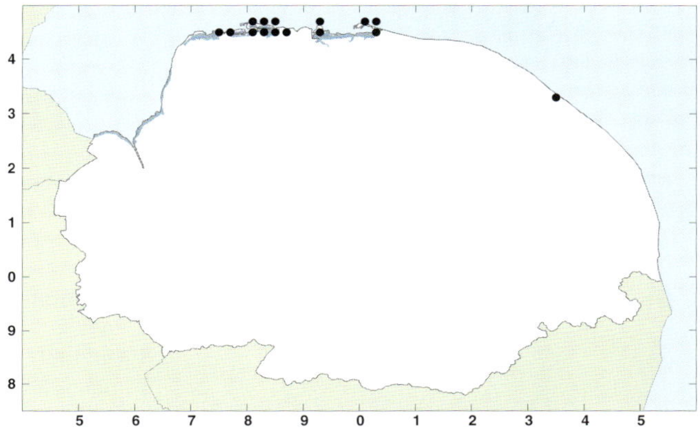

Winter green perennial confined to the drier, shingly edge of salt marshes where it is usually found in association with the small sea lavenders, *L. bellidifolium* and *L. binervosum*, but extends farther east than these at Blakeney. The record at Walcot, E33 is by a road which the sea spray reaches during storms, 1998 ALB. A nationally scarce species.

Frankenia laevis - Burnham Overy

CUCURBITACEAE

Bryonia dioica Jacq. **White Bryony**
Herbaceous perennial of hedgerows, thickets, and waste places. The only member of the cucurbit family native to the cool temperate latitudes of Great Britain. Noticeably quick to die down in autumn. W617. E500.

■ **Cucumis melo** L. **Melon**
Annual originating from Africa, now widely grown. Seeds on refuse tips and in sewage can germinate, but the plants do not survive the winter. Recorded at Harford Tip E20 in 1971, G.Hanson.

■ **C. sativa** L. **Cucumber**
Annual, native to India, widely grown in greenhouses and occasionally as a field crop. Seeds on refuse tips and in sewage will produce plants, but these do not survive the winter. Recorded at Beetley Pit W91 in 1971, ALB & ELS.

■ **Cucurbita pepo** L. **Marrow, Courgette, Gourd**
Annual from the Americas, grown in gardens and in fields. Another tender casual of refuse tips and sewage works. Last recorded at Beetley Pit W91 in 1971, ALB & ELS.

SALICACEAE

Populus alba L. **White Poplar, Abele**
Deciduous tree perhaps once native in the Fenland, but not obviously so in Norfolk today. It is found both as a tree and as a hedgerow shrub. Sometimes confused with the next species, especially sucker shoots, and possibly over-recorded. W78. E133.

P. x canescens (Aiton) Sm. **Grey Poplar**
(P. alba x P. tremula)
Deciduous tree possibly introduced in prehistoric times. Now found in wet woods and carr, also in hedges on moist soils. Suckers vigorously. W88. E156.

P. tremula L. **Aspen**

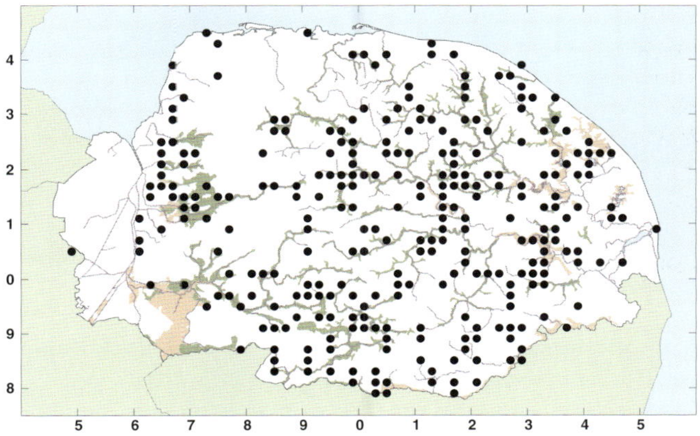

Deciduous tree native to woods and carr where it can sucker to form dense thickets, also in hedges.

P. nigra L. ssp. **betulifolia** — **Black Poplar**

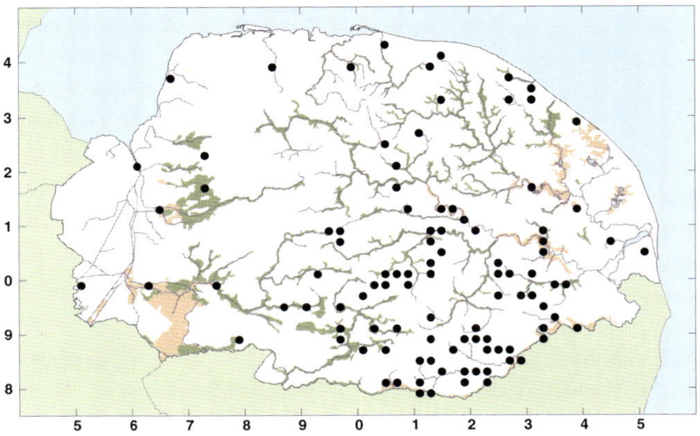

Deciduous tree confined to northern Europe, a scarce native of deep alluvial soils, but also widely planted in the past before the arrival of hybrid poplars involving American species. The tree is strangely scarce in north-east Norfolk. It has been the subject of long study by the late E.Milne-Redhead and work in Norfolk has been continued by EV.Rogers. Most trees are male, but there is a fine female at Old Buckenham E09. Trees have recently been planted in many suitable sites throughout the county, all from known local stock. Ssp. **nigra** is the subspecies found in southern Europe and from it several cultivars have been raised, the best known being 'Italica', the Lombardy Poplar which has been widely, though not commonly, planted in the county. No regeneration from seed has been observed.

Populus nigra - East Walton (felled a week later)

P. x canadensis Moench — **Hybrid Black Poplar**
Syn. *P. x euramericana* (Dode) Guinier. *(P. nigra x P. deltoides)*
Deciduous tree. Many cultivars have been selected for timber, screening and ornament, these are widely planted but do not regenerate except occasionally by suckering. W110. E312.

P. x jackii Sarg. — **Balm of Gilead**
Syns. *P. candicans* Aiton; *P. gileadensis* Roul. *(P. deltoides x P. balsamifera)*.
Deciduous tree, of uncertain origin but probably a complex hybrid, always planted. The cultivar 'Aurora' which has white blotched leaves and appears diseased, is sometimes planted. W0. E3.

P. trichocarpa Torr. & A. Gray ex Hook. — **Western Balsam Poplar**
Deciduous tree, native to western North America, planted for timber and for ornament, spreading by suckers. Its hybrid with balsam poplar, *P. balsamifera* L, is also planted for timber and as a windbreak, usually as 'Balsam Spire'. W3. E36.

Salix — **Willows**
One of the authors of the 1968 *Flora of Norfolk*, Eric Swann, made a particular study of willows, often in the company of RCL.Howitt of Newark. Unfortunately most willows are short lived and often found in somewhat ephemeral habitats such as gravel pits and dyke banks where they are often soon destroyed. Many of the old willow holts have also been grubbed up in the intervening years, so it is difficult to refind many of the specimens cited.

Salix pentandra L. — **Bay Willow**
Deciduous tree. Recent work suggests that this tree was probably a native of the Fenland. Existing trees may have been planted, but there is a possibility that some are remnants of a former population. Found in 1818 in Bardolph Fen, W50 by the Rev W Skrimshire. Recent records include Blackborough W61 in 1950; Old Lakenham E20 in 1976; Costessey E11 in 1978; Dereham, W91 in 1988 and Gt.Melton E10, D.Howlett. Coppiced plants occur in Hindolveston Wood E02 and at Shimpling E18. W1. E3.

S. x meyeriana Rostk. ex Willd. — **Shining Willow**
(S. pentandra x S. fragilis)
Deciduous tree first recorded at Croxton in 1969, HD.Wiard, more recently at Gaywood W62, Brinton and Sharrington E03, Sustead E13 and Roughton E23. W1. E4.

S. x ehrhartiana Sm. — **Ehrhart's Willow**
(S. pentandra x S. alba)
Deciduous tree found at Holkham W84 in 1993, KAB. W1. E0.

S. fragilis L. — **Crack Willow**

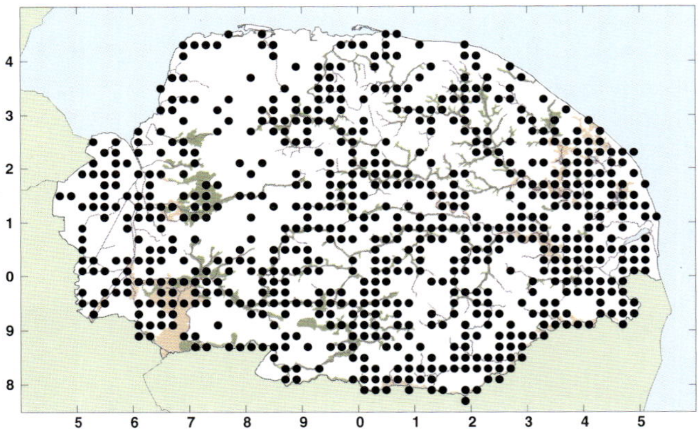

Deciduous tree frequent by rivers, streams and ditches where it was often planted in the past to try and stablise the banks. Most Norfolk specimens appear to be referable to var. **russelliana**, the **Bedford Willow**, as noted by EL.Swann. Var. **furcata** was seen in the churchyard at Larling, W98 in 1994, KAB while a specimen seen at Seamere, Hingham E00 in 1996 by ALB equals the largest girth ever recorded nationally for this species. We have been unable to refind var. **decipiens**, the **White-Varnished Willow** at Cranberry Rough, Hockham, W99, where it was frequent in the 1960s in a fairly open area now dense birchwood.

S. x rubens Schrank — **Hybrid Crack Willow**
Syn. *S. x basfordiana*. Scaling ex. Salter *(S.alba x S. fragilis)*.
Deciduous tree of which a number of cultivars are grown, our plants appear to be var. **basfordiana** which was a favourite basket willow. W6. E5.

S. alba L. **White Willow**

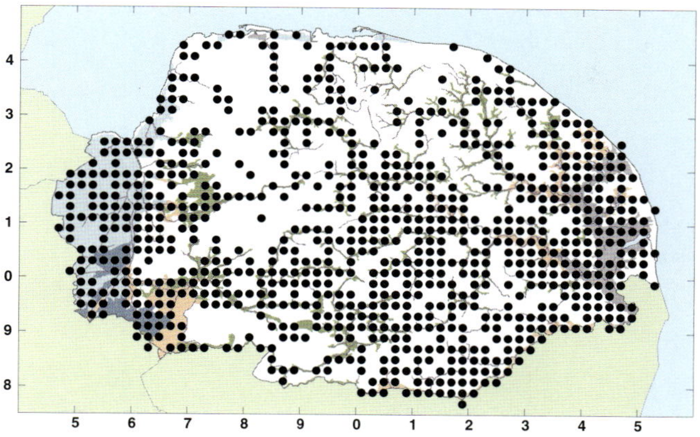

Deciduous tree of wet ground, found throughout suitable habitats. Most are var. **caerulea** (Sm.) Dumort **Cricket Bat Willow** which arose in Norfolk in the early 1700s and may still be native in some sites though it has also been widely planted.

S. triandra L. **Almond Willow**

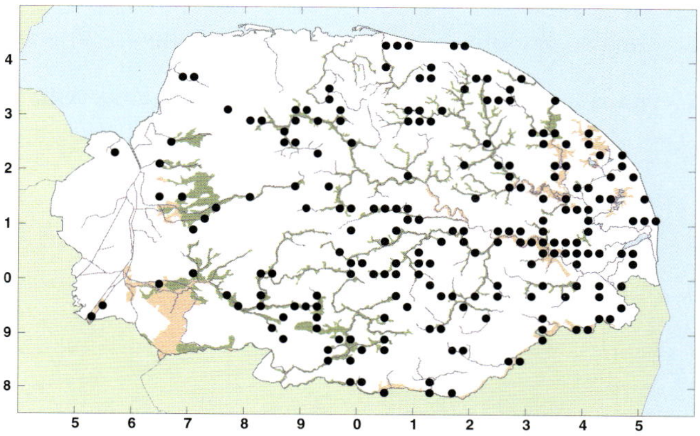

Deciduous tree native of damp places, especially in the Broadland. Much planted elsewhere for basketry and not infrequent. Var. **hoffmanniana** is found in a few sites as at Larling and Roudham, W98.

■ **S. x mollissima** Hoffm. ex Elwert **Sharp-stipuled Willow**
(S. triandra x S. viminalis)
Deciduous shrub or small tree with ten pre-1970 records, last seen at East Ruston E32 in 1972.

S. purpurea L. **Purple Willow**

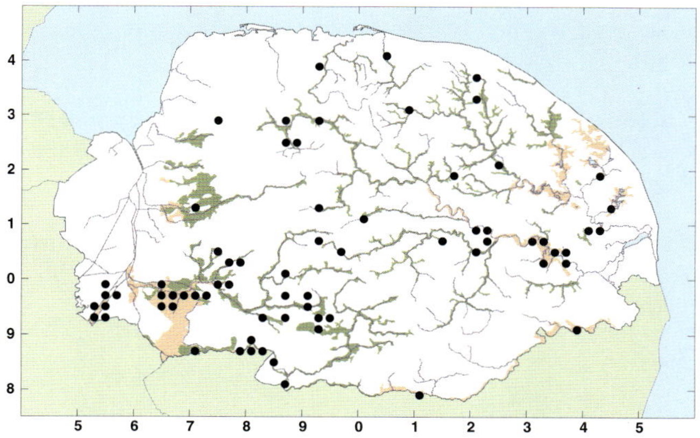

Deciduous shrub or small tree sometimes frequent in fens, carr and streamsides, also planted for windbreaks and basketry, commoner in west Norfolk.

■ **S. x pontederiana** Willd.
Syn. *S. x sordida* A. Kern. *(S. purpurea x. S. cinerea)*
Deciduous shrub recorded once, at Bromehill, Brandon W88, in 1969, HD.Wiard.

S. x rubra Huds. **Green-leaved Willow**
(S. purpurea x S. viminalis)
Deciduous tree, a scarce hybrid recorded at Guist W92 in 1979, ELS and again in 1996 by KAB. W1. E0.

S. viminalis L. **Osier**

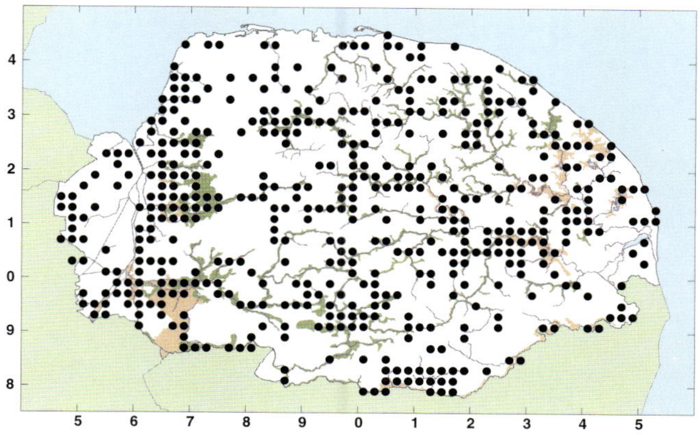

Deciduous shrub once grown throughout the county for making baskets, frequent on ditch and stream sides and in hedgerows.

S. elaeagnos Scop. **Olive Willow**
Found by the new King's Lynn bypass at Bawsey W62 in 1978, ELS. Of planted origin but still present. W1. E0.

S. x sericans Tausch ex A. Kern. **Broad-leaved Willow**
(S. viminalis x S. caprea)
Deciduous shrub of wet woodlands and gravel pits, scattered through the county. A fast growing willow now occasionally being cultivated for fuel. W7. E21.

S. x calodendron Wimm. **Holme Willow**
(S. viminalis x S. caprea x S. cinerea)
Deciduous shrub mostly found in marshy areas by streams and gravel pits, not uncommon in the Waveney Valley and may also prove to have a similar distribution in the Broads. W4. E41.

S. x stipularis Sm. **Eared Osier**
(S. viminalis x S. caprea x S. aurita)
Deciduous shrub, found on wet ground at Bradwell E50 in 1996, RWE & WM. W0. E1.

S. x smithiana Willd. **Silky-leaved Osier**
(S. viminalis x S. cinerea)
Deciduous shrub frequent in wet habitats, with or without its parents. W7. E5.

S. caprea L. **Goat Willow**

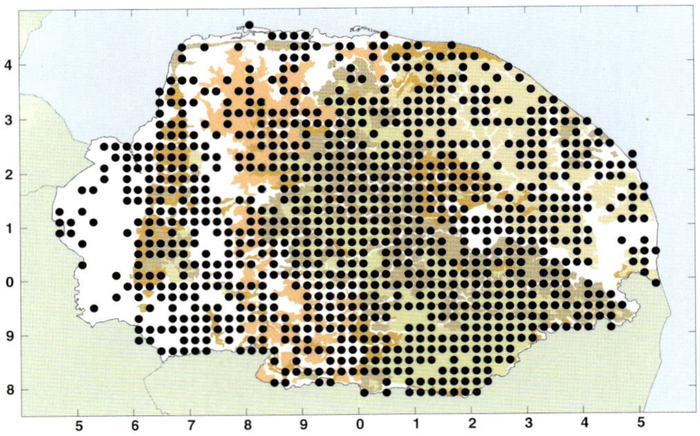

Deciduous shrub or small tree frequent in old woodlands and hedgerows especially on alkaline soils. One of the few willows to be at home away from water. Our plants are ssp. **caprea**.

S. x reichardtii A. Kern.
(S. caprea x S. cinerea)
Deciduous shrub or small tree frequently growing with one or both parents, also in the absence of either. W38. E2.

■ **S. x capreola** Jos.Kern. ex Andersson
(S. caprea x S. aurita)
Deciduous shrub not recorded since 1968.

S. cinerea L. **Grey Willow**

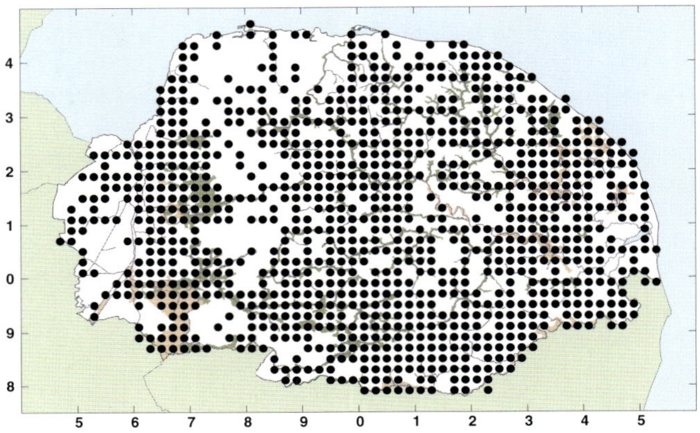

Deciduous shrub frequent throughout the county in damp situations in both natural and artificial habitats. The commonest subspecies in west Norfolk is ssp. **cinerea**, but ssp. **oleifolia** Macreight (syn. *S. atrocinerea* Brot. does occur as at Tottenhill Row, W61 CD.Preston, 1996. In east Norfolk ssp. **oleifolia** appears to be the commonest taxon.

S. x multinervis Döll
(S. cinerea x S. aurita)
Deciduous shrub, less frequent than *S. aurita*, but sometimes found in sites where that species no longer grows. Found at Syderstone W83 in 1991, KAB and Broome Heath E39 in 1990, GM & E. Beaumont. W1. E1.

S. x subsericea Döll
(S. cinerea x S. repens)
Deciduous shrub, an uncommon hybrid found at Litcham, W81 in 1992; Winterton Dunes, also in 1992, R.Southwood and Roydon W62 in 1993, KAB. W2. E1.

S. aurita L. **Eared Willow**

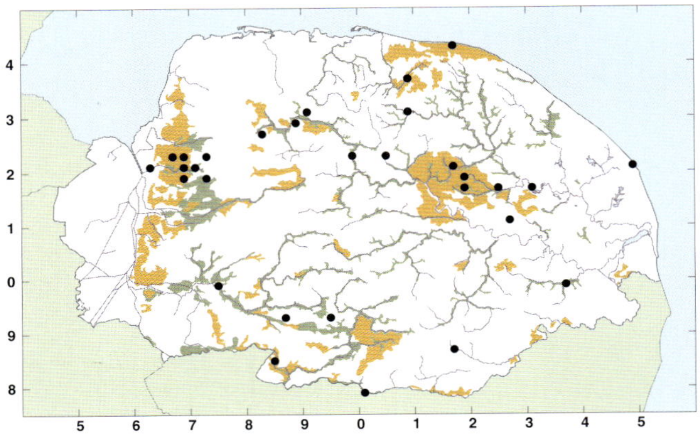

Deciduous shrub found chiefly on damp heaths, but always uncommon.

S. x ambigua Ehrh.
(S. aurita x S. repens)
Deciduous shrub recorded from Buxton Heath, E12 in 1977, ETD and Roydon Common W62 where it has been known for at least 30 years and still occurs. W1. E0.

S. repens L. **Creeping Willow**

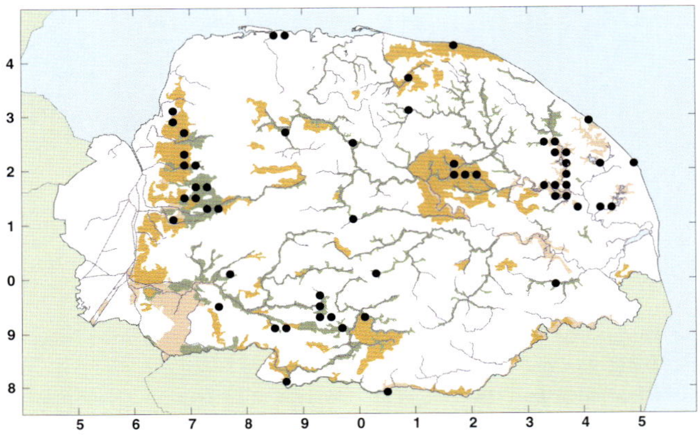

Deciduous shrub occurring in a wide range of damp habitats, most plants are var. **repens**. Var. **argentea** (Sm.) Wimm. & Grab. is found in dune slacks at a few coastal sites.

BRASSICACEAE (CRUCIFERAE)

■ **Sisymbrium irio** L. **London Rocket**
Annual from mainland Europe. Casual recorded twice, at Watton W90 in 1911 and near Cromer E24 in 1957.

■ **S. loeselii** L. **False London Rocket**
Annual from mainland Europe. Casual recorded once, near Harford Bridges Norwich E20 in 1963.

■ **S. volgense** M.Bieb ex E Fourn. **Russian Mustard**
Perennial originating in Russia, a casual found once, at Caister St Edmunds E20 in 1978, PG.Lawson.

S. altissimum L. **Tall Rocket, Tumbling Mustard**

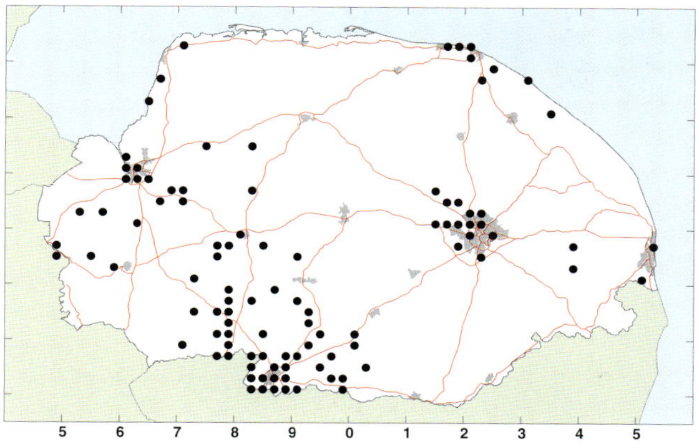

Annual, naturalised from mainland Europe and quite frequent in the Breckland where it was first observed on military camp sites between 1914 and 1918 by WG. Clarke. Otherwise mainly a plant of waste ground in urban areas.

■ **S. polyceratum** L.
Annual from the Mediterranean region. Casual recorded once, in King's Lynn W62 in 1949.

S. orientale L. **Eastern Rocket**
Annual, naturalised from mainland Europe, most frequent in the vicinity of towns but has also been recorded in several coastal sites.
W21. E22.

S. officinale (L.) Scop. **Hedge Mustard**
Annual or biennial, ubiquitous in field edges, waste ground and walls, absent only from continuously grazed grassland.
W701. E704.

Descurainia sophia (L.) Webb ex Prantl **Flixweed**

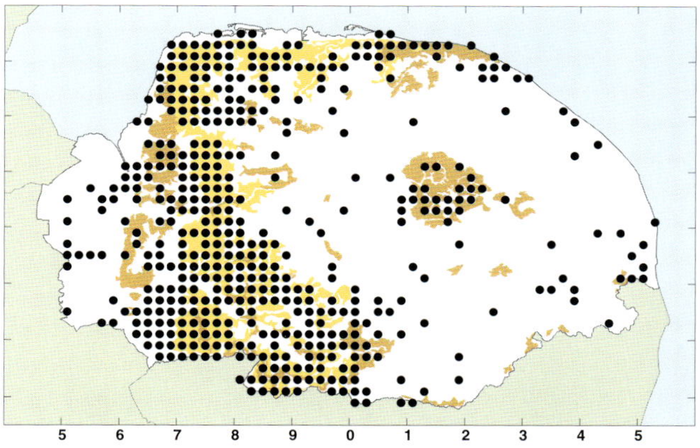

Annual or biennial, a frequent field weed of lighter soils where it can be abundant.

Alliaria petiolata (M. Bieb.) Cavara & Grande
Garlic Mustard, Jack by the Hedge
Biennial, once used for salads and for flavouring and still strongly associated with habitation, also often abundant along old hedges and wood margins. Absent from some parts of the fenland and from coastal marshes. W617. E681.

Arabidopsis thaliana (L.) Heynh. **Thale-cress**

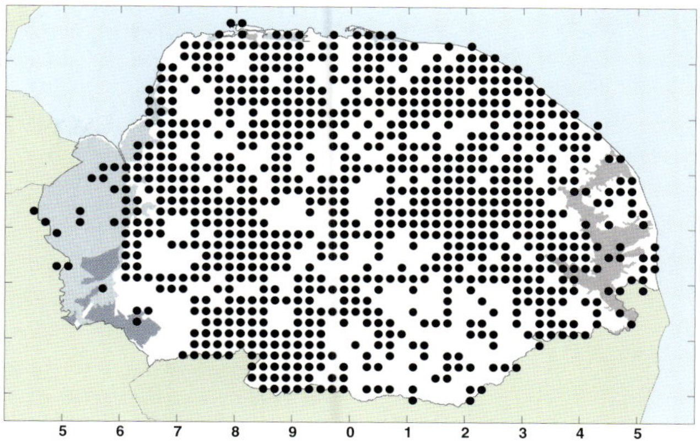

Summer and autumn germinating annual, abundant in open, dry soils, on waste ground and on walls. Those germinating following June rains may flower in September and October. On heavier soils and in Broadland, often confined to wall tops and gravel paths.

Arabidopsis thaliana - North Wootton

■ **Isatis tinctoria** L. **Woad**
Biennial grown widely as a dye plant until the arrival of cheap imported indigo, the last East Anglian factory, near Wisbech, closing in 1913. Last noted at Wroxham E21 in 1932 where it was a garden escape. Woad is, however, being grown again in the county, so it may yet recur as an escape.

Bunias orientalis L. **Warty Cabbage**
Biennial from mainland Europe, a casual recorded at Gayton W71 in 1945. During the present survey found scattered along roadsides on the former airfield at Weston Longville where it has been known for many years, E01, ALB and E11, J.Humphris. W0. E2.

Erysimum cheiranthoides L. **Treacle Mustard**

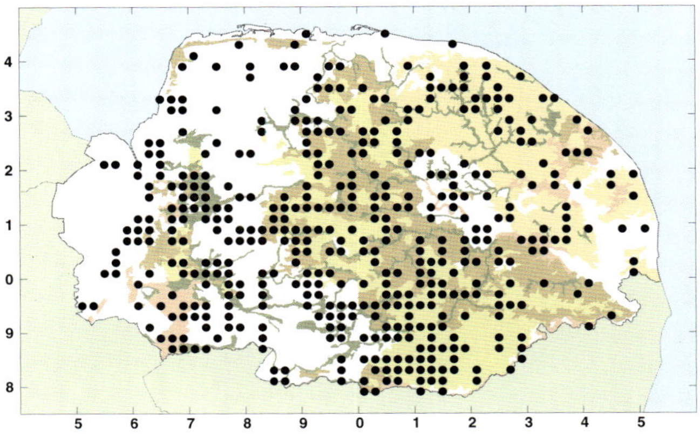

Annual to biennial, sometimes a short-lived perennial, introduced from mainland Europe, widespread but perhaps not as frequent as formerly apart from the clay soils of south Norfolk.

E. x marshalii (Henfr.) Bois **Siberian Wallflower**
(E. decumbens x E. perowskianum)
Hybrid of garden origin occasionally escaping, recorded from Norwich E29 in 1990, ETD and Heacham W63, in 1992, PRC. A fertile hybrid widely grown from seed. W1. E1.

E. cheiri (L.) Crantz **Wallflower**
Syn. *Cheiranthus cheiri* L.
Winter green perennial from the eastern Mediterranean, long persisting on old walls and cliffs. W65. E64.

Hesperis matronalis L. **Dame's Violet, Dame's Rocket**
Biennial to winter green perennial introduced to gardens from mainland Europe, sometimes thrown out and becoming naturalised, especially on moisture-retentive soils, thence spreading by seed. It has become well established by watersides along the Tas at Saxlingham Thorpe E29 and in the Waveney Valley, the seeds probably floating short distances along the water courses. W70. E101.

Malcolmia maritima (L.) W.T.Aiton **Virginian Stock**
Annual, in spite of its common name originating from coastal sands in Greece and Albania. Once a common garden annual, now less often grown and reported during the present survey only around Norwich E20 and Sheringham E14 in 1996. W0. E5.

Matthiola incana (L.) W.T.Aiton **Hoary Stock**
Annual or perennial introduced to gardens from mainland Europe, an occasional escape. W1. E1.

M. longipetala (Vent.) DC **Night-scented Stock**
Syn. *M. bicornis* (Sibth. & Sm.) DC
Annual originating from Greece, though now widely naturalised around the Mediterranean, once a garden favourite, now less frequently seen. Recorded during this survey as an escape in Norwich E20 by ETD. W0. E1.

Barbarea vulgaris W.T.Aiton **Common Winter-cress**

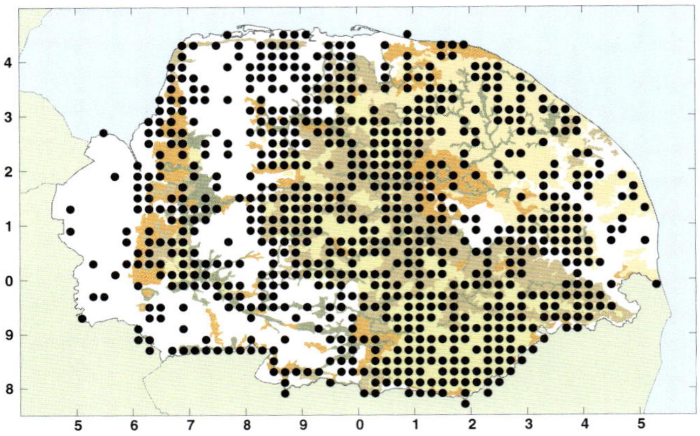

Biennial to winter green perennial scattered on roadsides, stream and dyke banks. Missing only from the driest areas and curiously also from parts of northern Broadland.

B. intermedia Boreau **Medium-flowered Winter-cress**
Biennial to winter green perennial introduced from mainland Europe, occasional on roadsides and waste ground. W4. E11.

■ **B. stricta** Andrz. **Small-flowered Winter-cress**
Biennial to winter green perennial rare on waste ground. ELS observed in 1967 that it was increasing, but it is no longer so as the last record was Whitlingham E20 in 1983, ETD.

B. verna (Mill.) Asch **American Winter-cress**
Biennial to winter green perennial introduced from mainland Europe (not American in spite of its common name) and grown as a winter salad plant. Occasional on roadsides and waste ground. W7. E8.

Rorippa nasturtium-aquaticum (L.) Hayek **Water-cress**
Syn. *Nasturtium officinale* W.T.Aiton

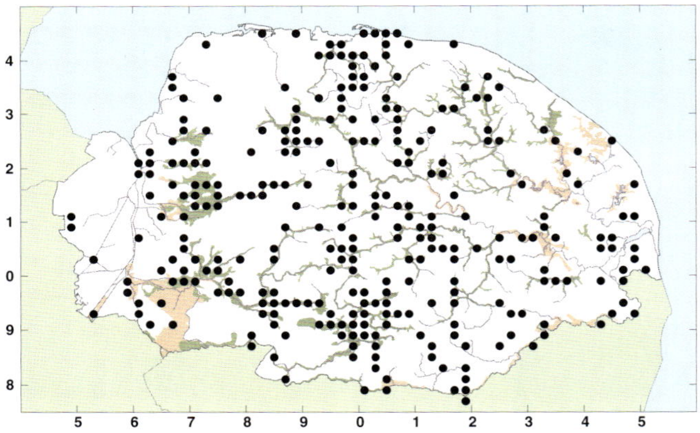

Winter green perennial, frequent in streams, wet ditches and marshes.

R. x sterilis Airy Shaw **Hybrid Water-cress**
(R. nasturtium-aquaticum x R. microphylla)
Winter green perennial recorded from Hockwold W68 in 1991 TCG.Rich; Castle Acre W81 in 1996, MK where it was noted in 1946 by ELS and Ditchingham E39 in 1997, ALB. W3. E1.

R. microphylla (Boenn.) Hyl. ex A. & D. Löve.
Narrow-fruited Water-cress
Syn. *Nasturtium microphyllum* (Boenn.) Rchb.

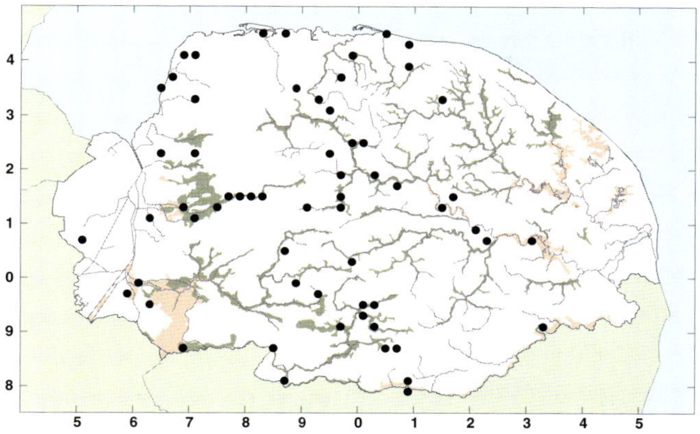

Winter green, or more accurately a winter-bronze perennial scattered in streams and ponds.

[R. islandica (Oeder ex Gunneras) Borbás
Northern Yellow-cress
This name was formerly used to include *R. palustris* and Norfolk records are referable to that species.**]**

R. palustris (L.) Besser **Marsh Yellow-cress**

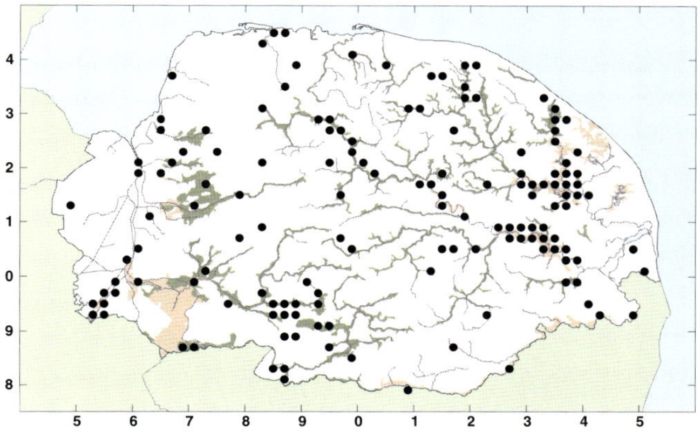

Annual or short-lived perennial often found by water courses, also frequently by drying ponds.

R. sylvestris (L.) Besser **Creeping Yellow-cress**

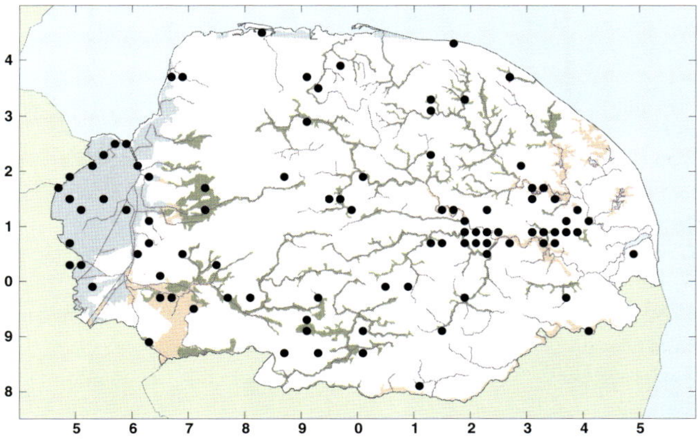

Perennial widely scattered near stream courses, occasionally in damp arable ground as at Cawston E12 and Old Catton E21 but also a garden weed even in dry places, observed flourishing in a gravel drive.

■ **R. x anceps** (Wahlenb.) Rchb.
(R. amphibia x R. sylvestris)
This hybrid was recorded at Welney W59 in 1954.

R. amphibia (L.) Besser **Great Yellow-cress**

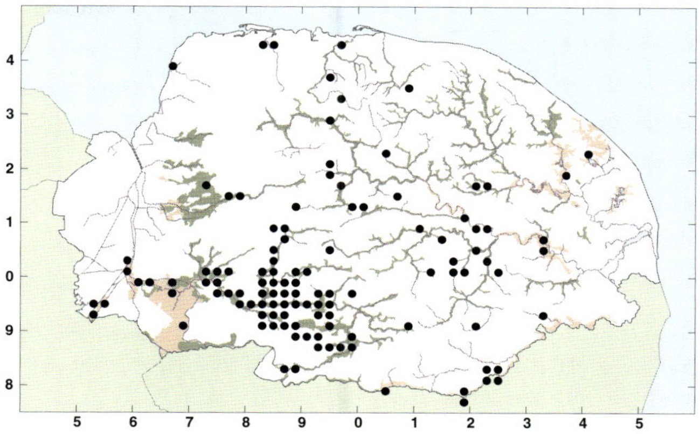

Perennial found in still water, most frequent in Breckland meres and in pools in that area.

Rorippa amphibia - Thompson

Armoracia rusticana P.Gaertn., B.Mey. & Scherb.
Horse-radish

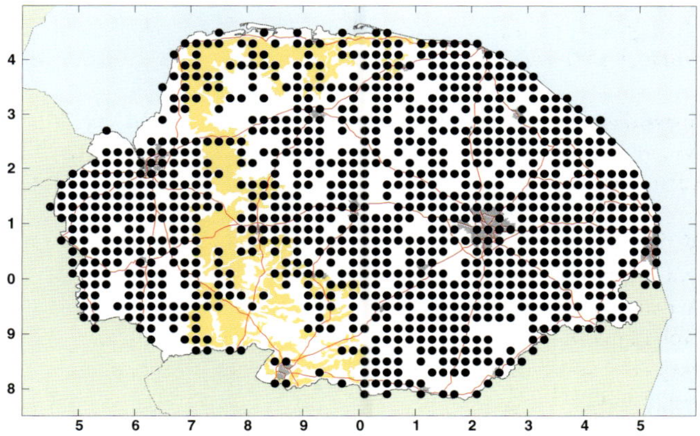

Herbaceous perennial, introduced, probably from western Asia, persistent as a throwout from gardens. Apparently always spreading vegetatively, the fruit being said to ripen rarely, if at all, but remarkably widespread if this is so.

Cardamine amara L. **Large Bitter-cress**

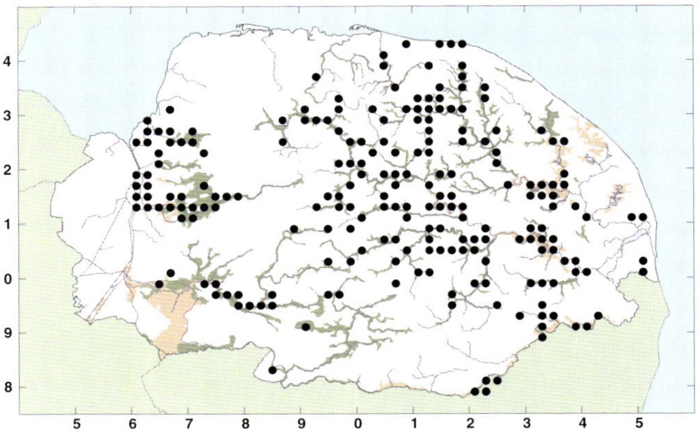

Winter green perennial of streamsides and marshes, thriving in shady sites, usually, but not exclusively, on acid soils.

C. pratensis L.
Cuckoo Flower, Ladies' Smock, locally **Milkmaids**

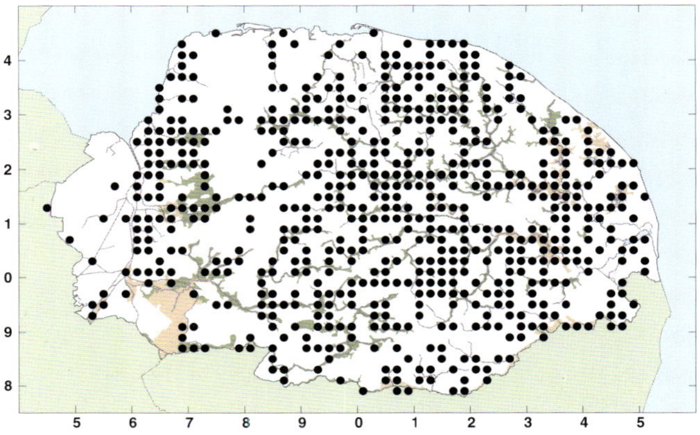

Winter green perennial of moist to wet grassy places, occasionally in damp churchyards, absent from dry sandy soils and from most of the Fenland.

C. flexuosa With. **Wavy Bitter-cress**

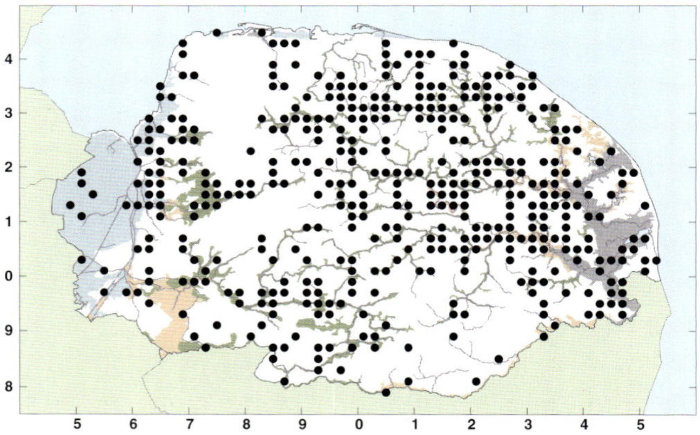

Autumn germinating annual of moist soils, often in shade, also occasionally a weed in cultivated ground.

C. hirsuta L. **Hairy Bitter-cress**
Annual of cultivated ground, capable of germinating whenever the weather is suitable, particularly common in nurseries and garden centres, hence also in most gardens, also on walls, waste ground but only rarely as a field weed. W554. E588.

C. corymbosa Hook.f.
Syn. *C. uniflora* (Hook.f.) Allan non Michx.
Introduced from New Zealand, now apparently spreading as a weed of cultivation, particularly in nurseries. One record, Stanhoe W83 in 1994, KAB. W1. E0.

Arabis glabra (L.) Bernh. **Tower Mustard**
Syn. *Turritis glabra* L.

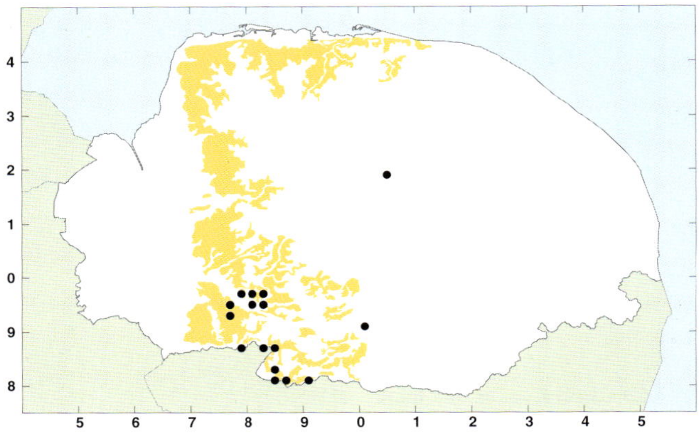

Biennial of sandy, grassy places formerly more widely known. Now almost confined to Breckland where it is to be found in seventeen sites, mostly in forest rides, in some of which it can be abundant. Also at Bylaugh E01, where it was recorded by Trimmer in his Flora of 1866 and is still present, ALB 1997. A nationally scarce species.

A. caucasica Willd. ex Schltdl. **Garden Arabis**
Winter green perennial introduced to gardens from south-west Europe to the Caucasus, an occasional escape, holding its own. W4. E1.

A. hirsuta (L.) Scop. **Hairy Rock-cress**

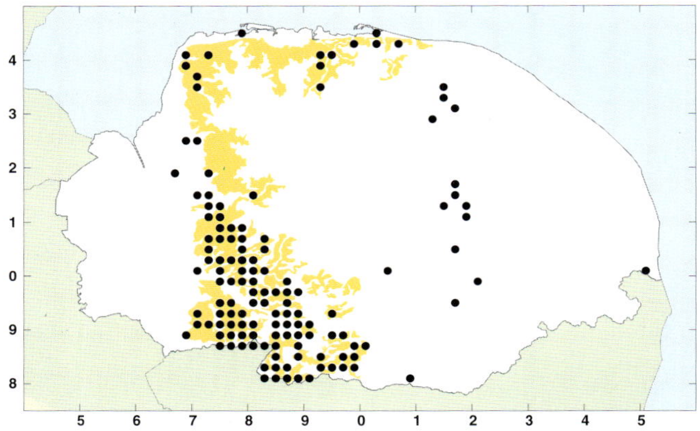

Biennial to winter green perennial, locally frequent in short grassland on chalky soils. In spite of the fact that walls are given in Petch & Swann, 1968 and in Stace, 1997 as one of its principal habitats, it is not so in Norfolk, only two sites on walls being recorded by RMP during his 1995-1997 wall survey of west Norfolk; at Sedgeford W73 and Swaffham W80.

Aubrieta deltoidea (L.) DC. **Aubretia**
Winter green perennial introduced from southern Europe to gardens and occasionally escaping and surviving, especially on walls and pathways. W20. E14.

Lunaria annua L. **Honesty**
Biennial, introduced to gardens from south-east Europe, not infrequently escaping, but rarely persisting. W137. E385.

Arabis glabra - Santon

L. rediviva L.
Perennial from mainland Europe, introduced to gardens. Recorded at Wroxham E31 in 1978, JA.Newbould and Watton W90 in 1989, M.Merson. W1. E0.

■ **Alyssum alyssoides** (L.) L. **Small Alison**
Winter germinating annual from mainland Europe, formerly known from four sites in west Norfolk, surviving at West Acre W71 until 1951.

A. saxatile L. **Golden Alison**
Syn. *Aurina saxatile* (L) Desv.
Winter green perennial introduced to gardens from south-east Europe and occasionally escaping on to adjacent walls and banks. W1. E0.

Berteroa incana (L.) DC. **Hoary Alison**
Perennial from mainland Europe, long established at Whittingham, W79 where it has been known since 1930. Also observed at Weeting W78 in 1994, R.Hoblyn. W2. E0.

Lobularia maritima (L.) Desv. **Sweet Alison**
Annual or short-lived perennial introduced to gardens from south-east Europe, escaping on to walls and dry ground but not surviving long. W20. E82.

Erophila majuscula Jord. **Hairy Whitlow-grass**
Annual of dry, sandy soils recorded in west Norfolk in Rich 1991; on a path by Eaton golf course E20 in 1996, M.Gurney and at Grimston W72 in 1998, GB. W1. E1.

E. verna (L.) DC. **Common Whitlow-grass**

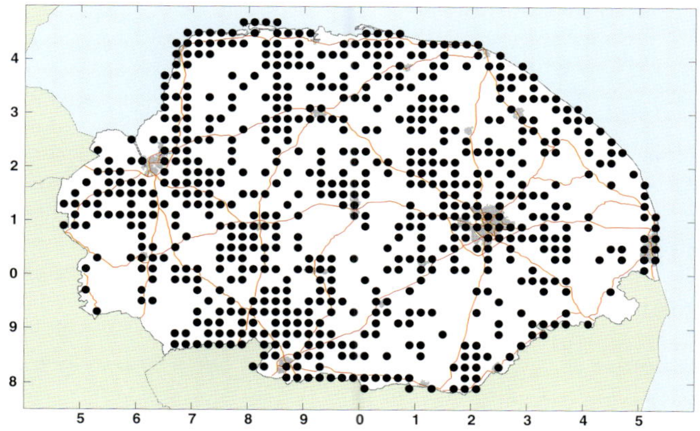

Autumn germinating annual, widespread on open, dry grassland, notably so on Breckland tracks, also on walls. Var. **praecox** (Steven) Diklic was recorded on the shingle bank at Heacham W63 in 1994, PRC.

E. glabrescens Jord. **Glabrous Whitlow Grass**
Autumn-germinating annual similar to, and confused with *Erophila verna*, but rare. Recorded for west Norfolk in Rich 1991, also at Roydon near Diss E17&18 in 1990, AC. W0. E2.

Cochlearia anglica L. **English Scurvy-grass**

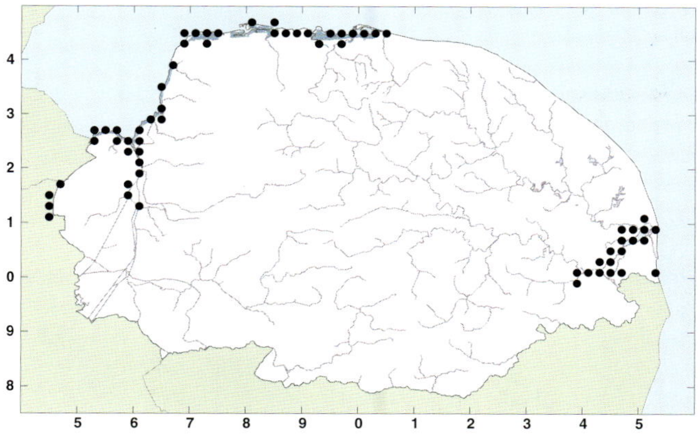

Biennial to perennial of salt marshes. Our commonest coastal scurvy-grass, found on the firmer areas of mud by estuaries and tidal rivers, also on coastal saltmarshes. Not yet recorded inland by roads as it has been in other parts of England.

C. officinalis L. **Common Scurvy-grass**

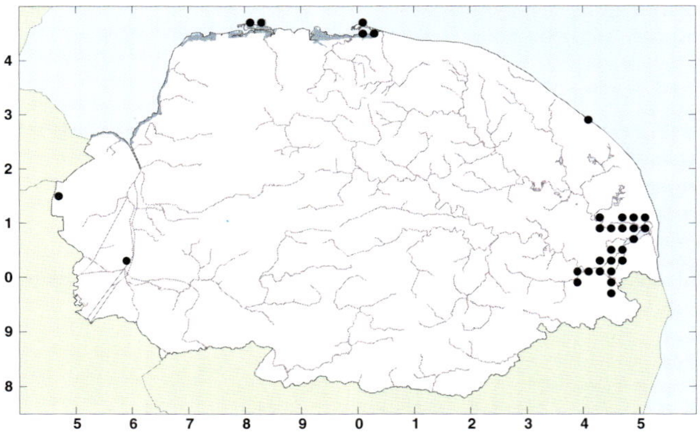

Winter green perennial of muddy areas on salt marshes, very scarce on the north coast and Wash marshes. Not infrequent by the rivers immediately above Breydon Water, especially in niches behind the steel piling of river banks where these are overtopped by occasional high tides as on the R.Chet at Hardley E39. Our plants belong to ssp. **officinalis**.

C. danica L. **Danish Scurvy-grass, Early Scurvy-grass**

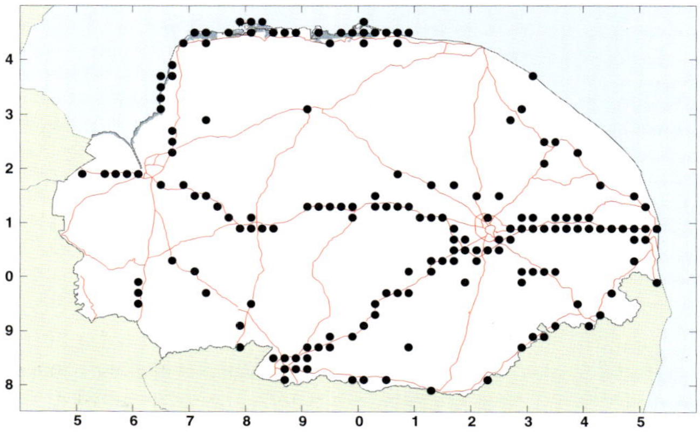

Autumn germinating annual to biennial, scattered and rather scarce on the drier parts of salt marshes, also on the shingle bank at Heacham W63. In the last twenty years it has become abundant by many inland roads apparently spreading into the county on three fronts, along the A17 from the west, the A11 from the south-west and the A12 from the south-east.

Camelina sativa (L.) Crantz **Gold of Pleasure**
Annual from southern Europe, formerly cultivated for oil and fibre, now an occasional casual, probably from bird seed. Reported as abundant around the Village Hall at Brundall E30 in 1990 SP and seen at Bressingham E08 in 1994, JH. W0. E2.

Neslia paniculata (L.) Desv. **Ball Mustard**
Annual, a casual recorded from Norwich E20 in 1915 and near the docks in King's Lynn W62 in 1993, RMP. W1. E0.

Capsella bursa-pastoris (L.) Medik. **Shepherd's Purse**
Annual, germinating whenever weather is suitable, abundant in arable fields, gardens, roadsides, waste ground and on walls. W703. E704.

Cochlearia danica - Roudham (A11)

Teesdalia nudicaulis (L.) W.T.Aiton **Shepherd's Cress**

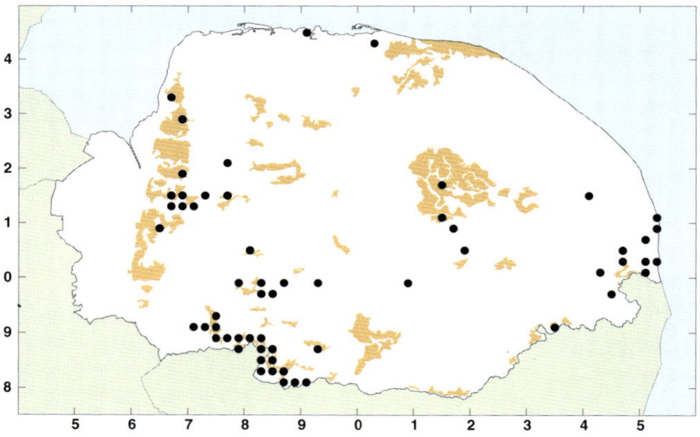

Autumn germinating annual, occurring wherever soils are light and sandy, commonest on tracks in the Breckland.

Thlaspi arvense L. **Field Penny-cress**

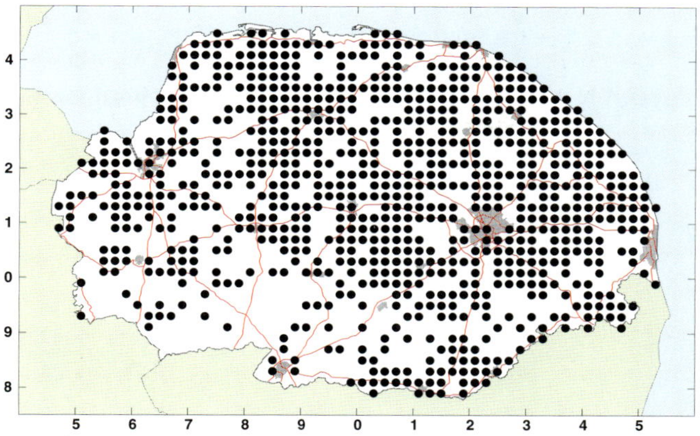

Mainly spring germinating annual of arable land, most frequent on heavier soils.

T. alliaceum L. **Garlic Penny-cress**
An introduction from mainland Europe, very similar to penny-cress, found as a casual in ornamental planters in Fakenham W93 and Wells W94 in 1991, MK, possibly from the same source. W2. E0.

Iberis sempervirens L. **Perennial Candytuft**
Evergreen sub-shrub from southern Europe, grown in gardens and occasionally self-sown as in the wall of Terrington St. Clement's church W52 in 1993, RMP. W1. E0.

■ **I. amara** L. **Wild Candytuft**
Autumn-germinating annual, last recorded at Two Mile Bottom, Thetford W79 in 1943.

I. umbellata L. **Garden Candytuft**
Annual introduced to gardens from southern Europe, found occasionally on walls and waste places. W6. E7.

Lepidium sativum L. **Garden Cress**
Annual introduced from western Asia as a salad crop, also as a casual in bird seed. Most records are from tips. W0. E1.

Teesdalia nudicaulis - Santon

L. campestre (L.) W.T.Aiton **Field Pepperwort**

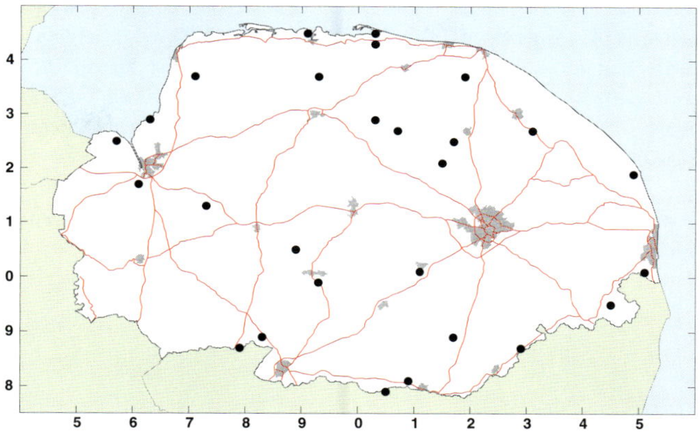

Annual to perennial scattered on roadsides throughout the county.

L. heterophyllum Benth. **Smith's Pepperwort**
Syn. *L. smithii* Hook.

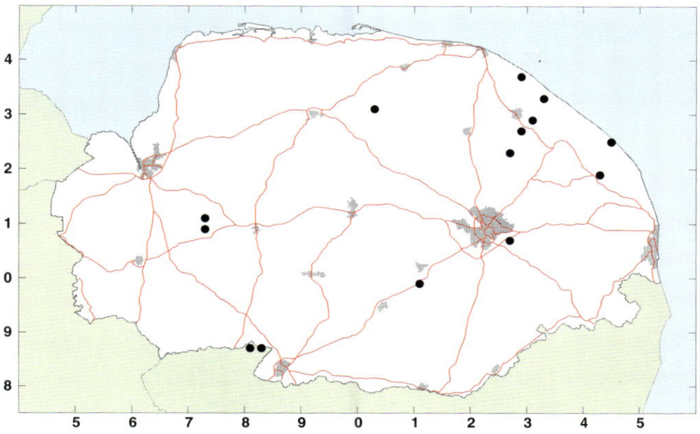

Annual, usually on dry banks, apparently most frequent in the North Walsham area, otherwise scattered through the county. Now rare as a field weed though several hundred plants were found growing together in a field at Marham W71 in 1990, J.Williamson.

L. ruderale L. **Narrow-leaved Pepperwort**

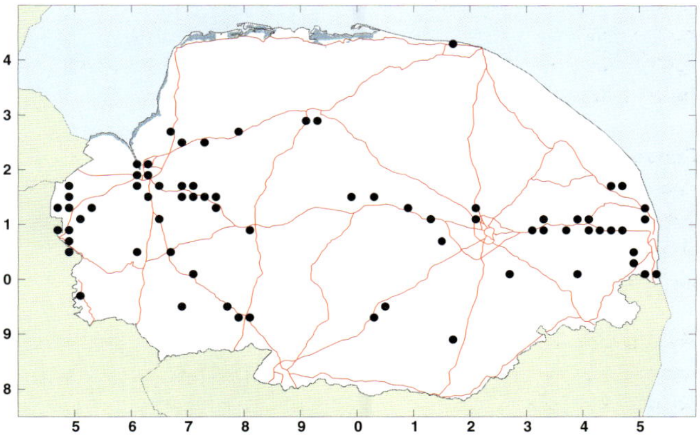

Winter green annual or biennial once considered a native coastal plant, but during this survey found chiefly by roadsides. It appears to be extending its range along the trunk roads in a similar fashion to Danish scurvy-grass.

L. latifolium L. **Dittander**

Deciduous perennial long established near the sea, a large colony between Sheringham and Beeston E14 has been known for many years; also on Haddiscoe Island E40 in 1992, J Harris; Yarmouth north denes E50 in 1994, B.Hancy and on a bank near the Gt.Ouse at King's Lynn W61 in 1996, JEG. Inland found on a roadside verge at South Raynham W82 in 1994, P.Oswald. W2. E4.

L. draba L. **Hoary Cress, Thanet Cress**

Syn. *Cardaria draba* (L.) Desv.

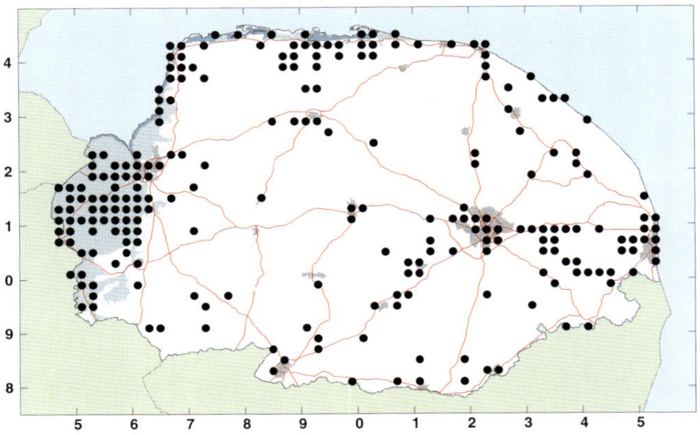

Herbaceous perennial introduced from Europe and established as a weed in scattered sites, particularly beside railway lines close to urban areas. All our plants belong to ssp. **draba**.

Coronopus squamatus (Forssk.) Asch. **Swine Cress**

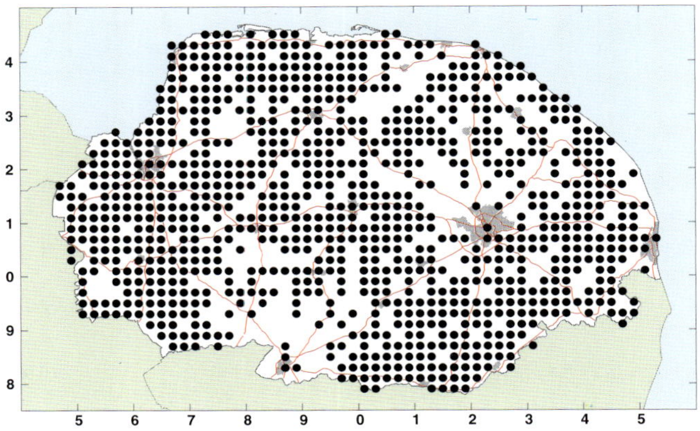

Spring germinating annual, most frequent in gateways, paths and field edges where soil is compacted.

C. didymus (L.) Sm. **Lesser Swine Cress**

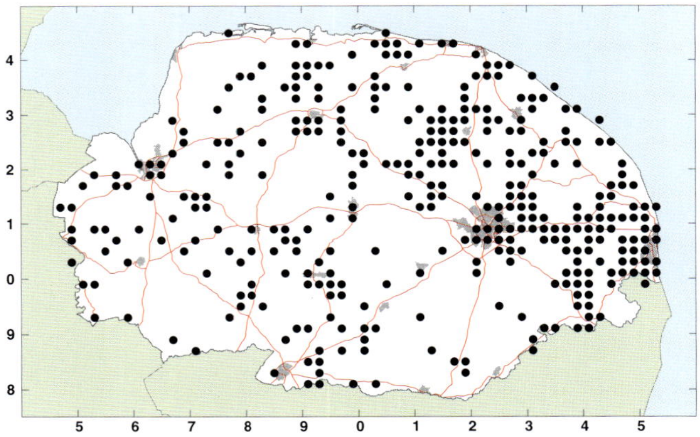

Spring germinating annual introduced from South America. It appears to be increasing both its range and frequency. Notably absent from some areas perhaps due to a lack of seed sources.

■ **Conringia orientalis** (L.) Dumort **Hare's-ear Mustard**

Annual of casual occurrence, probably chiefly a birdseed alien. Last recorded at Stanhoe W83 in 1966 and Norwich E20 in 1968.

Diplotaxis tenuifolia (L.) DC. **Perennial Wall Rocket**

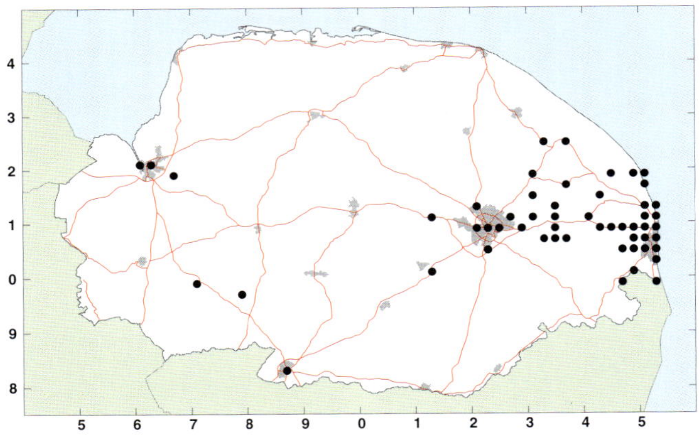

Winter green perennial, particularly frequent in the triangle Norwich E20 to Caister-on-Sea E51 and Hopton-on- Sea E59. Its irregular distribution is not easily explained. In spite of its vernacular name, rare on walls.

D. muralis (L.) DC. **Annual Wall Rocket**

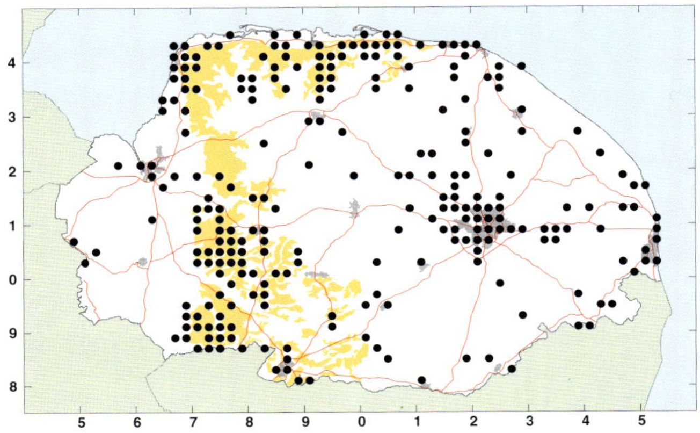

Annual or short-lived winter green perennial locally abundant with a largely urban-based distribution. The perennial form var. **babingtonii** Syme, was seen on a wall at Marham W70 in 1995, RMP.

D. erucoides (L.) DC **White Wall Rocket**

Annual of casual occurrence, found once only, at Hockwold W78 in 1990, TCG.Rich. W1. E0.

Brassica oleracea L. **Cabbage**

Biennial to perennial, occasionally found on roadsides and near allotments and gardens. Only ever a casual. W3. E6.

B. napus L. ssp. **oleifera** (DC) Metzg. **Oil-seed Rape**

Annual germinating when seed shed, abundant on roadsides wherever seed from the crop is carried. W431. E454.

B. rapa L. ssp. **rapa** **Turnip**

Annual germinating when seed shed, scattered on roadside verges, field edges and waste places. W30. E15.

B. juncea (L.) Czern. **Chinese Mustard**

Annual germinating when seed shed. Recorded in Norwich E20 in 1990, ETD; on Weston airfield E01, ALB and on the shingle bank at Heacham W63 in 1992, PRC. W1. E3.

B. nigra (L.) W.D.J.Koch **Black Mustard**

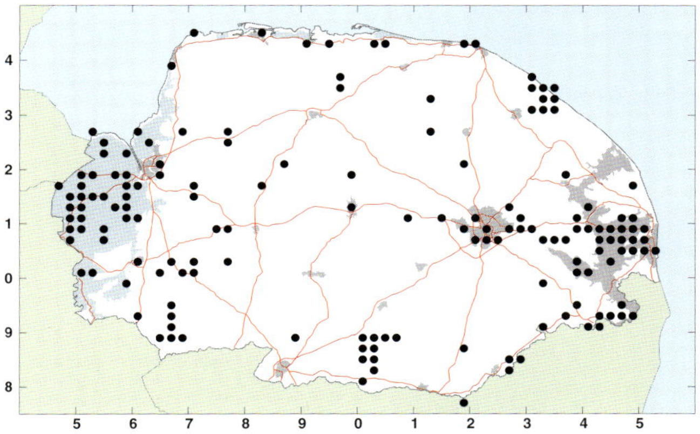

Annual germinating when seed shed. Probably native along river banks such as near the Waveney, by the dykes and droves on Haddiscoe Levels and perhaps in similar sites in the Fenland. Casual elsewhere.

Sinapis arvensis L. **Charlock**

Annual germinating when seed shed, an abundant weed of fields and waste ground though less frequent on acid, sandy soils where it is replaced by *Raphanus raphanistrum.* W562. E566.

S. alba L. **White Mustard**

Annual germinating when seed shed. Formerly grown commercially in parts of the county under contract to Colmans of Norwich. W39. E78.

Erucastrum gallicum (Willd.) O.E.Schulz **Hairy Rocket**

Annual or biennial from mainland Europe, a scarce casual found at Methwold Hythe W79 in 1966; at Brandon W78 in 1988, J.Partridge and Pentney W71 in 1991, KAB. W2. E0.

■ **Coincya monensis** (L.) Greuter & Burdet **Wallflower Cabbage**

Syn. *Rhynchosinapis cheiranthos* (Vill.) Dandy

Biennial from southern Europe. One record, from Harford tip E20 in 1970, CG.Hanson, is of ssp.**cheiranthos** (Vill.) Greuter & Burdet.

Hirschfeldia incana (L.) Lagr.-Foss. **Hoary Mustard**

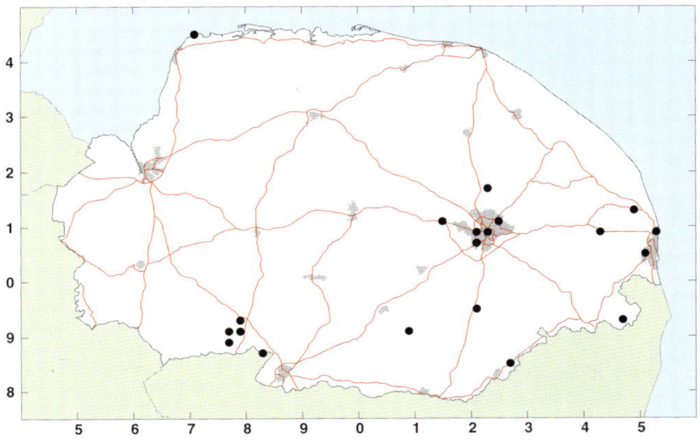

Over-wintering annual introduced from southern Europe, well established. Known to have been present at some sites in the Breckland for forty years, also frequent in south and west Norwich.

Cakile maritima Scop. **Sea Rocket**

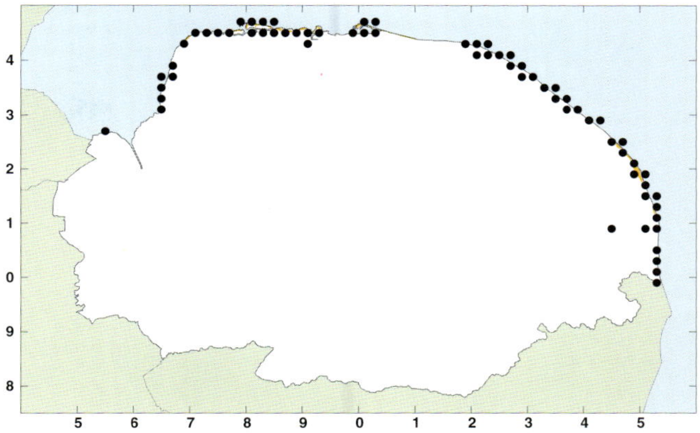

Spring-germinating annual frequent on the drift-line of sandy and shingly beaches.

Rapistrum rugosum (L.) J.P.Bergeret **Bastard Cabbage**

Syn. *R. orientale* (L.) Crantz.

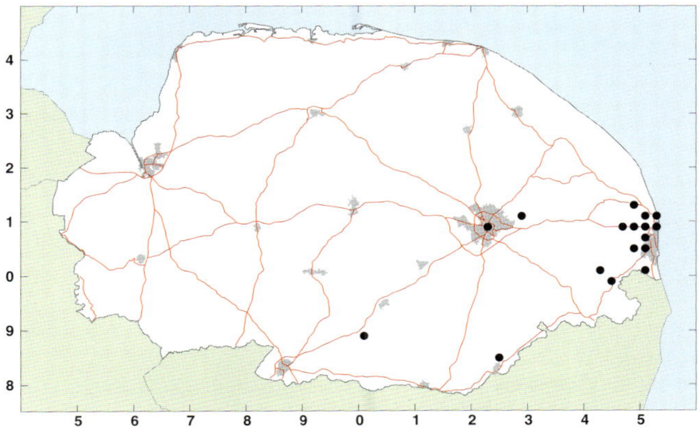

Annual from southern Europe. A casual recorded chiefly from tips as at Harford, Norwich E20 in 1968; King's Lynn W20 in 1979 and Beetley W91in 1980. During this survey found on waste ground and sea banks around Yarmouth and Breydon Water E40, E50 where it can be frequent.

■ **R. perenne** (L.) All. **Steppe Cabbage**

Biennial from southern Europe, last seen in a lay-by at Lynford W79 in 1978, RPL & ELS.

Crambe maritima L. **Sea Kale**

Herbaceous perennial of coastal shingle banks described in 1863 as abundant on the North Norfolk coast. Now scarce and surviving only between Snettisham and Heacham W62 and 63 and at Blakeney E04, though it was not seen there in 1997. Its hold is now always somewhat precarious because of increasing damage to its habitat by winter storms and the subsequent use of bulldozers to rebuild the shingle banks. Seed was introduced to Blakeney E04 in 1912. A nationally scarce species. W2. E1.

Crambe maritima - Blakeney (also illustrated overleaf)

Crambe maritima - Snettisham page 105

Raphanus raphanistrum L. — **Wild Radish**

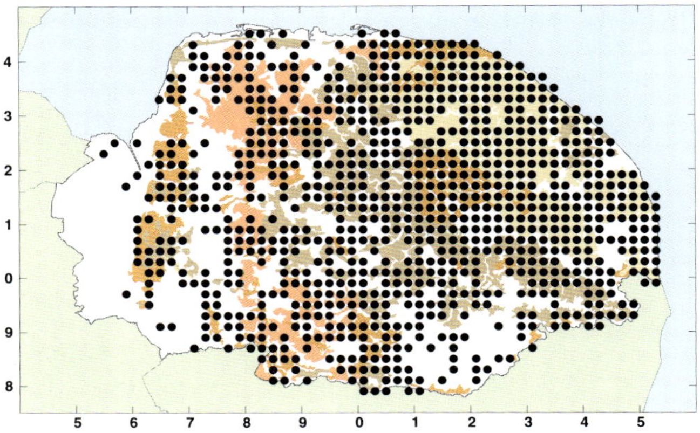

Annual or short-lived perennial widely distributed. Most plants belong to ssp. **raphanistrum** which is scarce on the clay and in Fenland. Ssp. **maritimus** (Sm.) Thell, occurs on muddy shores and sea-banks as by Breydon Water E50 and cliffs at Little Marl Point E23. Ssp. **landra** (Moretti ex. DC) Bonnier & Layens, was noted at Yarmouth E50 in 1964, JH. Silverwood and again in 1978, ETD.

R. sativus L. — **Cultivated Radish**

Annual with several old records from tips, more recently found where seed has been sown for game-bird feeding. W3. E6.

RESEDACEAE

Reseda luteola L. — **Weld, Dyer's Rocket**

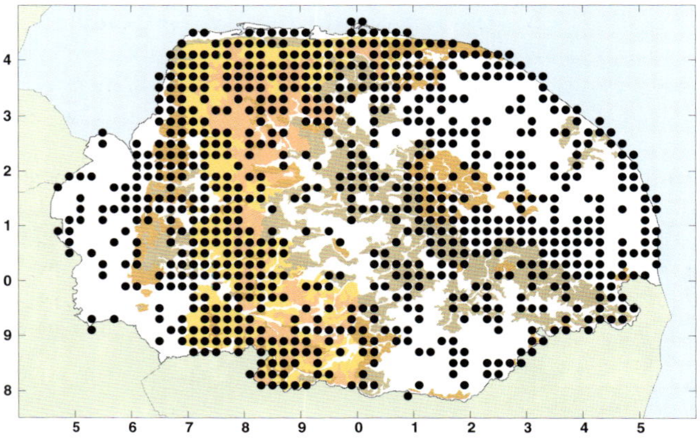

Biennial, most successful on the drier soils. Formerly cultivated for its fast, bright yellow dye and still perhaps surviving from that source on waste ground near villages.

■ **R. alba** L. — **White or Upright Mignonette**

Usually an annual, introduced from south-east Europe. Only three records, the last being from waste ground in Norwich E20 in 1966.

R. lutea L. — **Wild Mignonette**

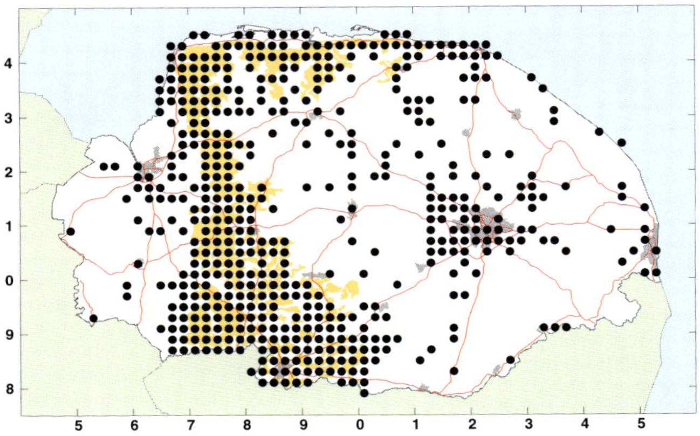

Biennial to perennial frequent in areas of light, calcareous soils, but rare on heavy or wet soils.

■ **R. odorata** L. — **Garden Mignonette**

Annual introduced to gardens from the Mediterranean. Found twice on Harford tip, Norwich E20 in 1969 and 1972, probably less frequently grown today.

■ **R. inodora** Rchb.

Biennial or perennial from southern Europe, the only record from a carrot field at Wootton W62 in 1984, CPP.

EMPETRACEAE

■ **Empetrum nigrum** L. — **Crowberry**

Evergreen shrub from upland Britain. Recorded on the cliff-top near Overstrand, E24 by CD.Raby and at Winterton E41 by JH.Silverwood, both in 1967. Presumed to be of bird-sown origin.

ERICACEAE

Rhododendron ponticum L. — **Rhododendron**

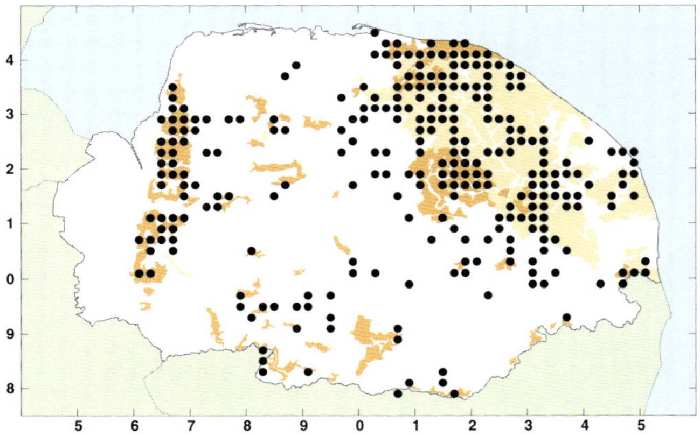

Evergreen shrub introduced to parks and gardens from south-east Europe and south-west Asia. Many of our plants show signs of hybridity with other species, mainly *R. maxima*. Whatever their origin, they can be pernicious woodland weeds with abundant self-sown seedlings on acid soils.

■ **Andromeda polifolia** L. **Bog Rosemary**
Evergreen shrub recorded early in the 19th century from Larlingford and Shropham. Possibly also a former native of raised bogs in the fenland, but long extinct from all sites.

Calluna vulgaris (L.) Hull **Heather, Ling**

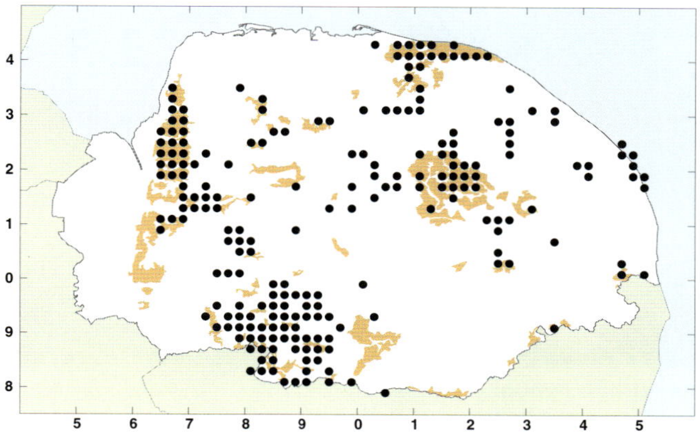

Evergreen shrub once frequent in north and west Norfolk and in the past a feature of many sheepwalks where it grew on leached soils. Now, apart from a number of remnant sites, confined to the main areas of heathland, apart from the acid dunes of the east coast where it thrives.

Calluna vulgaris - Winterton Dunes

Erica tetralix L. **Cross-leaved Heath**

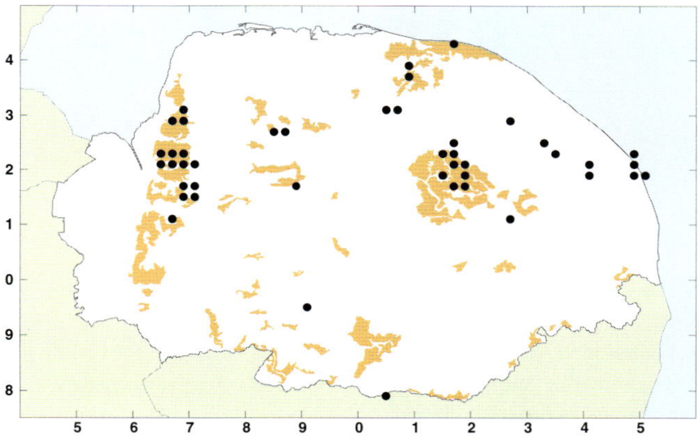

Evergreen shrub confined to moist soils in fens and bogs. Also a feature of acid dune slacks between Hemsby and Horsey, E41, 42, 51.

E. cinerea L. **Bell Heather**

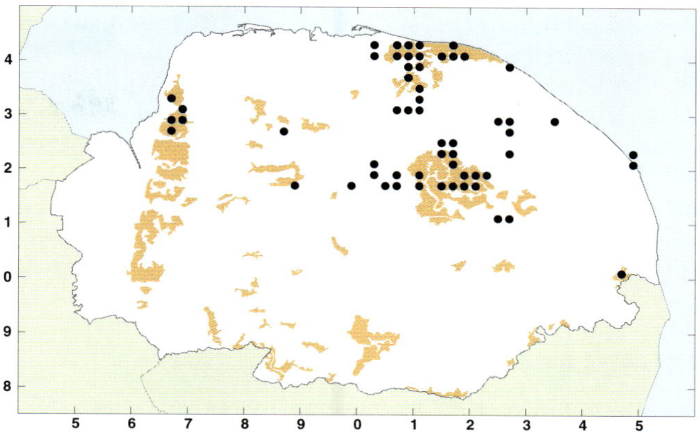

Evergreen shrub, notably absent from the Breckland where *Calluna* is frequent, needing soils of a greater acidity.

Vaccinium oxycoccos L. **Cranberry**
Syn. *Oxycoccus palustris* Pers.
Evergreen shrub now confined to bogs at Dersingham and Roydon with a few plants surviving at Sugar Fen, all W62. It also persists at Cranberry Rough, Hockham W99 where it was observed in 1975 by E.Campbell after long being thought extinct. From the frequency of the name Cranberry in minor place names such as Cranberry Farm and Cranberry Fen, this plant was once much more frequent. Until 1913 it was to be found at Horning E31 in Broadland, EAE. Now very rare in south and east England. W5. E0.

■ **V. vitis-idaea** L. **Cowberry**
Evergreen shrub, planted in 1908 on heathland at Attleborough but not surviving. A single plant found at Leziate W61 in 1984, CRS must also have been an introduction, though whether deliberate or bird-sown is uncertain.

V. myrtillus L. **Bilberry**
Evergreen shrub oddly absent from East Anglia. The plant recorded in Trimmer, 1866 on Mousehold Heath, Norwich was later found to have been planted as was a colony at Horsford E11 in 1908, while those found at Attlebridge E11 in 1968, ETD and still present, may well have derived from this last introduction. Whether the plant mentioned as growing wild on 'Paulin' (Poringland) Heath E21 in 1769 was truly native can never be known. W0. E1.

PYROLACEAE

Pyrola minor L. **Common Wintergreen**
Evergreen perennial for which only eight sites have ever been recorded, that from Roydon Common W62 in 1852 (as *P. rotundifolia*) the only west Norfolk site, all others in east Norfolk. Refound at Holt Lowes E03 in 1997 by F.Rose & PWL. W0. E1.

P. rotundifolia L. **Round-leaved Wintergreen**
Evergreen perennial formerly occurring quite widely in moist, acid sites, now confined to dune slacks at Winterton E42; Upton Fen E30 and Great and Little Fens at Catfield, E32. Although some records were listed as ssp. **maritima** (Kenyon) E.Warb. in the 1968 Flora, ELS later realised this was an error and confirmed that all our plants belong to ssp. **rotundifolia**. W0. E4.

MONOTROPACEAE

Monotropa hypopitys L. **Yellow Bird'snest**
Saprophytic perennial occurring occasionally and somewhat erratically in pine plantations both inland and on the coast. Most records are for ssp. **hypopitys** but ssp. **hypophegea** (Wallr.) Holmboe has also been recorded. Seen in seven sites between 1968 and 1984 and more recently at Weeting W78 in 1986; Santon W88 in 1987, N.Gibbons; Cawston E12 in 1994, ALB; Selbrigg E12 also in 1994, AR.Leech and on coastal dunes at Wells-next-the-Sea W94 in 1986, PRB and Holkham W84 in 1992. W4. E3.

Monotropa hypopitys - Holkham

PRIMULACEAE

Primula vulgaris Huds. **Primrose**
Syn. *P. acaulis* (L.) Hill

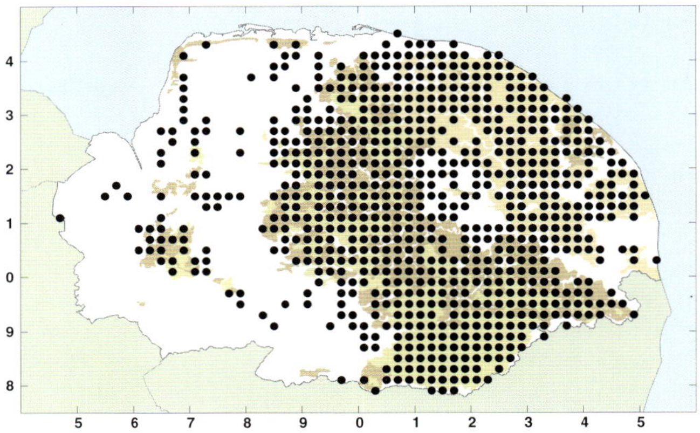

Winter green perennial often summer dormant in dry seasons, frequent in moist, shady hedgebanks and woods, also introduced to churchyards, parks and gardens which masks but does not completely disguise its natural distribution. Absent from Breckland as a native plant.

P. x polyantha Mill. **False Oxlip**
Syn. *P. x tommasinii* Gren & Godr., *P. x variabilis* Goupil non Bastard *(P. vulgaris x P. veris).*

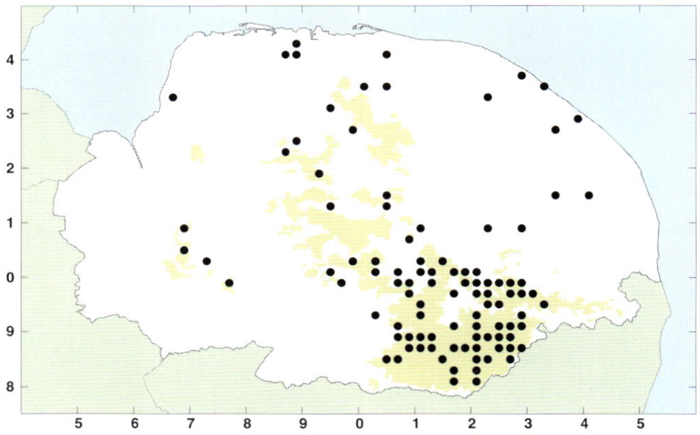

Winter green perennial, usually with or near the parent species. W16. E67.

■ **P. elatior** (L.) Hill **Oxlip**
Semi-winter green, summer dormant perennial reported at Dickleburgh Moor E18 in Galpin's *Flowering Plants of Harleston* (1888). The area was later visited by Miller Christy in 1912 who described a hummock 'about 150 yards in extent, rising to 2.5-3ft above the level of the surrounding boggy meadow'. On it were 'probably less than two score plants (of oxlips) growing in association with about the same number of cowslips and perhaps a dozen hybrids between the two'. In 1961, Dr Woodell, when studying hybrids in *Primula*, visited Dickleburgh. He found that a hummock, similar to that described by Miller Christy, had been ploughed when part of the Moor was reclaimed. The Norfolk Wildlife Trust's Dickleburgh Pightle Reserve now made up the rest of the site. In 1963, a thorough search of this area revealed 10 oxlip plants, but no cowslips or hybrids. The site was characterised by the presence of *Phragmites* which was not on Miller Christy's list of associated species which also suggests that the original site is lost. In the 1970s, PW.Lambley photographed a plant with deep yellow flowers which may have been the hybrid *P. elatior x P. veris*. Dr Woodell, visiting again in the 1980s, found an oxlip- primrose hybrid, but in 1997, a thorough search by ALB failed to find any primulas at all. It seems certain that oxlips are now lost from their only recorded site in Norfolk. More recent records refer to garden escapes. A nationally scarce species.

P. veris L. **Cowslip**

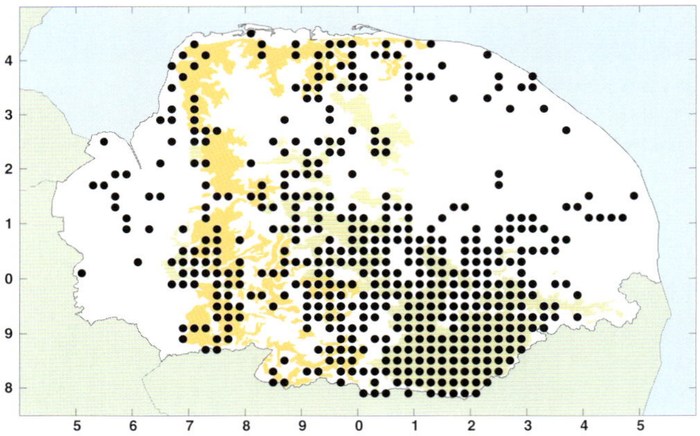

Winter green perennial, semi-dormant in summer, found in neutral to basic, moist and dry grassland and on ditch-sides and banks in reclaimed land, possibly the result of the past importation of chalk for their construction making a suitable seed bed. Also sown beside some recent bypasses.

P. pulverulenta Duthie
Herbaceous perennial introduced to gardens and moist places from the Himalayan region, holding its own as at Thursford W93. W1. E0.

P. japonica A. Gray **Japanese Cowslip**
Herbaceous perennial with the same status as the species above. W1. E0.

Hottonia palustris L. **Water Violet**

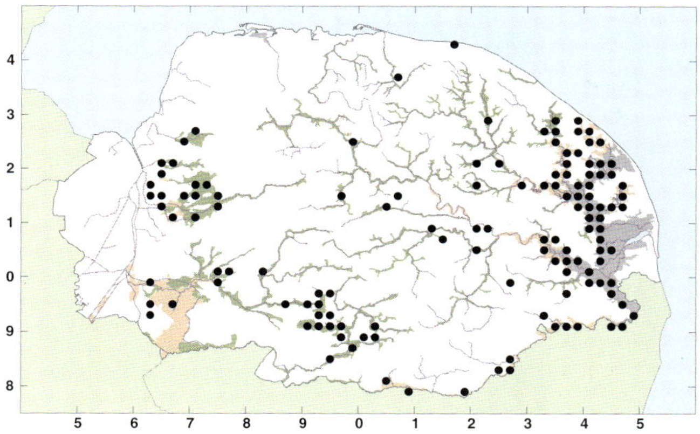

Perennial with condensed, leafy over-wintering shoots. Will survive short-term drying of its water supply, but is eventually killed. Much reduced in numbers in recent years for this reason. Nationally, the plant is almost confined to eastern England.

Hottonia palustris - Thompson

Cyclamen hederifolium Aiton **Cyclamen**

Syn. *C. neapolitanum* Ten.

Summer dormant, autumn-flowering tuberous-rooted perennial introduced to gardens from southern Europe, freely sowing and long persisting particularly in the shrubberies of old rectories and in churchyards. W1. E13.

C. coum

Summer dormant, spring flowering tuberous-rooted perennial introduced to gardens from southern Europe, well established in a shrubbery at Seething E39 in 1996, ALB. W0. E1.

Lysimachia nemorum L. **Yellow Pimpernel**

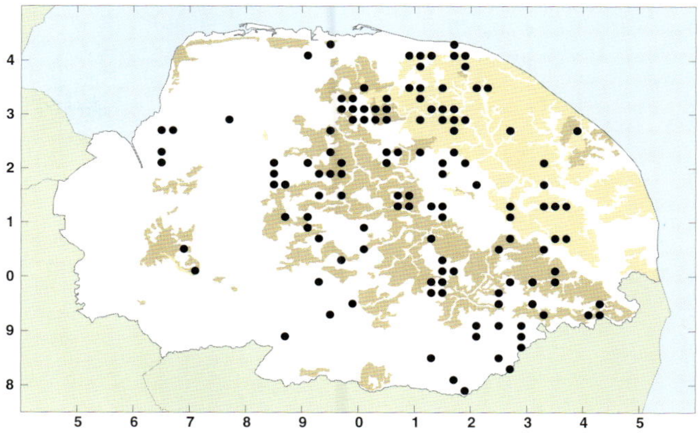

Winter green perennial of woods and moist, shady places.

L. nummularia L. **Creeping Jenny**

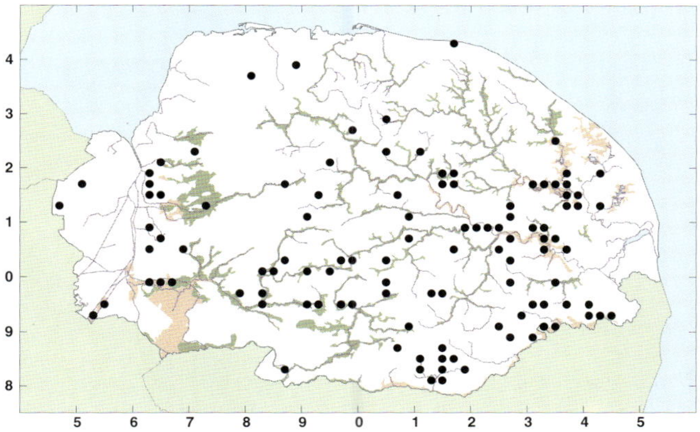

Winter green perennial of damp, shady grasslands, marshes and wet woods. Can occur as a garden escape, but appears native in most of its Norfolk sites.

L. vulgaris L. **Yellow Loosestrife**

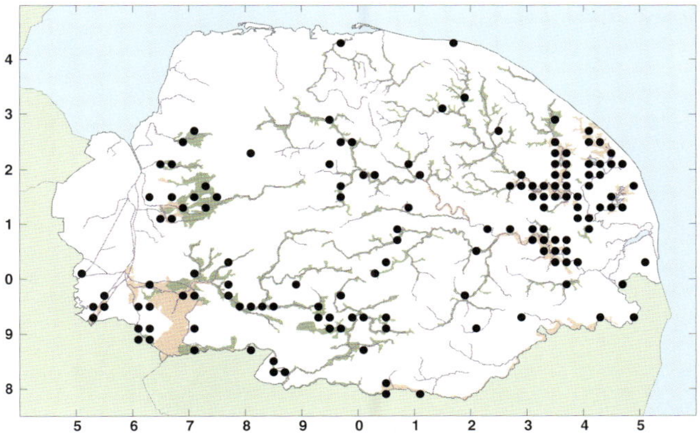

Herbaceous perennial scattered in ditches and streamsides, surprisingly scarce in some areas notably so in the north and west coastal marshes.

L. punctata L. **Dotted Loosestrife**

Herbaceous perennial introduced to gardens from south-east Europe, naturalised in damp soils, presumably from garden throwouts as it rarely self sows. W13. E35.

Anagallis tenella (L.) L. **Bog Pimpernel**

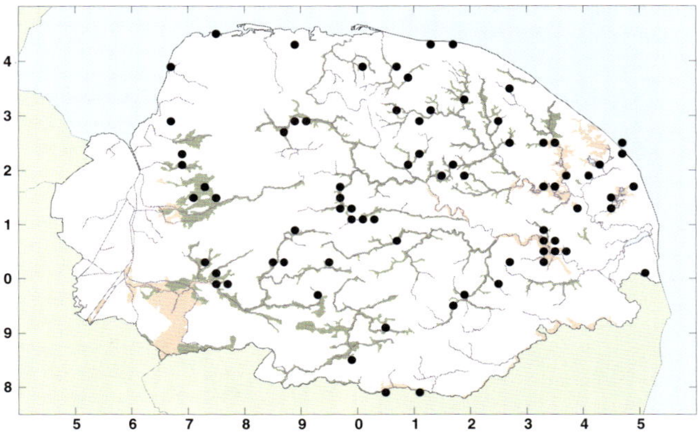

Semi-winter green perennial of open, damp sites on sandy and peaty ground, widely scattered and often abundant where it occurs, but never frequent. Persists abundantly in a mown lawn near one of the springs on the former Poringland Heath, E20.

Samolus valerandi - Brancaster

A. arvensis L. **Scarlet Pimpernel**

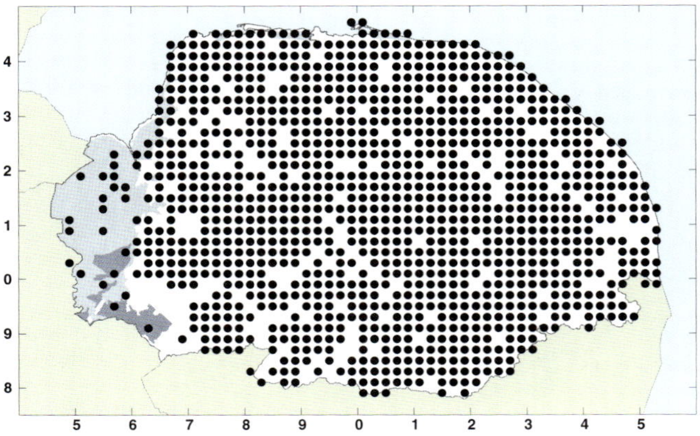

Annual, germinating spring or autumn, still frequent on the margins of arable fields, shingle banks and on waste ground, absent from ground reclaimed during the last two centuries. Ssp. **arvensis** normally has red flowers, but pink, mauve and blue flowered forms, both forma **azurea** and ssp. **foemina** (Mill.) Schinz & Thell. **Blue Pimpernel**, have been found during this survey.

A. minima (L.) E.H.L. Krause **Chaffweed**

Syn. *Centunculus minimus* L.

Annual formerly known from a few sites on damp, sandy ground on greensand. Last seen on Roydon Common W62 in 1975 and Wolferton W62 in 1983, both CPP. Found at Thorpe St Andrews E21 in 1997, SP & S.Henson. W0. E1.

Glaux maritima L. **Sea Milkwort**

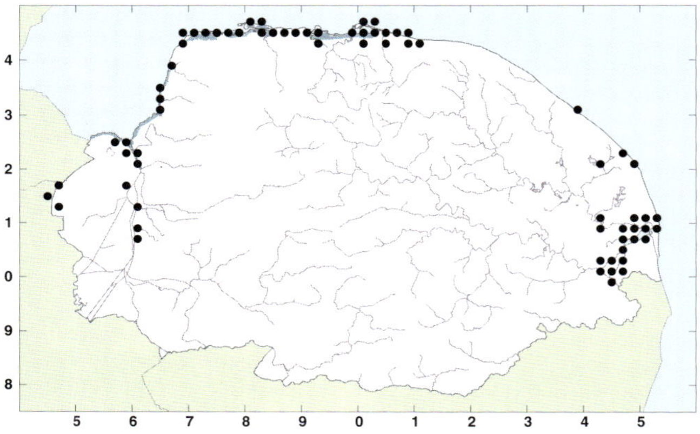

Herbaceous perennial of the sandy edges of salt marshes and damp ground in dunes, penetrating inland along the banks of tidal waters.

Samolus valerandi L. **Brookweed**

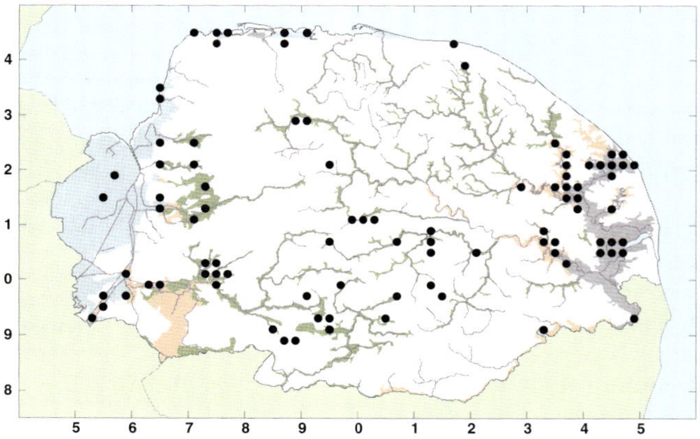

Winter green perennial of damp ground and thin grassland, largely avoiding calcareous soils. In most of Britain it grows within the coastal belt, but not so in Norfolk.

HYDRANGEACEAE

Philadelphus coronarius L. **Mock Orange, Syringa**
Deciduous shrub introduced to gardens from mainland Europe and sometimes found growing far from habitation as at Honing E32 where a three metre bush was found on the edge of an isolated fen in 1990, ALB. W0. E4.

GROSSULARIACEAE

Escallonia macrantha Hook.& Arn. **Escallonia**
Evergreen shrub introduced to gardens from Chile, much planted for hedging near the coast. An occasional relic of cultivation. W1. E2.

Ribes rubrum L. **Red Currant**

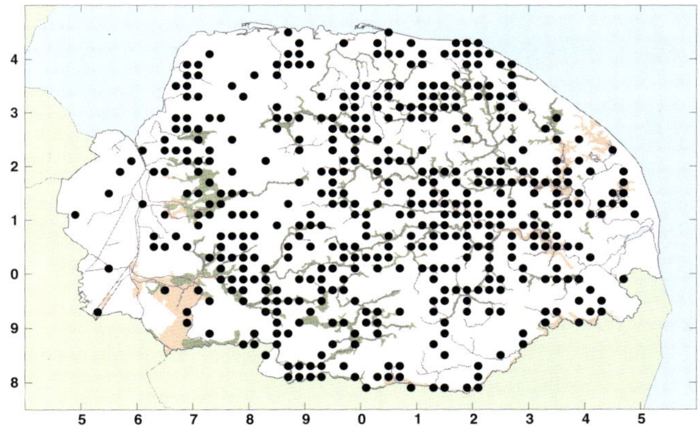

Deciduous shrub, not uncommon in woodland, also sometimes in hedgerows and then perhaps a garden escape.

R. nigrum L. **Black Currant**

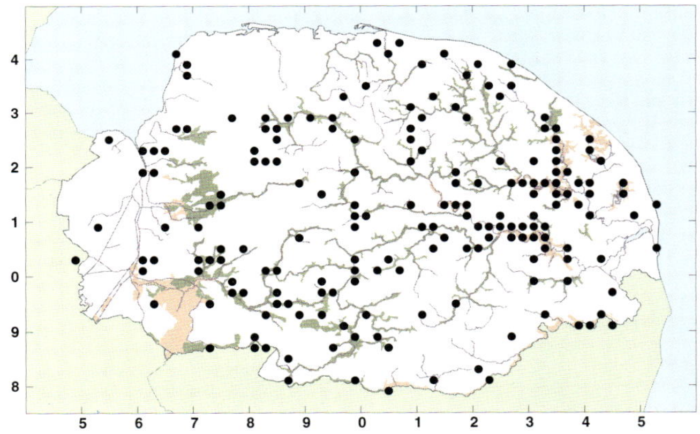

Deciduous shrub, less common than red currant. Truly native in fen carr and wet woodlands and seldom seen outside these habitats, only rarely found as a garden relic.

R. sanguineum Pursh **Flowering Currant**
Deciduous shrub introduced to gardens from North America, sometimes escaping, occasionally planted as at West Lexham W81where there is a 150m long roadside hedge. W26. E64.

R. alpinum L. **Mountain Currant**
Deciduous shrub native to northern Britain. Two bushes occur about 300m apart in hedgerows on either side of Skipping Block Corner, Kimberley E00, ALB. One appears contemporary with the enclosure-type hedge, the second is younger. Although the species is dioecious, the female does sometimes produce a few fruit, so self-seeding is not impossible. W0. E1.

R. uva-crispa L. **Gooseberry**

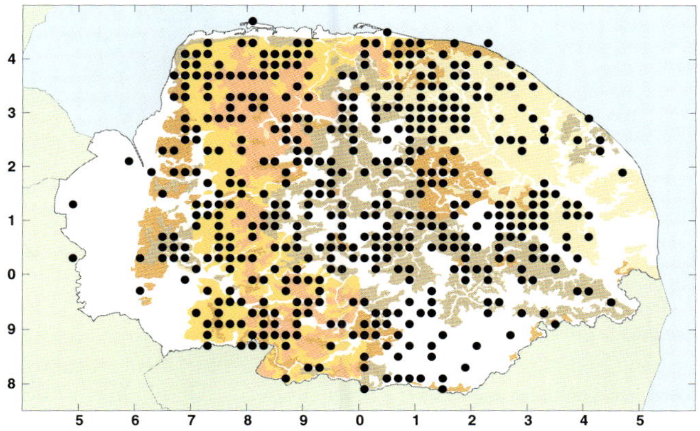

Deciduous shrub widespread in woods, scrub and hedgerows on drier soils. May sometimes be a garden escape.

CRASSULACEAE

Crassula tillaea Lest.-Garl. **Mossy Stonecrop**
Syn. *Tillaea muscosa* L.

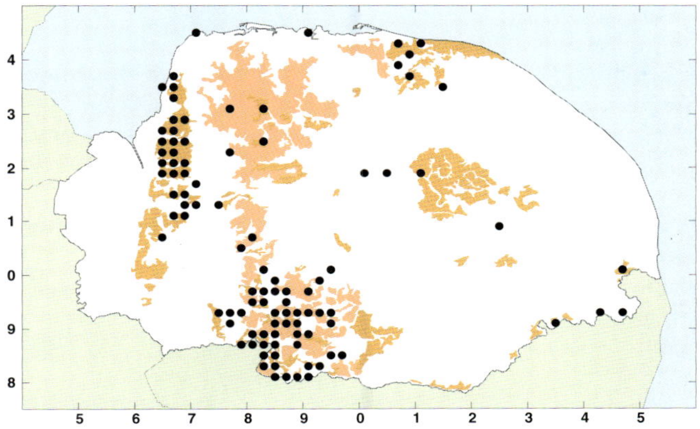

Autumn germinating annual, not infrequent on acid, sandy tracks especially on the greensand of west Norfolk and in Breckland. The Monthly Magazine of April 1824 states, 'Dry heaths in Norfolk and Suffolk and as yet not noticed in any other county in England,' this is not so now, but it is still a nationally scarce species.

Crassula tillaea - Bawsey Pits

C. helmsii (Kirk) Cockayne **New Zealand Pigmyweed**

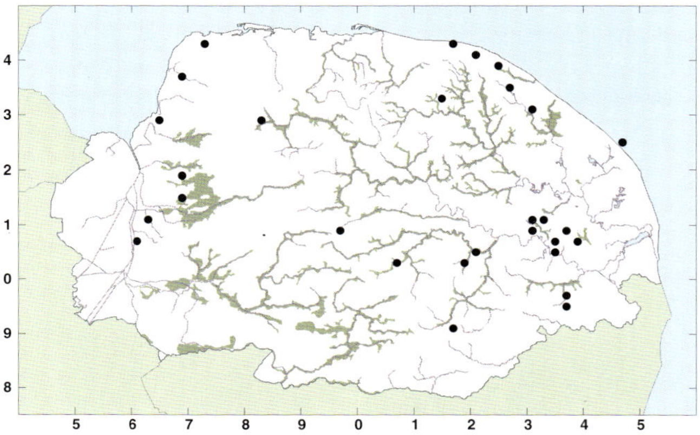

Evergreen perennial introduced as an aquarium oxygenator from Australia and New Zealand, thrown out or deliberately introduced to ponds and ditches, very invasive once it has a hold. At present most of its sites are in isolated ponds.

Umbilicus rupestris (Salisb.) Dandy **Navelwort**
Evergreen perennial native to western Britain. A single plant found in Sisland Carr E39, is believed to have been planted. W0. E1.

Sempervivum tectorum L. **Houseleek**
Evergreen perennial still surviving on house, outhouse and barn roofs, but less frequent than formerly due to re-roofing and generally 'tidying up'. W8. E10.

S. arachnoideum L. **Cobweb Houseleek**
Evergreen perennial introduced to gardens from central Europe, still thriving where it was found growing on a barn roof at Denver W50 in 1991 by RMP, the first British record. W1. E0.

Sedum spectabile Boreau **Butterfly Stonecrop, Ice Plant**
Herbaceous perennial introduced to gardens from China, occasionally thrown out and persisting. Well naturalised at several sites including Narborough W71 in 1995 and Bradfield E50 in 1996. W3. E13.

S. telephium L. **Orpine, Live-long**

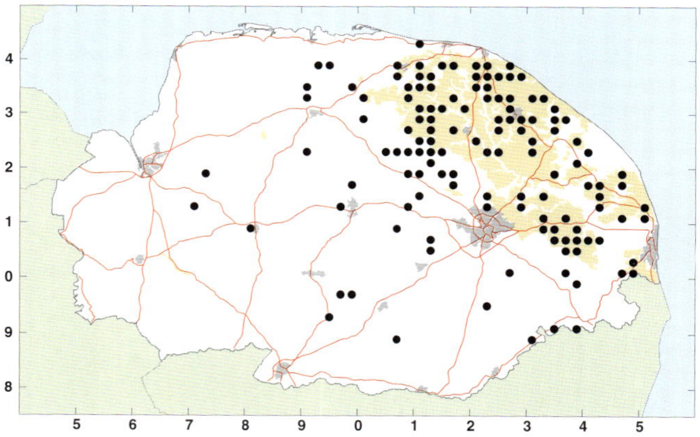

Herbaceous perennial of roadsides and grassy places, now much less common through mis-timed bank cutting, especially on the dry soils of north-west Norfolk. Our plants all appear to be ssp. **telephium**.

S. spurium M. Bieb. **Caucasian Stonecrop**
Evergreen succulent perennial introduced to gardens from the Caucasus, thrown out and sometimes persisting. W11. E23.

S. rupestre L. **Reflexed Stonecrop**
Syn. *S. reflexum* L.

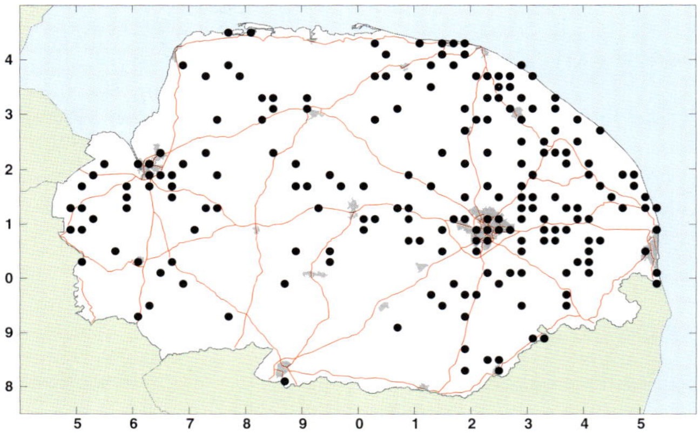

Evergreen succulent perennial introduced to gardens from the Caucasus, escaping and well established, mostly on walls and roadsides. See also comments under the following species.

S. forsterianum Sm. **Rock Stonecrop**
Syn. *S. rupestre* auct. non L..
Evergreen perennial, native to western Britain, introduced and occasionally naturalised in well-drained sites. Because of a past confusion of names, *S. rupestre*, used now for the former *S. reflexum*, is also synonymous with *S. forsterianum*. Swann, 1968, described *S. forsterianum* as 'locally common in hedgerows in north-east Norfolk' but it is worth noting that all but three of the records supporting this were made before 1920 as *S. rupestre*, and in his card idex is written 'care here' in capital letters suggesting he was himself doubtful about them. Because of this it does seem possible that some of these older records may refer to *S. rupestre* (syn. S. *reflexum*), see above. Recent confirmed records of *S. forsterianum* include Great Bircham churchyard, W73 KAB, and a roadside bank at Suffield E23 in 1997, ALB. W1. E6.

S. acre L. **Biting Stonecrop**

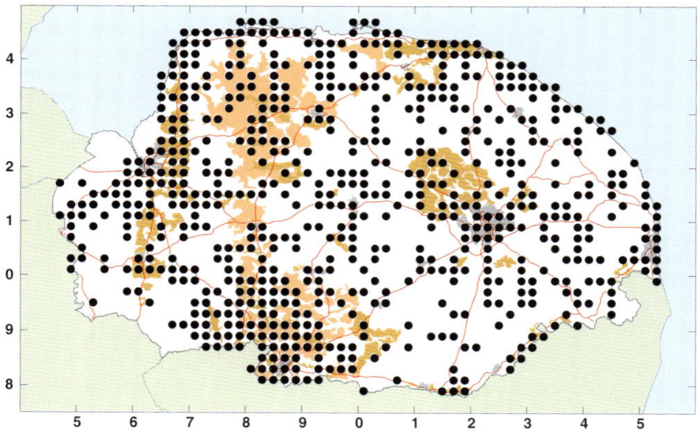

Evergreen, succulent perennial frequent in sandy places, short grassland and on walls.

S. album L. **White Stonecrop**

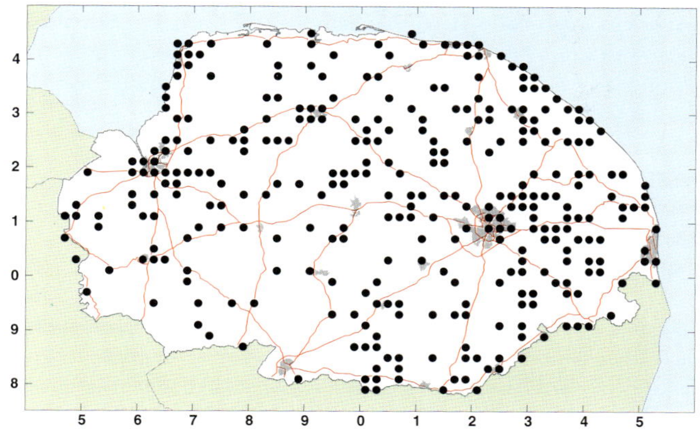

Evergreen, succulent perennial possibly native in western Britain but introduced to gardens and becoming established on walls and roadsides, particularly over concrete as on former air strips. Found in this habitat for example at Thorpe Abbots E18, where a strip a metre wide and several hundred metres long was observed in 1991, ALB, also in a similar but unusually shady site at Honeypot Wood, W91.

S. anglicum Huds. **English Stonecrop**

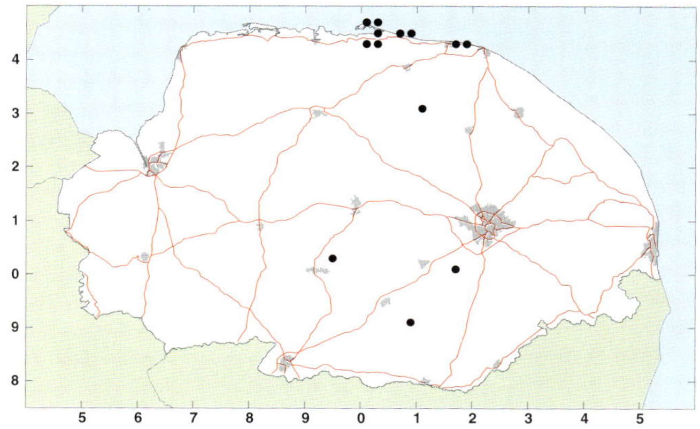

Annual or short-lived perennial of coastal cliffs and dunes, first recorded at Caister-on-Sea E51 in 1793 and last seen there in 1933, EAE. During this survey reported from a few sites between Blakeney E04 and Sheringham E14 but now becoming rare. Now also cultivated and other records are probably garden escapes.

S. dasyphyllum L. **Thick-leaved Stonecrop**
Evergreen succulent perennial introduced to gardens from mainland Europe and occasionally escaping as on walls in Norwich E20 and in Dersingham W63. W1. E4.

SAXIFRAGACEAE

Bergenia cordifolia (Haw.) Sternb. **Elephant Ears**
Evergreen perennial from Siberia and Mongolia, introduced to gardens and occasionally found on waste ground as at Wymondham tip E10 in 1978. Always the result of deliberate planting or the dumping of garden rubbish; very tenacious in old, abandoned gardens. W0. E1.

B. crassifolia (L.) Fritsch **Elephants' Ears**
Evergreen perennial introduced to gardens from Siberia and Mongolia, sometimes thrown out and very persistent when established. Seen at Saham Toney W90, and Whissonsett W92. W2. E0.

Saxifraga cymbalaria L. **Celandine Saxifrage**
Annual introduced to gardens from the eastern Mediterranean, germinating autumn and spring whenever weather is suitable. Established at Ovington and Saham Toney W90 where it has been since at least 1961. W2. E0.

S. x urbium D.Webb **London Pride**
(S. spathularis x S. umbrosa)
A perennial evergreen hybrid of garden origin, occasionally established on walls and waste ground. Recorded at Salthouse E04 and Overstrand E 24. W0. E2.

S. granulata L. **Meadow Saxifrage**

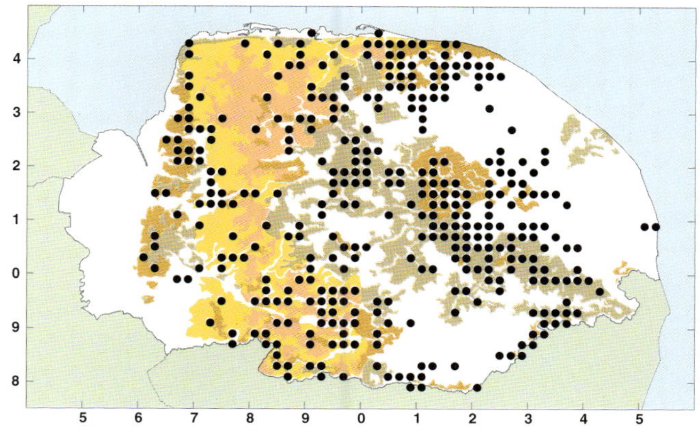

Winter green perennial, dormant in summer, occurring in established grassland. A feature of the short turf of some churchyards as at Attlebridge E11.

S. tridactylites L. **Rue-leaved Saxifrage**

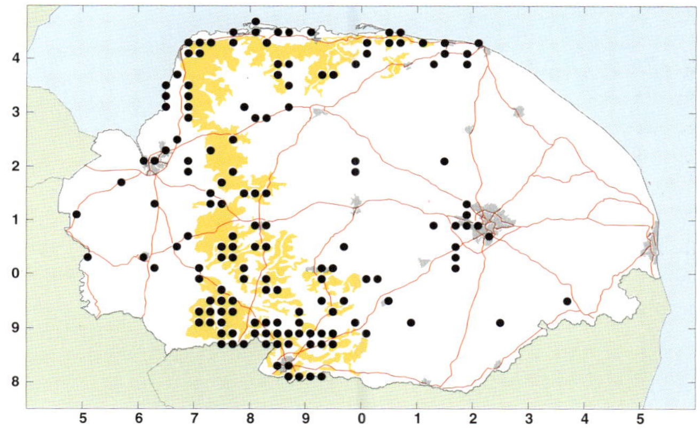

Autumn germinating annual, growing, sometimes in quantity, on bare sandy soils especially where the substrate is firm, frequently colonising abandoned concrete tracks, also on walls.

Tellima grandiflora (Pursh) Douglas ex Lindl.
Evergreen perennial introduced to gardens from North America, occasionally naturalised in parks, woodland and on roadsides. Recorded at Hunworth E03 in 1975, M.Wood and again in 1997 KKH, also in woodland at Barwick W83 in 1990, P.Tomlinson. W2. E3.

Chrysosplenium oppositifolium L. **Opposite-leaved Golden Saxifrage**

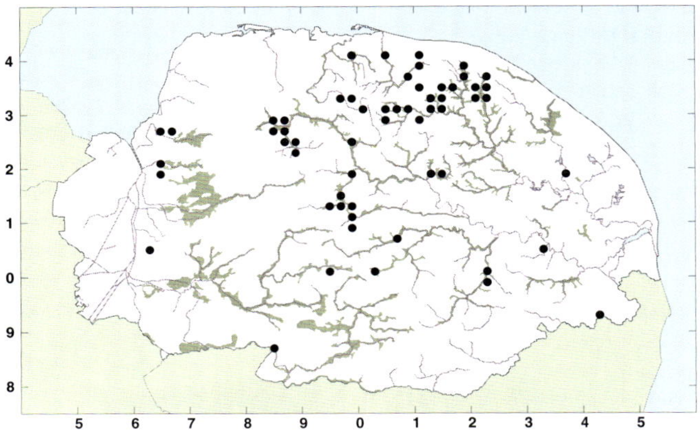

Evergreen perennial of moist, acid woods and streamsides, most frequent around the headwaters and upper tributaries of the Wensum, Bure and Yare. Introduced to a spring-fed pool and overflow at Santon W88 in the late 1940s and still thriving, also to the carr at Wheatfen E30 where it is spreading.

C. alternifolium L. **Alternate-leaved Golden Saxifrage**

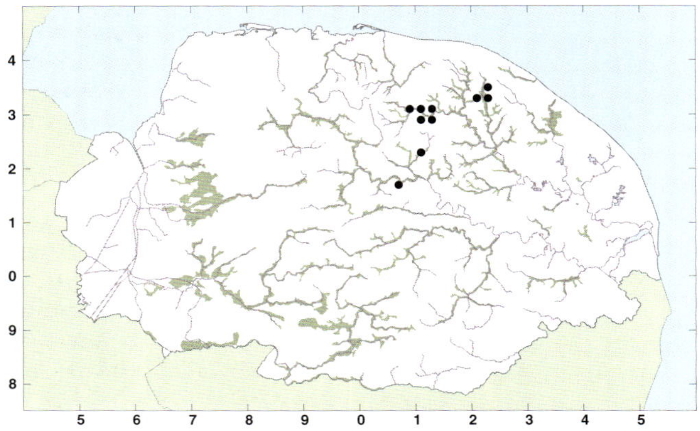

Evergreen perennial of moist woods and streamsides, much more restricted in distribution than *C. oppositifolium*.

Parnassia palustris L. **Grass of Parnassus**

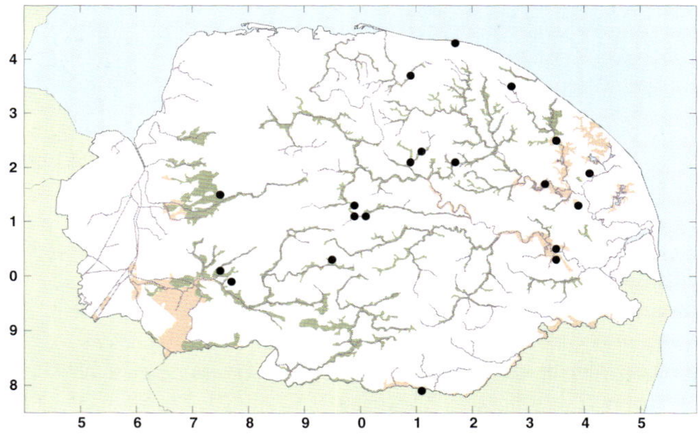

Herbaceous perennial, now scarce, needing a permanently moist site in grassland and marshes.

ROSACEAE

Sorbaria sorbifolia (L.) A. Braun **Sorbaria**
Deciduous shrub from Siberia, introduced to gardens and occasionally escaping. Planted more than 50 years ago at Thursford W93 and flourishing. W1. E0.

Spiraea
Although six species and hybrids have been recorded, a further 19 sites have not been identified beyond generic level. The whole genus needs more careful recording in the county.

Spiraea salicifolia L. **Bridewort**
Deciduous shrub introduced to gardens from North America, occasionally established in moist, acid woods and hedgerows as at Stanford W89 and Wood Green, Long Stratton, E29, but probably over-recorded in the past for *S. x pseudosalicifolia*. W3. E1.

S. x rosalba Dippel **Intermediate Bridewort**
(S. salicifolia x S. alba)
Deciduous shrub more frequently grown than its parents. Occasionally found as an escape from gardens as at Mundford W89. W1. E0.

S. x pseudosalicifolia Silverside **Confused Bridewort**
Syn. *S. x billardii* auct. non Herincq *(S. douglasii x S. salicifolia)*
Deciduous shrub, a garden hybrid well established in woodland at Hargham E09; Leziate W61 and near Snetterton E09. W1. E2.

S. douglasii Hook. **Steeple Bush**
Deciduous shrub introduced to gardens from North America, naturalised and surviving, especially in urban sites as at Swaffham W80. W1. E2.

S. tomentosa L. **Hardhack**
Deciduous shrub introduced to gardens from North America, recorded at Attleborough E09. W0. E1.

S. x vanhouttei (Briot) Carrière **Van Houtte's Spiraea**
(S. cantonensis x S. trilobata)
Deciduous garden shrub, occasionally escaping. Recorded from Watton airfield W90. W1. E0.

Filipendula vulgaris Moench **Dropwort**

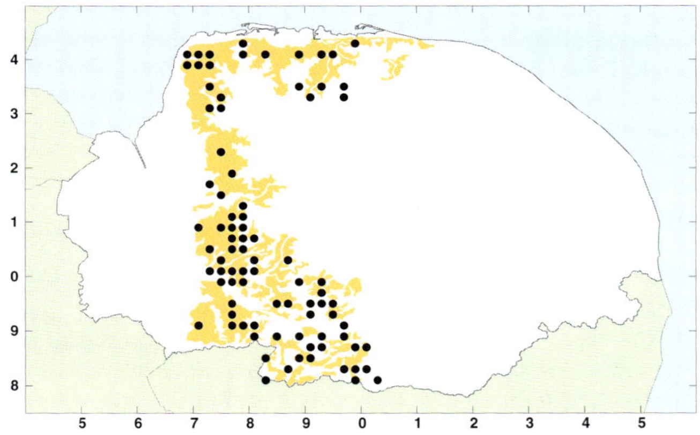

Winter green perennial confined to calcareous grassland, extending eastwards to East Harling W98 and the Lophams E08.

F. ulmaria (L.) Maxim. **Meadowsweet**

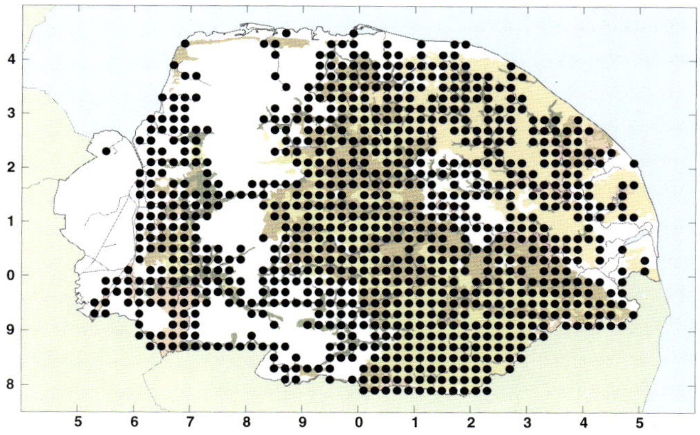

Herbaceous perennial of river, stream and dyke banks, marshes and any permanently wet site apart from bogs. Largely absent from the Fenland apart from the Welney Washes W59. Planted in W52.

Kerria japonica (L.) DC. **Kerria, Batchelors' Buttons**
Deciduous shrub from Japan usually grown in its double form, **'Pleniflora'** (Syn. 'Florapleno'). Found in scattered sites where garden rubbish has been dumped. W2. E3.

RUBUS

(by A.L.Bull)

A much neglected genus except in the autumn when the fruit are ripe. Blackberry gatherers all agree that some sorts have fruit which are excellent in pies while others are better for bramble jelly and some are worthless for either. They also seem to know at a glance which plants fall into which category, something which seems to have baffled and eluded all but a very few botanists down the ages. Norfolk has been poorly served since the genus first started to receive critical attention following the realisation some two hundred years ago that many clearly deliniated taxa could be found even within a relatively small area. Thus the study of literature relating to Rubus in the historical sense within the county is restricted to the ten years 1879-1889 during which time the Rev.E.F.Linton was rector at Sprowston. While many counties have benefitted by visits from such as the Rev. Augustine Ley in the late 19th and early 20th centuries and some of his contemporaries including the Rev W.Moyle-Rogers, followed later in the 20th century by such batalogical stalwarts as Linnaeus Cumming, W.C.R.Watson, W.C.Barton, H.J.Riddlesdell and B.A.Miles, Norfolk was not so fortunate. The county had to be content with a fleeting visit by Riddlesdell in 1925 and some records from E.S.Edees during the 1950s while visiting relations near Swaffham, an area amongst the poorest in the county for brambles.

The following account must be regarded as the first definitive study of the brambles in the county as a whole. The present author's interest was aroused by the publication of Petch and Swann's *Flora of Norfolk* which quoted Linton's gatherings for almost all taxa. Thirty years later there is still much ground to be covered, but at least a reasonable understanding of distributions is now possible and as a result of thirty years of effort, six new names have been added to the British list, three of which have already been found to be of regional or even national significance.

While it has obviously not been possible for one person to visit every tetrad, an effort has been made to visit all the potential important sites in West Norfolk while East Norfolk has mainly been covered during the course of ordinary recording, special visits having been made where it was felt that such visits were desirable. This being the case, it is likely that there are actually more good sites for Rubus in East Norfolk that have not been visited than there are in the West, given that East Norfolk has much more of the acid gravels preferred by brambles generally.

Apart from *Rubus caesius* and *R. ulmifolius*, all Rubus species are apomictic, producing seed without pollination, a characteristic shared by *Hieracium* and *Taraxacum*. Even so, interesting possible hybrids do occur and, following apomictic characteristics thereafter, can then develop into good microspecies in their own right.

Classification follows *Brambles of the British Isles* Edees & Newton, 1988.

<u>Genus Rubus</u>. L.
<u>Sub-genus 4. Idaeobatus</u> Focke.

Rubus idaeus L. **Raspberry**

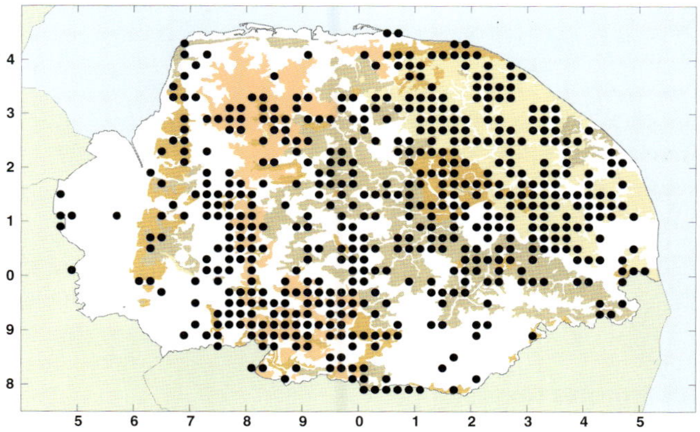

As with all members of the genus, the perennial roots give rise to biennial stems which produce only leaves during the first year and are known as primocanes. Flowers and fruits are produced in the second year and the stems then die. The present species is native on damp heaths and commons, especially around the margins of fens and boggy areas in the Broads and elsewhere. Distribution of native stock may be confused in some areas due to naturalisation of root stocks discarded from gardens and allotments or from seed.

<u>Sub-genus 5</u>.
<u>Section A. Rubus: Sub-section 1. Rubus</u>
<u>Sub-section 1</u>. is characterised by its affinity with *R. idaeus*. Stems suberect, not rooting at the tips, often suckering from the roots. The lack of tip-rooting restricts the spread of members of the group formerly known as the Suberecti. Because of this the presence of any of the following is often taken as the indicator of an undisturbed site.

R. arrheniiformis W.C.R.Wats.
A plant of damp commons on heathy soils which appears to have declined somewhat since work during the 1970s and early 1980s (Bull 1985) though new sites have been found during the present survey. It should be remembered, however, that all Rubus tend to disappear when sites become overgrown, only to reappear with renewed vigour following clearances. W1. E5.

■ **R. bertramii** G.Braun
A specimen in the herbarium of the South London Botanical Institute determined as this by BA.Miles was collected by EF.Linton at Beeston St.Andrew E21 in 1886.

R. fissus Lindley
Rare. Recorded by EF.Linton under *R. rogersii* at Westwick E22 and in the current survey at Southrepps Warren E23 and Honing Common E32. W0. E2.

R. nessensis W.Hall

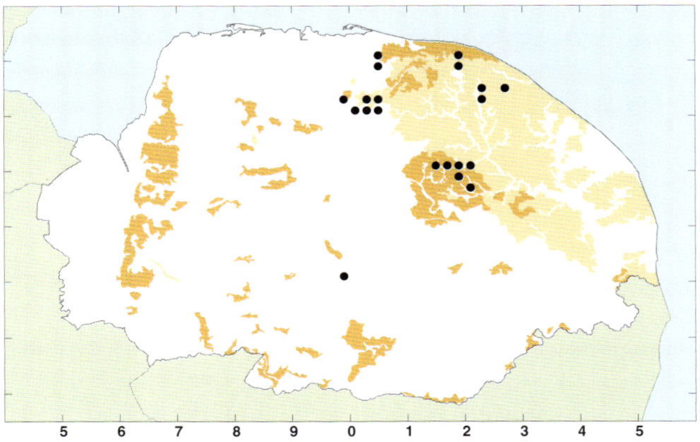

One of the best indicators of ancient woodland sites, confined in Norfolk mainly to such woods in the central-north of the county. At Swanton Novers Little Wood E03, the primocanes often grow to a height of 3m.

R. nitidiformis Sudre
Nationally scarce and at its most northerly in this county. It occurs at Billingford Wood E18; Westwick E22 and Skeyton E22 where it is at its best along some old meadows by the Stake Bridge Beck. W0. E3.

R. pergratus Blanchard
North American alien. This ally of *R. nessensis* is something of a curiousity with long narrow tips to its large leaves and a robust, deeply furrowed stem without prickles. Frequent in Common Plantation, Aylmerton E13. W0. E1.

R. plicatus Weihe & Nees.

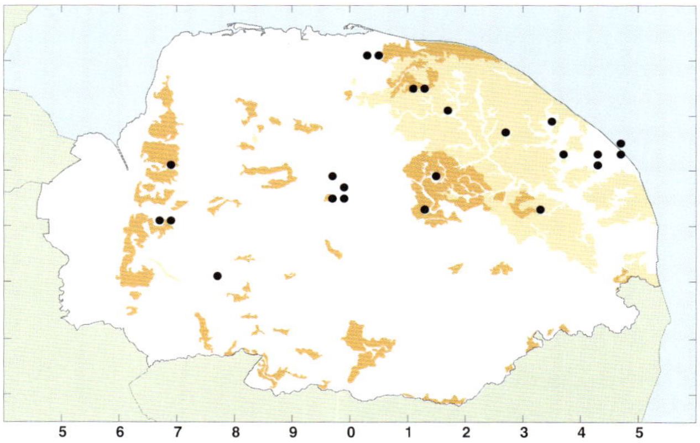

The most widespread member of the group, not infrequent on heaths and along the margins of fens. The flowers are usually white, but a deep-pink flowered variant, similar to the one that is frequent in Scotland, occurs at Sugar Fen, Leziate W62.

R. sulcatus Vest.
A rare plant nationally. In Norfolk it occurs only in Barney Wood, W93 and Hevingham Park woodlands E12. W1. E1.

R. vigorosus Mueller & Wirtgen

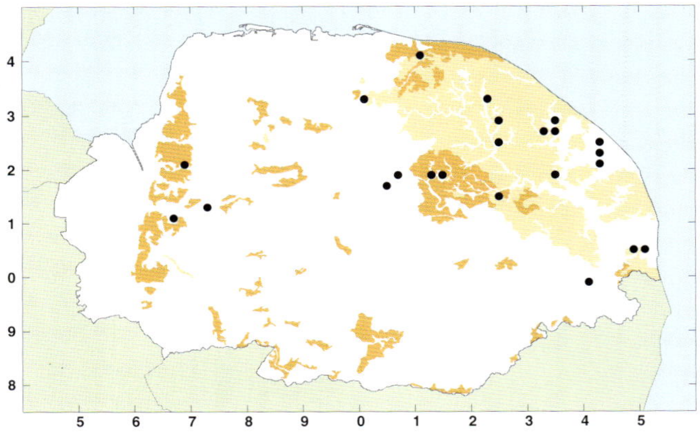

Scattered through Norfolk on damp heathland, though mainly in the East.

<u>Subsection. Hyemales</u> E.H.L.Krause. Plants able to proliferate by tip-rooting as well as seed.
<u>Series 1. Sylvatici</u> (Mueller) Focke. Stalked glands absent or rare and insignificant, leaflets green on both surfaces, sometimes softly-hairy below but never felted.

R. adspersus Weihe ex. H.E.Weber

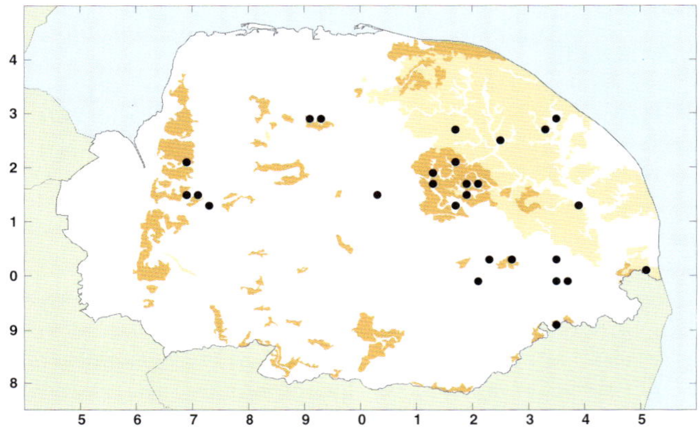

A plant of damp commons and heaths, sometimes frequent where these are still undisturbed. Scattered thinly throughout the county.

R. crudelis W.C.R.Wats.
Very rare. Restricted to a single known site in a green lane at Saxthorpe E13. W0. E1.

R. gariannensis A.L.Bull

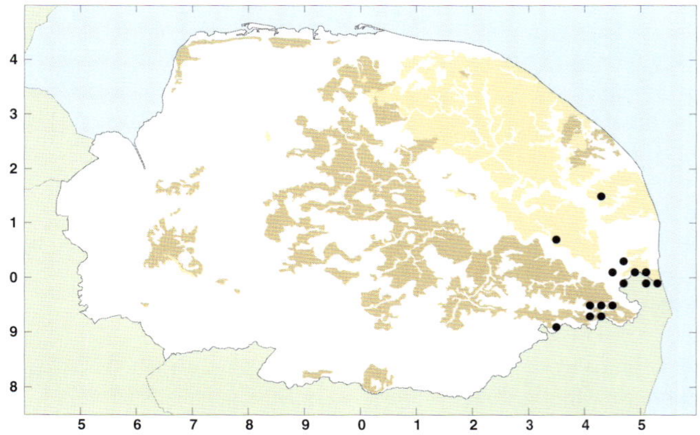

A plant of heaths and hedgebanks in north-east Suffolk and south-east Norfolk, also occurring on flood banks in the Waveney Valley and in at least one spot out on the grazing levels. Almost the only bramble found there except for *R. ulmifolius*; also found in one or two isolated wooded field margins. The holotype was taken from Howards Common, Belton E40 and is in the BM.

R. gratus Focke

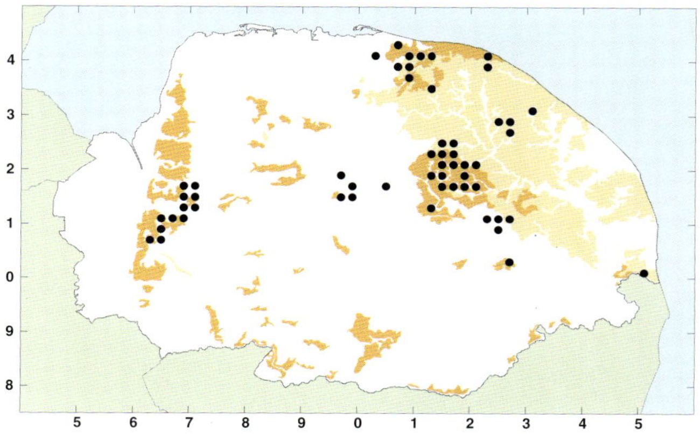

This plant with crab apple-like flowers and hairy anthers is typical of the margins of heaths and on former heathland sites.

R. laciniatus Willd.
Differentiated from all other British species by the dissected leaflets and fringed petals. In Norfolk it occurs where bird sown around woodland margins, or as a relic of cultivation, especially near urban areas such as Norwich E20 and Yarmouth E50. Not known for certain in the wild. W3. E20.

R. leptothyrsos G.Braun
One of the Norfolk plants of dubious origin occurring only in and around plantations on the Sandringham Estate, W62. Widespread only in Scotland. (See Bull 1997). W5. E0.

R. lindleianus Lees

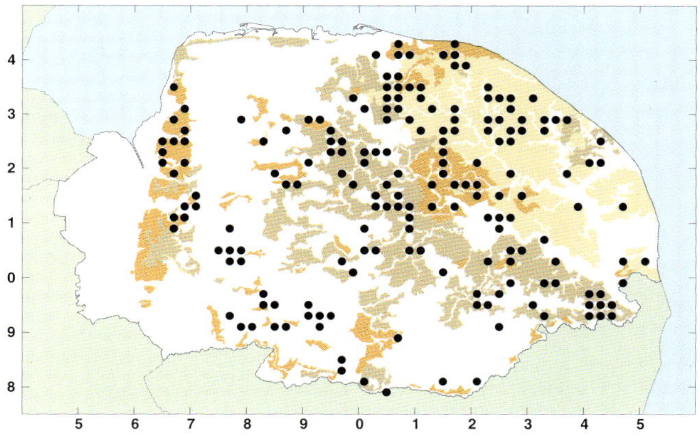

Frequent in many parts of the county, even occurring from time to time on the clay.

R. macrophyllus Weihe & Nees

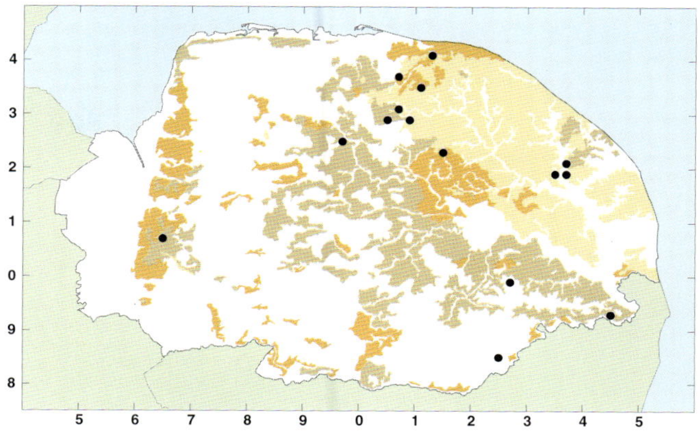

A scarce plant with us, almost confined to the sites of ancient woodland as at Hindolveston E02.

R. platyacanthus Mueller & Lef.

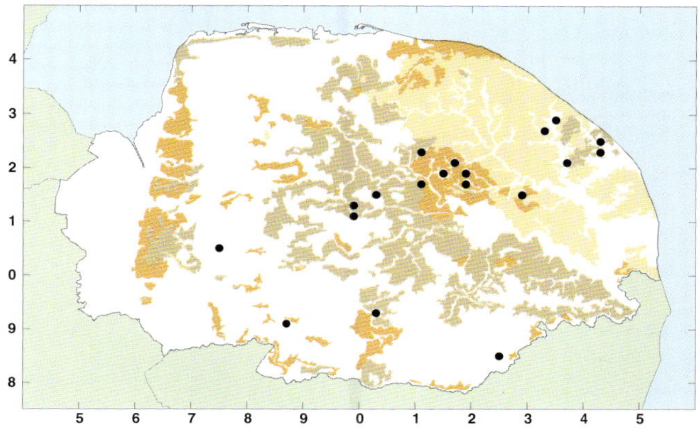

Rather scarce on heaths, fen margins, wood margins and hedgebanks. A strange plant collected at Scarning Fen W91 was described by A.Newton as being uncannily like a hybrid between this plant and *R. insectifolius* even though it was fully fertile. Both putative parents were present.

R. poliodes W.C.R.Wats.

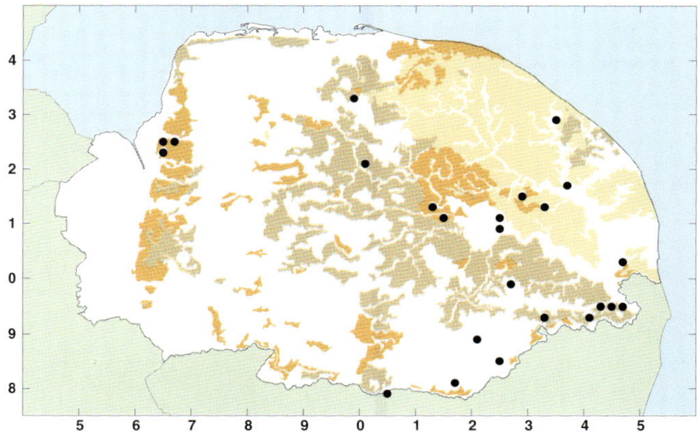

A regional endemic at its most northerly in Norfolk. Scattered through heathland sites to the north of Norwich and in slightly acid woodlands on the clay above the Waveney Valley. Also on one or two heathland sites in the far west of the county.

R. pyramidalis Kalt.

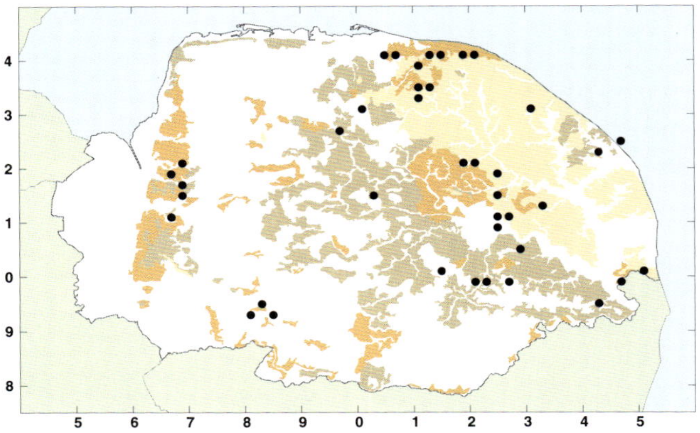

Scattered through the county. The 'dwarf' form which is much smaller in all its parts, var. **parviflorus**, has been found at Roman Camp, Aylmerton E14.

Series 2. Rhamnifolii (Bab.) Focke. Similar to Sylvatici, but leaves usually felted beneath.

R. amplificatus Lees in Steele

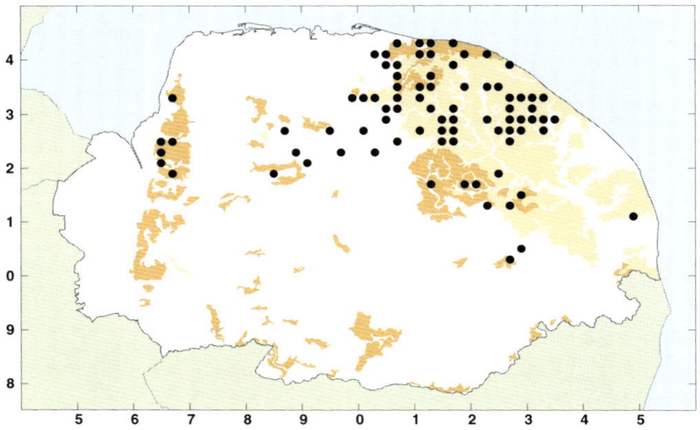

This plant seems to be typical of the acid, heathy ground across north Norfolk and is abundant and even dominant in suitable areas such as parts of the Cromer ridge. Not apparently penetrating as far south as such potentially suitable sites as Mousehold Heath E21.

R. boudiccae A.L.Bull & E.S.Edees

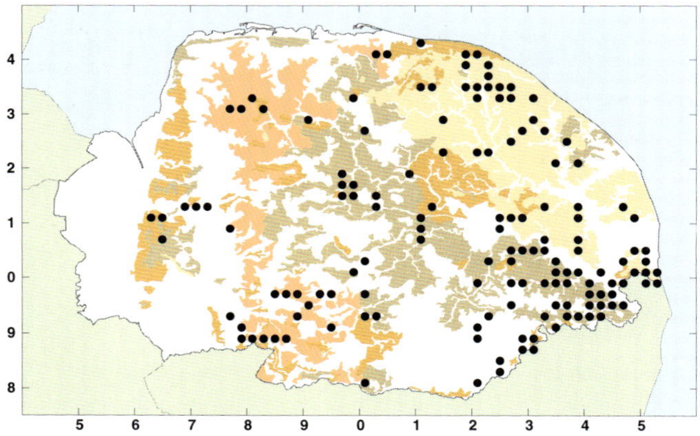

A thorny problem faced with fortitude by Bull & Edees and finally awarded to the Queen of the Iceni! From being known to be frequent in East Anglia, the plant has now been found as far afield as Warwickshire, Ireland, Anglesey and the Scottish Island of Aran. Edees collected the holotype on Ringland Hills E11, it is in the BM.

Rubus boudiccae - Stanta

R. cardiophyllus Lef. & Mueller
This is the plant which was confused with the much more frequent and at that time unnamed *R. boudiccae*. Rare. Alderford Common E11; Mousehold Heath E21; Abel Heath, Blickling E12; Crostwight Common E32; Litcham W92 and Holme W74. W2. E5.

R. cissburiensis W.C.Barton & Riddelsd.
First recorded in 1970 at Colney Hall Wood E10 but now shaded out, one of the southern English dominants that scarcely occurs north of the Thames. It has been found more recently in small quantity at Shotesham E29 and Pedham E31. W0. E2.

R. elegantispinosus (A.Schum.) H.E.Weber
A continental plant introduced into gardens, bird sown at Rushall, E18 and Furze Covert E28 in the same parish where it is dominant. One bush in Earlham Cemetary E20, also bird sown. The Rushall site may have been seeded from a similar site across the Waveney at Hoxne in Suffolk where the plant is also dominant and appears to have been longer established. W0. E3.

R. incurvatiformis E.S.Edees
This species and the next probably arrived in Norfolk by way of migrating birds. The present plant is a regional endemic of the west Midlands and on the west-facing slopes of the Pennines, so the discovery of a single bush beside Suffield Road at the south-western corner of Gunton Park woodlands, E23 in 1997 caused great surprise. W0. E1.

R. incurvatus Bab.
As with the previous species, mainly western and most frequent in North Wales and with a single locality in Denmark. This, paradoxically, could have been the origin of the seed which started a colony on the top of Bard Hill, Salthouse E04, a likely first stop for a flock of fieldfares newly arrived from across the North Sea. W0. E1.

R. londinensis (Rogers) W.C.R.Wats.
Only recently recognised as a Norfolk plant in a white-flowered form, this south-eastern endemic occurs at Hockering Wood, E01; Stody E03 and Bear's Grove, Rackheath E21 also at Swannington Upgate E11 where the flowers are pale pink. The deep pink flowered plant which Watson originally named is very scarce north of the Thames. W0. E4.

R. nemoralis Mueller

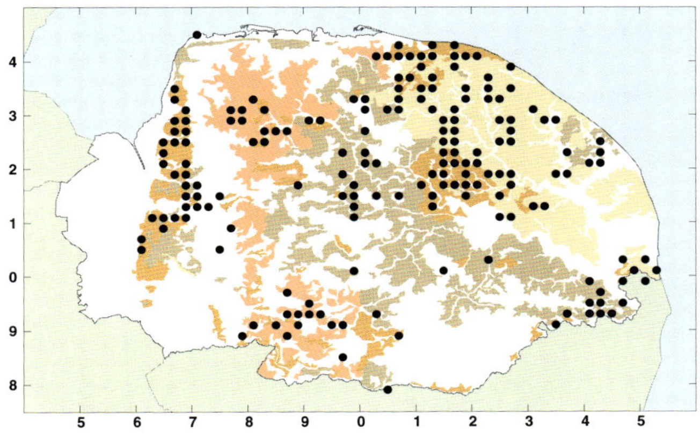

The common heathland blackberry whose strong, hooked prickles defend the often large and thimble- shaped fruits from the attentions of blackberry pickers. The showy pink flowers with notched petals are also worth more than a glance.

R. pistoris W.C.Barton & Riddelsd.
Several bushes were found at Horsford Woods E11 in 1998, ALB & A.Newton. This is the first official county record, but a specimen labelled *R. lentiginosus* in Herb. BM. collected in the adjacent Felthorpe Woods E11 in 1921 by ES.Todd has been found to be this species. W0. E1.

R. polyanthemus Lindeb.

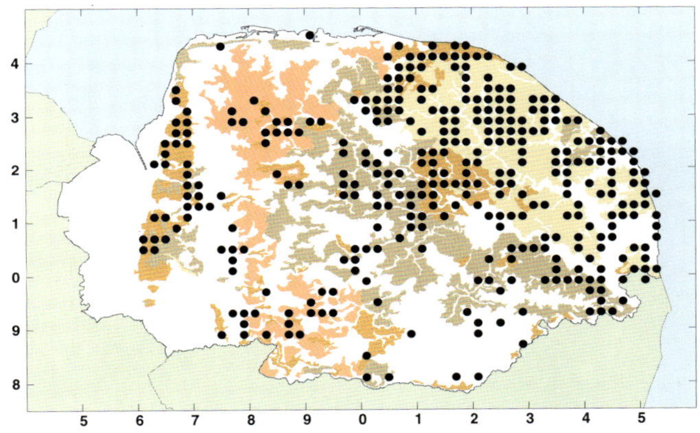

One of the two commonest species in the county after *R. ulmifolius* and one that can be found in almost every type of habitat. In some areas, such as parts of the Broads, it is even commoner than *R. ulmifolius.*

[R. rhombifolius Weihe ex Boenn.
A dot on the map in Edees & Newton 1988 for this plant south of King's Lynn has been investigated and its origin cannot be traced, so must be presumed to be an error. A.Newton pers.comm.**]**

R. septentrionalis W.C.R.Wats.
Another Scottish tree-planting alien which is abundant throughout all Forest Enterprise's plantings at Shouldham Warren W71 with a former single site at Weeting W79. There is evidence that some bird sowing is taking place outside the area of primary colonisation, a plant having been found at Beechamwell Cross W70 in 1997.
W9. E0.

[R. subinermoides Druce
Recorded in Edees and Newton at Ringland Hills E11. This was later proved to be an error, the plant concerned was found to be a shade-grown plant of *R. poliodes.***]**

Series 3. Sprengeliani Focke. Stems low arching and procumbent, stalked glands and acicles usually present, stamens shorter than the styles.

R. sprengelii Weihe

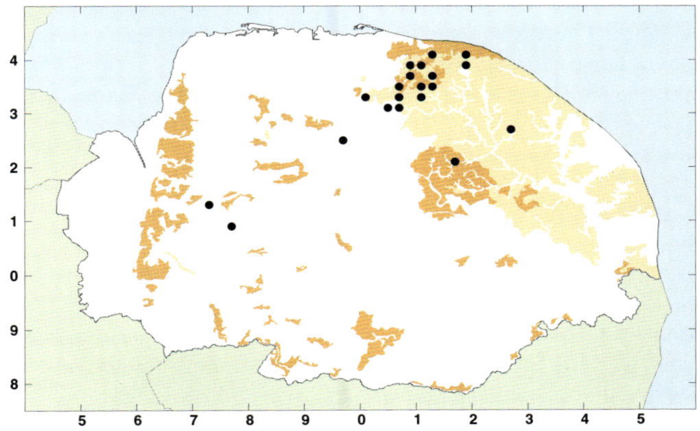

Apart from a population on the gravels around Narborough W71, this plant is almost confined to the Cromer glacial moraine. The pretty pink, starry flowers carried at very little above ground level are very characteristic, though there is a small, white-flowered population at Pond Hills, Hempstead E13.

Series 4. Discolores (Mueller) Focke. Armature strong. Stalked glands absent. Stems frequently waxy pruinose. Leaves coriaceous, white-felted beneath.

R. anglocandicans A.Newton.

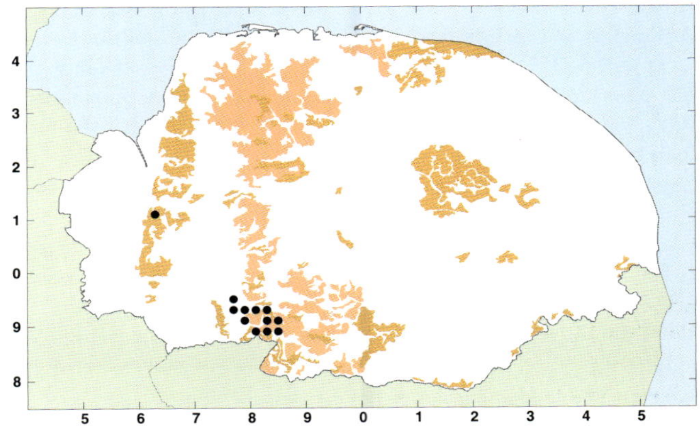

A midlands endemic which is not infrequent in the western part of Breckland and extends roughly from W.Tofts W89 to the greensand at Tottenhill W61.

■ **R. armipotens** W.C.Barton ex. A.Newton
This south-eastern endemic is a forest planting introduction which may have been shaded out from both the sites from which it was known in the 1970s, at Emily's Wood, Weeting W78 and Colney Hall Wood E10. It could well re-appear when the sites are cleared.

R. armeniacus Focke
Syn. *R. procerus* auct.brit. non P.J.Mueller ex Boulay
The 'Himalayan Giant' of horticulture, with massively spreading primocanes and large, insipid fruit which is becoming far too frequent in moist soils, especially south of King's Lynn and on some commons in east Norfolk. The plant was originally discovered growing in a neighbour's garden in Germany by the man for whom it was given its cultivar name, 'Theodore Reimers' in 1889. He grew it on from seed and it has been suggested that stock was then sent from Germany to nurseries in the USA where it was propagated in quantity and returned to Europe as 'Himalayan Giant'. W29. E97.

R. ulmifolius Schott.
The ubiquitous blackberry which has the best flavoured fruit for culinary purposes, though some find it too pippy. It is rare in woodland as it gets shaded out more quickly than most species. The beautiful double-flowered cultivar 'Bellidiflorus' has been seen on St John's Hill, Mousehold Heath E21. EA.Ellis had both this and an albino plant in cultivation, the prickles of which were yellowish instead of purple and the ripe fruit amber-coloured. W221. E488.

R. winteri P.J.Mueller ex Focke
Native in parts of the midlands and south-west. Its presence at Ketteringham E10 may be entirely natural, but the plant is occasionally cultivated having large fruits, so it could easily be an escape here though well established. W0. E1.

Series 5. Vestiti Focke & Focke. Conspicuously hairy with stalked glands and acicles present on the primocanes and more frequent among the densely pilose rhachis hairs.

[R. adscitus Genev.
This species was recorded by EF.Linton from Sprowston E21 under the synonym of *R. hypoleucos*. Further research has revealed that all such plants found during this survey belong to an undescribed taxon which looks superficially like *R. adscitus* when in full flower. The stems are more robust and more glandular, the leaves are similar in shape but much more irregularly toothed and are slightly plicate and undulate at the margins. The panicle branching, flower shape and colour are all like *R. adscitus* but the rhachis and pedicels are thicker and coarser and occasionally the flowers are of a markedly deeper pink with red styles. The present plant grows in more than seven 10km squares scattered between Barney Wood W93; Catfield E32 and Brooke Wood E29, with no doubt other populations awaiting discovery. A.Newton considers that the present plant should be placed in Series 7, Micantes.**]** W1. E1.

R. boraeanus Genev.

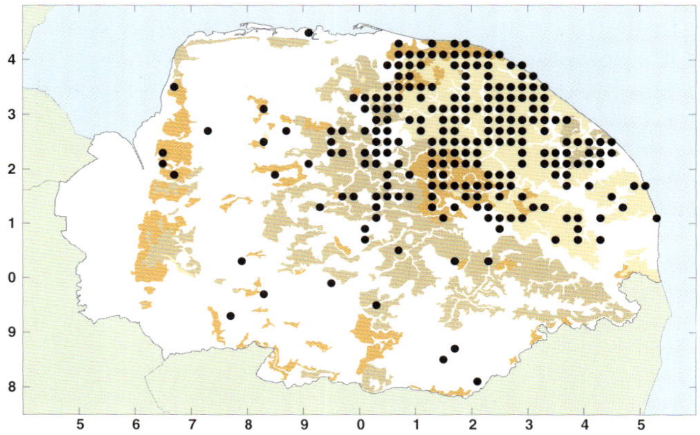

Thinly scattered across southern Britain and in west and north-west France. In Norfolk this plant has its greatest centre of population in the north-east where it is often abundant in all habitats and frequently out- performs *R. ulmifolius* in hedgerows. It rapidly becomes scarcer the further south and west one goes though it does reach the King's Lynn area.

R. criniger (E.F.Linton) Rogers

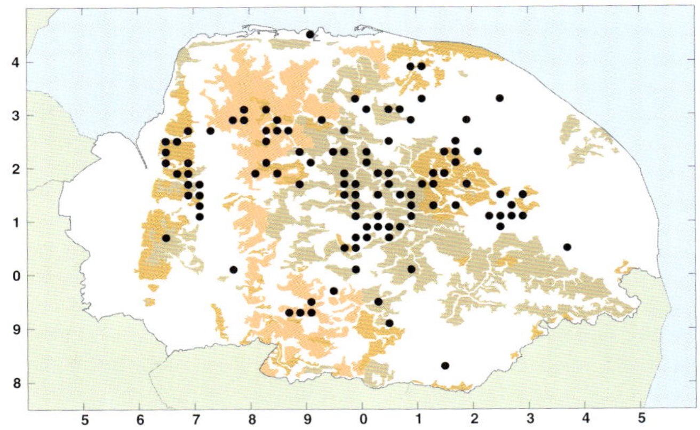

Most frequent in central Norfolk from whence EF.Linton collected the lectotype on Lexham Heath W81, on the unlikely date of October 22nd 1888. The plant is still there in good quantity. A ternate-leaved plant found abundantly around Colchester, in north Essex and extending into Suffolk along the Stour valley has been found at S.Runcton W60 and Stanford W89. This has been identifed as *R. criniger* var *trifolius* by A.Newton. The present author, however, finds this ternate-leaved plant to be less glandular than the nominate form having a very different terminal leaf shape, even in the rare examples which have 4-5 leaflets. The panicles are stiff and erect, while in *R. criniger* sensu stricto the panicles are curved and drooping and the anthers are also pilose which, in the ternate form, never occurs. This suggests that, although close to *R. criniger* in many respects, the ternate plant is more likely to be a separate taxon.

Rubus criniger - East Tuddenham

R. leucostachys Schleicher ex Sm.
A south-east regional endemic found only in woods at Pretty Corner and Sheringham Park E14. W0. E3.

R. surrejanus W.C.Barton & Riddelsd.
A widespread endemic of southern England which is abundantly successful when introduced into new areas by forestry operations as at Ken Hill, Snettisham W63, Weeting W79, Colney Hall Wood E10 and woodland near Loddon E39. W4. E4.

R. vestitus Weihe in Bluff & Fingerh.

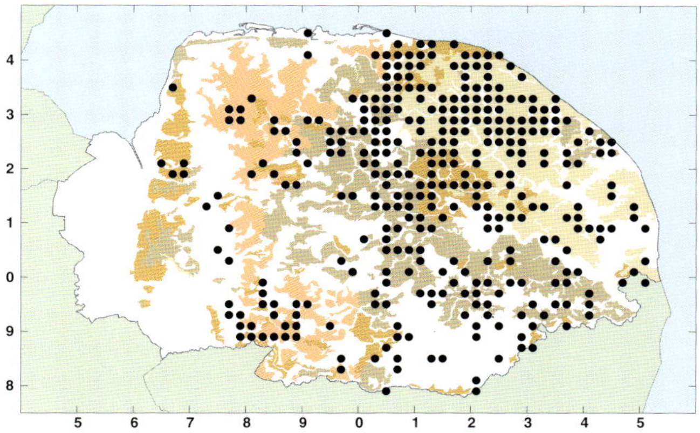

This common woodland blackberry also occurs on occasions in hedgerows and on commons and is one of the few species other than *R. ulmifolius* to tolerate both clay and chalk. It occurs in two colour forms, the commoner being pale pink to whitish and named var. **albiflorus**, the less common variety has almost red petals, filaments and styles. One of the best plants for bramble jelly.

Series 6. Mucronati (Focke) H.E.Weber. Sometimes hairy but often glabrous or glabrescent. Stalked glands and acicles present in varying quantities with more on the panicles. Terminal leaflets usually broadly obovate with an abrupt, cuspidate tip.

R. egregius Focke
Rare throughout its English range. Known since Linton's time from Mousehold Heath E21. Also discovered on North Tuddenham Common E01 and near Pedham Bridge E31. W0. E3.

R. mucronulatus Boreau

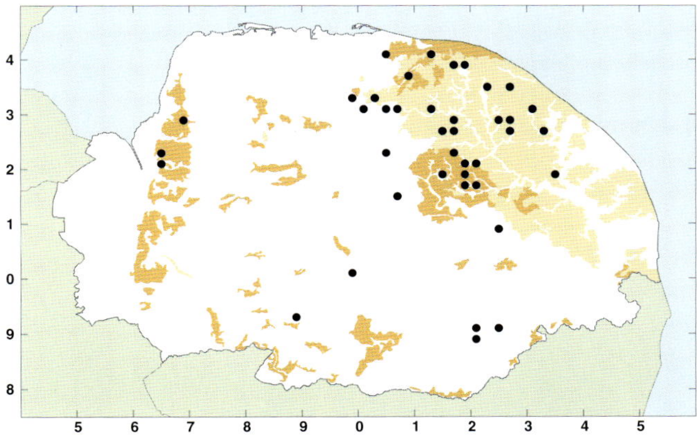

A northern plant in the county at the southern end of its range, its most southerly known station being at Tyrrell's Wood, Pulham Market E28. Widespread in woodlands in north Norfolk extending almost to the Wash.

R. villosior A.L.Bull

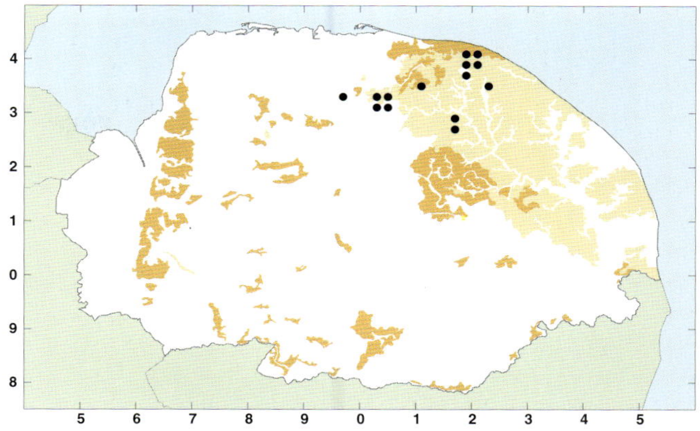

Abundant and often dominant in woods across north Norfolk from Thursford W93 to Gunton E23. Characterised by the very shaggy panicle, the long silky hairs becoming yellowish as the summer advances giving the impression that every bush is affected by felt disease, a condition caused by mites of the order *Acari*. The holotype was taken from Thursford Wood and is in the BM.

Series 7. Micantes Sudre. Stems glabrous or hairy with unequal acicles, pricklets and stalked glands. Prickles on angles of the stem usually distinct from the pricklets. Panicles similarly armed.

R. glareosus Rogers
Plentiful through the woods from Pretty Corner, Sheringham to the southern end of Beeston Common both in E14. W0. E3.

R. lintonii Focke ex Bab.

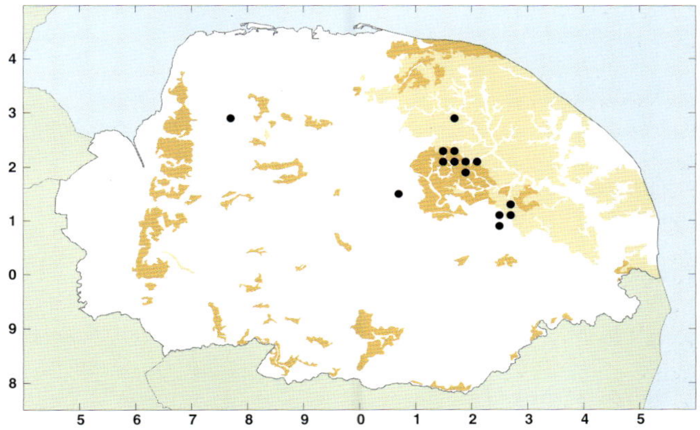

Locally frequent near Norwich, occurring also in a few woods farther north. In the west only at Houghton W72. Though classed as 'a widespread endemic', it is elsewhere only known to occur in Essex, Hertfordshire and Staffordshire. Lectotypes were collected by EF.Linton at Sprowston E21, in September 1884 and August 1886 and are in Babington's Herbarium at Cambridge.

R. moylei W.C.Barton & Riddelsd.
First found c.1972 in Colney Hall Wood E10 but now shaded out. Found during this survey near Emily's Wood, Weeting W78 and in Cherry Breck Plantation, Snettisham W63. This endemic of Wales and the south of England is thought to have arrived with nursery stock when the woods were planted. W4. E0.

R. norvicensis A.L.Bull & E.S.Edees

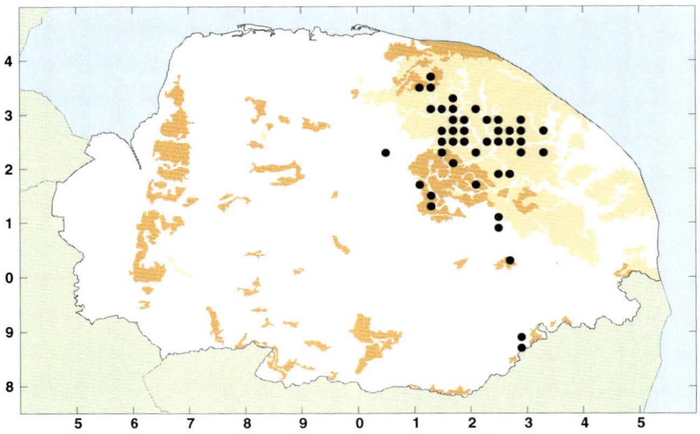

Most frequent in the area of the county around Wolterton E13 and Blickling E12, but widespread elsewhere, extending to the Suffolk border. The plant re-appears in south Suffolk, Essex and Hertfordshire and since being named has been found elsewhere in southern England. Edees took the holotype from Sloley E22 and it is in the BM.

R. raduloides (Rogers) Sudre
A plant of scattered distribution though never common, even though it is widespread in Britain and north- west Europe. W0. E10.

R. trichodes W.C.R.Wats.
This south-eastern endemic is closely related to several other species and is consequently sometimes difficult to determine. It grows abundantly in almost all the woods surrounding Gunton Park E23, and as these are even aged conifer plantations the plants were probably introduced with nursery stock. Material from this site has been approved by Edees and Newton. W0. E5.

<u>Section 8. Anisacanthi</u> H.E.Weber. Stems hairy or glabrescent. Prickles, pricklets and stalked glands very variable in quantity even on the same bush. Prickles subequal and confined to the angles as in <u>Micantes</u> or unequal and grading to pricklets, and scattered around the stem as in <u>Hystrices</u>.

R. adamsii Sudre
A widespread endemic reaching Norfolk at the Mayflower, Houghton W72 and Bacton Woods E33. W4. E1.

R. anisacanthos G. Braun
Rare in Norfolk. So far recognised only at Barney Wood W93; Hockering Wood E01 and Bush Wood, Bayfield E04. W1. E2.

R. cromerensis A.L.Bull

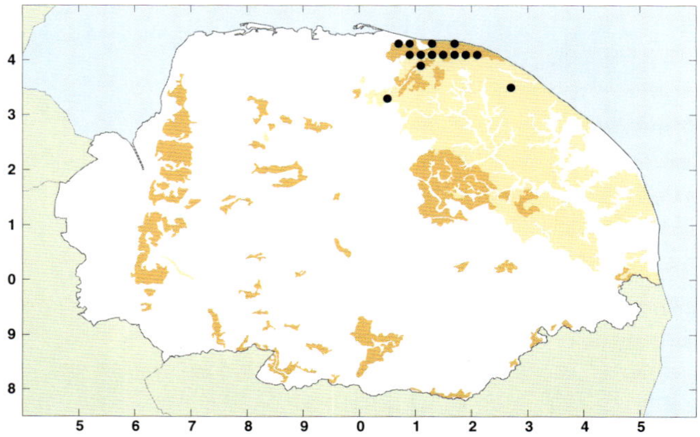

This plant may not belong in this section as it is not very glandular. The numerous sub-equal prickles are distributed around the stem. A frequent plant, mainly close to the sea on the north facing slopes of the Cromer ridge. Only two inland stations to date, at Southrepps E23 and Stody E03. Holotype from Salthouse Heath E04 and is in the BM.

R. infestus Weihe ex Boenn.
Mainly a northern species in Britain with only a single known Norfolk site at Booton Common E12. W0. E1.

R. leyanus Rogers

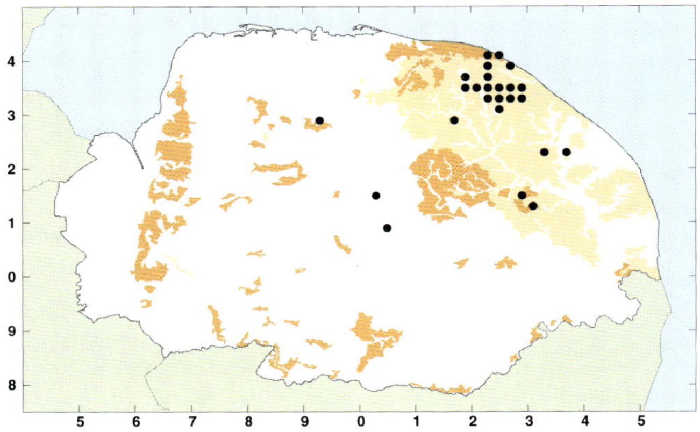

This is a plant of south-west England and Wales with an outlying population in Hampshire and an isolated site at Bourne Wood in Lincolnshire. In view of this, it is curious that it should be an abundant feature of hedgerows in old lanes and wood margins particularly towards the north-east coast, but extending inland as far as Sprowston E21; Mattishall and North Tuddenham E01.

<u>Series 9. Radulae</u> (Focke) Focke. Stems thinly hairy or glabrous, rough with numerous pricklets, acicles and stalked glands all short and roughly equal. Prickles on the angles not grading into pricklets.

R. cantianus (W.C.R.Wats.) E.S.Edees & A.Newton
Known as the 'Kentish Blackberry' with the centre of its distribution in that county, it has been found in post war coniferous plantings at Thorpe St Andrew E21 and Woodbastwick E31. Presumably another tree planting introduction. W0. E3.

■ **R. echinatoides** (Rogers) Dallman
Found in the 1970s in Colney Hall Wood E10 and on Mousehold Heath E21 but not seen during the present survey.

R. flexuosus Mueller & Lef.

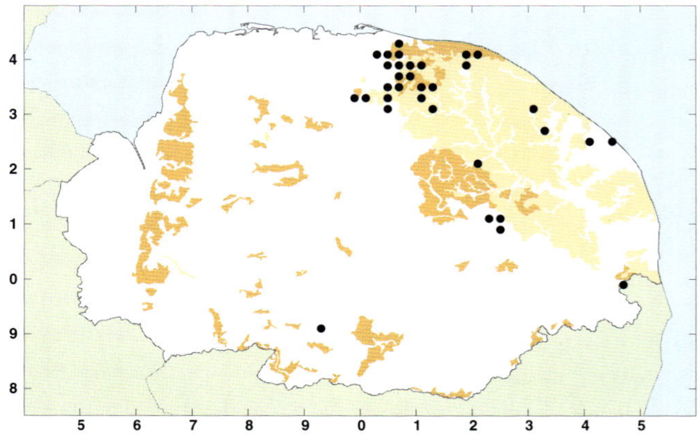

Widely scattered with the main concentration of records in the north of the county.

R. insectifolius Lef. & Mueller

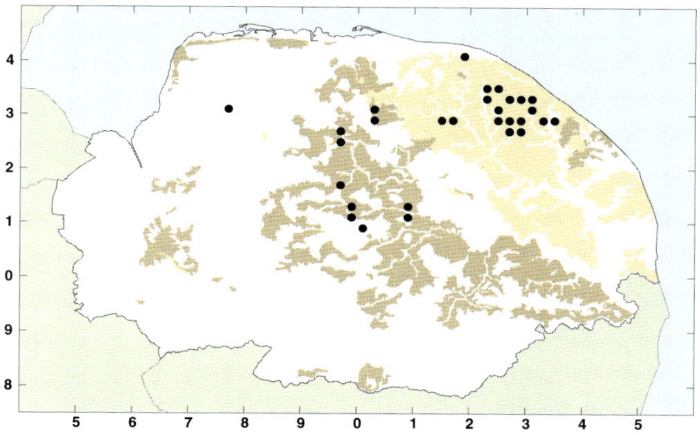

Another plant which is mainly northern in the county. Where it occurs it is frequently dominant, the long, trailing stems with dark green glossy leaves climbing over everything and the white flowers with bright red styles being particularly noticeable in mid-July.

R. malvernicus E.S.Edees
Not found outside the Malvern area until it was found in Race Course Plantation, Thorpe St Andrew E21 in 1993. W0. E1.

R. pallidus Weihe in Bluff & Fingerh.
Found in Sprowston Wood E21 since Linton's time and also during the 1970s at Langley E30. Still there in 1998. W0. E1.

R. radula Weihe ex Boenn.

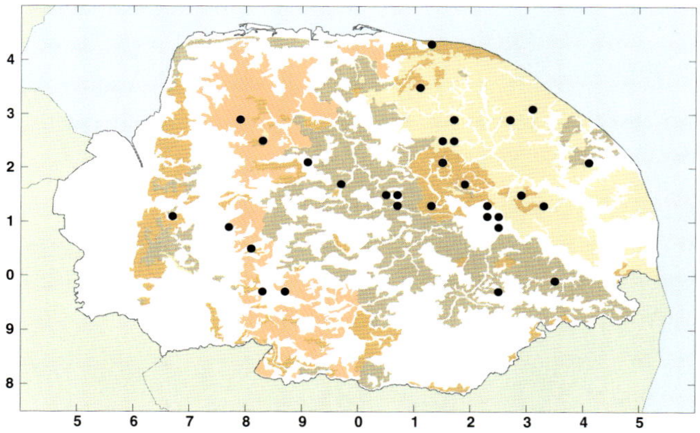

Scattered through Norfolk and one of the few brambles which may have been in Breckland before Forest Enterprise.

R. rudis Weihe in Bluff & Fingerh.
Only one known site in the middle of forestry plantings on Shouldham Warren W71 where it occupies a sizeable area. As it is plentiful in Bardney & Twyford Forests in Lincolnshire, this must be regarded as the probably source of the present colony. W2. E0.

R. subtercanens W.C.R.Wats.
New to Norfolk. Found at Hempnall, E29 in 1998, ALB, conf. A.Newton. W0. E1.

R. rufescens Lef. & Mueller

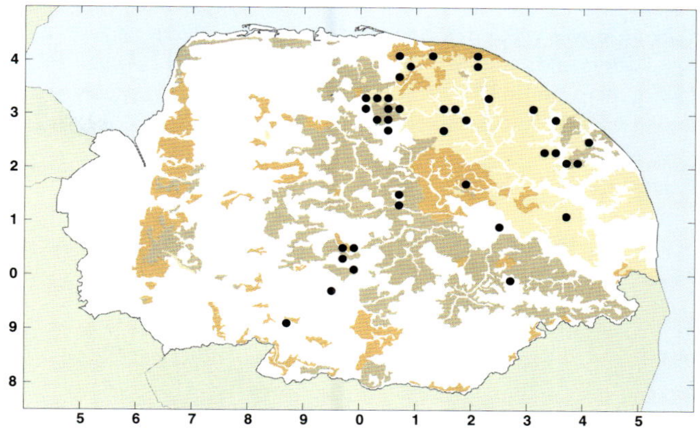

A typical plant of old woodland, often dominant where it occurs, the map, however, suggests that it avoids such woods if they are on clay.

Series 10. Hystrices Focke. Prickles, pricklets, acicles and stalked glands grading into each other and very numerous. Pricklets more obvious than stalked glands.

R. dasyphyllus (Rogers) E.S.Marshall

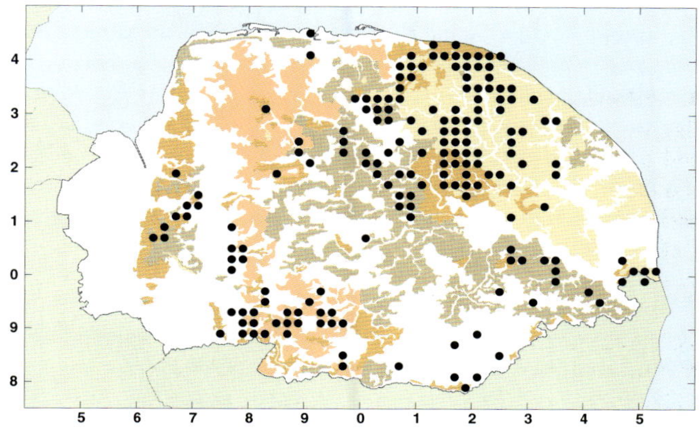

Edees and Newton say 'common and often abundant in many habitats. In England and Wales probably the most frequent bramble extending into many areas not reached by *R. ulmifolius*'. In Norfolk it is found mainly in woods and on heaths and is rare on clay.

R. hylocharis W.C.R.Wats.

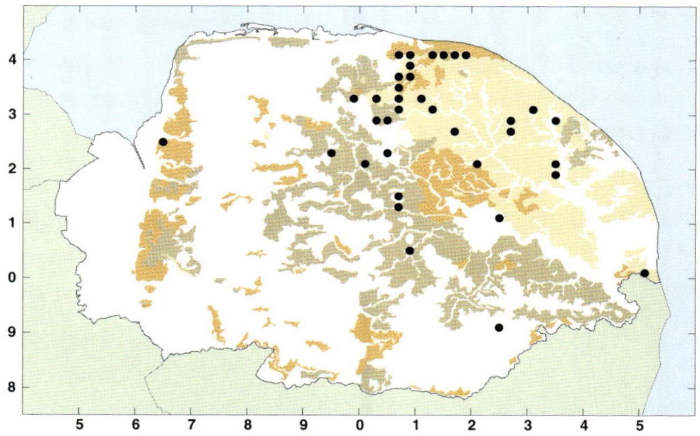

Scattered through Norfolk with the main concentration of records to the north and north-west of Norwich.

R. '<u>Superhystrix</u>'
A plant with even more intense armature than *R. hylocharis*, bluntly obovate leaves which are hard beneath and large panicles with neat pink flowers and red styles. It is frequent to abundant between Holt E03 and Saxthorpe E13 and is very prominent where it occurs. It is not yet sufficiently widespread to merit a name. W0. E3.

R. iceniensis A.Newton & H.E.Weber

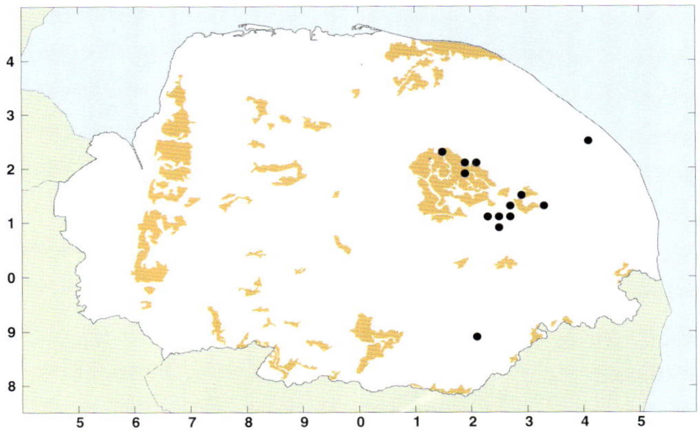

Listed in Petch & Swann as *R. menkei* Sudre, this plant was the subject of a visit by A.Newton and HE.Weber in 1977 in order that Weber could assess this putative outlier of a continental species. The plant was immediately rejected as being unlike *R. menkei* and awarded the name *R. iceniensis* in honour of the East Anglian tribe. The herbarium sheet nominated as the holotype was collected in July 1893 on the borders of Sprowston and Rackheath as *R. tereticaulis*. This name was given it by PJ.Mueller and EF.Linton and is in the herbarium at Manchester. But for a plant found at Mardley Heath, Herts in 1993, it would be an east Norfolk endemic.

■ **R. marshallii** Focke & Rogers
Two or three bushes formerly occurred in the centre of Colney Hall Wood E10 (about 1972-1978) but are now shaded out.

R. phaeocarpus W.C.R.Wats.
The occurrence of this plant in Norfolk is not as straight forward as some other forestry introductions. In the west it is a member of the Surrey bramble flora of Cherry Breck Plantation, Snettisham W63; of woodland at Raynham W82 and at Eastfield Wood at Tittleshall W92. In the east it occurs in hedgerows in old streamside water meadows at Itteringham E12 and by the Stake Bridge Beck, Skeyton E22 where it looks very like a native. W4. E2.

R. watsonii W.H.Mills
There is a single site in the Ash Plantation at Felbrigg E23 where the plant could be either bird sown or introduced. This is an east Midlands endemic so its occurrence here is not surprising. A.Newton has also recently confirmed an earlier gathering at Shouldham Warren W71, perhaps a forest planting introduction. W1. E1.

R. atrebatum A.Newton
Found in Little Wood, Edgefield E13 in 1998, ALB, conf. A.Newton. New to Norfolk. W0. E1.

<u>Series 11. Glandulosi</u> (Wimmer & Grab.) Focke. Stems prostrate and frequently pruinose with many slender prickles grading into gland-tipped acicles and stalked glands all round. Flowers usually small and white and sometimes with red styles.

R. scaber Weihe in Bluff & Fingerh.
Two sites are known, about 2km apart at Tea Plantation, Stanford and Widdowshill Plantation, Sturston; both in the Stanford Training area in W89. W2. E0.

R. sp.
A plant which is abundant in Little Wood, Edgefield E03 and nearby Pond Hills, Hempstead E13 has also been found in south Suffolk and possible also in north Hertfordshire. When ES.Edees first saw it in the field in 1977 he thought that it might be *R. leptadenes* Sudre. Subsequent research shows that it is a very different taxon to that species which occurs in Lee Clump, Buckinghamshire, so it must remain nameless for the time being.

<u>Section B. Corylifolii</u> Lindley. Stems low arching, often pruinose. Leaflets imbricate, the basal pair sessile or subsessile. Fruit often imperfect with few drupelets.

R. britannicus Rogers

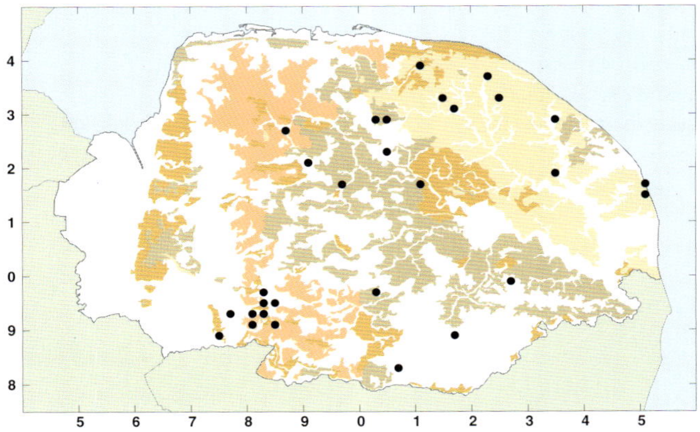

Scattered in Norfolk though nowhere common. A southern plant near its northernmost limit.

R. conjungens (Bab.) Rogers

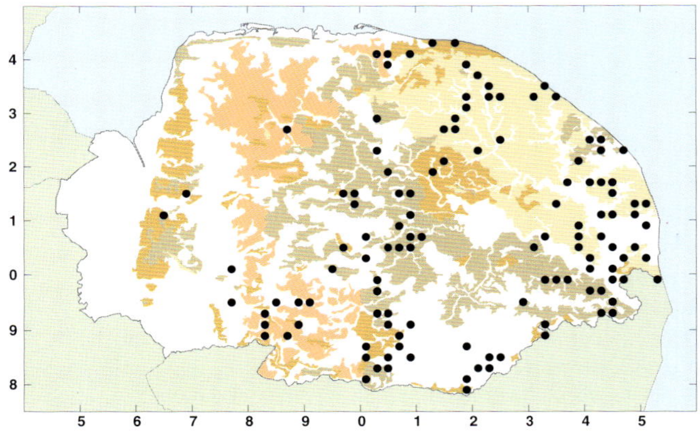

One of the more widespread members of the group occurring mainly in hedgerows, often found in parts of the Broads and occasionally on clay.

R. eboracensis W.C.R.Wats.

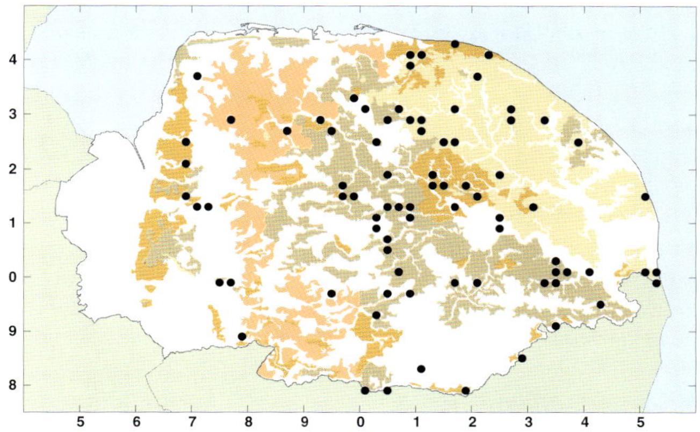

'The Yorkshire Blackberry' is widely scattered and needs to be differentiated from the last species. The flowers of this plant are smaller and paler pink, the stems are round and not angled, the prickles are small, especially on the panicles, the simple leaves of which are roughly rhomboid.

R. intensior E.S.Edees
A surprise find of this west Midlands endemic by Great Heath Road, North Elmham W92 was made and confirmed in 1996. W1. E0.

R. nemorosus Hayne & Willd.

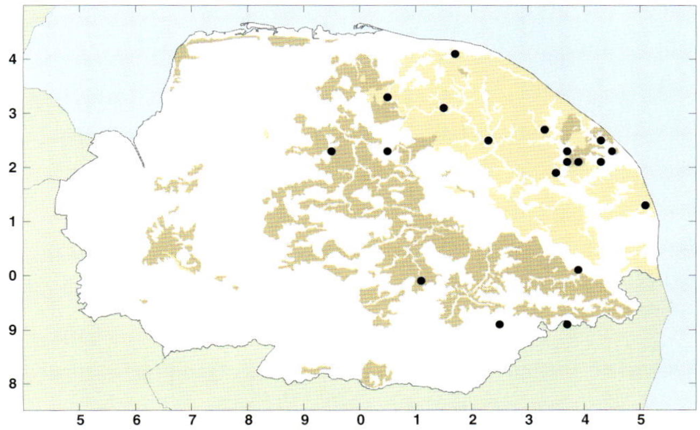

The large pink flowers with red styles are produced very early. It can be readily distinguished in this group by its hairy anthers. It likes a damp situation, so not surprisingly is most frequent in the Broads.

R. hindii A.L.Bull

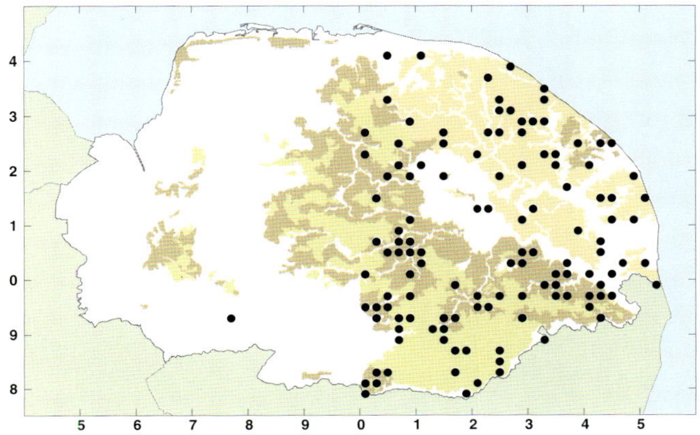

Frequent in the east of the county and extending into the west in a few places, also known from several other counties. It may eventually prove to be the plant errroneously attributed to the continental *R. scabrosus* Mueller, by WCR.Watson. The holotype was taken from Wordwell in Suffolk.

R. pruinosus Arrh.

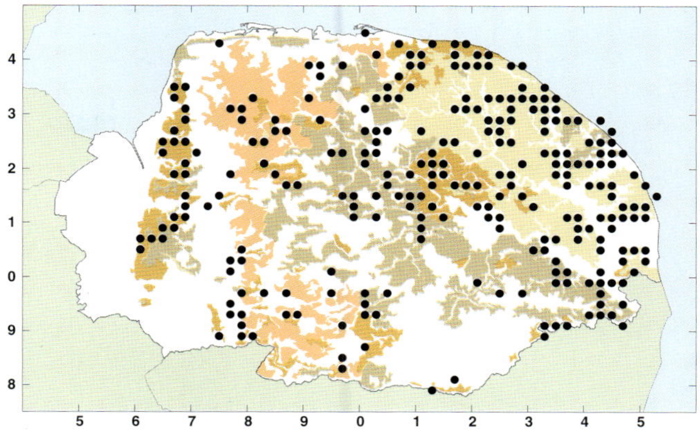

Frequent except on clay and like *R. nemorosus* most frequently met with in the Broads.

R. tuberculatus Bab.

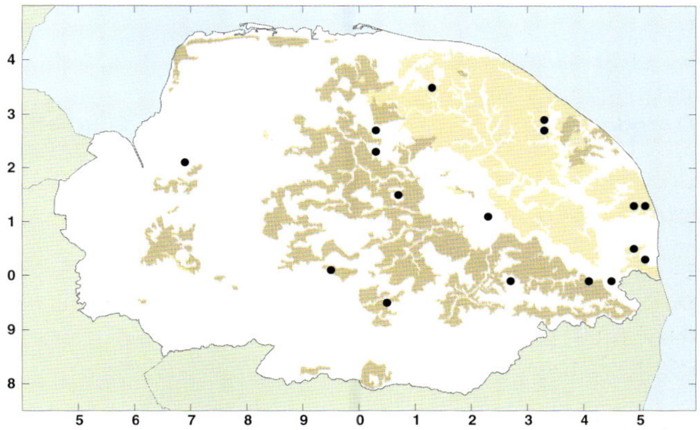

Not very common in the county and originally confused with *R. hindii* which with us usually has pink flowers. True *R. tuberculatus* in Norfolk has white flowers. There are other differences, namely the shape of the terminal leaflet, the frequency of stem hair and the distribution of stalked glands.

Section C. Caesii Lej. & Courtois.

R. caesius L. **Dewberry**
Much over-recorded in the present survey, there being confusion with what, for want of a better term, are called Caesian hybrids. True *R. caesius* is almost confined with us to hedgerows and woodlands on clay and chalk and shuns waterlogged situations and acid sands.
W277. E275.

In closing the above account, the author would like to record his gratitude firstly to the late ES.Edees whose help in the early days gave the incentive needed to persist with this difficult group, and latterly to Alan Newton whose continuing encouragement and advice has led to the naming of a number of new species and the discovery of other unexpected plants. All of this has resulted in Norfolk having a much more respectable total than could have been envisaged when the late EL.Swann was told by Watson 'You have no brambles in Norfolk because the county was under the last ice sheet!'

ALB

Potentilla palustris (L.) Scop. **Marsh Cinquefoil**
Syn. *Comarum palustre* L.

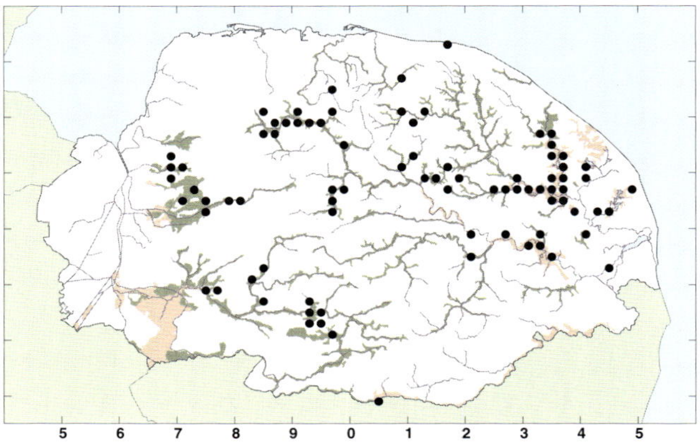

Herbaceous perennial, frequent in parts of the Broads but scarce elsewhere, usually in wet, peaty ground.

P. anserina L. **Silverweed**
Herbaceous perennial of roadsides, grassland, gardens and waste ground, absent only from very dry soils. W579. E621.

P. argentea L. **Hoary Cinquefoil**

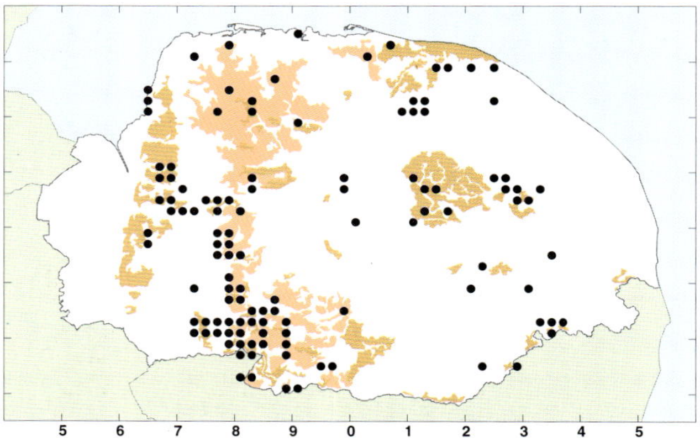

Winter green to semi-winter green perennial, frequent in west Norfolk on light, sandy soils, particularly in forest rides in Breckland.

Potentilla argentea - Syderstone

P. recta L. **Sulphur Cinquefoil**
Perennial introduced to gardens from southern Europe, surviving in dry, sandy soils for a number of years as at Sheringham E14 and a gravel pit at Docking W73 where it has been known for over 30 years. W15. E2.

■ **P. intermedia** L. **Russian Cinquefoil**
Biennial or perennial, a casual introduced from Russia. Recorded from Snake Wood, Weeting W89 in 1971, RPL.

■ **P. norvegica** L. **Ternate-leaved Cinquefoil**
Usually an annual, introduced from northern Europe and found once near High Kelling E03 in 1973, AT.Forest.

■ **P. neumanniana** Rchb. **Spring Cinquefoil**
Syns. *P. tabernaemontani* Asch.nom. illeg., *P.verna* auct. non L.
Winter green perennial of dry, basic grassland, two records, Garboldisham W98 in 1952 and Santon Downham W88 in 1955.

P. erecta (L.) Rausch. **Tormentil**

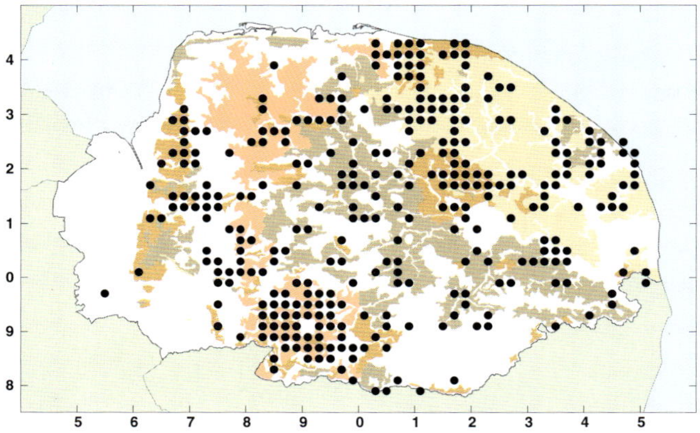

Herbaceous perennial, frequent in damp grasslands on acid soils and especially around the margins of fens and bogs. Our plants are ssp. **erecta**.

P. anglica Laichard. **Trailing Cinquefoil**

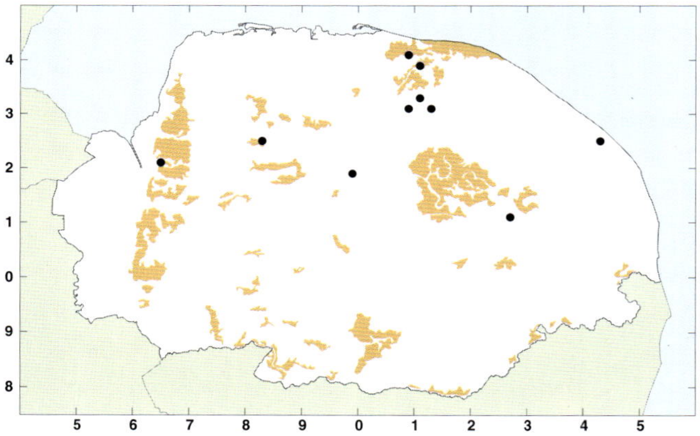

Evergreen perennial, rare on heaths and dry, acid grassland, but easily overlooked.

P. x mixta Nolte ex Rchb. **Hybrid Cinquefoil**
Syn. *P. x italica* Lehm. (*P.erecta x reptans*) and (*P. anglica x P. reptans*)
A scarce hybrid recorded from Sugar Fen W62 in 1972, RPL; Stow Bedon W99 in 1980, RPL & CRS; Wolferton W62 in 1985, CPP and Holt E03 in 1997, KAB. W1. E1.

P. reptans L. **Creeping Cinquefoil**
Herbaceous perennial of roadsides, waste ground, gardens and grassland, abundant on all but the wettest soils. W669. E655.

P. sterilis (L.) Garcke **Barren Strawberry**

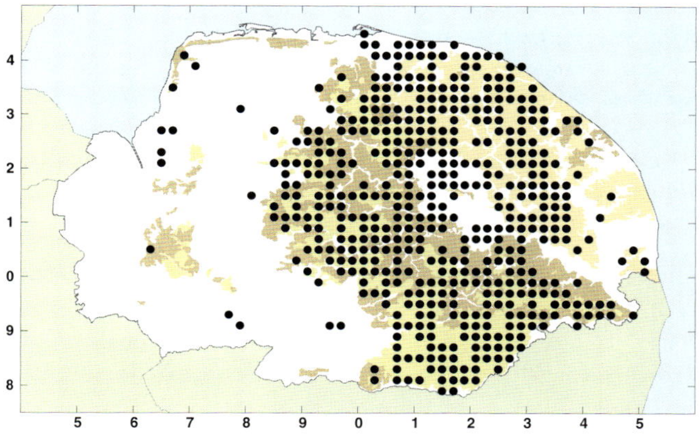

Evergreen perennial restricted to heavier soils where it can be abundant.

Fragaria vesca L. **Wild Strawberry**

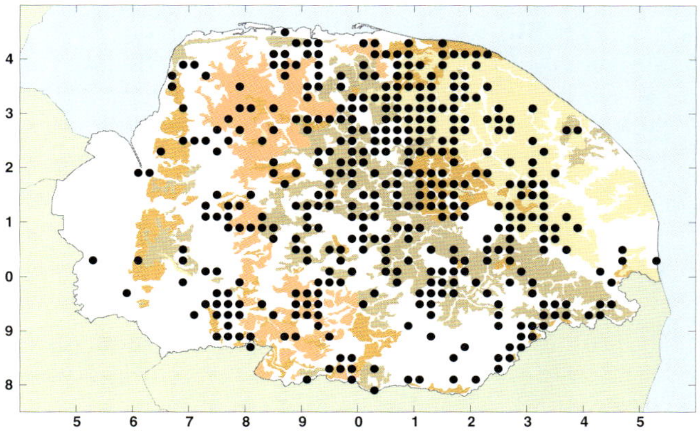

Semi-winter green perennial found in woodland and on shady places by roadsides. A white-fruited form grows on derelict heathland at Cawston E12, ALB.

■ **F. moschata** (Duchesne) Duchesne **Hautbois Strawberry**
Syn. *F.muricata* sensu D.H.Kent non L.
Semi-winter green perennial introduced from eastern Europe, last recorded in Nicholson 1914.

F. x ananassa (Duchesne) Duchesne **Garden Strawberry**
Winter green perennial of garden origin, sometimes found in waste places, once frequent by railway lines now rarely persisting, though one colony extends along the embankment of the former railway, now Marriott's Way, Cawston E12. W14. E23.

Duchesnea indica (Jacks.) Focke
Yellow-flowered Strawberry
Evergreen perennial introduced from south-east Asia to gardens where it can self-sow and become a weed. Occasionally escaping as at North Runcton and Wimbotsham in W61; Stanhoe W83; Norwich E20 and Ditchingham E39. W3. E2.

Geum rivale L. **Water Avens**

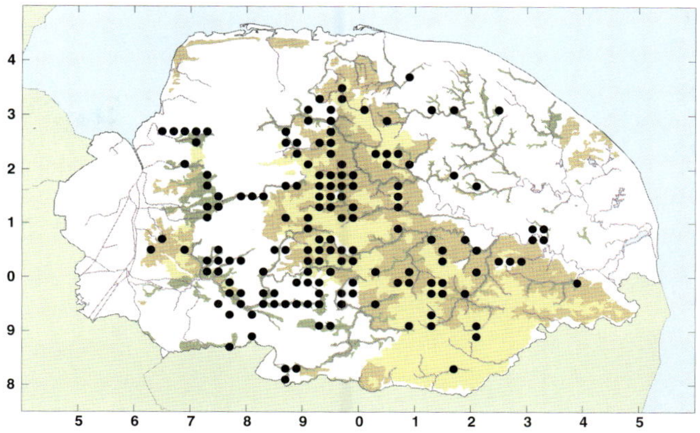

Winter green perennial of marshes, wet meadows and moist woods, most frequently associated with ancient woodland and notably absent from both Broadland and Fenland which lack such woods. Forms with double petals are not uncommon and are sometimes confused with the hybrid below.

G. x intermedium Ehrh. **Hybrid Avens**
(*G. rivale x G. urbanum*).
Frequent with the parents, usually in woodland sites. W23. E7.

G. urbanum L. **Wood Avens, Herb Bennet**

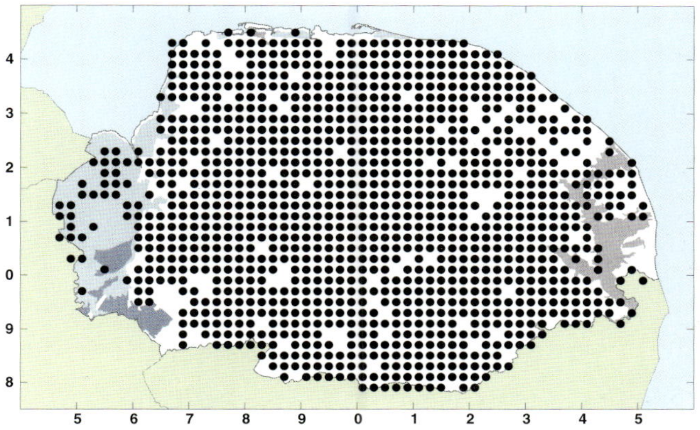

Winter green perennial, widespread in shady places, quick to colonise new plantations.

Agrimonia eupatoria L. **Agrimony**

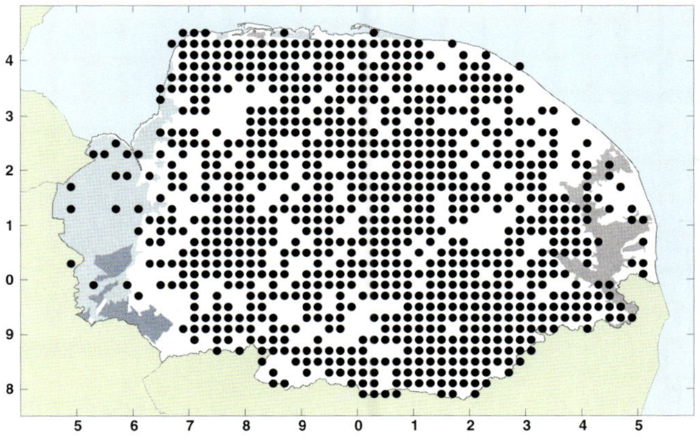

Herbaceous perennial of established grassland, especially on well-drained soils.

A. procera Wallr. **Fragrant Agrimony**
Syn. *A. odorata* auct. non (L.) Mill.

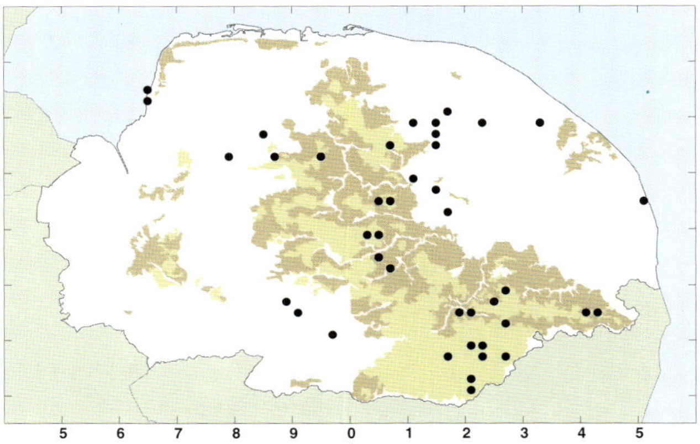

Herbaceous perennial of established grassland, far less common than *A. eupatoria* and usually on moister soils.

Sanguisorba officinalis L. **Great Burnet**
Syn. *Poterium officinalis* (L.) A.Gray
Herbaceous perennial of long established moist grassland, confined to the Waveney valley. Known for many years from its only sites at Blo Norton E07 and Roydon E17. W0. E2.

S. minor Scop. **Salad Burnet**
Syn. *Poterium sanguisorba* L.

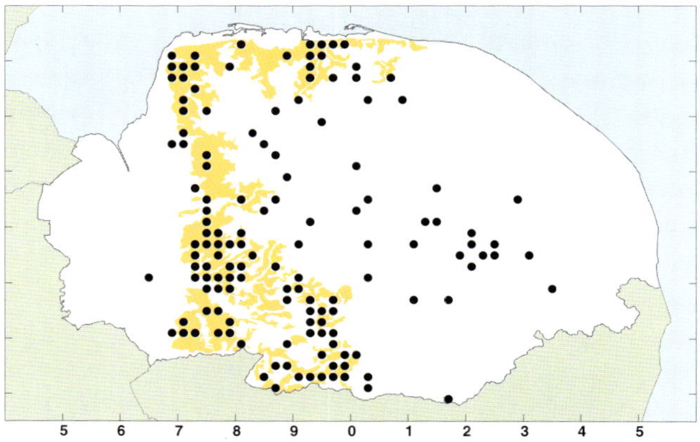

Herbaceous perennial with ssp. **minor** native to established, neutral to calcareous grassland and ssp. **muricata** (Gremli) Briq., **Fodder Burnet**, (syn. *P.polygamum* Waldst. & Kit.) introduced as a fodder plant from southern Europe. It is occasionally established on roadside verges often where it has been sown with so-called native seed mixes.

Acaena novae-zelandiae Kirk
New Zealand Bur, Bidi-bidi, Pirri-pirri
Winter green perennial introduced to gardens from New Zealand and long confused with *A. anserinifolia* (J.R.& G.Forst.) Druce, a separate species. Occasionally escaping. Known from Kelling Heath since 1928, CPP, still present, KKH; at Selbrigg in 1994, AR.Leech and in forestry near Holt, 1997, K&GB. All are E04 where the burs may be distributed by deer. Also at Heacham W63, PRC. W1. E3.

Alchemilla filicaulis Buser **Lady's Mantle**
Syn. *A. vestita* (Buser) Raunk.
Perennial recorded in 1966 in Hockering Wood E01, ALB & ELS and during the 1970s in a series of old meadows along the R.Tud from Hockering E01 to Honingham E11, where it became extinct by 1986 due to agricultural 'improvements' or neglect, ALB. Also at Aslacton E18, GD.Watts. All are ssp. **vestita** (Buser) M.E.Bradshaw. W0. E2.

■ **A. glabra** Neygenf. **Lady's Mantle**
Perennial recorded twice, at Swanton Novers E03 in 1956 and still present in 1968, also at Wolferton W62 in 1983, CPP.

A. mollis (Buser) Rothm. **Garden Lady's Mantle**
Herbaceous perennial introduced to gardens from the Carpathians and occasionally surviving on nearby roadsides. W5. E19.

Aphanes arvensis L. **Parsley Piert**

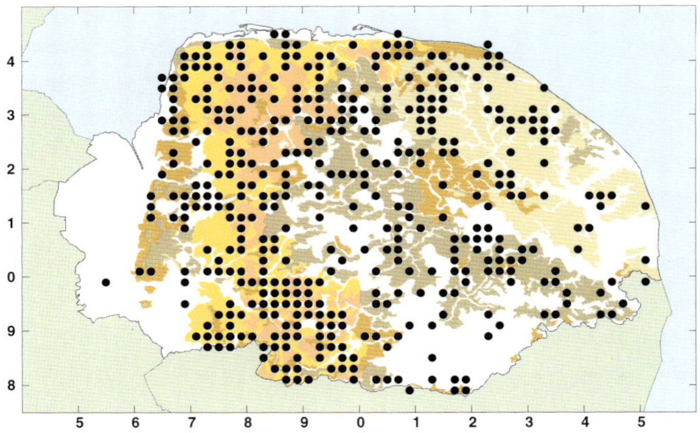

Autumn germinating annual of arable lands, scarce on moist soils.

A. australis Rydb. **Slender Parsley Piert**
Syn. *A. microcarpa* auct. non (Boiss & Reut.) Rothmaler, *A.inexpectata* Lippert

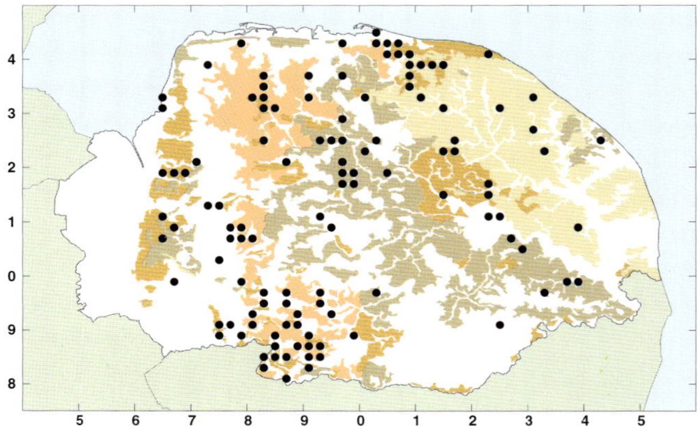

Autumn germinating annual of light, well drained soils, the commoner species on acid soils.

Rosa multiflora Thunb. ex Murray **Many-flowered Rose**
Deciduous shrub introduced to gardens from eastern Asia, recorded from North Elmham W92 in 1974 and during this survey at Whitwell E02 before 1990; Syderstone Common W83 in 1980, ALB and Great Witchingham E12 in 1983, ETD. W0. E1.

R. 'Dorothy Perkins'
A familiar garden rambler, well naturalised at Reedham E40, recorded in 1997, RWE & WM.. W0. E1.

R. arvensis Huds. **Field Rose**

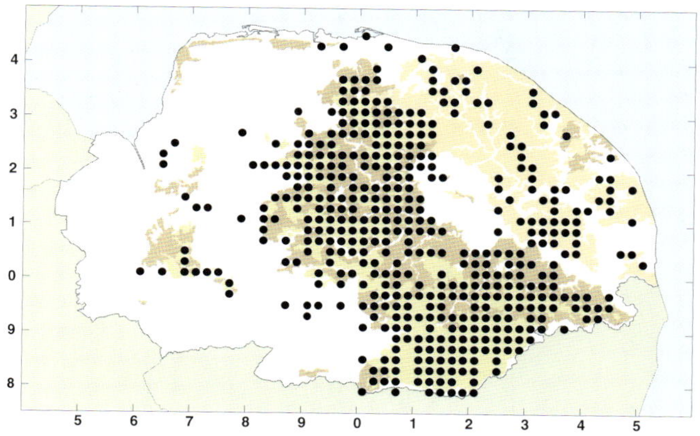

Deciduous shrub of hedges and scrub avoiding light soils.

Rosa arvensis - Gately

R. x verticillacantha Mérat
(*R. arvensis x R. canina*)
Deciduous shrub recorded at East Tuddenham E01, ALB where it is rampant in a hedge; SwaffhamW80 in 1995, R.Maskell & AL.Primavesi and Bawdeswell E02 in 1997, KAB. W1. E1.

R. pimpinellifolia L. **Burnet Rose**
Deciduous, suckering shrub only certainly native close to the north coast at Salthouse and Kelling E04. Introduced in a few inland sites and at Scolt Head, W84 where it was noted in 1975, EAE and is still present, SW. In these sites the presence of double, fragrant flowers indicate garden cultivar origin. W2. E7.

R. rugosa Thunb. ex Murray **Japanese Rose, Rugosa Rose**
Deciduous spreading shrub introduced from eastern Asia. Sometimes planted for game cover as at Weeting W79, or spread by birds as on Scolt Head W84 in 1993, SW. W17. E68.

R. x praegeri Wolley-Dod
(*R. rugosa x R. canina*)
A rare hybrid recorded twice in the British Isles, in Ireland in 1927 and at Scolt Head W84 in 1993, SW. W1. E0.

R. virginiana Herrm. **Virginian Rose**
Syn. *R. lucida*
Deciduous shrub introduced to gardens from eastern North America, planted beside new roads as by the Tuddenham bypass E01, ALB, and likely to spread. W0. E2.

R. ferruginea Vill. **Red-leaved Rose**
Syns. *R. rubrifolia* Vill. nom. illeg., *R. glauca* Pourr. non Vill. ex Loisel
Deciduous shrub found naturalised on dunes at Holme W74 in 1993, JEG. W1. E0.

R. stylosa Desv. **Short-styled Field-rose**
Deciduous shrub, scarce and usually on heavy soils. Found at Brooke Wood E29 in 1977, ETD and recently at Barney W93 in 1994, MK and Heacham W63, PRC. W2. E0.

■ **R. x andegavensis** Bastard
(*R. stylosa x R. canina*)
Deciduous shrub found once at Saham W90 in 1970, DMM det. Melville.

R. canina L. **Dog Rose**
Deciduous shrub abundant in hedges, wood edges and scrub. Different colour forms occur but appear to be no more than forms and varieties of this species. W664. E704.

R. x dumalis Bechst.
(*R. canina x R. caesia*)
Deciduous shrub found on four occasions even though one of its parents has yet to be found in the county. Hockwold W78 in 1992, CD.Preston; Stow Bedon W99 in 1994, Sugar Fen, Leziate W62 and Brancaster W74 in 1996, KAB. W4. E0.

R. x dumetorum Thuill.
(*R. canina x R. obtusifolia*)
Deciduous shrub found near Swaffham W80 in 1995 by R.Maskell and AL.Primavesi and East Harling E08; Roughton and Northrepps E23, all JH. W2. E2.

■ **R. x nitidula** Besser
(*R. canina x R. rubiginosa*)
Deciduous shrub recorded once, at E.Walton W71 in 1976, B.Burt det. Melville.

■ **R. x belnensis** Ozanon
(*R. canina x R. agrestis*)
Deciduous shrub found at Eaton in or near the Donkey Lane chalk pit E20 by M.Wigginton in 1975, still present in 1977, ETD.

■ **R. caesia** Sm. **Glaucous Dog-rose**
Deciduous shrub, one parent of *R. x dumalis* but not yet found in the county.

R. obtusifolia Desv. **Round-leaved Dog-rose**
Deciduous shrub recorded at East Bradenham W90 in 1988, ALB; Gooderstone Fen W70 in 1994, JEG and Swaffham W80 in 1995 by R.Maskew and AL.Primavesi. W3. E0.

R. tomentosa Sm. **Harsh Downy-rose**
Deciduous shrub occasional on heavy soils. Recorded at Shipdham and Wood Rising, both W90 in 1988, DMM and at Saxlingham Thorpe E29 in 1996, ALB, RWE and WM. W2. E2.

R. sherardii Davies **Sherard's Downy-rose**
Deciduous shrub with two pre-1970 records, more recently at Carbrooke W90, G.Nobes; Fritton E40 in 1995, P&I.Green and Yelverton E20 in 1998, ALB & J.Mott. W1. E2.

R. mollis Sm. **Soft Downy-rose**
Syn. *R. villosa* auct.non L.

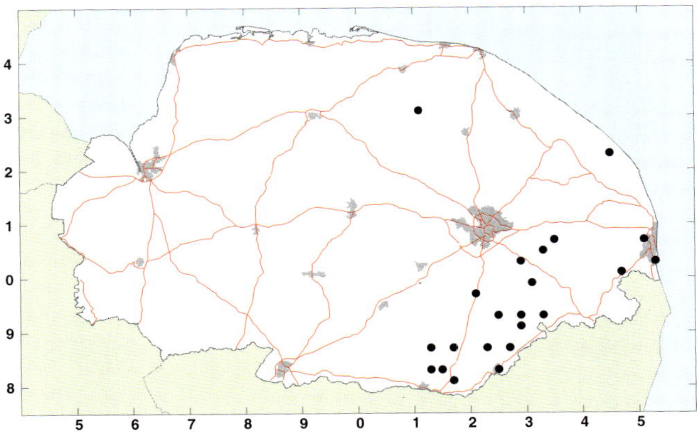

Deciduous shrub not uncommon on the claylands of south-east Norfolk, scarce elsewhere.

R. rubiginosa L. **Sweet Briar**

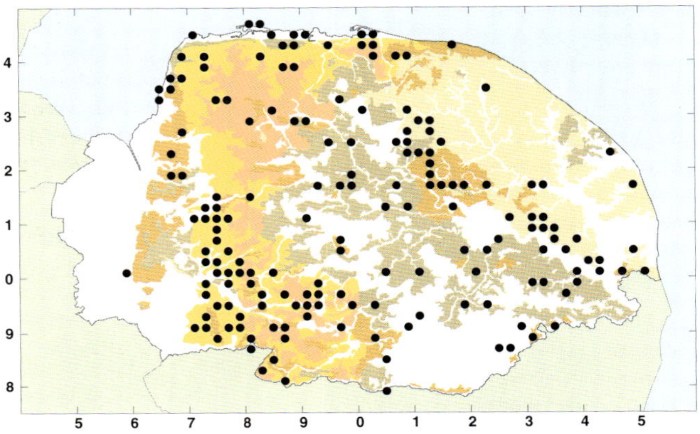

Deciduous shrub, by far the commonest of the sweet briars, occurring chiefly on chalky soils in scrub and rough pastures.

■ **R. x bigeneris** Duffort ex Rouy
(*R. rubiginosa x R. micrantha*)
Deciduous shrub recorded at Morston E04 in 1978 by RPL & ELS.

R. micrantha Borrer ex Sm. **Small-flowered Sweet Briar**
Deciduous shrub, rare, found during this survey at Whinburgh E00; East Tuddenham E01 and Corpusty E12. W0. E3.

R. agrestis Savi **Small-leaved Sweet Briar**
Deciduous shrub recorded at Eaton E20 near the Donkey Lane chalkpit, five bushes still present in 1985, ETD. W0. E1.

Prunus cerasifera Ehrh. **Cherry plum, Myrobalan**

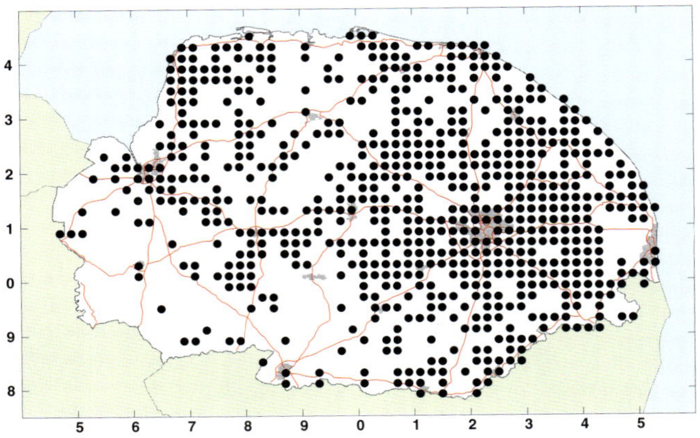

Deciduous tree or shrub introduced from western Asia and locally frequent in hedges, especially around villages. Much planted in the past for its fruit. The earliest hedgerow *Prunus* to flower and often confused by the layman with blackthorn. A very large tree at Hindringham W93.

P. spinosa L. **Blackthorn, Sloe**
Deciduous suckering shrub or small tree frequent in hedges and scrub, a rapid coloniser of grassland and then very difficult to eradicate. Often hybridising with the following species, adding to the problems of identification within this group. Many of these hybrids are sterile producing no fruit. W583. E687.

P. domestica. L. **Plum, Bullace, Damson**
Syn. *P. insititia* L.

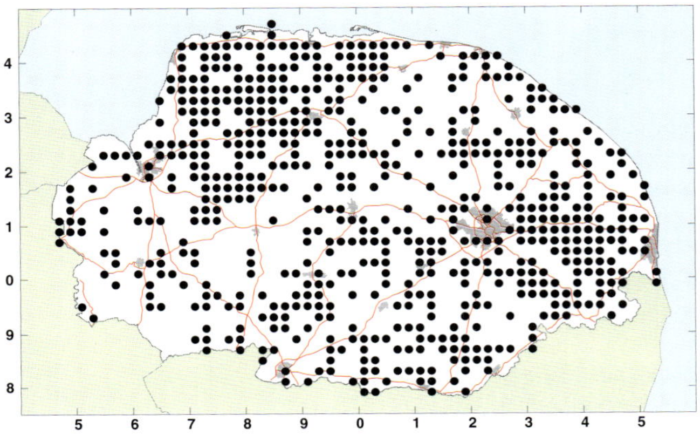

Deciduous shrub or tree widespread in hedges, perhaps particularly a feature of those planted in the 18th and 19th century. Trees of true ssp. **insititia** (L.) Bonnier & Layens occur in hedges, including both bullace with round, amber-coloured fruit and damson which is oval and blue, but they are much hybridised with ssp. **domestica**, the cultivated plum, and for mapping purposes it has not been found feasible to separate them.

P. avium (L.) L. **Wild Cherry**
Deciduous shrub or tree widely scattered, but its native distribution much masked by recent planting. W179. E330.

P. cerasus L. **Dwarf Cherry**

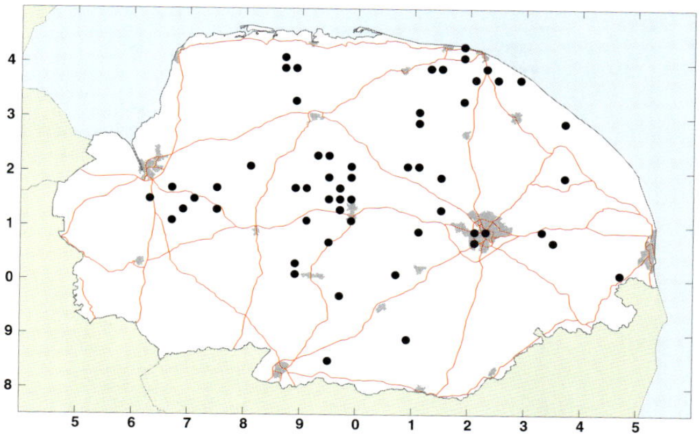

Deciduous suckering shrub or small tree, most frequent on heavy soils, but perhaps over-recorded.

P. padus L. **Bird Cherry**

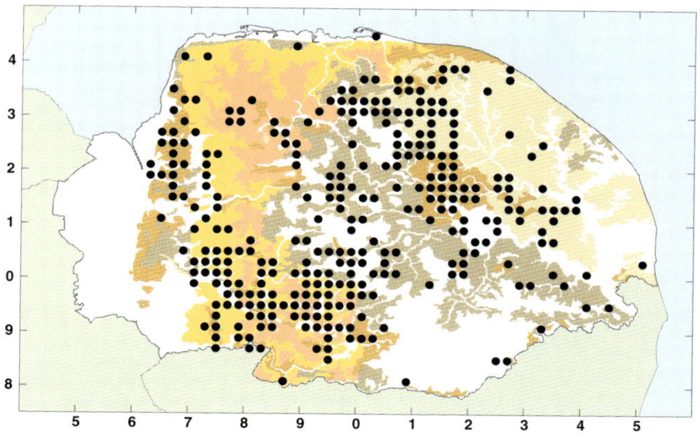

Deciduous shrub or small tree abundant in some areas of fen carr and in moist woodland as at Wayland Wood W99. In Britain common only in the north but undoubtedly native with us.

Prunus padus - Beechamwell

P. serotina Ehrh. **Rum Cherry**
Deciduous shrub or tree introduced from eastern North America, recorded at Heacham W63 in 1991, PRC; Barnham Cross Common, Thetford W88 in 1992, KAB and Roydon E18 in 1990, AC. W2. E1.

P. lusitanica L. **Portugal Laurel**
Winter green shrub or tree introduced from south-west Europe to parks for ornament and game cover. W0. E21.

P. laurocerasus L. **Cherry Laurel**
Winter green shrub or tree from south-east Europe, much planted in parks and woodlands as cover and food for pheasants but unlike the previous species, self sowing freely. W28. E160.

Cydonia oblonga Mill **Quince**
Deciduous shrub or small tree introduced from Asia, occasionally found in hedges or rough ground where it may have originated as stock for pear trees. Recorded at Thetford W88 and at Wiveton E04 where two plants were found in a poplar plantation, KKH. W1. E1.

Chaenomeles speciosa (Sweet) Nakai **Chinese Quince**
Deciduous shrub introduced to gardens from China. A single bush, presumably self-sown, was found flowering on waste ground near the Loddon bypass E39, ALB, also at Heacham W63, PRC. W1. E1.

Pyrus communis L. **Domestic Pear**
Deciduous shrub or tree mostly originating as seed from cores or from the remains of cultivated trees which until recently were grafted onto domestic stock, or as today, on quince. Young plants can be very thorny leading to confusion with **P. pyraster** (L.) Burgsd., **Wild Pear** which can only be separated with certainty by its small, hard, round fruits. It has been confirmed at Mattishall, E01. W15. E13.

Malus sylvestris (L.) Mill. **Crab Apple**

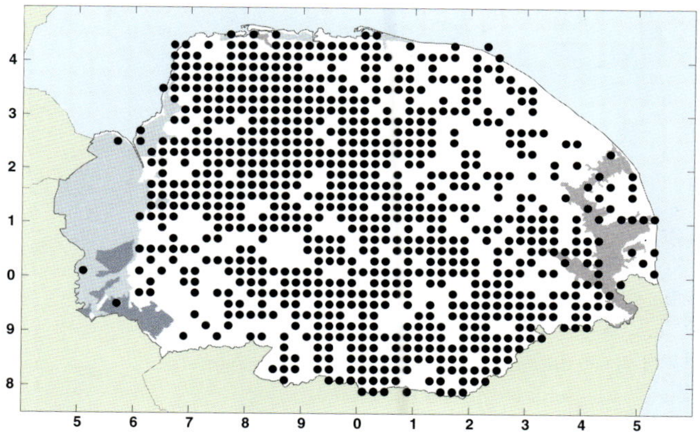

Deciduous shrub or tree frequent in hedges especially where they mark the edge of old commons, also around villages.

M. domestica Borkh. **Apple**
Deciduous shrub or tree of complicated origin. Not uncommon in roadside hedges where it is most likely to be derived from cores thrown from passing cars. Particularly frequent along new bypasses. W271. E435.

Sorbus aucuparia L. **Rowan**

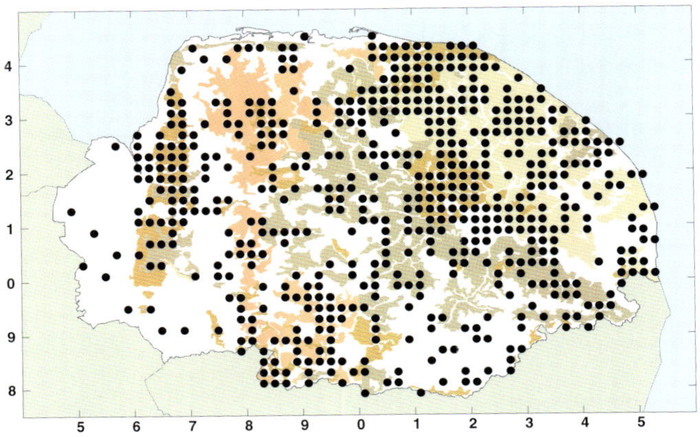

Deciduous shrub or tree most frequent on sandy, acid soils, but its present distribution is not always clear because of recent plantings.

S. intermedia (Ehrh.) Pers. **Swedish Whitebeam**
Deciduous shrub or tree from the Baltic region. Sometimes planted, but also spread by seed and found far from habitation. Known from an isolated hedge on Feltwell Fen W69 since the 1960s, E & J.Secker and frequent on parts of Mousehold Heath E21, ALB where it is bird-sown. W11. E27.

S. aria (L.) Crantz **Whitebeam**
Deciduous shrub or tree native to south-west England. In Norfolk, always introduced but self-sowing once established. W7. E42.

S. torminalis (L.) Crantz **Wild Service Tree**

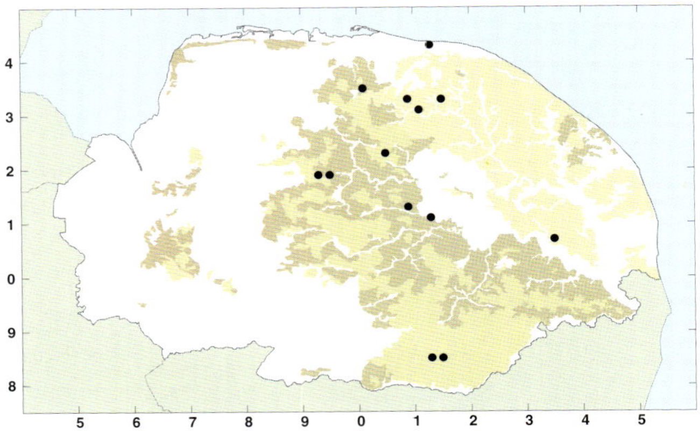

Deciduous shrub or tree native to ancient woodlands and occasionally also found in nearby hedges.

Amelanchier lamarckii F.G.Schroed. **Juneberry**
Deciduous shrub introduced from North America, found at Sheringham E14, FF and in woodland at Hempnall E29, ALB. W0. E2.

Cotoneaster frigidus Wall. ex Lindl. **Tree Cotoneaster**
Deciduous shrub or small tree introduced from the Himalayan region. One plant found in a hedgerow at Marlingford E10 and one in a marl pit at Garboldisham W98. W1. E1.

C. x watereri Exell **Waterer's Cotoneaster**
(*C. frigidus x C. salicifolius*)
Semi-evergreen shrub or small tree of garden origin often planted in cemeteries and churchyards and self-sowing as in Norwich E20 and King's Lynn W62. W2. E2.

C. salicifolius Franch. **Willow-leaved Cotoneaster**
Evergreen to semi-evergreen shrub from western China recorded on walls in King's Lynn W61 in 1996, RMP and at Didlington W79 in 1994, Flora Group. W3. E0.

C. lacteus W. W.Sm. **Late Cotoneaster**
Evergreen shrub from China recorded on two occasions, at Southery W69 in 1997, RMP and Sculthorpe W83 where it was self-sowing on the former air-base in 1994, Flora Group. W2. E0.

C. integrifolius (Roxb.) G.Klotz **Entire-leaved Cotoneaster**
Evergreen shrub usually grown as *C. microphyllus*, long established on sand dunes at Burnham Overy Staithe W84. W1. E0.

C. horizontalis Decne. **Wall Cotoneaster**
Deciduous shrub by far the commonest naturalised cotoneaster self-sown on walls and into pavement cracks. W56. E50.

C. hjelmqvistii Flinck & B.Hylmö **Hjelmqvist's Cotoneaster**
Deciduous shrub from China found at Heacham W63 in 1993, PRC. W1. E0.

C. divaricatus Rehder & E.H.Wilson **Spreading Cotoneaster**
Deciduous shrub from China recorded on the railway bank at Wighton W94 in 1981, KAB and at Heacham W63 PRC. W2. E0.

C. simonsii Baker **Himalayan Cotoneaster**
Deciduous shrub from the Himalayan region, frequently self-sowing on to walls and waste ground, mostly in urban areas. W18. E10.

C. bullatus Bois **Hollyberry Cotoneaster**
Deciduous shrub from China found naturalised at Anmer W72 in 1989, KAB; Cranwich W79 in 1992; Thetford W88 in 1997, Flora Group; Beeston Common E14 in 1997, FF; Old Buckenham E09 and Hickling E42, both KAB, 1998. W3. E3.

C. franchetii Bois **Franchet's Cotoneaster**
Evergreen to semi-evergreen shrub from China, found on a wall in Fakenham W93 in 1997, RMP and in Norwich E20, ETD. W1. E3.

C. sternianus (Turrill) Boom **Stern's Cotoneaster**
Evergreen to semi-evergreen shrub from China found on walls in King's Lynn W61 and W62; Wisbech W41 and Downham Market W60 all RMP and Norwich E20, ETD. W5. E3.

C. dielsianuus E.Pritz. ex Diels **Diel's Cotoneaster**
Deciduous shrub from China, found self-sown on the banks of the New Cut at Stoke Ferry W69 in 1993, Flora Group; near Marsham E12 in 1996, K&GB and on walls at Thetford W88 and King's Lynn W61, RMP. W3. E1.

Pyracantha coccinea M.Roem. **Firethorn**
Evergreen shrub introduced from southern Europe, self-seeding on a wall in King's Lynn W61 in 1996, RMP and at Keswick E20. W1. E1.

Crataegus persimilis Sarg. **Broad-leaved Cockspur-thorn**
Syn. *C. prunifolia* Pers. non (Marshall) Baumg.
Deciduous tree or shrub from North America being planted increasingly frequently beside bypasses and soon likely to be found self-sown. Also occasionally around new plantations as at Hardingham E00. W0. E3.

C. monogyna Jacq. **Hawthorn**
Deciduous shrub or small tree long planted as hedging and also frequent in scrub. Pink and double-flowered cultivars are met with in cemeteries and churchyards and a specimen of the Glastonbury

Thorn, which retains its leaves until well into the winter and flowers around Christmas, grows beside a former layby at Etling Green E01. Fruits produced by it in spring contain no seeds, while fruit ripening in autumn is normal, ALB obsv. W690. E716.

C. x media Bechst.
Syn. *C. x macrocarpa* auct. non Hegetschw. (*C. monogyna x C. laevigata*)
Deciduous shrub or small tree sometimes found without one parent or well away from the normal range of *C. laevigata*. In this case it seems likely that young plants have at some time been received from nurseries in a batch of *C. monogyna*. W3. E12.

C. laevigata (Poir.) DC. **Midland Hawthorn**
Syn. *C. oxyacanthoides* Thuill.

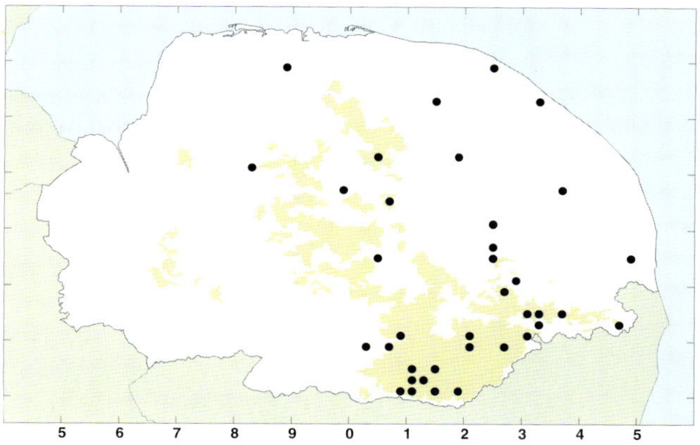

Deciduous shrub or small tree scattered in woodland and hedgerows on heavy, clay soils. Occasionally planted with native shrub mixes as on the Swaffham bypass W80 in 1994, where young plants were noted growing in plastic tubes, JEG.

Other *Crataegus* species have been planted in hedges and used in recent planting schemes, but none has been noted as regenerating.

FABACEAE (LEGUMINOSAE)

Robinia pseudoacacia L. **False Acacia**
Deciduous tree introduced from North America, regenerating freely from suckers and occasionally by seed. W22. E54.

Vigna radiata (L.) Wilczek **Mung Bean**
Syn. *Phaseolus aureus* Roxb.
Annual, a casual used by Chinese restaurants as bean sprouts and found at King's Lynn, in W61 and 62 in 1993 by RMP. W2. E0.

■ **Psoralea americana** L. **Scurfy Pea**
Syn. *P.dentata* DC
Herbaceous perennial introduced from southern Europe, a casual found at Binham W94 in 1959 and Harford, Norwich E20 in 1961.

Galega officinalis L. **Goat's Rue**
Herbaceous perennial introduced from mainland Europe to gardens for ornament or as a medicinal herb, also grown as a fodder crop. Recorded on a bank at Hoveton E31 in 1988, PA.Smith and by a bridleway at Old Buckenham E09 in 1997, ALB & WM. W0. E2.

Colutea arborescens L. **Bladder-senna**
Deciduous shrub introduced to gardens from mainland Europe, formerly naturalised on Mousehold Heath E21. Scarce on a few road verges and hedgerows. W2. E4.

Astragalus danicus Retz. **Purple Milk-vetch**

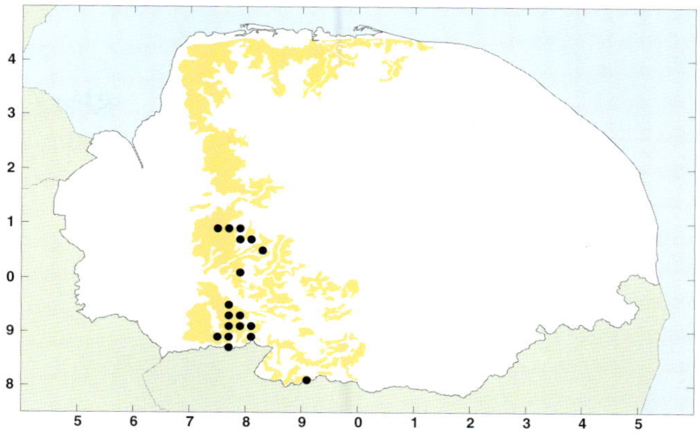

Winter green perennial confined to well-drained, chalky soils including open rides in coniferous forest. Curiously absent from the chalk grasslands north of the River Nar.

Astragalus danicus - Cranwich

A. glycyphyllos L. **Wild Liquorice**
Herbaceous perennial with five old records. Found during this survey by a footpath over the North Norfolk Railway, leading to the cliff west of Sheringham E14 in 1995, BSBI meeting and on a roadside verge near Downham Market W60 in 1996, RMP. W1. E1.

Onobrychis viciifolia Scop. **Sainfoin**

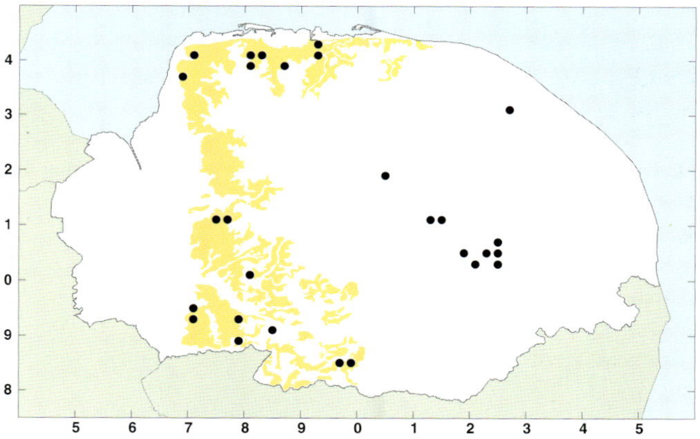

Herbaceous perennial, native on chalk, and perhaps so in most of its west Norfolk sites, also a relic of cultivation. Also sown on some new road verges as along the Norwich southern bypass E10, 11 & 20.

■ **O. caput-galli** (L.) Lam.
This annual occurred at East Tuddenham E01 as a garden weed in 1979, ALB.

Anthyllis vulneraria L. **Kidney Vetch, Ladies' Fingers**

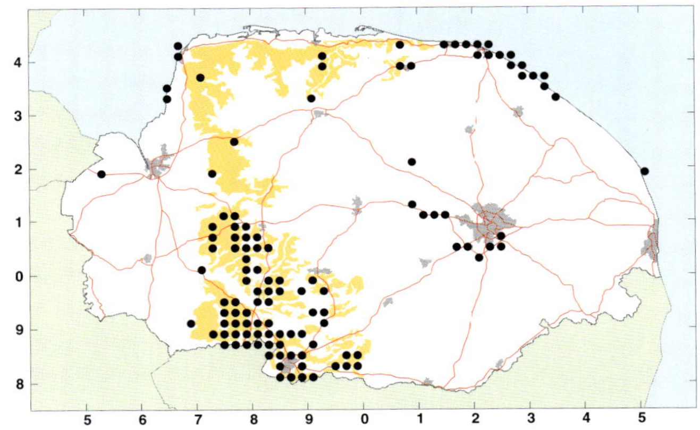

Winter green perennial of light sandy, calcareous soils. The native plant is ssp. **vulneraria** but the continental ssp. **polyphylla** (Ser.) Nyman has been sown on some new road verges. The records immediately south of Norwich are of introductions on the southern bypass.

Anthyllis vulneraria - Cranwich

Lotus glaber Mill. **Narrow-leaved Bird'sfoot Trefoil**
Syns. *L. tenuis* Waldst. & Kit. ex Willd. *L. corniculatus* ssp. *tenuis* (Waldst. & Kit. ex Willd.) Syme

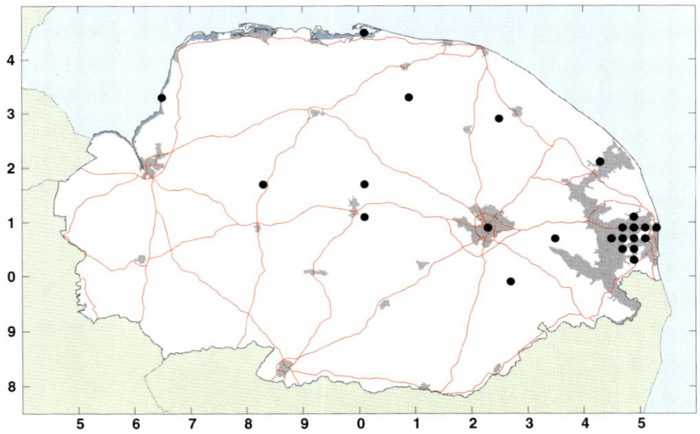

Herbaceous perennial of seabanks and pastures with a saline influence, a few scattered records inland, many on disused railway embankments and in sand pits. Perhaps sometimes overlooked for *L. corniculatus.*

L. corniculatus L. **Common Bird'sfoot Trefoil, Bacon and Eggs, Lady's Slipper**

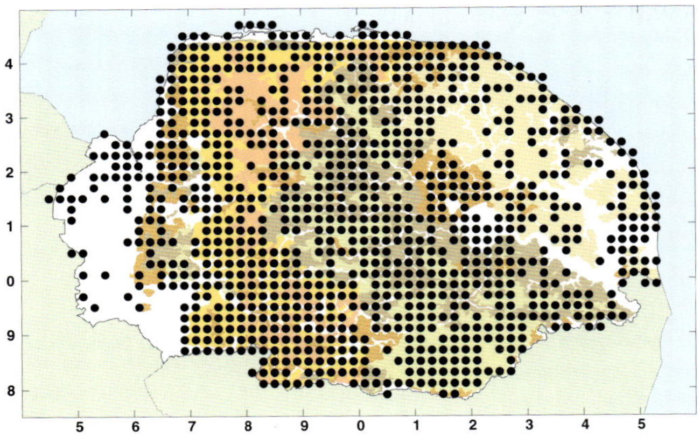

Herbaceous perennial frequent in well-drained grassland. Taller plants with larger leaves especially on the verges of new roads, may be the result of the introduction of ssp. *sativa* from mainland Europe.

L. pedunculatus Cav. **Greater Bird'sfoot Trefoil**
Syn. *L. uliginosus* Schkuhr.

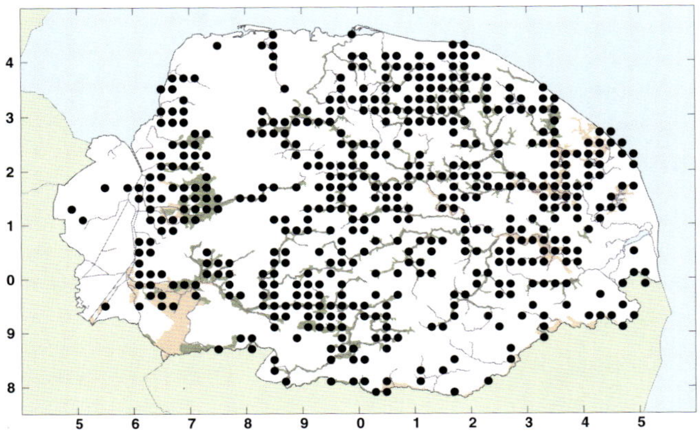

Herbaceous perennial frequent in moist grassland, fens and streamsides.

Ornithopus perpusillus L. **Bird's foot**

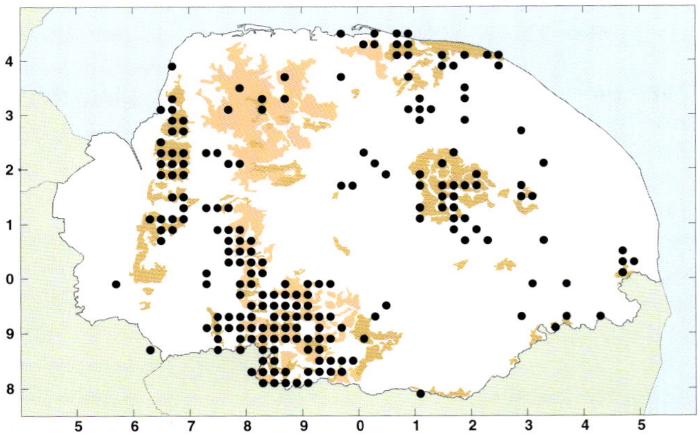

Autumn germinating annual of sandy and gravelly places, common only on acid soils.

■ **Coronilla scorpioides** (L.) W.D.J.Koch
Annual Scorpion-vetch
Spring-germinating annual from southern Europe, a casual recorded at Shotesham E29 in 1969 and at Kilverstone W88 in 1978, EAE.

Hippocrepis comosa L. **Horseshoe Vetch**

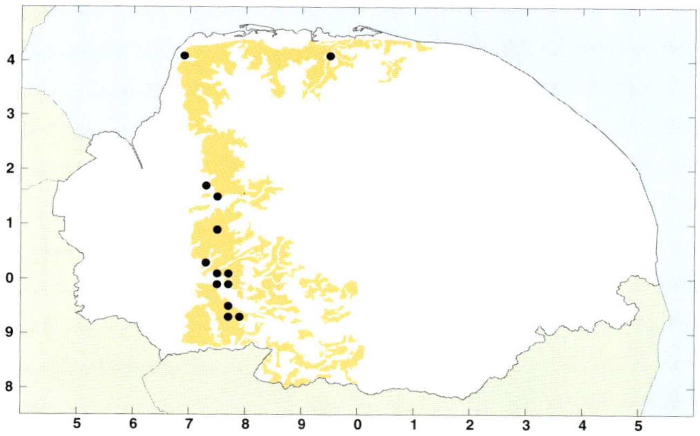

Winter green perennial restricted to undisturbed calcareous grassland. A nationally scarce species.

Hippocrepis comosa - Foulden Common

Securigera varia (L.) Lassen **Crown Vetch**
Syn. *Coronilla varia* L.
Herbaceous perennial introduced from mainland Europe, recorded from Thetford W88 in 1996, NG and Norwich E20, ETD. W1. E1.

■ **Scorpiurus muricatus** L. **Caterpillar Plant**
Spring-germinating annual from southern Europe, an occasional casual, recorded from Loddon E39 and Ludham E31 in 1968.

■ **Cicer arietinum** L. **Chick Pea**
Annual, probably originating from south-west Asia, a casual recorded once, from King's Lynn, W62 in 1942.

Vicia cracca L. **Tufted Vetch**

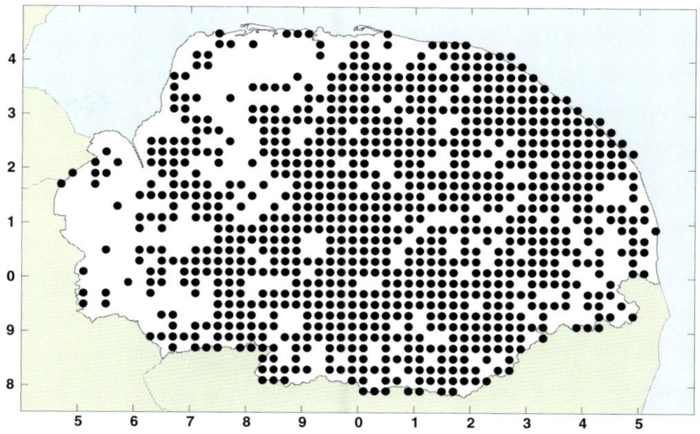

Herbaceous or somewhat winter green perennial of hedgerows and rough grassland, a feature of Broadland fen margins, but absent from drier areas where sheep grazing may have been an important factor in the past, also from recently reclaimed land. Colonises new areas slowly.

V. tenuifolia Roth **Fine-leaved Vetch**
Herbaceous perennial from mainland Europe found at Coltishall E21 in 1965; East Wretham W98 in 1978, ALB and in a forest ride in the adjacent parish of Roudham W98 in 1997, K&GB. W1. E0.

V. villosa Roth **Fodder Vetch**
Annual from mainland Europe with three records, the most recent at King's Lynn W62 in 1993, RMP. W1. E0.

■ **V. benghalensis** L. **Purple Vetch**
Annual from the Mediterranean recorded as a garden weed at East Tuddenham E00 in 1982, ALB.

V. hirsuta (L.) Gray **Hairy Tare**

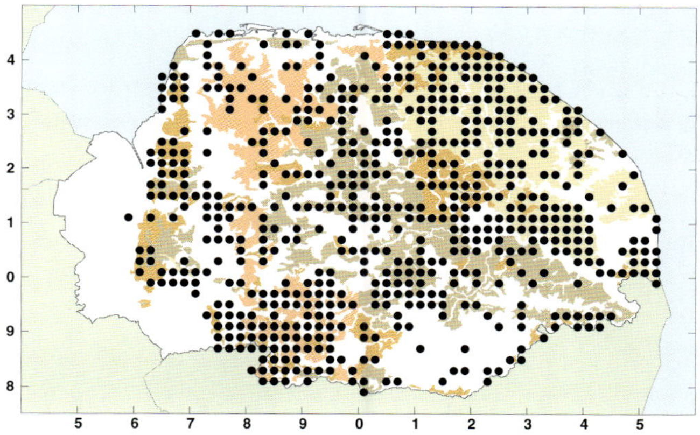

Annual of disturbed, dry grassland and waste places.

V. parviflora Cav.
Syn. *V. tenuissima* auct. non (M.Bieb) Schinz & Thell.
Native to southern England, but probably no more than a casual in Norfolk, recorded at Alburgh E28 in 1969 and during this survey at Brundall E30 in 1990, SP. W0. E1.

V. tetrasperma (L.) Schreb. **Smooth Tare**

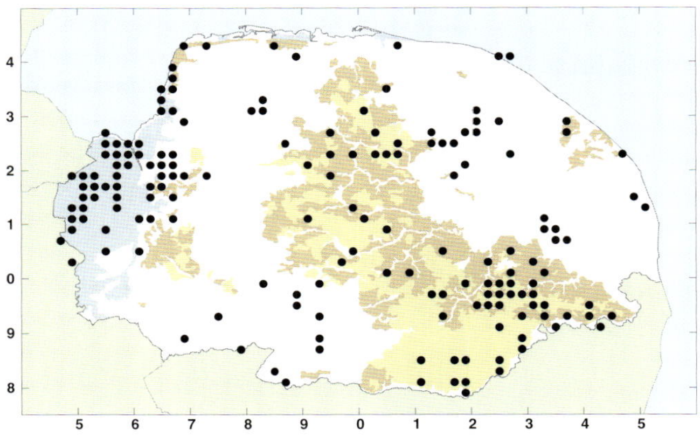

Annual of rough grassland, avoiding sandy soils both acid and calcareous. Its distribution is almost complimentary to that of *V. hirsuta*.

V. sepium L. **Bush Vetch**

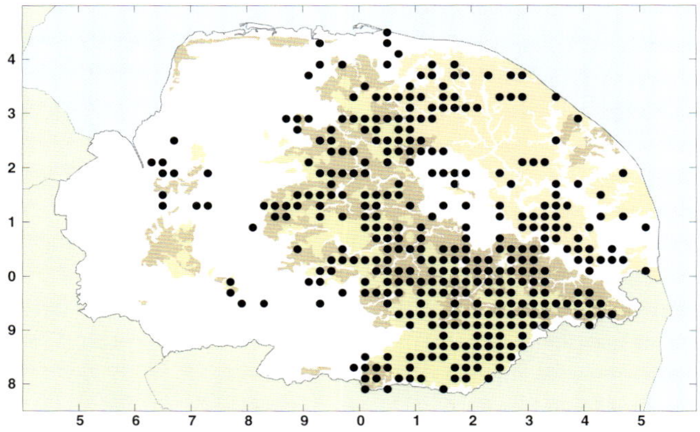

Perennial, frequent in hedgerows and wood edges on moist, heavy soils, absent elsewhere. Var. *ochroleuca* Bast. appeared annually in Foxley Wood E02 in the 1970s, ALB.

V. sativa L. ssp. **nigra** (L.) Ehrh. **Slender Vetch**
syn. *V. angustifolia* L.

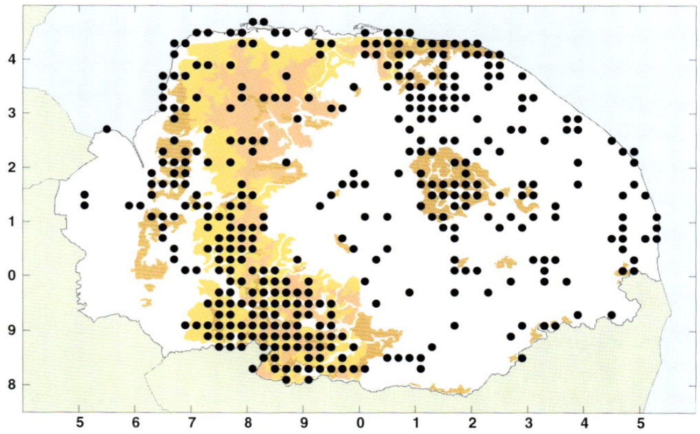

Short-lived perennial found on light, dry soils in open areas and short grassland.

V. sativa L. ssp. **segetalis** (Thuill.) Gaudin **Common Vetch**

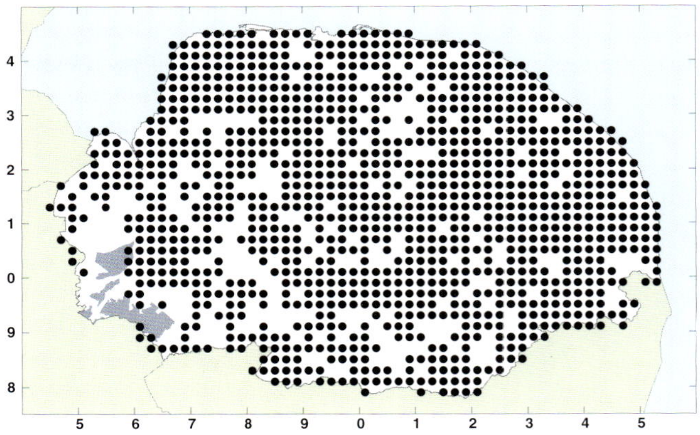

Short-lived perennial widespread in grassland, on roadside verges and in waste places on all soils including clay.

V. sativa L. ssp. **sativa** **Cultivated Vetch**
Annual introduced from mainland Europe for fodder. Rare, surviving occasionally on field edges and roadsides as at Welney W59 in 1990, MK. W1. E0.

V. lathyroides L. **Spring Vetch**

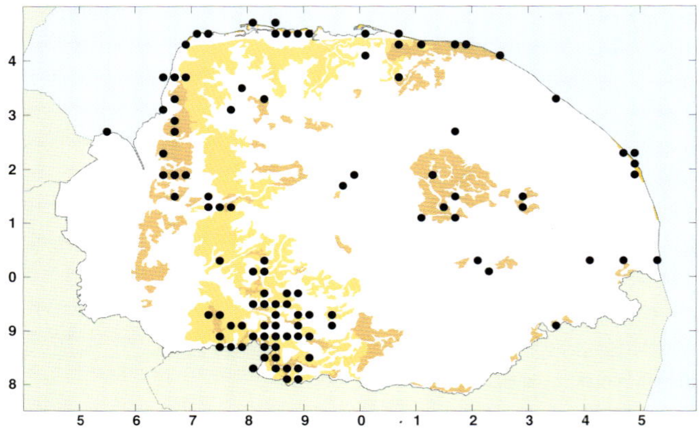

Annual found on dry heaths, coastal dunes and sandy places.

V. lutea L. **Yellow Vetch**
Annual native in parts of England but a casual here. Known at Brisley E92, between 1960 and 1981, DMM and Hardingham E00 in 1992, J.Dawson. W0. E1.

■ **V. bithynica** (L.) L. **Bithynian Vetch**
Annual perhaps native in parts of England, recorded as a casual in Norwich E20 in 1934.

■ **V. narbonensis** L. **Narbonne Vetch**
Annual from southern Europe recorded at Lynford W89 in 1970, VM.Leather.

V. faba L. **Horse Bean, Field Bean**
Annual of uncertain origin. Our plants are var. *equina* Pers. It has been in cultivation since prehistoric times and is still grown as a field crop, plants surviving briefly from spilt seed by roadsides and in field edges. W33. E40.

Lens culinaris Medik. **Lentil**
Annual from South-west Asia seen on Harford tip, Norwich E20 in 1958 and 1969, also from Norwich E20 in 1990, ETD. W0. E1.

Lathyrus japonicus Willd. **Sea Pea**
Herbaceous perennial, its main area of distribution in Britain from Norfolk to Dorset. Recorded from Holkham in 1935. In 1996 it was still to be found at Blakeney Point and Cley E04 where it was known to have been sown in 1912 and 1953, but could not be found in 1997 after the shingle ridge was overrun by the sea in spring storms. A nationally scarce species. W0. E2.

L. pratensis L. **Meadow Vetchling**

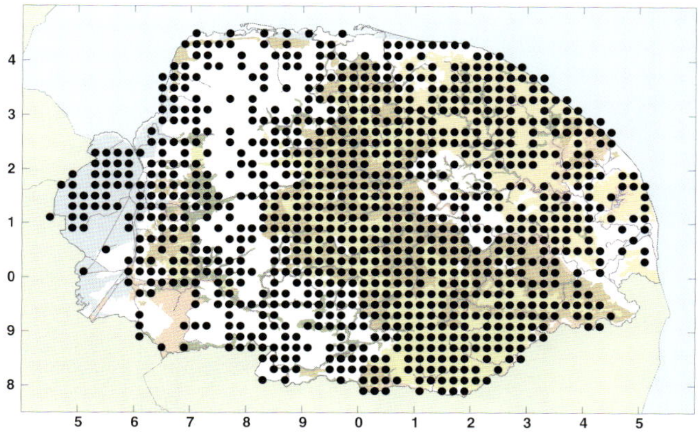

Herbaceous perennial absent only from very dry or peaty soils.

L. palustris L. **Marsh Pea**

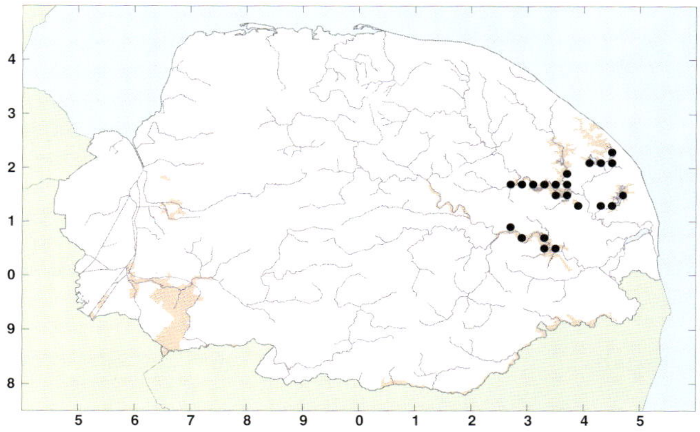

Herbaceous perennial still well distributed in Broadland. Formerly also in the Fenland and adjacent valleys. Last seen outside the Broads at Blackborough End W62 in 1970, CPP. A nationally scarce species.

L. tuberosus L. **Tuberous Pea, Fyfield Pea**
Herbaceous perennial from mainland Europe, established for some years at Feltwell W79 where it was reported in 1991, R.Tofts and in 1997 J.Secker, also at Sedgeford W73 in 1994, SF, and Blofield E30 in 1998, SP. W2. E1.

L. grandiflorus Sm. **Two-flowered Everlasting Pea**
Herbaceous perennial from the Mediterranean region well established at Langford in the Stanford Training area W89 where a colony which has been undisturbed for 45 years covers almost 35sq.m., ALB. W2. E5.

Lathyrus palustris - Wheatfen

L. sylvestris L. **Narrow-leaved Everlasting Pea**

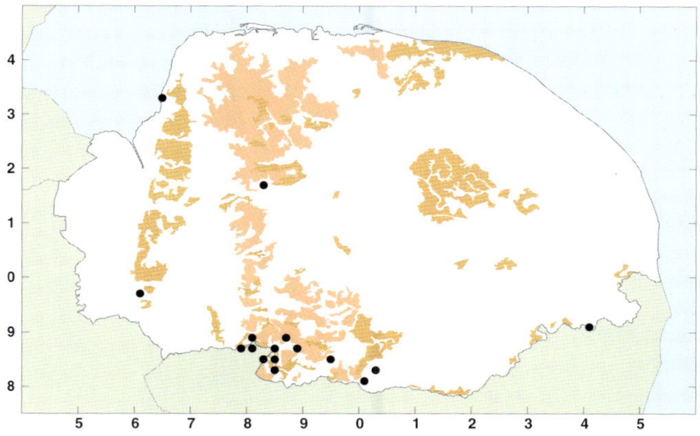

Herbaceous perennial scattered in shrubby places on light soils, particularly in the Brandon-Thetford area, W78- 88.

L. latifolius L. **Broad-leaved Everlasting Pea**
Herbaceous perennial from mainland Europe, a very conspicuous and long lasting garden escape. W23. E43.

L. heterophyllus L. **Norfolk Everlasting Pea**
Herbaceous perennial from mainland Europe, a survivor at Burnham Overy Staithe W84 from a garden on the dunes made before 1940, this plant is var **unijugos** W.D.J.Koch, with one pair of leaflets. W1. E0.

■ **L. odoratus** L. **Sweet Pea**
Annual introduced to gardens from southern Italy, occasionally found on waste places, not lasting long. Harford tip, Norwich E20, 1971, AC & PG.Lawson.

■ **L. sativus** L. **Indian Pea**
Annual from the Mediterranean found on a quayside at King's Lynn W62 in 1942.

■ **L. hirsutus** L. **Hairy Vetchling**
Annual from mainland Europe recorded from near Harford tip, Norwich E20 in 1972.

L. nissolia L. **Grass Vetchling**

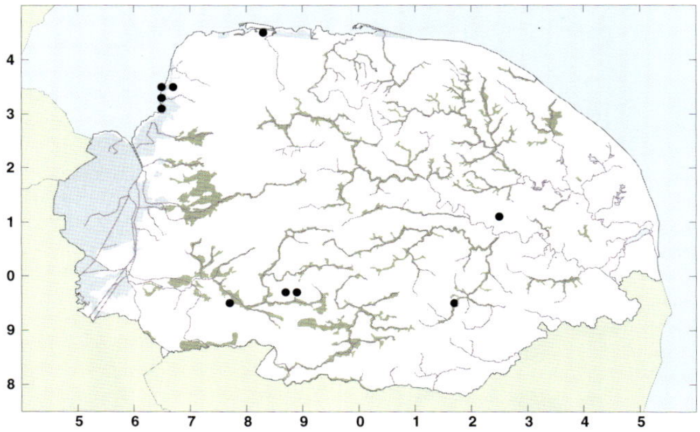

Autumn germinating annual of coastal grassland, now scarce. Also occasionally inland in the Breckland and at Tharston E19, M.Cooper where it was first recorded in 1963, and as a casual on an industrial estate in Norwich E20 in 1998, C.Bennett.

■ **L. aphaca** L. **Yellow Vetchling**
Annual from mainland Europe, recorded five times before 1960.

Pisum sativum L. **Garden Pea**
Annual in its wild form native to southern Europe and occasionally met with as an escape from gardens. The variety **arvense**, the former field pea, known in the trade as 'Dun Pea' has bi-coloured purple flowers and also occurs as a birdseed alien. Found in King's Lynn W62 in 1991, RMP. W1. E0.

Ononis spinosa L. **Spiny Restharrow**

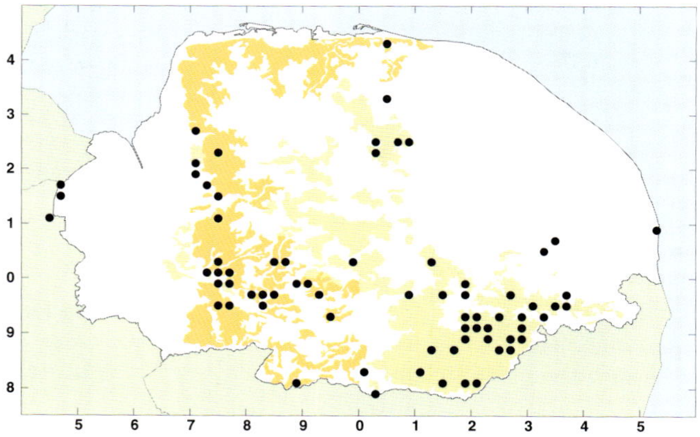

Deciduous shrub typical of old grassland and green lanes on chalk and on heavy, moist soils, particularly on the boulder clay where it was described as common by Galpin, 1888.

Ononis spinosa - East Walton

O. repens L. **Common Restharrow**

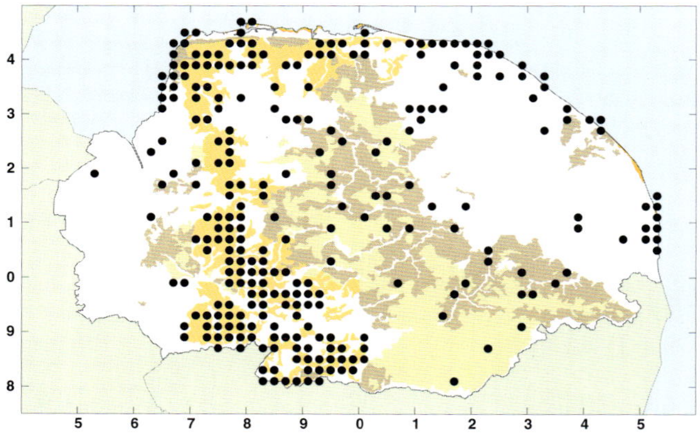

Herbaceous perennial frequent in well-drained grassland and on established dunes.

Melilotus altissimus Thuill. **Tall Melilot**

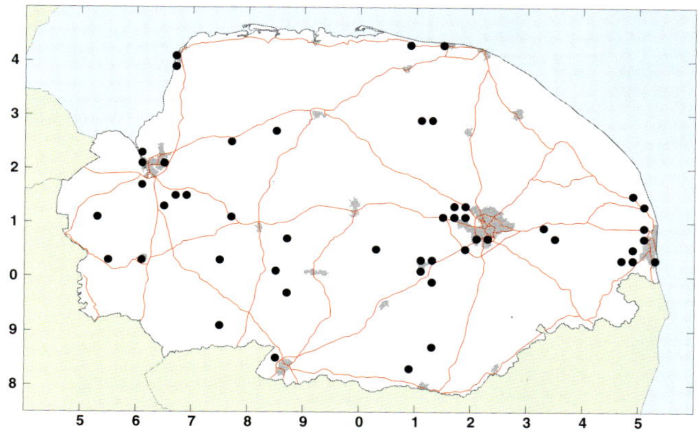

Biennial to perennial from mainland Europe formerly grown as a fodder crop, now scattered on lighter soils.

M. albus Medik **White Melilot**

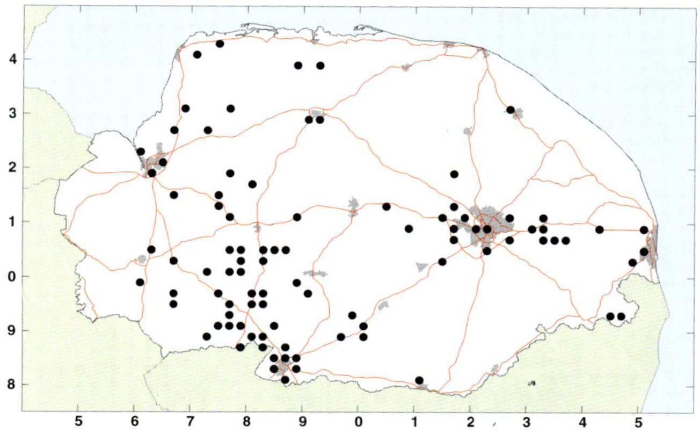

Herbaceous annual to biennial from mainland Europe, formerly grown on lighter soils for sheep feed and, like other members of the clover family, to improve the fertility of the soil. Scattered, mainly on lighter soils.

M. officinalis (L.) Pall. **Ribbed Melilot**

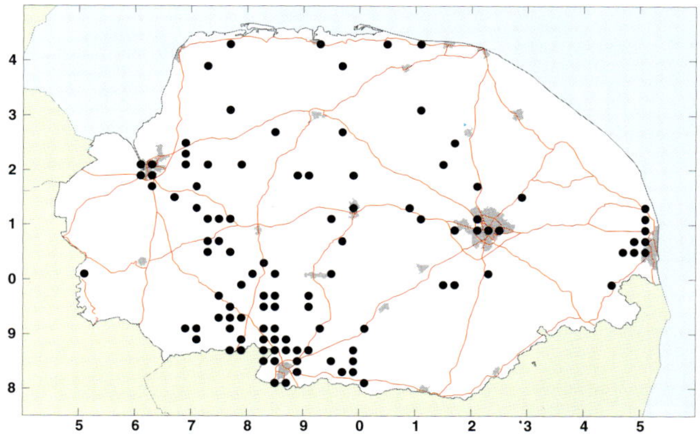

Biennial from mainland Europe introduced as a fodder crop, now usually found by roadsides and in waste places.

M. indicus (L.) All. **Small Melilot**
Annual from southern Europe, recorded fifteen times before 1980, but only once during this survey, at Snetterton Heath E08 in 1996, JH.
W0. E1.

■ **Trigonella foenum-graecum** L. **Fenugreek**
Annual from the eastern Mediterranean, usually introduced from birdseed, two records, in Norwich E20 in 1915 and again in 1974.

Medicago lupulina L. **Black Medick**
Annual to short-lived perennial frequent on all but the wettest soils.
W609. E578.

M. sativa L. ssp. **sativa** **Lucerne, Alfalfa**

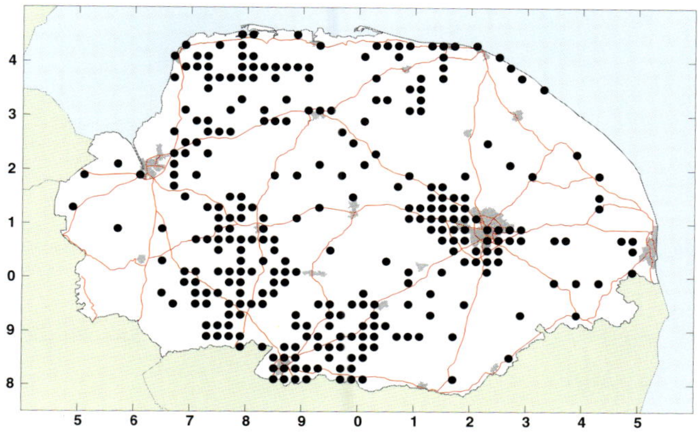

Winter green perennial or annual introduced as a fodder and green manure crop. It has escaped widely onto roadside verges, field margins and waste places, especially on light soils.

M. sativa L. ssp. **falcata** (L.) Arcang. **Sickle Medick**
Syn. *M. falcata* L.

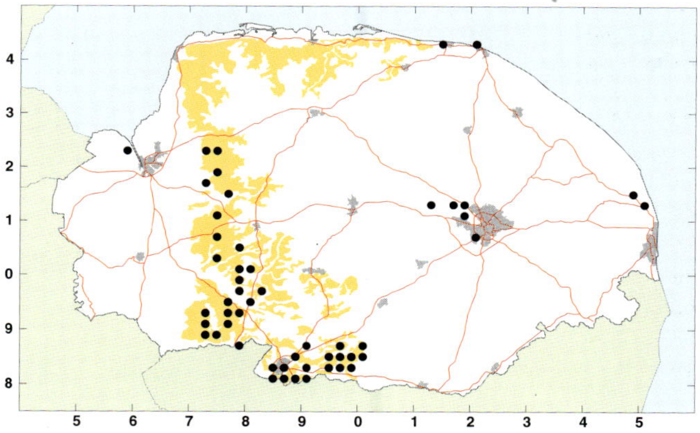

Winter green perennial, native in grassy places and waste ground, chiefly on calcareous soils in the Breckland. It is becoming much less frequent because of hybridization with lucerne A nationally scarce species

Medicago sativa ssp. **falcata** - **Feltwell**

M. sativa L. ssp. **varia** (Martyn) Arcang. **Sand Lucerne**
Syn. *M.x varia* Martyn

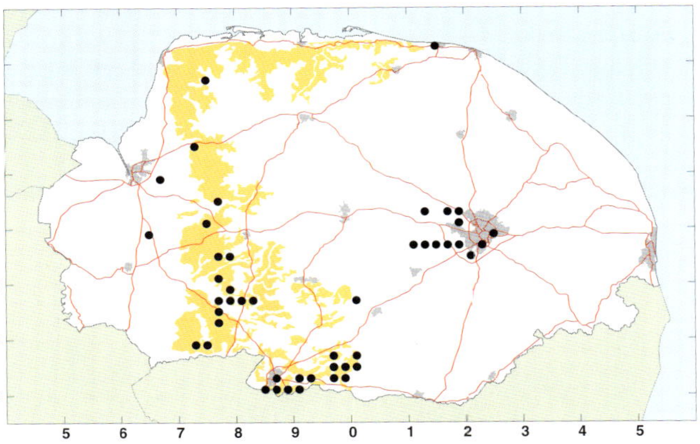

This perennial hybrid occurs spontaneously, but in the past was also introduced from Germany for fodder as 'Grimm' lucerne.

M. minima (L.) Bartal. **Bur Medick**

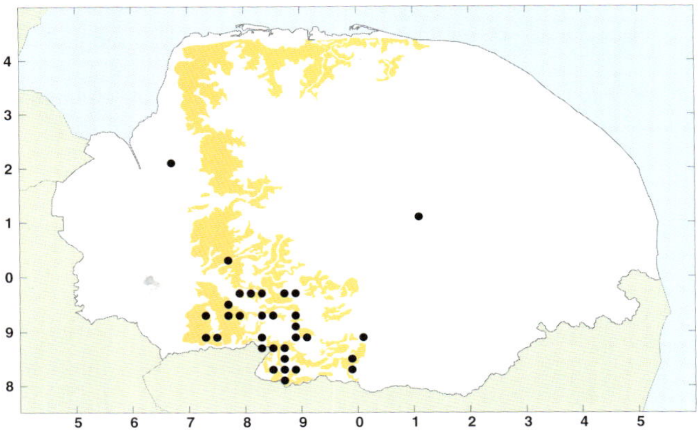

Annual, locally frequent in the Breckland on dry, sandy soils, rare elsewhere. A nationally scarce species.

M. polymorpha L. **Toothed Medick**
Syn. *M. hispida* Gaertn. nom. illeg.

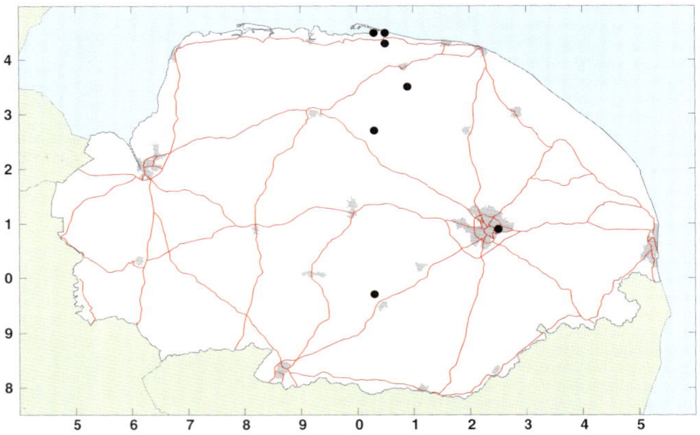

Annual, usually found in sandy ground near the sea, but during this survey also found inland where it has more the status of a casual.

M. arabica (L.) Huds. **Spotted Medick, Calvary Clover**

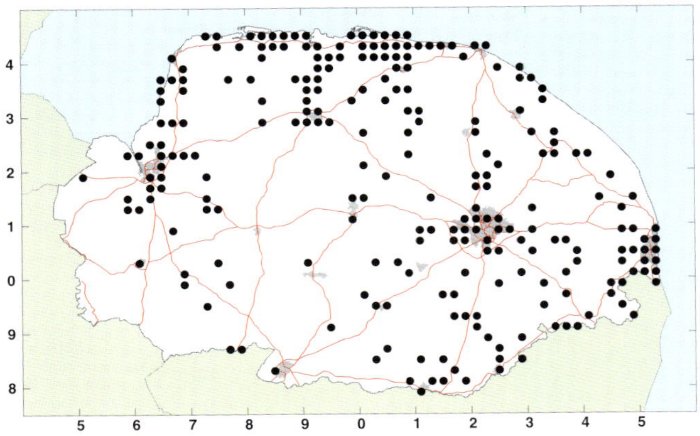

Autumn germinating annual growing in short turf including lawns and open sandy places, commonest near the sea, but found increasingly inland.

Medicago minima - Mundford

Trifolium ornithopodioides L. **Bird's-foot Clover**

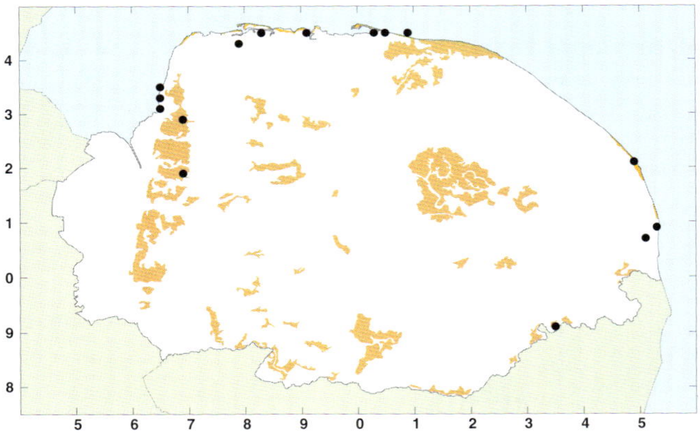

Autumn germinating annual, rare in dry grassland, mostly near the sea though the first Norfolk record was from Mousehold Heath, Norwich E21 in 1824.

T. repens L. **White Clover**
Winter green perennial, ubiquitous in grassland and waste places.
W692. E711.

T. hybridum L. **Alsike Clover**
Winter green perennial introduced from mainland Europe as a fodder plant, well established by roadsides where it was formerly sown in a standard verge mix, also in waste places. All plants belong to ssp. **hybridum**. W52. E76.

T. glomeratum L. **Clustered Clover**

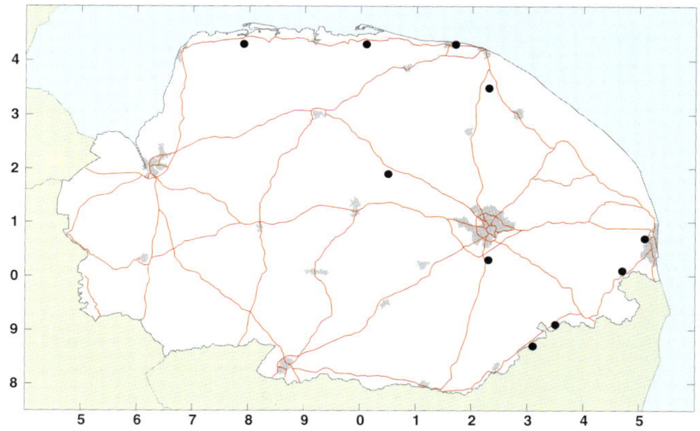

Annual found in bare places on sandy ground near the sea, also on heaths and tracks inland. A nationally scarce species.

T. suffocatum L. **Suffocated Clover**
Annual found on the Denes at Yarmouth W50 in 1932 and again in 1995, P&I.Green, and on Broome Heath in 1862 and again in 1992, GM & E.Beaumont. A nationally scarce species. W0. E2.

T. fragiferum L. **Strawberry Clover**

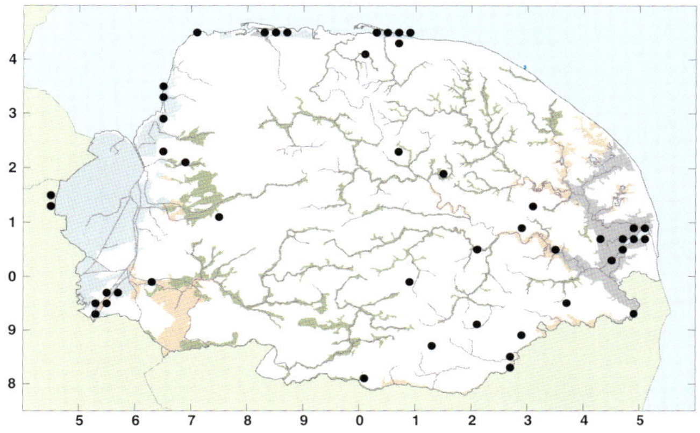

Perennial of moist grassland, tolerant of slightly saline conditions, now most frequent on and behind sea banks as around Breydon Water E40. Also inland, especially on the boulder clay where it was described by Galpin, 1888 as 'frequent in meadows and pastures' and is still to be found on green lanes and roadside verges.

■ **T. resupinatum** L. **Reversed Clover**
Annual from southern Europe, one record as a casual in Hellesdon E10 in 1945.

T. campestre Schreb. **Hop Trefoil**

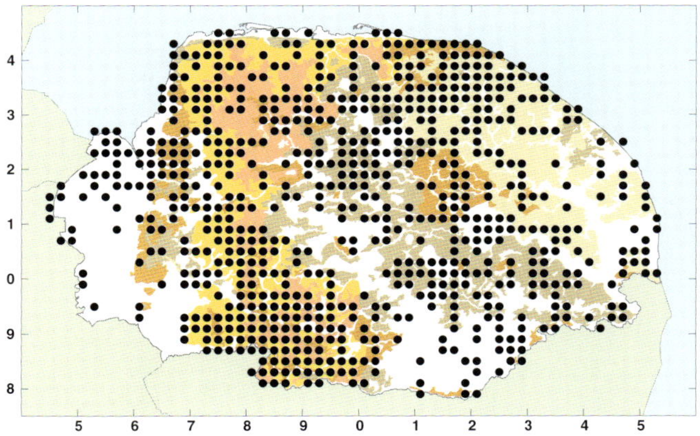

Annual of grassy places, markedly scarce along the watershed and in Broadland.

T. dubium Sibth. **Lesser Trefoil**

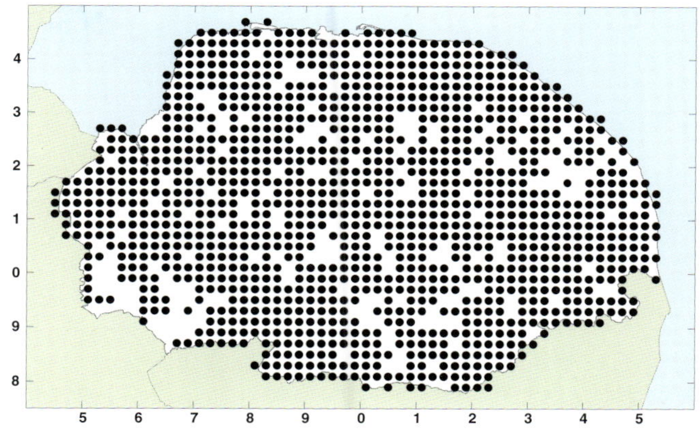

Annual of grassy places, sometimes a weed of lawns on thin soils. Scarce on peaty soils.

T. micranthum Viv. **Slender Trefoil**

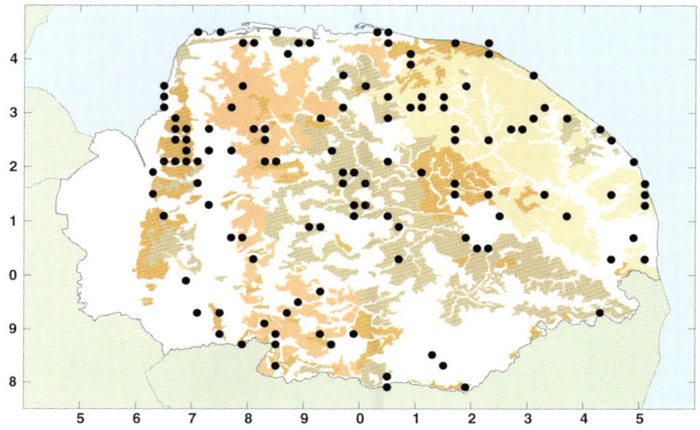

Annual of dry, gravelly places, scattered throughout the county where the habitat is suitable.

T. pratense L. **Red Clover**
Partially winter green perennial of roadside verges and grassy places, widely grown in sown pasture. Some authorities place the cultivated plant in var. **sativum**. W609. E570.

T. medium L. **Zig-zag Clover**

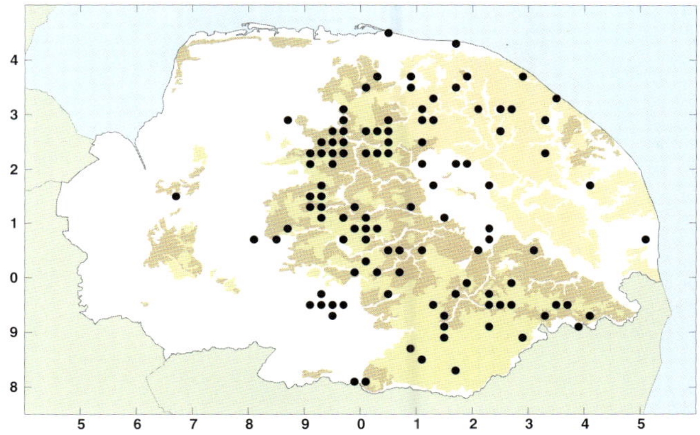

Perennial found on roadside verges and grassy places on heavy soils.

T. ochroleucon Huds. **Sulphur Clover**

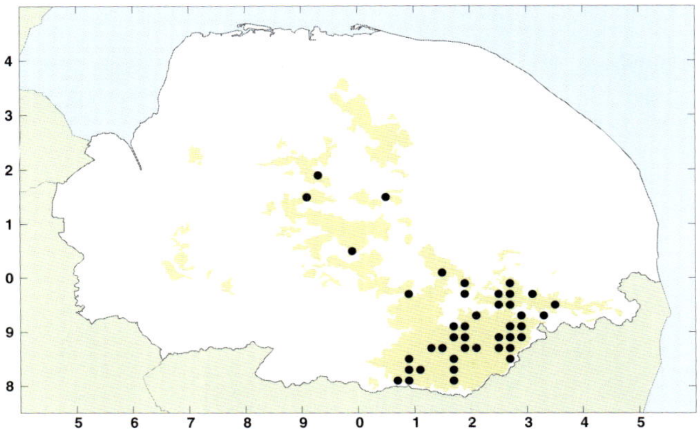

Perennial confined to heavy, calcareous clays in East Anglia, described by Galpin, 1888, as frequent in pastures. Now because of the destruction of this habitat, largely confined to roadside verges and becoming rare on these because of increasing rankness through fertiliser drift and frequent traffic damage. A nationally scarce species.

Trifolium ochroleucon - Topcroft

T. incarnatum L. **Crimson Clover**
Annual from southern Europe formerly grown for fodder and found in five sites before 1960. During this survey seen at Shotesham E29 in 1991, M.Cooper. This plant belonged to ssp. **incarnatum**. W0. E1.

T. striatum L. **Knotted Clover**

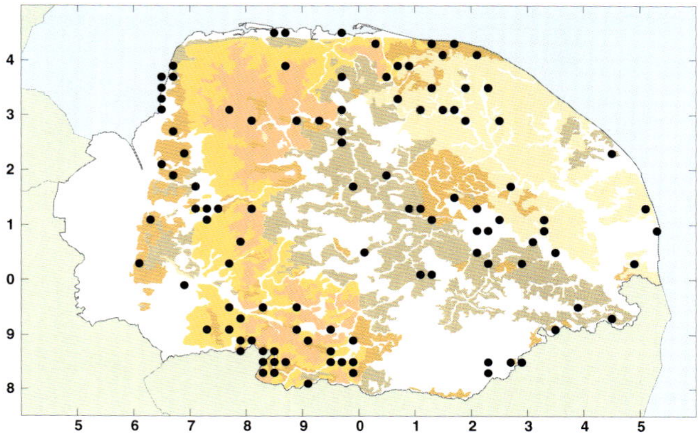

Annual of short grassland and open places on light soils.

T. scabrum L. **Rough Clover**

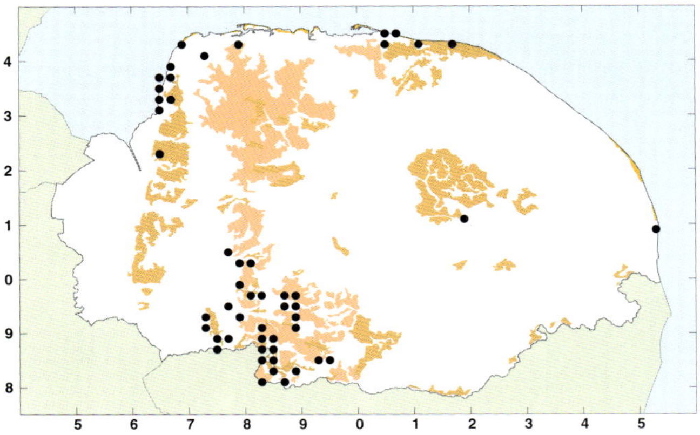

Annual of short grassland and sandy places, commonest in Breckland but also found near the sea.

T. arvense L. **Hare'sfoot Clover**

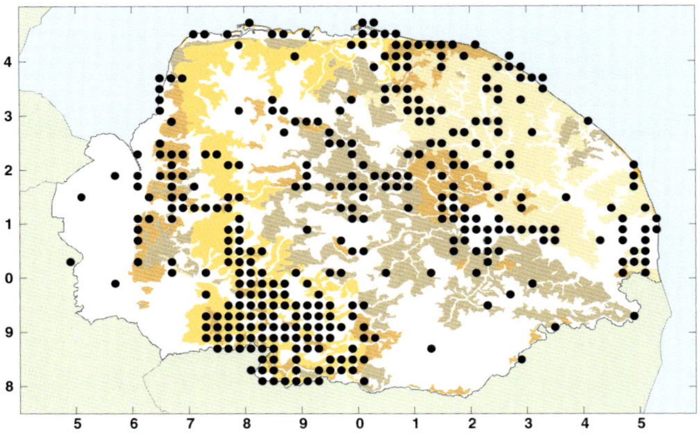

Annual, frequent on tracks and open areas on light, sandy soils, increasingly also in fields managed as long-term setaside.

■ **T. angustifolium** L. **Narrow Clover**
Annual from southern Europe, recorded once, in a chicken-run at West Acre W71 in 1952.

■ **T. squamosum** L. **Sea Clover**
Annual last recorded at Great Yarmouth E50 in 1830.

T. subterraneum L.
Subterranean Clover, Burrowing Clover

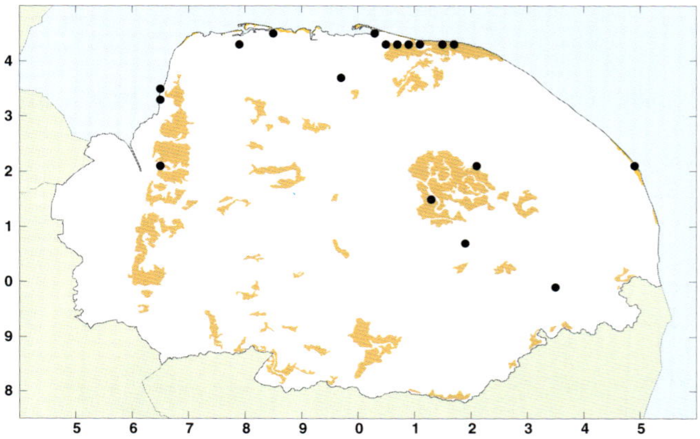

Annual of short turf and sandy places, mostly near the sea but also inland on dry soils.

Lupinus arboreus Sims — **Tree Lupin**

Partially evergreen shrub introduced to gardens from California and seeding onto waste ground. The plant is not totally hardy so persists mainly near the sea, also on the embankments of some bypasses where continuous traffic movement through cuttings meliorates the climate, obs. ALB. W16. E89.

L. x regalis Bergmans — **Russell Lupin**

(*L. arboreus x L. polyphyllus*)

Herbaceous perennial, a hybrid of garden origin, persisting on waste ground. Records of *L. polyphyllus* almost certainly belong here as the species is rarely grown. W6. E23.

L. angustifolius L. — **Narrow-leaved Lupin**

Annual from the Mediterranean, grown occasionally as a field crop, more often a casual of waste places. Recorded from King's Lynn W62 in 1994 and Wimbotsham W60 in 1995, both RMP. W2. E0.

■ **L. luteus** L. — **Annual Yellow Lupin**

Annual from the Mediterranean, a scarce arable weed found once only, at Gt Cressingham W80 in 1954.

Laburnum anagyroides Medik. — **Laburnum**

Deciduous tree introduced to gardens from the mountains of southern Europe, sometimes planted as a roadside ornamental and seeding on to waste ground. W33. E90.

Cytisus striatus (Hill) Rothm. — **Hairy-fruited Broom, Woolly-podded Broom**

Green-stemmed shrub introduced to gardens and parks from Spain and sometimes found planted in error for the native broom, also self-sowing. Noted at Leziate sandpits W61 in 1995, KAB and by the Aylsham by-pass E22, D.McClintock 1975, still present 1995, KAB. W1. E1.

C. scoparius (L.) Link — **Broom**

Syn. *Sarothamnus scoparius* (L.) W.D.J.Koch

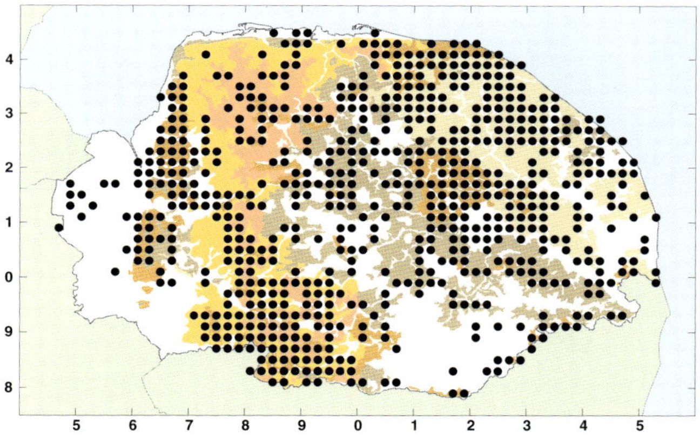

Green-stemmed shrub, a frequent free-seeding component of scrub on neutral to acid land.

Spartium junceum L. — **Spanish Broom**

Shrub introduced from the Mediterranean, planted beside the Loddon bypass E39 and spreading freely from seed. W0. E2.

Genista monspessulana (L.) L.A.S.Johnson — **Montpellier Broom**

Shrub introduced to gardens from the Mediterranean and occasionally naturalised. Seedlings found in pavement cracks in Norwich E20 in 1992, ETD. W0. E1.

G. tinctoria L. — **Dyer's Greenweed**

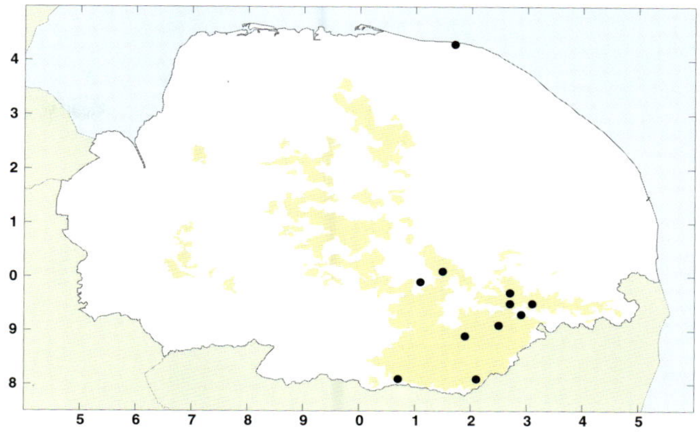

Deciduous shrub, a native on some commons and green lanes on the boulder clay in south-east Norfolk where it was described as frequent by Galpin, 1888 but is now scarce. Once widely used as a dye plant. Recorded at Wacton Common by K.Trimmer in 1866, still a few plants there in 1997, R.Maidstone. Also at Gayton Thorpe W71 on chalky clay where it was last seen 1966. Plants at Beeston E14 are likely to be an introduction.

Genista tinctoria - Wacton

G. anglica L. — **Petty Whin, Needle Furze**

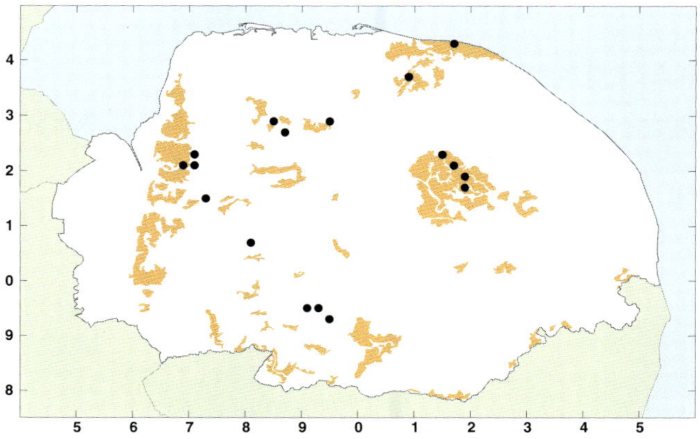

Shrub confined to moist, acid sites, very local.

Ulex europaeus L. **Furze, Gorse, Whin**

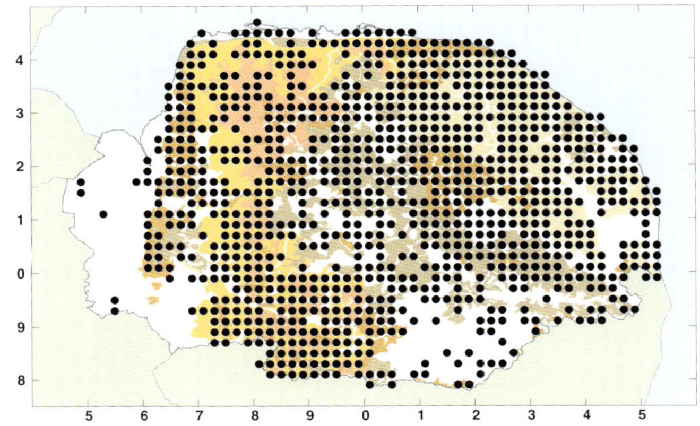

Green-stemmed shrub absent only from heaviest and wettest soils.

U. europaeus L. x **U. gallii**
Green-stemmed shrub occasionally found where the two parents grow together. W3. E0.

U. gallii Planch. **Western Gorse**

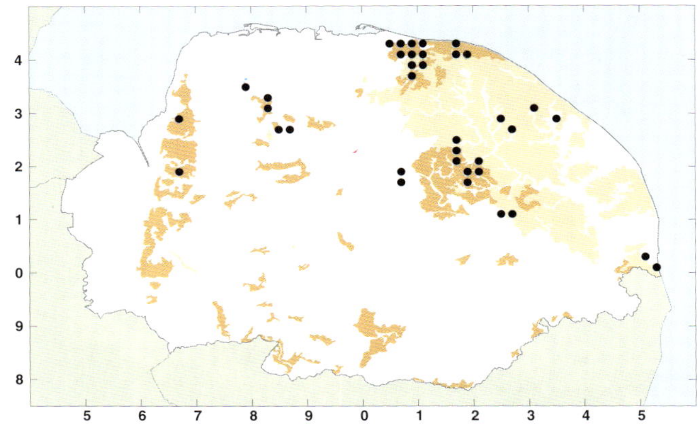

Green-stemmed shrub found on the acid, glacial soils of north-east Norfolk and on the greensand.

U. minor Roth **Dwarf Gorse**
Green-stemmed shrub formerly known from only one site, at Snetterton E09, not seen since 1964. Planted and surviving beside the Aylsham bypass, E22. W0. E1.

ELAEAGNACEAE

Hippophae rhamnoides L. **Sea Buckthorn**

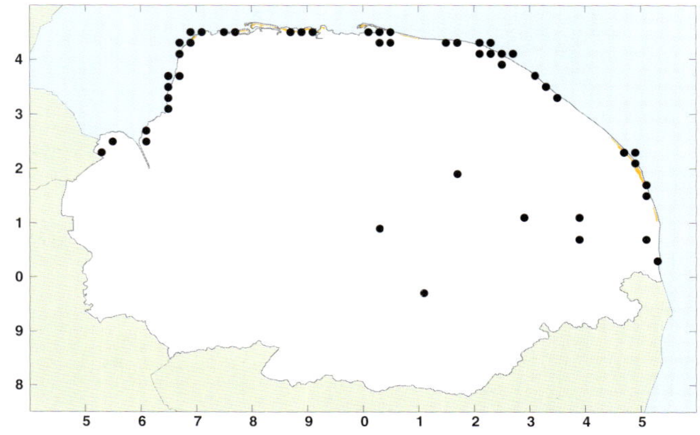

Deciduous shrub, considered native around the Norfolk coast from Snettisham eastwards. Introduced to the Wash coast west of Lynn, also found inland, often from roadside shrub plantings and soon forming dense thickets.

HALAGORACEAE

Myriophyllum verticillatum L. **Whorled Water-milfoil**

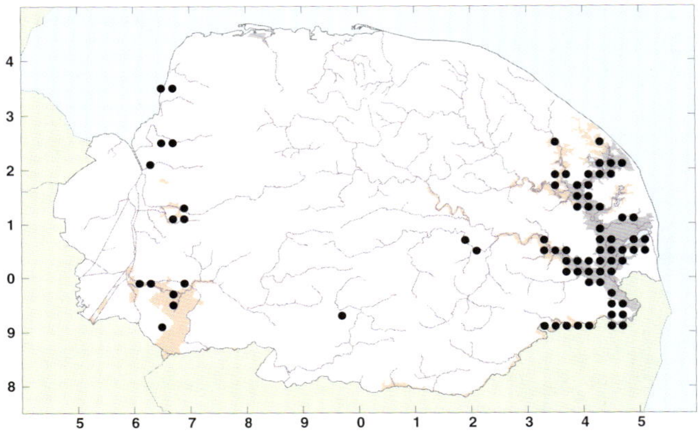

Winter green perennial, widely distributed but never common. A nationally scarce species.

M. aquaticum (Vell.) Verdc. **Parrot's Feather**
Syn. *M. brasiliense* Cambess.
Winter green perennial introduced to aquaria from South America and thrown into ponds and dykes where it can become established for a while, but not surviving overwinter if the water is deeply frozen. In a pond at Burnham Market W84 from 1989-1994, GB and on Southrepps Common E23 in 1995, PWL associated with the similarly introduced *Crassula helmsii.* W2. E11.

M. spicatum L. **Spiked Water-milfoil**

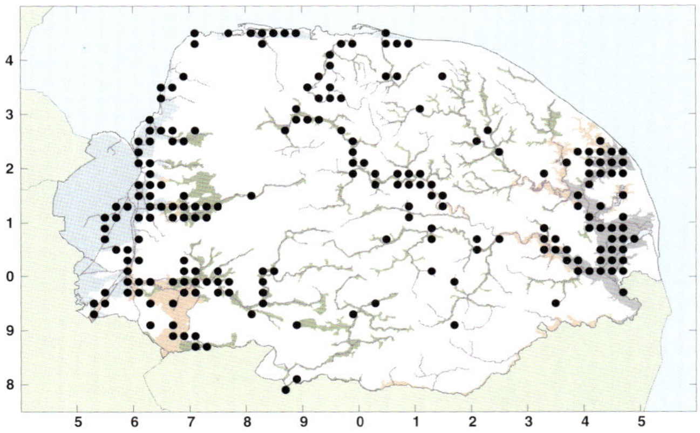

Winter green perennial found in still and slow moving water, often with hornwort, *Ceratophyllum demersum*. Not in very shallow water.

M. alterniflorum DC. **Alternate Water-milfoil**
Winter green perennial found in the Hickling to Sea Palling area in 1972-1974, RJ.Driscoll. Seen in Reservoir Pond in Melton Constable Park E03 in 1996, ALB and at Catfield Dyke E42 in 1997, J.Heath. W0. E2.

GUNNERACEAE

Gunnera tinctoria (Molina) Mirbel. **Giant Rhubarb**
Perenial occasionally planted in damp sites. Well grown self-sown seedling plants were noted with the parents by the lake at Seamere, Hingham E00 in 1996, ALB. W0. E2.

LYTHRACEAE

Lythrum salicaria L. **Purple Loosestrife**

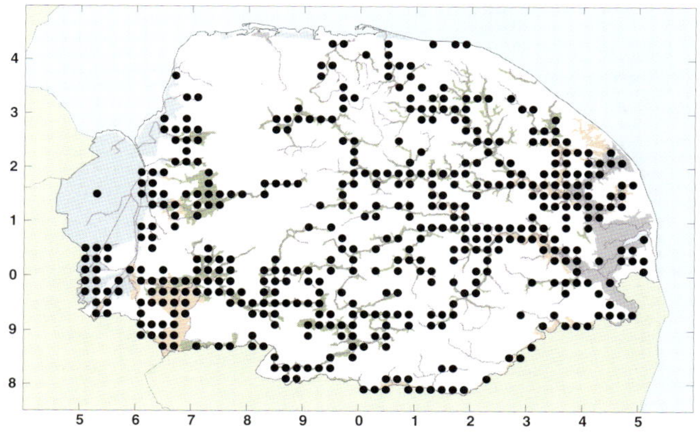

Herbaceous perennial frequent by streams and rivers, in marsh and fen.

Lythrum salicaria - How Hill

L. junceum Banks & Sol. **False Grass-poly**
Perennial but not surviving the winter in Britain so behaving as an annual. Introduced in birdseed from the Mediterranean. Four records have been made, most recently at Strumpshaw E30 in 1983, ETD and Norwich E20, P.Westley in 1998.

L. portula (L.) D.A.Webb **Water Purslane**
Syn. *Peplis portula* L.

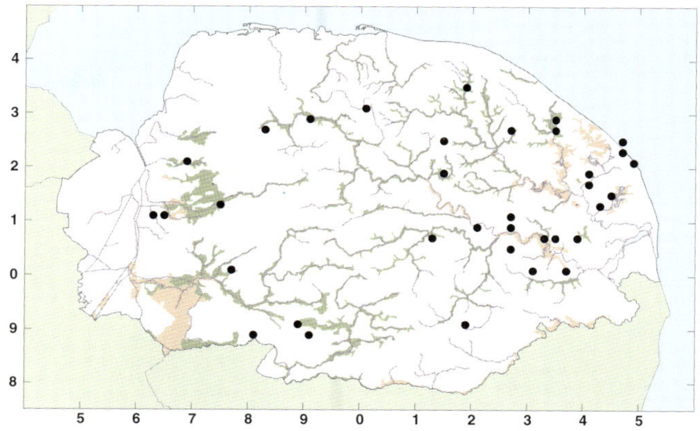

Annual scattered on bare, damp ground or the edges of pools, always where the soil is acid. Perhaps sometimes overlooked. Abundant along a recently cut flooded ditch at Horsey Gap E42 in 1997, ALB.

THYMELAEACEAE

Daphne mezereum L. **Mezereon**
Deciduous shrub, perhaps native in some sites but widely bird-sown from gardens, occurring in a wide range of habitats from dry ground to fen carr. W10. E3.

D. laureola L. **Spurge Laurel**

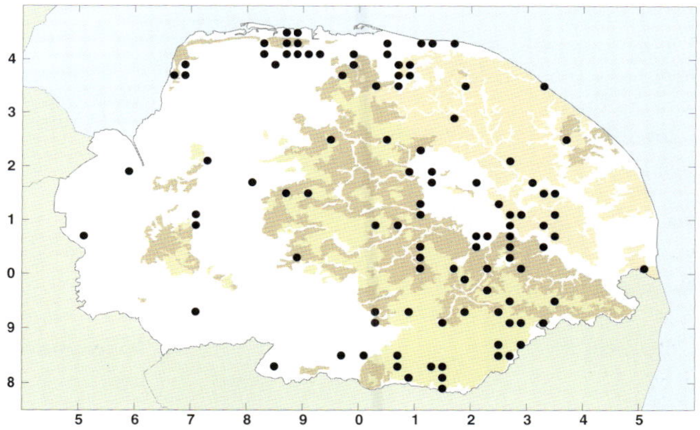

Evergreen shrub, native in woodland on chalky soils and notably in all the woods on the chalk ridge between Heacham W63 and Wells-next-the-Sea W84. Also on the boulder clay where Galpin, 1888, described it as frequent. Sometimes found bird-sown from gardens.

ONAGRACEAE

Epilobium hirsutum L.
Great Willowherb, Codlins and Cream
Deciduous perennial, the commonest of all the willowherbs, found in all moist areas except bogs, but also capable of colonising surprisingly dry sites. W642. E695.

E. x subhirsutum Gennari
(*E. hirsutum x E. parviflorum*)
One record during this survey, Castle Acre W71 in 1988, TCG.Rich. W1. E0.

■ **E. x novae-civitatis** Smejkal
(*E. hirsutum x E. ciliatum*)
Has been recorded for west Norfolk, Stace 1975.

E. parviflorum Schreb. **Hoary Willowherb**

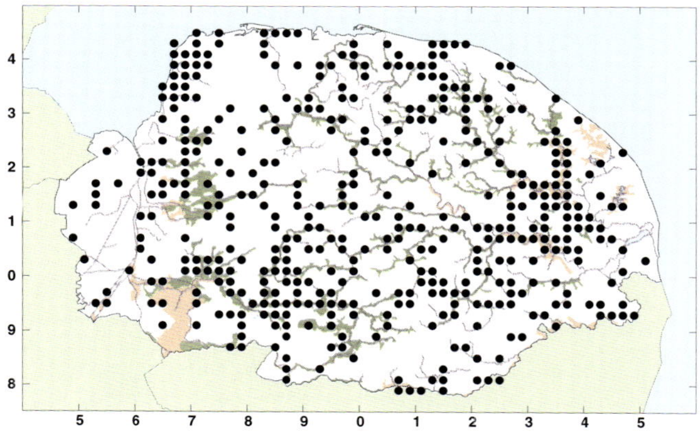

Herbaceous perennial found mainly on streambanks and in marshy places, but also on arable land and walls.

■ **E. x limosum** Schur
(*E. parviflorum x E. montanum*)
One record, at Thursford W93 in 1954, ELS.

■ **E. x palatinum** F.W.Schultz
(*E. parviflorum x E. montanum*)
Has been recorded from east Norfolk, Stace 1975.

■ **E. x dacicum** Borbás
(*E. parviflorum x E. obscurum*)
Three records before 1960.

E. parviflorum x E. ciliatum
Found during this survey on set aside fields in east Norfolk. W0. E6.

■ **E. x rivulare** Wahlenb.
(*E. parviflorum x E. palustre*)
One record, at Hickling E42 in 1936.

E. montanum L. **Broad-leaved Willowherb**

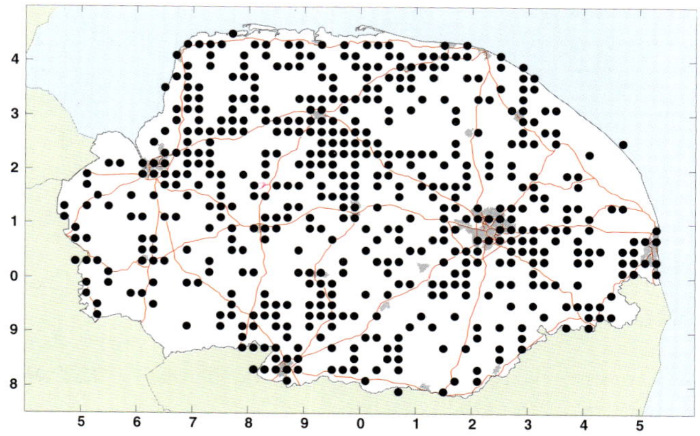

Partially evergreen perennial frequent in shady places in woods, hedgebanks and gardens.

E. montanum x E. ciliatum
Ocasional, especially in gardens. W2. E0.

E. lanceolatum Sebast. & Mauri **Spear-leaved Willowherb**
Local, first found in Norfolk at Beetley W91 in 1971, ELS & ALB. During this survey at Stradsett W60 in 1992 in gravel paths and Helhoughton churchyard W82 in 1995, both RMP. W2. E0.

E. tetragonum L. **Square-stemmed Willowherb**
Syn. *E. adnatum* Griseb.

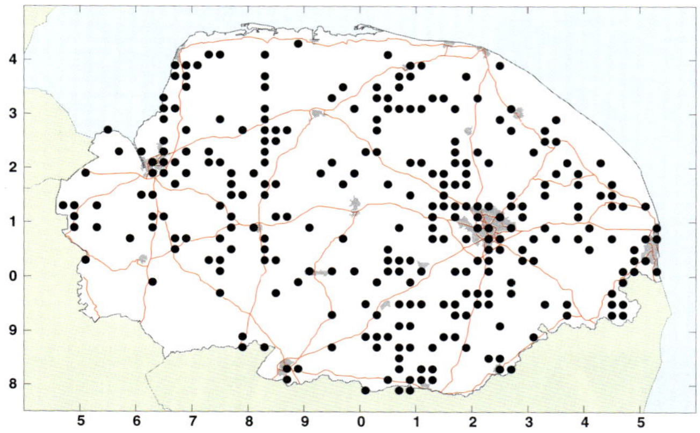

Our plant is mostly ssp. **tetragonum**, a perennial locally frequent in a wide range of habitats but particularly on soils which remain moist in winter. Ssp. **lamyi** (F.W.Schultz) Nyman, syn. *E. lamyi* F.W.Schultz, has been recorded in five sites, the last in dune slacks at Old Hunstanton in 1952.

■ **E. x semiobscurum** Borbás
(*E. tetragonum x E. obscurum*)
One record, Runcton Holme W61 in 1954.

E. obscurum Schreb. **Short-fruited Willowherb**

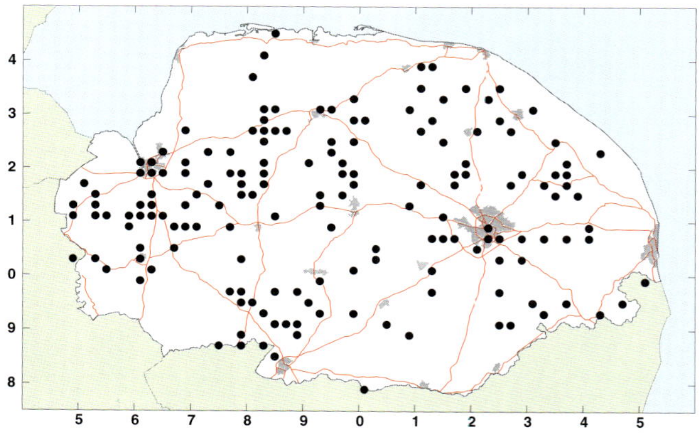

Perennial, occasional, mainly in damp, shady places, also on walls.

■ **E. obscurum x E. ciliatum**
Three records, the last at Wretham W88 in 1972, RPL & ELS.

■ **E. x schmidtianum** Rostk.
(*E. obscurum x E. palustre*)
Has been recorded for east Norfolk, Stace 1975.

E. roseum Schreb. **Pale Willowherb**
Rare in damp places. During this survey recorded at sites scattered from King's Lynn W61 to Hoveton E31. W1. E6.

E. ciliatum Raf. **American Willowherb**
Syn. *E. adenocaulon* Hausskn.

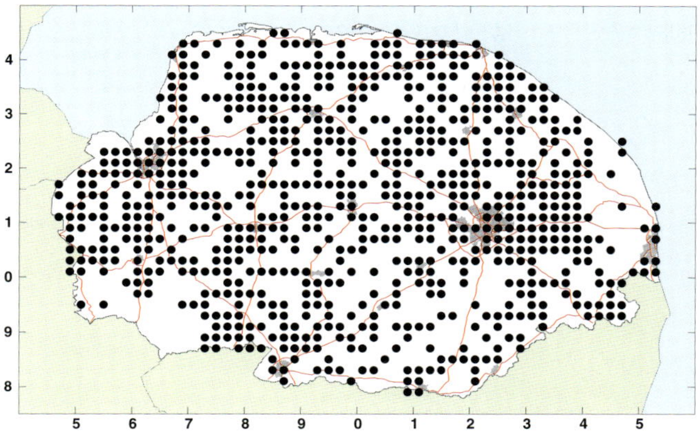

Perennial not recorded in Norfolk until 1952, now much the commonest of the smaller willowherbs and found in almost all habitats except on wet soils.

E. palustre L. **Marsh Willowherb**

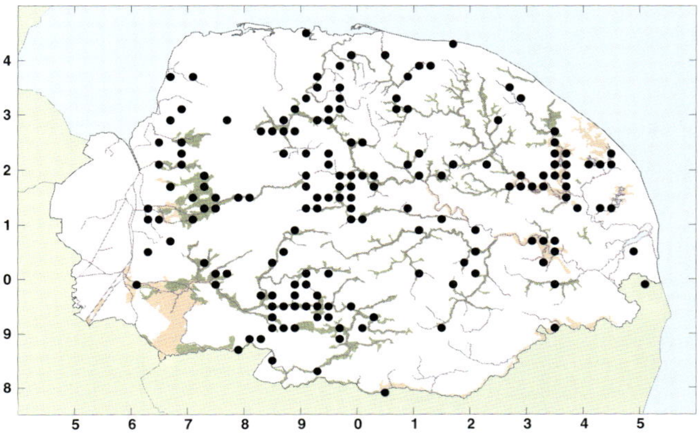

Perennial frequent in marshy places and by streams.

Chamerion angustifolium (L.) Holub **Rose-bay Willowherb**
Syns. *Epilobium angustifolium* L. *Chamaenerion angustifolium* (L.) Scop. nom. illeg.
Herbaceous perennial, described as rare by Trimmer in 1866, now almost ubiquitous except in water- logged ground. Least common on clay. W622. E614.

Oenothera
Oenothera species and hybrids are now being grown on a commercial basis for their oil for cosmetic and herbal uses. They appear as volunteers in succeeding crops and spread locally on to roadsides and waste ground. As the species hybridise freely, identification is likely to become increasingly difficult.

O. glazioviana P.Micheli ex Mart. **Large-flowered Evening Primrose**
Syn. *O. erythrosepala* Borbás
Winter green biennial introduced from North America, first recorded in 1954 and now scattered in waste places, particularly by roads and on waste ground. W56. E145.

O. fallax Renner **Intermediate Evening Primrose**
(*O. glazoviana x O. biennis*)
Found in a cornfield at Loddon E39 in 1975 by PG.Lawson, det. CA.Stace; as a relic of a previous crop at Garboldisham W98 in 1994, JH det. J.Bowra and in several sites in the same parish, E08 in 1997 by RWE. W1. E2.

O. biennis L. **Common Evening Primrose**
Introduction of European origin, occasional in waste places, far less common than *O. glazoviana* and formerly much confused with it. W18. E21.

O. cambrica Rostanski **Small-flowered Evening Primrose**
A large colony was seen beside the Loddon bypass E39 in 1995 by RWE and ALB, also at Dickleburgh E18 in 1995, JH. W0. E2.

Oenothera cambrica - Loddon

O. stricta Ledeb. ex Link **Fragrant Evening Primrose**
Annual introduced from Chile, occasionally occurring on sandy soils. First recorded from Yarmouth in 1880 (as *O. odorata*) more recently from Sidestrand E23 in 1980 ETD & RM.Leaney and Pentney W71 in 1989, HW. W1. E0.

■ **O. perangusta** R.R.Gates
One record, at Bergh Apton tip E30 in 1982, ETD.

■ **Clarkia unguiculata** Lindl. **Clarkia**
Annual from North America a short-lived garden escape.

■ **C. amoena** (Lehm.) A.Nelson & J.F.Macbr. **Godetia**
Annual from western North America a short-lived garden escape.

Fuchsia magellanica Lam. **Fuchsia**
A deciduous shrub introduced to gardens from South America. Surviving as a garden throwout in Norwich E20, ETD; Bradfield E23, ALB and Overstrand E24, FF. W0. E3.

Circaea lutetiana L. **Enchanter's Nightshade**

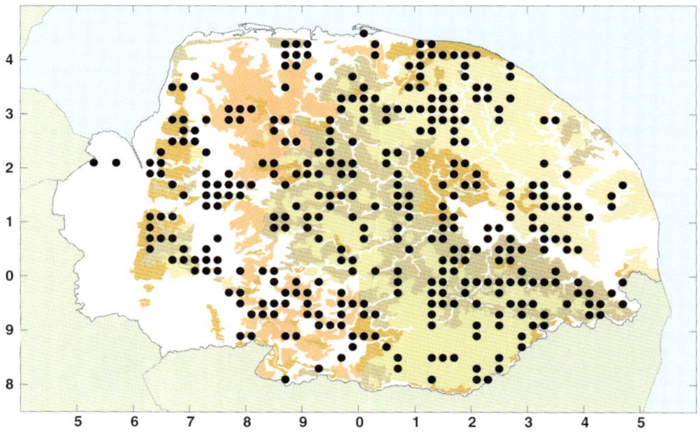

Herbaceous perennial of shady places, found both in ancient woods and in established plantations.

CORNACEAE

Cornus sanguinea L. **Dogwood**
Syn. *Thelycrania sanguinea* (L.) Fourr.

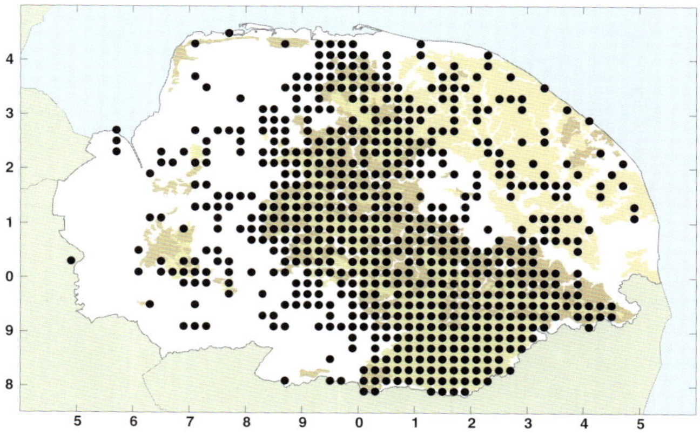

Deciduous shrub found in woods and hedgerows on clay soils. Modern plantings are beginning to obscure this pattern.

C. sericea L. **Red Osier Dogwood**
Syn. *C. stolonifera* Michx.
Deciduous shrub introduced to gardens and parks from North America, forming thickets where no longer managed. W3. E8.

C. alba L. **White Dogwood**
Deciduous shrub introduced from eastern Asia to gardens and parks, often well established away from habitation. W5. E4.

C. mas L. **Cornelian Cherry**
Deciduous shrub introduced to gardens from mainland Europe, occasionally in hedgerows, either from plantings or bird-sown. At Hargham E09 it was noted in 1967 and still remains in quantity in a wood and adjacent hedge and an old tree occurs in Westwick churchyard E22. W0. E2.

Aucuba japonica Thunb. **Spotted Laurel**
Evergreen shrub introduced to gardens from Japan, reported as self-sown in Norwich E20 by ETD. W0. E4.

SANTALACEAE

Thesium humifusum DC. **Bastard Toadflax**
■ Herbaceous perennial once widespread on chalk grassland, surviving until 1953 on the Devil's Dyke at Cranwich W79 and at Gooderstone W70 until 1965. A nationally scarce species.

VISCACEAE

Viscum album L. **Mistletoe**

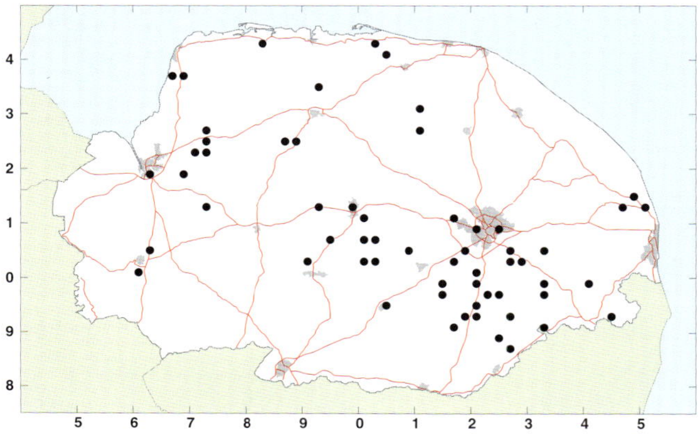

Evergreen shrub parasitic on a variety of hosts, the commonest of which in Norfolk appears to be lime. Also recorded during this survey on poplar, apple, almond, hawthorn, field maple and willow. Trimmer's 1866 records on ash and oak have not been refound.

CELASTRACEAE

Euonymus europaeus L. **Spindle**

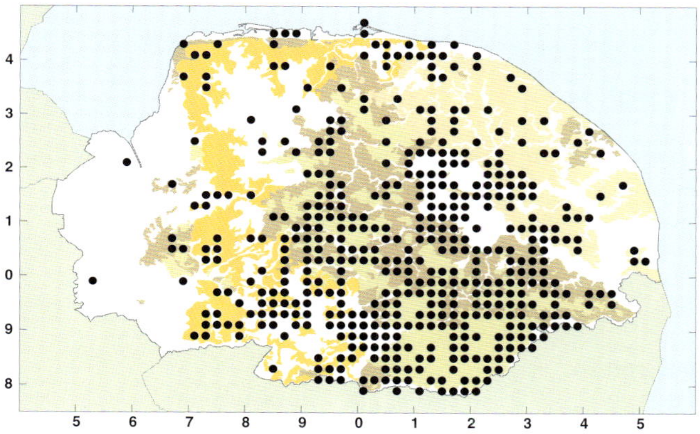

Deciduous shrub not infrequent on base rich soils and occasionally elsewhere. Bushes with white arils occur a few yards over the county boundary with Suffolk at Knettishall W98 N & MG.

E. japonicus L.f. **Evergreen Spindle**
Evergreen shrub introduced to gardens from Japan, established at Overstrand E24, 1994, FF. W0. E1.

AQUIFOLIACEAE

Ilex aquifolium L. **Holly, Hulver**

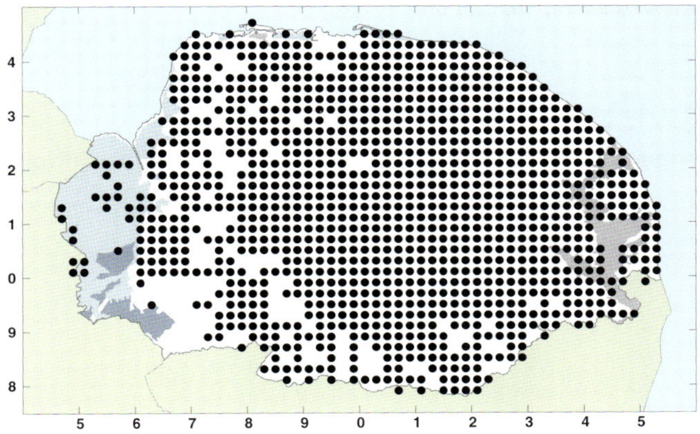

Evergreen shrub frequent in woods and hedges in all but the peat fens and dry soils away from habitation. Trees were often left standing when hedges were cleared away in the 1950s and 1960s as it was long considered unlucky to fell a holly. Only a few of these survive as, if not protected by brushwood, their thin bark soon splits in severe winters and they then die. In some parts of the county holly was formerly cut and fed to sheep in winter (Mabey 1996). It is also palatable to rabbits who kill many young trees by ring barking them in winter. Yellow-fruited specimens are occasionally met with as near Hopton Point W89 in the Stanford Training Area, ALB.

Ilex x altaclerensis (Loudon) Dallim. **Highclere Holly**
(*I aquifolium x I. perado*)
Evergreen shrub, a hybrid of garden origin widely planted in parks and occasionally self-sown. W4. E0.

BUXACEAE

Buxus sempervirens L. **Box**
Evergreen shrub presumably native in the county as the name Bixley (E20) means a clearing in a box wood, but also widely introduced to game coverts and plantations and sometimes forming dense thickets. The cultivar 'Latifolia Bullata' is frequently met with, especially in Breckland. W80. E93.

EUPHORBIACEAE

Mercurialis perennis L. **Dog's Mercury**

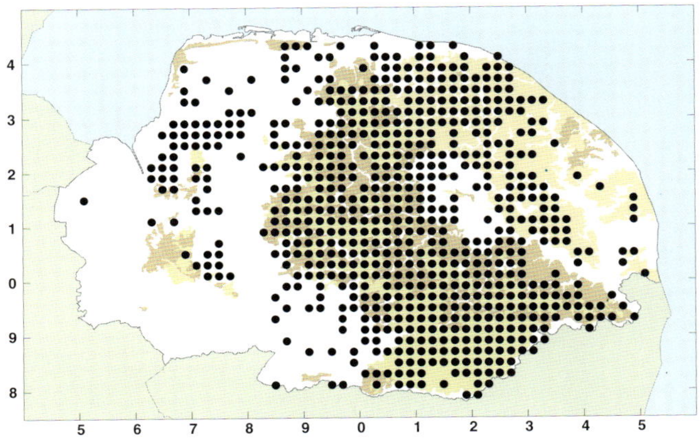

Herbaceous perennial closely associated with past or present areas of ancient woodland where it can carpet the ground. Very slow to colonise new areas.

Mercurialis perennis - Heacham

M. annua L. **Annual Mercury**

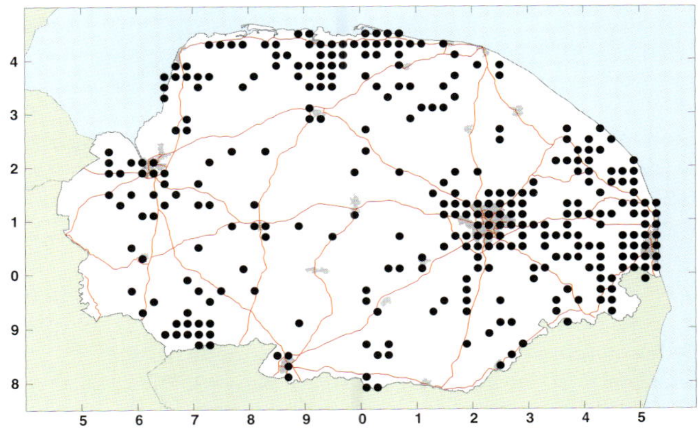

Largely spring germinating annual but occasionally surviving through mild winters. Native in some areas of Britain, its largely urban-based distribution suggests that it is not so with us.

■ **Euphorbia platyphyllos** L.
Broad-leaved Spurge, Warted spurge
Annual, rare and found only once, at High Kelling E04 in 1938.

E. helioscopia L. **Sun Spurge**

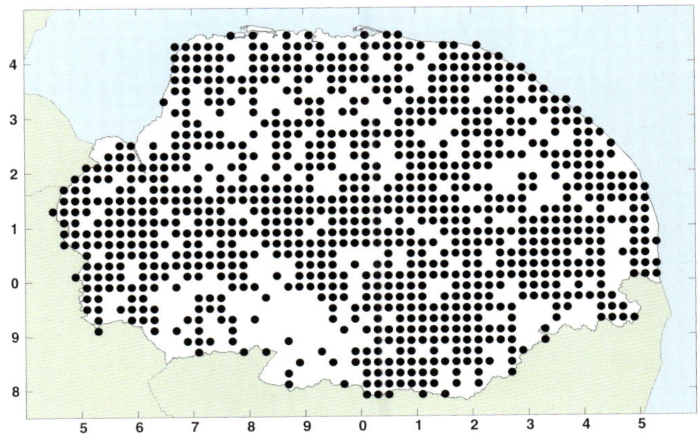

Annual, germinating whenever it is warm enough, a frequent arable weed.

E. lathyris L. **Caper Spurge**
Biennial, introduced to gardens from mainland Europe rarely found far from them though widespread through the county. W61. E108.

E. exigua L. **Dwarf Spurge**

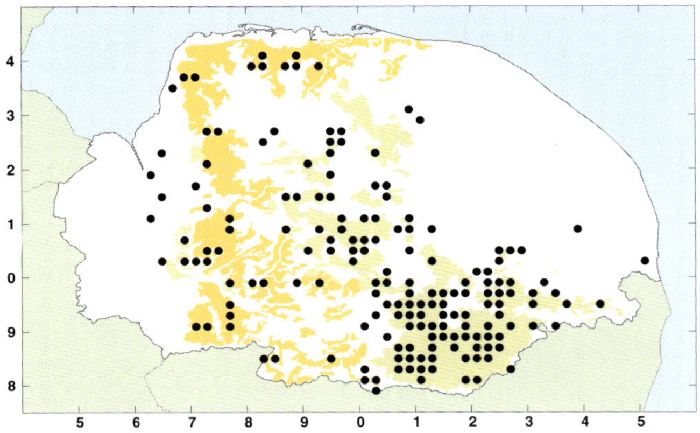

Annual, germinating whenever it is warm enough, widely scattered and almost always an arable weed.

E. peplus L. **Petty Spurge**
Annual, germinating whenever it is warm enough, most frequent in gardens and waste places, less common in fields. W409. E517.

E. paralias L. **Sea Spurge**
Winter green perennial, scarce on coastal sand dunes, recorded between Old Hunstanton and Wells-next-the-Sea W64, 74 and 84 and near Yarmouth E50. W8. E1.

E. x pseudovirgata (Schur) Soó **Twiggy Spurge**
(*E. esula x E. cyparissias*)
Herbaceous perennial long established on a roadside near Swaffham W80 where it has been known since 1957 and still thrives, JEG and at Downham Market W60 in 1991, RMP. W2. E0.

■ **E. esula** L. **Leafy Spurge**
Herbaceous perennial introduced to gardens from mainland Europe. Four old records, but much confused in the past with the previous taxon to which some may belong. A plant found at Kilverstone W98 by HD.Wiard in 1972 was det. PD.Sell.

E. cyparissias L. **Cypress Spurge**
Herbaceous perennial introduced into gardens from mainland Europe and perhaps south-east England. A scarce but long established weed of light, sandy soils. W7. E33.

E. amygdaloides L. ssp. **amygdaloides** **Wood Spurge**

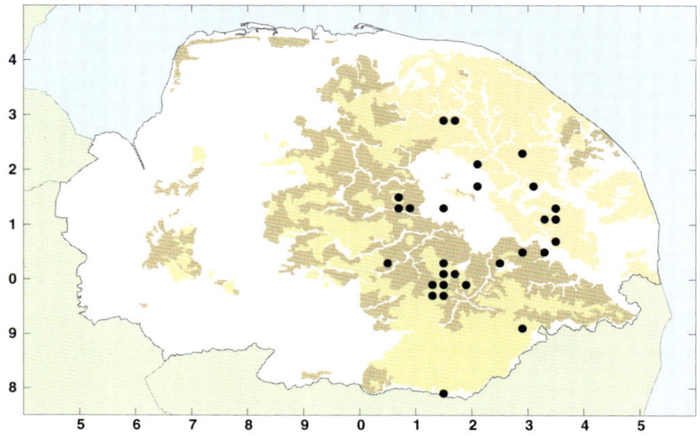

Evergreen perennial confined to ancient woodlands on boulder clay. Away from overhead cover, the emerging leaves in spring are tender. Records on other soils and away from woods may well refer to ssp. **robbiae** (Turrill) Stace, (syn. *E. robbiae* Stace), introduced to gardens from Turkey and sometimes thrown out onto verges. It can be a very dominant weed.

E. characias L. **Mediterranean Spurge**
Evergreen, semi-shrubby perennial introduced to gardens from the eastern Mediterranean, recorded on a wall at Rushford W98 in 1993, RMP. W1. E0.

RHAMNACEAE

Rhamnus cathartica L. **Purging Buckthorn**

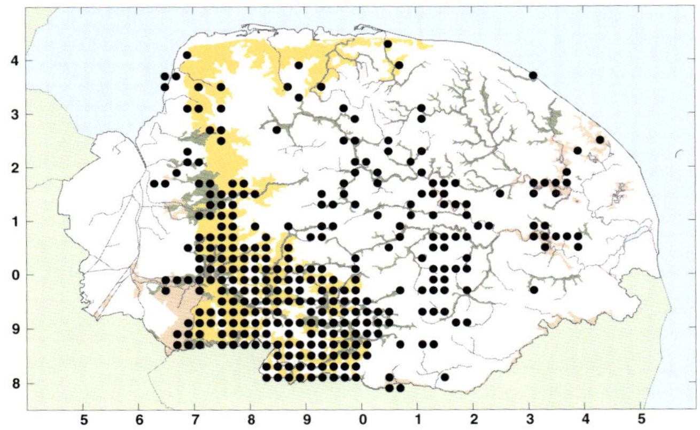

Deciduous shrub commonest on moist, calcareous soils. Abundant in fen and carr in and around the Breckland where it can form dense thickets, also on drier soils in association with blackthorn, *P. spinosa*.

Frangula alnus Mill. **Alder Buckthorn, Berry Alder**

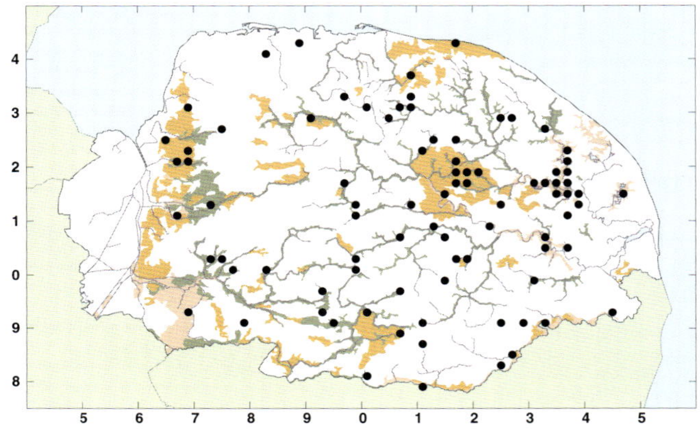

Deciduous shrub of moist, neutral to acid soils, most frequent in Broadland. Once cultivated in a few areas for its use as high quality charcoal especially for making gunpowder.

Frangula alnus - Roydon

VITACEAE

Vitis vinifera L. **Grape Vine**

Deciduous climbing shrub introduced to gardens from mainland Europe, occasionally naturalised as at Snettisham and Heacham, W63 where large plants sprawl across the shingle bank, also on abandoned allotments at Downham Market W60, RMP and the disused airbase at Raynham W82. W3. E0.

Parthenocissus quinquefolia (L.) Planch. **Virginia Creeper**

Deciduous climbing shrub introduced to gardens from North America and producing good seed. It was recorded as a mature plant on the shingle bank at Snettisham in 1989 and Heacham in 1993, both W63 PRC; at Langford W89 where it has been established since 1942 and also in a few roadside and urban sites. Although **P. tricuspidata** (Siebold & Zucc.) Planch. (**Boston Ivy**) from China and Japan is by far the commonest species in cultivation, it has not been recorded in the county as an escape. W5. E6.

P. inserta (A. Kern.) Fritsch **False Virginia Creeper**

Deciduous climbing shrub, introduced to gardens from North America and found on waste ground in King's Lynn W62 in 1994, RMP and Norwich E20, ETD. W1. E2.

LINACEAE

Linum bienne Mill. **Pale Flax**

Herbaceous biennial or perennial known from only two sites. Refound in 1997 on the sea bank at Wells-next-the-Sea W94 PRB, where it was first seen in 1922 and still flourishes after being thought lost when the area was flooded by the sea in 1963. Also at Wingland W52 in 1996, SF only yards from the Lincolnshire border where it was recorded in 1962. W2. E0.

L. usitatissimum L. **Flax, Linseed**

Annual introduced to cultivation formerly for linen, today the variety grown is mostly for the production of oil. A frequent short-term casual by roadsides from spilt seed, but not persisting. W72. E52.

■ **L. perenne** L. **Perennial Flax**

Syn. *L. anglicum* Mill.

A rare perennial of calcareous grassland for which eight sites were listed in Nicholson, 1914. Last seen at Aylmerton E14 in 1920. A nationally scarce species.

L. catharticum L. **Fairy Flax**

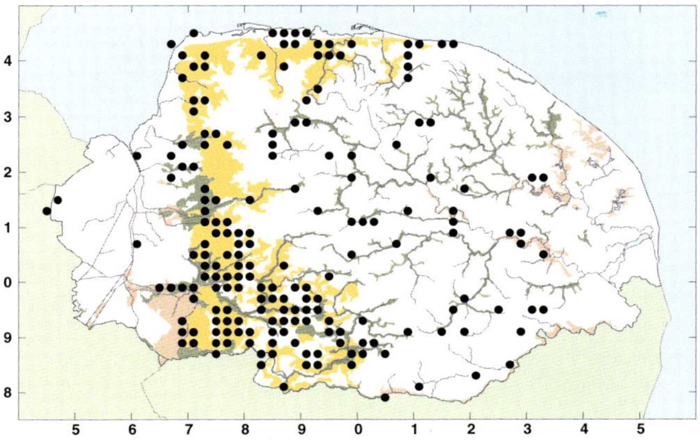

Annual of dry, calcareous soils, occasionally elsewhere, especially on old railway banks.

Radiola linoides Roth **Allseed**

Minute annual typically growing on bare, acid, sandy ground where water stands in winter in woods and heaths. Seen at Litcham W81 in 1976, ALB and Wolferton W62 in 1977, CPP. During this survey at Thorpe St Andrew E21 in 1997, ALB and Winterton Dunes E42, Flora Group. W0. E2.

POLYGALACEAE

Polygala vulgaris L. **Common Milkwort**

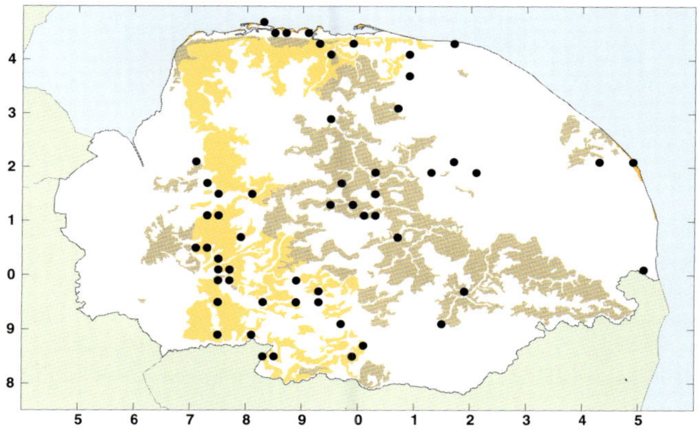

Winter green perennial, chiefly in undisturbed calcareous grassland and dunes.

P. serpyllifolia Hosé **Heath Milkwort**

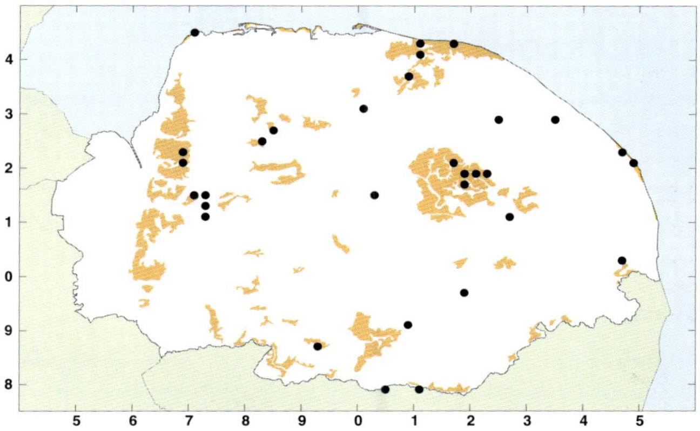

Winter green perennial scattered on light, acid soils. ELS in 1950 considered it the commonest of the milkworts, but this does not appear to be so now, presumably because of the loss of its heathy habitat.

HIPPOCASTANACEAE

Aesculus hippocastanum L. **Horse Chestnut**

Deciduous tree introduced from south-east Europe, widely planted and regenerating from seed. W228. E548.

A. x carnea J.Zeyh. **Red Horse Chestnut**

(*A. hippocastanum x A. pavia*)

Deciduous tree mostly of garden origin but planted in woodland at Woodrising W90, ALB. Although more or less fertile, it rarely produces self-sown seedlings. W1. E11.

ACERACEAE

Acer platanoides L. **Norway Maple**

Deciduous tree introduced to gardens and parks from mainland Europe and freely regenerating in some sites, also planted as an ornamental by roadsides. W69. E136.

A. campestre L. **Field Maple**

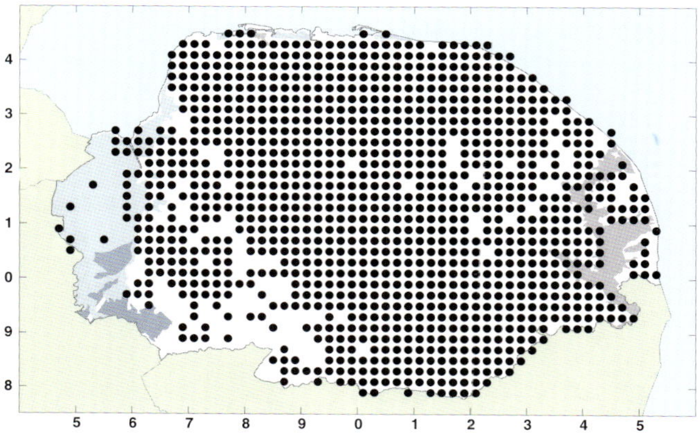

Deciduous tree found in ancient woods, providing most of the canopy at Yaxham Wood E00, also in long-established hedges. Absent from Fenland and scarce in Breckland and Broadland. As with many native trees, this pattern is being obscured by recent planting. A tree at Alburgh E28 is host to mistletoe while a fine standard in Gawdy Hall Wood E28 has a girth of 290cm, close to the country's largest.

A. pseudoplatanus L. **Sycamore**
Deciduous tree introduced from central Europe, according to some authorities, by the Romans. Very free seeding and often becoming a weed in plantations and waste ground on all soils. W637. E688.

ANACARDIACEAE

Rhus hirta L. **Stag'shorn Sumach**
Syn. *Rhus typhina* (L.) Sudw.
Deciduous shrub introduced to gardens and parks from North America, sometimes planted by roadsides and forming thickets by suckering. W4. E46.

SIMAROUBIACEAE

Ailanthus altissima (Mill.) Swingle **Tree of Heaven**
Deciduous tree introduced to gardens and parks from China, often producing root-suckers, but has also been found self-sown on walls in King's Lynn W61 RMP and at West Tofts W89, ALB. W2. E6.

OXALIDACEAE

Oxalis corniculata L. **Procumbent Yellow Sorrel**
Winter green perennial introduced to gardens from warm temperate areas and now an almost ineradicable weed in many gardens, on paths and waste places, also on walls. Often seen in its more ornamental, but equally invasive form, *purpurea*. W64. E129.

O. exilis A.Cunn. **Least Yellow Sorrel**
Winter green perennial introduced to gardens from New Zealand and Tasmania, occasionally escaping and found on walls, paths, pavements and waste ground scattered across the county. W4. E14.

O. stricta L. **Upright Yellow Sorrel**
Syn. *O. europaea* Jord.
Winter green perennial introduced to gardens from North America, recorded on a wall in King's Lynn W61 in 1995, RMP and in Norwich E20, ETD. W1. E2.

O. articulata Savigny **Pink Sorrel**
Syns. *O. floribunda* auct. non Lehm., *O. rosea* Jacq.
Winter green perennial introduced to gardens from South America, frequently naturalised by house walls, on pavements and roadside banks. W18. E115.

O. acetosella L. **Wood Sorrel**

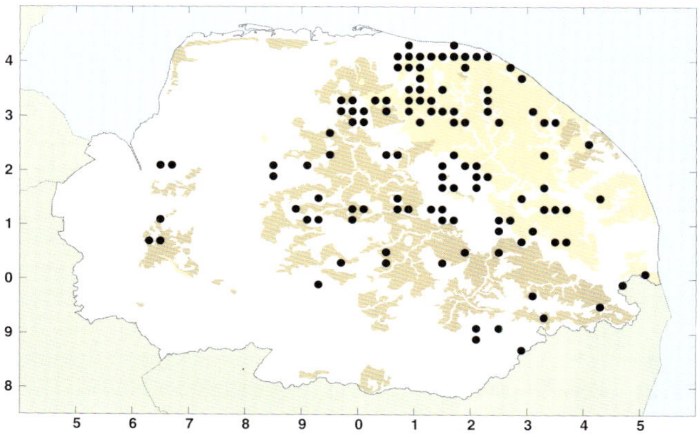

Winter green perennial, our only native *Oxalis*, widespread but only locally common in woods on acid soils.

Oxalis acetosella - Blickling

O. debilis Kunth **Large-flowered Pink Sorrel**
Syn. *O. corymbosa* DC
Winter green perennial introduced from South America, an occasional and perhaps overlooked weed of nurseries and gardens, walls and waste places, spreading by means of bulblets. W3. E6.

O incarnata L. **Pale Pink Sorrel**
Winter green perennial introduced to gardens from South Africa, not frost hardy but surviving by means of deep seated bulblets. Occasionally established in waste places and on walls as by the church of Tilney All Saints W51 in 1992, RMP. W2. E2.

GERANIACEAE

Geranium endressii J.Gay — **French Cranesbill**

Semi-winter green perennial introduced to gardens from the Pyrenees and occasionally surviving in rough grass, particularly in churchyards where it was probably once planted. W1. E4.

G. x oxonianum Yeo — **Druce's Cranesbill**

(*G. endressii x G. versicolor*)

Winter green perennial, a hybrid of garden origin occasionally naturalised but not persisting. W2. E24.

G. versicolor L. — **Pencilled Cranesbill**

Semi-winter green perennial introduced to gardens from the Mediterranean, occasionally naturalised. W6. E17.

G. nodosum L. — **Knotted Cranesbill**

Semi-winter green perennial introduced to gardens from southern Europe and occasionally thrown out on to roadsides. W1. E0.

G. rotundifolium L. — **Round-leaved Cranesbill**

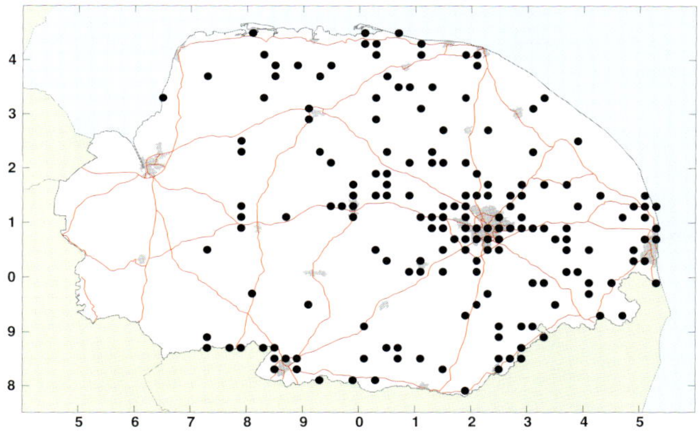

Autumn-germinating annual usually considered a southern species, its status doubtful with us. Found on dry banks and in short grass it is frequent where it occurs but quite absent from some areas of the county. It appears to be increasing.

G. pratense L. — **Meadow Cranesbill**

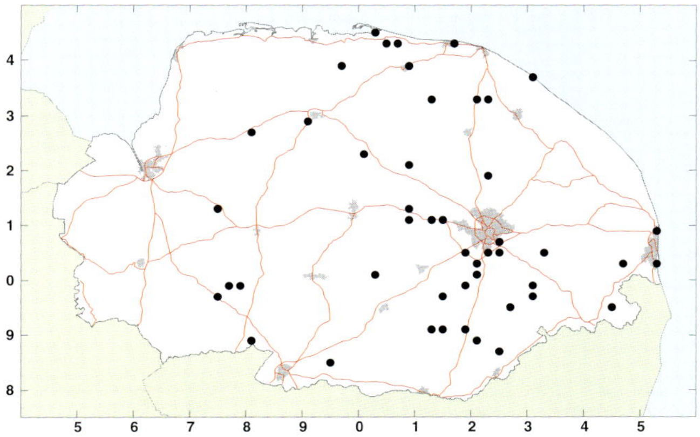

Herbaceous perennial perhaps native in Norfolk in a few sites on the boulder clay as at Hardwick E29. Most other records are considered garden throwouts; also sown by new roads as part of 'wild flower' mixes.

G. sanguineum L. — **Bloody Cranesbill**

Semi-evergreen perennial introduced to gardens from other parts of Britain, not native in Norfolk. An occasional garden throwout, notably well established on the shingle bank at Snettisham W63. W4. E5.

G. columbinum L. — **Long-stalked Cranesbill**

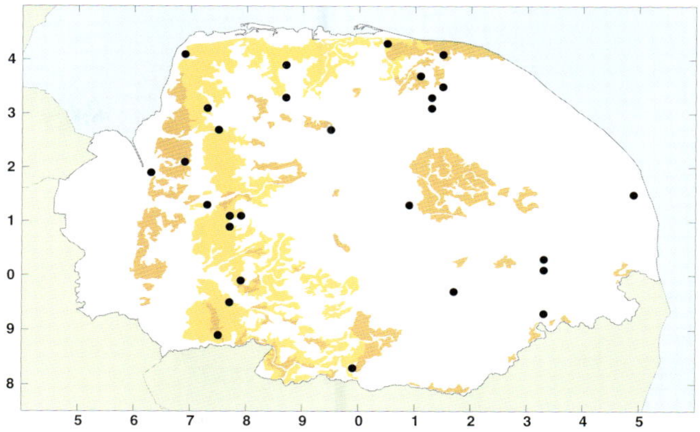

Spring germinating annual usually found in small numbers. Normally a plant of chalky soils, both on light soils and on the boulder clay.

Geranium columbinum - Ringstead

G. dissectum L. — **Cut-leaved Cranesbill**

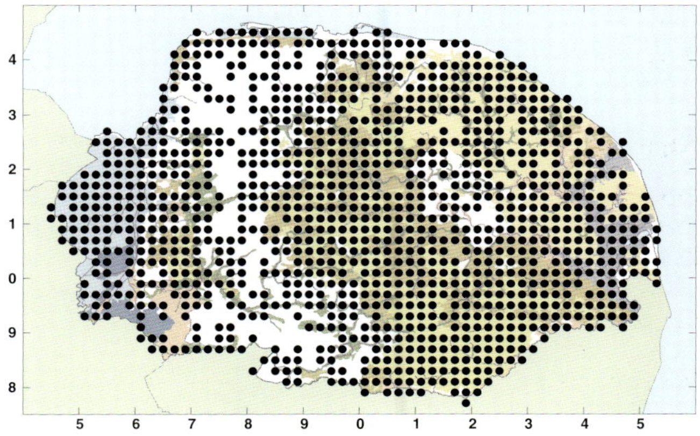

Autumn germinating annual, a weed of arable land especially on heavy soils.

G. ibericum Cav. — **Caucasian Cranesbill**

Herbaceous perennial introduced to gardens from the Caucasus and occasionally naturalised. W1. E0.

G. x magnificum Hyl. — **Purple Cranesbill**

(*G. ibericum x G. platypetalum*)

Herbaceous perennial of garden origin, an occasional garden throwout. W4. E6.

G. pyrenaicum Burm.f. **Hedgerow Cranesbill**

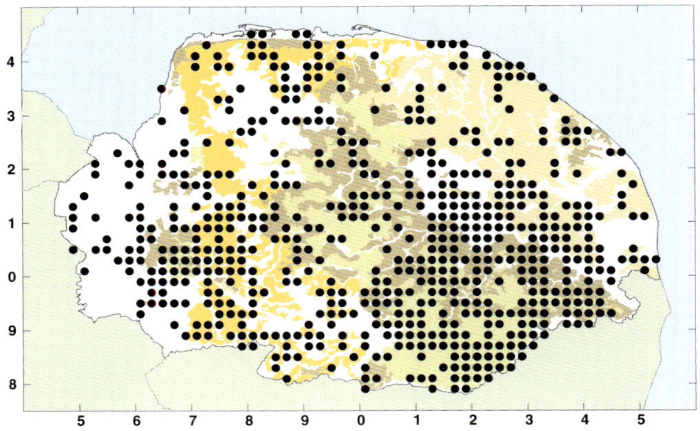

Winter green perennial introduced from southern Europe and first recorded in Britain in the early years of the eighteenth century. Still considered rare by Nicholson in 1914, but now widespread throughout the county. Most frequent on heavier soils. A pale-flowered form sometimes occurs.

G. pusillum L. **Small-flowered Cranesbill**

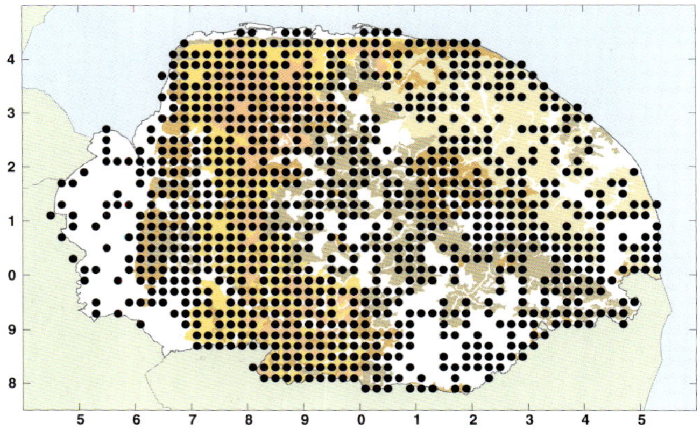

Autumn germinating annual most frequent on light soils.

G. molle L. **Dove's-foot Cranesbill**
Autumn germinating annual, our commonest cranesbill, found in arable land and waste places throughout the county. Quite large colonies of a white-flowered form occur sporadically. W661. E666.

G. macrorrhizum L. **Rock Cranesbill**
Semi-winter green perennial, introduced to gardens from southern Europe, an occasional garden throwout. W0. E2.

G. lucidum L. **Shining Cranesbill**

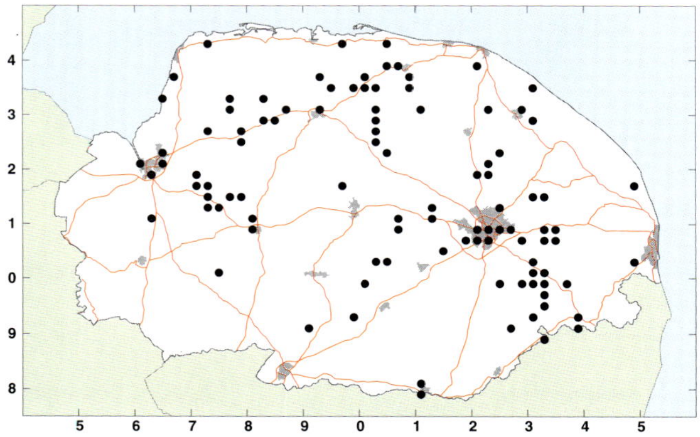

Autumn germinating annual occasional on roadside banks and walls, usually near habitation and doubtfully native in many of its sites.

G. robertianum L. **Herb Robert**

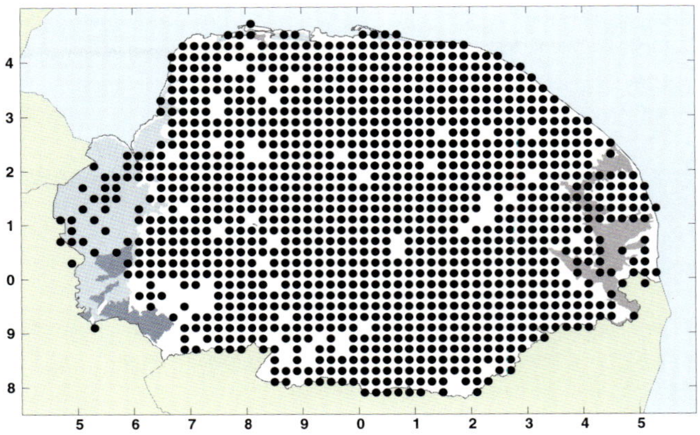

Winter green biennial or annual of hedgerows and waste ground, usually in shady sites. Most of our plants are ssp **robertianum**. In dry, sunny sites such as walls it can be very dwarf and deep-red in colour. White-flowered forms are not uncommon and, like most white-flowered plants, come true from seed. Ssp **maritimum**, a compact, but distinct subspecies has been known on the shingle bank at Wolferton and Snettisham W63 since at least 1939, ELS and is still present, SF.

G. phaeum L. **Dusky Cranesbill**
Semi-winter green perennial introduced to gardens from southern Europe and escaping, many old records exist but few plants survive long, the exception being at Saham Hills W90 where it was first recorded in 1953 and again in 1988, ALB. W5. E6.

Erodium moschatum (L.) L'Hér. **Musk Stork's-bill**
Annual, found during this survey at Kirby Cane E39; Mundesley E33 and Aldeby E49, all ALB and carpeting the churchyard at Gorleston E50 in 1998, Flora Group. Although considered an introduction in Norfolk, its previous four records were all coastal which would add to the possibility that it is native along this stretch of coast. W0. E5.

E. cicutarium (L.) L'Hér. **Common Stork's-bill**

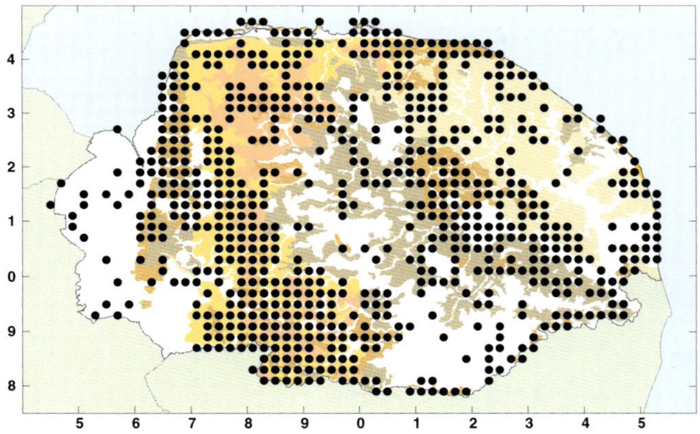

Autumn germinating annual frequent in light, sandy soils, in the dry spring of 1997 turning areas of dry grassland in Stanford Training Area pink (ALB). Ssp. **dunense** Andreas, a condensed form with glandular hairs, is found on coastal dunes and is sometimes recorded as *E. lebelii*, syn. *E. glutinosum*, but the presence of that species in the county has never been substantiated.

Erodium cicutarium - Stanta

LIMNANTHACEAE

Limnanthes douglasii R.Br. **Meadow-foam**
Annual from California often planted in gardens and escaping, though rarely persisting. W1. E19.

TROPAEOLACEAE

Tropaeolum majus L. **Nasturtium**
Annual introduced from South America, an occasional throwout but does not survive long. W2. E10.

BALSAMINACEAE

Impatiens capensis Meerb. **Orange Balsam, Jewel-weed**

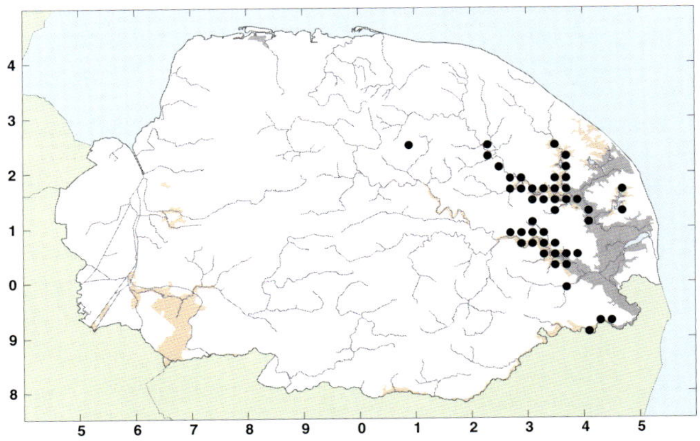

Spring germinating annual from North America, first seen in Britain in 1822. Apparently introduced to the Mermaid stream at Bolwick, Marsham E12 in 1927 whence it has spread throughout the Broads reaching the Yare in 1959 and most other rivers a few years later.

I. parviflora DC. **Small Balsam**

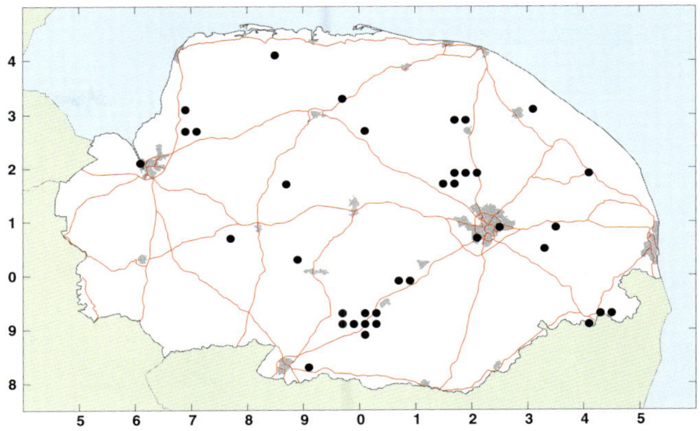

Spring germinating annual introduced to gardens from Asia, found occasionally on damp, shady ground but still with a curiously discontinous distribution as was noted by EAE in 1963.

I. glandulifera Royle
Indian Balsam, Himalayan Balsam, Policeman's Helmet

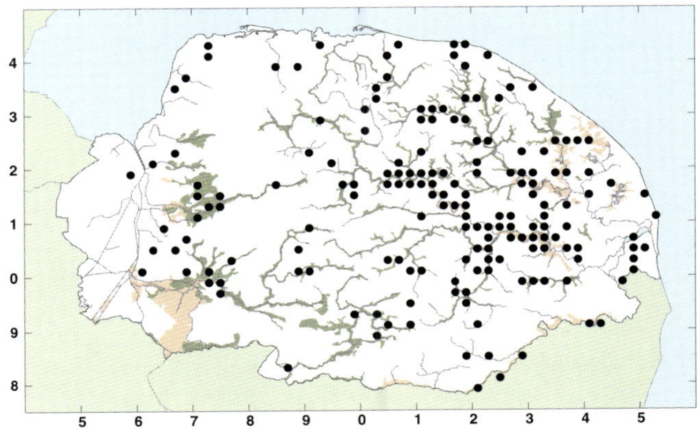

Spring germinating annual introduced to gardens from the Himalayas early in the 19th century, escaping and sometimes a pest by rivers. Never as dominant in Norfolk rivers as elsewhere, perhaps because of fringing reed beds. A large stand occurs on the Wensum near Lyng E01 where there is little reed.

ARALIACEAE

Hedera helix L. **Ivy**
A winter green climbing shrub ubiquitous where there is shade. Our native ivy is ssp. **helix**. The cultivar '**Hibernica**' with larger, thicker leaves, occurs frequently in parks, shrubberies and large gardens where it regenerates. W643. E696.

APIACEAE (UMBELLIFERAE)

■ **Caucalis platycarpos** L. **Small Bur-parsley**
Syn. *C. lappula* Grande
Annual found in Norwich E20 in 1915, also on a rubbish tip on Mousehold Heath E20 in 1959.

■ **Turgenia latifolia** (L.) Hoff. **Greater Bur-parsley**
Syn. *Caucalis latifolia* L.
Annual from southern Europe found in Norwich E20 in 1915.

Hydrocotyle vulgaris L. **Pennywort**

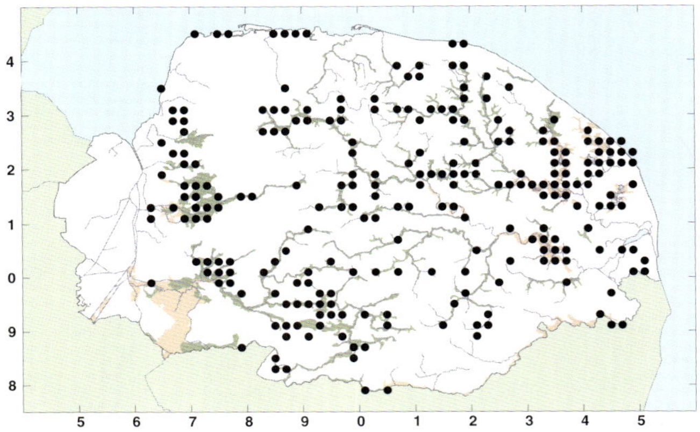

Winter green perennial frequent in marshy ground.

Sanicula europaea L. **Wood Sanicle**

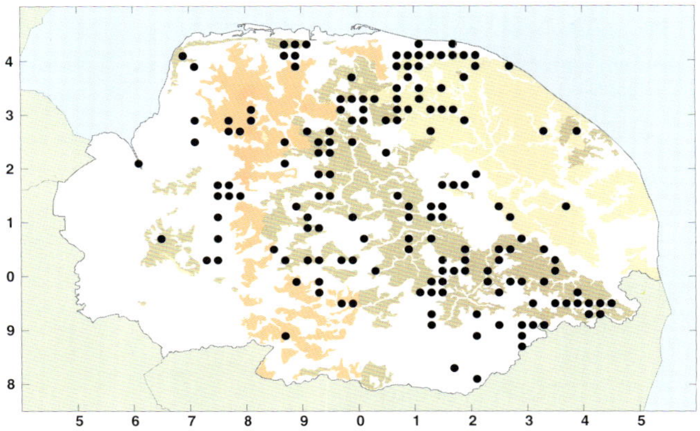

Winter green perennial found in shady places and deciduous woods where soils are well-drained and neutral to alkaline. Absent from many of the wet acid woods of central Norfolk.

Astrantia major L. **Masterwort**
Herbaceous perennial introduced to gardens from mainland Europe and occasionally escaping or surviving from plants thrown out on to waste ground and roadsides. Three records before 1970, noted during this survey on a hedgebank at Massingham W82 in 1995, MK. W1. E0.

Eryngium maritimum L. **Sea Holly, Eryngo**

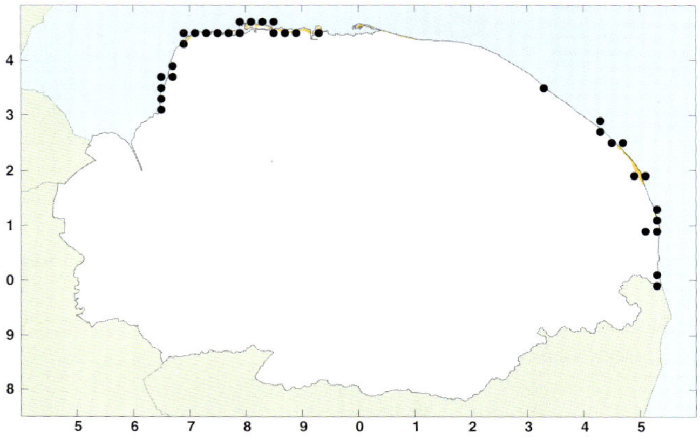

Herbaceous perennial confined to sand dunes and shingle by the sea. First recorded when Sir Thomas Browne noted that he had received information from the 'eryngo diggers' at Yarmouth in 1668. At that time it was plainly more common and its roots were collected, candied and sold as a sweet-meat.

Eryngium maritimum - Burnham Overy

E. planum L. **Blue Eryngo**
Evergreen perennial introduced to gardens from mainland Europe. Found growing in Heacham chalk pit W63 in 1990, PRC. W1. E0.

Chaerophyllum temulum L.
Rough Chervil, Queen Anne's Lace

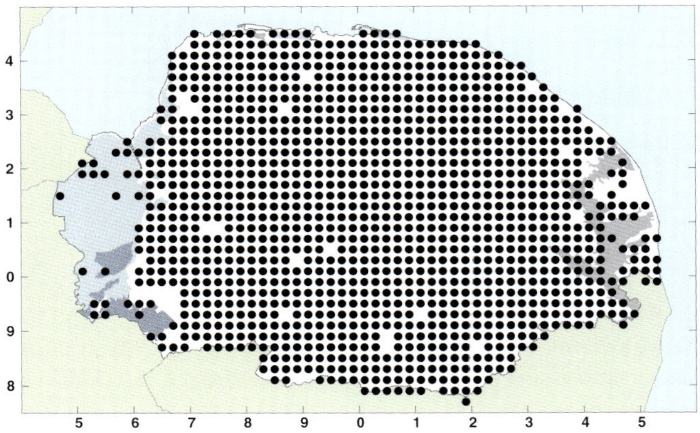

Winter green perennial abundant in grassy places and hedgebanks over most of the county, but sparse where there is little shade.

Anthriscus sylvestris (L.) Hoffm. **Cow Parsley, Queen Anne's Lace, Sheep's Parsley, Mother Die**
Winter green perennial abundant throughout the county. Locally less common on shallow soils over chalk. W694. E711.

A. cerefolium (L.) Hoffm. **Garden Chervil**
Introduced to gardens from south-east Europe as a herb, found on roadsides at Roydon E08 in 1990, AC. W0. E1.

A. caucalis M. Bieb. **Bur Chervil**

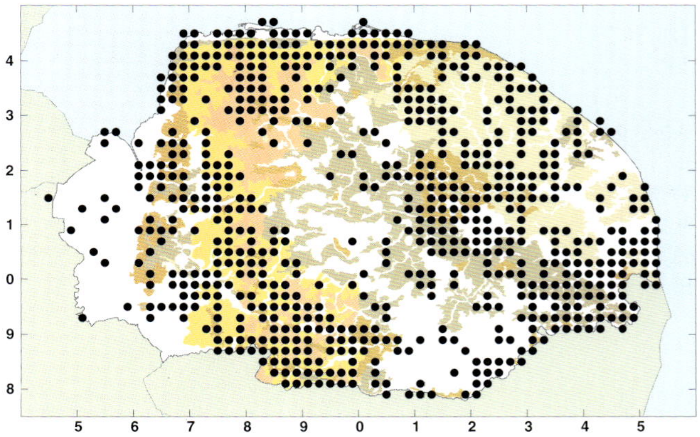

Autumn-germinating annual frequent in field edges and hedgerows on light, sandy soils. Absent from the wetter clays and loams.

Scandix pecten-veneris L. **Shepherd's Needle**

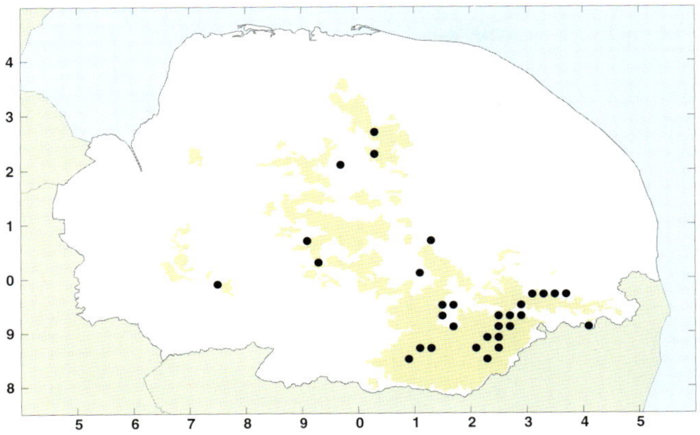

Spring germinating annual, once considered almost lost, but making a come-back on heavier soils. It has been suggested that its fruits are being carried to new sites on the fleece of the increasing number of sheep flocks which are moved around the county. A nationally scarce species.

■ **S. stellata** Banks and Sol.
Annual from the eastern Mediterranean, recorded at Shotesham E29 in 1980, D.Hedges.

Myrrhis odorata (L.) Scop. **Sweet Cicely**
Native to northern England and only casual in Norfolk. Noted as escaping in Norwich E20 in 1990, ETD and by woodland edges at Seamere, Hingham E00 in 1996, ALB. W0. E3.

Coriandrum sativum L. **Coriander**
Spring germinating annual introduced from the Mediterranean as a flavouring and in birdseed. Recorded at Welney W59 on a field edge in 1990, MK and on road verges at Thetford W88 in 1997, DM and 1998, NG. W5. E0.

Smyrnium olusatrum L. **Alexanders**

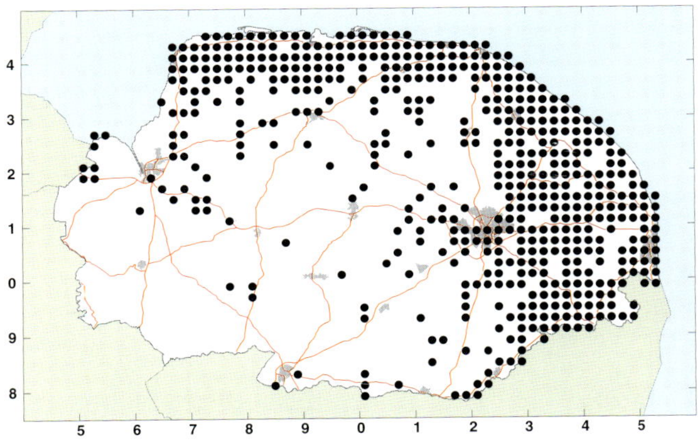

Winter green biennial introduced from south-east Europe as a pot herb, perhaps by the Romans. Well established along the county's milder coastal belt, also frequent on lower land in east Norfolk. Scarce where winter temperatures are lower, both inland and on the higher ground of the north of the county, but has been rapidly spreading inland in the last five years.

Conopodium majus (Gouan) Loret **Pignut, Earthnut**

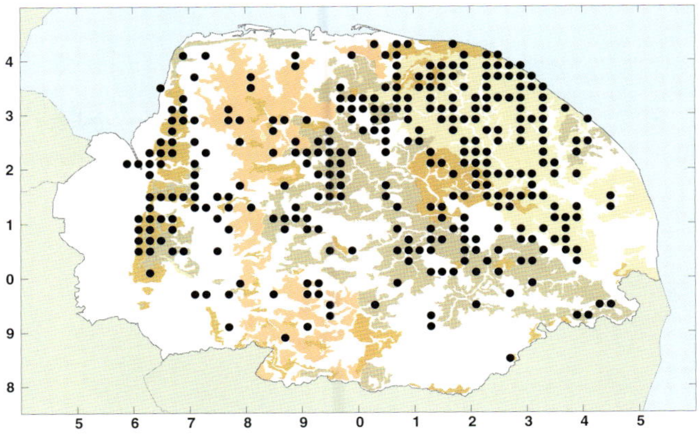

Herbaceous perennial of woods, meadows and established grassland, frequent in churchyards. Commonest on neutral to acid soils.

Pimpinella major (L.) Huds. **Greater Burnet-saxifrage**
Herbaceous perennial, not native in Norfolk but introduced many years ago to the garden at Wheatfen E30 by EAE and now well established by fen dikes nearby. W0. E1.

P. saxifraga L. **Burnet-saxifrage**

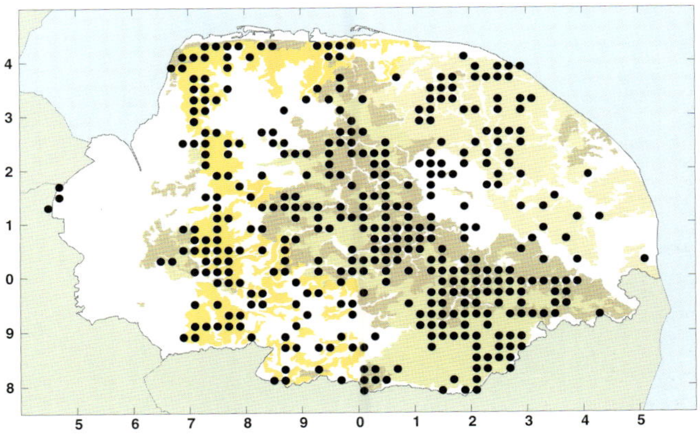

Herbaceous perennial of long established grassland, most frequent on base rich soils.

Smyrnium olusatrum - Burnham Overy

Aegopodium podagraria L. **Goutweed, Ground Elder**
Herbaceous perennial probably originally introduced to gardens from mainland Europe as a pot and medicinal herb, now widely naturalised, usually where there has been habitation. A pernicious weed of many gardens. W579. E656.

Sium latifolium L. **Greater Water Parsnip**

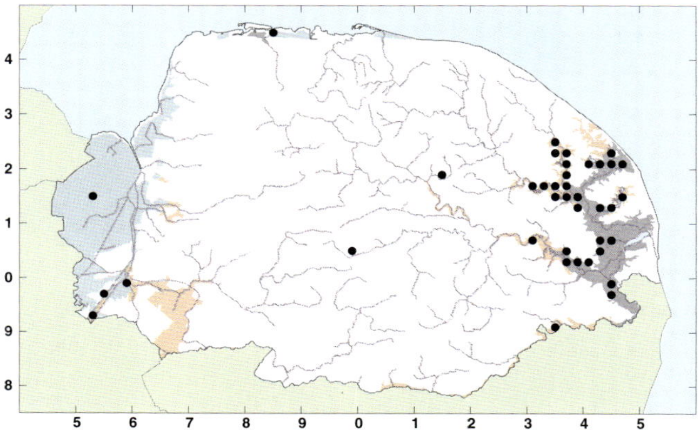

Herbaceous perennial found chiefly along ditch-sides in fens, least scarce in Broadland but always in small colonies. A nationally scarce species.

Sium latifolium - How Hill

Berula erecta (Huds.) Coville **Lesser Water Parsnip**

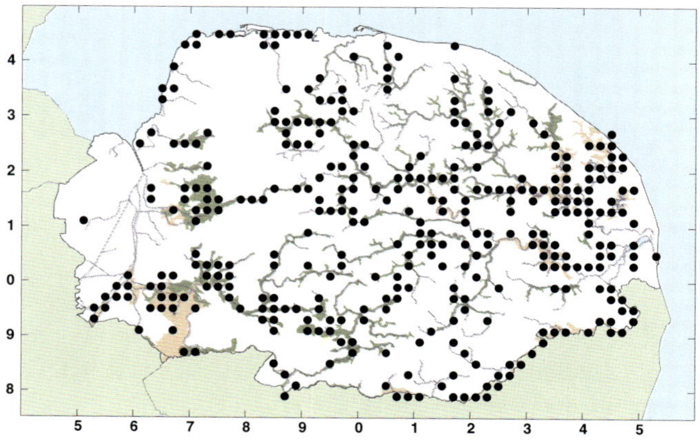

Herbaceous perennial frequent along ditches, streams and rivers, often with *Apium nodiflorum*, but never as abundant.

[**Crithmum maritimum** L. **Rock Samphire**
Native to rocky coasts, not in Norfolk. A 19th century record for Hemsby E41 is thought to have been an error.]

Oenanthe fistulosa L. **Tubular Water-dropwort**

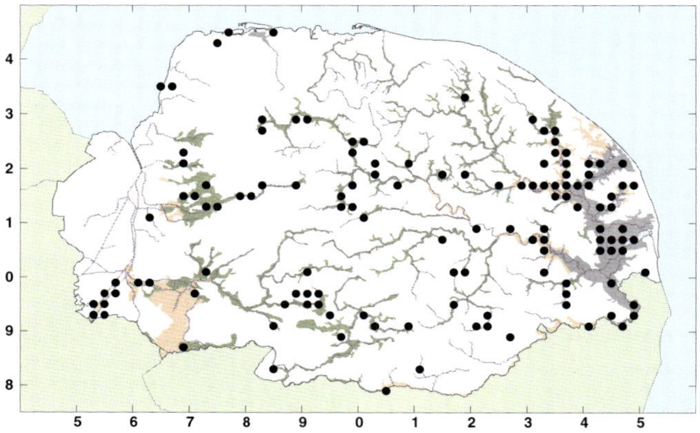

Herbaceous perennial found in ditches, streams and marshes, but much less common where the water is brackish.

■ **O. silaifolia** M. Bieb. **Narrow-leaved Water-dropwort**
This perennial species of ditches and wet marshes was recorded in eleven sites by Nicholson, 1914. A.Bennett, who wrote about the species in *Trans.NNNS* Vol.9 pp190-1 of 1910, considered it a native (Nicholson's quotation is in error) and noted that it was often overlooked as it 'flowers early and is often cut with the grass', presumably for marsh hay.

O. lachenalii C.C.Gmel. **Parsley Water-dropwort**

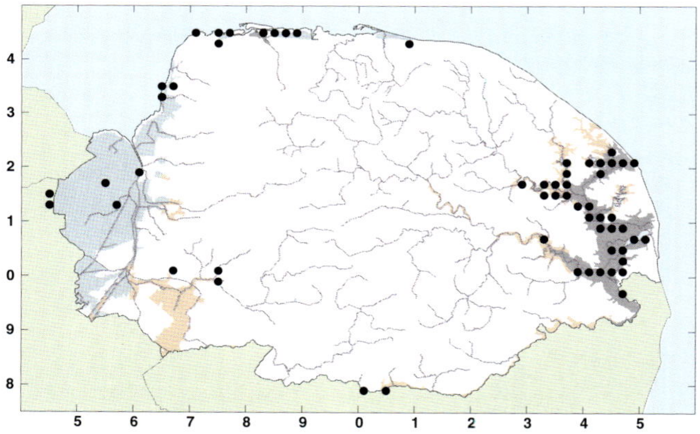

Herbaceous perennial chiefly in ditches and streams near the sea including those which are brackish, but also occasionally found inland as at Foulden W70 in 1990.

O. crocata L. **Hemlock Water-dropwort**
Recorded from five Broadland sites by Trimmer in 1866, and for Rockland Broad E30 by Nicholson 1914. Rediscovered during scrub clearance at Wheatfen, E30 in 1998, D Nobbs, where first seen by EAE in 1934 (P.Ellis pers. comm.). W0. E2.

O. fluviatilis (Bab.) Coleman **River Water-dropwort**

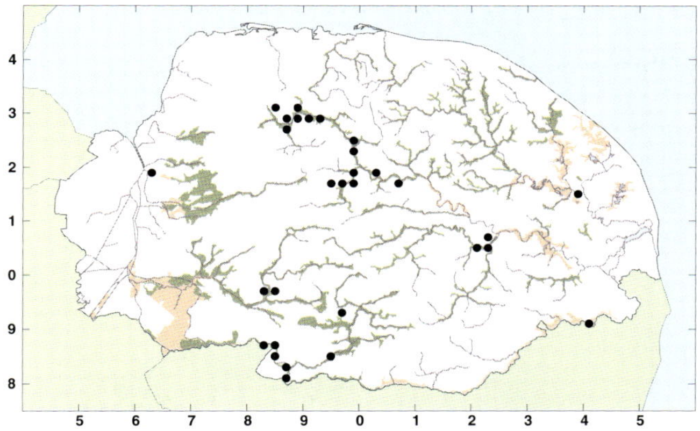

Herbaceous perennial with an oddly patchy distribution, normally confined to slow-flowing rivers.

O. aquatica (L.) Poir. **Fine-leaved Water-dropwort**

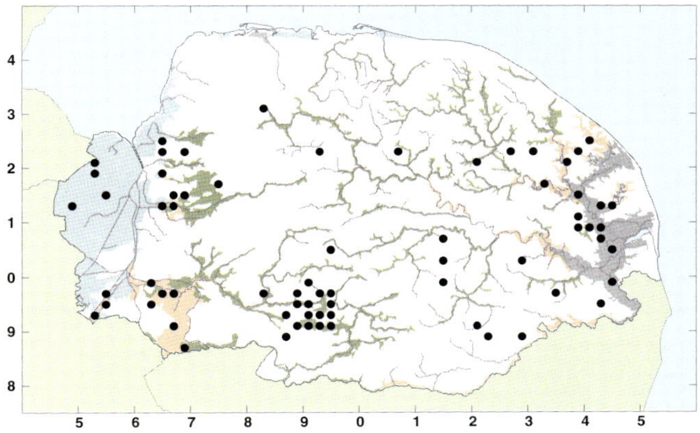

Herbaceous perennial usually found in still water, especially pools in marsh and fen, sometimes dominant in pingoes as on Thompson Common W99.

Aethusa cynapium L. **Fool's Parsley**

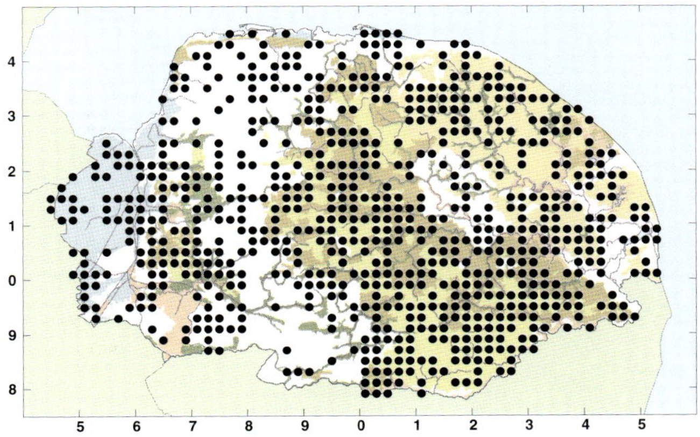

Annual or winter green biennial found largely as a weed in cultivated ground, particularly gardens. Our plant is ssp. **cynapium**.

Foeniculum vulgare Mill. **Fennel**

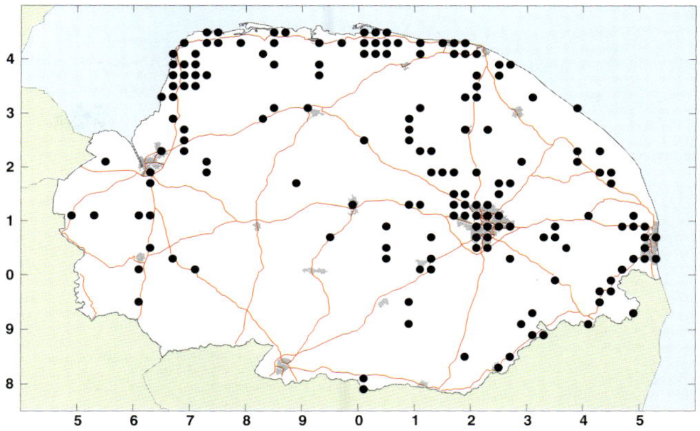

Herbaceous perennial probably introduced to gardens from mainland Europe, well naturalised on road verges, grassy banks and on shingle, especially near the coast. Also on waste ground in and around Norwich.

Oenanthe aquatica - Thompson

Silaum silaus (L.) Schinz & Thell. **Pepper Saxifrage**

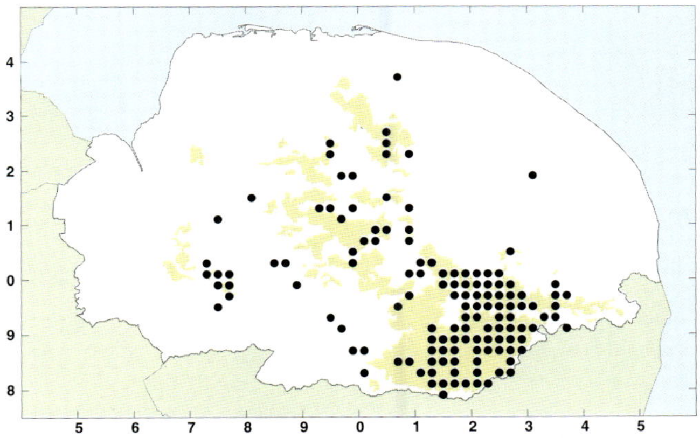

Perennial found in permanent, unimproved grassland on moist, base rich soils, frequent only on the chalky boulder clay.

Conium maculatum L. **Hemlock**

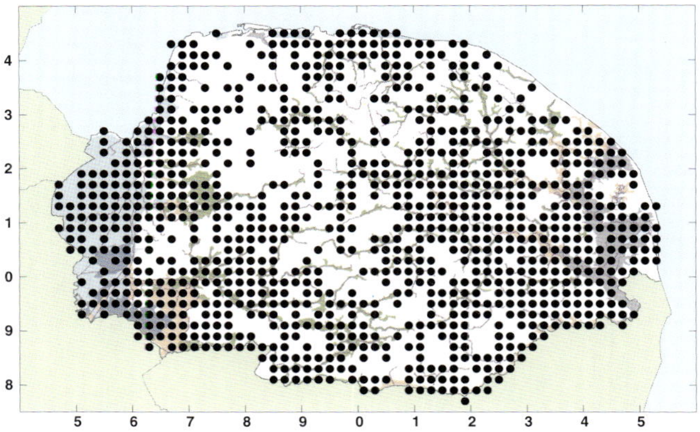

Winter green biennial, common on damp ground by roadsides, ditch banks, field margins and waste places.

Bupleurum tenuissimum L. **Slender Hare's ear**
Annual, once found on banks and grassland adjacent to the sea from Wolferton W62 to Blakeney E04 and around the estuary of Breydon Water E40. Always uncommon, now rare. During this survey seen at Wolferton W62 in 1987, CPP; Berney and Halvergate E40 in 1990, L.Street; Belton E40 in 1994, J.Child & G.Taylor and Thornham W74 in 1998, SP. A nationally scarce species. W2. E3.

■ **B. rotundifolium** L. **Thorow-wax**
Annual, once a cornfield weed in much of England, now a scarce casual mainly in gardens from bird-seed. Last seen at Harleston E28 in 1971, C.Forrest.

■ **B. subovatum** Link. ex Spreng. **False Thorow-wax**
Syn. *B. lancifolium* auct. non Hornem.
Annual introduced from the Mediterranean in birdseed and sometimes found in gardens. Eight records were made between 1950 and 1980.

Apium graveolens L. **Wild Celery**

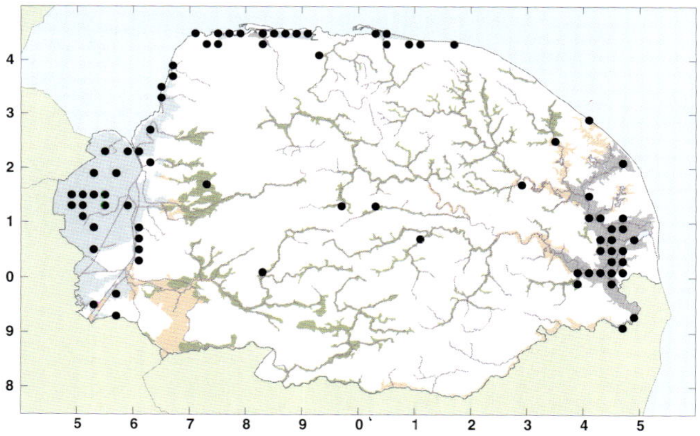

Winter green biennial not infrequent in ditches where there is or has been maritime influence, often detectable from some distance by its strong celery smell. Rare inland.

A. nodiflorum (L.) Lag. **Fool's Water-cress**

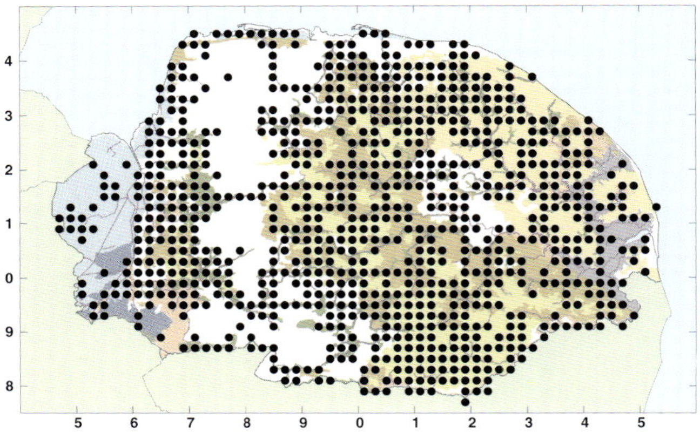

Winter green perennial present in all but the most polluted watercourses, often growing with water- cress. Not poisonous, just rather tasteless.

A. inundatum (L.) Rchb.f. **Lesser Marshwort**

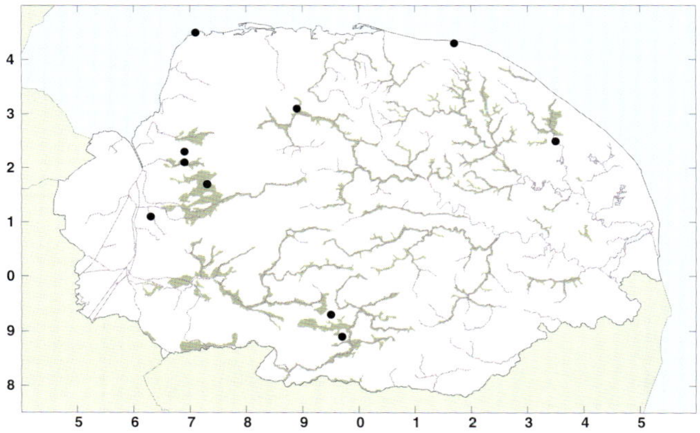

Winter green perennial of shallow pools, always very scarce.

■ **Petroselinum crispum** (Mill.) Nyman ex A.W. Hill **Garden Parsley**
Winter green biennial occasionally found as an escape from cultivation.

P. segetum (L.) W.D.J.Koch **Corn Parsley**

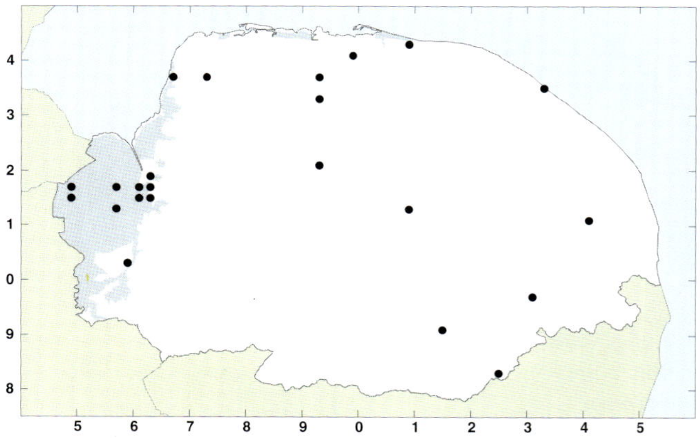

Winter green biennial, a scarce plant of open hedgebanks and dry ground, often alongside estuaries. A nationally scarce species.

Sison amomum L. **Stone Parsley**

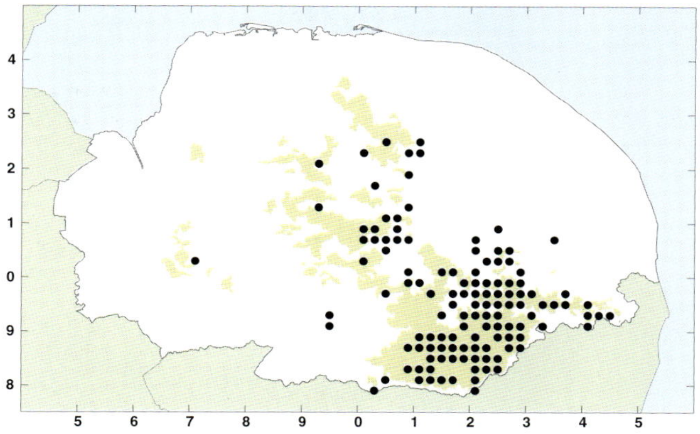

Winter green biennial with a distribution remarkably similar to that of *Silaum silaus* but confined to calcareous clay.

Cicuta virosa L. **Cowbane**

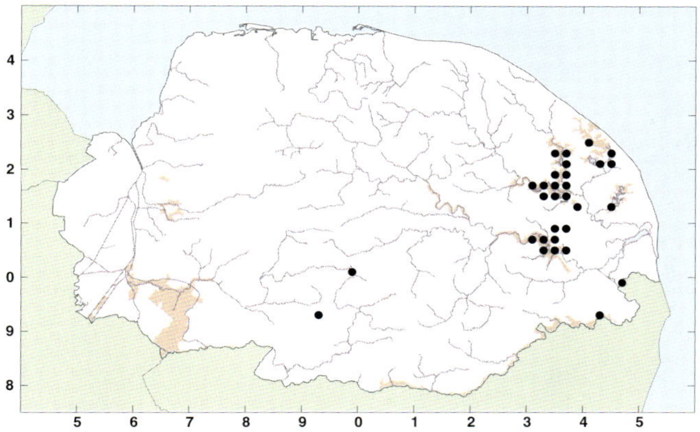

Herbaceous perennial, almost confined to banks of streams and dykes in the Broadland, elsewhere last seen at Cranberry Rough, Hockham W99 in 1983 but still at Scoulton W90 in 1996, KAB & EVR.

Cicuta virosa - Catfield

Ammi majus L. **Bullwort**

Annual introduced from southern Europe, found on arable land at Bircham W73 in 1989, ALB; in gardens at Carbrooke W90, G.Nobes and Dereham W91 in 1991, B.Griffin. W3. E0.

Falcaria vulgaris Bernh. **Longleaf**

Evergreen perennial from mainland Europe, a rare casual with three records before 1980 and one since, at North Elmham W91 in 1989, D&BM. W1. E0.

■ **Carum carvi** L. **Caraway**

Annual or biennial from mainland Europe introduced as a herb. Described as 'plentiful near Lynn' at the end of the 17th century and found a number of times in the 18th and 19th centuries.

Angelica sylvestris L. **Wild Angelica**

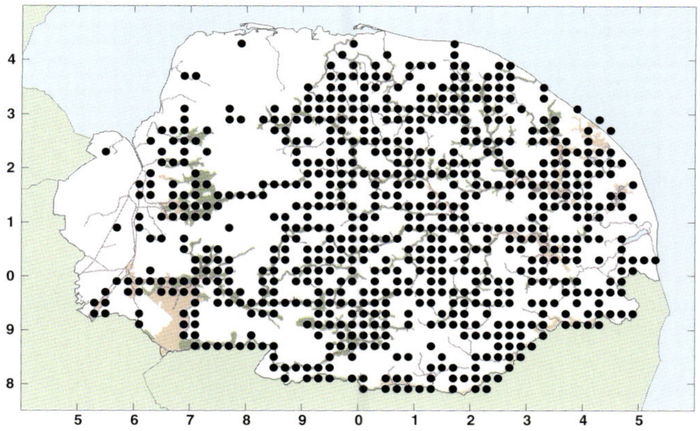

Herbaceous perennial frequent by streamsides, in marshes and wet ground generally, occasionally also in woodland.

■ **A. archangelica** L. **Garden Angelica**

Herbaceous perennial introduced to gardens from northern Europe, formerly cultivated for its herbal uses and for candied confectionery. An occasional escape on waste ground as at Bramerton E20 in 1982, ETD.

Peucedanum palustre (L.) Moench **Milk Parsley**

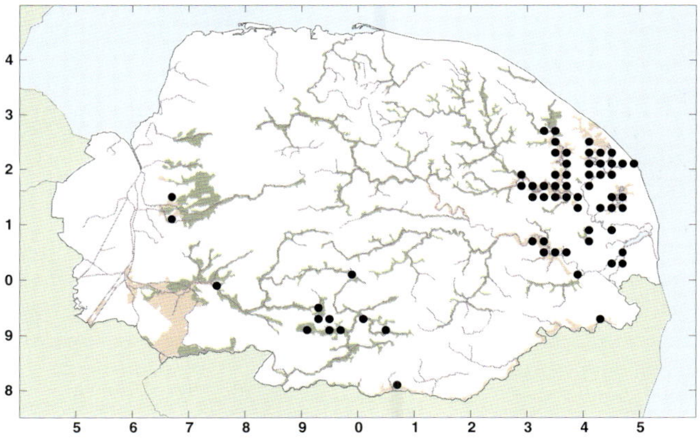

Herbaceous biennial well known as the foodplant of the swallowtail butterfly, still frequent in some areas of Broadland, needing an open habitat to thrive. Also surviving in a few isolated fens elsewhere, but much reduced in recent years because of the invasion of its sites by scrub. A nationally scarce species.

Pastinaca sativa L. **Parsnip**

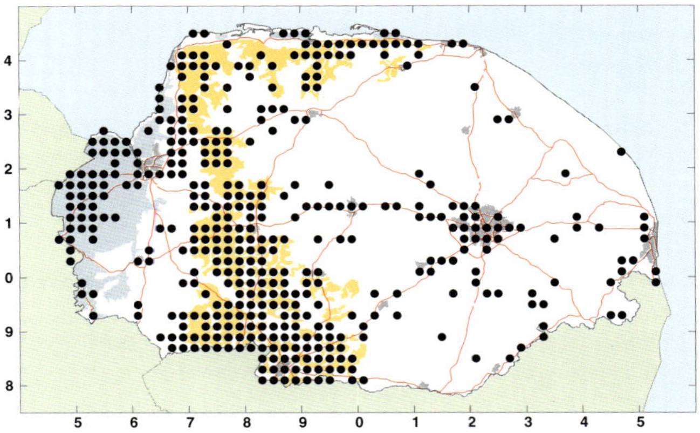

Winter green biennial of chalky soils described by HC.Watson in 1835 as rare in Norfolk, now quite abundant in west Norfolk. It is also notably frequent on the silty soils of the northern Fenland and it is always possible that some plants may have reverted from crops as it is now widely grown.

Heracleum sphondylium L.
Hogweed, Cow Parsnip, locally **Cow Mumble**
Winter green perennial, one of the truly ubiquitous species. Our plant is ssp. **sphondylium**. Ssp. **sibiricum** (L.) Simonk., with greenish-white flowers which have equal-sized petals was collected in the 1950s from several sites in the north-east of the county by TG.Tutin. While searching at one of these sites at Waxham E42 in 1997, one plant was found with this characteristic whilst var. **angustifolia** was found in several sites in this area, such as at Barton Turf E31, ALB. W694. E714.

H. mantegazzianum Sommier & Levier **Giant Hogweed**

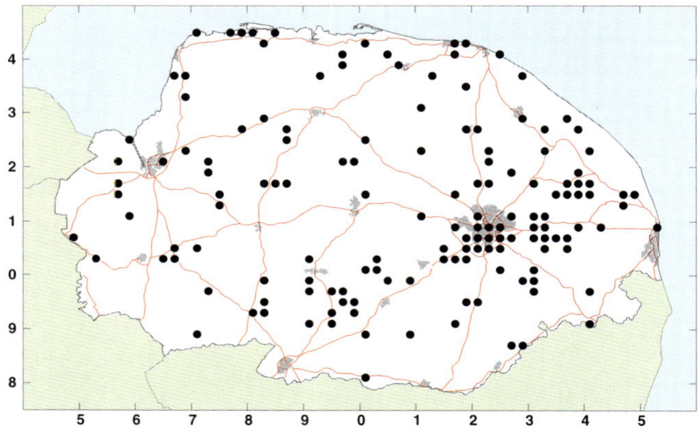

Winter green perennial introduced to gardens from south-west Asia, seeding freely and often naturalised. Not poisonous, but the subject of much press coverage owing to the presence in its sap of a chemical which, if spilled on to the skin, can temporarily remove its natural protection from sunburn.

■ **H. villosum** Fisch.
Winter green perennial recorded once, from Potter Heigham E41 in 1923.

Torilis japonica (Houtt.) DC. **Upright Hedge Parsley**

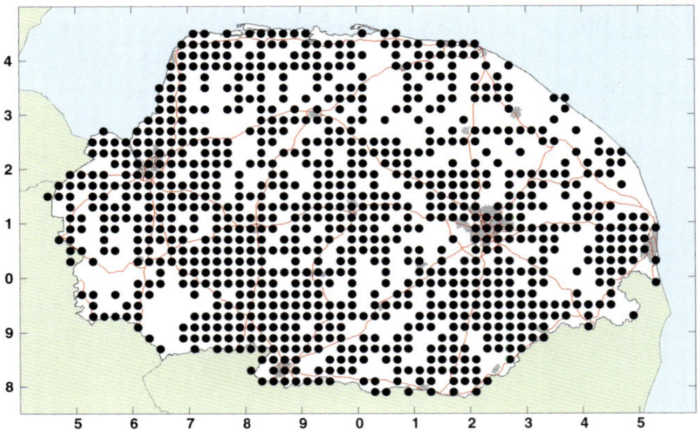

Annual of hedgebanks, woody margins and shady sites on dry soils.

T. arvensis (Huds.) Link **Spreading Hedge Parsley**
Annual of field edges and waste places, noted as 'locally abundant in hedgerows on the heaths' (around Swaffham W80) in the 1950s by E.S.Edees but now very rare and seen once only during this survey, a long standing garden weed at Tuddenham E01, ALB. A nationally scarce species. W0. E1.

T. nodosa (L.) Gaertn. **Knotted Hedge Parsley**

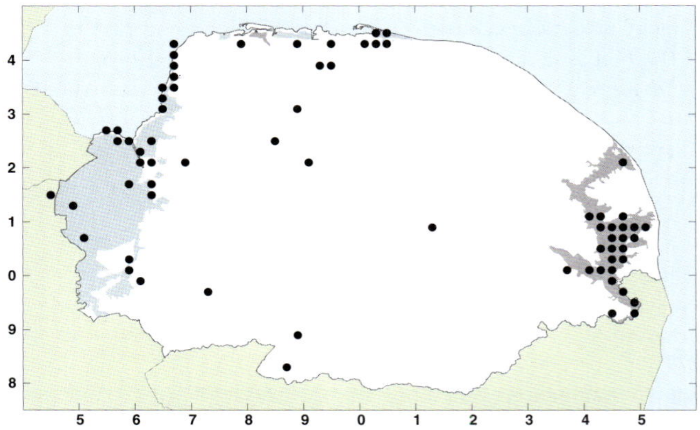

Annual found in short grassland and dry places chiefly near the sea. Particularly frequent on estuarine grazing levels in both east and west, but also occasionally as a casual from birdseed.

Daucus carota L. ssp **carota** **Wild Carrot**

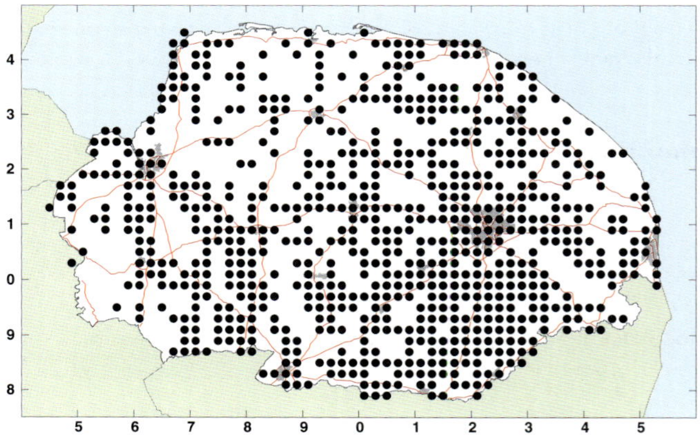

Winter green biennial, frequent. Often, but not always, on chalky soils.

GENTIANACEAE

■ **Cicendia filiformis** (L.) Delarbre **Yellow Centaury**
Spring-germinating annual formerly seen at Roydon W63 and Brancaster W74 but last recorded in 1928 and presumed extinct.

Centaurium erythraea Rafn **Common Centaury**

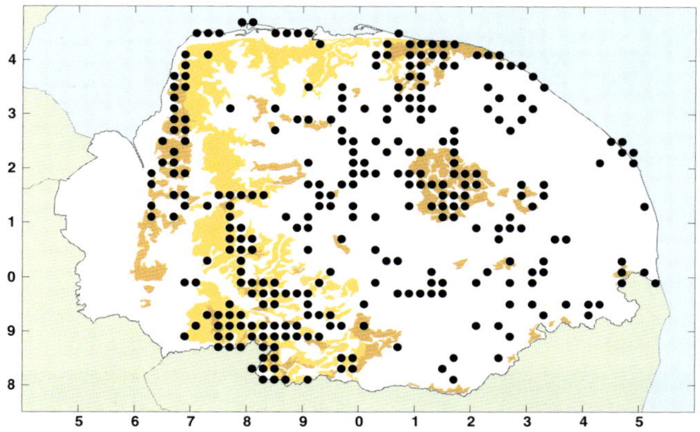

Annual not uncommon on light soils. Sometimes found in sunny woodland rides on chalky boulder clay.

C. littorale (Turner ex Sm.) Gilmour
Syn. *C. minus* Moench.
Biennial recorded once between Wells-next-the-Sea and Holkham in 1900. Not seen since. Recorded near Bacton E33 in 1996, T.Harwood, but not refound in 1997. W0. E1.

C. pulchellum (Sw.) Druce **Lesser Centaury**

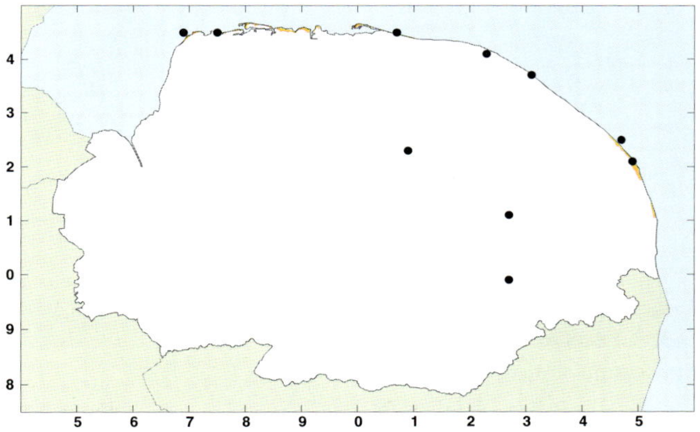

Annual of damp grassy areas, especially near the sea, also on the boulder clay where it has been found in grassland and on woodland rides.

Blackstonia perfoliata (L.) Huds. **Yellow-wort**
Autumn-germinating annual formerly more widespread, especially on the heavier, alkaline soils of central Norfolk with twenty sites listed between 1860 and 1960, but now apparently reduced to two large colonies, one on the banks of the Cut-off Channel at Feltwell,W68,W69 J.Secker, the second in recently cleared woodland at Framingham Earl E20, ALB & J.Mott. The reason for this widespread loss is not certain. W2. E2.

Gentianella campestris (L.) Börner **Field Gentian**
Autumn-germinating annual recorded from eight sites in the last hundred years, surviving at Derby Fen until drainage in the 1960s, now apparently confined to East Walton W71 where it was seen in 1988, DE.Coombe. W1. E0.

G. amarella (L.) Börner **Autumn Gentian**

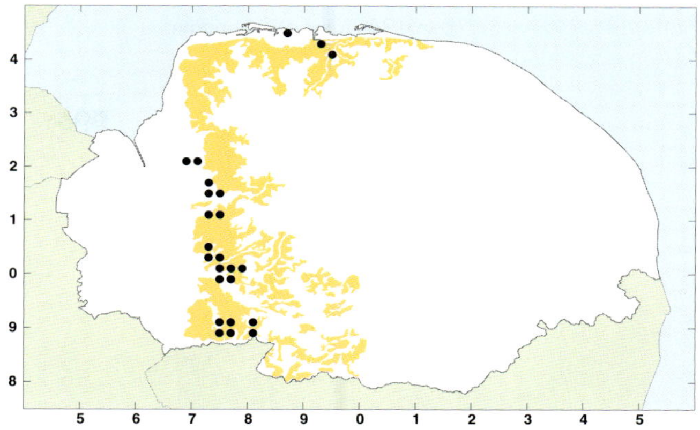

Autumn-germinating annual confined to chalk grassland with one colony on calcareous dunes at Burnham Overy W84. Much reduced since the 1968 Flora, probably because of the increasing rankness of such grassland especially after myxomatosis.

Gentiana pneumonanthe L. **Marsh Gentian**
Herbaceous perennial formerly a feature of the wet heaths north of Norwich and near King's Lynn. Now confined to four sites in E11, 12 and 14 and W71. A nationally scarce species. W2. E4.

Gentiana pneumonanthe - Buxton Heath

APOCYNACEAE

Vinca minor L. **Lesser Periwinkle**

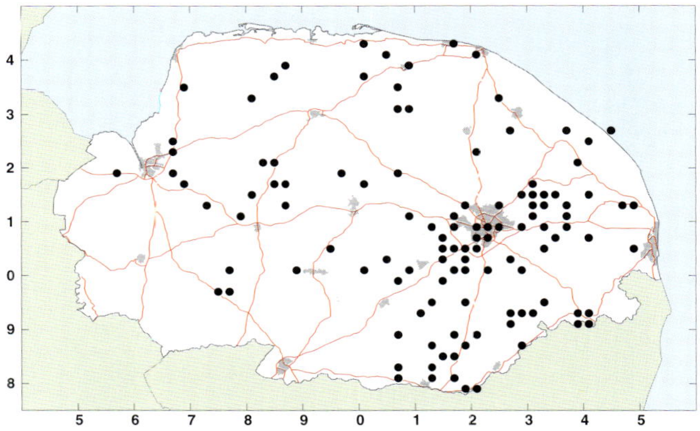

Winter green perennial introduced to gardens from mainland Europe and western Asia, naturalised in shrubberies and on hedgebanks usually near gardens, most frequent on the chalky boulder clay where it has been suggested it is native (ALB).

V. major L. **Greater Periwinkle**
Winter green perennial, introduced to gardens from central Europe to North Africa, often naturalised in similar sites to *V. minor* but far commoner. W146. E498.

SOLANACEAE

Nicandra physalodes (L.) Gaertn. **Apple of Peru**
Spring-germinating annual introduced from South America, occasional on tips and waste ground. W6. E8.

Lycium barbarum L.
Duke of Argyll's Tea-plant, FalseTea-tree
Syn. *L. halimifolium* Mill.
Deciduous shrub introduced to gardens from China, formerly much grown as cottage hedging and escaping. W159. E216.

L. chinense Mill. **Chinese Teaplant**
Similar to and not always separable from *L. barbarum* and far less frequent. W2. E45.

Atropa belladonna L. **Deadly Nightshade, Devil's Cherry**

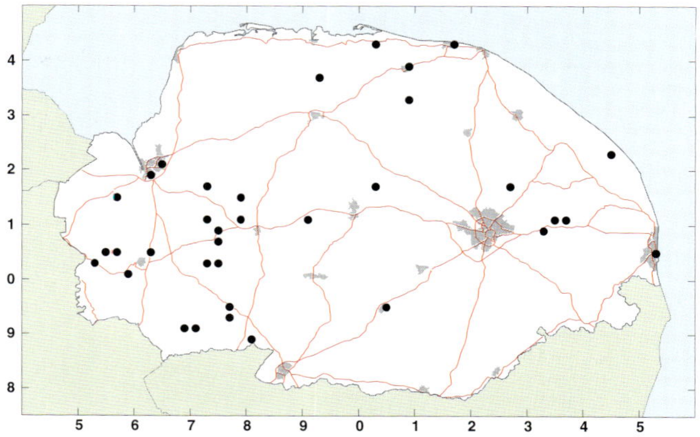

Herbaceous perennial, perhaps native on the chalk though often an escape from cultivation. Very persistent when established and sometimes associated with old monastic sites as at Carrow Abbey E20 where it was first noted in 1759.

Hyoscyamus niger L. **Henbane**

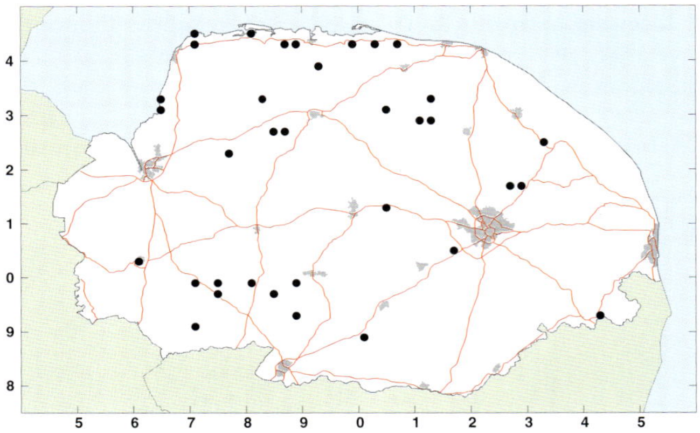

Annual or biennial native near the coast, also as a casual on waste ground and in arable fields, particularly on sandy soils. Sometimes abundant.

Hyoscyamus niger - Corpusty

Physalis alkekengi L. **Japanese Lantern**
Perennial introduced to cultivation from southern Europe to Asia. Sometimes bird sown but not long lasting. Recorded from Needham E28 ALB, also at Caister E51, ALB, DM and WM. W0. E3.

P. peruviana L. **Cape Gooseberry, Chinese Lanterns**
Deciduous perennial from South America, several plants occurred on land treated with sewage, Brettenham W98 in 1993, JEG&GB and Stow Bedon W99 in 1996, EVR. W2. E0.

■ **P. philadelphica** Lam.
Annual from the Americas introduced as a fruit, usually accidentally in batches of the closely related *P. ixocarpa* Brot. ex Hornem., Tomatillo. Found once, at Stacksford near Old Buckenham E09 in 1981, ALB.

Lycopersicon esculentum Mill. **Tomato**
Spring germinating annual to short-lived perennial from Central and South America, occasionally found on tips and waste places and in arable land where sewage has been spread. Plants do not survive the winter. W9. E12.

Solanum nigrum L. **Black Nightshade**

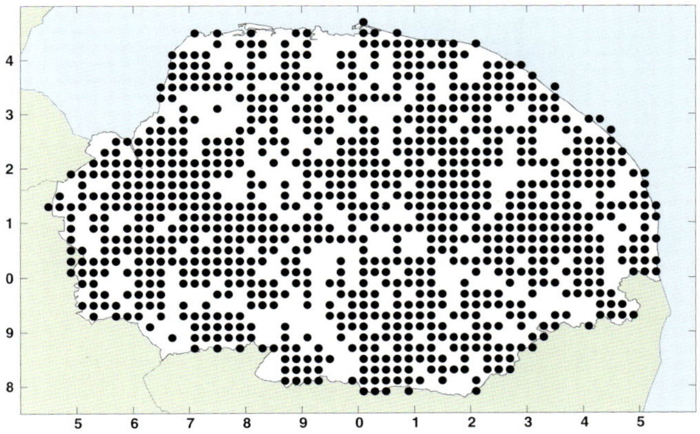

Spring germinating annual frequent in arable land, though its patchy distribution is not easy to explain. Most of our plants are ssp. **nigrum.** The much rarer ssp. **schultesii** (Opiz) Wessely, a casual introduced from mainland Europe, was found at Thornham W74 in 1979, DM.Donald and during this survey in King's Lynn W62 in 1992, RMP and Stow Bedon W99 in 1997, EVR.

S. physalifolium Rusby **Green Nightshade**
Syns. *S. nitidibaccatum* Bitter, *S. sarachoides* auct. non Sendtn.

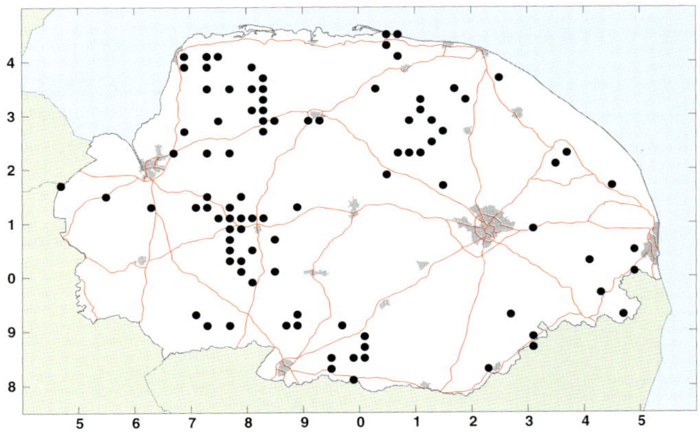

Spring germinating annual introduced as a casual from South America. First recorded at Bircham W73 in 1949 in carrots, but not becoming frequent until the 1970s, now well established especially in sugar beet.

■ **S. sarachoides** Sendtn. **Leafy-fruited Nightshade**
Spring germinating annual, not certainly found within the county because of past confusion over naming. Most earlier finds are probably referable to *S. physalifolium.*

■ **S. triflorum** Nutt. **Small Nightshade**
Spring germinating annual introduced as a casual from North America and formerly known at Wolferton where it was first seen in 1938 and last in 1984, both by CPP. Recorded also from Bagthorpe W73; Flitcham W72 and Bawdeswell E02 all before 1980. Still in the Suffolk Breckland.

S. dulcamara L. **Bittersweet, Woody Nightshade**

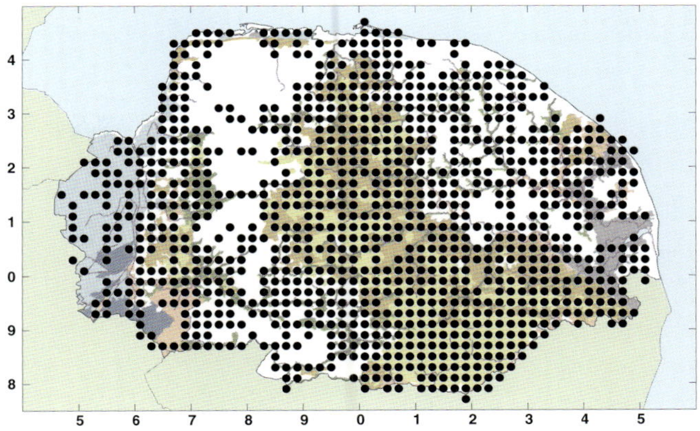

Herbaceous perennial capable of twining through shrubs both clockwise and anti-clockwise. Frequent in ponds, shallow ditches, wet ground and hedgerows in all moisture-retentive soils.

S. tuberosum L. **Potato**
Tender herbaceous perennial introduced as a crop from South America, sometimes surviving in fields and on roadsides, but not persistent. W54. E99.

■ **S. sisymbriifolium** Lam. **Red Buffalo-bur**
Annual from South America, recorded once, in sugar beet at Great Hockham W99 in 1975 by E.Campbell.

S. rostratum Dunal **Buffalo-bur**
Syn. *S. cornutum* auct. non Lam.
Annual from North America found first at Upton E31 in 1959 and on seven occasions during the 1970s. During this survey, noted in potatoes at Shouldham W60 in 1989, GM and again in 1994, Lady Rose Hare; in a former linseed field at Ryston W60 in 1995, RMP and on Beeston Common E14 in 1997, P.Buckle. W2. E1.

Datura stramonium L. **Thorn-apple**
Spring germinating annual introduced from America, not infrequent in waste places, sometimes persisting for several years. Especially associated with former smithies from the habit of some blacksmiths of giving a horse a leaf to chew to keep it calm (EAE) while a pinch of the crushed seeds was added to their food to make their coats shine (ALB). Today usually found where poultry and game feeding has taken place. The variety **chalybaea** (*D.tatula*) with coloured flowers is not infrequent. W24. E13.

Nicotiana alata Link & Otto **Sweet Tobacco**
Annual introduced to gardens from South America and occasionally escaping. W1. E0.

N. x sanderae W.Watson
(*N. alata x N. forgetiana*)
A hybrid of garden origin occasionally found as an escape. W2. E1.

Petunia x hybrida (Hook.) Vilm.
Annual of garden origin occasionally escaping and self-sowing. W0. E1.

CONVOLVULACEAE

Convolvulus arvensis L. **Field Bindweed**
Deciduous perennial very variable in flower markings, often producing no fruit, but still a common weed. W690. E694.

■ **C. tricolor** L. **Trailing Convolvulus**
Annual, a common cornfield weed in the Mediterranean introduced to gardens and occasionally escaping as at Horning E32 in 1979, ETD.

Calystegia soldanella (L.) R. Br. **Sea Bindweed**

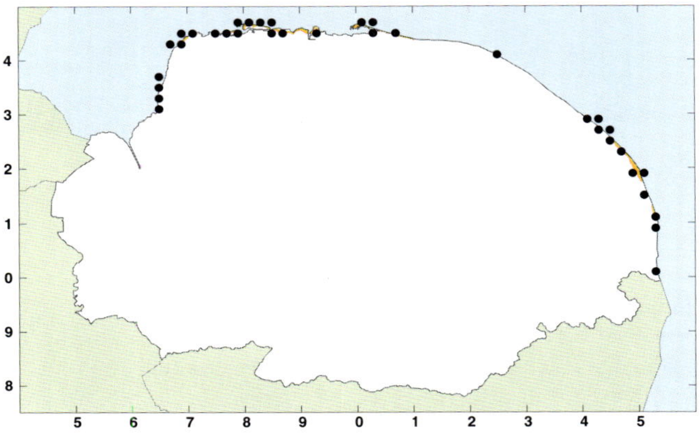

Deciduous perennial confined to sand and shingle by the sea where it can be abundant.

C. sepium (L.) R. Br. **Hedge Bindweed**

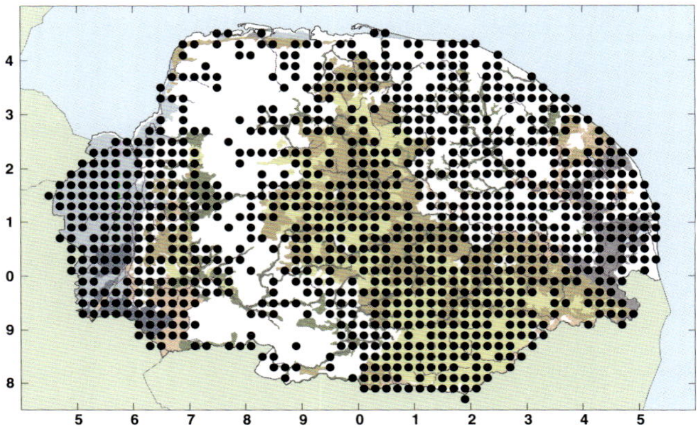

Deciduous perennial, a natural inhabitant of fen carr and frequent in moist soils. Most plants belong to ssp. **sepium**. The pink-flowered ssp. **roseata** Brummitt, occurs occasionally and forma **colorata** which is white with five deep pink stripes is frequent around Wacton E19, ALB.

C. pulchra Brummitt & Heywood **Hairy Bindweed**
Syn. *C. dahurica* auct. non (Herb.) G.Don.
Deciduous perennial probably always introduced and occasional in moist places as at Litcham W81 and Thorpe next Haddiscoe E49.
W3. E16.

Cuscuta epithymum - Runton

C. silvatica (Kit.) Griseb. **Large Bindweed**

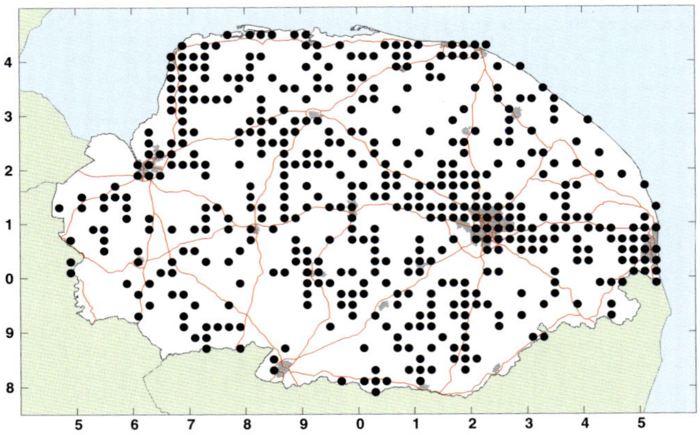

Deciduous perennial introduced from mainland Europe and now the commonest of the large bindweeds in urban areas and waste places, equally at home on wet and dry soils.

Ipomoea hederacea Jacq. **Ivy-leaved Morning-glory**
Annual scrambler introduced to gardens from the Americas, found in an arable field at Shouldham W60 in 1989, GM together with other casuals, presumably from pheasant feed. W1. E0.

CUSCUTACEAE

■ **Cuscuta campestris** Yunck. **Yellow Dodder**
Spring germinating parasitic annual from North America, recorded on carrots at Sprowston E21 in 1930 and Tivetshall E18 in 1973 and on chrysanthemums at Dereham W91 in 1971, KC.Durrant.

C. epithymum (L.) L. **Common Dodder**

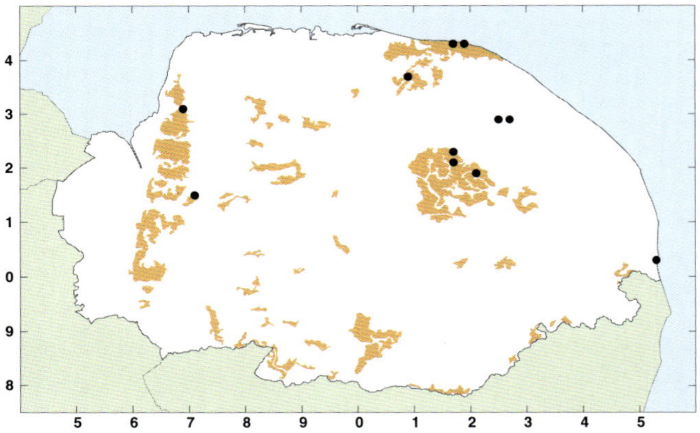

Spring germinating annual most commonly parasitic on heather, but sometimes also on gorse. Noted on *Lotus corniculatus* and *Teucrium scorodonia* by ELS, 1968. Increasingly scarce with the shading of its habitats by scrub.

■ **C. trifolii** Bab. **Clover Dodder**
Recorded ten times in the 19th century on clover.

■ **C. epilinum** Weihe **Flax Dodder**
Found growing on flax on three occasions in the 19th century.

■ **C. europaea** L. **Great Dodder**
Recorded from twelve sites on herbaceous plants and gorse by Nicholson, 1914, mostly in the 19th century.

MENYANTHACEAE

Menyanthes trifoliata L. **Bog-bean**

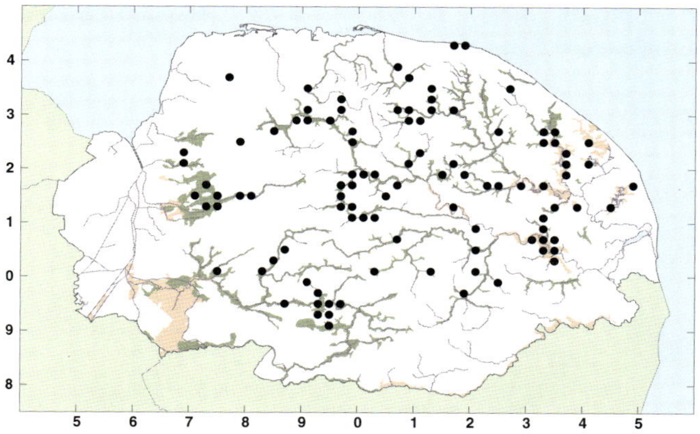

Herbaceous perennial of ponds and ditches where the water is at least moderately acid in reaction, absent from many of the calcareous streams of west Norfolk and from the Fenland.

Menyanthes trifoliata - Thompson

Nymphoides peltata Kuntze **Fringed Waterlily**

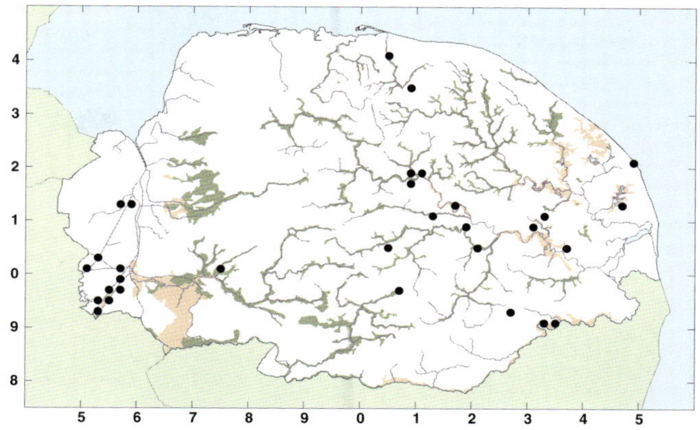

Herbaceous perennial native in some of the slow-moving rivers and streams of the Fenland. Recorded by W.Skrimshire near Outwell W50 in 1819 where it still grows, B&DM. Also introduced in some sites as in a number of gravel pits, now fishing lakes, in east Norfolk.

POLEMONIACEAE

■ **Polemonium caeruleum** L. **Jacob's Ladder**

Herbaceous perennial native in northern England, in Norfolk introduced to gardens and escaping. Found at Bawdeswell Heath E02 in 1971, ALB.

HYDROPHYLLACEAE

Phacelia tanacetifolia Benth. **Phacelia**

Spring germinating annual introduced from south-west USA and grown as a rich source of nectar for bees, occasionally self-sowing. Also from birdseed. First record at Cawston E12 in 1957. W6. E3.

BORAGINACEAE

Lithospermum officinale L. **Common Gromwell**

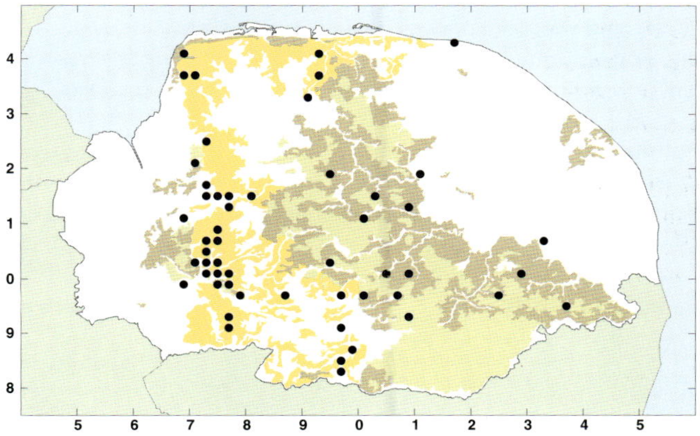

Herbaceous perennial, not infrequent in scrubby sites on calcareous soils, but also occasionally found at the edges of woods on neutral to acid soils.

L. arvense L. **Field Gromwell**
Syn. *Buglossoides arvensis* (L.) I.M.Johnst.

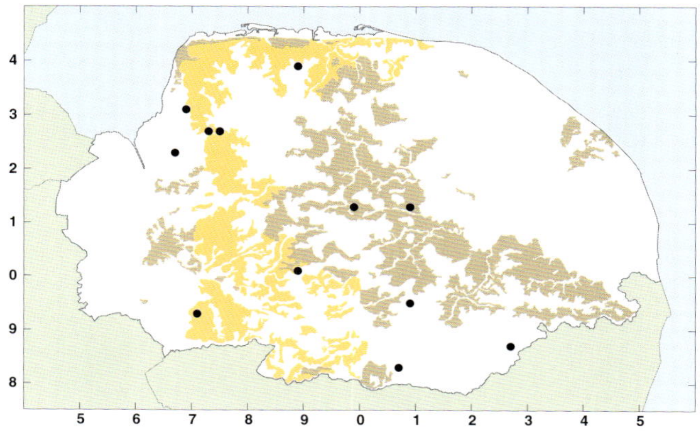

Spring or autumn germinating annual found mostly at the edges of arable fields on base rich soils, always scarce.

Cerinthe major L. **Greater Honeywort**
Annual introduced to gardens from the Mediterranean region. The variety **purpurata** was found on a gravel parking area in Dereham W91 in 1998, ALB, the first Norfolk record. W1. E0.

Echium vulgare L. **Viper's Bugloss,** locally **Devil's Guts**

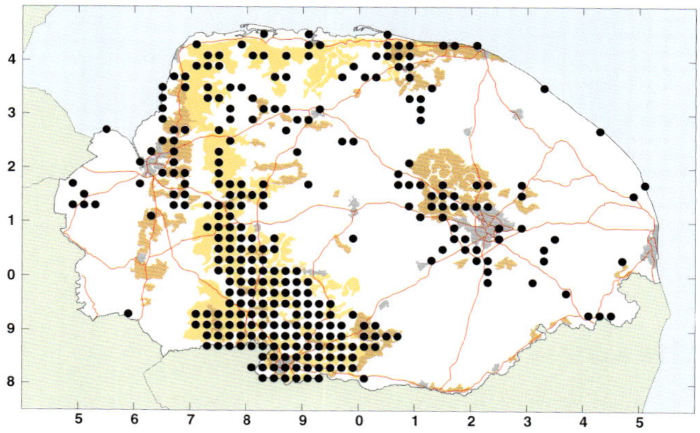

Winter green biennial or short-lived perennial commonest on poor dry, acid soils and frequent only in the Breckland. Formerly a serious arable weed and hindrance to the plough, hence its local name.

Pulmonaria officinalis L. **Lungwort**
Winter green perennial introduced to gardens from mainland Europe and occasionally escaping on to road verges as at Egmere W93 in 1995, K&GB. W1. E0.

Symphytum officinale L. **Common Comfrey**

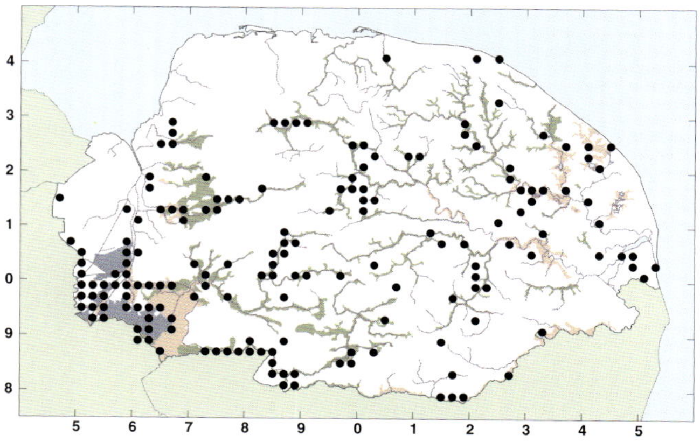

Winter green perennial of river banks and damp, grassy places, almost always creamy-flowered.

S. x uplandicum Nyman **Russian Comfrey**
(S. officinale x S. asperum)

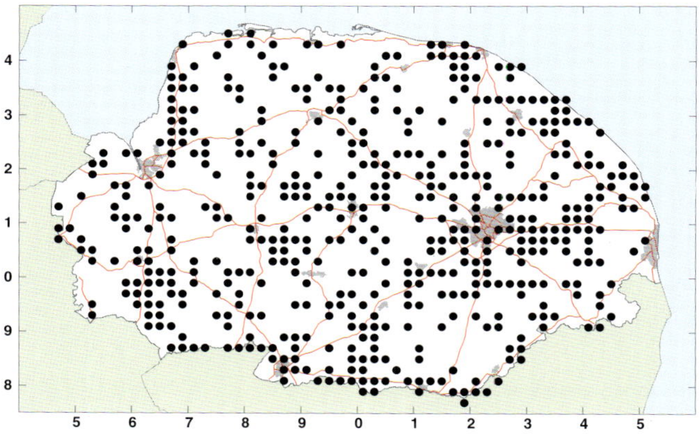

Winter green perennial introduced from mainland Europe as a fodder plant, naturalised on roadsides and rough ground where it can form large, dominant patches.

Echium vulgare - East Wretham Heath

S. asperum Lepech. **Rough Comfrey**
Introduced from south-west Asia as a fodder plant, soon out-performed by its offspring, *S. x uplandicum*, with which it is often confused. Found at Eccles, E08 in 1987 and Yarmouth E50 in 1995, I&P.Green. W0. E2.

S. tuberosum L. **Tuberous Comfrey**
Winter green perennial probably introduced and a garden escape in Norfolk. Most frequent within Broadland. W1. E8.

S. 'Hidcote Blue' **Hidcote Comfrey**
(Hybrid of uncertain parentage, possibly S. grandiflorum x S. x uplandicum)
Winter green perennial found established from garden refuse in a pit at Scarning W91 in 1995, KAB; along a woodland edge at Hevingham E12 in 1997, ALB and at Woodton E29 in 1998, Flora Group. W1. E2.

S. grandiflorum DC **Creeping Comfrey**
Winter green perennial introduced to gardens from the Caucasus and occasionally escaping on to roadside verges, usually close to gardens. W3. E6.

S. orientale L. **White Comfrey, Turkish Comfrey**

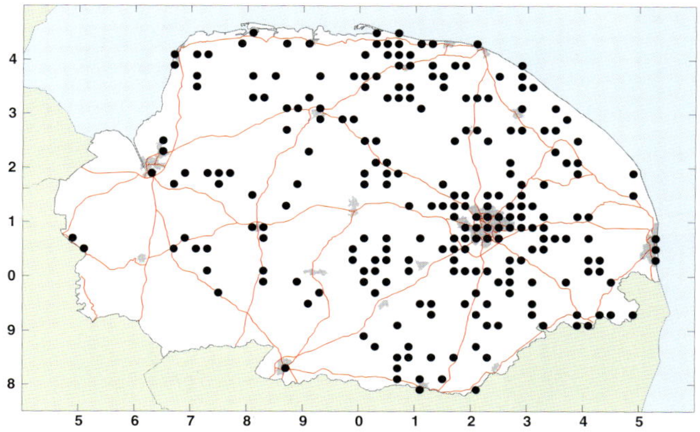

Winter green perennial introduced to gardens from south-eastern Europe and Turkey and naturalised on shady hedgebanks and in churchyards. Spreading rapidly once established, a small colony noted in a chalk-pit at Barton Bendish W70 in 1958 was dominant over a large area in 1997, RMP.

Brunnera macrophylla (Adams) I.M.Johnst. **Giant Forget-me-not**
Herbaceous perennial introduced from the Caucasus and sometimes self-sown. Found on a woodland edge at Seamere, Hingham E00 in 1996, ALB. W0. E1.

■ **Anchusa ochroleuca** M.Bieb. **Yellow Alkanet**
Perennial from eastern Europe, recorded at King's Lynn docks W62 in 1913.

A. arvensis (L.) M. Bieb. **Bugloss**
Syn. *Lycopsis arvensis* L.

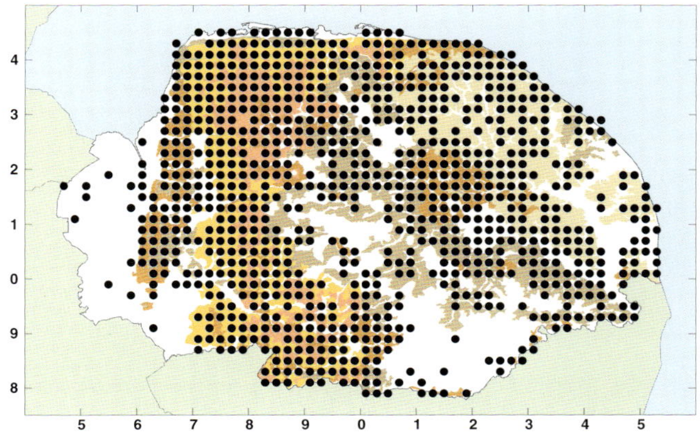

Annual, a weed of field margins, frequent except on heavy and moisture-retentive soils.

Pentaglottis sempervirens (L.) Tausch ex L.H.Bailey **Green Alkanet**
Winter green perennial introduced to gardens from south-western Europe, perhaps originally for the red dye produced by its roots. Well naturalised in shady sites. W122. E306.

Borago officinalis L. **Borage**
Mostly spring germinating annual introduced as a herb from southern Europe and sometimes persisting outside gardens. W27. E58.

Trachystemon orientalis (L.) G.Don **Abraham, Isaac and Jacob**
Perennial from the Caucasus introduced to gardens and occasionally established on shady banks. Found at Gt Snoring W93 in1971, T.Baker and still surviving, also at Strumpshaw E30 in 1990, SP and long established by the A140 at Tasburgh E29 SR.Martin. W1. E3.

■ **Mertensia maritima** (L.) Gray **Oysterplant**
Herbaceous perennial recorded on Blakeney Point E04 in 1910, last seen as a single plant in 1921.

Amsinckia lycopsoides (Lehm.) Lehm. **Scarce Fiddleneck**
Spring germinating annual introduced as a casual from western North America, recorded on several occasions before 1954. Also at Stiffkey W94 in 1991, C.Pogson, comm E.Hyde. W1. E0.

A. micrantha Suksd. **Common Fiddleneck**

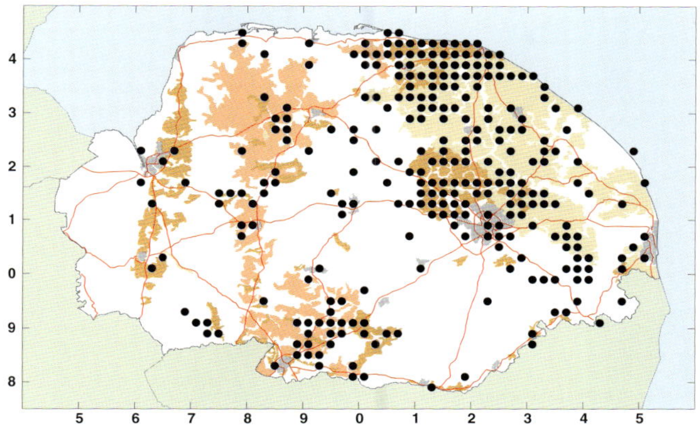

Spring germinating annual from western North America. The earliest record is from Wells-next-the-Sea W94 in 1916, but it became frequent only in the 1950s when it was recorded in the Weybourne area E14 and north of Norwich E11. It is now scattered on light soils throughout the county. At Larling W98 it has become a pernicious weed capable of flowering and seeding at 6cm high.

Myosotis scorpioides L. **Water Forget-me-not, Scorpion-grass**

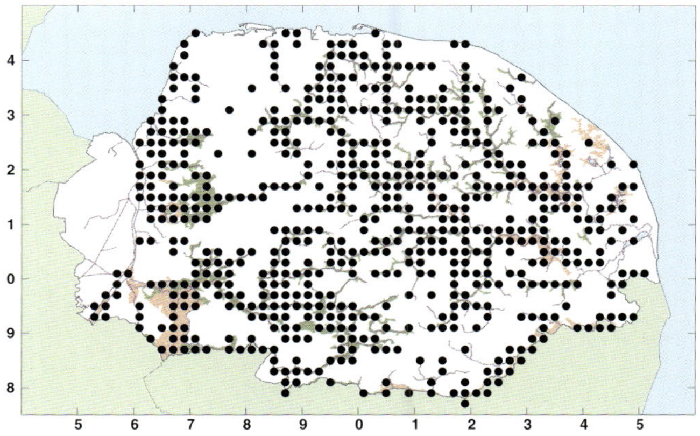

Winter green perennial frequent in ditches and by watercourses.

M. secunda A. Murray **Creeping Forget-me-not**

Annual to perennial of wet, acid places, known only from Holt E03. First recorded in 1981, CRS, still present in 1998, ALB. W0. E1.

M. laxa Lehm. **Tufted Forget-me-not**

Syn. *M. caespitosa* Schultz.

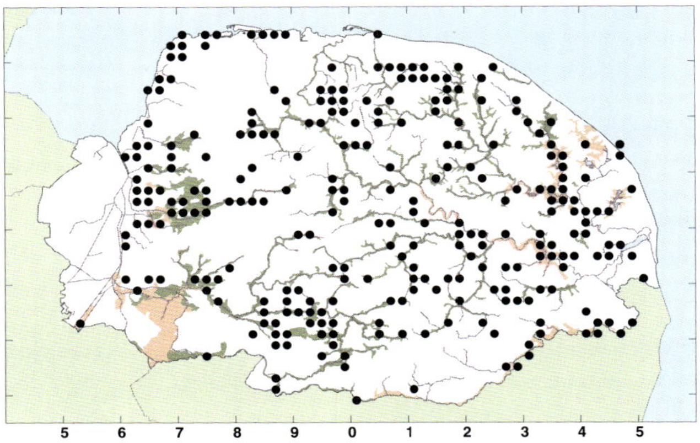

Winter green perennial in similar sites to *M. scorpioides* but not as common.

M. sylvatica Hoffm. **Wood Forget-me-not**

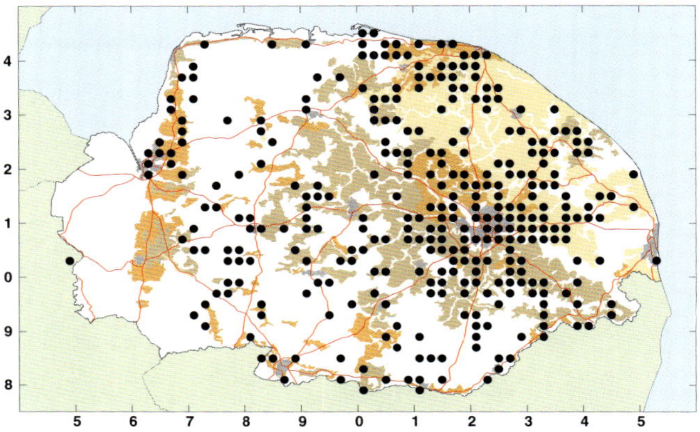

Winter green perennial, possibly native in some woodlands as at Cranworth W90 and Foxley E02, but usually occurring as a garden escape. The smaller, more compact, brighter blue-flowered forms found in gardens and also escaping, are derived from hybridization between *M. sylvatica* and *M. alpestris*.

M. arvensis (L.) Hill **Field Forget-me-not**

Winter green annual to perennial frequent on arable land, also in waste places and roadsides. W589. E598.

M. ramosissima Rochel **Early Forget-me-not**

Syn. *M. hispida* Schltr.

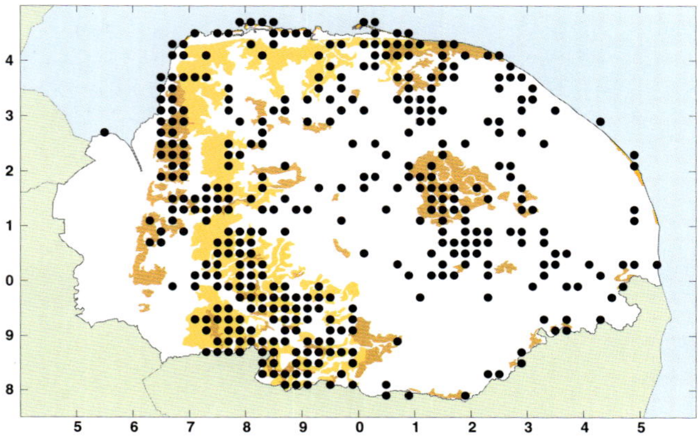

Autumn germinating annual frequent on light, sandy soils, both inland and on dunes.

M. discolor Pers. **Changing Forget-me-not**

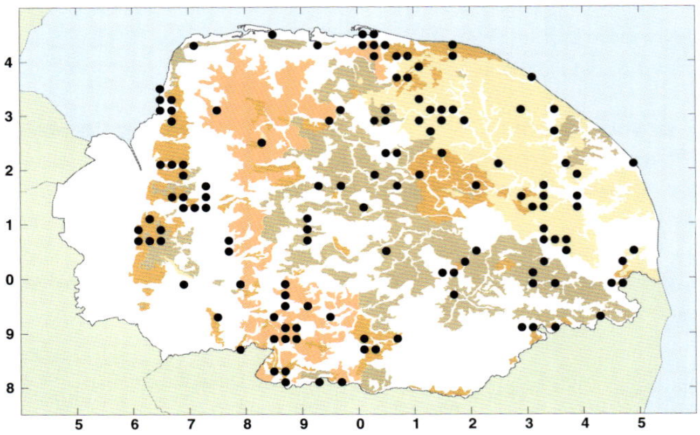

Autumn germinating annual found in similar sites to the last species, but much less frequent.

Cynoglossum officinale L. **Hound's-tongue**

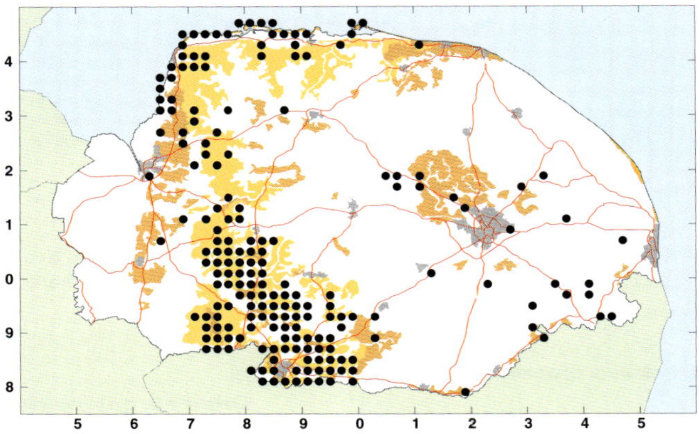

Winter green biennial found on light, sandy soils, common only in Breckland and on the north coast.

VERBENACEAE

Verbena officinalis L. **Vervain**

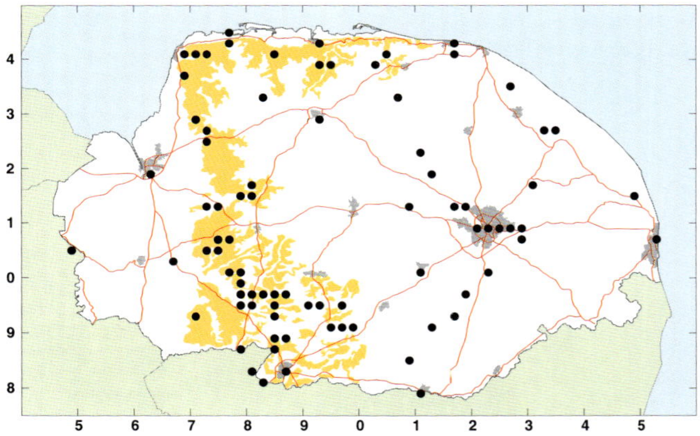

Perennial, scattered on roadside verges and permanent grasslands, usually on base rich soils.

LAMIACEAE (LABIATAE)

Stachys officinalis (L.) Trevis **Betony**
Syn. *Betonica officinalis* L.

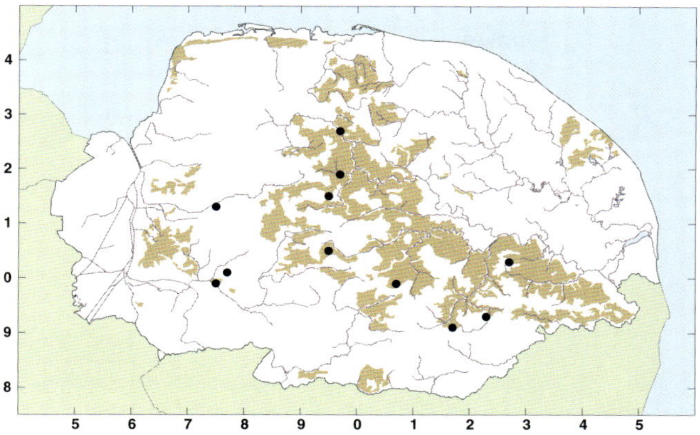

Perennial of moist, somewhat acid grassland. A plant with a mainly northern distribution and scarce in the county.

S. byzantina K.Koch. **Lamb's-ear**
Syn. *S. lanata* Jacq. non Crantz
Winter green perennial from south-western Asia, an occasional escape, seen in Norwich E20 in 1990, ETD. W0. E1.

Verbena officinalis - Stanford

S. sylvatica L. **Hedge Woundwort**
Herbaceous perennial abundant in hedgerows, roadsides and waste places, scarce only in the peat Fens. W571. E632.

S. palustris L. **Marsh Woundwort**

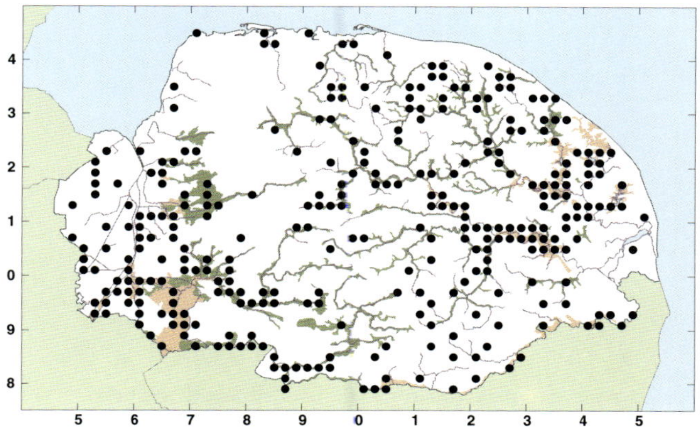

Herbaceous perennial usually found in moist soils in grassland and by ditches, also in drier ground as an arable weed.

■ **S. annua** (L.) L. **Annual Yellow Woundwort**
Annual introduced from mainland Europe, a rare casual recorded at Pulham St Mary E28 in 1952 and Harford tip, Norwich E20 in 1963.

S. arvensis (L.) L. **Field Woundwort**

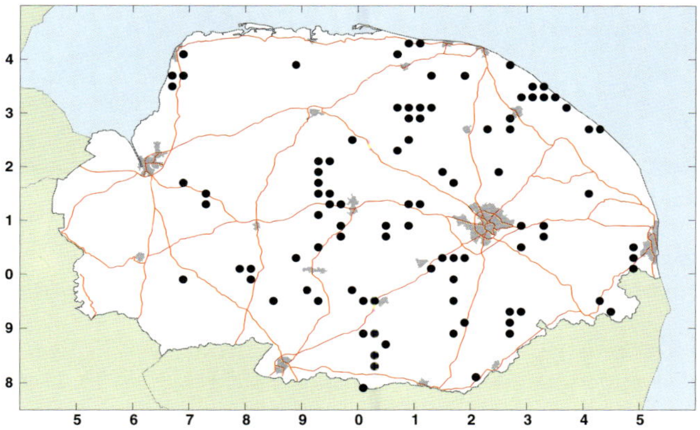

Annual, widely scattered on moist soils, but nowhere common, now found on waste ground as well as arable fields.

Ballota nigra L. **Black Horehound**
Winter green perennial, very common and surprisingly frequent on walls, RMP pers.obs. A white flowered form at Loddon E39 was very striking. W590. E655.

Leonurus cardiaca L. **Motherwort**
Winter green perennial introduced from mainland Europe as a herbal plant. Formerly a not infrequent escape with many sites noted in Nicholson,1914, but not seen again until 1993 when a plant was found on a riverside bank at Earsham E38, W.Kennison. W0. E1.

Lamiastrum galeobdolon (L.) Ehrend. & Polatschek
ssp. **montanum** (Pers.) Ehrend. & Polatschek. **Yellow Archangel**
Syn. *Galeobdolon luteum* Huds.

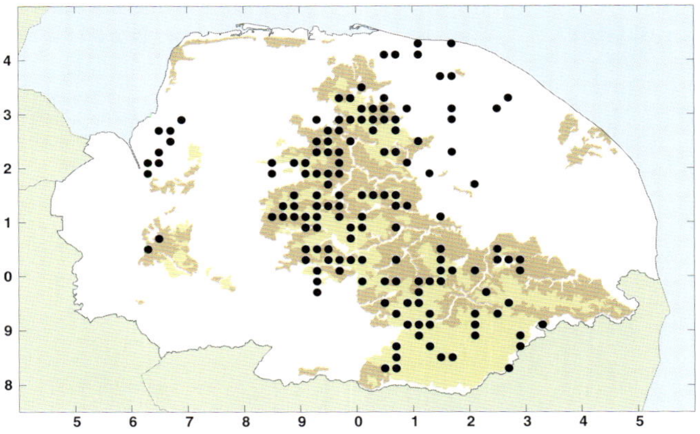

Winter green perennial found only in ancient woodland, also hedgebanks and lanes where such woodland once occurred.

Lamiastrum galeobdolon - Wayland Wood

Lamiastrum galeobdolon (L.) Ehrend. & Polatschek
ssp. **argentatum** (Smejkal) Stace **Variegated Archangel**
Syn. *Galeobdolon argentatum* Smejkal
Winter green perennial with silver-variegated leaves, widely grown in gardens, very persistent when it escapes. W38. E145.

Lamium album L. **White Dead-nettle**
Winter green perennial, a ubiquitous weed. W690. E691.

L. maculatum (L.) L. **Spotted Dead-nettle**
Winter green perennial introduced to gardens from mainland Europe, found as an escape but not usually long- lasting. W22. E57.

L. purpureum L. **Red Dead-nettle**
Autumn or spring germinating annual abundant throughout the county. A white-flowered form is sometimes found. W652. E702.

L. hybridum Vill. **Cut-leaved Dead-nettle**

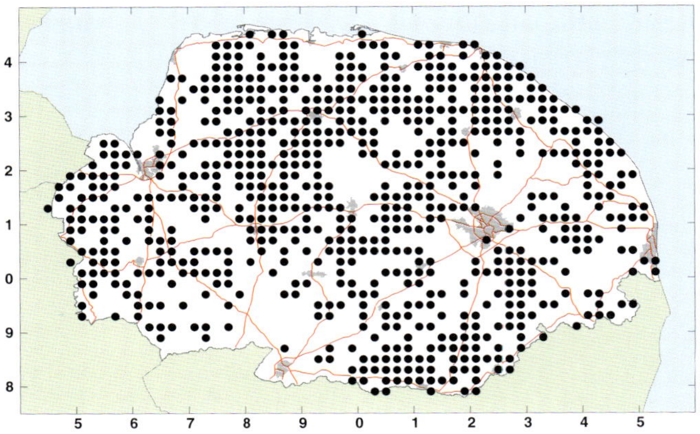

Autumn or spring germinating annual, a widely distributed arable and garden weed, found on all soils but never as frequent as *L. purpureum*. In Petch & Swann, 1968 it is noted as being less common than *L. amplexicaule*, but the reverse is true today.

L. amplexicaule L. **Henbit**

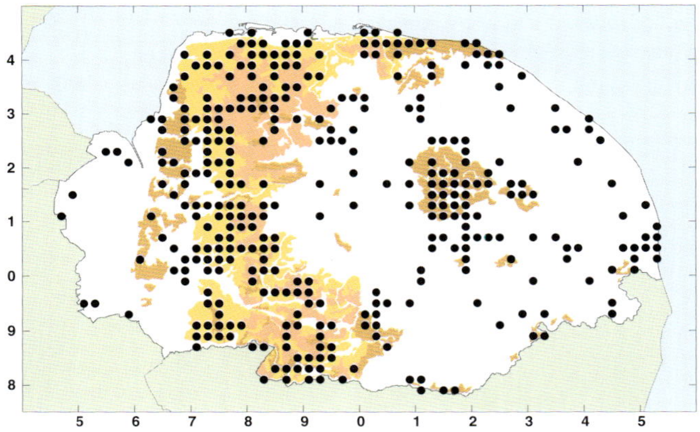

Autumn or spring germinating annual, usually, but not always found on lighter soils. Perhaps decreasing.

Galeopsis angustifolia Ehrh. ex Hoffm. **Red Hemp-nettle**
Spring germinating annual of light soils and shingle, now very scarce and recently seen only at Snettisham W63 where it was recorded from 1951 to 1970, then re-found in 1995 by SF. A nationally scarce species. W1. E0.

G. speciosa Mill. **Greater Hemp nettle**

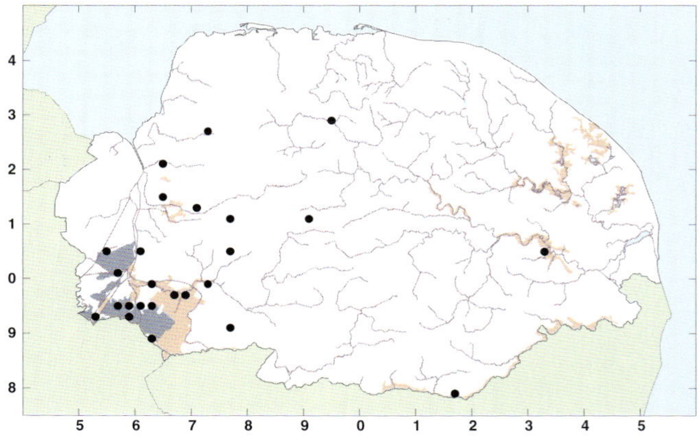

Spring germinating annual most frequent on peaty soils.

G. tetrahit **Common Hemp-nettle**

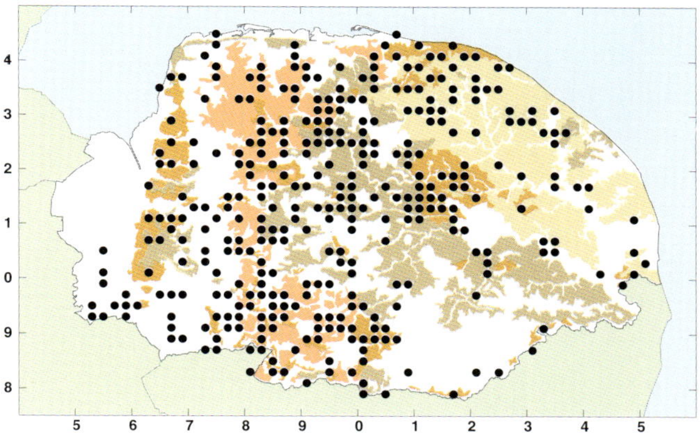

Spring germinating annual frequent in fields, rough grassland and waste places, avoiding calcareous soils.

Galeopsis speciosa - Billingford Common

G. bifida Boenn. **Bifid Hemp-nettle**

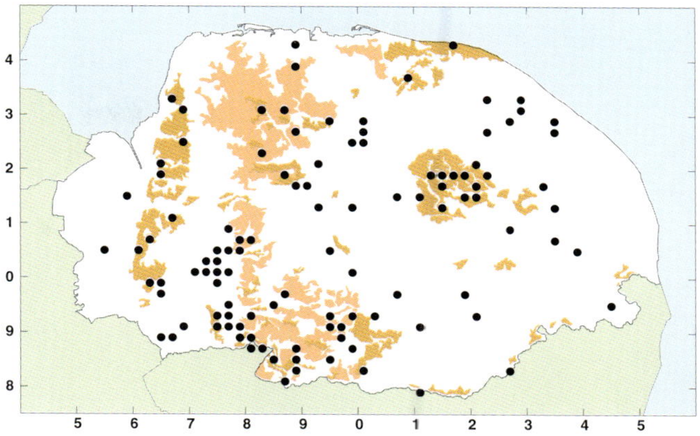

Spring germinating annual, with a preference for well-drained soils.

Marrubium vulgare L. **White Horehound**

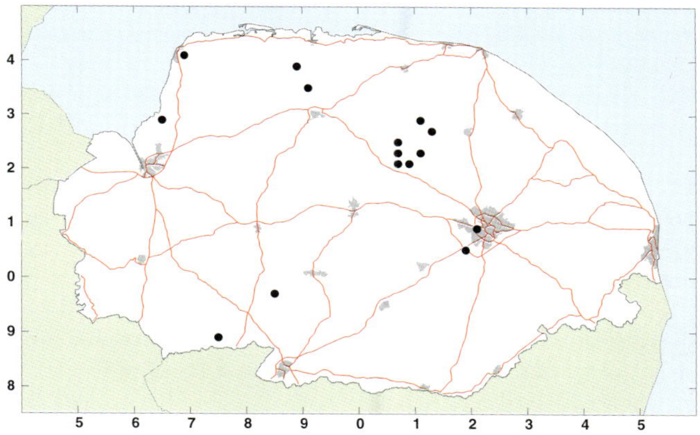

Semi-winter green perennial, once grown as a herb and perhaps sometimes a garden escape, very persistent around Bodney Lodge in the Stanford Training Area W89, ALB. Likely to be native at Ringstead, W64 & 74 in chalk grassland near the sea. W5. E9.

Scutellaria galericulata L. **Common Skullcap**

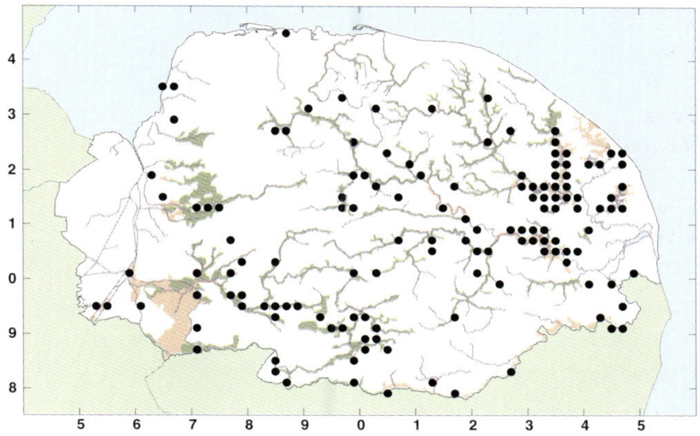

Herbaceous perennial, widely distributed on moist soils, but common only in Broadland.

■ **S. hastifolia** L. **Norfolk Skullcap**
Herbaceous perennial introduced from mainland Europe to a wood near Brandon. First recorded in 1948, last seen in 1980.

S. minor Huds. **Lesser Skullcap**
Herbaceous perennial of wet heaths formerly recorded from Blickling and Buxton Heath E12; Stalham E32 and Edgefield Heath E03. During this survey found only at Holt Lowes E03 in 1994, ALB where it was recorded by Trimmer, 1866. W0. E1.

Teucrium scorodonia L. **Wood Sage**

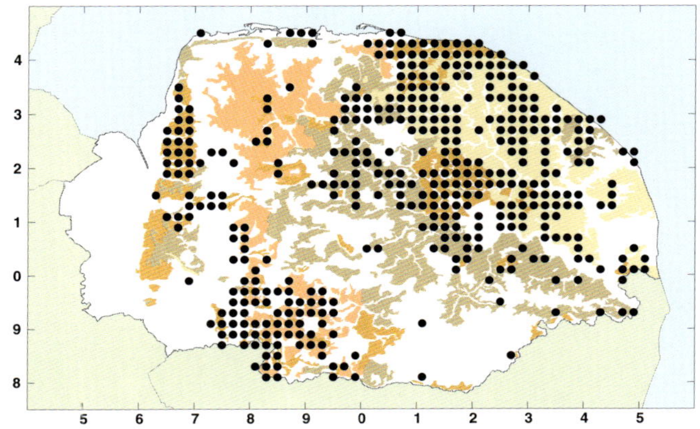

Winter green perennial restricted to well-drained, acid soils where it can be abundant.

T. chamaedrys L. **Wall Germander**
Winter green shrubby perennial introduced from mainland Europe, particularly as a herb. Noted on old walls in Norwich as early as 1723, a rare casual since. Established on walls at Castle Acre W81 near a herb garden, 1996, RMP. W1. E0.

■ **T. scordium** L. **Water Germander**
Perennial native to the Fenland and formerly recorded from Welney W59; Stow Bridge W60 and Horning E31 in the 19th century. Persisting now only in Cambridgeshire.

Ajuga chamaepitys (L.) Schreb. **Ground Pine**
Annual, native of southern Britain, recorded as a casual at East Tuddenham E01 in 1993, ALB. W0. E1.

A. reptans L. **Bugle**

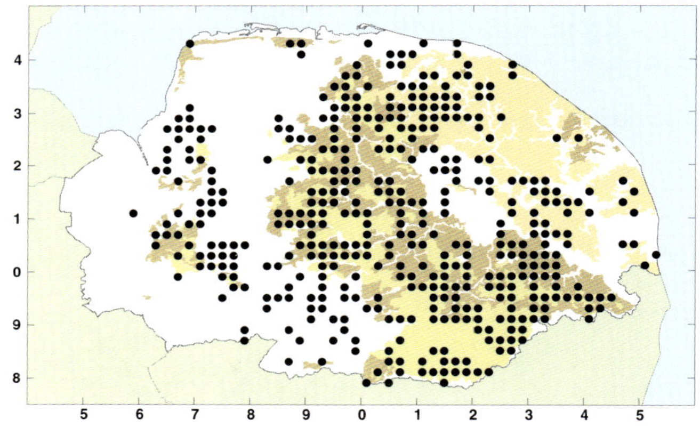

Winter green perennial found in ancient damp woodland as well as in long-established grassy meadows and fens. Also an occasional throw-out from gardens, often in its coloured leaved forms.

Nepeta cataria L. **Catmint**

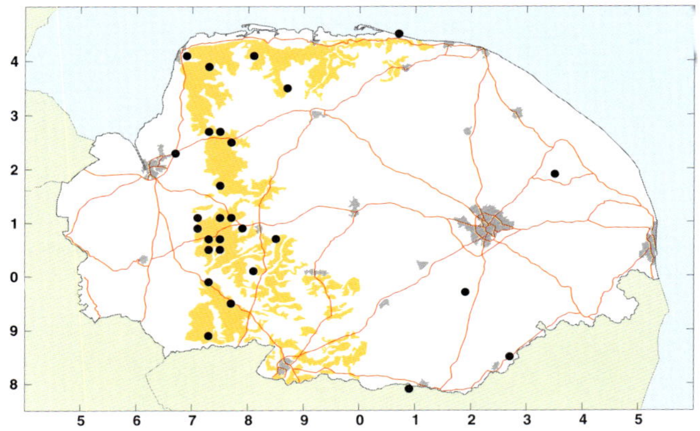

Winter green perennial of rough, chalky grassland and banks, now mostly surviving on roadside refuges where it is being rapidly reduced through unsuitable verge cutting.

N. x faassenii Bergmans ex Stearn **Garden Catmint**
(*N. mussinii x N. nepetella*)
A garden hybrid widely grown and occasionally established near gardens. W3. E0.

Glechoma hederacea L. **Ground Ivy, Ale-hoof**
Winter green perennial. Reputedly used in brewing before hops were introduced, ubiquitous. W671. E699.

Prunella vulgaris L. **Self-heal**

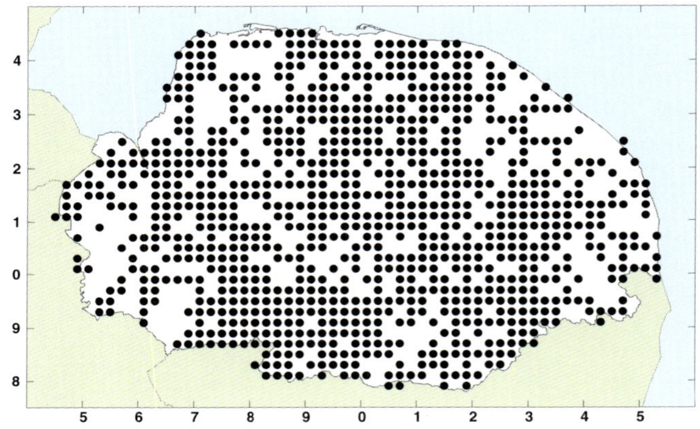

Winter green perennial abundant in shady sites. One of the few plants which thrives on acid and alkaline soils whether they are wet or dry.

Melissa officinalis L. **Lemon Balm, Bee Balm**
Herbaceous perennial introduced to gardens from southern Europe as a culinary herb, an occasional escape. W7. E17.

Satureja montana L. **Winter Savory**
Winter green sub-shrub introduced to gardens from southern Europe and found on waste ground at Upwell W50 in 1994, RMP. W1. E0.

Clinopodium ascendens (Jord.) Samp. **Common Calamint**
Syn. *Calamintha ascendens* Jord.

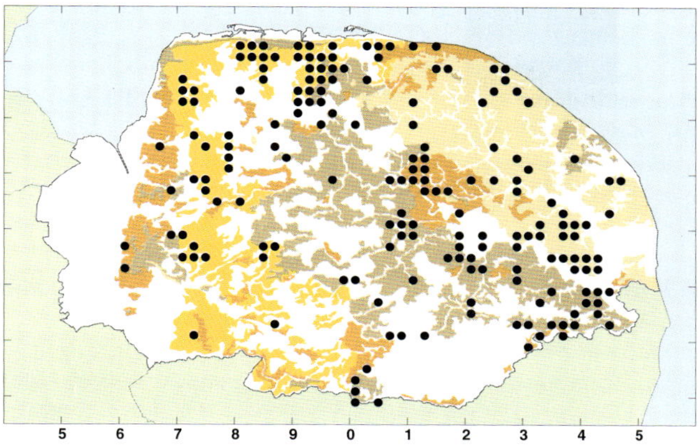

Semi-winter green perennial occurring widely but never commonly in well-established, dry grassland especially steep roadside banks; usually, but not always, on calcareous soils.

C. calamintha (L.) Stace **Lesser Calamint**
Syn. *Calamintha nepeta* (L.) Savi
Semi-winter green perennial first recorded from Hillington W72 in 1800 by W.Skrimshire. Always considered rare and now known only from a few localities in and around the village of Castle Acre W81 where it grows on walls, road verges and in short turf. A nationally scarce species. W4. E0.

C. vulgare L. **Wild Basil**

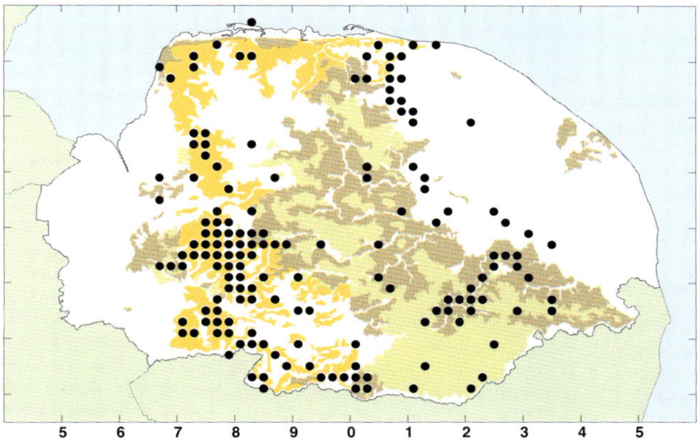

Semi-winter green perennial found chiefly in established grassland on base rich soils.

C. acinos (L.) Kuntze. **Basil Thyme**
Syn. *Acinos arvensis* (Lam.) Dandy

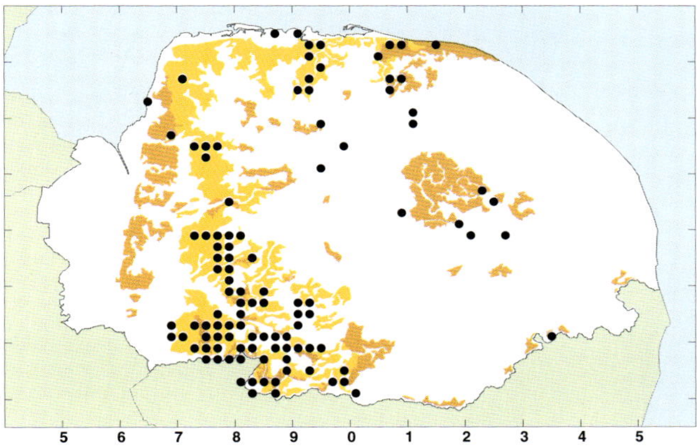

Annual, described in Petch & Swann, 1968 as 'Common. Dry places, equally on chalk or sand.' Today it is frequent only in rides in the Breckland forests, where upon occasions as many as 60% of the plants can be white-flowered, elsewhere it is scarce.

Hyssopus officinalis L. **Hyssop**
Evergreen sub-shrub, introduced to gardens from southern Europe and occasionally found as a garden escape as at Blo' Norton E07 where one plant was seen on a roadside verge in 1997, Flora Group. W0. E1.

Origanum vulgare L. **Marjoram, Oregano**

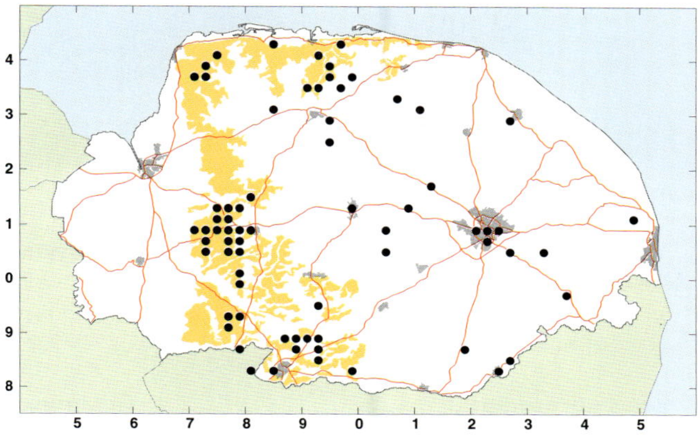

Herbaceous perennial of chalk grassland, now lost from many of its former roadside verge habitats as, like catmint, its stems are leafy so repeated cutting weakens the plants. Abundant in some forest rides.

Thymus pulegioides L. **Greater Wild Thyme**

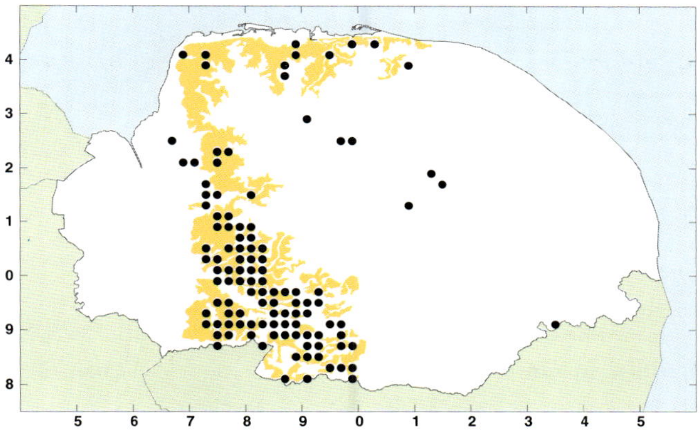

Winter green sub-shrub occurring in short turf on chalky to neutral soils but commonest on the former.

T. polytrichus A. Kern. ex Borbàs **Wild Thyme**
ssp. **britannicus** (Ronninger) Kergúelen.
Syn. *T. drucei* Ronn, *T. praecox* Opiz.
Winter green sub-shrub confined to a few chalky grasslands in the south of the county where it seems to be increasing, mainly at the expense of the rarer Breckland thyme. Recorded for W78, 88 and 98 and E18. W6. E3.

T. serpyllum L. **Breckland Thyme**
Winter green sub-shrub confined to the Breckland of Norfolk and Suffolk. Of the sites recorded by Petch & Swann, 1968, that at Brandon W78 is lost and the record for Weeting Heath was an error (ELS notes). Now known only from Santon and Thetford, W88. Red Data Book species. Illustrated on the following page. W4. E0.

Thymus serpyllum - Thetford page 175

T. vulgaris L. **Garden Thyme**

Winter green sub-shrub from the western Mediterranean introduced to gardens as a herb and found as an escape in Downham Market W60 in 1996, RMP and Norwich E20, SP. W1. E1.

Lycopus europaeus L. **Gipsywort**

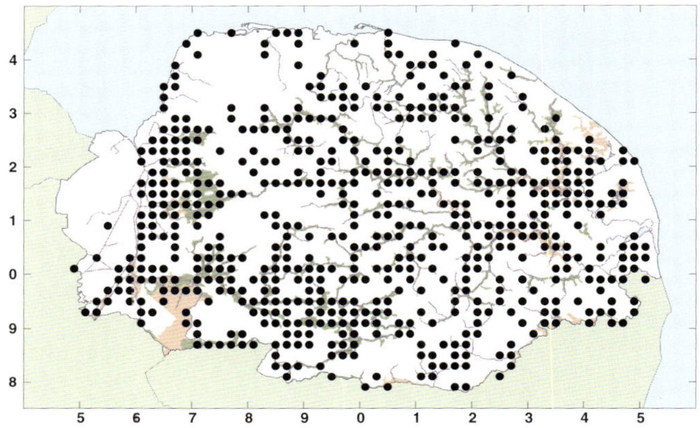

Herbaceous perennial found on moist soils and ditch sides throughout the county.

Mentha arvensis L. **Corn Mint**

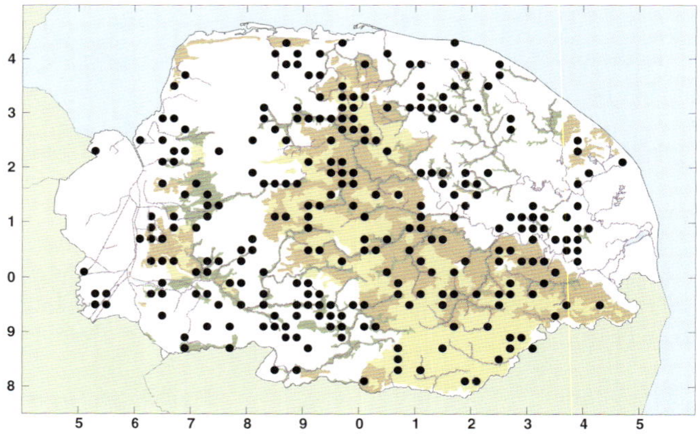

Herbaceous perennial found in moist soil in arable fields and in bare places in woodland.

M. x verticillata L. **Whorled Mint**

(*M. arvensis x M. aquatica*)

Herbaceous perennial found in similar habitats to *M. arvensis* but not as frequent. W33. E23.

M. x smithiana R.A.Graham **Tall Mint, Black Peppermint**

(*M. arvensis x M. aquatica x M. spicata*)

Herbaceous perennial of waste ground, usually of garden origin, found at Wendling W91 in 1990, DMM and Brandon W78 in 1993, KAB. W2. E0.

M x gracilis Sole **Bushy Mint**

Syn. *M. gentilis* auct. non L. (*M. arvensis x M. spicata*)

Herbaceous perennial, an occasional escape from gardens as on Litcham Common W81 in 1992, Flora Group and in Norwich E20 in 1990, ETD. W1. E2.

M. aquatica L. **Water Mint**

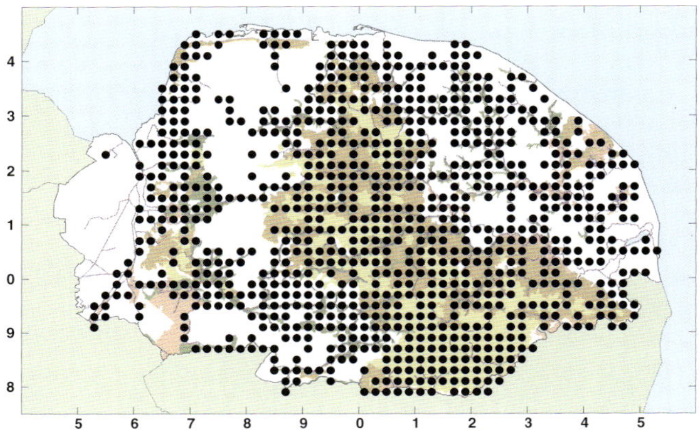

Herbaceous perennial abundant alongside all waterways except the steep-sided Fenland dykes.

M. x piperita L. **Peppermint**

Syn. *M. x citrata* Ehrh., *M. dumetorum* auct. non Schult. (*M. aquatica x M. spicata*)

Herbaceous perennial, usually an escape from gardens but populations of the pubescent form at Blo Norton E07 and Pond Hills E13 may be native, ALB. W4. E5.

M. spicata L. **Spear Mint**

Herbaceous perennial widely grown as a herb and escaping. Plants formerly listed as *M. longifolia* belong either here or with the following hybrid. W25. E30.

M. x villosonervata Opiz **Sharp-toothed Mint.**

(*M. spicata x M. longifolia*)

Herbaceous perennial occasionally escaping from gardens, recorded once each in W63, 70, 74, 78, 80 and E30. W5. E1.

M. x villosa Huds. **Apple-Mint, Bowles' Mint**
(*M. spicata x M. suaveolens*)

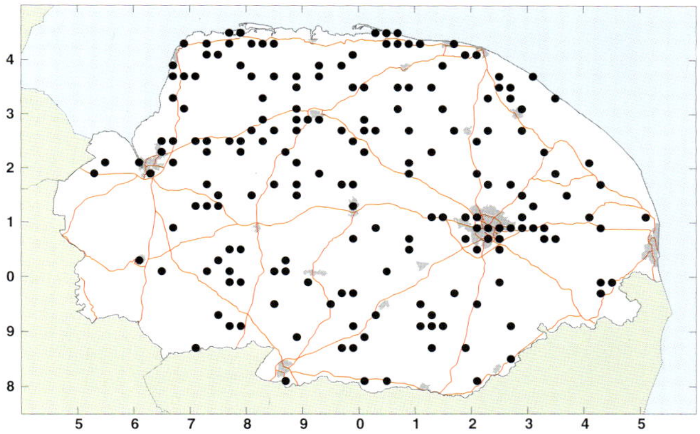

Herbaceous perennial, the most commonly grown garden mint, especially in the hairy form known as Bowles' Mint. The name *M. villosa* includes plants formerly listed as *M. x cordifolia* and *M. x niliaca* in part. Widely escaping and often well established.

M. x rotundifolia (L.) Huds. **Lamb Mint**
Syn. *M. x niliaca* Juss. ex Jacq. (*M. longifolia x M. suaveolens*)
Herbaceous perennial occasionally escaping from gardens, recorded at Cranworth W90 in 1990, ALB. W1. E0.

■ **M. suaveolens** Ehrh. **Round-leaved Mint**
Syn. *M. rotundifolia* auct. non (L.) Huds.
Herbaceous perennial, recorded once only, at Didlington W79 in 1968.

■ **M. pulegium** L. **Pennyroyal**
Herbaceous perennial, the only records in the 20th century being for Ditchingham E39 in 1957 and Tottenhill W61 in 1961. Rapidly becoming an endangered species nationally.

Lavandula angustifolia Loisel. **Lavender**
Evergreen shrub from the Mediterranean region. Self-sown plants have been found near gardens and survive all but the coldest winters. W3. E3.

Rosmarinus officinalis L. **Rosemary**
Evergreen shrub introduced from the Mediterranean, not completely hardy in severe winters, but self-sowing and surviving on the former Sculthorpe Airbase W83 in 1996, KAB. W1. E0.

■ **Salvia officinalis** L. **Garden Sage**
Evergreen sub-shrub introduced from southern Europe to gardens and found on Harford tip, Norwich E20 in 1969.

■ **S. pratensis** L. **Meadow Clary**
Perennial native to southern England but not considered native in Norfolk. Five old records, most recently at Kilverstone W88 where it was recorded from 1957 to 1977.

S. x sylvestris L.
(*S. nemorosa x S. pratensis*)
Winter green perennial found at Methwold W79 in 1954 and on the dunes at Burnham Overy Staithe W84 in 1953 where it was still present in 1997. W1. E0.

S. verbenaca L. **Wild Clary**
Syn. *S. horminoides* Pourr.

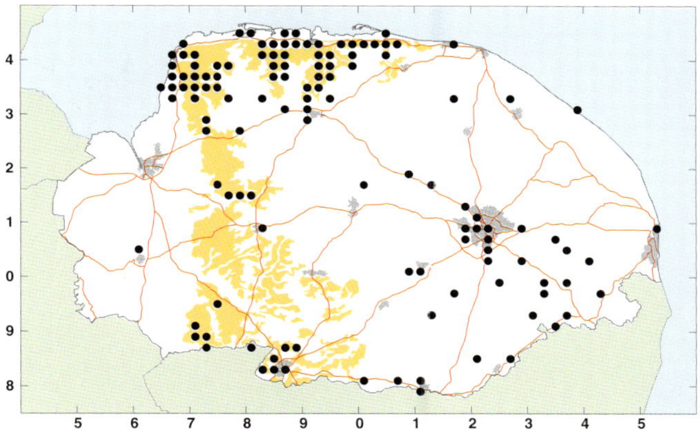

Winter green perennial found in dry, long-established neutral to calcareous grassland. A rosetted plant which loses only flowering stems when roadside verges are cut and perhaps increasing in some sites especially in the north-west of the county.

Salvia verbenaca - Shernborne

■ **S. reflexa** Hornem. **Mintweed**
Perennial introduced to gardens from North America. Three old records, the last for Harford tip, Norwich E20 in 1961.

■ **S. viridis** L. **Annual Clary**
Syn. *S. horminum* L.
Annual recorded once, surviving on a refuse tip at Beetley W91 in 1980, RPL & ELS.

■ **S. verticillata** L. **Whorled Clary**
Winter green perennial recorded once, at the entrance to the Harford Tip, Norwich E20 in 1960.

HIPPURIDACEAE

Hippuris vulgaris L. **Mare's-tail, Fox-tail** (local)

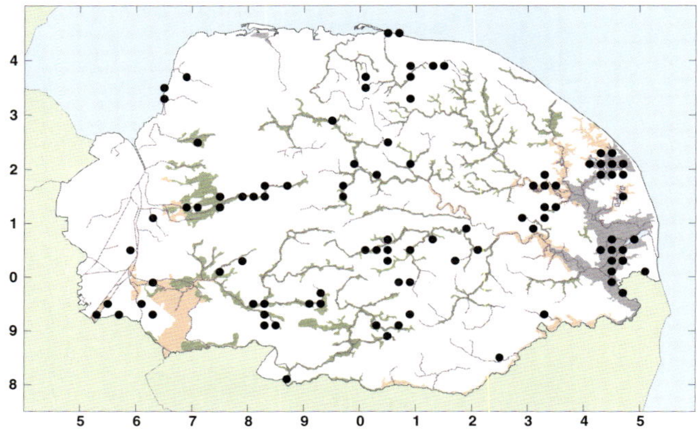

Herbaceous perennial described in 1968 as frequent in ponds, streams and rivers, now no longer so, though it can be plentiful where it occurs. Many of our waterways are, however, without it, notably the northern rivers of the Broads.

Hippurus vulgaris with **Potamogeton** leaves - **Holt**

CALLITRICHACEAE

Callitriche stagnalis Scop. **Common Water-starwort**

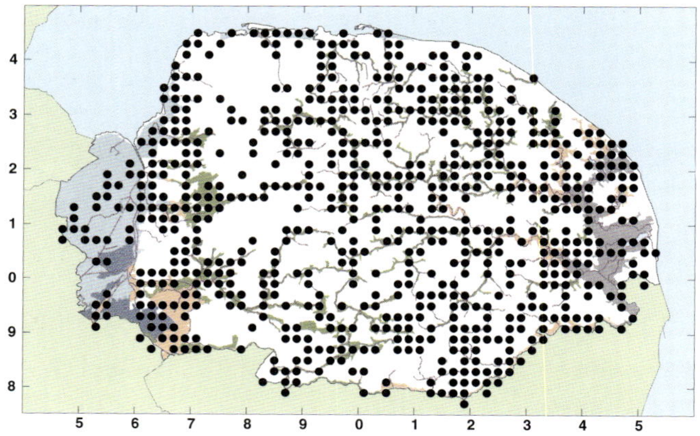

Annual or perennial, normally found on seasonally wet ground, in water collecting in wheel-ruts and the bare edges of muddy ditches and ponds. Present in both sun and shade. Its distribution appears to be random throughout the county, requiring only a suitable habitat.

C. platycarpa Kütz. **Various-leaved Water-starwort**

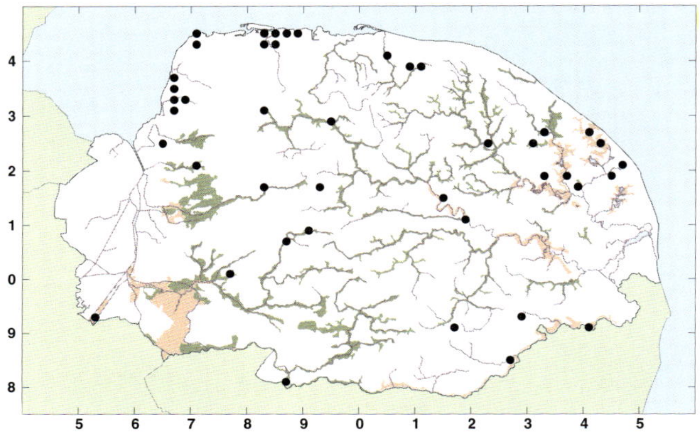

Annual or perennial found in a wide range of habitats, the commonest species in coastal dykes and pools, but also inland.

C. obtusangula Le Gall **Blunt-fruited Water-starwort**

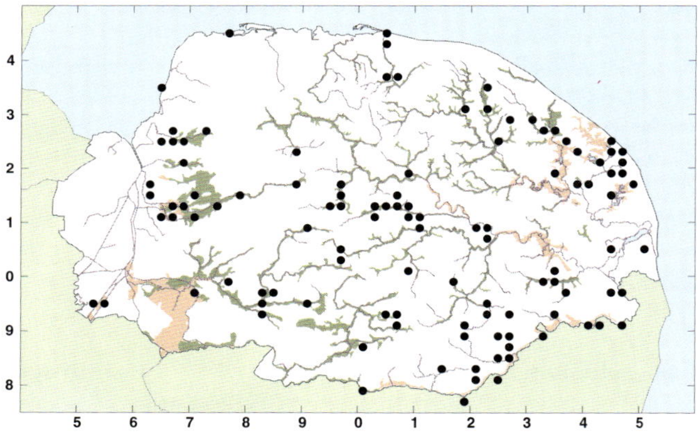

Perennial, frequent in ditches, ponds and rivers, the concentration on the chalk noted in Petch & Swann, 1968 is no longer obvious.

C. hamulata Kütz. ex W.D.J.Koch
Intermediate Water-starwort

Syn. *C. intermedia* ssp *hamulata* (Kütz.ex W.D.J.Koch) A.R.Clapham
Annual or perennial, always scarce and during this survey found on coastal fresh water marshes at Heacham W63 and Wootton W62, inland at Narborough W71; Rockland St. Mary E30; Salhouse E31 and in a pond at Bacton Woods E33. W3. E3.

PLANTAGINACEAE

Plantago coronopus L. **Buck's-horn Plantain**

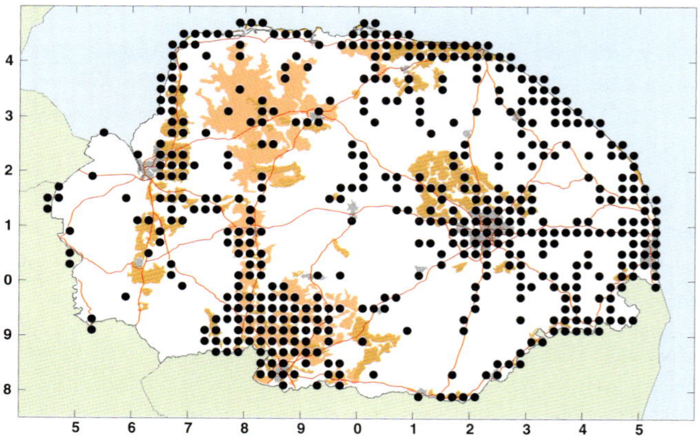

Winter green perennial confined to light, sandy soils but now increasingly at the edge of roads. Thriving 6m above the ground on the buttresses of Wiggenhall St German's church W51 in 1989, GB.

P. maritima L. **Sea Plantain**

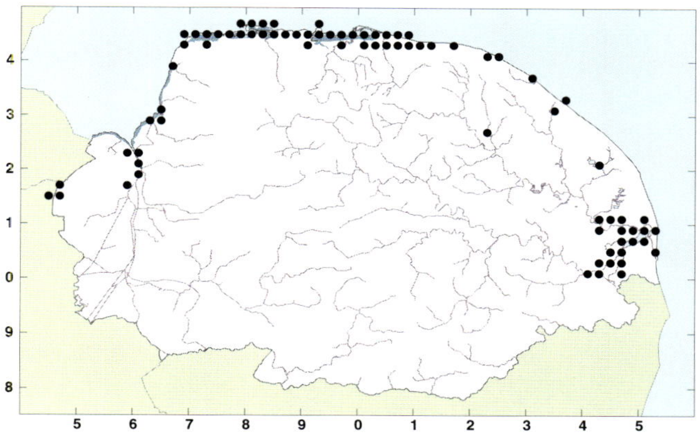

Perennial still found largely in salt marshes, though plants are occasionally turning up inland on the edge of roads.

P. major L. **Greater Plantain, Rat's-tail Plantain**
Deciduous perennial, an early coloniser of waste ground, road verges, heavily-cut lawns and sports fields. A camp follower spread all over the world with European settlers. W705. E713.

P. media L. **Hoary Plantain**

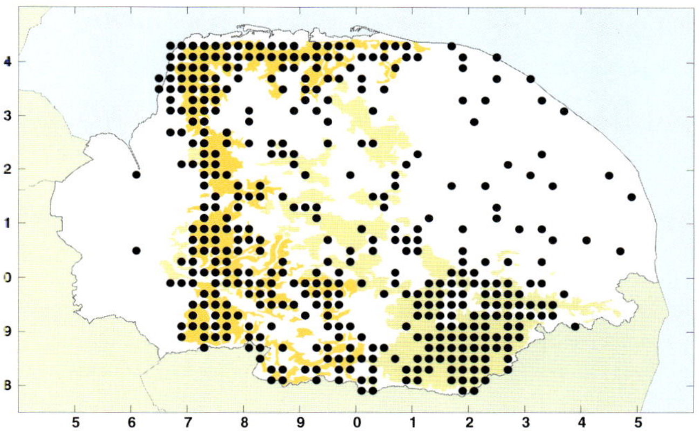

Winter green perennial usually indicating some lime in the soil, a plant chiefly of long-established grassland. In many areas almost confined to road verges or churchyards.

P. lanceolata L. **Ribwort Plantain**
Winter green perennial found in grassland of all sorts, also common on walls. In well-established, dry turf a form with small, pill-shaped flowers occurs while in more ruderal situations the heads tend to be much longer. Even more of a world wide weed than *P. major*, found even in the tropics. W697. E707.

■ **P. arenaria** Waldst. & Kit.
Syn. *P. indica* L. nom. illeg.
Recorded three times, on the rubbish tip at Harford, Norwich and on Mousehold Heath, both E20 in the 1950s and at Wolferton W62 in 1983, where it had apparently been grown as a crop, CPP.

Littorella uniflora (L.) Asch. **Shoreweed**
Winter green perennial formerly known from five sites, the only survivors of which are at Home Mere W98 within the Stanford Training Area where it was observed flowering in 1991, DE.Coombe and at Westwick E22 also in 1991, PWL. W1. E1.

BUDDLEJACEAE

Buddleja davidii Franch. **Butterfly bush**
More-or-less deciduous shrub introduced to gardens from China, widely planted and freely seeding. First recorded in large urban centres, but now likely to be found on any piece of waste ground and on walls in towns and villages. A valuable butterfly plant in late summer. W108. E209.

OLEACEAE

Forsythia x intermedia Zabel. **Forsythia**
(*F. suspensa x F. viridissima*)
Deciduous shrub, a garden hybrid sometimes rooting and surviving from dumped prunings. W2. E6.

Jasminum nudiflorum Lindl. **Winter Jasmine**
Deciduous shrub, self-layered on the church wall at Anmer W72. Presumably its origin was a shrub once planted by the church wall but now removed. Observed in 1991 independently by KAB and RMP. W1. E0.

Fraxinus excelsior L. **Ash**
Deciduous tree absent only from a few newly reclaimed areas, mostly near the coast, also from parts of the peat fens. W653. E689.

Syringa vulgaris L. **Lilac**
Deciduous shrub introduced to gardens from south-eastern Europe, probably during the Middle Ages, found on old house sites, in parks, plantations and hedges, regenerating from suckers. W133. E211.

Ligustrum vulgare L. **Common Privet**

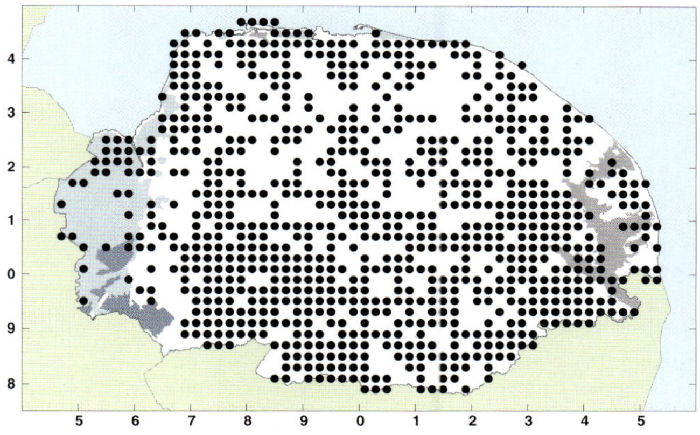

Semi winter green shrub most frequent on chalky soils, its natural distribution muddled by its use until the late nineteenth century as a hedging plant. Plants with greenish-yellow ripe berries were found at Narborough W71 in 1995, HW.

L. ovalifolium Hassk. **Garden Privet**
Winter green shrub introduced to gardens from Japan in the late nineteenth century. As it is almost winter green it soon supplanted the last species as a hedging plant and survives in relic hedges and shelter belts, also occurring as a gap filler in more recent times, sometimes marking the sites of old airfields. W76. E232.

SCROPHULARIACEAE

■ **Verbascum sinuatum** L.
Herbaceous perennial from southern Europe recorded near Thetford W88 in 1943.

■ **V. blattaria** L. **Moth Mullein**
Winter green annual or biennial introduced from mainland Europe and last recorded at Eaton chalkpit E20 in 1973, R.Jones and Corpusty E03 in 1974, Rev.R.Hooper.

V. virgatum Stokes **Twiggy Mullein**
Winter green biennial native to south-west England, a casual in Norfolk. Recorded during this survey at Snetterton & Kenninghall E08 in 1990, R.Parker. W0. E2.

V. pyramidatum M.Bieb. **Caucasian Mullein**
Winter green perennial introduced from the Caucasus and found at Wretham W89 in 1997, ALB. W1. E0.

V. phlomoides L. **Orange Mullein**
Winter green biennial introduced from mainland Europe, recorded in W70, 72, 93 and 99 as a casual. W4. E0.

V. densiflorum Bertol. **Dense-flowered Mullein**
Winter green biennial introduced from mainland Europe, found once, near houses at Choseley W73 in 1995, KAB, probably from rubbish thrown out from gardens. W1. E0.

V. thapsus L. **Great Mullein**

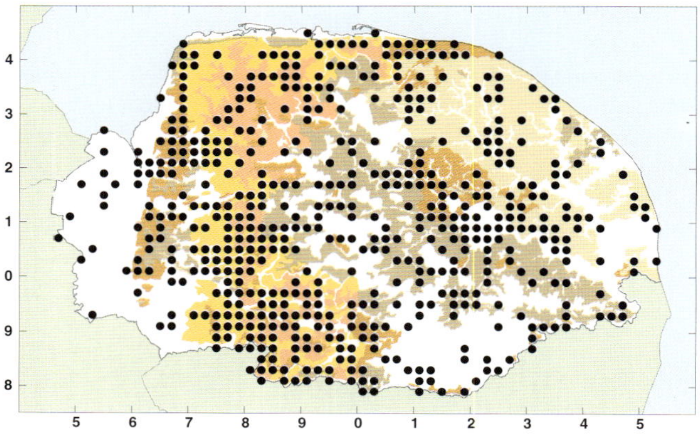

Winter green biennial found on light soils. Its commonest habitat in the county appears to be roadside banks, but it is also on waste ground and rough grassland.

V. x semialbum Chaub.
(*V. thapsus x V. nigrum*)
Winter green biennial. During this survey recorded at Harling Road W98 in 1994; Grimes Graves, Weeting W88 in 1994; Santon W88 and East Barsham W93 in 1995. W4. E0.

Verbascum pulverulentum - Whitlingham Marsh

V. x duernsteinense Teyber
(*V. thapsus x V. speciosum*)
The first British record for this hybrid was of a plant at Didlington W89 in 1976, VML. It was found again in the adjacent parish of Foulden W79 in 1993, KAB. W1. E0

V. x godronii Boreau
(*V. thapsus x V. pulverulentum*)
Winter green biennial seen at Ringstead W73 in 1992, PRC and at two sites in Hillington W72 in 1994, K&GB. W3. E0

V. nigrum L. **Dark Mullein**

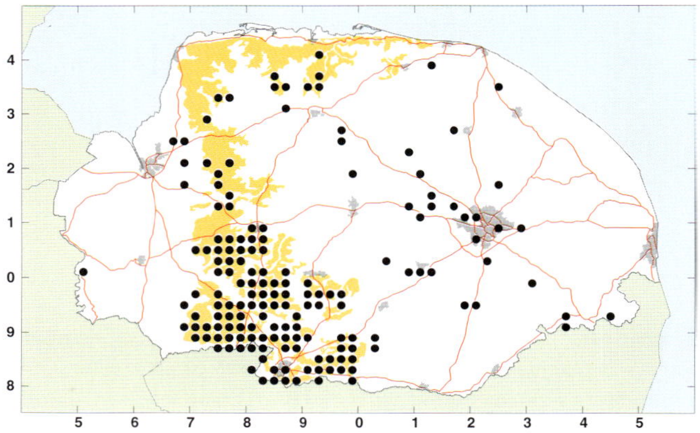

Herbaceous perennial of lime rich soils, outside the main chalky areas growing on steep slopes above the river valleys where erosion has cut through to the underlying soils.

V. x mixtum Ramond ex. DC.
Syn. *V x wirtgenii* Franch. (*V. nigrum x V. pulverulentum*)
Winter green biennial introduced from mainland Europe, found with its parents at Guist W92 in 1996, JEG & RMP. W1. E0.

V. speciosum Schrad. **Hungarian Mullein**
Winter green biennial introduced to gardens from south-east Europe, long established in the Didlington to Foulden area W79 and still surviving. First recorded there as *V. longifolium* in 1958 and later as *V. olympicum* but finally confirmed as *V. speciosum*. Also near Smokers' Hole in the Stanford Training Area W89 in 1997, ALB. W4. E0.

V. pulverulentum Vill. **Hoary Mullein**

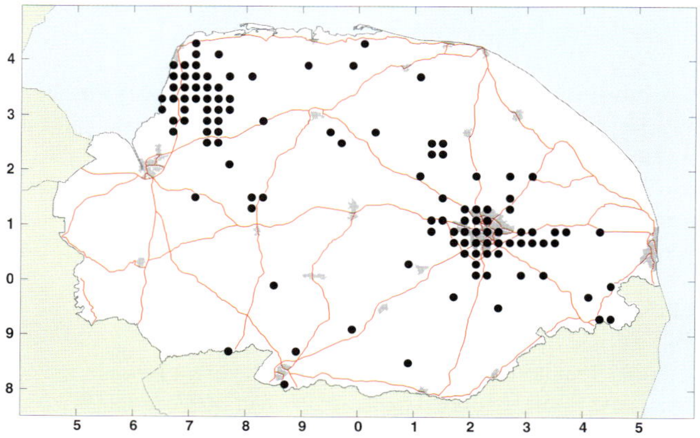

Winter green biennial or monocarpic perennial with two centres of distribution, one on the shingle bank between Snettisham and Heacham W63, nearby dry grasslands and road verges, the second on the river gravels near Norwich, E20. Very rare outside East Anglia. A nationally scarce species.

V. lychnitis L. **White Mullein**

Winter green biennial native to southern England. Found during this survey in Norwich E20, ETD and Saxlingham Thorpe E29, W.Mitchell. W0. E2.

Scrophularia nodosa L. **Common Figwort**

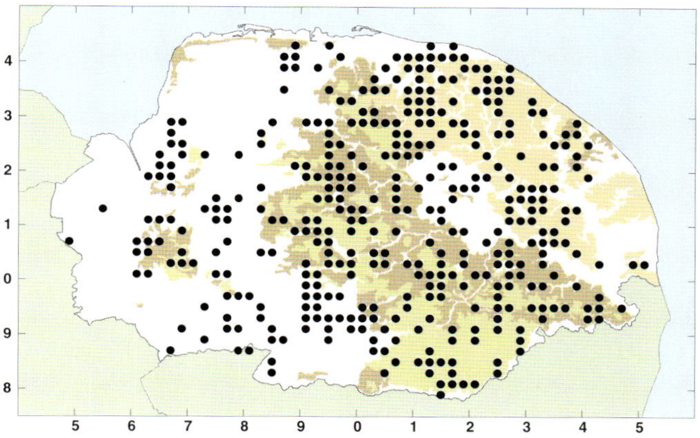

Semi-winter green perennial scattered in shady places on moist soils.

S. auriculata L. **Water Figwort, Water Betony**

Syn. *S. aquatica* auct. non L.

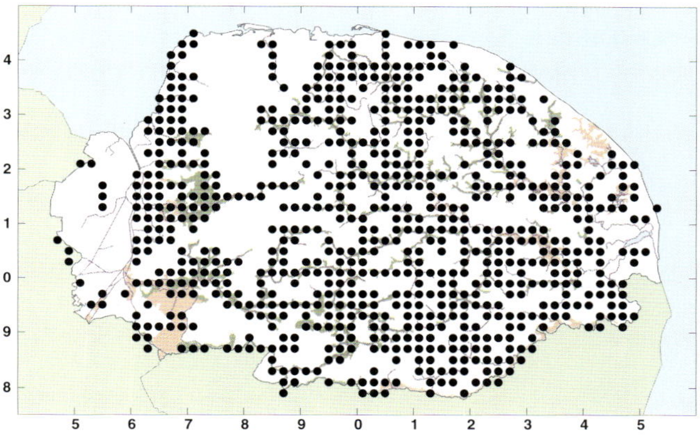

Semi-winter green perennial frequent by watercourses.

S. umbrosa Dumort. **Green Figwort**

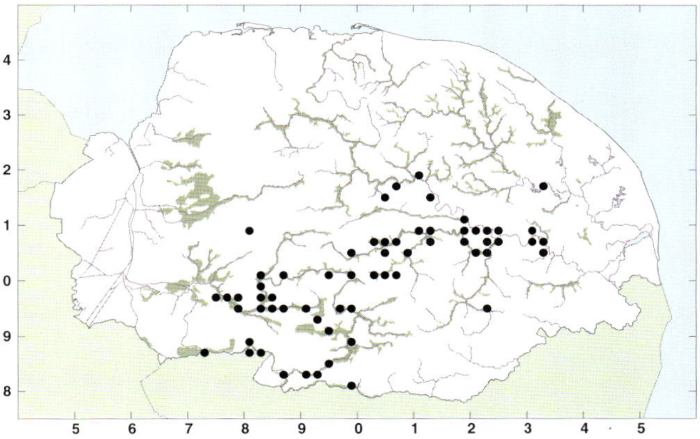

Semi-winter green perennial recorded from two sites at Scoulton W00 and Watton W90 by Nicholson, 1914, and in the 1920s at Bridgham W98 and Norwich E30. The species has spread since that time, especially along the Yare valley where it has now been recorded from Wood Rising W90 to Whitlingham E20 and also in some of the tributaries of the Wensum. Curiously it seems to be spreading upstream. A nationally scarce species.

S. scorodonia L. **Balm-leaved Figwort**

Semi-winter green perennial native to south-west England, a casual recorded at Marston marshes E20 in 1990, R.Land. W0. E1.

S. vernalis L. **Yellow Figwort**

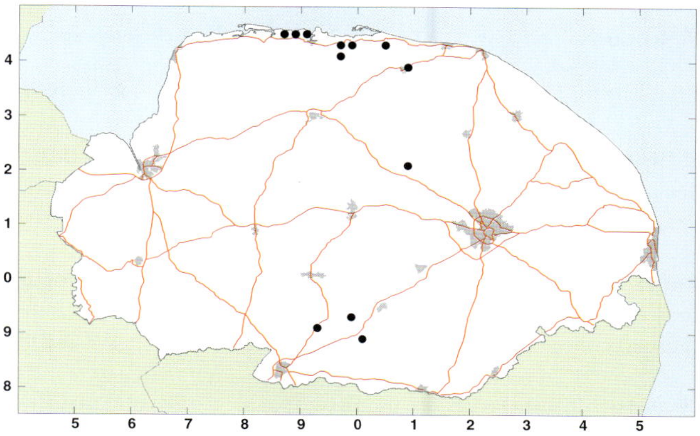

Winter green perennial introduced to gardens from mainland Europe and escaping. Very persistent, the colony first recorded at the end of the 18th century by Sir J.E.Smith is still at Stiffkey W94.

Mimulus guttatus DC. **Monkeyflower**

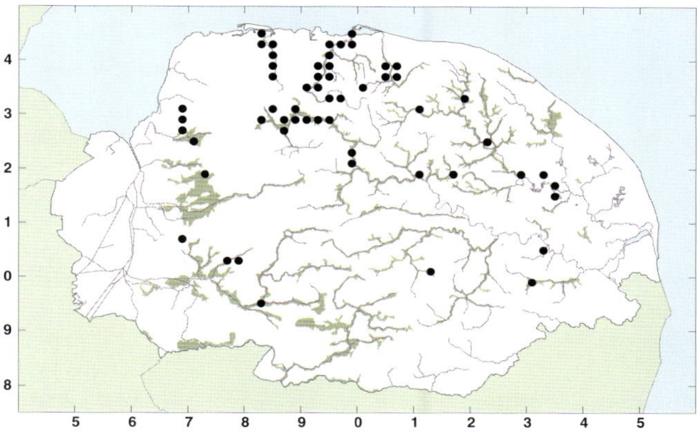

Mainly deciduous perennial introduced to gardens from western North America, frequently escaping by watersides. A feature of the north coast rivers and their tributaries.

Scrophularia umbrosa - Langford

M. x robertsii **Hybrid Monkeyflower**
(*M. moschatus x M. luteus*)
Deciduous perennial, a hybrid between two North American species which arose in gardens. An occasional escape. Recorded in 1996 at Bayfield E04, Flora Group and Beeston Common E13 FF. W0. E2.

M. luteus **Blood-drop-emlets**
Mainly deciduous perennial introduced to gardens from Chile. Recorded from Burntfen Broad E31 by PA.Smith 1988. W0. E1.

■ **Limosella aquatica** L. **Mudwort**
Annual now very rare nationally, recorded at Foulden W70, F.Newton and Scoulton Mere W98, F.Robinson both in 1914.

Antirrhinum majus L. **Snapdragon**
Semi-winter green perennial introduced to gardens from south-west Europe, scattered on old walls throughout the county. W60. E58.

Chaenorhinum minus (L.) Lange **Small Toadflax**

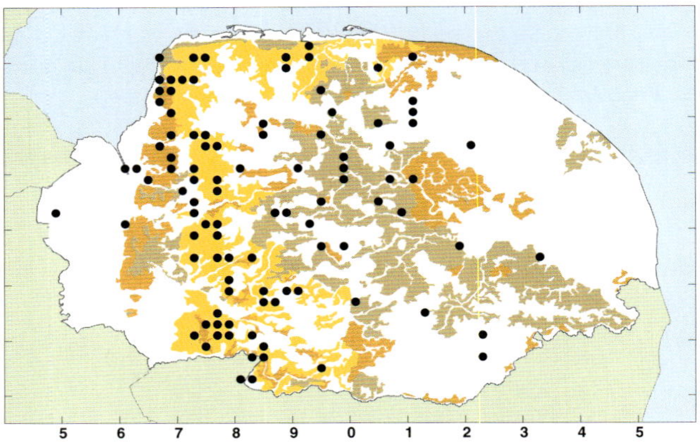

Spring germinating annual of well-drained soils, widely scattered but never frequent. Often found on disused railway lines and forest rides and in east Norfolk in arable fields on clay.

Misopates orontium (L.) Raf. **Weasel's-snout**
Syn. *Antirrhinum orontium* L.

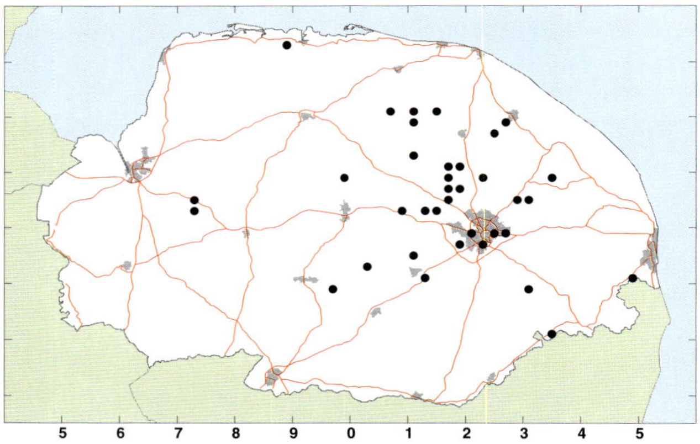

Spring germinating annual, a scarce arable and garden weed, usually on moist soils.

■ **M. calycinum** Rothm. **Pale Weasel's-snout**
Annual recorded once as a casual from birdseed, Didlington W79 in 1978, VML.

Asarina procumbens Mill. **Trailing Snapdragon**
Syn. *Antirrhinum asarina* L.
Winter green perennial introduced from south-west Europe to gardens and turning up in waste places. Found during this survey at Fincham W60 in 1992, MK and on a wall at Cawston E12 in 1995, PG.Lawson, still present in 1998. W1. E1.

Cymbalaria muralis P. Gaertn., B.Mey. & Scherb.
Ivy-leaved Toadflax, Kenilworth Ivy

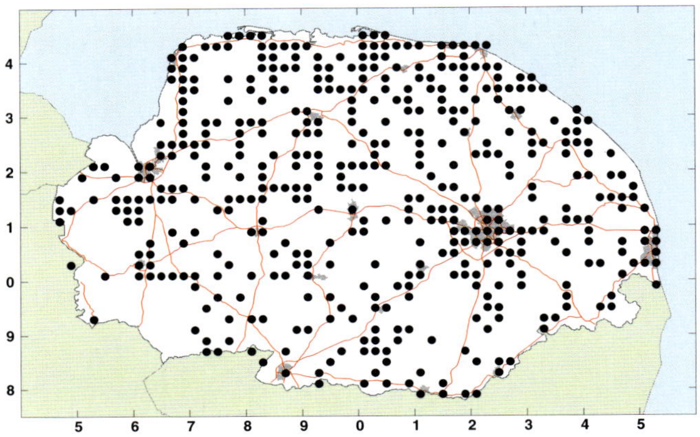

Winter green perennial introduced to gardens from mainland Europe in the 16th century, now widely naturalised throughout the county on walls in towns and villages but not reported in a natural habitat. The record from the middle of the forest at Cranwich W79 in 1997 came from some dumped rubble used as hardcore.

Kickxia elatine (L.) Dumort. **Sharp-leaved Fluellen**

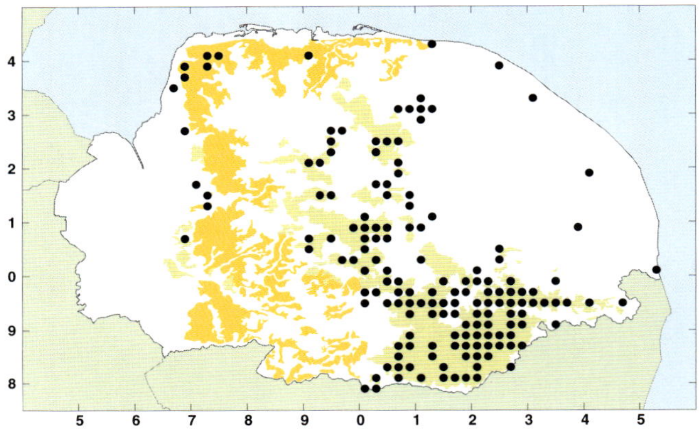

Spring germinating annual found mostly but not exclusively on calcareous soils, notably on the boulder clay where it can be frequent.

K. spuria (L.) Dumort. **Round-leaved Fluellen**

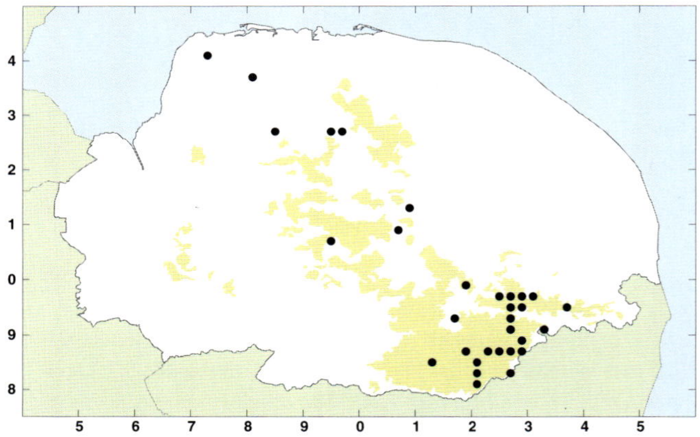

Spring germinating annual with a similar distibution to the previous species with which it sometimes grows, but far less common.

Linaria vulgaris Mill. **Common Toadflax, Butter and Eggs**

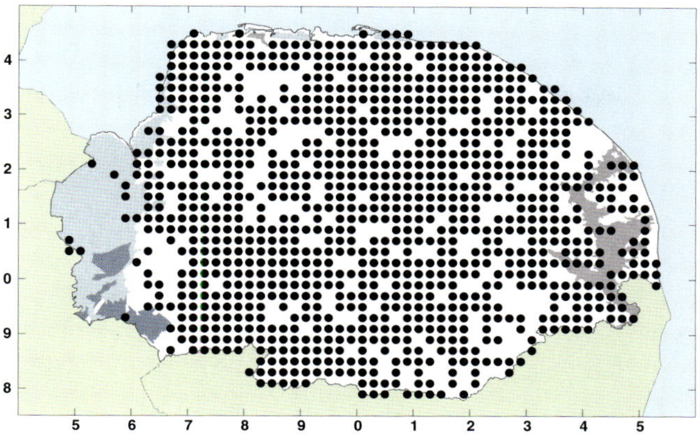

Herbaceous perennial, generally frequent but with some gaps in its distribution which are difficult to explain. The peloric form has been found on at least four occasions, but not during this survey.

L. purpurea (L.) Mill. **Purple Toadflax**
Semi-winter green perennial introduced to gardens from Italy, well naturalised in towns and villages, most often on walls. W100. E143.

L. repens (L.) Mill. **Pale Toadflax**
Perennial native to southern and western Britain, a garden escape in Norfolk occurring rarely. Most recently found in King's Lynn W62 in 1992, RMP; Wymondham E10, ALB and Scarning W91 on the edge of an old rail track in 1996, D&BM. W2. E1.

L. maroccana Hook. f. **Annual Toadflax**
Annual introduced from Morocco to gardens and found as an escape on sandy cliffs at Mundesley E33 in 1959, also at Cley E04 in 1993, KKH. W0. E1.

Digitalis purpurea L. **Foxglove**

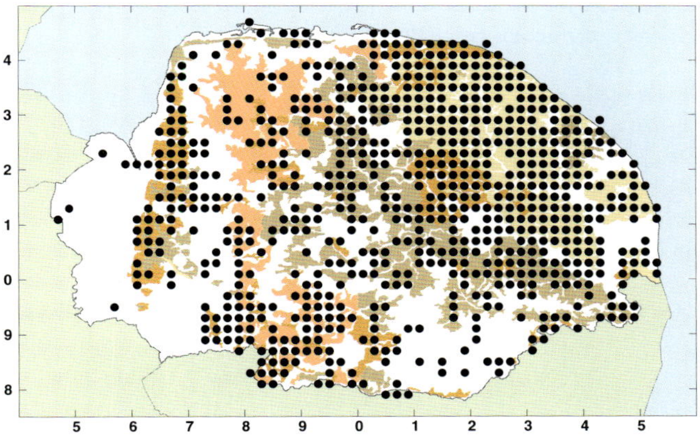

Winter green biennial occurring as a native on neutral to acid soils. Noted by William Turner in 1548 as being 'Much in England and especially in Norfolk'. Perhaps occasionally as an introduction from gardens though most modern cultivated plants are distinct strains.

Erinus alpinus L. **Fairy Foxglove**
Winter green perennial introduced to gardens from southern Europe and occasionally escaping. Known from walls at Cawston E12 since 1943 and still present and in a similar habitat at West Acre W71 in 1996, K&GB. W1. E2.

Veronica serpyllifolia L. **Thyme-leaved Speedwell**

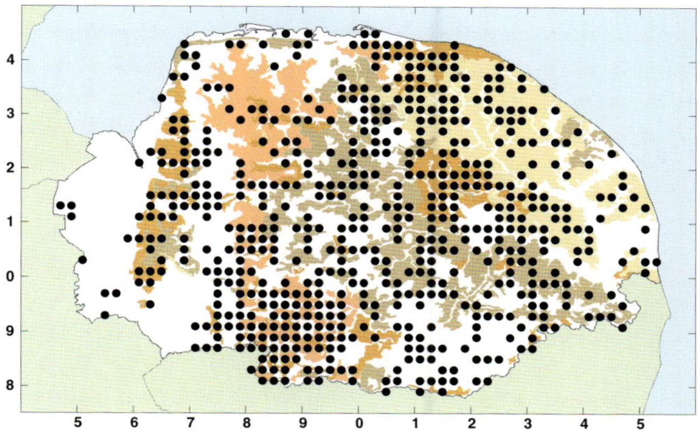

Winter green perennial frequent in grassland on neutral to acid soils, especially in open woodland rides.

V. officinalis L. **Heath Speedwell**

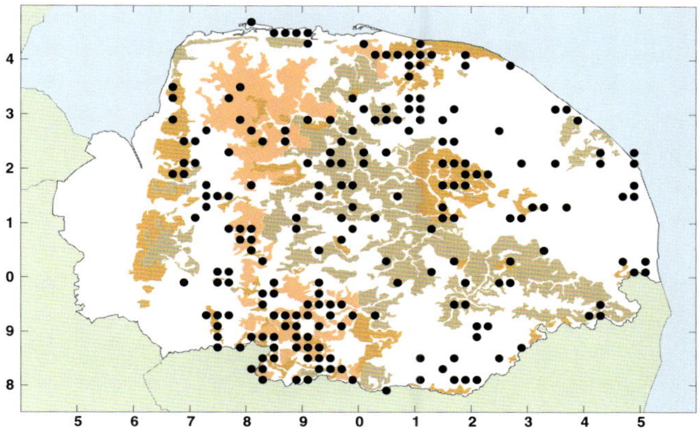

Winter green perennial of grasslands and heaths on acid soil, less widespread than the preceding species. Frequent on rides in the Breckland forests.

V. chamaedrys L. **Germander or Bird's-eye Speedwell**
Winter green perennial common in established grassland, especially abundant in churchyards and on roadside verges which have not become too rank. W605. E662.

V. montana L. **Wood Speedwell**

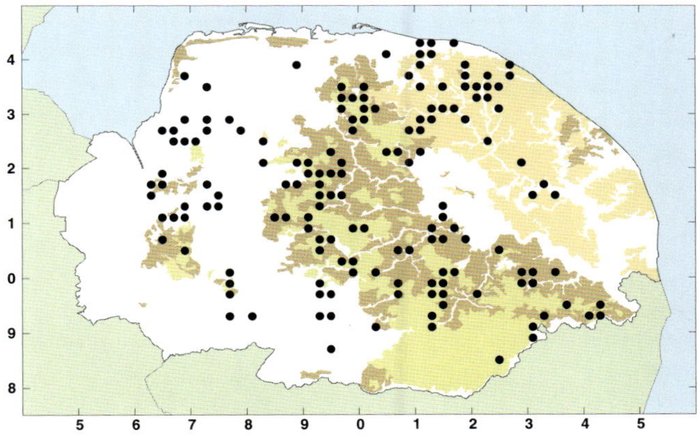

Winter green perennial occasional in long-established woodland but also in damp, shady places.

V. scutellata L. **Marsh Speedwell**

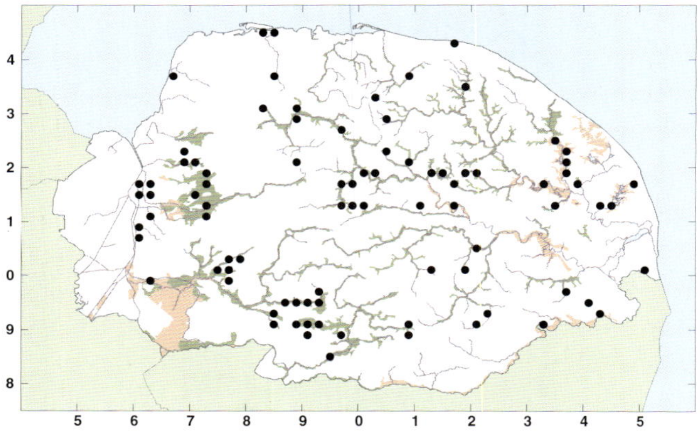

Herbaceous perennial uncommon but widespread in damp, marshy ground.

V. beccabunga L. **Brooklime**

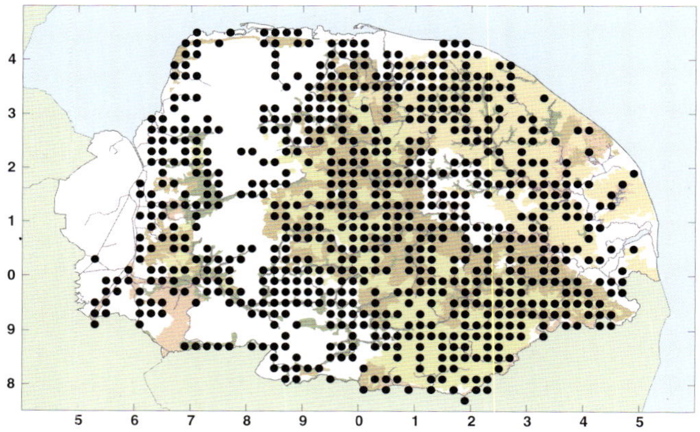

Winter green perennial widely distributed throughout the county by water and in marshy ground. Absent from the Fenland where water courses are closely confined, except for the Ouse Washes where it is common.

V. anagallis-aquatica L. **Blue Water Speedwell**

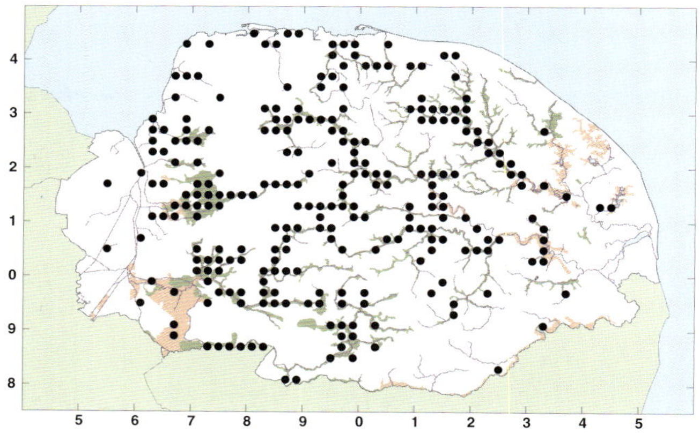

Herbaceous perennial found growing in water, in ponds, rivers and streams.

■ **V. x lackschewitzii** J.B.Keller **Hybrid Water Speedwell**
(*V. anagallis-aquatica x V. catenata*)
Recorded at Burnham Overy W84 in 1972, RPL and Castle Acre W81 in 1973 and 1979, ELS.

V. catenata Pennell **Pink Water Speedwell**

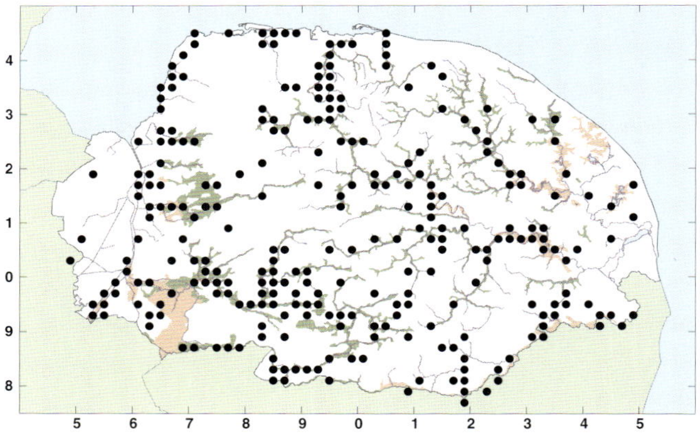

Herbaceous perennial usually growing in still or slow-moving water.

V. praecox All. **Breckland Speedwell**
Autumn germinating annual confined to Norfolk and Suffolk. Once recorded from five sites in the southern Breckland but since 1990 found only at Feltwell W79; Thetford W88 and at Weeting W78 where it was originally introduced. A Red Data Book species.
W3. E0.

Veronica praecox - Feltwell

V. triphyllos L. **Fingered Speedwell**
Autumn germinating annual now confined to Norfolk and Suffolk and found at Thetford W88 where, together with *V. praecox*, it struggles to survive as its original cornfield habitat is now a modern housing estate. Also sown at Weeting W78 where it has become well established. A Red Data Book species. W2. E0.

V. arvensis L. **Wall Speedwell**

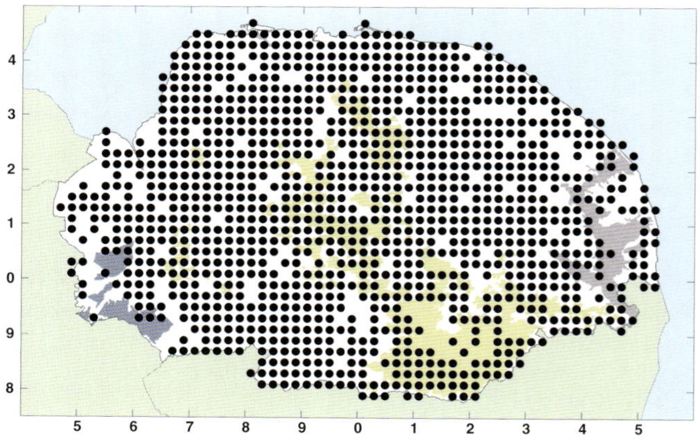

Autumn germinating annual abundant throughout the county wherever there are suitably open, dry soils, also commonly on walls.

V. verna L. **Spring Speedwell**

Autumn germinating annual first recorded at Thetford in 1850 by JD.Salmon. Recorded at Santon Warren between 1918, when it was described as abundant, and 1946 but not seen since. Also introduced at Weeting W78 together with *V. praecox* and *V. triphyllos* where it is well established though not as successful as the other two species. A Red Data Book species. W1. E0.

V. agrestis L. **Green Field Speedwell**

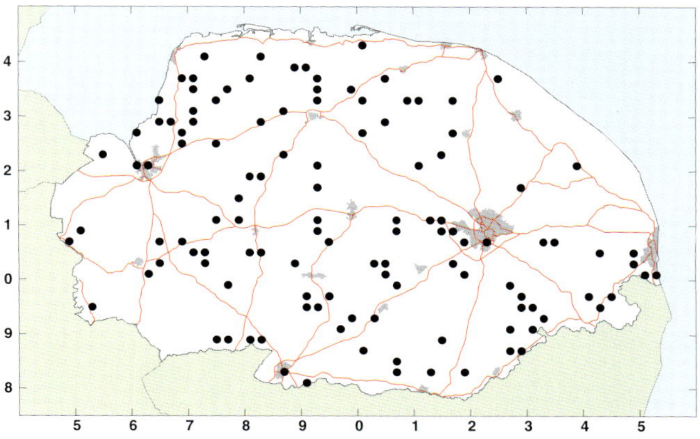

Autumn germinating annual, a weed of arable land, uncommon but widely distributed.

V. polita Fr. **Grey Field Speedwell**

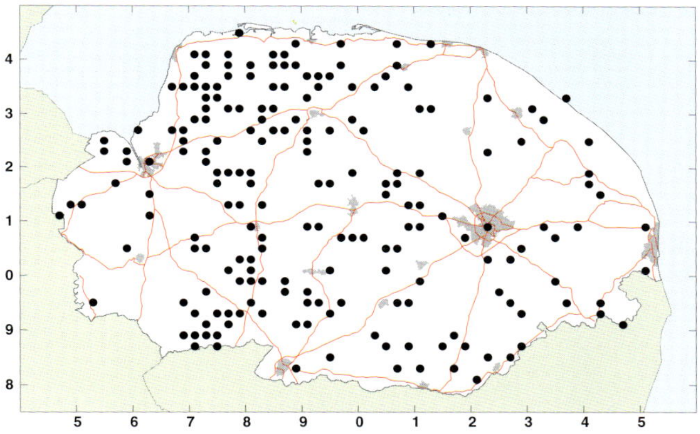

Autumn germinating annual, a weed of arable land, gardens and tracks, uncommon.

Veronica spicata - Weeting

V. persica Poir. **Common Field Speedwell**

Annual, introduced from mainland Europe in the 19th century (Galpin, 1888, suggests 1829) and first recorded in Norfolk in 1860. It will germinate throughout the year and has largely replaced the previous two species. It is now by far the commonest speedwell as a field and garden weed. W673. E685.

V. filiformis Sm. **Slender Speedwell**

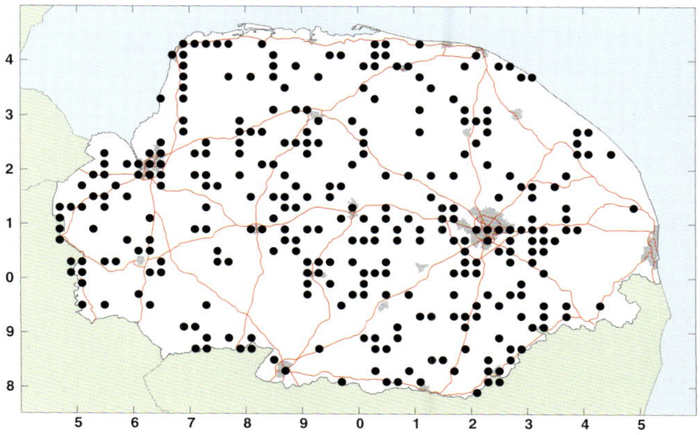

Winter green perennial introduced to gardens from Turkey and the Caucasus, first noted in this country in Surrey in 1808. Now widely naturalised in short and particularly mown grassland in spite of the fact that it is sterile in Britain.

V. hederifolia L. **Ivy-leaved Speedwell**

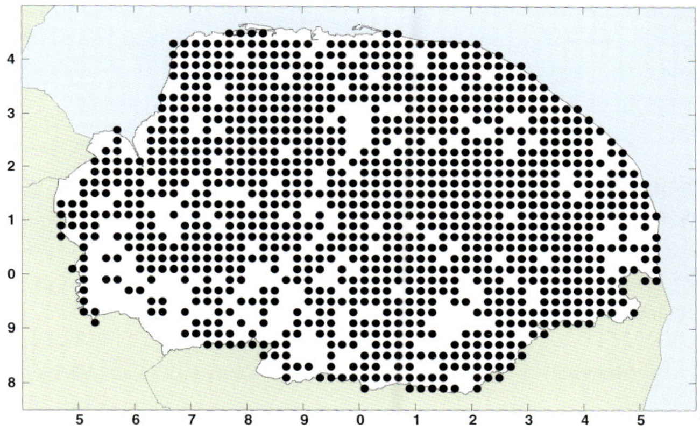

Autumn germinating annual, an abundant weed of arable and waste ground, walls and gardens. Ssp. **hederifolia** appears to be the most frequent in semi-natural sites while ssp. **lucorum** (Klett & Richt.) Hartl. (*V. sublobata* M.A.Fisch.) is found in more ruderal ones, especially in gardens and in shade.

■ **V. longifolia** L. **Garden Speedwell**

Perennial introduced to gardens from mainland Europe. Found once on Harford Tip, Norwich E20 in 1961.

V. spicata L. **Spiked Speedwell**

Semi-winter green perennial, the subspecies **spicata** confined to Norfolk, Suffolk and Cambridgeshire. It was formerly known from Garboldisham and West Harling Heaths W98 which were ploughed up in the 1950s and last seen on the adjacent Devil's Dike in 1978. Material from this stock was introduced to forest rides at Harling in the late 1980s and still survives. Good native colonies are thriving on Weeting Heath W78. Red Data Book species. W2. E0.

Hebe

Seedlings of this large and popular cultivated genus of evergreen shrubs from New Zealand often turn up outside gardens, especially in urban areas. Most plants grown in gardens are cultivars of hybrid origin and this, combined with the fact that all hebes, both species and hybrids, readily cross with each other, means that such chance seedlings will almost certainly be hybrids of involved parentage, making identification difficult if not impossible. Often the probable parentage can only be guessed at, the exception being the occasional seedling derived from an isolated plant of a true species. The following species have been identified with some degree of confidence.

H. salicifolia (G.Forst.) Pennell

A well established escape recorded on waste ground at Thetford W88 in 1997, KAB and Norwich E20 in 1990, ETD. W1. E1.

H. parviflora (Vahl.) Ckn. & Allan

One small plant was found established on a wall in King's Lynn W62 in 1996, RMP, det. KAB. W1. E0.

H. parviflora x H. salicifolia

A plant of or close to this parentage was found on a wall in King's Lynn W62 in 1996, RMP, det. KAB. W1. E0.

H. brachysiphon Summerh. **Hooker's Hebe**

Self sown on waste ground at Walsoken W41 in 1995, KAB. W1. E0.

Melampyrum cristatum L. **Crested Cow-wheat**

Spring germinating annual found in scrub and along wood-edges on acid soils. First recorded in Norfolk in 1832 in Godwick Wood; from Horningtoft W92 until 1960 and Beetley W91 where it survives precariously on a road verge. A nationally scarce species. W1. E0.

■ **M. arvense** L. **Field Cow-wheat**

Spring germinating annual of arable and field edges. First recorded in England 'between Lynn and Norwich in a cornfield' in 1716 by a Mr Petiver, seen again 'near Litcham' in 1724 by J.Sherard and at Costessey in 1792, not seen since.

M. pratense L. **Common Cow-wheat**

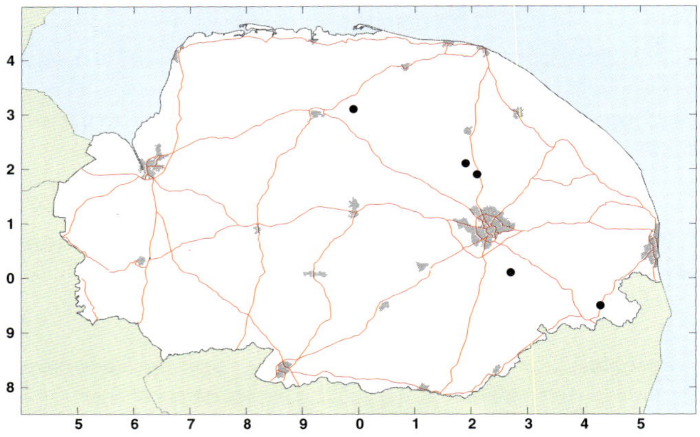

Spring germinating annual of heaths, woods and scrub on acid soils, known from nine sites in the county before 1978, now from only five.

Euphrasia officinalis agg. **Eyebright**

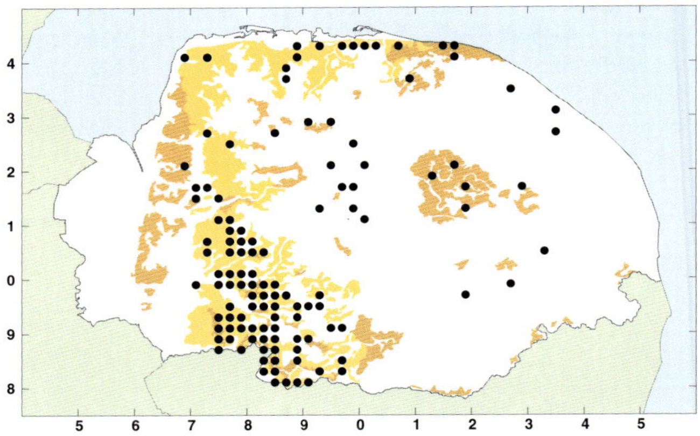

Spring germinating annual divided into a number of microspecies. All are found in undisturbed turf and, except perhaps for *E. confusa*, appear to have been much reduced in abundance since 1968, probably as a result of the loss of many of their habitats to rough grass and scrub.

E. anglica Pugsley

Found in damp, acid turf. Formerly recorded from 12 sites in west Norfolk. During this survey seen at Sugar Fen, Leziate W62 in 1994, JEG; East Walton W71 in 1993, HW and Hoe Rough W91 Flora Group in 1997. W5. E0.

E. nemorosa (Pers.) Wallr.

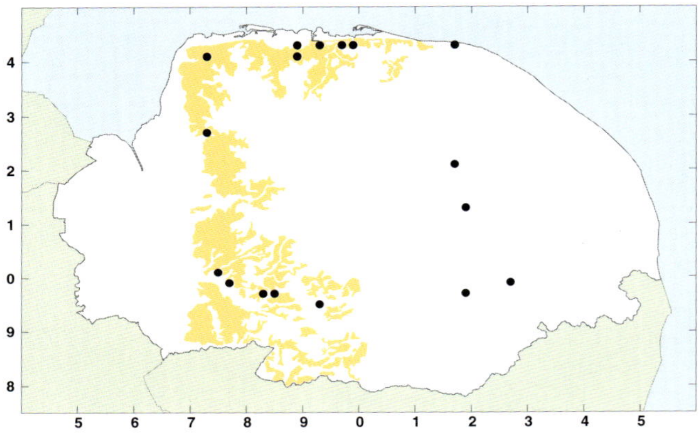

Recorded as the most frequent species in Norfolk, by CPP and ELS in 1968, but most specimens were later corrected by ELS to *E. confusa* and the note '<u>not</u> the commonest species in Norfolk' is against *E. nemorosa* in his card index.

E. pseudokerneri Pugsley

A scarce species of calcareous grassland, most recently seen at Pentney W71 in 1988, ALB; Ringstead W74 in 1995 and Hilborough W80 in 1996, K&GB and by a chalky stream on Beeston Common E14 in 1995, FF. W3. E1.

Parentucellia viscosa - Stow Bedon

E. confusa Pugsley

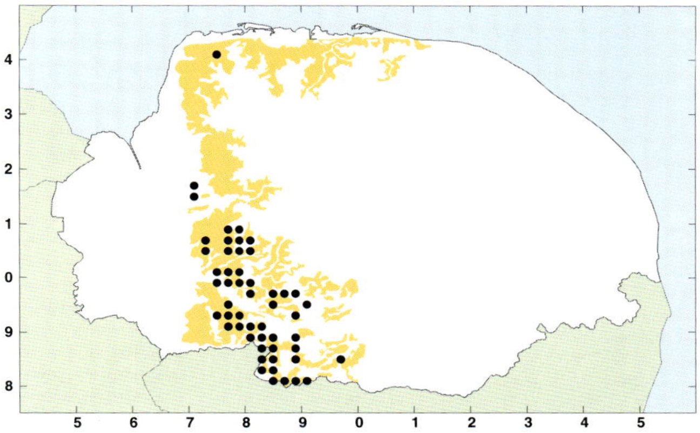

Frequent in neutral to basic grassland, especially so within the rides of the Breckland forest.

■ **E. micrantha** Rchb.
■ **E. rostkoviana** Hayne
Two heathland species occurring largely in upland areas. Both were reported from calcareous sites in Norfolk by Nicholson, 1914 but must be considered very doubtful. Neither has been seen again.

Odontites vernus (Bellardi) Dumort. **Red Bartsia**

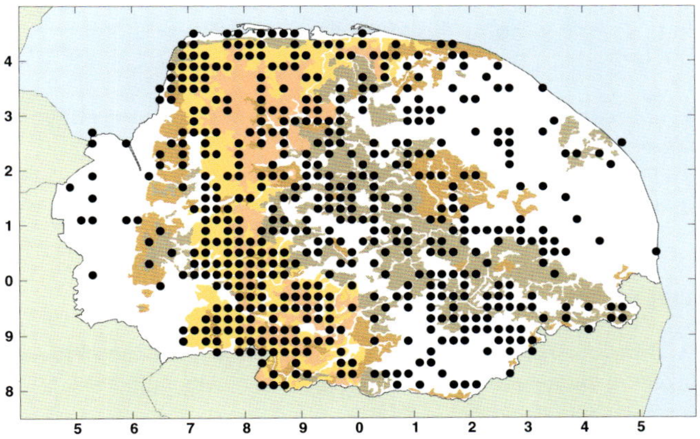

Spring germinating annual widespread in grassland, our plants belong to ssp. **serotinus** (Syme) Corb.

Parentucellia viscosa (L.) Caruel **Yellow Bartsia**
Spring germinating annual, native to south and west Britain and an introduction in Norfolk, probably originally as an impurity in grass seed for sports fields. Recorded first on a cricket field at East Winch W71 in 1949, also on golf courses at Hunstanton W64 in 1956 and Wells W94, 1954. More recently at Stow Bedon W99 in 1972; Ellingham E39 in 1973, R.Hadman; East Dereham W91 in 1991 and Leziate W61 from 1962, all still present. W3. E0.

Rhinanthus minor L. **Yellow-rattle, Rattle Baskets** (local)

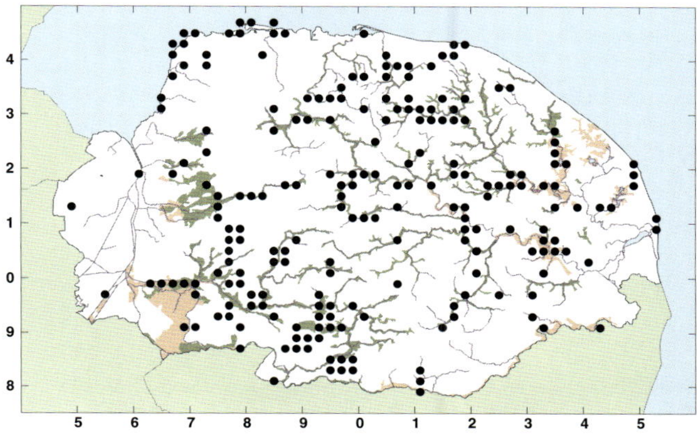

Spring germinating annual not infrequent in permanent grassland, that on sand dunes and dry, base rich grasslands is ssp. **minor** while ssp. **stenophyllus** (Schur) O.Schwarz occurs on moist grassland and in fens.

Pedicularis palustris L. **Marsh Lousewort**

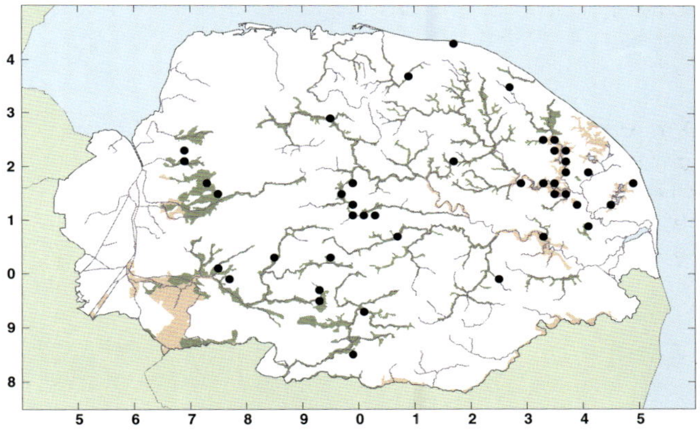

Annual to biennial occasionally found on wet heaths and in bogs, often growing in water. Scarce in south-east England.

P. sylvatica L. **Lousewort**

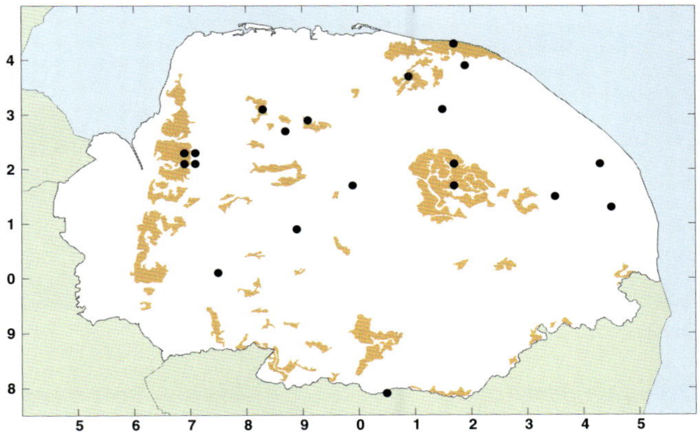

Biennial or perennial, found in moist, usually acid soils, now scarce.

OROBANCHACEAE

Lathraea clandestina L. **Purple Toothwort**
Herbaceous perennial introduced from south-west Europe, well established on the roots of poplar trees on an estate at Fincham W70. W1. E0.

Orobanche purpurea Jacq. **Purple Broomrape**

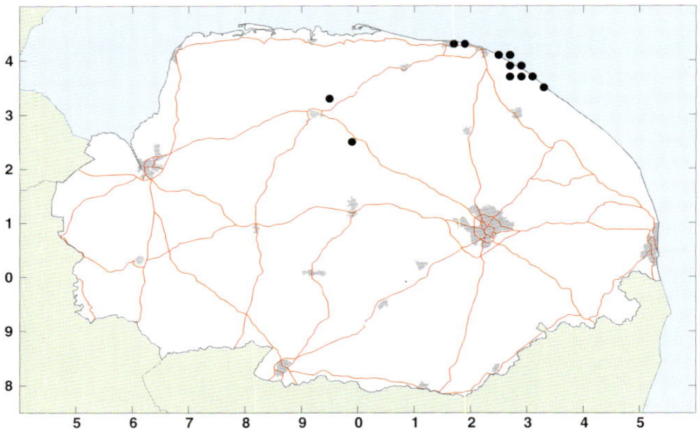

Herbaceous perennial parasitic on yarrow. Rare away from the north-east coast but seen at Little Snoring W93 and at Gt.Ryburgh W92 during this survey. A Red Data Book species.

Orobanche purpurea - Mundesley

O. rapum-genistae Thuill. **Greater Broomrape**
Herbaceous perennial parasitic on gorse and broom. Recorded at Heacham W63 in 1920 and at Great Hockham W99 from 1983, B.Campbell. Last seen in 1986, EVR. W1. E0.

O. elatior Sutton **Tall Broomrape**

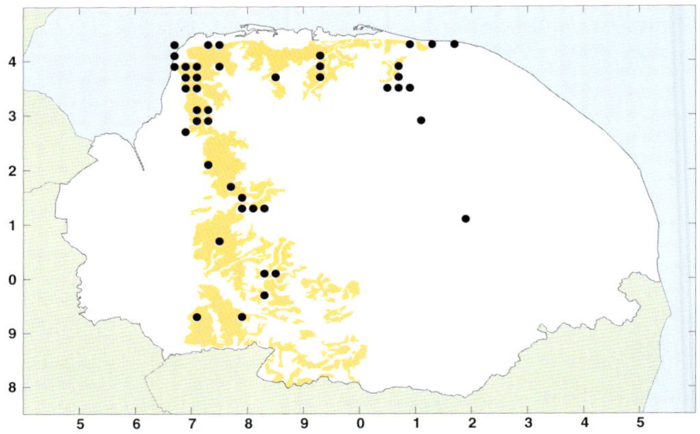

Herbaceous perennial found in calcareous grassland, parasitic on *Centaurea scabiosa*, rarely on *C. nigra*, widely distributed on the chalk, but never frequent.

■ **O. ramosa** L. **Hemp Broomrape**
Herbaceous perennial parasitic on hemp and introduced from southern Europe. Recorded at Starston E28 and Lakenham E20, both in 1880.

O. minor Sm. **Common Broomrape**

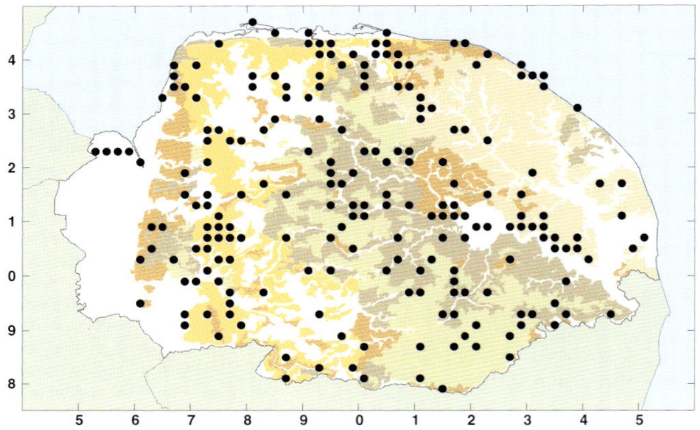

Herbaceous perennial parasitic on a wide range of hosts including *Tagetes* on a roundabout at Hunstanton W64 in 1991, I.&F.Simper and on *Verbena bonariensis* in a garden at Snettisham W63 in 1993, P.Reynolds. Widely scattered, usually on drier soils.

ACANTHACEAE

Acanthus mollis L. **Bear's-breech**
Winter green perennial introduced to gardens from southern Europe and occasionally surviving as a throwout as at Harpley W72 in 1993, MK. W1. E0.

A. spinosus L. **Spiny Bear's-breech**
Winter green perennial introduced to gardens from the Mediterranean, a survivor of a long-abandoned garden at Oxborough W70 in 1992, JEG. W1. E0.

BIGNONIACEAE

Eccremocarpus scaber Ruiz & Pav. **Chilean Glory Flower**
Self-sown seedlings, some flowering, between wall and pavement at Kenninghall E08 in 1996, K&GB. W0. E1.

LENTIBULARIACEAE

Pinguicula vulgaris L. **Common Butterwort**

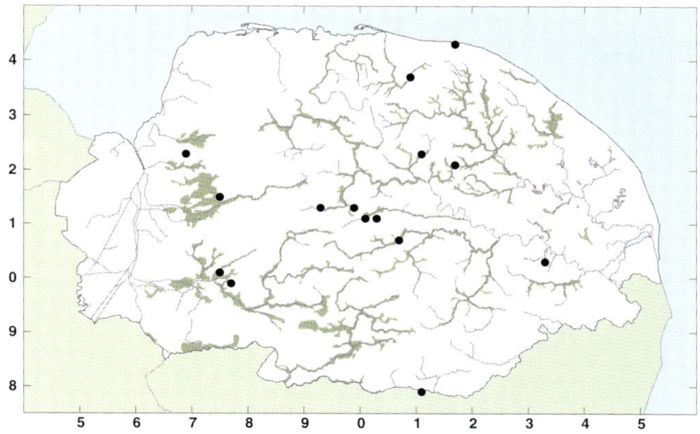

Herbaceous perennial confined to fens and wet heaths where there is some alkalinity present. Now much less common than formerly.

Utricularia vulgaris L. **Greater Bladderwort**

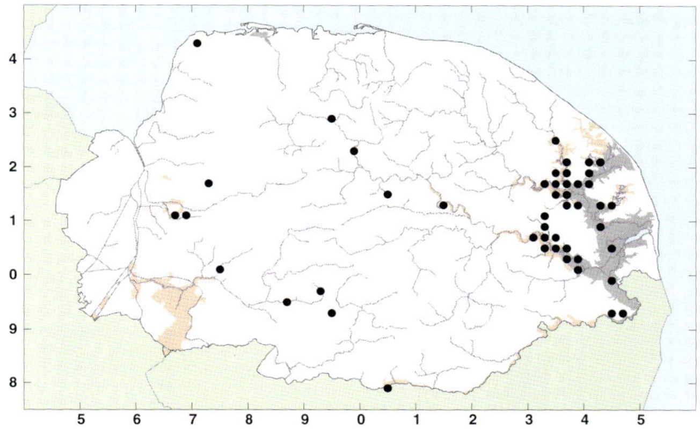

Perennial, once well distributed in pools, fens and dykes, but quick to die out in polluted water. Making a welcome return in a few recently cleaned Broadland waterways.

U. australis R.Br. **Bladderwort**
Syn. *U. neglecta* Lehm.
Perennial, old records for Martham E41 and Stow Bedon W99 in 1917, Wheatfen E30 in 1937 and Catfield E32 in 1975 have not been refound. Now only known to survive at Shouldham W60, H&JW. W1. E0.

■ **U. intermedia** Hayne **Intermediate Bladderwort**
Perennial, a plant of peaty pools with 12 sites known in 1968. In 1910 it was described at Roydon W62 by WG.Clarke and WH.Burrell as 'covering hundreds of square yards, [we were] treading on it at every step'. With decreasing water levels, it now appears to be very rare if not lost and no records have been made since 1979.

U. minor L. **Lesser Bladderwort**
Perennial of peaty pools and ditches, recorded in 1968 from 22 sites of which only one has been seen during this survey, at Dilham E32 in 1992, G.Kennison. W0. E1.

CAMPANULACEAE

Campanula persicifolia L. **Peach-leaved Bellflower**
Winter green perennial introduced to gardens from mainland Europe, an occasional escape to walls and waste ground. W12. E30.

C. medium L. **Canterbury Bells**
Winter green biennial introduced to gardens from southern Europe, recorded at Harford tip, Norwich E20 in 1969 and at Downham Market W60 in 1995, RMP. W1. E0.

C. glomerata L. **Clustered Bellflower**

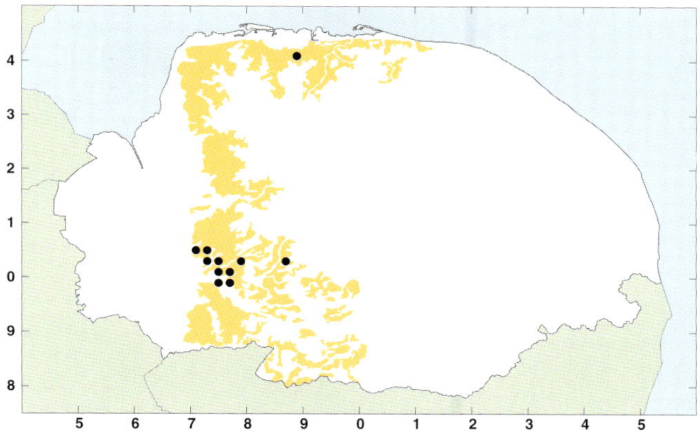

Herbaceous perennial confined to a few areas of calcareous grassland and markedly absent from many apparently suitable sites. Its Swaffham W80 site on a former firing range was ploughed in the winter of 1992-3. White-flowered forms occur.

Campanula glomerata - Foulden

C. portenschlagiana Schult. **Adriatic Bellflower**
Semi-winter green perennial introduced to gardens from Jugoslavia, occasionally established on walls. W3. E4.

C. poscharskyana Degen **Trailing Bellflower**
Semi-winter green perennial introduced to gardens from Jugoslavia, an escape to walls and waste ground. W15. E30.

C. latifolia L. — **Giant Bellflower**

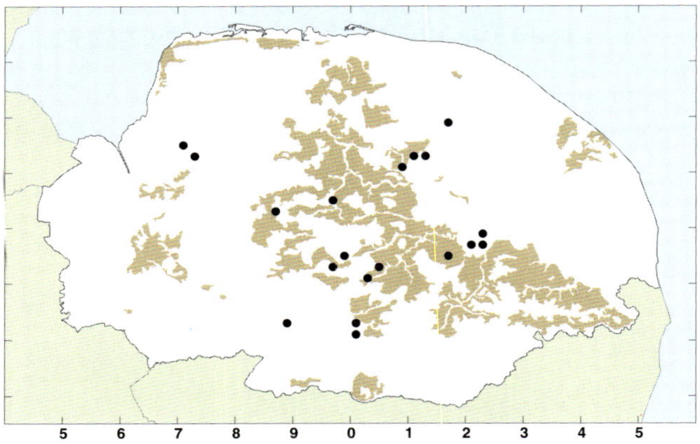

Herbaceous perennial a rare native of ancient, wet woodlands, a plant of largely northern distribution in Britain. It also occurs as an occasional garden escape.

C. trachelium L. — **Nettle-leaved Bellflower**

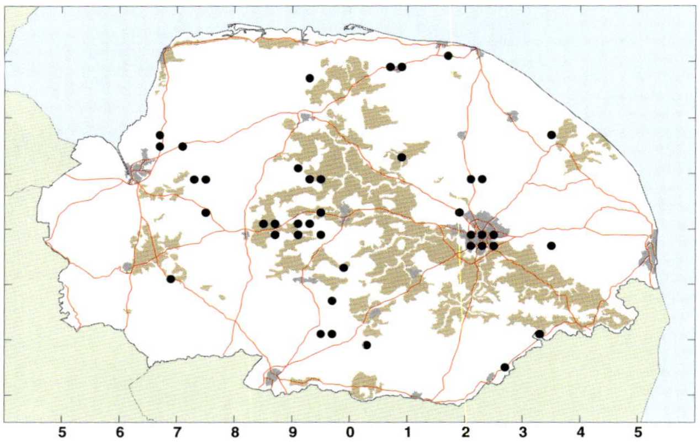

Herbaceous perennial, sometimes behaving as a biennial, not uncommon in old woods and as a native plant largely confined to them, a plant with a southerly distribution in Britain. Also an occasional garden escape.

C. rapunculoides L. — **Creeping Bellflower**

Herbaceous perennial introduced to gardens from mainland Europe, very invasive in gardens but curiously short-lived outside. W13. E30.

C. rotundifolia L. — **Harebell**

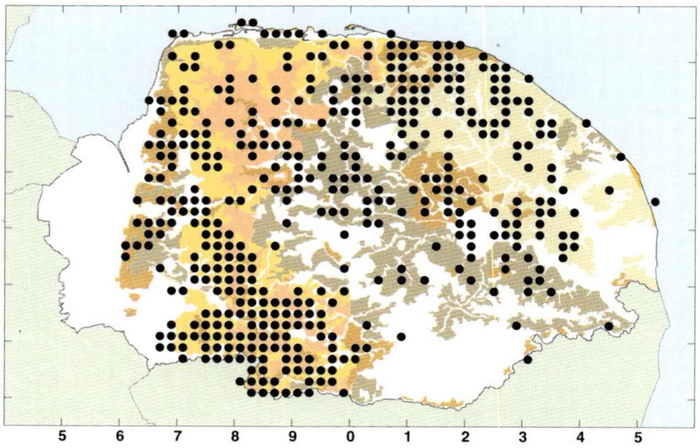

Herbaceous perennial frequent in established grassland on sandy, neutral to acid soils. It needs very sharp drainage and on heavier soils is sometimes found on ant hills or the shallow soils at the base of a hedgerow tree.

Legousia hybrida (L.) Delarbre — **Venus's Looking glass**

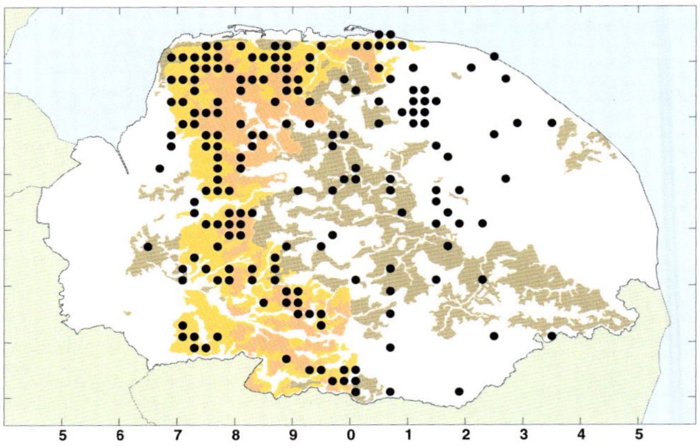

Spring germinating annual, an arable weed which appears to be increasing in frequency, especially on light soils.

Jasione montana L. — **Sheep's bit**

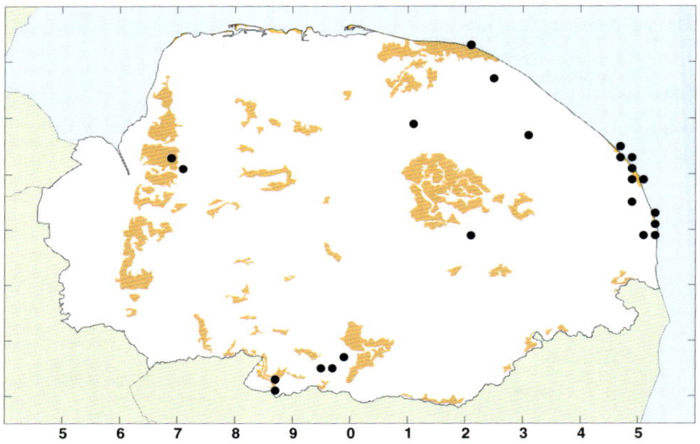

Semi-winter green perennial scarce in south-east England. In Norfolk confined to short, acid turf, most frequent on the acid dunes of the east coast.

Lobelia siphilitica L.

Herbaceous perennial introduced to gardens from North America, recorded on waste ground at Upwell W50 in 1994, RMP. W1. E0.

L. erinus L. — **Garden Lobelia**

Tender winter green perennial introduced to gardens from South Africa and grown as an annual. Occasionally escaping on to waste ground, in pavement cracks and at the base of walls. W8. E17.

RUBIACEAE

Sherardia arvensis L. — **Field Madder**

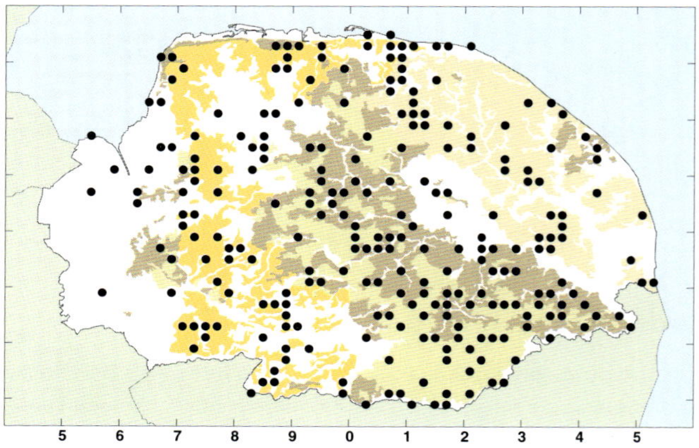

Spring germinating annual, widely scattered, usually on light soils. Now more frequent on tracks, in grassland and sandpits than in arable fields.

Asperula cynanchica L. **Squinancywort**

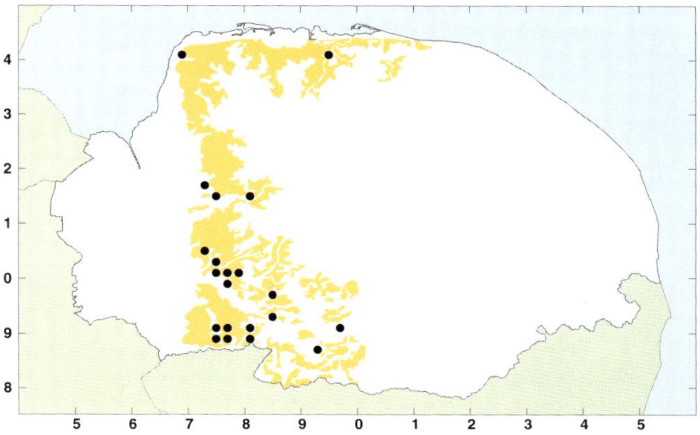

Perennial confined to chalk grassland. In established pastures sometimes found on old ant hills.

■ **A. taurina** L. **Pink Woodruff**

Perennial from southern Europe, recorded as naturalised in damp woodland at Hardingham E00 in 1949.

Galium odoratum (L.) Scop. **Sweet Woodruff**

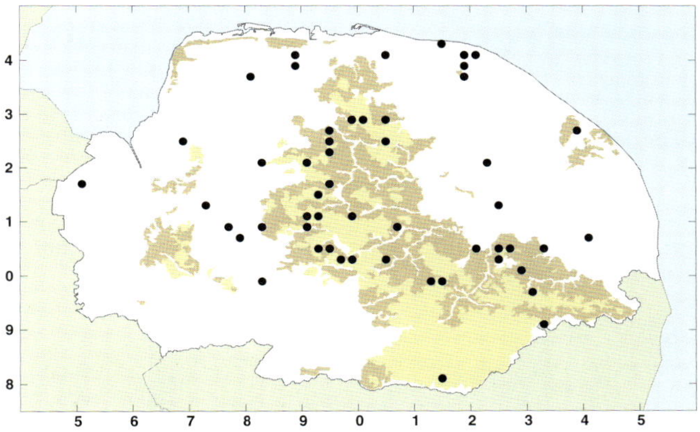

More-or-less winter green perennial, native to moist woodlands and hedgebanks, also occasionally as a garden escape.

G. uliginosum L. **Fen Bedstraw**

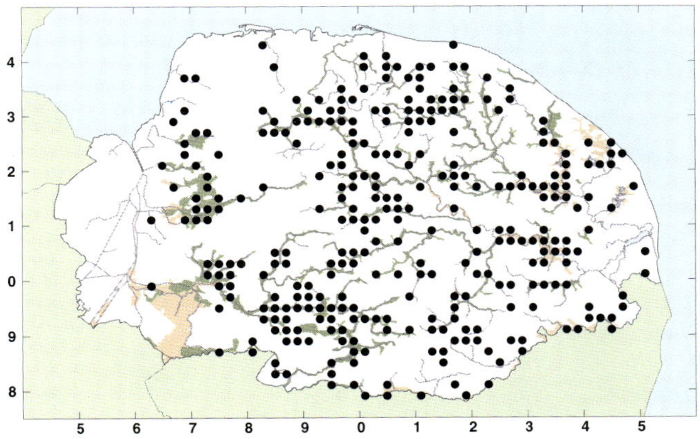

Herbaceous perennial, frequent in marshy ground, chiefly on base rich soils.

G. palustre L. **Marsh Bedstraw**

Herbaceous perennial, frequent in marshy ground, chiefly on neutral to mildly acid soils. Ssp. **palustre** appears to be the commonest subspecies with us though it is not so in the rest of the country. Ssp. **elongatum** (C.Presl) Arcang. is much less common, except for the Ouse Washes W59. W163. E216.

Galium odoratum - Cranworth

G. verum L. **Lady's Bedstraw**

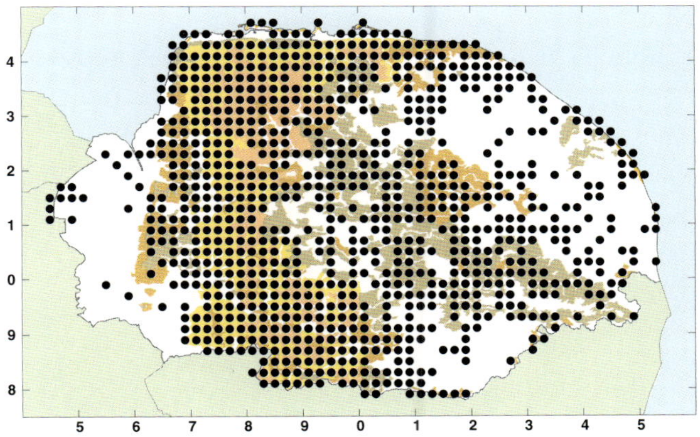

Semi-winter green perennial, frequent on light, calcareous to neutral soils and on coastal dunes. Illustrated on the following page.

Galium verum - Burnham Overy page 191

G. x pomeranicum Retz.
(*G. verum x G. mollugo*)
Herbaceous perennial recorded at West Rudham W82 in 1994, SF; the banks of Castle Acre castle W81 in 1997, KAB and a roadside at Cockley Cley W70 in 1997, JEG. W3. E0.

G. mollugo L. **Hedge Bedstraw**

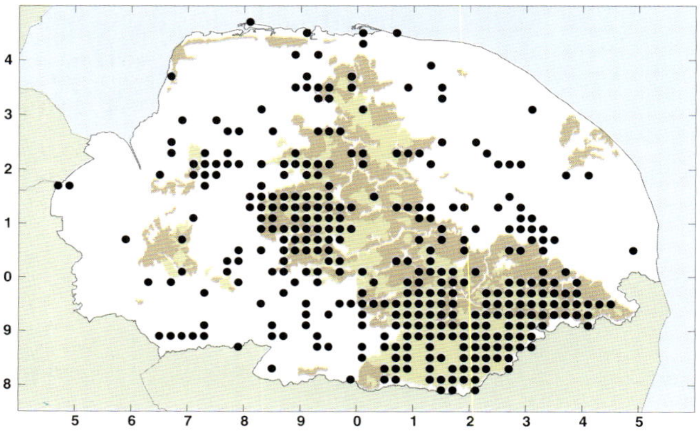

Herbaceous perennial most frequent on the calcareous boulder clays of south-east Norfolk where ssp. **mollugo** is usually found. Ssp. **erectum** Syme (Syn. *G. album* Mill.) occurs on lighter soils.

G. saxatile L. **Heath Bedstraw**

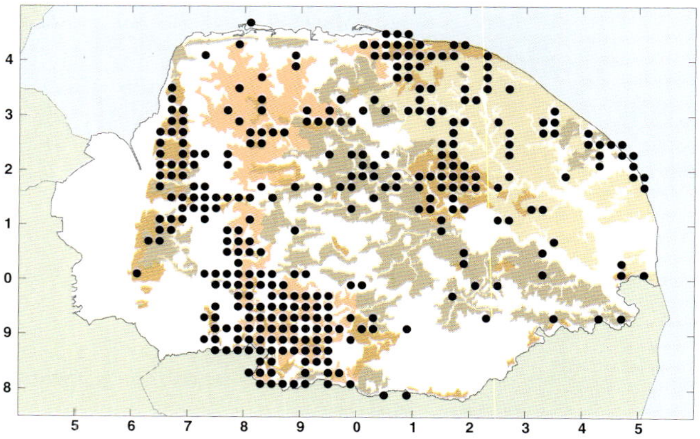

Semi-winter green perennial, an immediate indicator of dry, acid soils.

G. aparine L. **Goosegrass, Cleavers**
Annual, germinating throughout the year, a ubiquitous weed. W706. E712.

G. tricornutum Dandy **Corn Cleavers**
Annual introduced from mainland Europe, recorded in Norwich E20 in 1915, and at Heacham W63 in a corn field on chalky soils in 1992, G. Hirons conf., ALB. W1. E0.

G. parisiense L. **Wall Bedstraw**

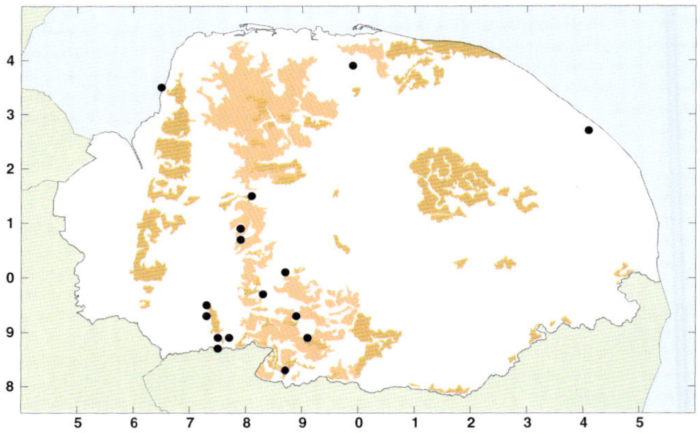

Spring and autumn germinating annual, very rare on sandy ground in the Breckland, elsewhere on long- established, mostly mediaeval walls where it is very persistent. A nationally scarce species.

Cruciata laevipes Opiz **Crosswort**
Syn. *Galium cruciata* (L.) Scop., *Cruciata chersonensis* auct. non (Willd.) Ehrend.

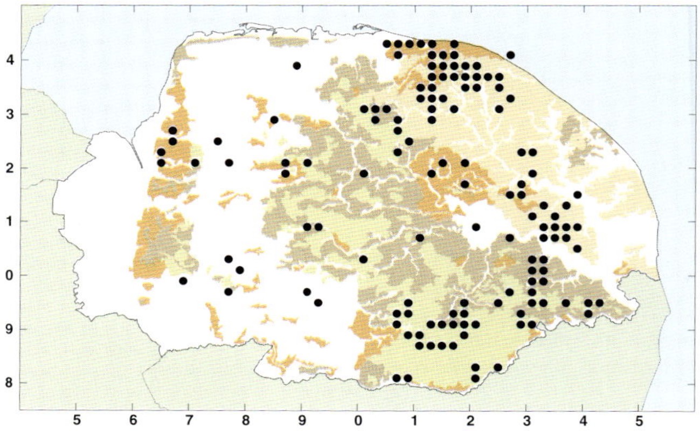

Semi-winter green perennial of rough grassland and verges, usually on heavier soils. Although often described as a plant of calcareous soils, its distribution does not bear this out in Norfolk.

CAPRIFOLIACEAE

Sambucus racemosa L. **Red-berried Elder**
Deciduous shrub introduced to gardens from Europe, western Asia and North America, sometimes planted for game cover and feed, occasionally naturalised as at How Hill and Alderfen, E31 where the plants are the American var. *pubens* (Michx.) Koehne. W0. E5.

S. nigra L. **Elder**
Deciduous shrub, leafing early. Abundant throughout the county. The variety *laciniata* with deeply dissected leaves is occasionally found in the wild; a weeping form was noted at Stow Bedon W99, by Dr C.Nelson and one with white fruits at Bagthorpe W83 by D.King. W700. E716.

S. ebulus L. **Dwarf Elder, Danewort**

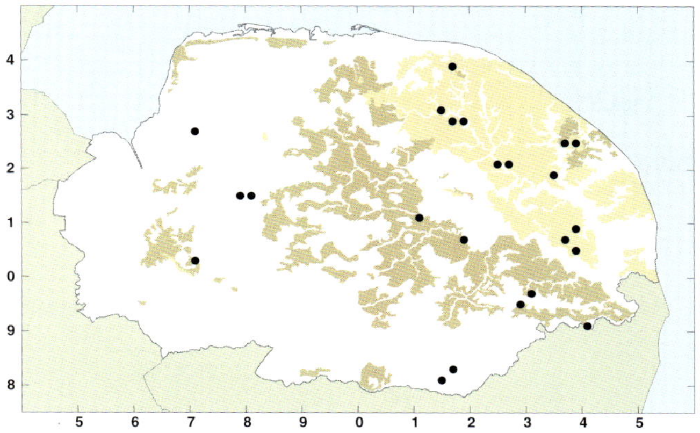

Herbaceous perennial found in hedgerows and on verges. It is unable to withstand annual summer cutting so is now lost from many of its verge sites and much reduced in frequency everywhere.

Viburnum opulus L. **Guelder Rose**

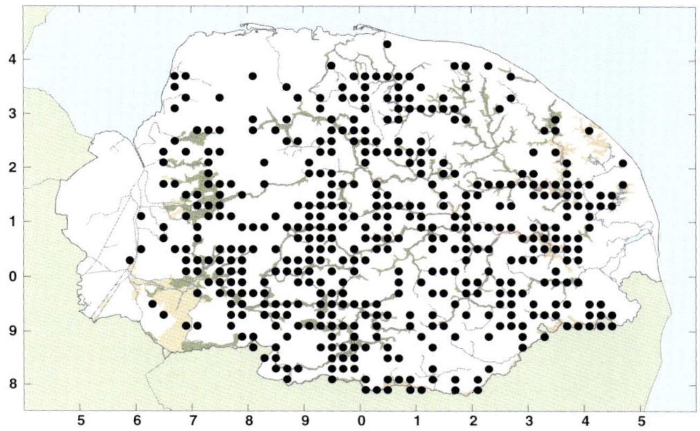

Deciduous shrub found in woods, hedgerows and scrub on moist soil throughout the county.

V. lantana L. **Wayfaring Tree, Whitten Tree**
Deciduous shrub recorded by Trimmer in five sites, considered extinct by 1968, but since then recorded in two sites which could be relics of its former distribution, at Gawdy Hall Wood, Hedenham E28 in 1974, O.Rackham, still present in 1998 and at Snettisham W63 in 1994, J.Austen. Frequently seen in recently planted hedgerows. W3. E24.

V. tinus L. **Laurestinus**
Evergreen shrub introduced from southern Europe to gardens and shrubberies, occasionally established on waste ground as at Fakenham W93 in 1992, MK. W1. E0.

V. rhytidophyllum Helms.
Evergreen shrub introduced from China, surviving from a garden throwout in Norwich E20 in 1992, ETD. W0. E1.

Symphoricarpos albus (L.) S.F.Blake **Snowberry**
Syn. *S. rivularis* Suksd.
Deciduous shrub introduced from western North America in 1817, widely planted as game cover and feed, also for hedging and often forming dense patches through suckering. W215. E426.

S. x chenaultii Rehder **Pink Snowberry**
(*S. microphyllus x S. orbiculatus*)
Deciduous shrub, a garden hybrid found on waste ground at Foulden W70 in 1994, KAB; Ashill W80 in 1997, R.Barrett; Freethorpe E40 in 1998, Flora Group and planted for game cover in several places as at Hardingham E00, ALB. W2. E2.

Leycesteria formosa Wall. **Himalayan Honeysuckle**
Deciduous shrub introduced to gardens from the Himalaya. Early leafing, free-seeding and occasionally naturalised as on walls in King's Lynn W61 in 1996, RMP. W4. E7.

Lonicera nitida E.H.Wilson **Wilson's Honeysuckle**
Evergreen shrub introduced to gardens and parks from China, particularly for hedging. Occasionally self-sowing. W12. E47.

■ **L. xylosteum** L. **Fly Honeysuckle**
Deciduous shrub perhaps native in southern Britain. Found in scrub at Wretham W99 in 1967, ALB.

L. japonica Thunb. ex Murray **Japanese Honeysuckle**
Semi-evergreen climber introduced to gardens from eastern Asia. A large colony was recorded on the sea-defence bank at Cley E04 in 1994, KKH. W0. E1.

L. periclymenum L. **Honeysuckle, Woodbine**

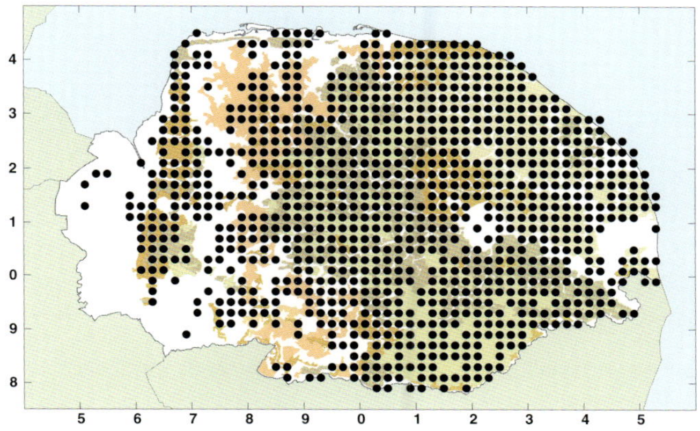

Deciduous climber frequent in woods and hedgerows, avoiding dry, calcareous soils.

L. caprifolium L. **Perfoliate Honeysuckle**
Deciduous climber introduced to gardens from southern Europe. Found recently at Norwich E20 in 1990, ETD; Gissing E18 in 1994, ALB; Denton E28 in 1994, ALB; Woodton E29 in 1996, Flora Group and Horsey E42 in 1997, ALB. W0. E5.

L. x italica Schmidt ex Tausch **Garden Honeysuckle**
Syn. *L. x americana.* auct., non (Mill.) K.Koch (*L. caprifolium x L. etrusca*)
Deciduous climbing hybrid established from gardens. Found on Barnham Cross Common, Thetford W88 in 1992, N.Gibbons. W1. E0.

ADOXACEAE

Adoxa moschatellina L. **Town Hall Clock, Moschatel**

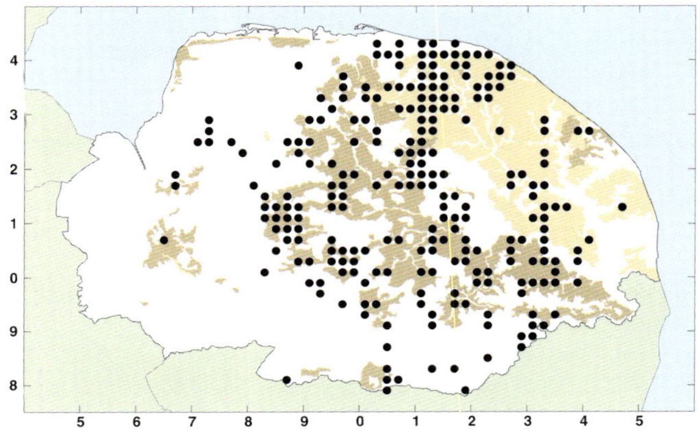

Herbaceous perennial found in shady places, usually on moist soils.

VALERIANACEAE

Valerianella locusta (L.) Laterr. **Common Cornsalad**

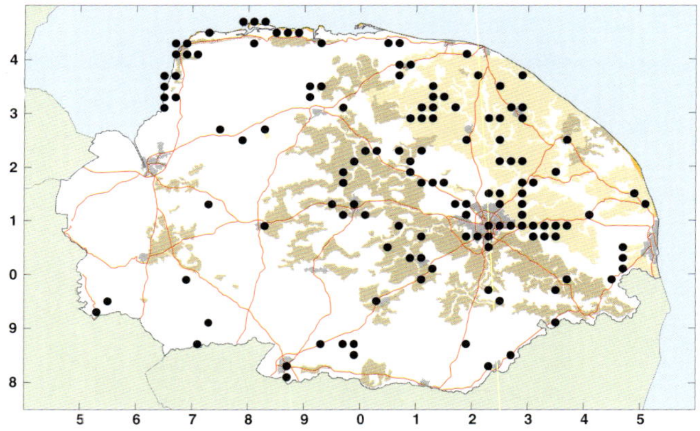

Autumn germinating annual found in open, dry places including arable land and coastal sand dunes.

Valerianella locusta - Holkham Dunes

V. carinata Loisel. **Keeled-fruited Cornsalad**

Annual found at Hardwick E29 in 1977, MAB and recently on the top and base of a wall at Kenninghall E08 in 1994, JH; on a disused railway track at Geldeston E39 in 1995, ALB and as a garden weed in Swaffham W80 in 1996, JEG. Possibly increasing. W1. E2.

■ **V. rimosa** Bastard. **Broad-fruited Cornsalad**

Spring germinating annual. Found in a cornfield between Felbrigg and Cromer E24 in 1891 and at High Kelling E13 annually between 1938 and 1941, S.Gomersal.

V. dentata (L.) Pollich **Narrow-fruited Cornsalad**

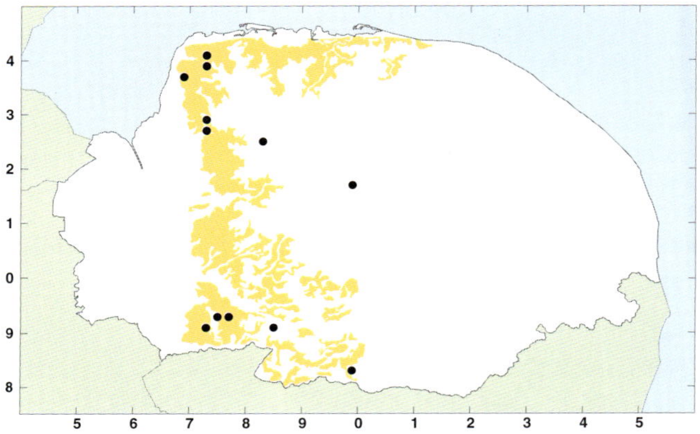

Spring germinating annual occasional in open places and arable fields on light calcareous soils.

Valeriana officinalis L. **Common Valerian**

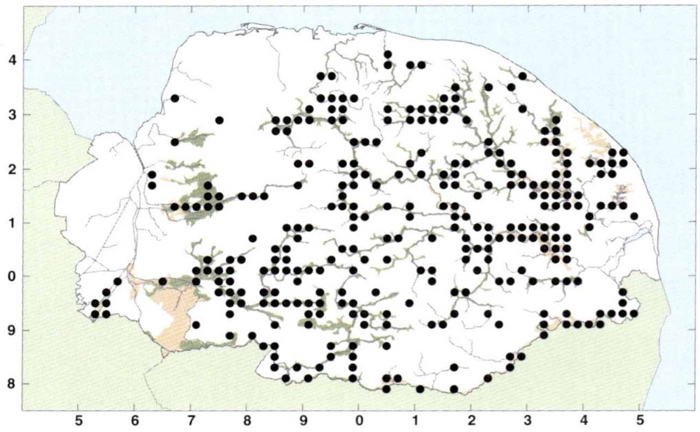

Semi-winter green perennial with two distinct habitats, in wet ground and along streamsides, and less frequently in rough grassland on drier soils.

V. dioica L. **Marsh Valerian**

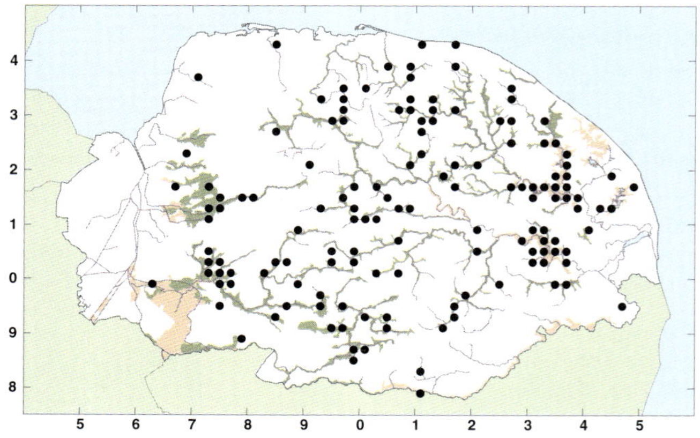

Semi-winter green perennial with a northern distribution in Britain but not uncommon in marshy places on both acid and calcareous soils.

Centranthus ruber (L.) DC. **Red Valerian**

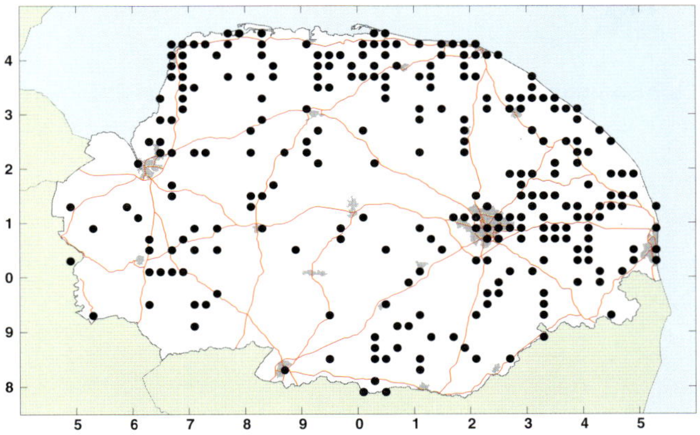

Semi-winter green perennial introduced long ago from the Mediterranean and thoroughly established on walls. Most common near the sea where its foliage over-winters more successfully.

DIPSACACEAE

Dipsacus fullonum L. **Wild Teasel**
Syn. *D. sylvestris* Huds.

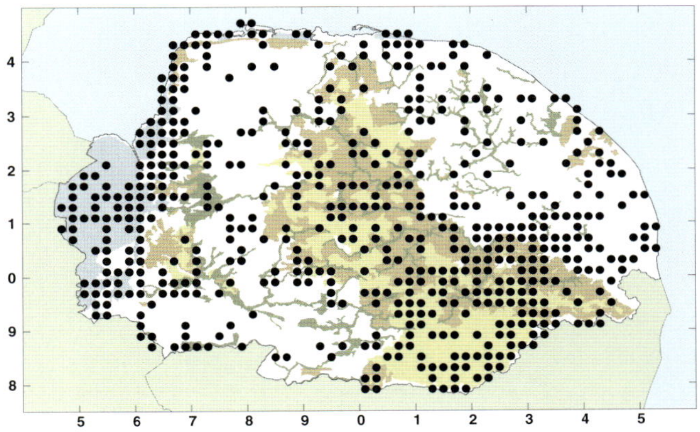

Winter green biennial widespread in moist soils. In the Fenland it is frequent on river and dyke banks.

Dipsacus pilosus L. **Small Teasel**

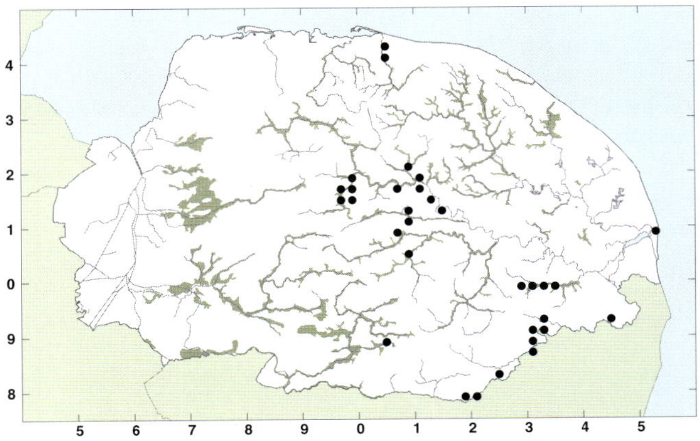

Spring germinating biennial confined to moist and alluvial soils, often in open woodland. The seeds are distributed by marsh tits and goldfinches.

Knautia arvensis (L.) Coult. **Field Scabious**

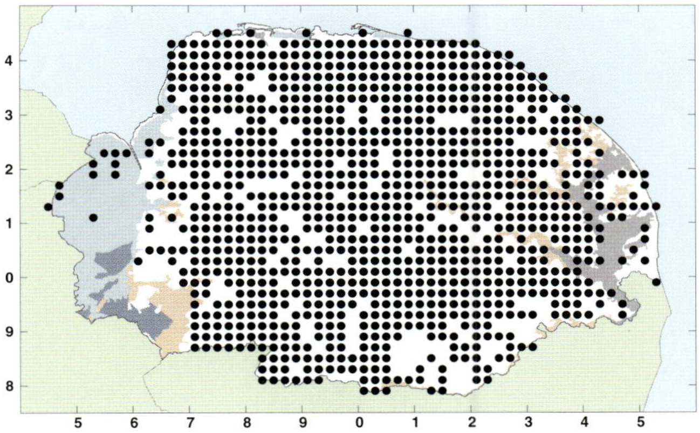

Semi-winter green perennial frequent in well-drained soils. White-flowered plants sometimes occur.

Succisa pratensis Moench **Devil's-bit Scabious**

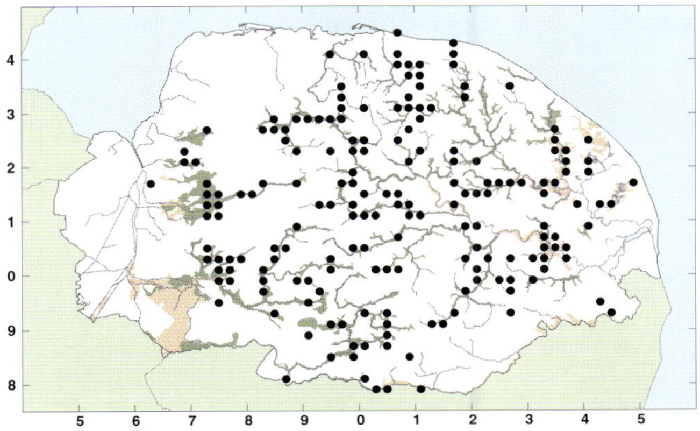

Herbaceous perennial of moist, established grassland on good loams. Flowers can show a wide colour range from white to very dark purple-red.

Dipsacus pilosus - Kirstead

Scabiosa columbaria L. **Small Scabious**

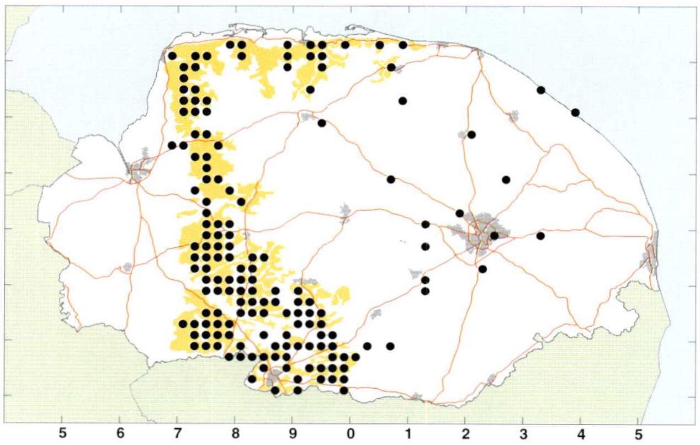

Semi-winter green perennial frequent only on light calcareous soils.

Scabiosa columbaria - Castle Acre

ASTERACEAE (COMPOSITAE)

Echinops sphaerocephalus L. **Glandular Globe Thistle**
Herbaceous perennial introduced to gardens from southern Europe. Garden escapes within this genus have been the source of much confusion in the past with many records made for *E. sphaerocephalus*, it is, however, scarce and occurs as a throwout only occasionally as at Heacham W63 in 1991, PRC and Blakeney E04 1992, K.Ferroussat. W1. E1.

E. exaltatus Schrad. **Tall Globe Thistle**
Syn. *E. commutatus* Jur.
Herbaceous perennial introduced to gardens from eastern Europe. A number of colonies of this species occur in the Foulden-Didlington-Ickburgh area W79 and 89, probably all derived from one original source. W6. E0.

Carlina vulgaris L. **Carline Thistle**

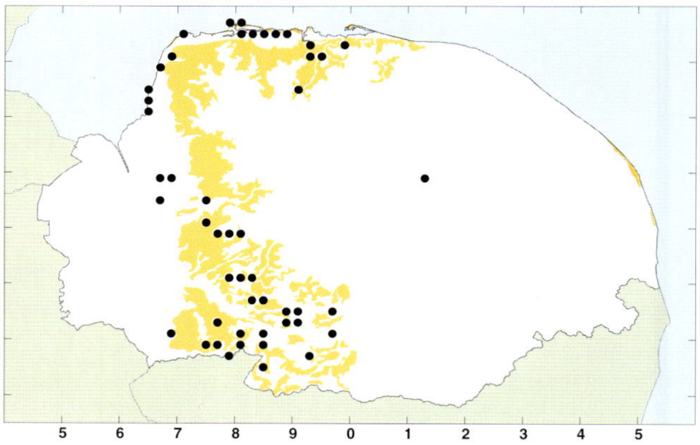

Herbaceous perennial found in established grassland on light, sandy soils usually over-lying chalk, also on calcareous dunes.

Arctium lappa L. **Greater Burdock**

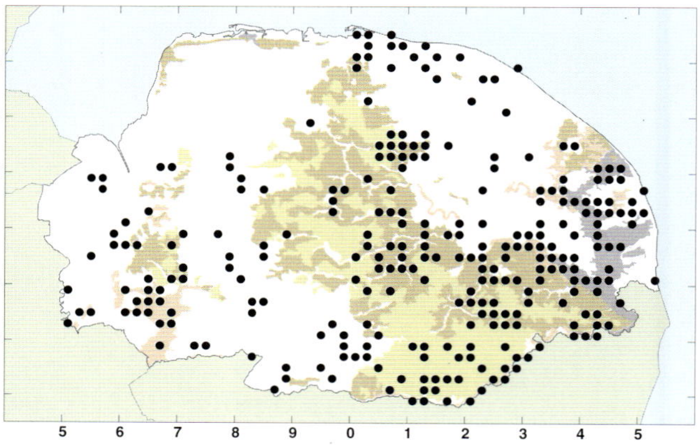

Winter green biennial of disturbed places on heavy, usually moist soils.

A. minus (Hill) Bernh. **Lesser Burdock**
Biennial which is widespread throughout the county. The nomenclature of this group has been the subject of much study and has been changed twice during the progress of this survey. *A. minus* and *A. nemorosum* were formerly given specific rank, then both were reduced to subspecies of *A. minus*. To these were added ssp. *pubens*, but more recently this has been demoted to varietal rank. Throughout the county there has been little difficulty in separating and naming ssp. **minus** which is chiefly found on lighter soils. Petch & Swann 1968 confined the use of this name to the very small-headed form found in the Breckland but recent taxonomy does not support this view. Ssp. **nemorosum** is a plant of heavier soils and has been found on relatively few occasions, chiefly in the east and extreme north-west of the county. W636. E658.

Carduus tenuiflorus Curtis **Slender Thistle**

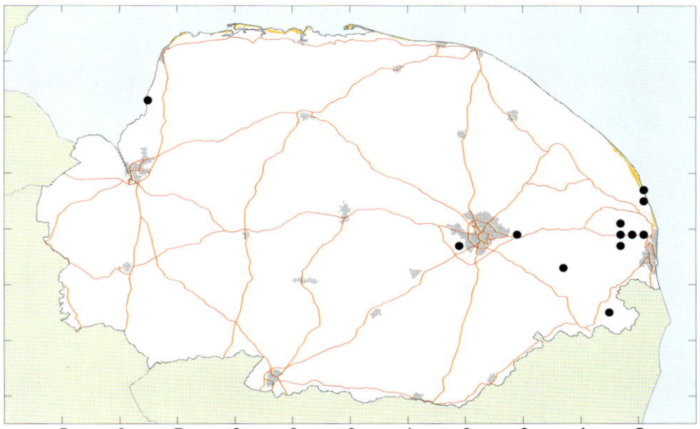

Winter green biennial of waste places especially near the coast.

C. crispus L. **Welted Thistle**
Syn. *C. acanthoides* auct. non L.

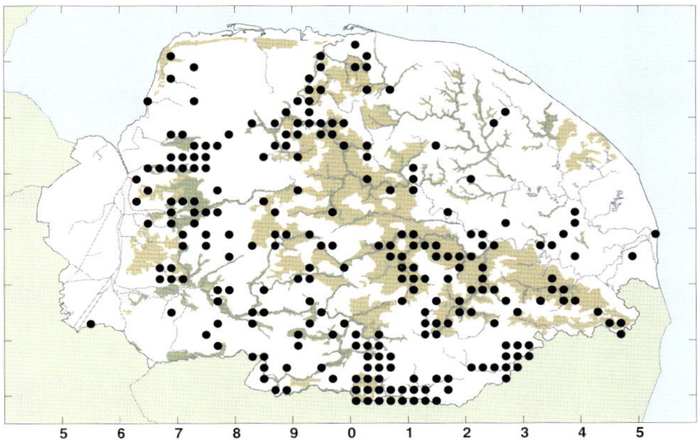

Widely scattered on waste ground and roadsides, usually on fertile soils.

■ **C. x dubius** Balbis
C. crispus x C. nutans
Recorded once at Loddon E39 in 1937.

C. nutans L. **Musk Thistle**

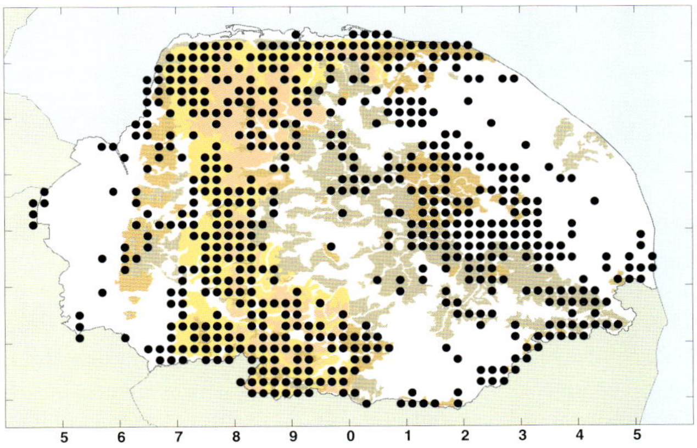

Winter green biennial frequent on light soils, especially where they are base rich, occurring both in grassland and waste places as well as a weed of arable fields .

■ **Cirsium eriophorum** (L.) Scop. **Woolly Thistle**
Winter green biennial, a casual recorded twice, at Longham W91 in 1919 and Scarning W91 in 1955.

C. vulgare (Savi) Ten. **Spear Thistle**
Winter green biennial, ubiquitous. W708. E717.

C. dissectum (L.) Hill **Meadow Thistle**

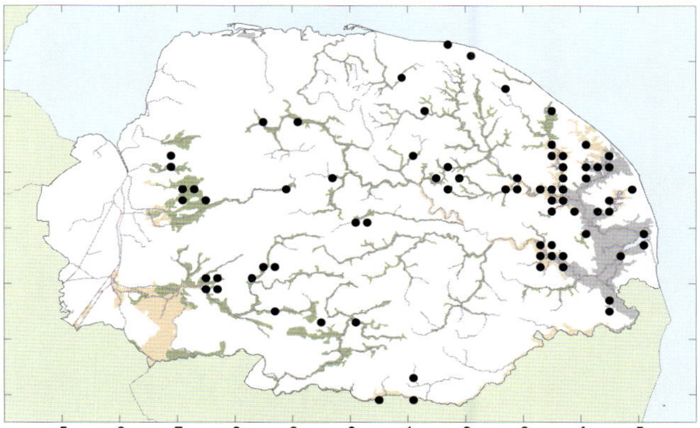

Winter green perennial, very local in undisturbed, moist grassland and fen.

■ **C. x woodwardii** (H.C.Watson) Nyman
(*C. dissectum x C. acaule*)
One record, at South Lopham W08 in 1953.

■ **C. x forsteri** (Sm.) Loudon
(*C. dissectum x C. palustre*)
One record at Southrepps E23 in 1975, PWL & F.Rose.

C. acaule (L.) Scop. **Dwarf Thistle**

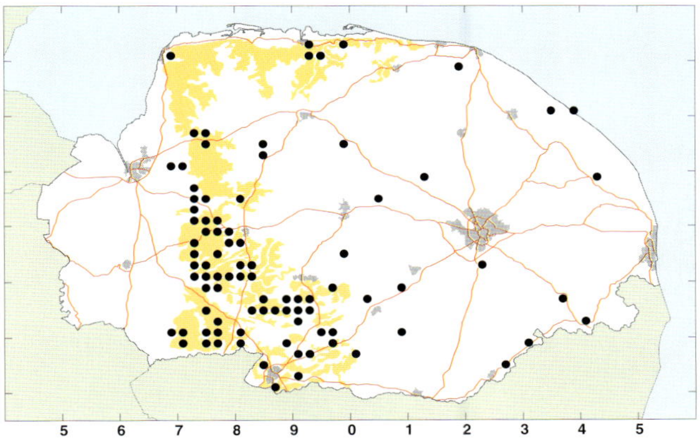

Winter green perennial local in dry, base rich grassland, occasional elsewhere.

Cirsium acaule - Narborough

C. palustre (L.) Scop. **Marsh Thistle**

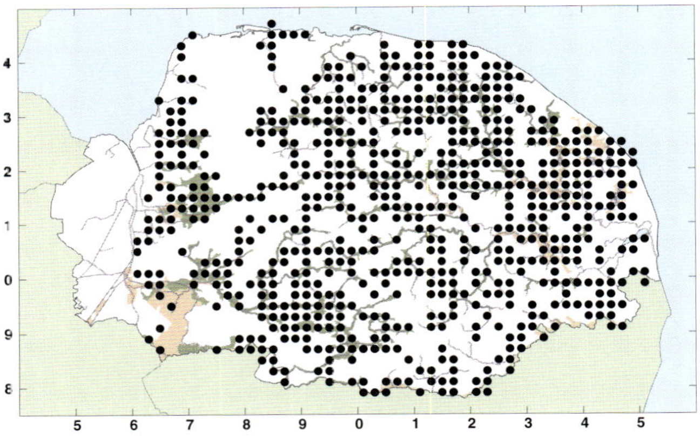

Winter green biennial found wherever the ground is wet; white-flowered plants are frequent.

C. arvense (L.) Scop. **Creeping Thistle**
Herbaceous perennial, ubiquitous, the most frequently recorded plant in the county. W712. E718.

Onopordum acanthium L. **Cotton Thistle, Scotch Thistle**

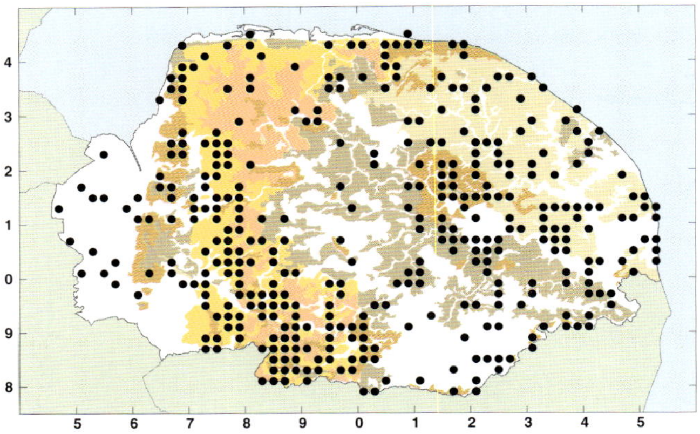

Winter green biennial thoroughly established on drier soils and perhaps native.

Cynara cardunculus L. **Globe Artichoke**
Perennial introduced from the Mediterranean and found as an escape in Norwich E20 in 1990, ETD. W0. E1.

Silybum marianum (L.) Gaertn. **Milk Thistle**

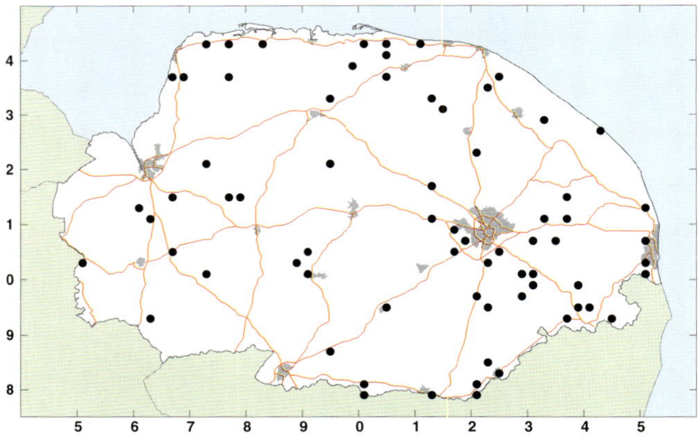

Annual or winter green biennial introduced from Europe, western Asia and North Africa. Scattered throughout the county on waste ground and hedgerows, it is particularly abundant on the banks of the Roman town at Caistor St Edmunds E20 and along the Waveney valley.

■ **Serratula tinctoria** L. **Saw-wort**
Herbaceous perennial native to heavy grassland in Suffolk, found by a deep dyke on heavy soil on the Norfolk side of the border at North Lopham E08 in 1970, ELS, and also Trist, 1979, probably the same site.

Centaurea scabiosa L. **Greater Knapweed**

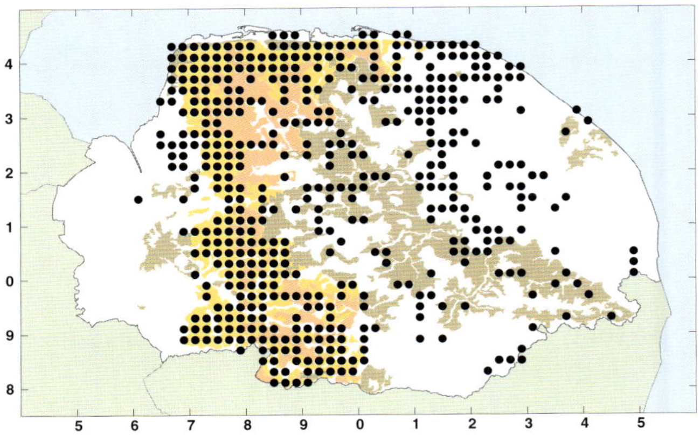

Herbaceous perennial occurring on well-drained, usually base rich soils. White-flowered forms can sometimes be found.

C. montana L. **Perennial Cornflower**
Herbaceous perennial introduced to gardens from mainland Europe, occasionally found as an escape but not long lasting. W2. E16.

C. cyanus L. **Cornflower**
Autumn and spring germinating annual, once an arable weed but now probably lost from this source. Recent records are mostly from deliberate sowings, on roadside verges, set aside or near gardens so it is difficult to be sure of the origin of single plants found in fields as at Pott Row W72 in 1998, HW. W6. E35.

■ **C. calcitrapa** L. **Red Star-thistle**
Winter green biennial introduced from mainland Europe. Recorded three times, the last at Stanhoe W83 in lucerne, in 1960.

■ **C. aspera** L. **Rough Star-thistle**
Perennial introduced from mainland Europe, found twice, the last on Weybourne Cliffs E14 in 1974.

■ **C. solstitialis** L. **Yellow Star-thistle**
Annual or winter green biennial introduced from southern Europe, occurring as an arable weed, chiefly in lucerne or carrots on twenty occasions, all before 1974.

■ **C. melitensis** L. **Maltese Star-thistle**
Annual or winter green biennial introduced from southern Europe, a casual found twice, the last at Thetford W88 in 1961.

■ **C. diluta** Aiton **Lesser Star-thistle**
Evergreen perennial introduced from the Mediterranean and found on five occasions between 1969 and 1982. Usually derived from birdseed.

C. nigra L. **Common Knapweed, Hardheads**
Syn. *C. nemoralis* Jordan
Herbaceous perennial abundant through the county except on very acid or peaty soils. W609. E690.

■ **C. dealbata** Willd.
Herbaceous perennial widely cultivated, recorded once, on cliff-tops at Mundesley E33 in 1973, ETD.

Carthamus tinctorius L. **Safflower**
Spring germinating annual introduced to waste places from south-west Asia, probably as a casual from birdseed, recorded as a birdseed alien at East Tuddenham E01 in 1997, ALB and on allotments at Cobholm E50 in 1996, C.Ley. W0. E2.

Cichorium intybus L. **Chicory, Succory**

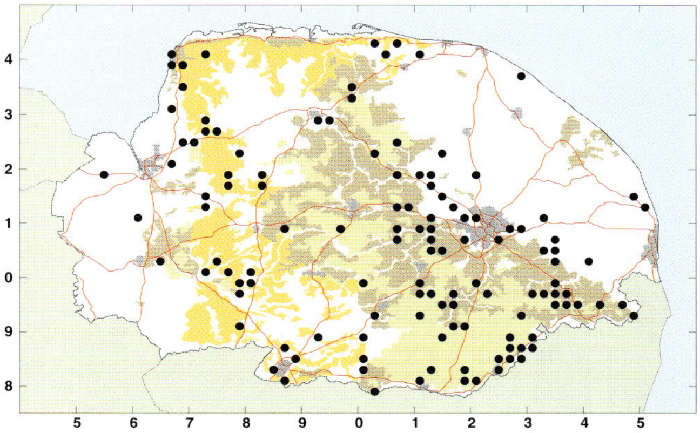

Winter green perennial introduced from mainland Europe, probably in the Middle Ages. In the time of Petch & Swann's 1968 Flora it was still widely grown as a crop and frequently survived on roadsides and field edges. Now it is usually found on waste ground.

■ **Arnoseris minima** (L.) Schweigg. & Körte **Lamb's Succory**
Spring germinating annual once found in arable fields on sandy soils but now considered extinct nationally. Last seen in a rye-field at Pentney W71 in 1953.

Lapsana communis L. **Nipplewort**
Autumn and spring germinating annual, ubiquitous. W643. E684.

Hypochaeris radicata L. **Cat's-ear**
Winter green perennial widespread in grassy places, roadsides and waste places. W580. E651.

H. x intermedia Richt.
(*H. radicata x H. glabra*)
A small colony of this hybrid was found growing amongst the parents at Langford W89 in the Stanford Training Area in 1998, ALB. W1. E0.

H. glabra L. **Smooth Cat's-ear**

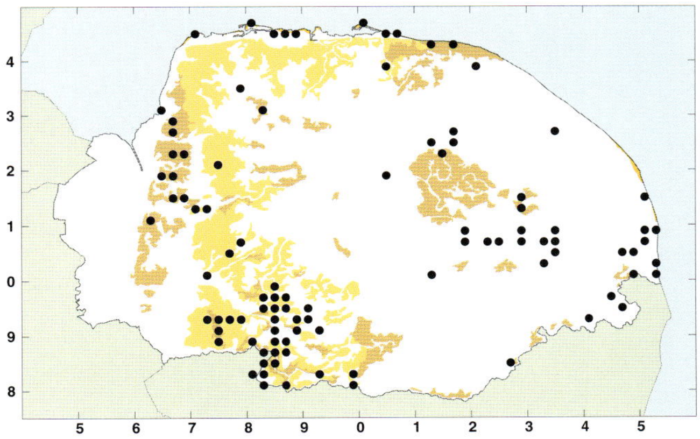

Winter green perennial, an uncommon plant of light, neutral to acid soils, generally decreasing, but still quite frequent in the Breckland on forest rides, in sandpits and occasionally on set aside. A nationally scarce species.

■ **H. maculata** L. **Spotted Cat's-ear**
Perennial, recorded twice before 1914, but neither record certain.

Leontodon autumnalis L. **Autumn Hawkbit**
Winter green perennial widespread in open grassland, a feature of late summer roadsides. W603. E652.

L. hispidus L. **Rough Hawkbit**

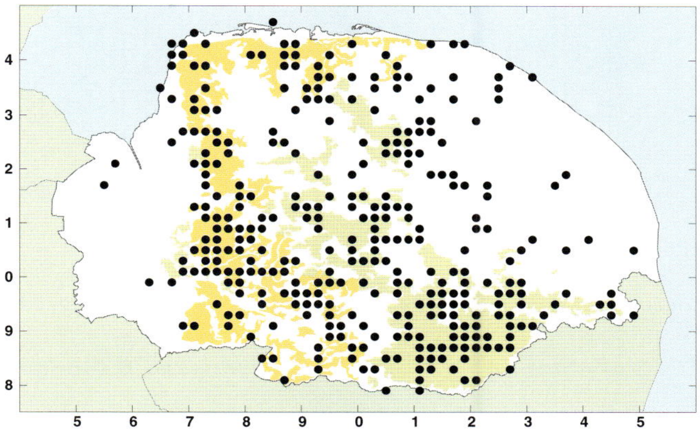

Winter green perennial found in established grassland especially on base rich soils.

L. saxatilis Lam. **Lesser Hawkbit**
Syn. *L. taraxacoides* (Vill.) Mérat. nom. illeg.

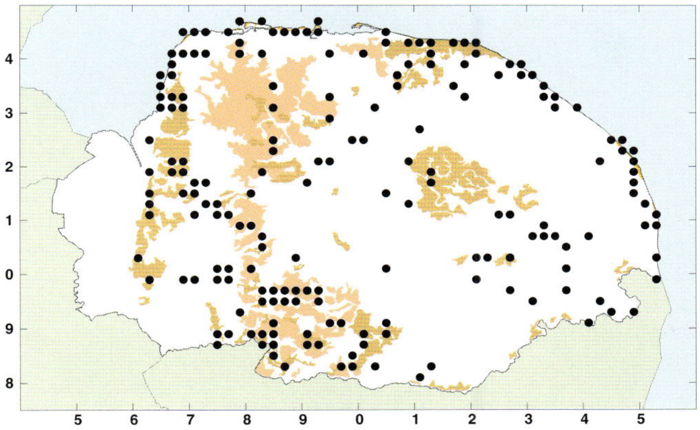

Winter green perennial found on dry, sandy soils in open sites.

Hypochaeris glabra - Bawsey

Picris echioides L. **Bristly Ox-tongue**

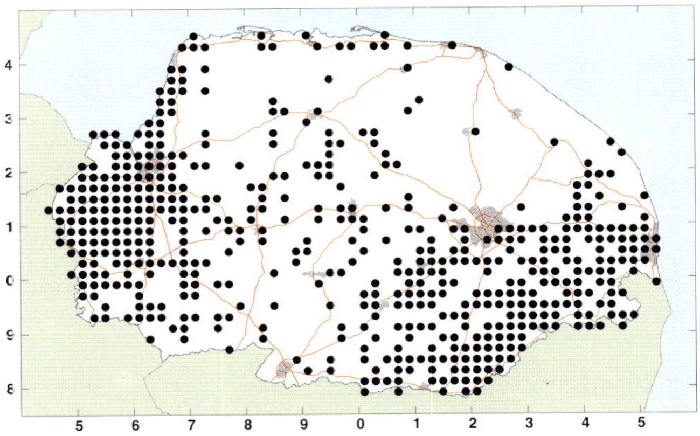

Spring germinating annual commonest on moist soils It has increased in frequency rapidly during the last twenty years especially in the south-east and west of the county.

P. hieracioides L. **Hawkweed Ox-tongue**

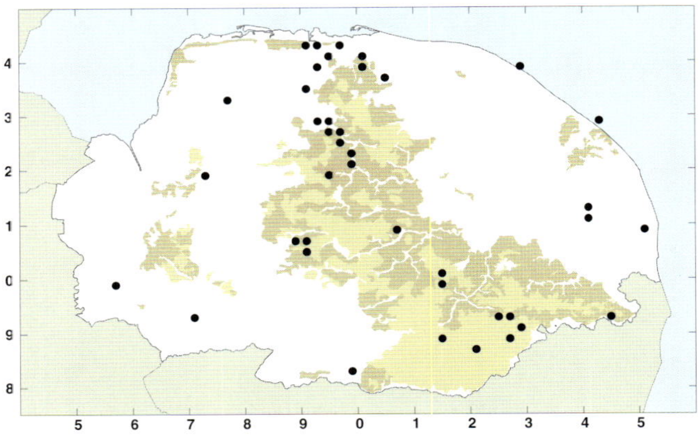

Winter green biennial occasional on chalky banks, particularly associated with old railways. Decreasing.

Tragopogon pratensis L. **Goat's-beard**

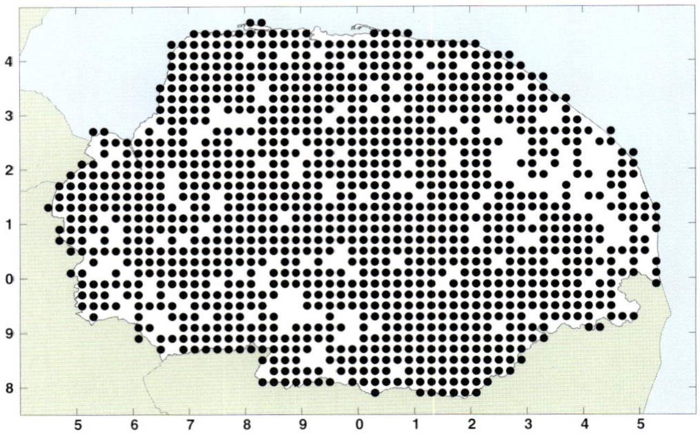

Annual to deciduous perennial generally frequent by roadsides, in grassland and on dunes. Our plants mostly belong to ssp. **minor** (Mill.) Wahlenb., but ssp. **pratensis** was found at Denton E28 in 1994, ALB where it was the common species, also seen at Langley E30 in 1998, R.Chancellor & DM.

■ **T. x mirabilis** Rouy
(*T. pratensis x T. porrifolius*)
Recorded in a railway yard at Yarmouth E50 in 1929.

T. porrifolius L. **Salsify**

Deciduous biennial from the Mediterranean, an occasional escape from cultivation, a notably thriving colony was found beside the Stalham bypass E32 in 1990, SP, it now extends from Potter Heigham to Smallburgh, FF in 1998. W4. E10.

■ **T. hybridus** L. **Slender Salsify**

Annual introduced from southern Europe, found as a casual in a garden at Swaffham W80 in 1977, AC.Wilson.

Sonchus palustris L. **Marsh Sow-thistle**

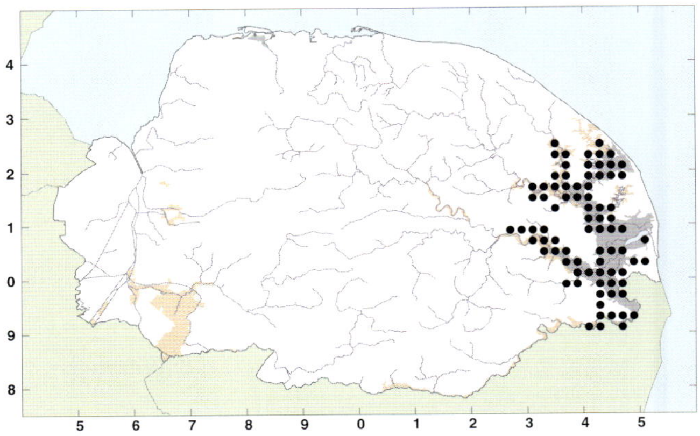

Perennial confined to Broadland where its main habitat is the banks and the flood land which lies between banks and water (ronds) alongside the lower reaches of rivers. Apparently much reduced by the early years of the twentieth century, it has made a great come-back in recent years, perhaps because of the increasing mineral richness of the river water. A remarkably robust colony was found in a dry woodland clearing at Toft Monks E49 in 1996, ALB.

Sonchus palustris - Wheatfen

S. arvensis L. **Perennial** or **Corn Sow-thistle**
Herbaceous perennial, a frequent weed of roadsides and waste places on heavy soils, often in drainage cuts (grups) by roads, also an arable weed. W578. E664.

S. oleraceus L. **Smooth Sow-thistle**
Annual or winter green biennial, ubiquitous. ELS in 1967 observed that this species had become less frequent than *S. asper* but this trend has not continued into the 1990s. W675. E697.

S. asper (L.) Hill **Prickly Sow-thistle**
Annual or winter green biennial, widespread. W692. E702.

Lactuca serriola L. **Prickly Lettuce**

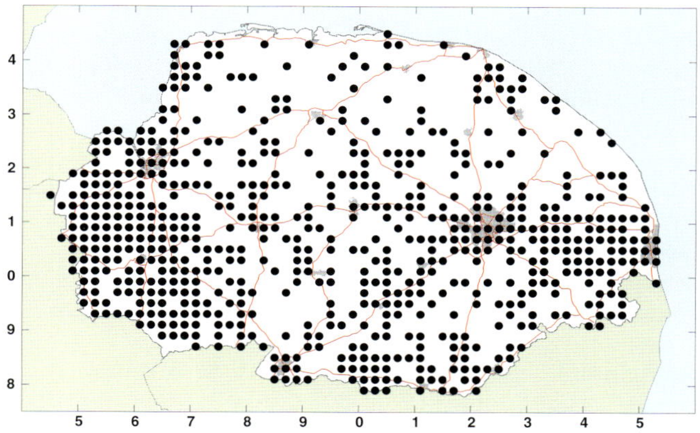

Spring germinating annual or biennial present in two forms. The form with deeply dissected leaves, forma **serriola** is native to the Fenland, being quite frequent around Ely in Cambridgeshire. During this survey it was found at Marshland St James W51 in 1993 and Barroway Drove W50 in 1994, both by RMP and at Feltwell W69 in 1998, EVR. It has also been reported from east Norfolk. The form with almost entire leaves, forma **integrifolia,** has spread widely through the county since Petch & Swann, 1968 when it was regarded as a rare casual and is now to be found on waste ground, on roadside verges and often abundantly on stump-rows in the Breckland forest.

L. sativa L. **Garden Lettuce**
Annual found once as a casual. W1. E0.

L. virosa L. **Greater Prickly Lettuce**

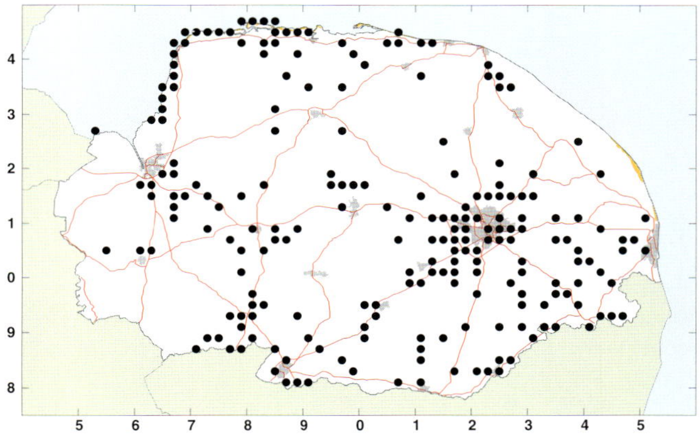

Annual or biennial long known from the north coast where it grows on sea banks, but now becoming increasingly common on road verges and field edges inland.

■ **L. saligna** L. **Least Lettuce**
Annual or biennial, always rare on partly saline soils along sea banks. Last observed at Tilney St Lawrence, W51 in 1953.

Cicerbita bourgaei (Boiss.) Beauverd **Pontic Blue Sow-thistle**
Herbaceous perennial introduced to gardens from Turkey and occasionally escaping as at Great Walsingham W93, where it has been known for 20 years; Emneth W40 and Snettisham W63. W4. E0.

Mycelis muralis (L.) Dumort. **Wall Lettuce**

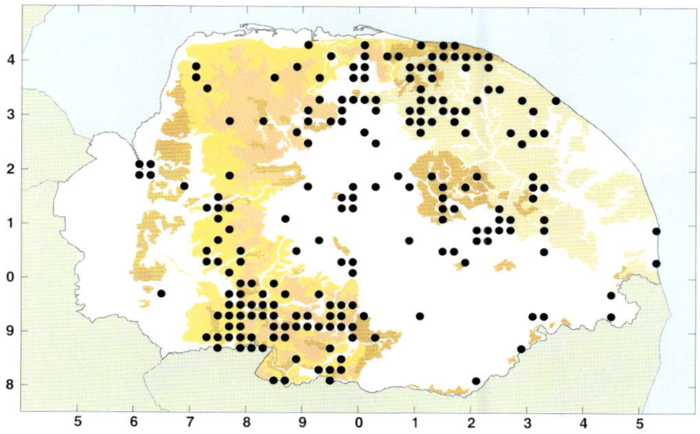

Winter green perennial, a plant of shady woodlands and walls, local in its distribution.

Lactuca virosa - Burnham Overy

Taraxacum. spp. **Dandelion**
Winter green perennial, one of the most familiar of all wild plants and ubiquitous in its distribution. Modern study recognises at least 235 microspecies in Britain, some very limited in their distribution, others widespread throughout the country. E.L.Swann studied the species in Norfolk, particularly in west Norfolk, from 1973 to 1983, for much of the time in collaboration with R.P.Libbey. Since that date, however, there has been no botanist with time and expertise to follow up their work and the publication of the monograph by Dudman and Richards in 1997 came too late for those working on this Flora.

Nine sections are recognised within *Taraxacum* and of these seven are represented in Norfolk. They are Erythrosperma, Palustria, Spectabilia, Naevosa, Celtica, Hamata and Ruderalia.

The following notes make use of records made since 1970 adjusted for recent changes in nomenclature and have been read and commented upon by Andrew Dudman

Section *Erythrosperma*
30 species have been recorded in Britain. These were once all included under *T. laevigatum*. They are perhaps the most interesting group as the 18 species found in the county are all plants of wild habitats, occurring on light soils, both sandy heaths and chalk grassland, also on dunes. They are very typical of the Breckland. The most frequent are **T. lacistophyllum** (Dahlst.) Raunk. which is frequent in chalk grassland from Ringstead W64 to Thetford W88; **T. brachyglossum** (Dahlst.) Raunk.; **T. rubicundum** (Dahlst.) Dahlst.; **T. oxoniense** Dahlst. recorded on the coast and in Breckland and **T. fulviforme** Dahlst. The following species have all been recorded at E.Wretham W88, 89 amongst other sites; **T. scanicum** Dahlst. which is confined to the Breckland; **T. arenastrum** A.J.Richards; **T. proximum** (Dahlst.) Raunk; **T. proxiforme** Soest in Lambinon & Soest; **T. falcatum**, Brenner; **T. retzii** Soest and **T. glauciniforme** Dahlst. **T. commixtum** G.E.Haglund has been recorded at Reffley W62; **T. parnassicum** Dahlst. at Ringstead W74; **T. dunense** Soest at Snettisham W63; **T. haworthianum** Dudman & Richards; **T. disseminatum** G.E.Haglund has been noted in W62 and **T. acutum** A.J.Richards has also been recorded.

Section *Palustria*
T. palustre (Lyons) Symons is the only species, a plant of old meadows. It has been noted at Scarning W91 and Carbrooke W90.

Section *Spectabilia*
T. faeroense (Dahlst.) Dahlst. is widespread through the country but scarce in the south-east. It has been recorded from a number of fens in the county.

Section *Naevosa*
A chiefly northern section with **T. euryphyllum** (Dalhst.) Hjelt and **T. maculosum** A.J.Richards the only species recorded.

Section *Celtica*
A section with a mainly western distribution, found in moist soils, mostly in natural habitats. Ten species occur, **T. gelertii** Raunk. recorded at Scarning W91 and South Lopham E08; **T. bracteatum** Dahlst. which is widespread; **T. britannicum** Dahlst. at Flitcham W72; **T. subbracteatum** A.J.Richards in A.J.Richards & C.C. Haw. at West Newton W62; **T. duplidentifrons** Dahlst. on both chalk grassland and sand dunes; **T. fulgidum** G.E.Haglund at Denver W60; **T. landmarkii** Dahlst. a plant of gullies and wet places in the north, curiously present at Snettisham W63; **T. nordstedtii** Dahlst. in wet sites from Santon W88 to Booton E12; **T. palustrisquameum** A.J.Richards at Foulden W79 and Beeston E14 and **T. ostenfeldii** Raunk. in the King's Lynn area W62.

Section *Hamata*
Many of the plants in this section can be described as weedy and many of the early flowering dandelions have proved to belong here, mostly to **T. pseudohamatum** Dahlst. The other species found are **T. hamatum** Raunk. which is widely distributed; **T. marklundii** Palmgr. found at Thetford W88; **T. quadrans** H.Ollg. recorded at Mundford W89; **T. atactum** Sahlin & Soest and **T. lamprophyllum** M.P.Christ.

Section *Ruderalia*
Most of the over 120 species in this section are weedy and many of them have been introduced from Europe. The following species occur in Norfolk of which those with a * are believed to have been introduced.:- ***T. laeticolor** Dahlst.; ***T. pannucium** Dahlst.; ***T. subexpallidum** Dahlst.; **T. alatum** H.Lindb.; **T. insigne** Ekman ex M.P.Christ. & Wiinst. in Raunk.; ***T. pallescens** Dalhst.; **T. sublaeticolor** Dahlst.; **T. expallidiforme** Dahlst.; **T. subcyanolepis** Dahlst.; **T. croceiflorum** Dahlst.; **T. stenacrum** Dahlst. an endemic species;***T. undulatiflorum** M.P.Christ.; **T. cyanolepis** Dahlst.; **T. ancistrolobum** Dahlst.; ***T. aequisectum** M.P.Christ.; ***T. adiantifrons** Ekman ex Dahlst.; ***T. aequilobum** Dahlst.; ***T. latissimum** Palmgr.; ***T. exsertum** Hagend., Soest & Zevenb.; ***T. exacutum** Markl.; ***T. pannulatum** Dahlst.;. **T. lingulatum** Markl.; **T. cordatum** Palmgr.; **T. sagittipotens** Dalhst. & R. Ohlsen ex G.E.Haglund; **T. ekmanii** Dahlst.; ***T. aurosulum** H. Lindb.; **T. oblongatum** Dahlst.; **T. cophocentrum** Dahlst. an endemic species; ***T. pachymerum** G.E.Haglund; ***T. tanyphyllum** Dahlst.; **T. laciniosifrons** Wiinst. in Raunk.; **T. dahlstedtii** H. Lindb.; ***T. obliquilobum** Dahlst.; **T. latisectum** H.Lindb.; ***T. huelphersianum** G.E.Haglund; **T. subundulatum** Dahlst.; ***T. pectinatiforme** H.Lindb.; ***T. trilobatum** Palmgr.; ***T. planum** Raunk; **T. polyodon** Dahlst.; ***T. xanthostigma** H.Lindb.; **T. longisquameum** H. Lindb. and **T. fasciatum** Dahlst.

Crepis biennis L. **Rough Hawk's-beard**
Winter green annual or biennial, a scarce plant of waste places. Recorded in 1961 at Cranworth W90 and present in 1997, ALB; Mattishall E01 where it grows along a quarter of a mile of road; Newton St. Faith's E21 in 1996, Flora Group Overstrand E24 in 1997, FF and a second colony in 1998, M.Crewe, also at Great Witchingham E01 in 1998, ALB. W1. E6.

■ **C. tectorum** L. **Narrow-leaved Hawk's-beard**
Annual introduced from Europe, a casual noted in Norwich E20 in 1971, ETD.

■ **C. nicaeensis** Balb. **French Hawk's-beard**
Annual introduced from Europe, a casual recorded in Norwich E20 in 1969.

C. capillaris (L.) Wallr. **Smooth Hawk's-beard**
Annual or winter green biennial frequent on all but the heaviest soils, also on walls. W645. E623.

C. vesicaria L. **Beaked Hawk's-beard**
Syn. *C. taraxifolia* Thuill.

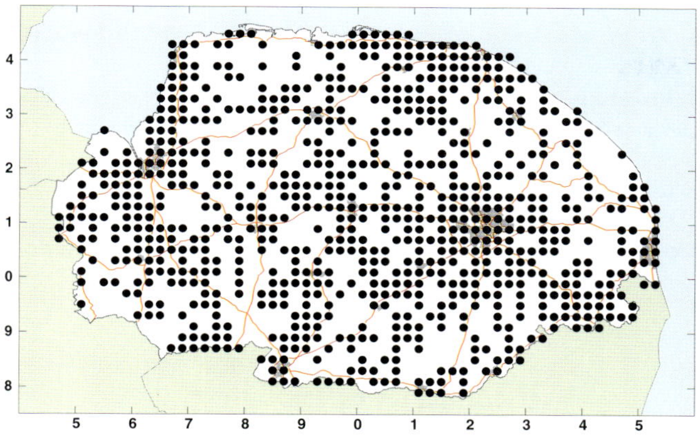

Deciduous perennial, very much a ruderal species and most frequent in urban areas, by main roads and on waste ground but not on walls, (pers. obsv. RMP).

■ **C. setosa** Haller f. **Bristly Hawk's-beard**
Annual introduced from southern Europe, found at Cringleford E10 in 1973, ETD.

■ **C. foetida** L. **Stinking Hawk's-beard**
Annual introduced from Europe, described as rare by Nicholson, 1914 and not seen since.

Pilosella officinarum F.W.Schultz & Sch.Bip.
Syn. *Hieracium pilosella* L. **Mouse-ear Hawkweed**

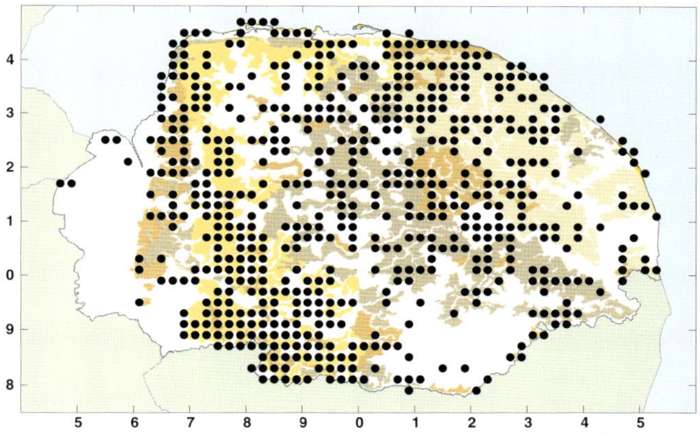

Winter green perennial widespread in permanent grassland on well-drained soils.

P. aurantiaca (L.) F. Schultz & Sch.Bip. **Fox and Cubs**
Syn. *Hieracium aurantiacum* L. (including *H. brunnocroceum* Pugsley) Winter green perennial introduced to gardens from mainland Europe, sometimes well established by roadsides. Our plants are ssp. *carpathicola* (Nägeli & Peter) Soják. W23. E68.

Hieracium.
The genus *Hieracium* is not well represented in Norfolk. Voucher specimens of records made between 1968 and 1985 are in the Herbaria of EL.Swann (NCM)or RP.Libbey (Univ.Leics.). Help with identification of records made during this survey has been given by J.Bevan and by D.McCosh who also read and commented on the text.

H. sabaudum L. **Shrubby Hawkweed**
Syn. *H. perpropinquum* (Zahn) Druce

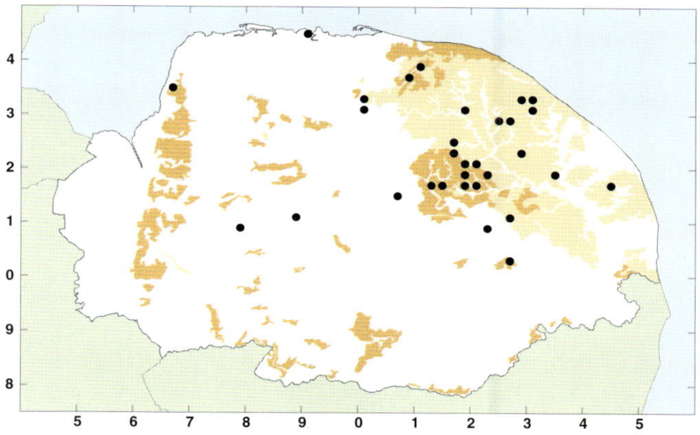

A plant of woods and hedgerows occuring chiefly on acid soils.

H. vagum Jord.
Recorded in the pinewoods at Holkham and Wells-next-the-Sea W84 & 94; at Swaffham W70 in 1995 JEG; Holt Country Park E03 in 1994, D.McCosh and Norwich E20, ETD. W3. E2.

Hieracium vagum - Holkham

H. umbellatum L.

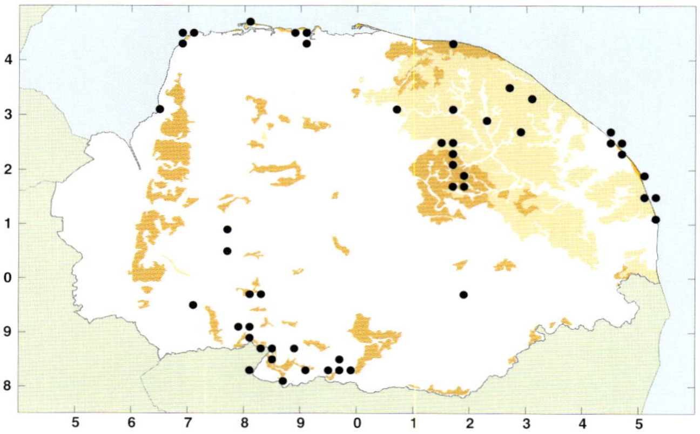

One of our most frequent hawkweeds chiefly found on light, acid soils particularly in shade, also on the acid dunes of East Norfolk.

■ **H. eboracense** Pugsley
Syn. *H. tridentatum* auct. non Fries
A plant of hedgebanks, usually in shade, seen at Whissonsett W92 in 1957 and Brooke E29 in 1974 ETD.

H. trichocaulon (Dahlst) Johannson
Syn. *H. tridentatum* auct. non Fries
Found at Holt Country Park E03 in 1994, D.McCosh; Briston E03 in 1995; Stody E13 in 1986 and Waxham E42 in 1997, all by ALB.
W0. E4.

■ **H. cheriense** Jord. ex Boreau
Syn. *H. lachenalii* C.C.Gmel.
Found once only, in a pine plantation at Ickburgh W89 in 1973 ELS det. PD.Sell.

H. acuminatum Jord.
Syns. *H. strumosum* (Ley ex W.R.Linton) Ley, *H. tunbridgensis* Pugsl.
Occasional on woodland banks and in shade, usually on acid soils. Recorded at Foulden W70 in 1990, R.Tofts; Holt E03 in 1991, D.McCosh; North Wootton W62 in 1995, FS and Fakenham W92 in 1995, RMP.
W3. E1.

H. diaphanum Lindeb.
Syn. *H. anglorum* (Ley) Pugsley
Far commonest on the acid soils of the north Norfolk glacial ridge, but also found at Mundford W79 in 1967, along the Waveney valley at Aldeby E49 and at Spooner Row E09 in 1997, ALB and on sand dunes at Wells-next-the-Sea in 1996, Flora Group.
W2. E7.

■ **H. maculatum** Sm.
Known to have been introduced to Norwich in 1781. A number of records were made on dry banks and walls around the city, E20, the last in 1970.

H. scotostictum N.Hyl.
Introduced from southern and central Europe, recorded in Norwich E20 in 1991, ETD and Taverham E11 in 1992, J.Humphris.
W0. E2.

H. exotericum agg.
Distinctive species found between 1964 and 1979 at six sites on shady banks on the glacial gravels of north Norfolk, recently only at Attlebridge E11 in 1992, J.Humphris det. J.Bevan.
W0. E1.

■ **H. grandidens** Dahlst.
Recorded near Brandon W78 in 1956, J.Rishbeth det. PD.Sell.

H. speluncarum Arv.-Touv.
Introduced from central Europe and long known from a wall in Flitcham W72 where it still grows, R&E.Cross. Its origin is uncertain.
W1. E0.

H. lanatum (L.) Villars
Introduced from mainland Europe and found on Yarmouth North Denes E50, in 1980, A.Lark, still present in 1996, C.Ley det. E.J.Clement.
W0. E1.

Filago vulgaris Lam. — **Common Cudweed**
Syn. *Filago germanica* L. non Huds.

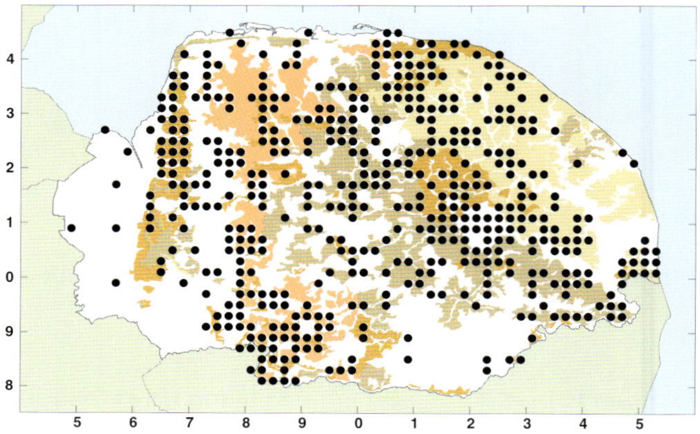

Autumn germinating annual well distributed on well-drained soils particularly common on some set aside fields on dry soils where it can be almost the dominant plant in early summer.

F. lutescens Jord. — **Red-tipped Cudweed**
Syn. *F. apiculata* G.E.Sm. ex Bab.
Chiefly autumn germinating annual of dry, sandy soils, known from five sites since 1950, now only where it was found in W63 in 1997. A Red Data Book species.
W1. E0.

■ **F. pyramidata** L. — **Broad-leaved Cudweeed**
Syn. *F. spathulata* auct. non C.Presl
Spring germinating annual considered rare by Nicholson in 1914, last seen at Narborough Field W70 in 1958.

F. minima (Sm.) Pers. — **Small Cudweed**

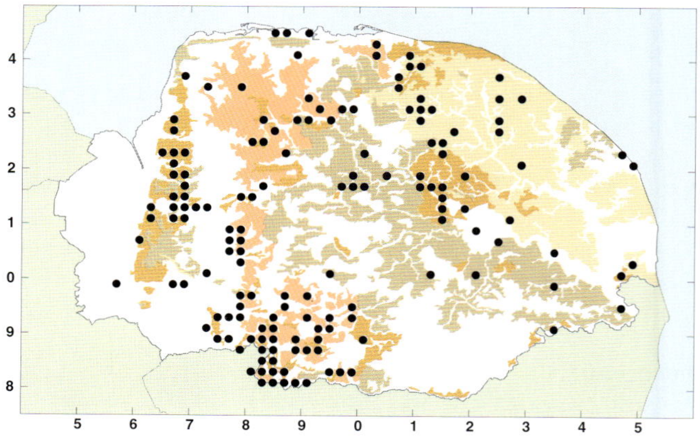

Spring germinating annual more completely confined to dry, sandy soils than *F. vulgaris* and far less frequent.

■ **Antennaria dioica** (L.) Gaertn. — **Mountain Everlasting**
Recorded from heathland at Stratton Strawless E22; Roughton E23, and Swaffham W80, in the 19th century, almost certainly native in all sites but now lost. Much reduced in extent throughout lowland England.

Gnaphalium sylvaticum L. **Heath Cudweed**

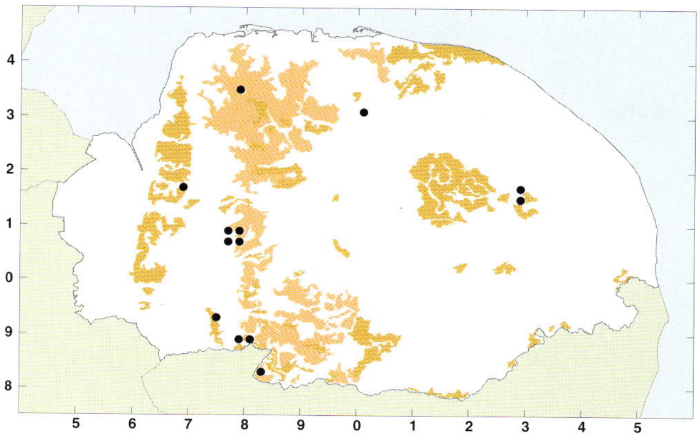

Perennial, confined to a few heathland and afforested sites while absent from many others which appear just as suitable.

G. uliginosum L. **Marsh Cudweed**

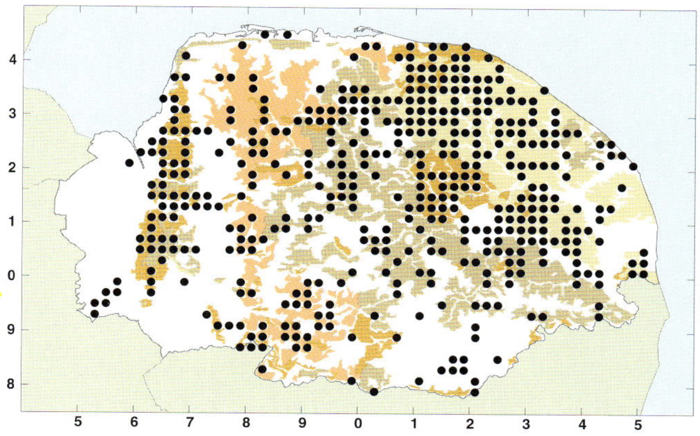

Autumn-germinating annual not infrequent in seasonally damp ground, surprisingly scarce on the boulder clay.

G. luteoalbum L. **Jersey Cudweed**
Autumn-germinating annual recorded from a number of open, sandy sites in the Breckland until 1918, since then only in dune slacks on the north Norfolk coast where it survives in one locality in W84. Red Data Book species. W1. E0.

Gnaphalium luteoalbum - Burnham Overy

Helichrysum bracteatum (Vent.) Andrews **Everlasting Flower**
Annual introduced to gardens from Australia found on dumped topsoil at Browston Green E40 in 1995, I&P.Green. W0. E1.

■ **Inula helenium** L. **Elecampane**
Herbaceous perennial introduced as a herb from western and central Asia, occasionally more or less naturalised but not long lasting. Last recorded at West Runton E14 in 1975, KH.Bull.

I. conyzae (Griess.) Meikle **Ploughman's Spikenard**
Syn. *I. conyza* DC

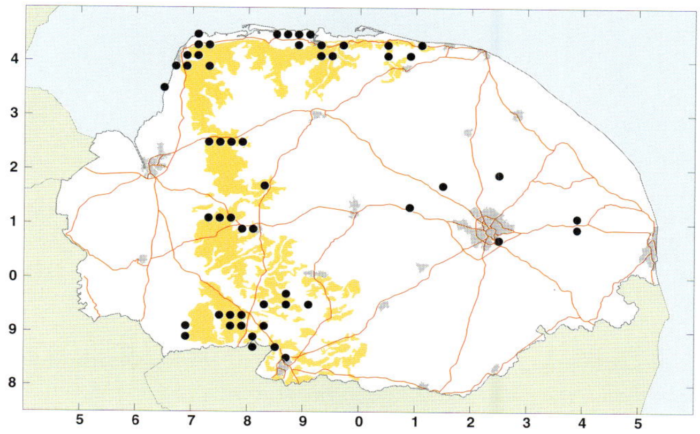

Herbaceous perennial scattered in bare places and short grassland on base rich soils. Has recently colonised newly made road banks which have provided a suitable habitat for the seeds carried along the roads by the draughts from passing traffic. It may have spread similarly in the 19th century along the banks of railway lines where it can still be found.

Pulicaria dysenterica (L.) Bernh. **Common Fleabane**

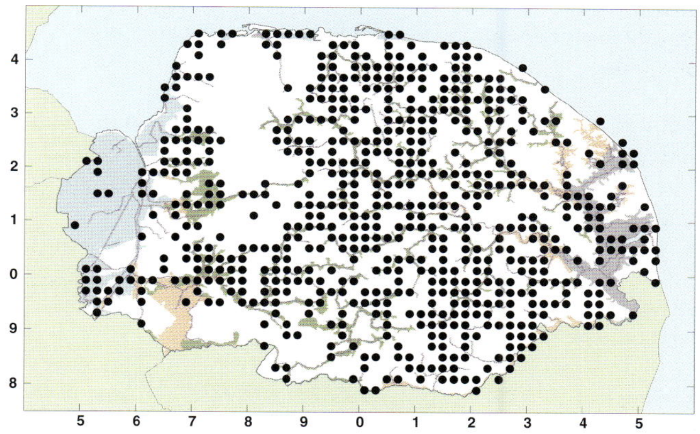

Herbaceous perennial widespread in wet places, ditches, marshes and streamsides.

■ **P. vulgaris** Gaertn. **Small Fleabane**
A nationally rare annual of sandy ground recorded twice in east Norfolk before 1918.

Telekia speciosa (Schreb.) Baumg. **Yellow Ox-eye**
Herbaceous perennial introduced to gardens from mainland Europe. Found near the entrance to Upton Fen E30 in 1992, M.Helliwell and established at Wheatfen E30 for fifty years, RWE. W0. E2.

Solidago virgaurea L. **Goldenrod**

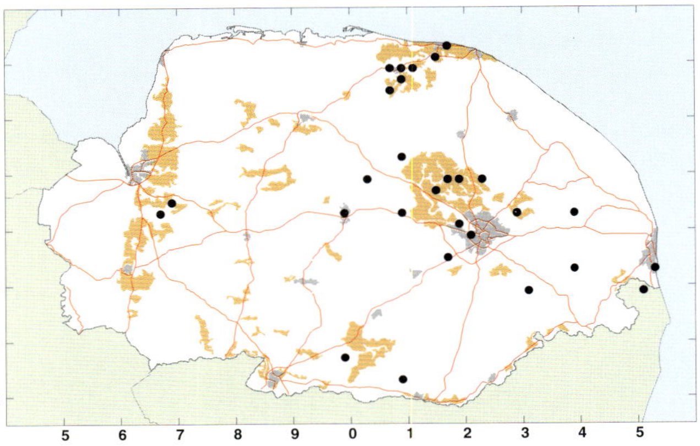

Herbaceous perennial, a scarce plant of moist grassland and woodland edges on acid soils.

S. canadensis L. **Canadian Goldenrod**
Herbaceous perennial introduced to gardens from North America, like *S. gigantea* now well established on waste ground and roadside verges, usually from discarded garden plants. W37. E69.

S. gigantea Aiton **Early Goldenrod**
Herbaceous perennial introduced to gardens from North America, now well established on waste ground and roadside verges, usually from discarded garden plants but also occurring well away from public access and then perhaps by seed. W36. E10.

Aster ssp
For many years most naturalised Michaelmas daisies were included in identification keys as either *A. novi-belgii* or *A. nova-angliae.* Recently, more close study has shown most to be of hybrid origin. One problem during the collection of records for this Flora has been that the record cards used were still listing *A. novi-belgii* as if it was the only escaped taxon, so that this species may still have been over-recorded.

■ **Aster novae-angliae** L. **Hairy Michaelmas Daisy**
Herbaceous perennial introduced to gardens from North America, three records between 1968 and 1981, two on refuse tips.

A. x versicolor Willd. **Late Michaelmas Daisy**
A. laevis x A. novi-belgii
Semi-winter green perennial introduced to gardens from North America, well established in some sites. W8. E7.

A. novi-belgii L. **Confused Michaelmas Daisy**
Semi-winter green perennial introduced to gardens from North America and frequently confused with its hybrids with *A. laevis* and *A. lanceolatus.* Many old records were from refuse tips.
W0. E13.

A. x salignus Willd. **Common Michaelmas Daisy**
Willow-leaved Michaelmas Daisy
(A. novi-belgii x A. lanceolatus)
Semi-winter green perennial introduced to gardens from North America and, together with the following species which is one of its parents, the commonest of the escaped Michaelmas Daisies. First noted at Horstead E11 in 1971 by EAE and recorded regularly since then. W16. E3.

A. lanceolatus Willd. **Narrow-leaved Michaelmas Daisy**
Semi-winter green perennial introduced to gardens from North America now well established by roadsides and on waste ground.
W6. E4.

A. tripolium L. **Sea Aster**

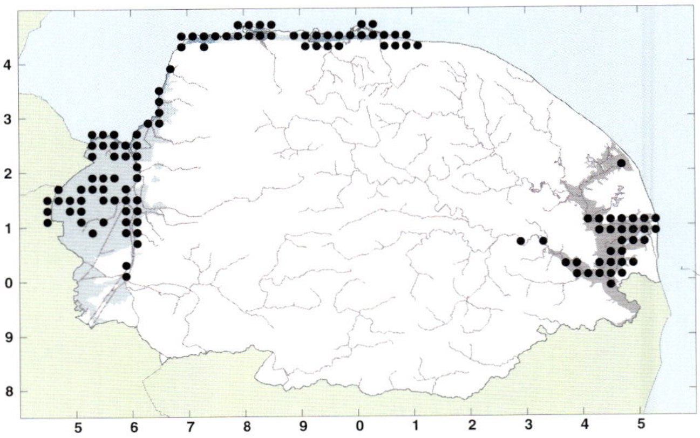

Perennial confined to salt marshes and muddy soils on sea banks. Found frequently on land which has or has had saline influence in the past. Forms on salt marshes are frequently the rayless form, var. *discoidea* while by ditches and banks away from regular inundation, the rayed var. *tripolium* is usual. This appears to have been so for many years as a note to this effect was made by ELS in 1937, though it does seem that var. *discoidea* may have become more frequent in recent years.

Aster tripolium var. **discoidea - Cley**

Felicia petiolata (Harv.) N.E.Br.
Evergreen sub-shrub introduced to gardens from South Africa. Long established at Saham Toney W90 RMP, 1995. W1. E0.

Aster tripolium var. **tripolium - Breydon North Bank**

Erigeron glaucus Ker Gawl. **Seaside Daisy**
Herbaceous perennial introduced from North America, and occurring as a garden escape at Burnham Market W84; Norwich E20 in 1990, ETD; Scratby E51 in 1995, Flora Group and Swanton Novers E03 in 1997, K&GB. W1. E4.

E. karvinskianus DC. **Mexican Fleabane**
Syn. *E. mucronatus* DC
Herbaceous perennial introduced to gardens from Mexico in the 19th century and often well naturalised on old walls particularly near the coast. Known on the churchyard wall at Burnham Deepdale W84 for at least 40 years. W3. E4.

E. acer L. **Blue Fleabane**

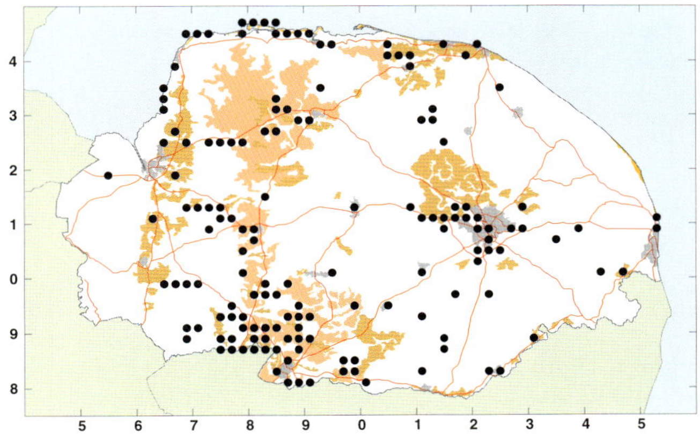

Herbaceous perennial of well-drained, sandy soils, both inland and on coastal sand dunes, usually where conditions are not too acid.

Conyza canadensis (L.) Cronquist **Canadian Fleabane**
Syn. *Erigeron canadensis* L.

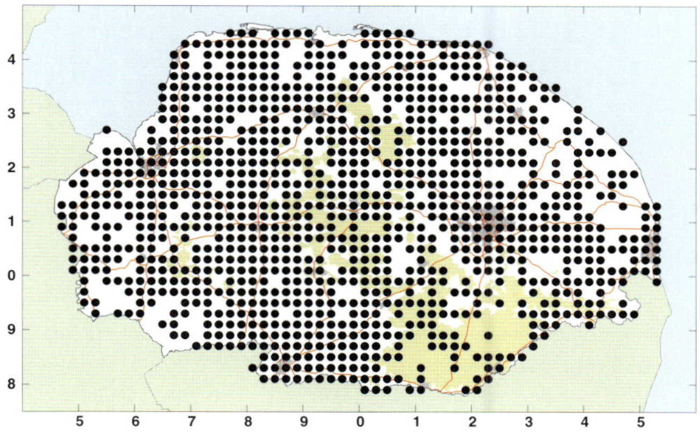

Annual, often overwintering, introduced from North America. A widespread weed first recorded in Norfolk in 1882 near Thetford W88 and now frequent on well-drained soils. Conspicuous on some set aside fields and forest clear-fells, it is also common on walls, especially in towns. Has only recently reached many inland villages where it is being noted as a new garden weed.

C. sumatrensis (Retz.) E. Walker **Guernsey Fleabane**
Annual from South America, recorded from King's Lynn W62 in 1993, RMP and Gt Yarmouth E50 in 1995. W1. E1.

■ **C. bonariensis** (L.) Cronquist **Argentinian Fleabane**
Annual from South America, recorded once, on the refuse-tip at Harford, Norwich in 1959.

■ **Callistephus chinensis** (L.) Nees **Chinese Aster**
Annual introduced to gardens from China, an occasional escape recorded twice on refuse tips, at Harford, Norwich E20 in 1970s and Beetley W91 in 1971 and 1974.

Olearia x haastii Hook. f. **Daisy-bush**
Evergreen shrub from New Zealand, a rare throwout from gardens noted on Beeston Common E14 in 1997, FF. W0. E1.

O. macrodonta Baker **New Zealand Holly**
Evergreen shrub introduced to gardens from New Zealand. One plant was found in a hedgerow at Friary Hills, Blakeney E04 in 1998, KKH. W0. E1.

Bellis perennis L. **Common Daisy**
Winter green perennial almost ubiquitous where there is habitation. W656. E695.

Tanacetum parthenium (L.) Sch. Bip. **Feverfew**
Syn. *Chrysanthemum parthenium* (L) Bernh.
Semi-winter green perennial introduced as a herb from the Balkans. In 1597 Gerarde described it as being 'very good for them that are giddie in the head', today it is used by some for migraine. It is widely naturalised throughout the county, usually close to habitation, and is almost a weed in some old gardens. W211. E378.

T. vulgare L. **Tansy**

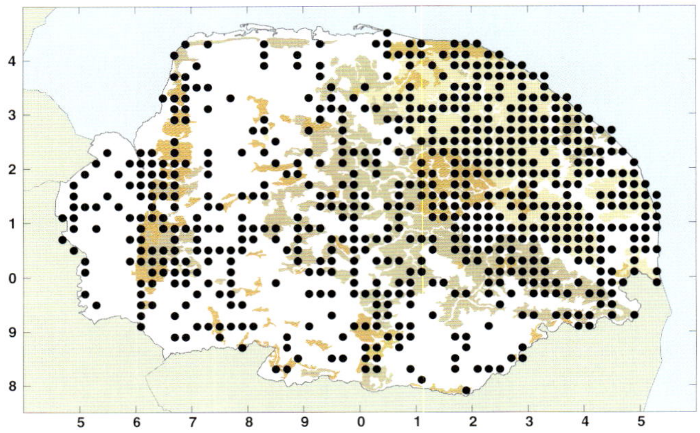

Winter green perennial, widely distributed on heavier, but not wet soils, often by roadsides.

Seriphidium maritimum (L.) Polj. **Sea Wormwood**
Syn. *Artemisia maritima* L.

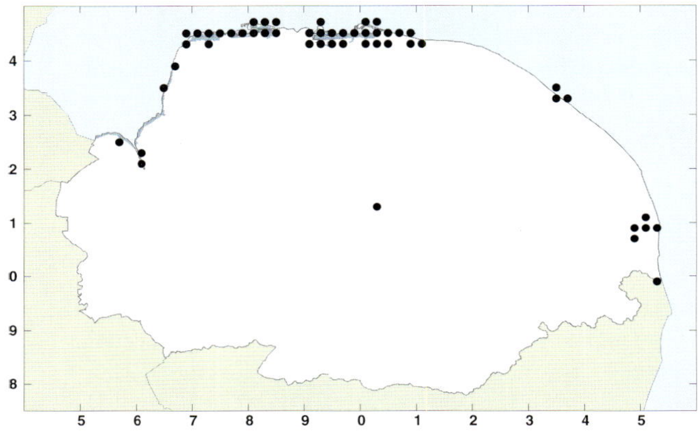

Herbaceous perennial frequent on raised areas in salt marshes and also on sea-walls and nearby embankments. Recorded on a roadside, by the A12 at Hopton-on-Sea E59 in 1997 and on the North Tuddenham bypass, E01 in 1998, ALB.

Seriphidium maritimum - Burnham Overy

Artemisia vulgaris L. **Mugwort**
Herbaceous perennial abundant through the county on road verges and waste ground. W673. E698.

■ **A. verlotiorum** Lamotte **Chinese Mugwort**
Herbaceous perennial, a casual from China, found at Old Hunstanton W64 between 1950 and 1954, then its most northerly station in the world.

A. absinthium L. **Wormwood**
Semi-winter green perennial once grown as a vermifuge, its more silvery-leaved forms now widely grown as foliage plants. A rare escape on rough ground and waste places. W3. E9.

■ **A. biennis** Willd. **Slender Southernwood**
Annual to biennial introduced from Asia and America, a rare casual found twice in 1920 and not since.

A. campestris L. **Field Wormwood, Breckland Mugwort**
Semi-winter green perennial, one of the specialities of Breckland, known from Cranwich W79 in the 1950s; Lynford W89 until 1981; Ickburgh W89 until 1989; at Santon W88 in 1987 where it may have been introduced and also at Weeting W78, a well recorded introduction. A plant in a lay-by at East Tuddenham E01 in 1992 and 1993, ALB can have been no more than a casual though the source of the seed must be a mystery. As a native plant now only at Thetford W88. W1. E1.

Artemisia campestris - Thetford

Achillea filipendulina Lam.
Herbaceous perennial introduced to gardens from Asia and well established on a churchyard wall at Rushford W98 in 1995, KAB and on waste ground at Hingham E00 in 1996, ALB. W1. E1.

A. ptarmica L. **Sneezewort**

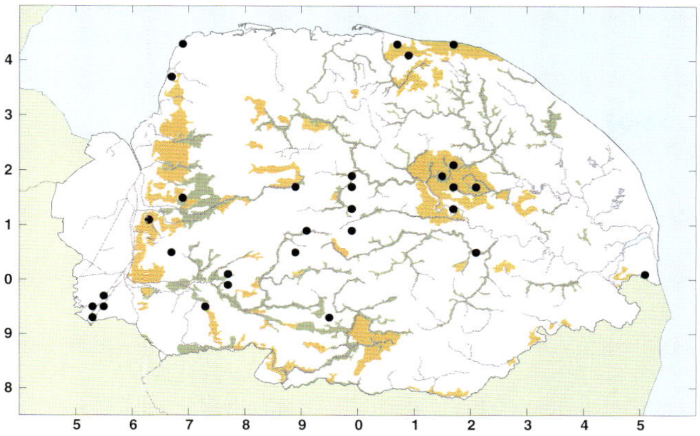

Herbaceous perennial found in moist, neutral to acid grassland, never common.

A. millefolium L. **Yarrow**
Winter green perennial of grassland, ubiquitous. Plants with bright pink flowers are occasionally found. W695. E707.

■ **Chamaemelum nobile** (L.) All. **Chamomile**
Winter green perennial recorded at Reymerston E00 in 1964 by ALB, not seen since. No later records have been confirmed although it may occasionally occur as a garden throwout.

Anthemis arvensis L. **Corn Chamomile**

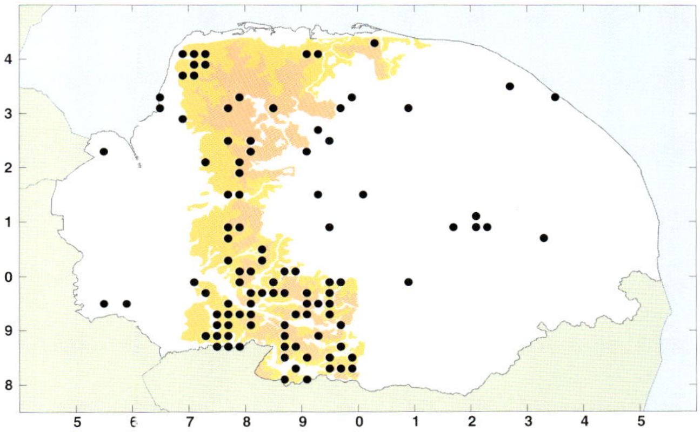

Autumn and spring germinating annual frequent only on light, well-drained soils, rapidly decreasing nationally.

A. cotula L. **Stinking Chamomile**

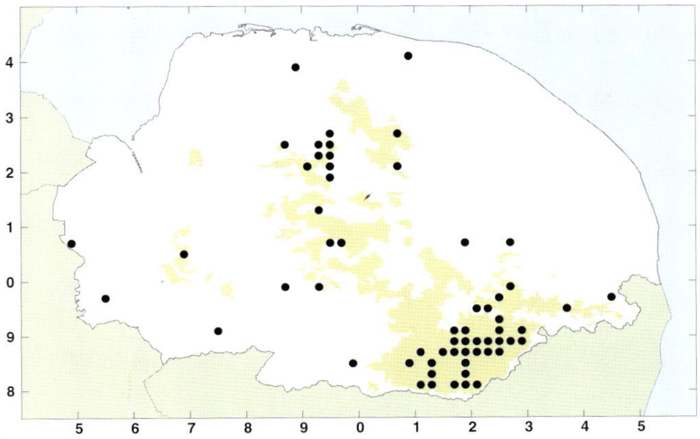

Autumn and spring germinating annual commonest on heavier soils, particularly the boulder clay.

A. tinctoria L. **Yellow Chamomile**
Semi-winter green biennial to perennial from mainland Europe, occasionally found as an escape as at Bradenham W91 in 1993, D&BM; Heacham W63 in 1993, PRC and Hempton W92 in 1994, SF. W3. E0.

Chrysanthemum segetum L. **Corn Marigold, Golds**

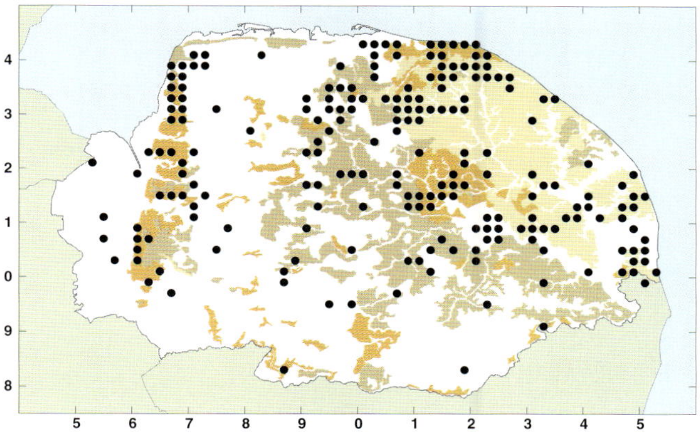

Annual weed of acid arable land, local in its occurrence, but sometimes frequent within restricted areas. Considered a sign of poor land.

Chrysanthemum segetum - Grimston

■ **C. coronarium** L. **Annual Chrysanthemum**
Annual introduced to gardens from mainland Europe, found as a casual near Norwich E20.

Leucanthemella serotina (L.) Tzelev **Autumn Oxeye**
Syn. *Chrysanthemum uliginosum* (Waldst. & Kit. ex Willd.) Pers.
Herbaceous perennial introduced to gardens from southern Europe noted at Whitlingham E20 in 1995, A.Lewis and Scottow Row E22 in 1997, ALB where it was growing with dumped garden rubbish, also at Bacton E33 in 1998, ALB. W0. E3.

Leucanthemum vulgare Lam.
Ox-eye Daisy, Wild Marguerite
Syn. *Chrysanthemum leucanthemum* L

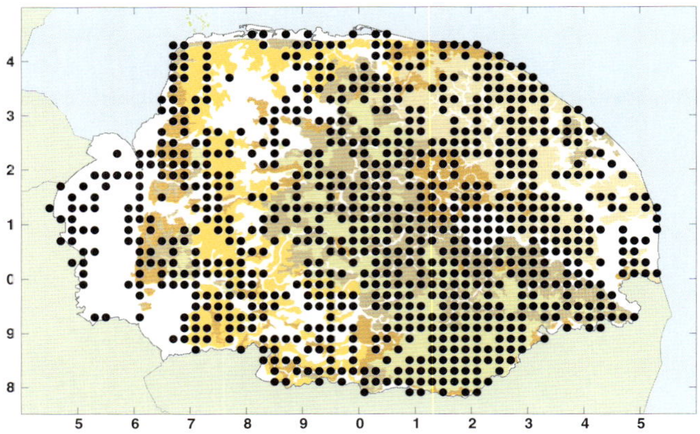

Semi-winter green perennial frequent on base rich and fertile soils, also introduced recently along some new roads.

L. x superbum (Bergmans ex J.W.Ingram) DH.Kent.
Shasta Daisy
Syn. *Chrysanthemem maximum* auct. non Ramond.
Semi-winter green perennial of garden origin, fertile in spite of its hybrid origin and established in waste places and roadsides, notably on earthy cliffs. W4. E40.

Matricaria recutita L. **Scented Mayweed**

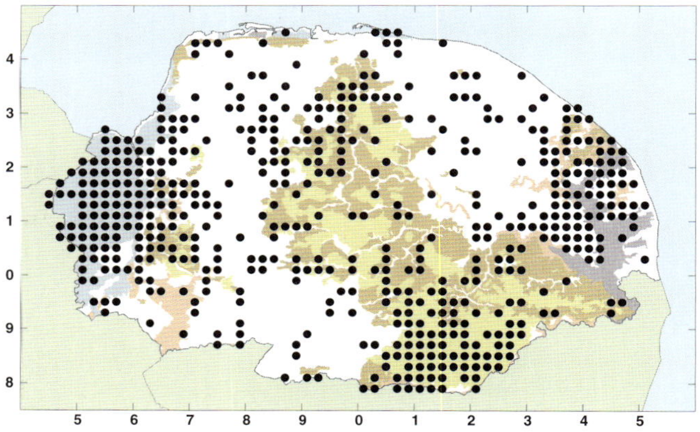

Autumn and spring germinating annual, a weed of heavier and wet soils, abundant where it occurs, particularly in Fenland, but local in its distribution.

M. discoidea DC. **Pineapple Weed**
Syn. *M. matricarioides* (Less.) Porter, per parte, nom. illeg.
Spring germinating annual probably originally from Asia but which appears to have reached Britain from Oregon in the middle of the 19th century. First recorded in Norfolk in 1904, and by 1923 already described as 'all over the county'. W696. E695.

Tripleurospermum inodorum (L.) Schultz-Bip.
Scentless Mayweed
Syn. *T. maritimum* ssp. *inodorum* (L.) Hyl. ex Vaar.
Autumn-germinating annual, ubiquitous. A rayless form has been observed and a double-flowered form appeared at Binham W93 in 1992, R.Scott. W690. E695.

T. maritimum (L.) W.D.J.Koch **Sea Mayweed**
Biennial to perennial found on beaches and waste ground near the sea. Recorded at sites scattered around the coast from Snettisham, W63 to Breydon Water E40. W3. E9.

Cotula coronipifolia L. **Buttonweed**
Annual to evergreen perennial introduced from South Africa or New Zealand. Found on the cattle-trampled margins of brackish ditches at Cobholm E50 in 1996, D.Green. W0. E1.

Senecio cineraria DC **Silver Ragwort**
Syns. *S. bicolor* (Willd.) Tod, *S. maritimus* Koidz.
Evergreen perennial introduced from the central and western Mediterranean, an occasional escape on to walls and waste ground, especially near the sea. W11. E11.

S. x albescens Burb. & Colgan
(*S. cineraria x S. jacobaea*)
A naturally occurring hybrid noted on a wall at Fring W73, in 1989, KAB; Snettisham W63 in 1989, PRC; Yarmouth E50 in 1995, I&P.Green; Glandford E04 in 1996, R.Barrett and at Marham W70 in 1996, RMP. W3. E2.

S. fluviatilis Wallr. **Broad-leaved Ragwort**
Herbaceous perennial introduced to gardens from mainland Europe, planted but now abundantly naturalised at Wheatfen E30 and also recorded from Hoveton E31 and Martham E42 in 1994, J.Parmenter. W0. E3.

■ **S. paludosus** L. **Great Fen Ragwort**
This herbaceous perennial, now almost extinct in Great Britain, was formerly a Norfolk plant. A.Bennet in 1911-12, elucidated much of its history. The plant was described in 1714 as in great plenty along the Old Podike (the dyke running across the Fenland from Outwell to Stowbridge W50), but these colonies were presumably soon lost through drainage. Druce, 1919, noted that it was found near Filby Broad E41 in 1876, but no record after that date can be verified.

S. jacobaea L. **Ragwort, Cankerweed, Gantweed.**
Semi-winter green perennial, abundant in grassland and road verges throughout the county. W699. E702.

■ **S. x ostenfeldii** Druce
(*S. jacobaea x S. aquaticus*)
Found once, at Castle Acre W81 in 1969, growing with both parents, ELS.

S. aquaticus Hill **Marsh Ragwort**

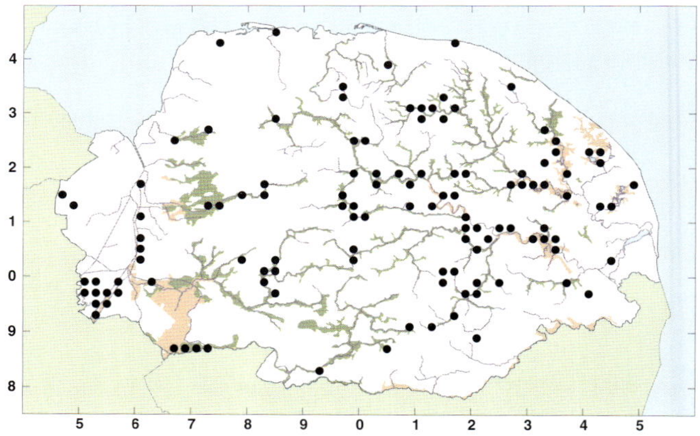

Semi-evergreen biennial to perennial locally frequent in seasonally flooded and wet pastures.

S. erucifolius L. **Hoary Ragwort**

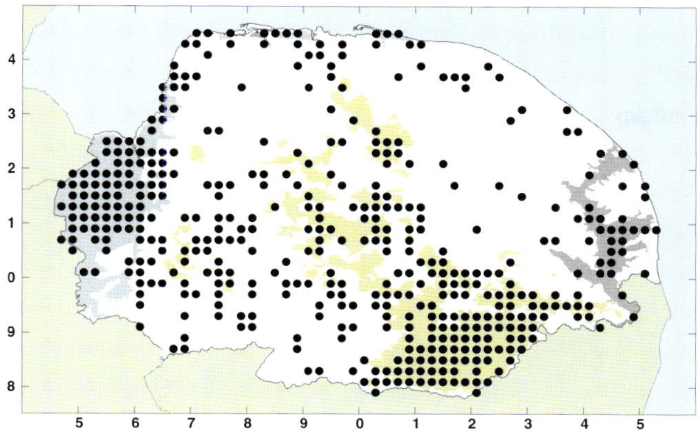

Semi-evergreen perennial found chiefly on heavy, fertile soils.

S. squalidus L. **Oxford Ragwort**

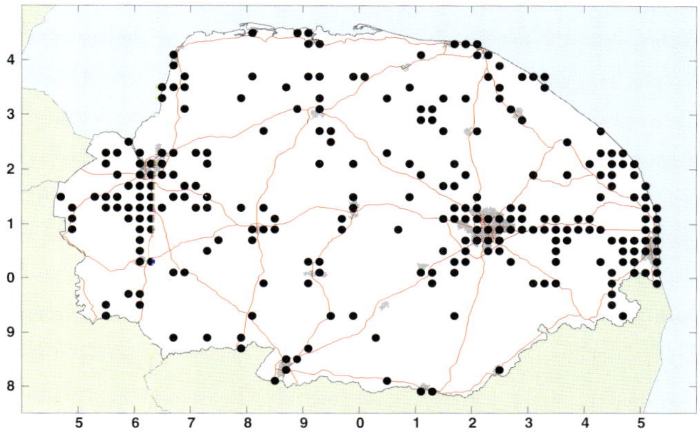

Winter green annual to perennial introduced from southern Europe. It was first recorded in Britain in 1794 and apparently introduced to Norfolk in 1836 (EAE). It is still spreading away from urban areas where it is a frequent weed, but is slow to establish far from habitation. Frequent on urban walls, one plant grew on the battlements at the top of St Margaret's church tower in King's Lynn W61 in 1990, GB.

Senecio squalidus - Norwich

S. x baxteri Druce
(*S. squalidus x S. vulgaris*)
An occasional hybrid usually occurring in urban areas, recorded at Hunstanton W64 in 1973, PRC; Norwich E20 in 1990, ETD; Strumpshaw E31 in 1991, SP; Dereham W91 in 1993, ALB and Rackheath E21 in 1997, RM.Leaney. W2. E3.

■ **S x subnebrodensis** Simonk.
(*S. squalidus x S. viscosus*)
A hybrid recorded once, at Snettisham W63 in 1979 with both parents, ELS.

S. vulgaris L. **Groundsel**
Annual of cultivated ground and waste places germinating throughout the year when temperatures are suitable, ubiquitous. Var. *hibernicus* Syme with ray florets is occasionally found. W700. E704.

S. sylvaticus L. **Heath Groundsel**

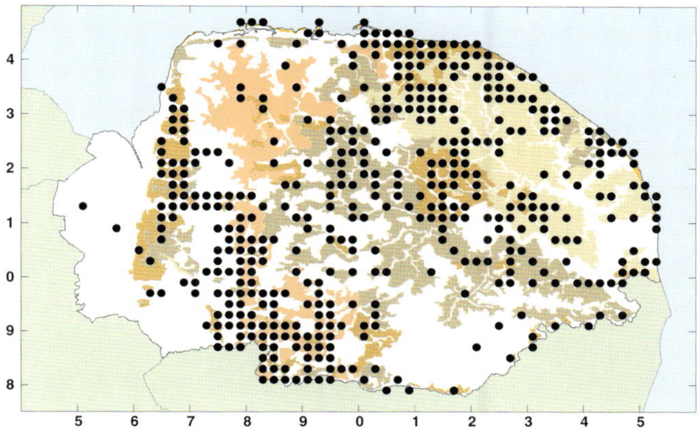

Annual growing on light, sandy soils, abundant in some newly planted pine plantations.

S. viscosus L. **Sticky Groundsel**

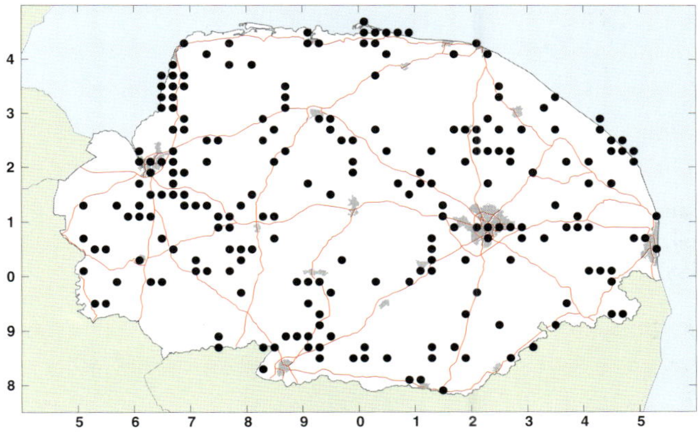

Annual of light soils, perhaps native to shingle beaches but now found chiefly on waste ground in urban areas and on disused railway tracks.

■ **Tephroseris palustris** (L.) Fourr.
Marsh Fleabane, Trumpets
Syn. *Senecio palustris*
Biennial to perennial now extinct in Great Britain. Once frequent in both Fenland and the Broads. Both EL.Swann and EA.Ellis studied the past records of this species and found a great deal of confusion, both with *S. aquaticus* and with the sources of both herbarium specimens and their labels. The Fenland stations may all have been lost before 1800, but in the Broads, where it seems to have been commonest near the coast, it was known to have survived at Fleggburgh, E41 until 1898.

Brachyglottis 'Sunshine' **Shrubby Ragwort**
(Formerly grown as *Senecio greyi*)
A garden cultivar derived by hybridisation between two New Zealand species. Increasingly planted and a persistent shrub on waste ground at Mundesley E33 in 1992, ALB and at Brooke E29 in 1998, Flora Group. W0. E3.

Doronicum pardalianches L. **Leopard's Bane**
Herbaceous perennial from western Europe, sometimes escaping or surviving from material thrown out from gardens, thriving in semi-shaded sites as at Shropham W99 where it has been known for almost 50 years. W7. E24.

D. columnae Ten. **Eastern Leopard's Bane**
Herbaceous perennial introduced from south-east Europe, recorded as an escape at Sedgeford W73 in 1993, PRC. W1. E0.

Tussilago farfara L. **Coltsfoot**

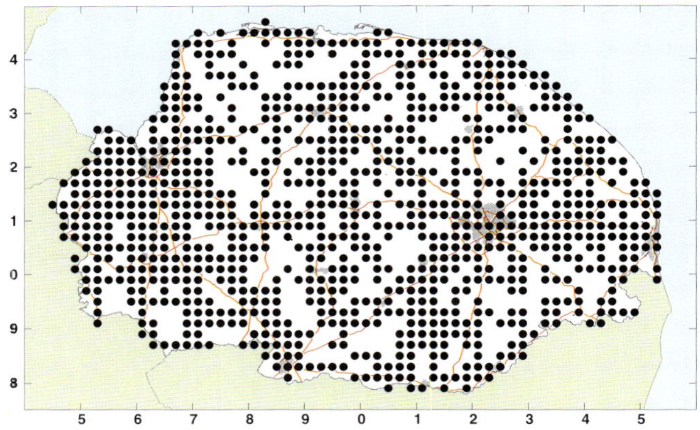

Herbaceous perennial commonest on moist and fertile soils, hence its abundance in the silt fens.

Petasites hybridus (L.) P.Gaertn., B.Mey. & Scherb.
Butterbur, Wild Rhubarb

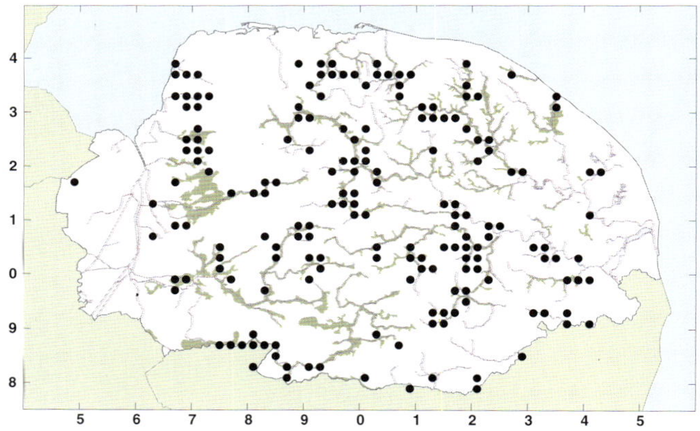

Herbaceous perennial possibly native in moist places where it can become dominant, but behaving more as an introduced species.

P. fragrans (Vill.) C. Presl **Winter Heliotrope, Cherry Pie**

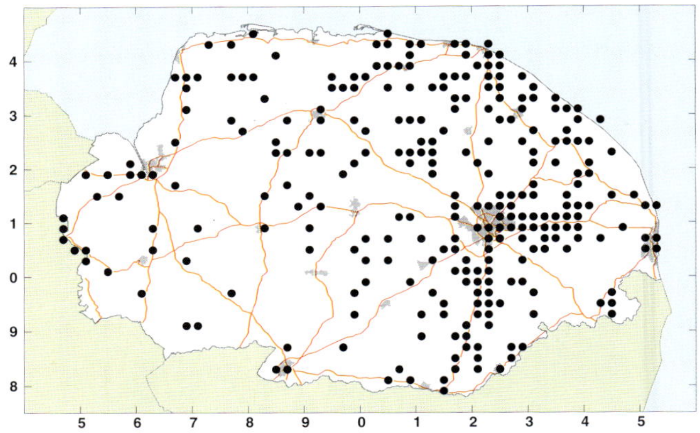

Herbaceous perennial introduced to gardens from North Africa, an occasional and very invasive throwout on to roadside verges, most frequent on heavier soils.

■ **Hemizonia pungens** Torr. & Gray **Spikeweed**
Annual from North America, a rare casual with grain recorded twice, at Salthouse E04 and at Ickburgh W89, both in 1967.

■ **Madia sativa** Mol.
Annual, native to Chile and cultivated in Europe for its oil-rich seeds which provide a substitute for olive-oil. Several plants were found on Weybourne rifle-range in 1950.

Calendula officinalis L. **Pot Marigold**
Annual or short-lived evergreen perennial from mainland Europe, an escape from gardens and sometimes over-wintering but not long-lasting. In the past most frequently seen on refuse tips. W21. E117.

Ambrosia artemisiifolia L. **Ragweed**
Spring-germinating annual from North America where it is one of the main causes of hayfever. A casual found in Foxley Wood E02 in 1970, ALB where pheasants were fed; in a potato field at Shouldham W90 in 1989, GM; at Watlington W61 in 1995, RMP and Stanhoe W83 in 1997, KB. W3. E1.

A. trifida L. **Great Ragweed, Buffalo Weed**
Spring-germinating annual from North America, a casual found at Cringleford E10 in 1983, ETD and Marham W70 in 1990, HW. W1. E0.

■ **Iva xanthifolia** Nutt. **Marsh-elder**
Spring-germinating annual from North America, recorded four times between 1955 and 1970.

■ **Xanthium spinosum** L. **Spiny Cocklebur**
Spring-germinating annual from South America recorded four times between 1960 and 1980.

■ **Guizotia abyssinica** (L.f.) Cass. **Niger**
Spring-germinating annual from eastern Africa recorded four times as a birdseed alien or on refuse tips, the last time at Blackborough End W61 in 1981, RPL & ELS.

Rudbeckia laciniata L. **Black-eyed Susan, Golden Glow**
Herbaceous perennial introduced from North America, an occasional throwout with garden rubbish. At West Tofts, W89 in the Stanford Training Area where it is presumably a relic of a garden deserted when the military occupation began in 1942, it formed a clump 3 x 4m, by 1997, ALB. Also at Watlington W61 in 1993, RMP and Lenwade E11, J.Humphris. W2. E1.

■ **R. hirta** L. **Bristly Coneflower**
(including *R.bicolor* Nutt.)
Annual noted on Harford Tip, Norwich E20 in 1978, ETD.

Helianthus annuus L. **Sunflower**
Spring-germinating annual from North America grown as a crop for oil and for pheasant feed, sometimes regenerating naturally. Also found as a birdseed alien, particularly on waste ground and refuse tips. W12. E11.

■ **H. decapetalus** L.
Herbaceous perennial from North America, recorded once, on Harford tip, Norwich E20 in 1978, ETD.

H. x laetiflorus Pers. **Perennial Sunflower**
(*H. pauciflorus x H. rigidus*)
Herbaceous perennial derived from North American parents, found in Norwich E20 in 1992, ETD; on the disused railway line at Hillington W72 in 1996, JEG & HW and at Redenhall E28 in 1994, ALB. This includes records of its parents which are rarely if ever found in Britain. W1. E2.

H. tuberosus L. **Jerusalem Artichoke**
Herbaceous perennial. Not infrequent on old and disused allotments and house sites, but not self-sowing so always as a relic of cultivation. W8. E16.

Galinsoga parviflora Cav. **Gallant Soldier**

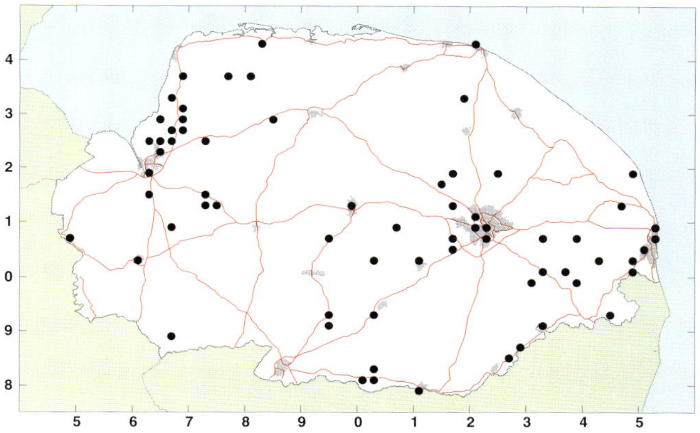

Spring-germinating annual from South America, in west Norfolk still a largely urban-based weed, scarcely increased in range since Petch & Swann's Flora of 1968, but in the eastern half found as a field weed in horticultural crops, potatoes and on setaside.

G. quadriradiata Ruiz & Pav. **Shaggy Soldier**
Syn. *G. ciliata* (Raf.) S.F.Blake

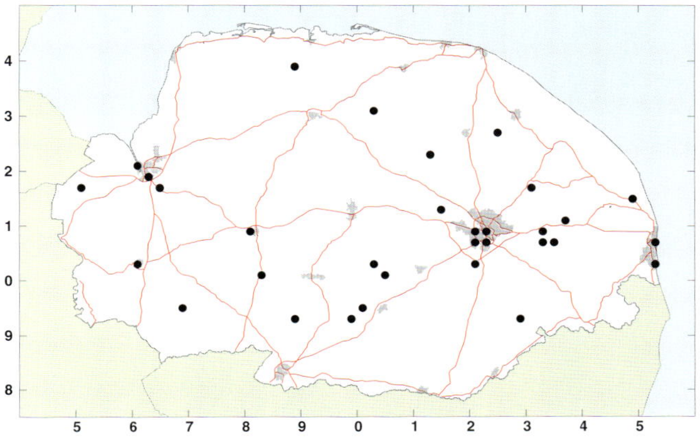

Spring-germinating annual from South America, widely scattered in fields and gardens, but scarce. Several recent records are from nurseries which suggests it may soon be found over a wider area!

Bidens cernua L. **Nodding Bur-marigold**

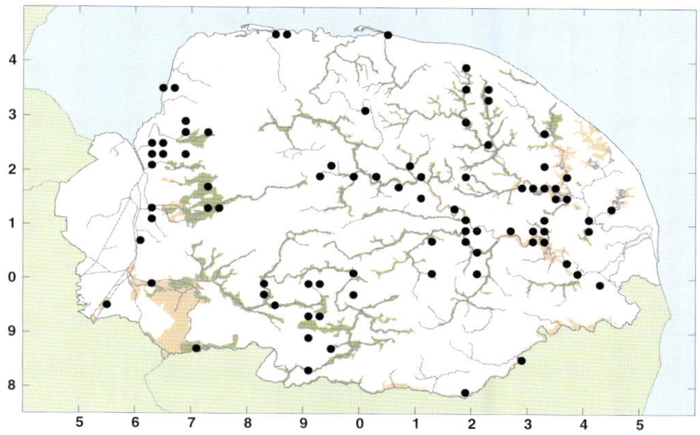

Spring germinating annual found in a wide range of damp habitats. The rayless form is normal while var. *radiata* D.C. with large ray florets is much more attractive but far scarcer.

Bidens cernua var. **radiata - Sparham**

B. tripartita L. **Trifid Bur-marigold**

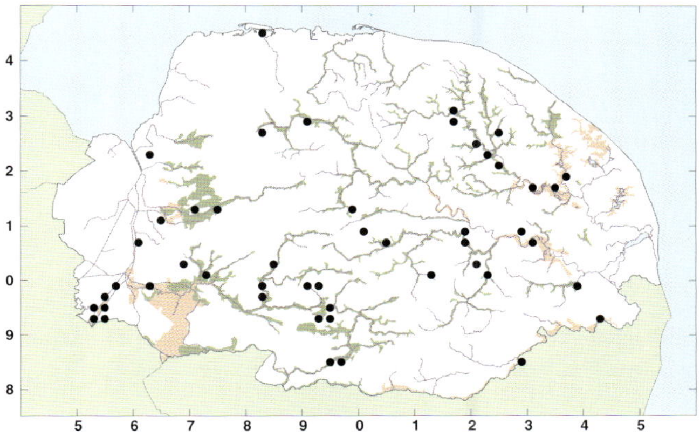

Spring germinating annual in a similar range of habitats but less frequent than *B. cernua* except in the Ouse Washes where it is the commoner of the two species.

■ **B. pilosa** L. **Black-jack**

A native of South America found at Northwold W79, in 1984 in a box of bedding plants from a nursery.

Cosmos bipinnatus Cav. **Cosmea, Cosmos**

Spring-germinating annual introduced to gardens from Mexico to southern USA. An occasional escape on refuse tips and waste ground. Two old records, more recently on the shingle bank at Heacham W63, PRC and Norwich E20, ETD, both in 1990. W1. E1.

Tagetes patula L. **French Marigold**

Spring-germinating annual introduced to gardens from Central America, occasional on refuse tips and waste ground. W2. E1.

Gaillardia x grandiflora Van Houtte

(*G. aristata x G. pulchella*)

Annual or short-lived winter green perennial derived in gardens from two North American parents. A number of plants were seen in the rough on the golf course at Thetford W88 in 1995, DM. W1. E0.

Eupatorium cannabinum L. **Hemp Agrimony**

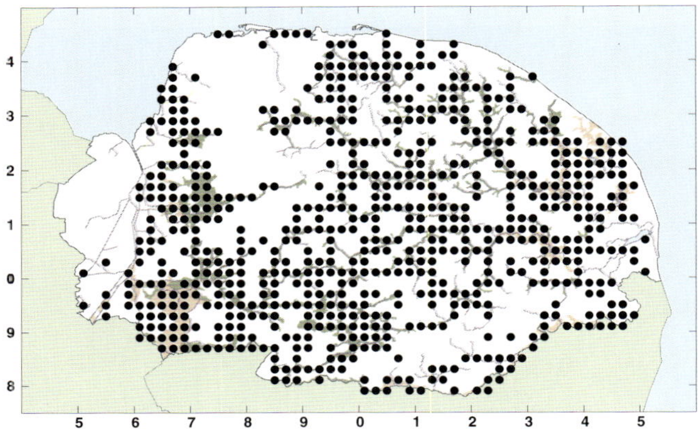

Herbaceous perennial found in moist ground and on stream and river banks.

BUTOMACEAE

Butomus umbellatus L. **Flowering Rush**

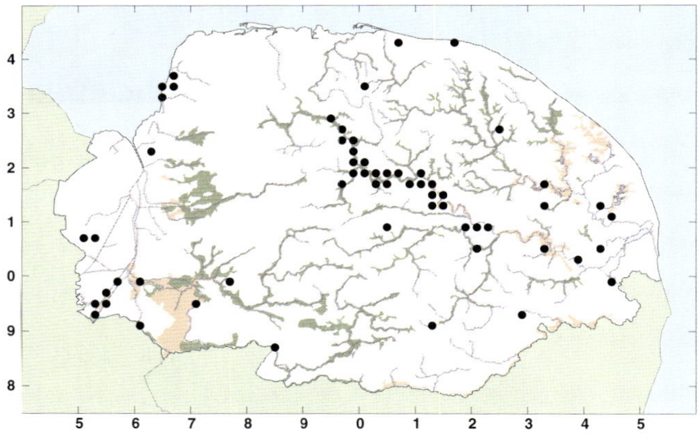

Herbaceous perennial found in ponds, ditches and slow-moving rivers, not present in strongly calcareous waters, frequent only in the upper reaches of the Wensum. Generally a shy flowerer so perhaps sometimes overlooked. May be introduced in a few ponds.

ALISMATACEAE

Sagittaria sagittifolia L.

Arrowhead, Serpent's Tongue, Broad Arrow

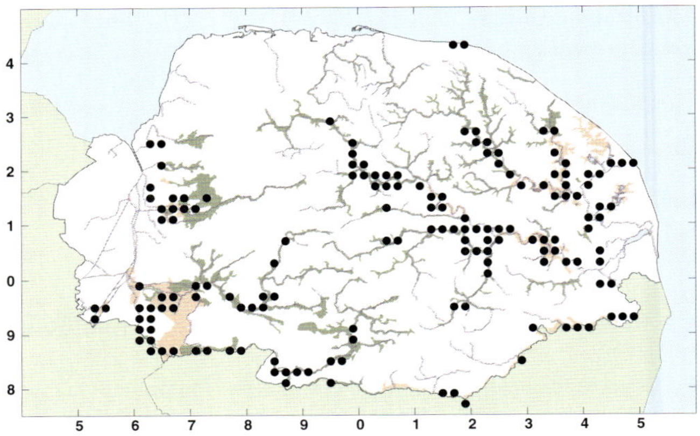

Herbaceous perennial frequent in rivers and dykes, absent from shallow streams and brackish water.

Butomus umbellatus - Snettisham

Baldellia ranunculoides (L.) Parl. **Lesser Water-plantain**

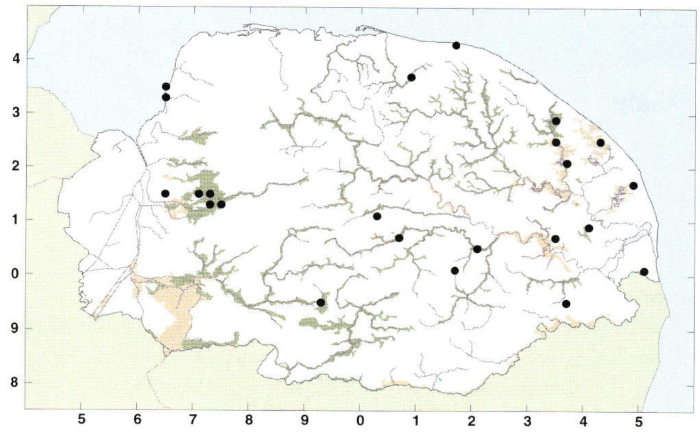

A scarce plant described by Nicholson, 1914 as rather common giving wet heaths as its habitat. Now scarce and found scattered in pools on marshes and in ditches, usually in small numbers though virtually dominant in a ditch at Blackborough End W61 in 1992, HW.

■ **Luronium natans** (L). Raf. **Floating Water-plantain**
Herbaceous perennial, not considered native in Norfolk. It was found in Calthorpe Broad E32 in 1974, R.Driscoll and at Potter Heigham E42 in 1979, RPL, ELS & ETD.

Alisma plantago-aquatica L. **Water-plantain**

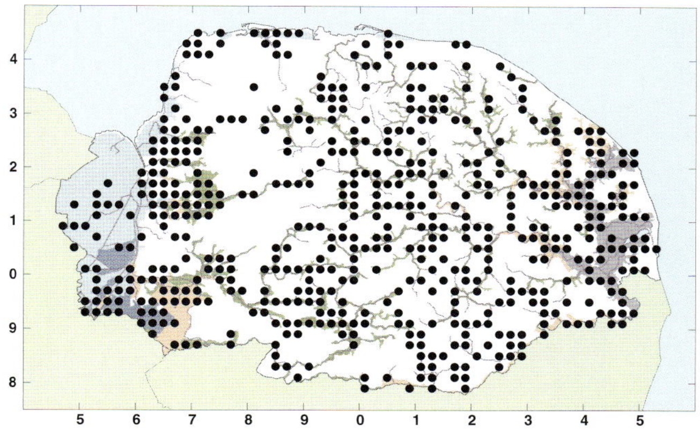

Herbaceous perennial frequent in ponds, streams, dykes and by rivers.

A. lanceolatum With. **Narrow-leaved Water-plantain**

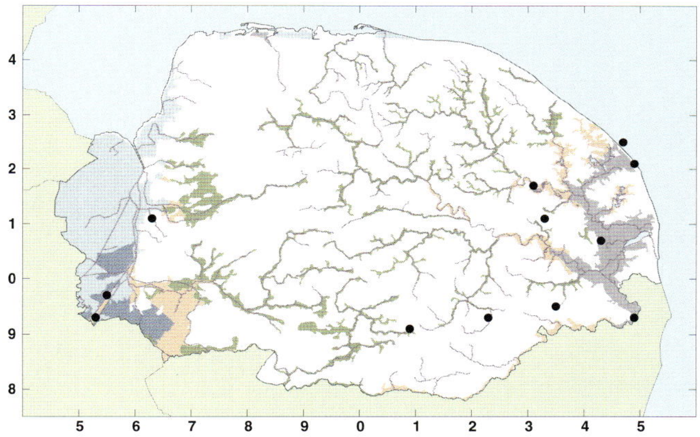

Herbaceous perennial first recorded in the county in 1947 and still rare in marshy places and dykes.

■ **A. gramineum** Lej. **Ribbon-leaved Water-plantain**
Herbaceous perennial, found on the shores of Langmere W98 in 1972, ELS & RPL and thought to have been brought there on the feet of migrating wildfowl.

HYDROCHARITACEAE

Hydrocharis morsus-ranae L. **Frogbit**

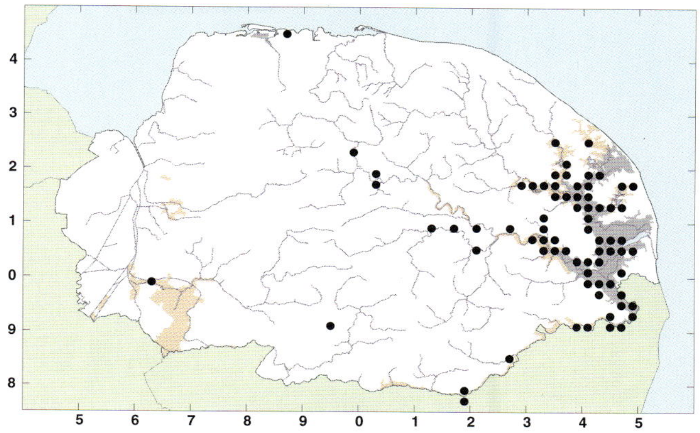

Herbaceous perennial of ponds and ditches, now very scarce outside the Broads. Lost from many former sites since 1968.

Hydrocharis morsus-ranae - Thompson

Stratiotes aloides L. **Water Soldier, Pickerel Weed** (local)

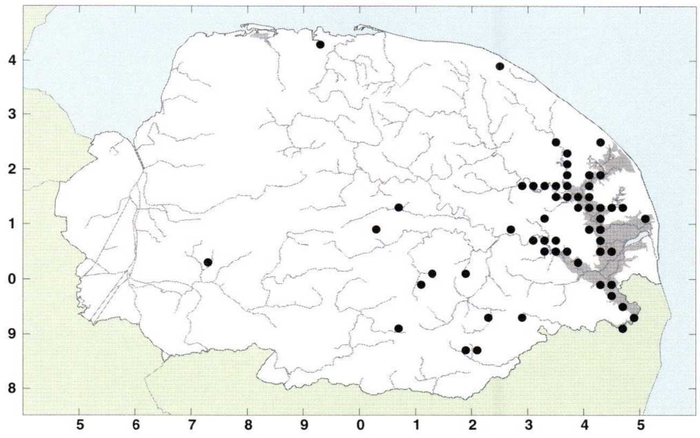

Winter green perennial, native and sometimes abundant in ditches and ponds in the Broads, formerly in Fenland and described in 1769 as being 'plentiful in ditches between Outwell (W50) and Downham Market (W60)'. Elsewhere, an introduction from gardens. Illustrated on the following page.

Stratiotes aloides - How Hill page 215

Elodea canadensis Michx.
Canadian Pondweed, Waterweed, American Weed

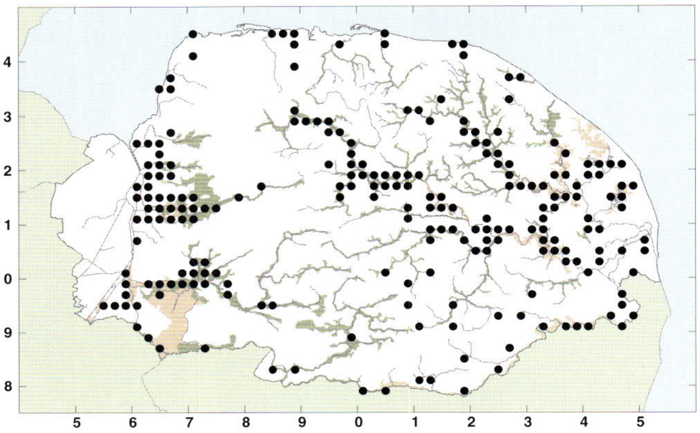

Winter green perennial from North America, first recorded in Britain in 1836 and already described as common by Nicholson, 1914. Now abundant in slow moving water in streams, dykes and ponds.

E. nuttallii (Planch.) H. St. John **Nuttall's Waterweed**

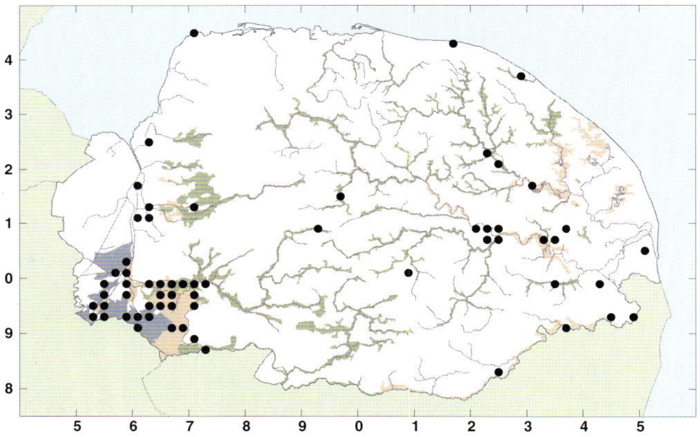

Winter green perennial from North America, first recorded in Britain in 1966 and in Norfolk in 1978 when it was found at Hilgay W69 by JM.Lock & J.Huntley. Still commonest in the southern Fenland, but already rapidly advancing in the Broads and some other waterways.

Lagarosiphon major (Ridl.) Moss **Curly Waterweed**

Winter green perennial from South Africa, an occasional throwout from aquaria. First recorded in Norfolk in 1982 at Brooke E29, ETD. During this survey also at Brewers Green, Diss E18, 1990, AC; King's Lynn W62, in 1991, RMP; Wells-next-the-Sea W94 in 1992, PRB; Wymondham E10 in 1995, D.Showler and Morston E04 in 1998, KKH. W2. E3.

APONOGETONACEAE

Aponogeton distachyos L.f.
Cape Pondweed, Water Hyacinth

Herbaceous perennial from South Africa, occasionally planted or thrown out from gardens, first noted in Norfolk at Westwick E22 in 1973, ETD and still present in 1995, I&P.Green, also in a pond at Irstead E31 in 1998, SP and at Suffield E23 where it has been since at least 1983 and was still present in 1998, G.Morris. W0. E3.

JUNCAGINACEAE

Triglochin palustre L. **Marsh Arrowgrass**

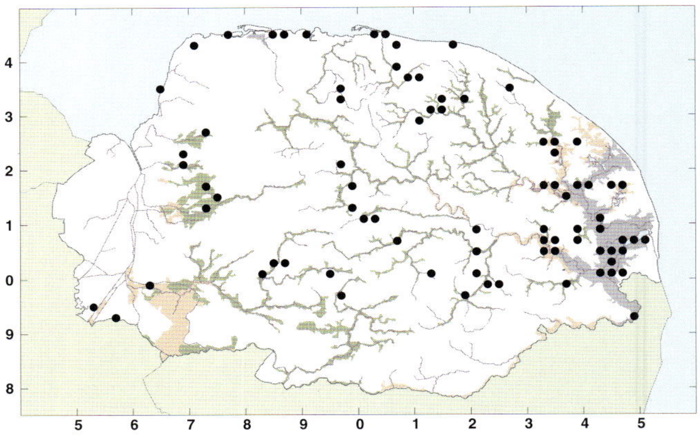

Evergreen perennial, widely scattered but local in unpolluted and 'unimproved' fens and marshes.

T. maritimum L. **Sea Arrowgrass**

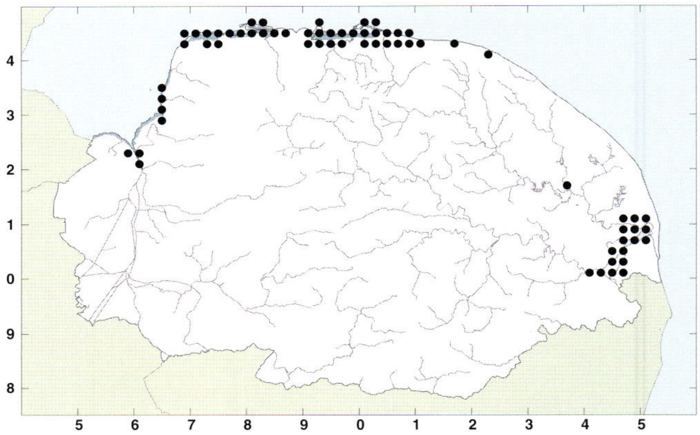

Perennial of salt marshes and damp ground by sea walls and tidal estuaries. Inland by the bridge at Ludham E31.

POTAMOGETONACEAE

All species are perennial.

Potamogeton natans L.
Broad-leaved Pondweed, Tench-weed (local)

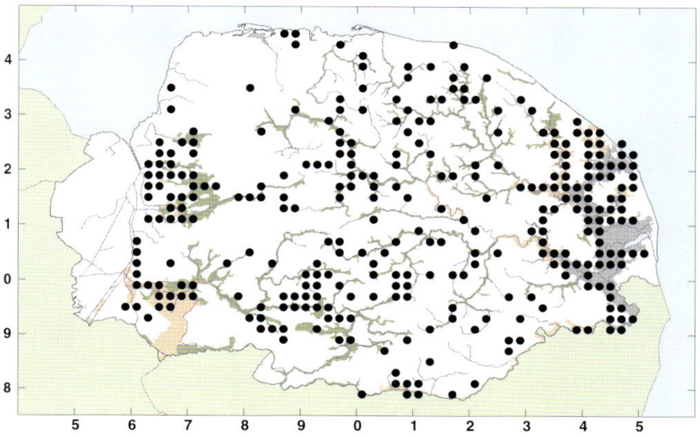

Widespread in slow-moving and still water, sometimes in quite isolated ponds.

Potamogeton natans - Narborough

P. x fluitans Roth
(*P. natans x P. lucens*)
Found at Limpenhoe E30 in 1989, CD.Preston. W0. E1.

P. polygonifolius Pourr. **Bog Pondweed**

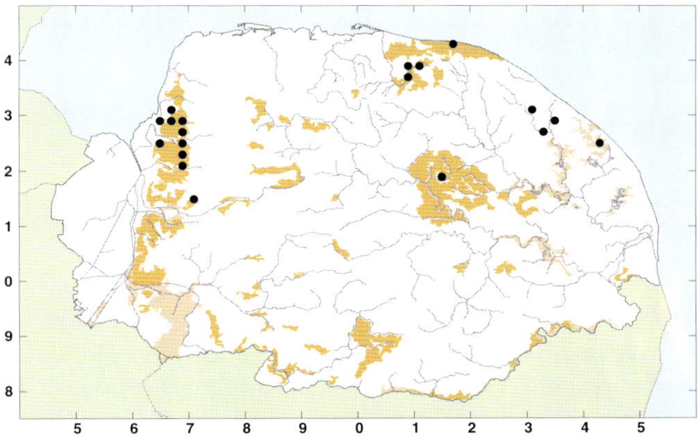

Confined to pools and ditches with acid, peaty waters, most frequent on the greensand in W62.

P. coloratus Hornem. **Fen Pondweed**

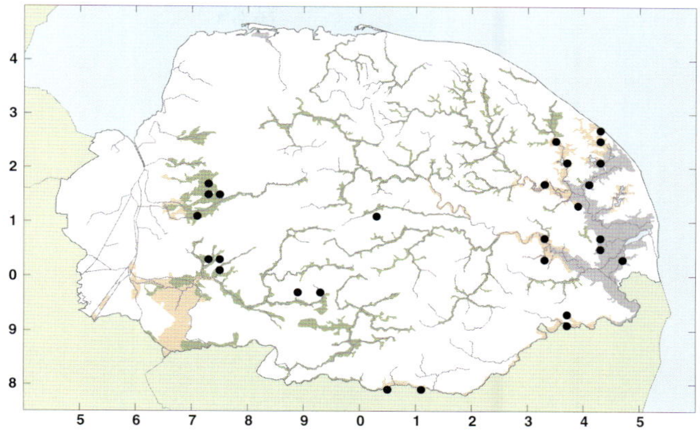

Confined to shallow pools, ditches and calcareous flushes, much reduced in recent years. A nationally scarce species.

P. lucens L. **Shining Pondweed, Bean-weed** (local)

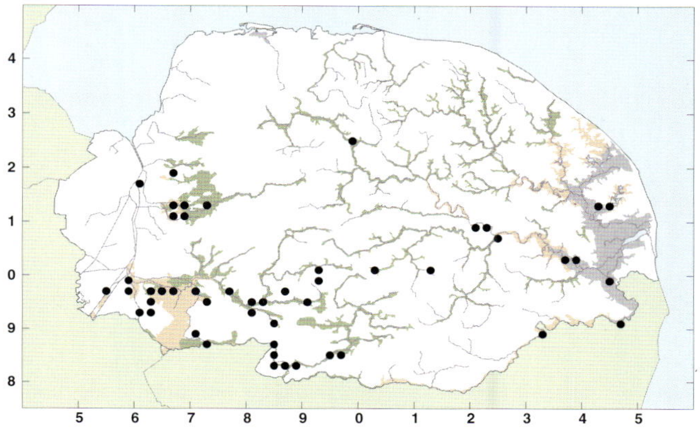

Frequent only in still and slow-moving water draining from base rich soils, commonest in the Wissey and Little Ouse and their tributaries

■ **P. x zizii** W.D.J.Koch ex Roth **Long-leaved Pondweed**
(*P. lucens x P. gramineus*)
Rare, recorded since Petch & Swann, 1968 only at East Wretham W98 in 1972, RPL&ELS and at Horning E31 in 1979, ALB.

P. x salicifolius Wolfg. **Willow-leaved Pondweed**
(*P. lucens x P. perfoliatus*)
Rare, recorded twice before 1920, recently only at Somerton E42 in 1990, G.Kennison and at Hilgay W69 in 1992, HW (both det. CD.Preston). W1. E1.

P. gramineus L. **Various-leaved Pondweed**
Rare, in meres, ponds and dykes. Seen in Langmere W98 in 1972, and at Cantley E30 in 1996, MW.George. W0. E1.

P. alpinus Balb. **Red Pondweed**
Scarce, in dykes and streams on peaty soils, described as locally common by Nicholson, 1914 but in only nine sites by Petch & Swann, 1968. Since then seen in five sites before 1985 and more recently noted at Wolferton W62 in 1990 and Shouldham and Wormegay boyh in W61 in 1994, all HW. W3. E0.

■ **P. praelongus** Wulfen **Long-stalked Pondweed**
Once frequent, especially in the Broads, last recorded in the river Waveney at Bungay E38 in 1976, PG.Lawson.

P. perfoliatus L. **Perfoliate Pondweed**

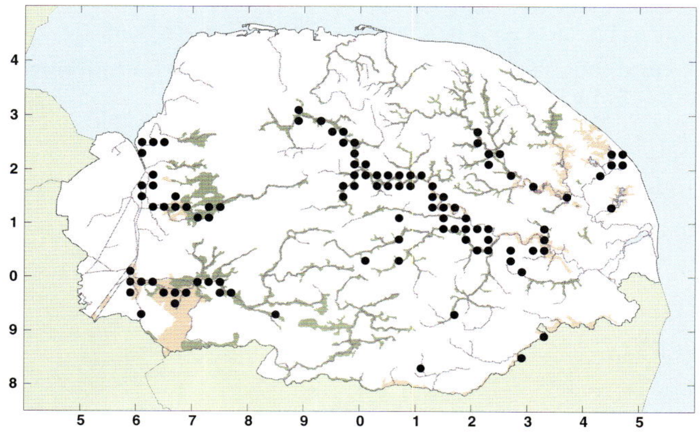

Present in most of the larger rivers and dykes, but few records made in the Waveney and Little Ouse valleys along the county boundary with Suffolk during this survey.

P. friesii Rupr. **Flat-stalked Pondweed**

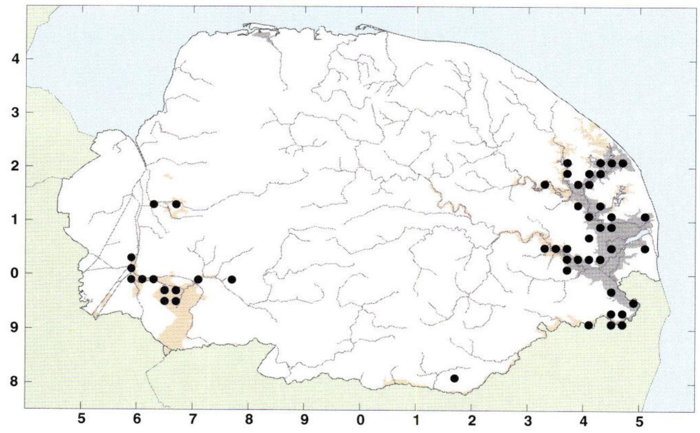

Local in ponds, ditches and streams, avoiding very acid conditions.

■ **P. x pseudofriesii** Dandy & G. Taylor
(*P. friesii x P.acutifolius*)
Recorded once, at Strumpshaw E30 in 1952. The first British record of this hybrid.

P. pusillus L. **Lesser Pondweed**

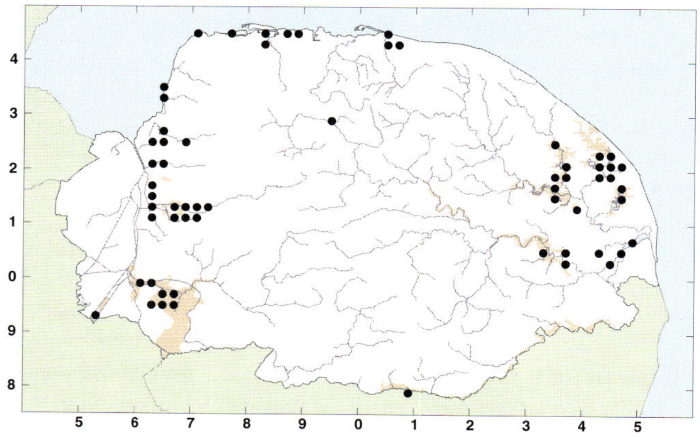

Found in ponds and ditches in both fresh and slightly brackish waters, most frequent in areas with a maritime influence.

■ **P. x grovesii** Dandy & G. Taylor
(*P. pusillus x P. trichoides*)
Recorded from one Broadland site before 1897.

P. obtusifolius Mert. & W.D.J.Koch **Blunt-leaved Pondweed**

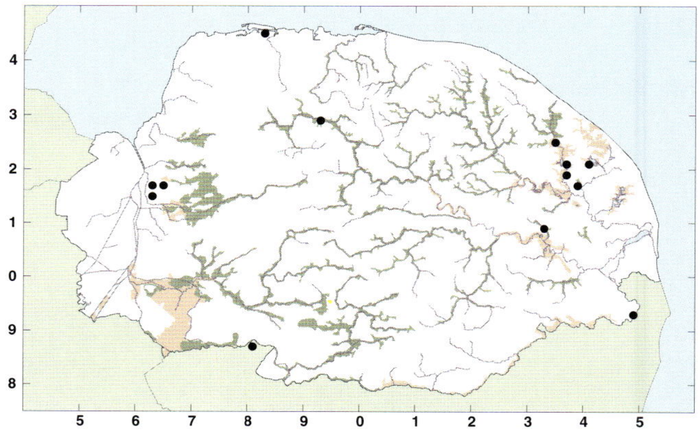

Scarce but widely scattered in dykes, ponds and shallow pools.

P. berchtoldii Fieber **Small Pondweed**

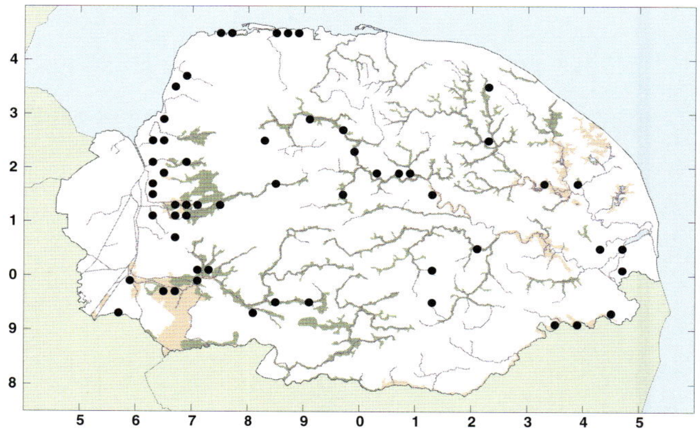

Not infrequent in ponds, dykes and streams, most frequent in alkaline waters but largely avoiding brackish conditions.

P. trichoides Cham. & Schltdl. **Hair-like Pondweed**

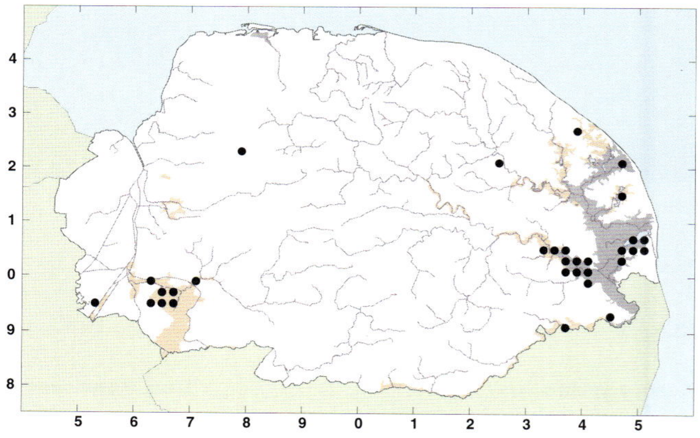

First found in Britain at Framingham Earl E20 in 1848, K.Trimmer. Very local, chiefly in ditches on peaty soils but found in an isolated pond near Massingham W72 in 1995, FS. A nationally scarce species.

P. compressus L. **Grass-wrack Pondweed**
A plant of ditches on grazing marshes with five old records from Broadland. Now found at one site on Upton marshes E31 in 1989, CD.Preston. A nationally scarce species. W0. E1.

P. acutifolius Link **Sharp-leaved Pondweed**
Characteristic of ditches on grazing marshes by the lower reaches of the Yare where it has been known since the 1880s. Recorded at twelve sites during this survey, all within this area. W0. E12.

P. crispus L. **Curled Pondweed**

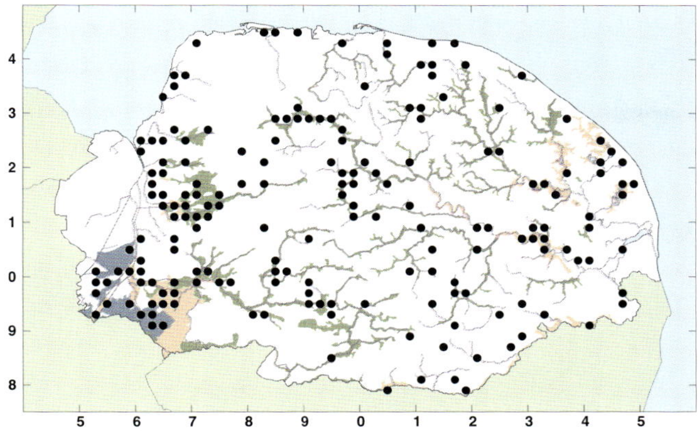

Widespread in ponds, dykes and streams throughout the county though never in abundance.

P. pectinatus L.
Fennel Pondweed, Grass-weed, Lace-weed (local)

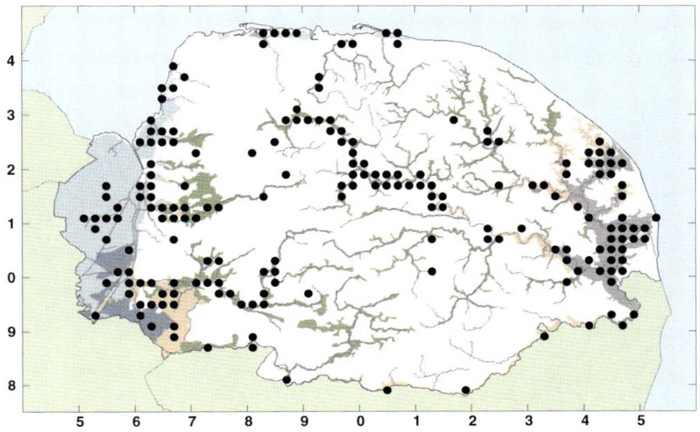

Frequent in streams and rivers, also in brackish dykes close to the sea, tolerant of pollution and tellingly now our commonest pondweed.

Groenlandia densa (L.) Fourr. **Opposite-leaved Pondweed**

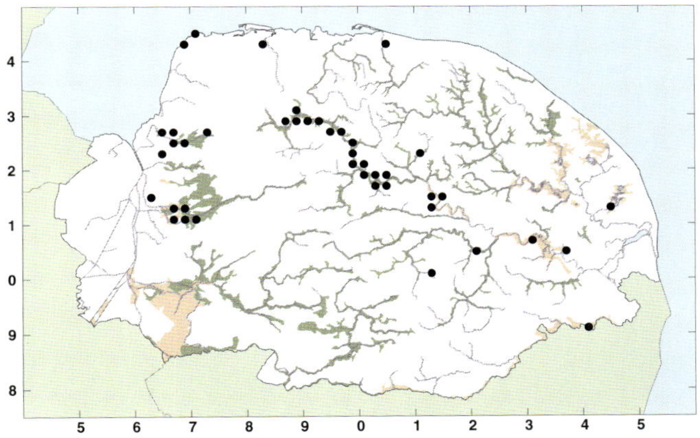

Always limited in its distribution, normally present in the upper reaches of rivers, especially those with calcareous waters.

RUPPIACEAE

Ruppia maritima L. **Beaked Tasselweed**
Perennial of brackish waters, once characteristic of ditches on reclaimed land, but many of these areas are now used for arable crops and the ditches too nitrogen-rich for its survival. Seen during this survey only at Cley and Salthouse E04 in 1992, KKH. W0. E2.

R. cirrhosa (Petagna) Grande **Spiral Tasselweed**
Syn. *R. spiralis* L. ex. Dumort
Perennial of brackish waters, formerly recorded from Snettisham W63 to Weybourne E04, recently at Cley and Salthouse E04 in 1997, PWL. A nationally scarce species. W0. E2.

NAJADACEAE

Najas marina L. **Holly-leaved Naiad**
Annual known only from partly brackish waters on the Broads, discovered in 1883 in Hickling Broad E42 and still found there, also during this survey at Cockshoot and Upton E31; Barton Broad E32; Heigham Sound and Martham Broad both E42, all by G.Kennison. W0. E7.

ZANNICHELLIACEAE

Zannichellia palustris L. **Horned Pondweed**

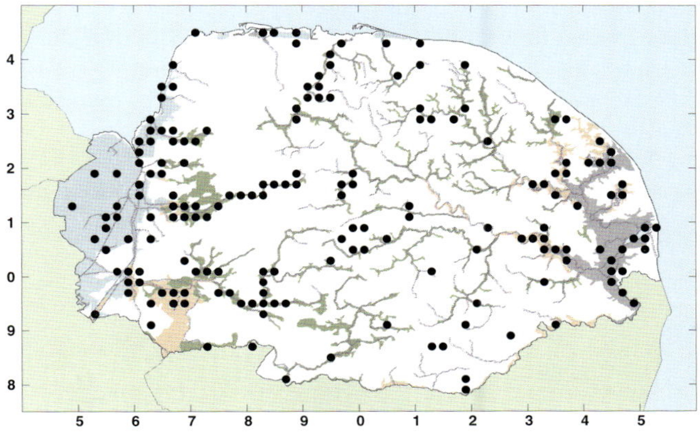

Frequent in rivers, streams and ditches, including those with brackish waters.

ZOSTERACEAE

Zostera marina L. **Eelgrass**
Winter green perennial found below the low-tide line, only partially recovered from its loss from disease. Seen recently at Breydon Water E50 in 1990, L.Street and Wells-next-the-Sea W94 in 1994, PRB. A nationally scarce species. W1. E1.

Z. angustifolia (Hornem.) Rchb.
Narrow-leaved Eelgrass, Wigeon Grass
Winter green perennial found near the low-tide line, recorded at Morston W94 and Blakeney Harbour E04 in 1995, A.Lewis; at Scolt Head W74 in 1993, SW and between Wells-next-the-Sea W94 and Blakeney E04 in 1997, J.McCallum. A nationally scarce species. W5. E2.

Z. noltii Hornem. **Dwarf Eelgrass**
Winter green perennial found near the low-tide line, now the most frequent of the eel-grasses, found in Breydon Water E50 in 1990, L.Smart; Scolt Head W74 in 1993, SW; Wells-next-the-Sea W94 in 1994, PRB and between that site and Cley E04 in 1997, J.McCallum. W4. E2.

ARACEAE

Acorus calamus L. **Sweet Flag**

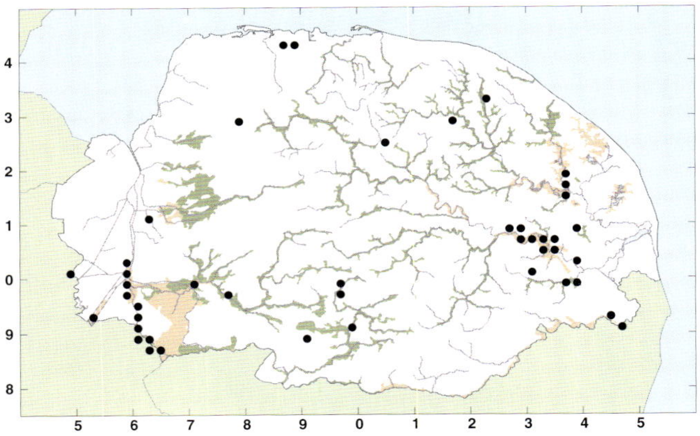

Semi-winter green to herbaceous perennial introduced from Asia in or before mediaeval times for strewing on floors, the plant's aroma persisting when it is dry. It was planted by suitable still or very slow-moving waters for a continuous supply of leaves and, in spite of the fact that it does not appear to set seed in Britain, spread downstream. Much reduced in quantity in Broadland by coypu but since their eradication, recovering slowly.

Arum maculatum L. **Lords and Ladies, Cuckoo-pint**

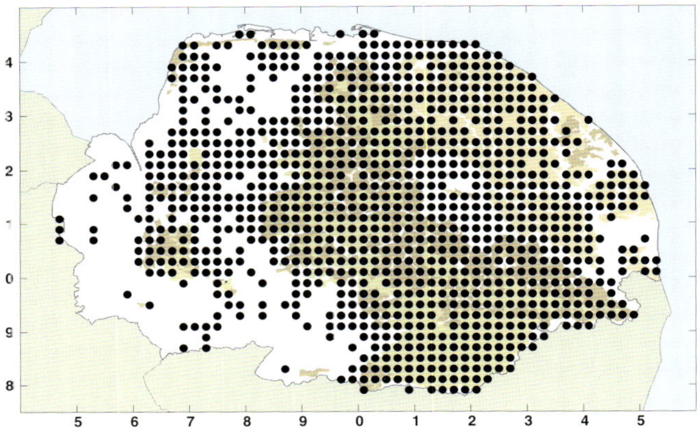

Summer dormant perennial leafing in late winter. A plant which is common on heavier, former woodland soils, but away from these is only near villages. As it was formerly used as the main source of starch, it may have a planted origin in these areas. In the majority of plants the spadix is purple, but colonies in which it is greenish-yellow can be found.

A. italicum Mill. **Italian Lords and Ladies**
Summer dormant perennial leafing in autumn. All plants recorded during this survey belong to the cultivar 'Pictum', a garden form of ssp. **italicum**, with white-marked leaves. W5. E2.

LEMNACEAE

All evergreen perennials producing smaller, rather atypical leaves in winter.

Spirodela polyrhiza (L.) Schleid. **Greater Duckweed**
Syn. *Lemna polyrhiza* L.

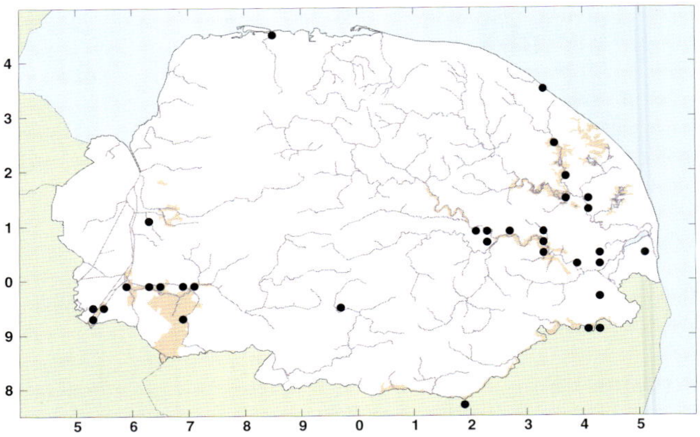

Widely scattered but scarce in dykes and streams, much reduced since the 1968 Flora especially in the north of the county.

Arum maculatum - Ashwellthorpe Wood

Lemna gibba L. — **Fat Duckweed**

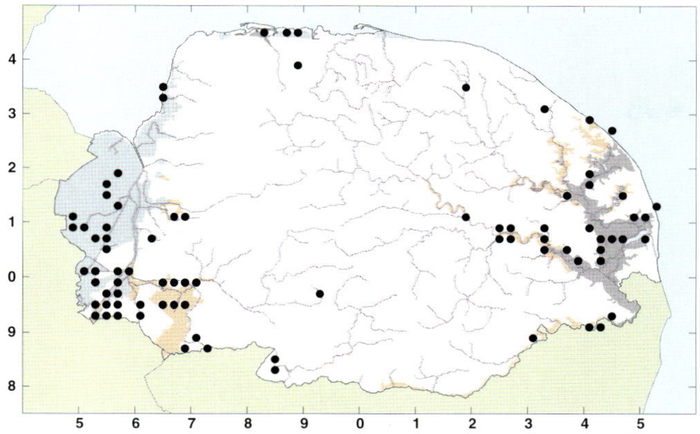

Widespread, but frequent only in the peat fens where it can sometimes be abundant.

L. minor L. — **Common Duckweed**

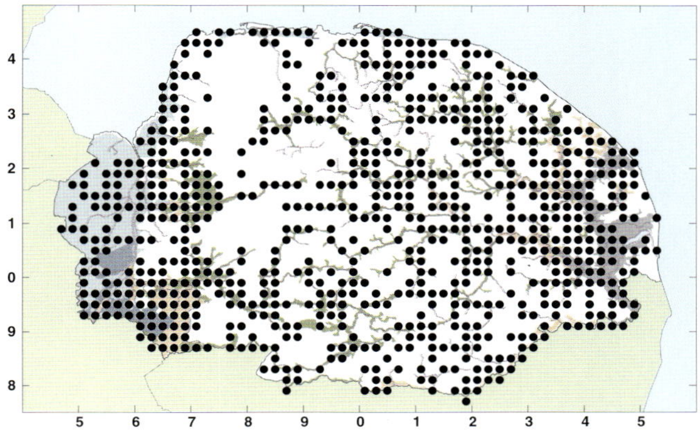

Frequent in all sorts of watercourses, commonest in ponds and dykes but also in all rivers, in backwaters if the flow is fast.

L. trisulca L. — **Ivy-leaved Duckweed**

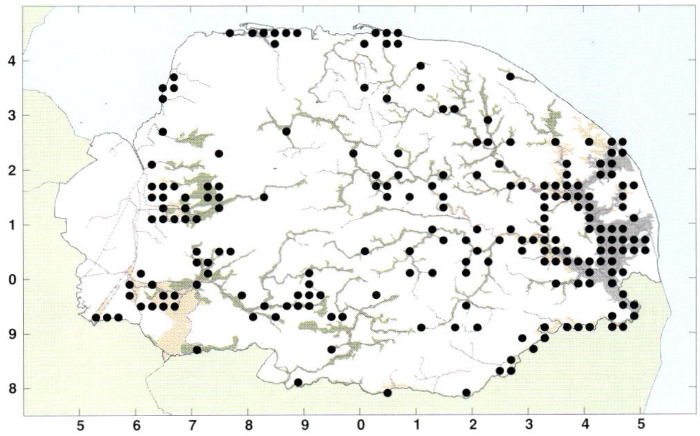

Widely distributed in ponds and dykes.

L. minuta Kunth — **Least Duckweed**

Syn. *L. minuscula* Herter nom.illeg.

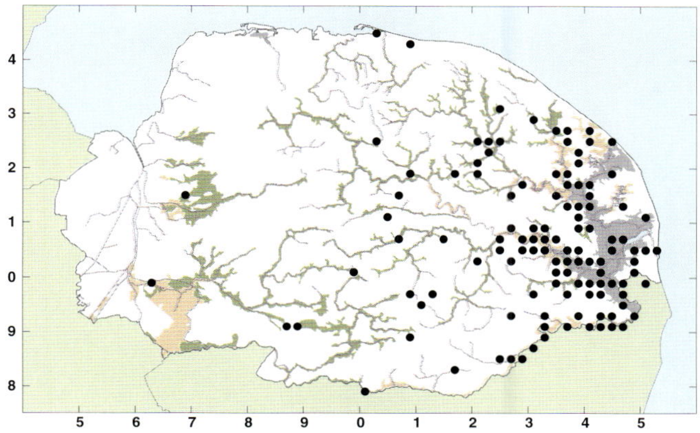

Introduced from North or South America, almost certainly as a throw-out from aquaria. First recorded in Britain in 1977 and in Norfolk in 1988 in both Barton Broad E32 by FJ.Rumsey and at Woodbastwick E31 by TCG.Rich. Now scattered throughout the county in ponds, dykes and pools but abundant only in parts of Broadland.

JUNCACEAE

Juncus squarrosus L. — **Heath Rush**

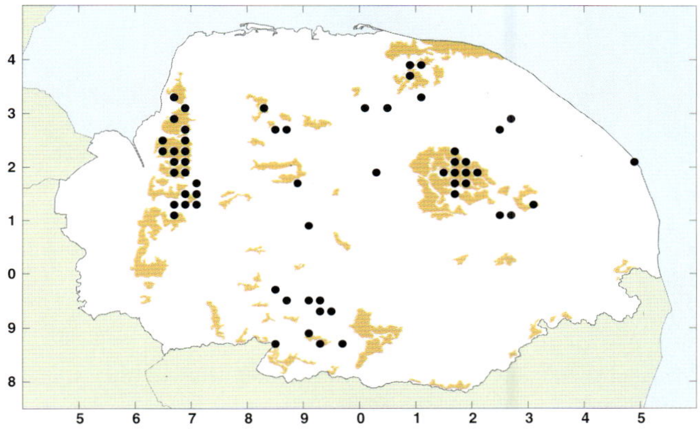

Evergreen perennial frequent only on moist, acid, often peaty soils.

J. tenuis Willd. — **Slender Rush**

Syn. *J. dudleyi* Wiegand

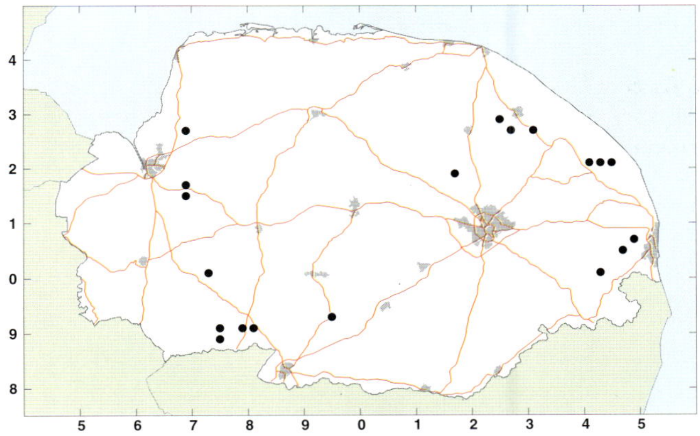

Evergreen perennial introduced from North America. First recorded in Norfolk in 1920, now scattered and well established on damp tracks, forest rides and old railway lines, though still scarce.

J. compressus Jacq. **Round-fruited Rush**

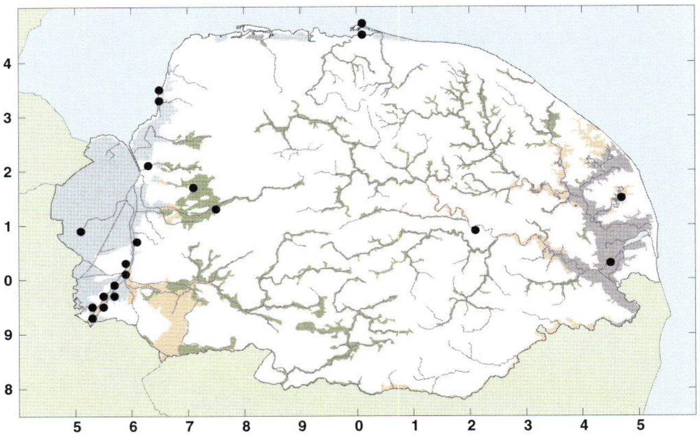

Evergreen perennial, scarce in pastures near the sea, frequent only in the Ouse Washes W59, 1994, J.Smith.

■ **J. compressus x J. gerardii**
Recorded once, on the banks of the Gt.Ouse near King's Lynn W61 in 1974, RPL.

J. gerardii Loisel **Saltmarsh Rush**

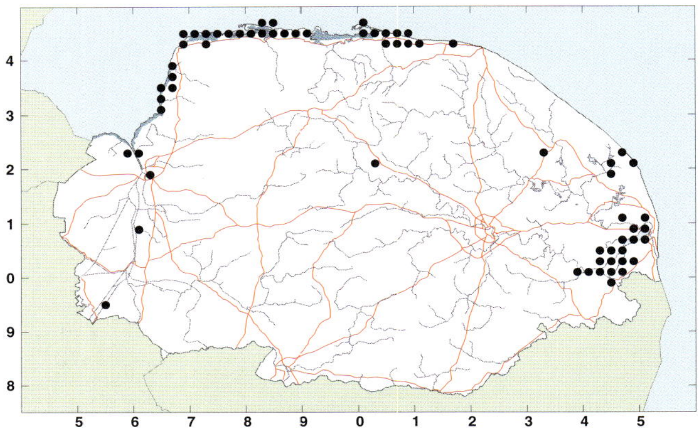

Perennial of salt marshes and brackish ground near the sea extending inland along tidal estuaries. Recorded on roadsides at Foxley E02 in 1996 and Smallburgh E32 in 1998, both by members of the Flora Group.

J. bufonius L. **Toad Rush**

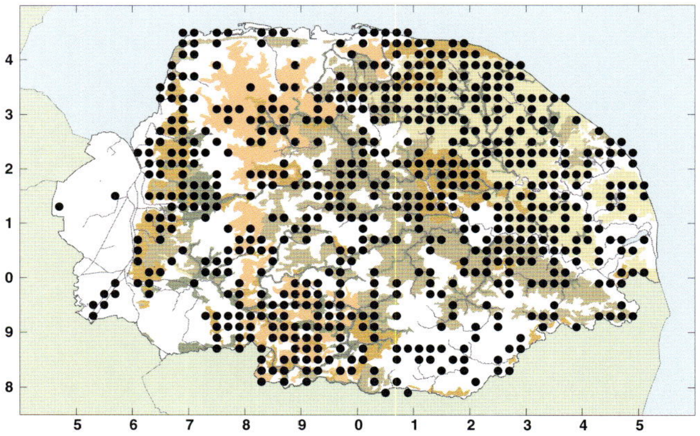

Annual, frequent in damp, bare places from open tracks to woodland rides, almost all records refer to **J.bufonius** s.s.

■ **J. ambiguus** Gusst.
Annual of coastal mud flats and brackish places. Found at Snettisham W63 in 1978, RPL & ELS.

J. subnodulosus Schrank **Blunt-flowered Rush**

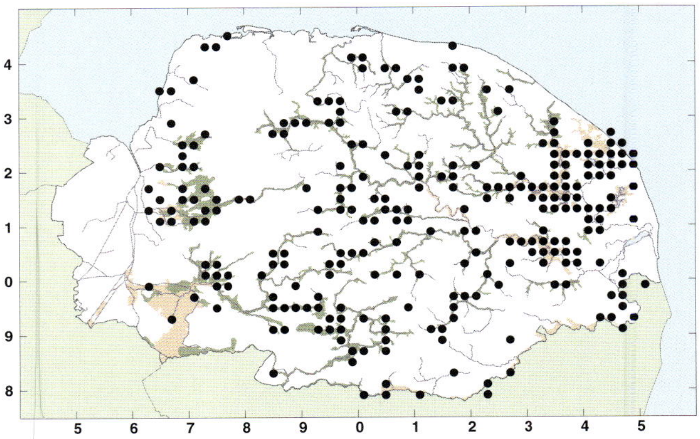

Perennial found on wet, base rich, peaty soils in fens and typical of mowing marshes. In suitable ground it can form large colonies which exclude virtually everything else.

Juncus subnodulosus - How Hill

J. articulatus L. **Jointed Rush**

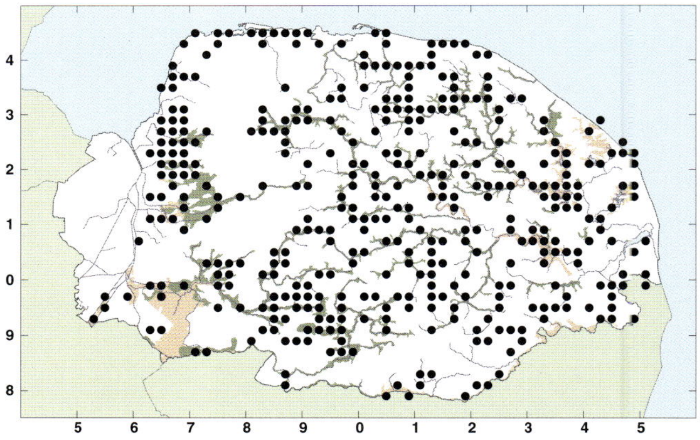

Perennial found in damp to wet places, especially on neutral to acid soils.

J. acutiflorus Ehrh. ex Hoffm. **Sharp-flowered Rush**

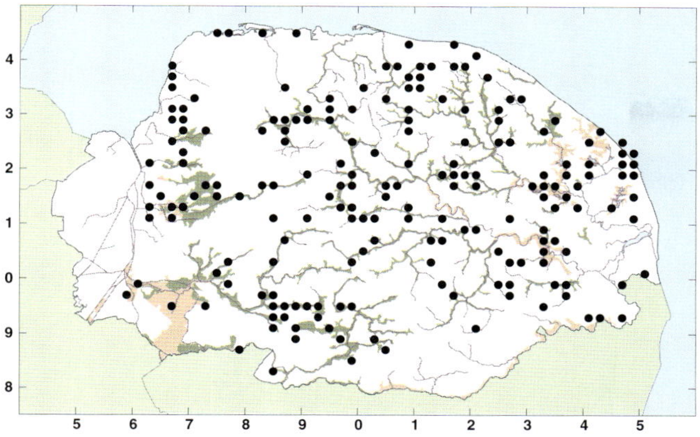

Perennial found in damp to wet acid, peaty bogs and marshes and by ponds and rivers.

■ **J. x surrejanus** Druce ex Stace & Lambinon
(*J. acutiflorus x J. articulatus*)
Recorded three times, from Derby Fen W61, East Walton W71 and Congham W62, all in 1946.

J. bulbosus L. **Bulbous Rush**

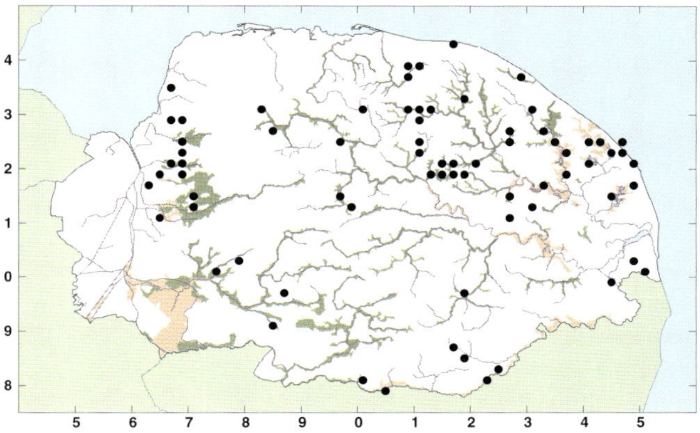

Perennial scattered in wet places on heaths and in bogs. **J.kochii** F.W.Schultz (*J. bulbosus ssp. kochii*) was recorded 1968 at Grimston W61, but is no longer recognised as a separate entity. Var. **uliginosus** with viviparous flowers was noted at East Winch W61 and South Wootton W62 in 1980, ELS.

J. maritimus Lam. **Sea Rush**

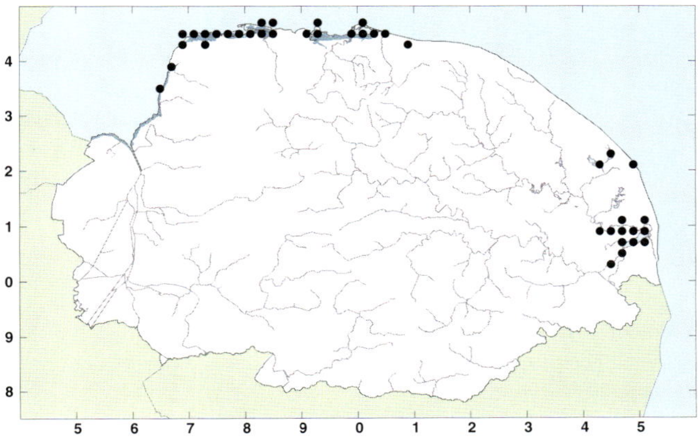

Perennial found in the upper parts of salt marshes and most commonly by dykes near the sea. Found inland at Hickling Broad E42 after the 1938 sea flood and still there.

J. acutus L. **Great Sharp Rush**
Winter green perennial, rare on the drier parts of salt marshes. First recorded at Brancaster W74 and W84 in 1828 where it still grows, also at Burnham Deepdale W84 and Thornham W74. A nationally scarce species. W3. E0.

J. inflexus L. **Hard Rush**

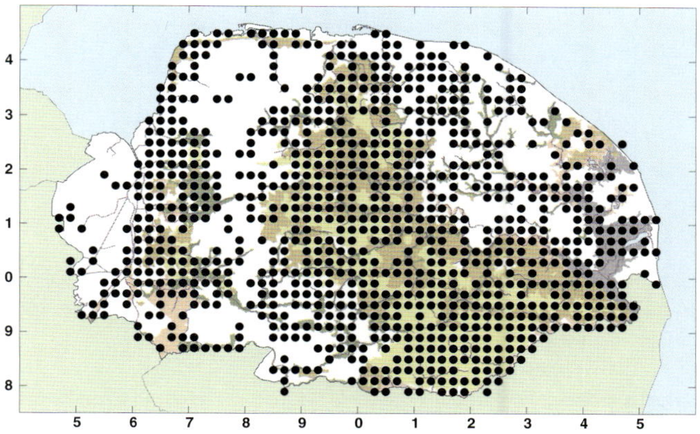

Winter green perennial, frequent in damp ground, said to need base rich soils, but does not appear to do so in Norfolk, though it is often on deeper soils.

J. effusus L. **Soft Rush**

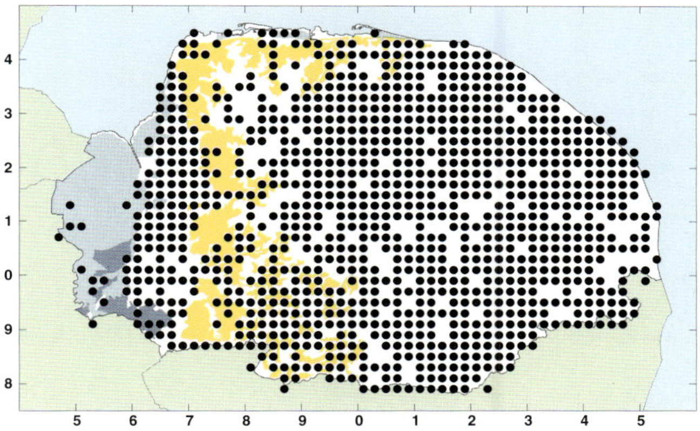

Winter green perennial, abundant in damp to wet habitats, often together with the previous species, surprisingly absent from most of the dykes of the silt fens. Var. **subglomeratus** D.C. (syn. *compactus* Lej. & Cortois ex Reichg.) is common and was used in the past for making rush lights.

J. x diffusus Hoppe.
(*J. inflexus x J. effusus*)
Six old records for this hybrid, found most recently at Leziate W62 in 1996, SF. W1. E1.

J. conglomeratus L. **Compact Rush**

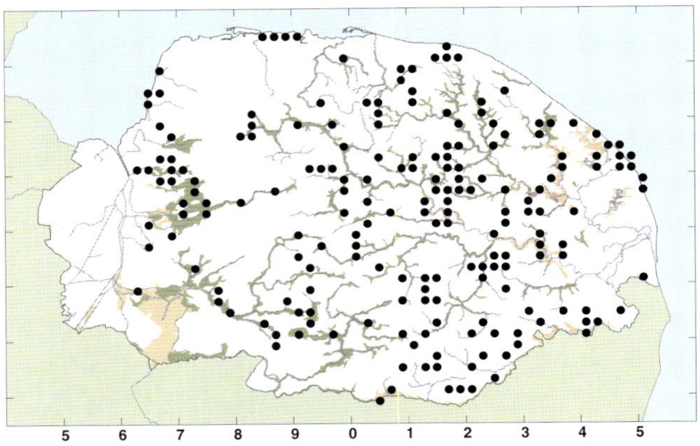

Perennial frequent on damp ground, chiefly on neutral to acid soils, perhaps over-recorded for *J. effusus* var. *subglomeratus* and like that species also used for making rush lights.

Luzula pilosa (L.) Willd. **Hairy Woodrush**

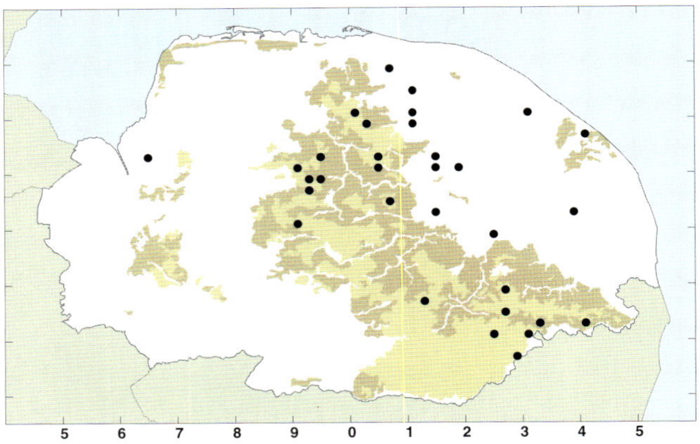

Perennial, uncommon in old woods and hedgerows on clay soil.

L. sylvatica (Huds.) Gaudin **Great Woodrush**

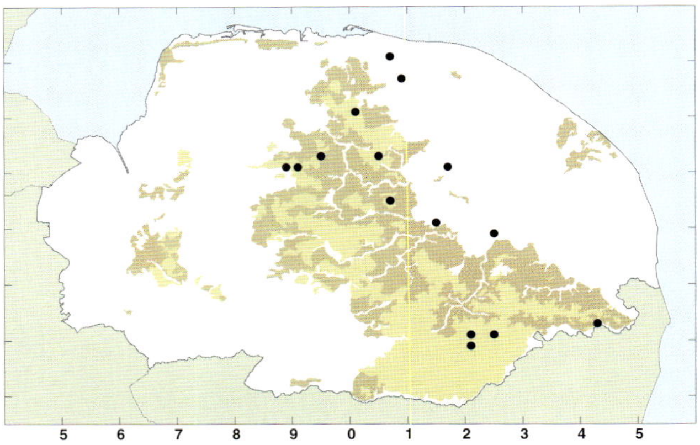

Perennial, rare in damp woods and adjacent hedgerows.

L. campestris (L.) DC. **Field Woodrush**

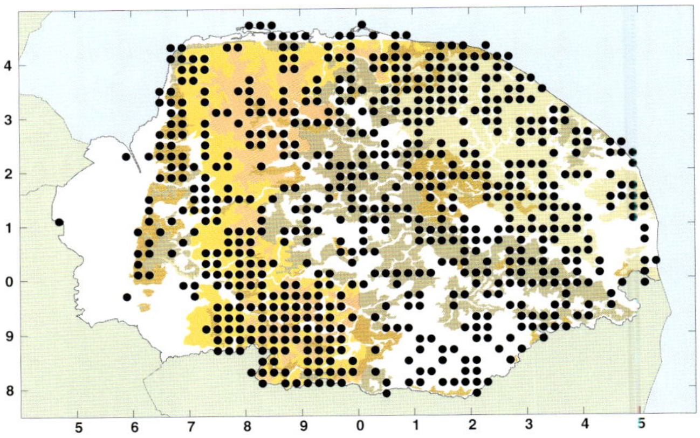

Perennial, frequent in established, short grassland on well-drained soils, often in churchyards.

L. multiflora (Ehrh.) Lej. **Heath Woodrush**

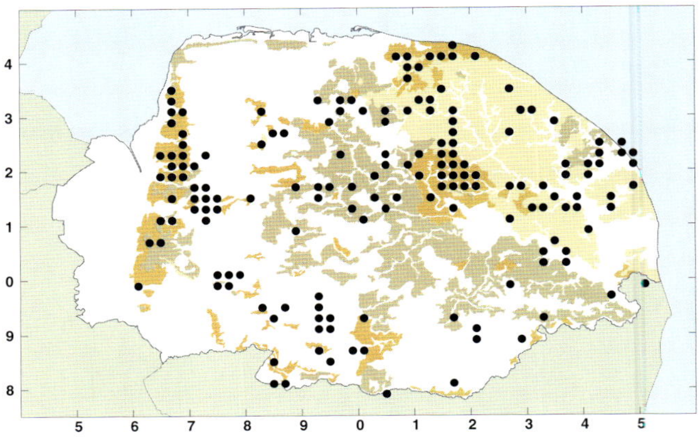

Perennial found on heaths and damp grassland on acid soils. Ssp. **congesta** (Thuill.) Arcang. is far the commoner of the subspecies throughout Norfolk.

CYPERACEAE

Eriophorum angustifolium Honck. **Common Cotton-grass**

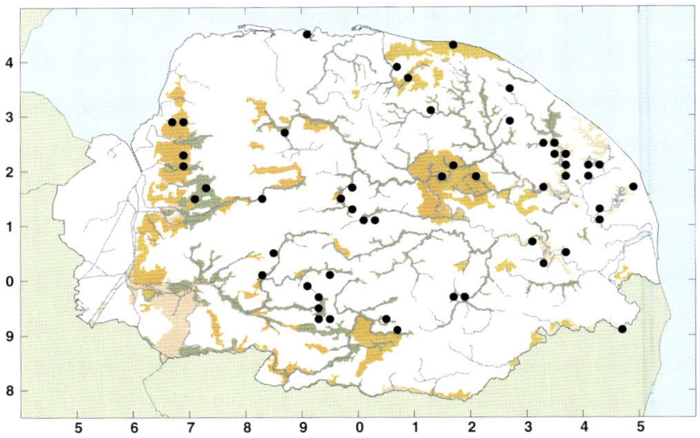

Perennial widely scattered in wet, acid peaty soils and bogs, but like all the Cotton-grass species, much at risk as pools dry out because of the lowering water table.

E. latifolium Hoppe **Broad-leaved Cotton-grass**

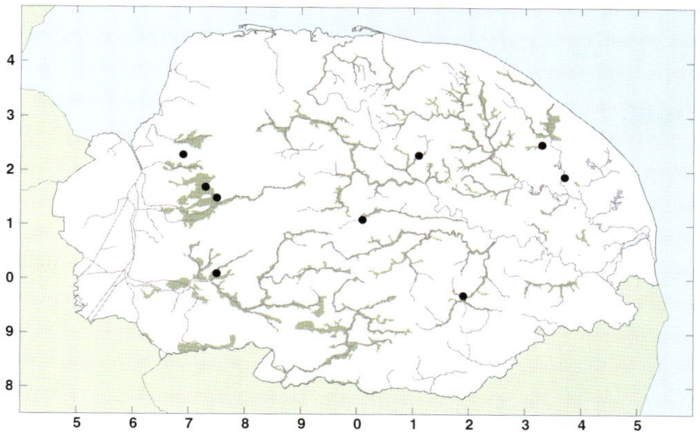

Perennial scattered on peaty soils, usually where the ground-water is lime rich and often associated with spring lines.

Eriophorum latifolium - East Walton

■ **E. gracile** W.J.D.Koch ex Roth. **Slender Cotton-grass**
Perennial found in acid, peaty areas, recorded at Stow Bedon W99 in 1916 and Acle E41 in 1959. Red Data Book species.

E. vaginatum L. **Hare's-tail Cotton-grass**
Perennial, the common cotton-grass of northern moors, rare in the lowlands. In Norfolk known only at Dersingham W62 in 1985, CPP; West Rudham W82 in 1994, E.Duffey and Roydon W62 in 1995, WAB. W3. E0.

Trichophorum cespitosum (L.) Hartm. **Deer-grass**
Syn. *Scirpus caespitosus* L.

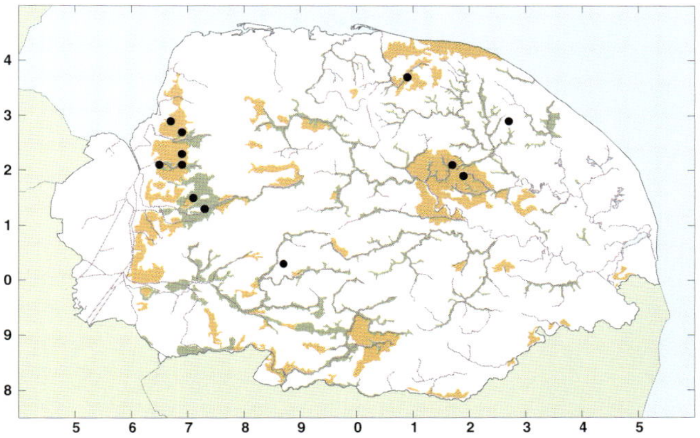

Perennial confined to damp, acid bogs and heaths, common in upland Britain, scarce in the lowlands.

Eleocharis palustris (L.) Roem. & Schult. **Common Spike-rush**

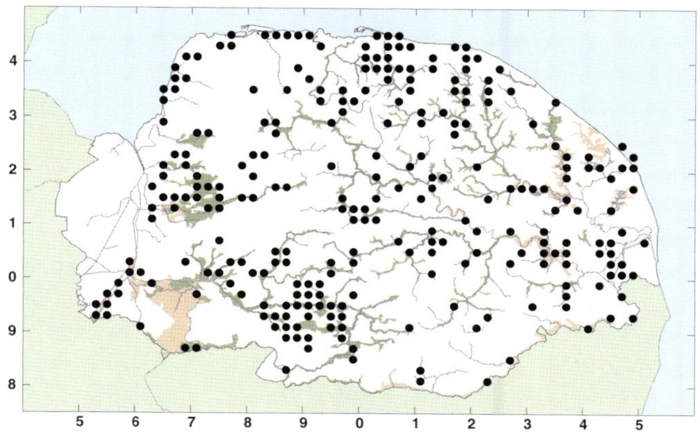

Evergreen perennial, widespread in marshes, ditches and ponds, often in standing water.

E. uniglumis (Link) Schult. **Slender Spike-rush**

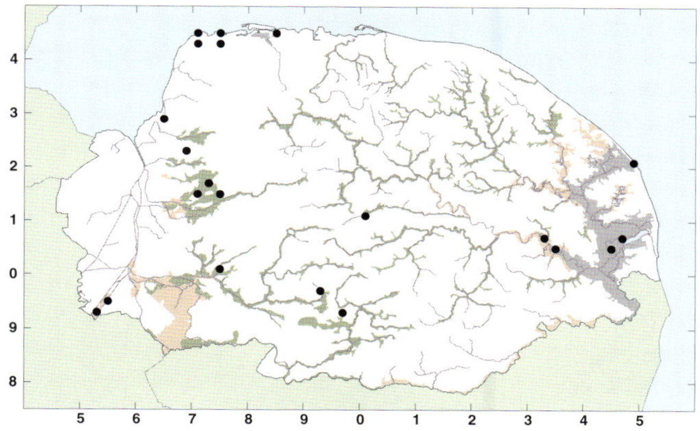

Perennial of marshy places, often in coastal habitats, but also inland.

E. multicaulis (Sm.) Desv. **Many-stalked Spike-rush**

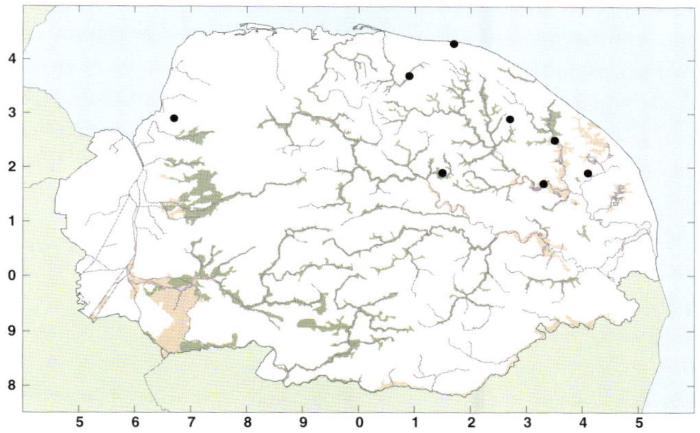

Perennial, a chiefly western species scarce in Norfolk, occurring in neutral to acid water-logged marshes and bogs. Known in 14 sites in the 1960s, now much reduced, especially in the west of the county.

E. quinqueflora (Hartmann) O. Schwarz
Few-flowered Spike-rush
Perennial, a chiefly north-western species found occasionally in water-logged marshes and fens. W1. E6.

■ **Scirpus sylvaticus** L. **Wood Club-rush**
Perennial, recorded in 'damp woods' before 1805. Not seen since.

Bolboschoenus maritimus (L.) Palla **Sea Club-rush**
Syn. *Scirpus maritimus* L.

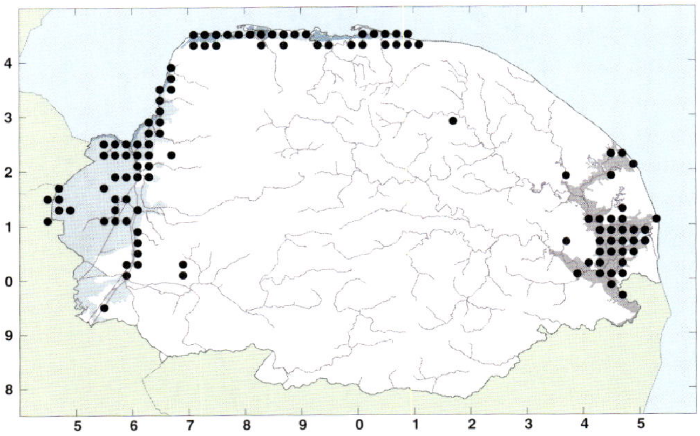

Frequent in estuaries, extending up tidal rivers, also by coastal dykes. Seen in a high roof gutter in South Lynn W61 in 1998, RMP. Introduced at some ornamental lakes such as that at Blickling E12, MK.

Schoenoplectus lacustris (L.) Palla
Common Club-rush, True Bulrush, Bolder (local)
Syn. *Scirpus lacustris* L.

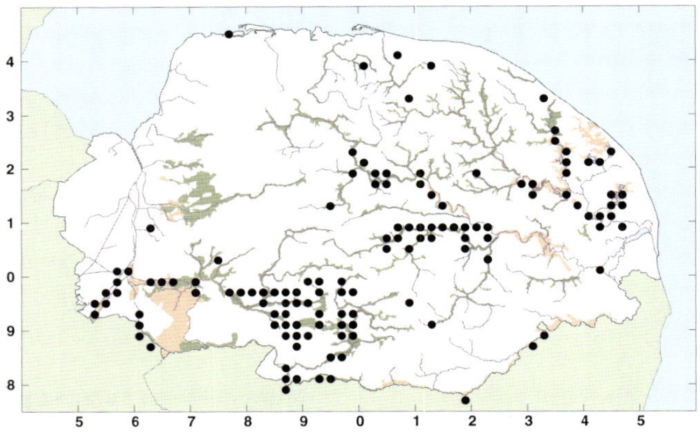

Perennial, scattered in shallow water, largely replaced in coastal areas by *S. tabernaemontani*. Probably introduced to a pond in W74. This is the true bulrush which has had its name usurped because of the 19th century painting of Moses in the Bulrushes by Alma-Tadema in which he used *Typha* (reed-mace) for his bulrushes.

S. tabernaemontani (C.C.Gmel.) Palla
Grey Club-rush, Grey Bulrush
Syn. *Scirpus tabernaemontani* C.C.Gmel.

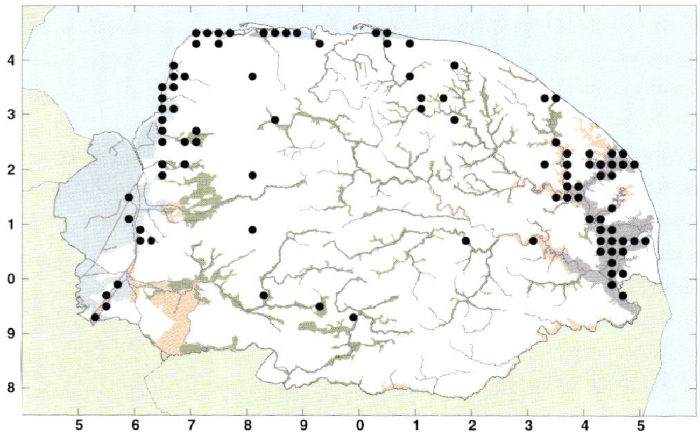

Perennial scattered in shallow water, chiefly in areas near the sea. In the Broads it was once gathered and used for mat making.

Isolepis setacea (L.) R. Br. **Bristle Club-rush, Bristle Scirpus**
Syn. *Scirpus setaceus* L.

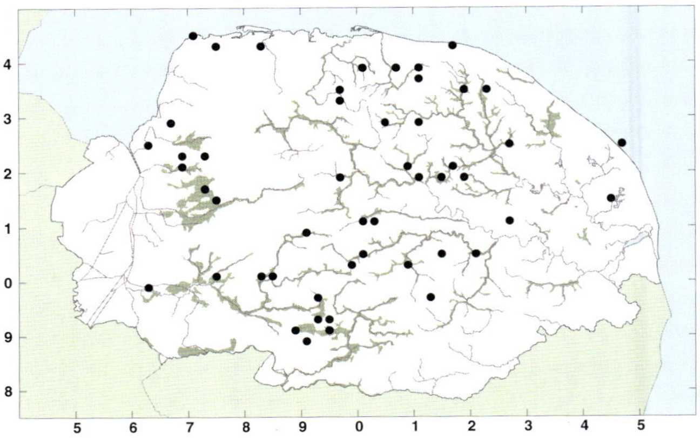

Over-wintering annual, widespread but always scarce in damp, sandy ground and open areas in meadows.

Isolepis setacea - Hockham

I. cernua (Vahl) Roem. & Schult.
Slender Club-rush, Nodding Scirpus
Syn. *Scirpus cernuus* Vahl.
Annual, rare in marshy ground, recorded from only nine sites this century, still present at Beeston E14 and Buxton Heath E12. W0. E2.

Eleogiton fluitans (L.) Link
Syn. *Scirpus fluitans* L. **Floating Club-rush, Floating Scirpus**

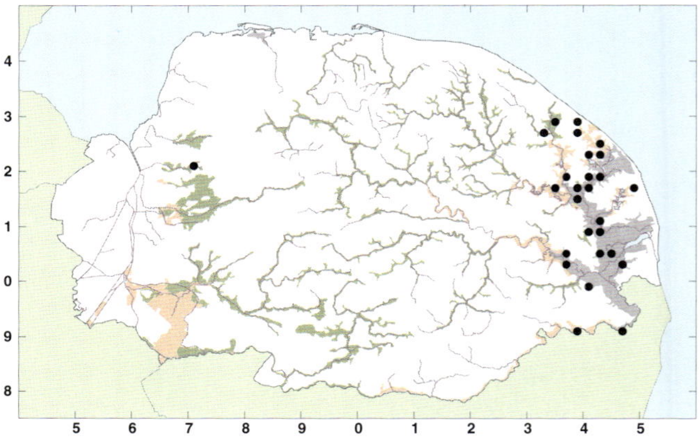

Perennial, occasional in dykes and pools in acid conditions. Much reduced in frequency since Petch & Swann, 1968 when ground water levels were far higher.

Blysmus compressus (L.) Panz. ex Link **Flat Sedge, Broad Blysmus**

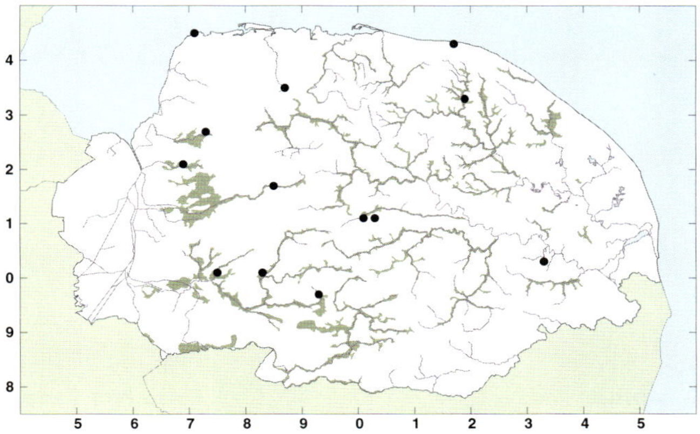

Scarce in long-established, damp meadows, always uncommon.

Cyperus longus L. **Galingale**
Evergreen perennial, native to southern England. Found in three sites where it was almost certainly introduced, Scoulton W90 in 1994, ER; Saxlingham E03 in 1994, Flora Group and Fakenham W92 in 1997 K&GB. W2. E1.

C. eragrostis Lam. **Pale Galingale**
Perennial introduced from tropical America and seen at Witton E30 in 1990, SP and Haddiscoe E49 in 1991, Flora Group. W0. E2.

Cladium mariscus - Foulden

Schoenus nigricans L. **Black Bog-rush**

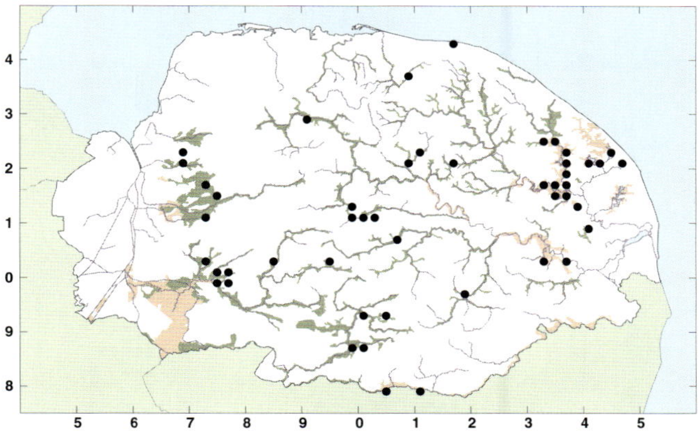

Perennial scattered in base rich fens and moist grassland.

Rhynchospora alba (L.) Vahl **White Beak-sedge**

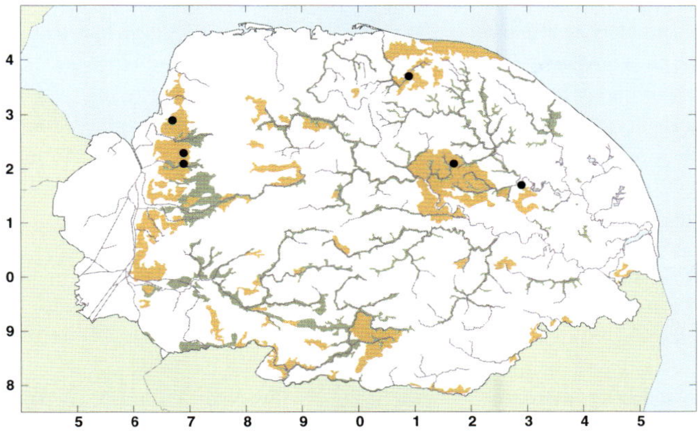

Perennial rare in water-logged, peaty soils, found in valley bogs where the ground water is acid.

Cladium mariscus (L.) Pohl **Sedge, Great Fen-sedge**

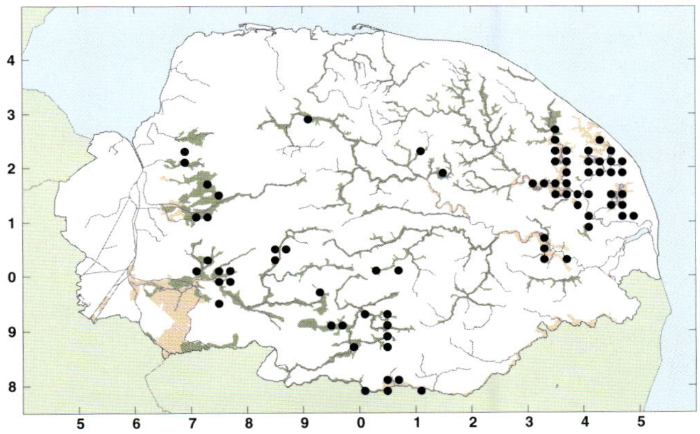

Winter green perennial found in base rich fens and by pools and streams. In the Broads long managed commercially for cutting in summer, a sedge marsh being locally known as a thack marsh. Still a valued crop for the ridges of roofs in thatching.

Carex
All Carex species are perennial and most are winter green.

Carex paniculata L. **Greater Tussock Sedge**

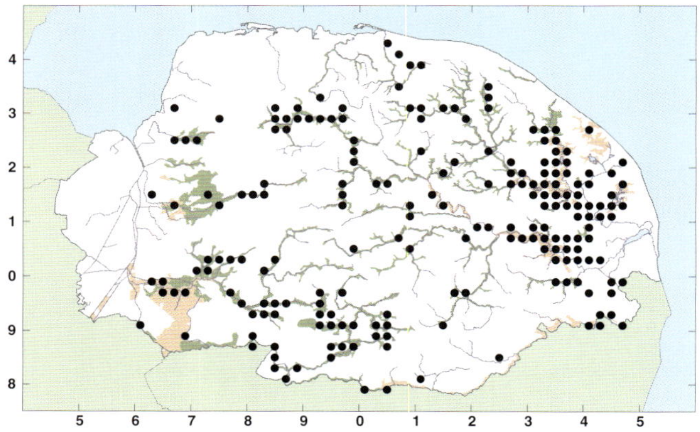

Frequent in marshes and fen carr, by broads and streams. Capable of growing in standing water and eventually forming high tussocks which were once cut and used as hassocks in the churches (now called kneelers). As the tussocks increase in size, other plants germinate and grow on them, supported above water level. When the invaders are trees, they will in time shade out the original sedge and form swampy woodland. This is the origin of much semi-floating fen carr.

■ **C. x rotae** De Not.
(*C. paniculata x C. appropinquata*)
Recorded at Upton Broad E31 and at Wheatfen E30 before 1940.

C. appropinquata Schumach. **Fibrous Tussock Sedge**

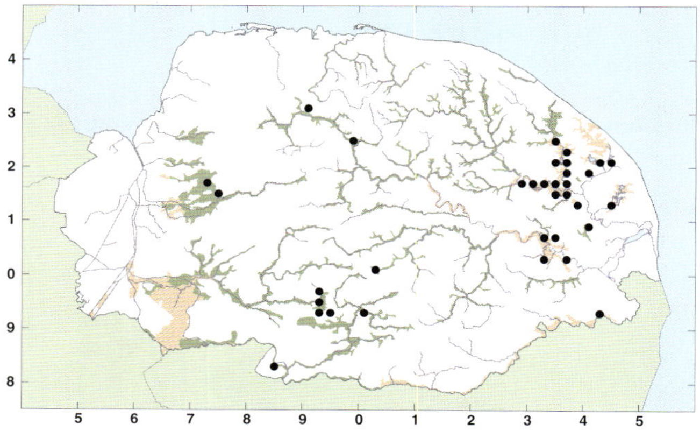

Scarce, found in damp, usually base rich, peaty soils. A colony of many hundred plants was found at Guist Fen W92 in 1985 by RW.David. A nationally scarce species.

Carex appropinquata - Thompson

C. diandra Schrank **Lesser Tussock Sedge**

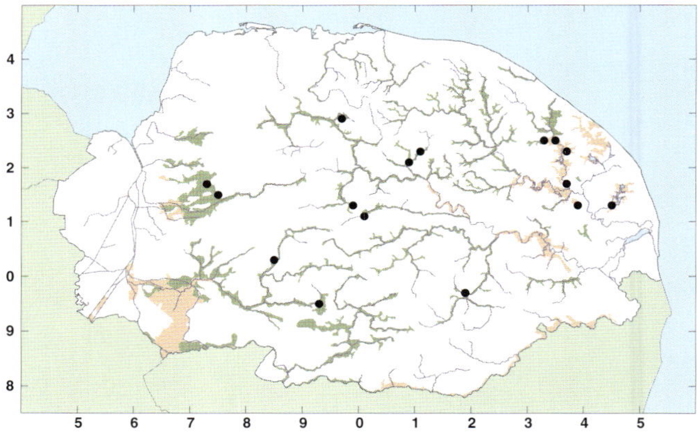

A scarce plant of fens, marshes and meadows.

C. otrubae Podp. **False Fox Sedge**

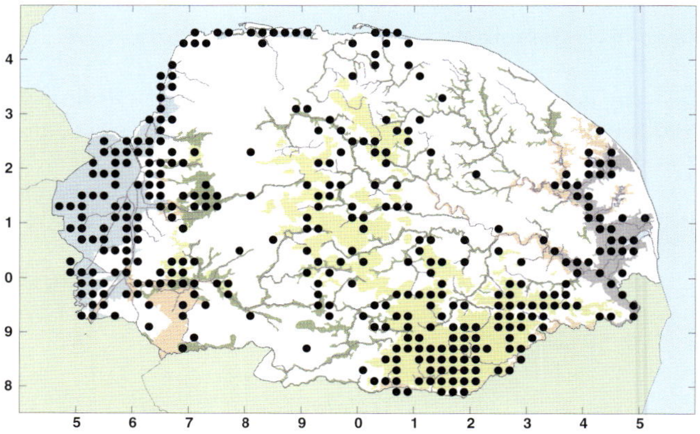

Frequent by dykes, streams and ponds especially on heavy, fertile soils, often found growing in association with rushes. The commonest sedge by fenland dykes.

C. x pseudoaxillaris K. Richt.
(*C. otrubae x C. remota*)
A hybrid found at Tyrrel's Wood, Pulham Market E28 in 1995, ALB conf. AC.Jermy and at Carleton Rode E19 in 1996; Stoke Holy Cross E20 in 1997 and Langley E30 in 1998, all Flora Group. W0. E4.

C. spicata Huds. **Spiked Sedge**
Syn. *C. contigua* Hoppe

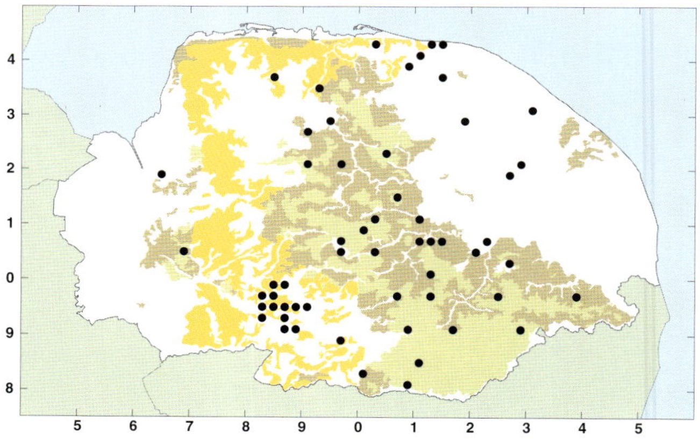

Scarce in grassland and woodland rides on moist, heavy soils.

C. muricata L. **Prickly Sedge**

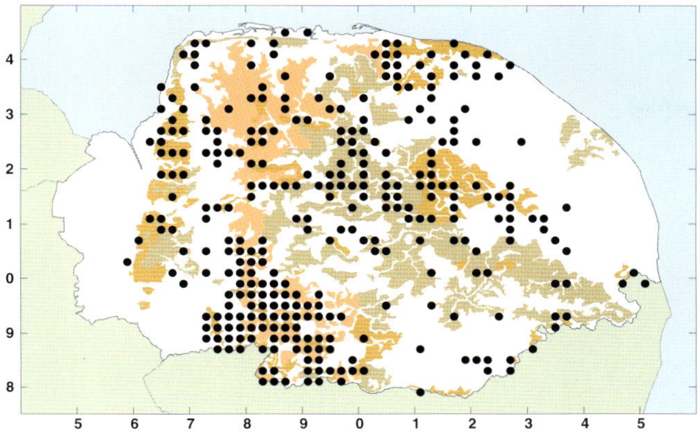

Frequent in grassland on light, sandy soils. Our plant is ssp. **lamprocarpa** Celak., (syn. *C. pairei* F.W.Schultz).

C. divulsa Stokes ssp. **divulsa** **Grey Sedge**

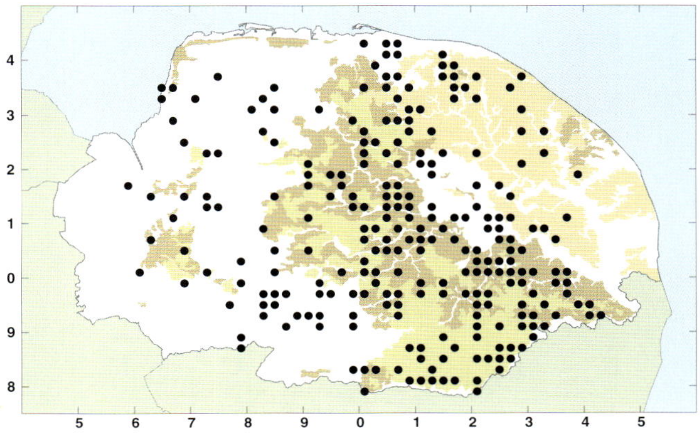

Frequent on neutral, dry soils, usually in the open though it can also be found in shady sites.

C. divulsa Stokes ssp. **leersii** (Kneuck.) W.Koch **Grey Sedge**
Syn. *C. polyphylla* Kar. & Kir.

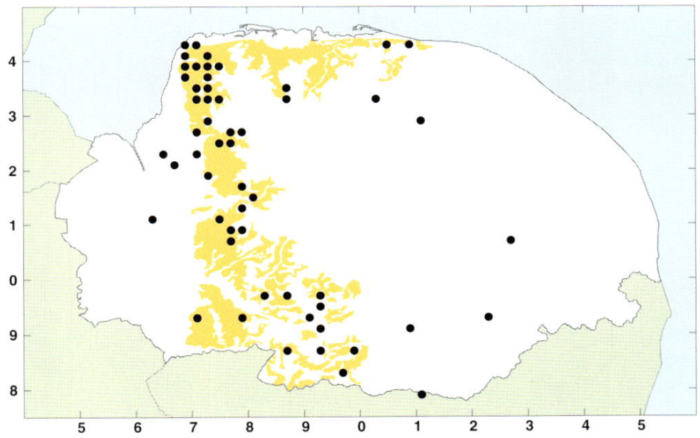

A plant of well-drained, chalky banks and calcareous grassland.

C. spicata and the two subspecies of *C. divulsa* can sometimes be very difficult to place, especially on roadside verges when they often grow atypically after being cut when in full growth.

C. arenaria L. **Sand Sedge, Net-rein** (local in Breckland)

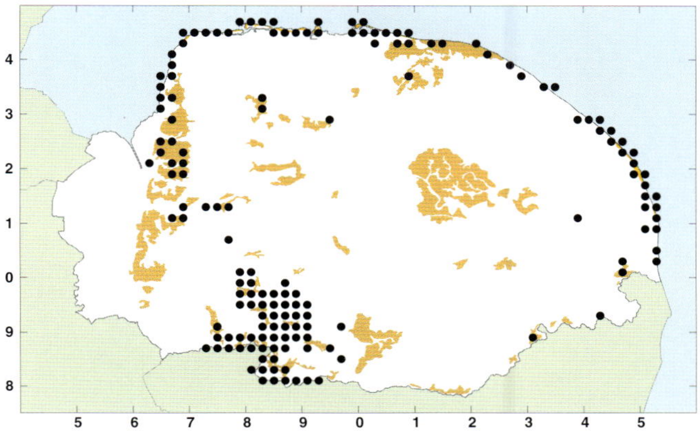

Frequent in dry, sandy places, both on coastal dunes and inland. A valuable sand binder, rhizomes rooting and continuing in the same direction forming typical lines of advance in sand pits as well as in more natural habitats.

C. disticha Huds. **Brown Sedge**

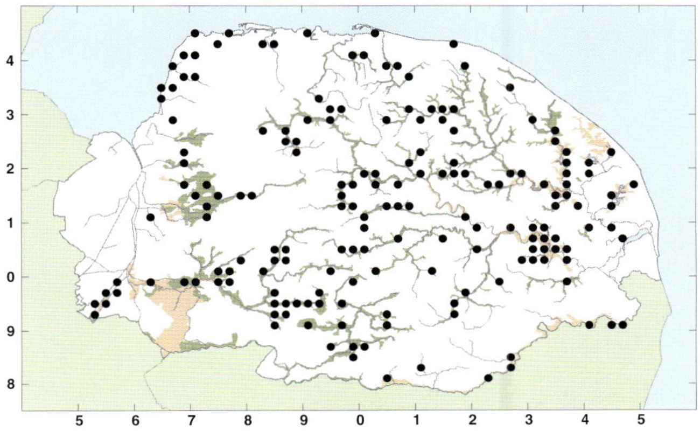

Not uncommon in marshes, fens and wet meadows, but avoiding water-logged soils.

C. divisa Huds. **Divided Sedge**

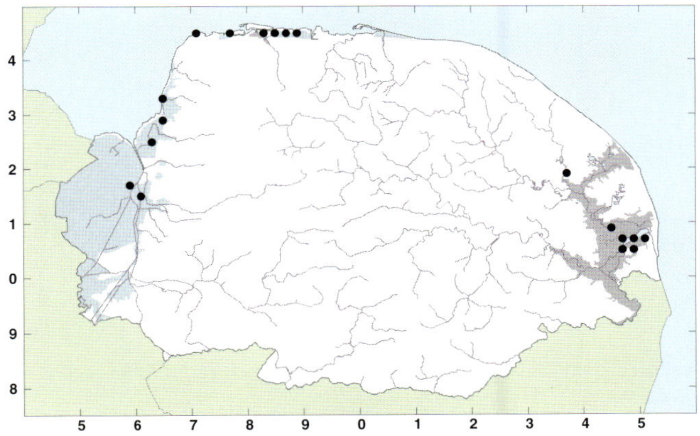

A scarce, though often locally frequent plant, found in coastal and estuarine marshes.

C. remota L. **Remote Sedge**

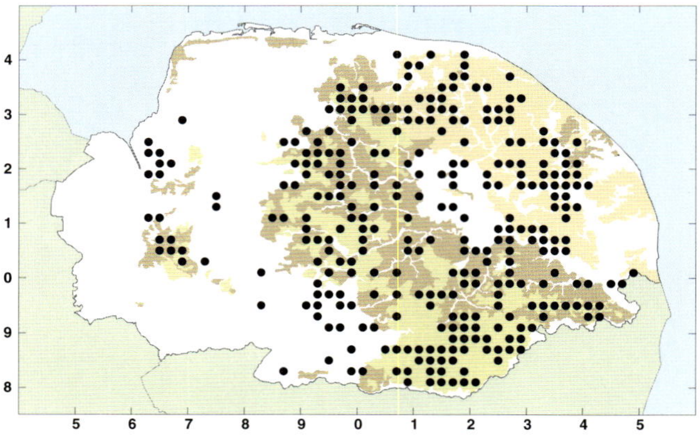

Frequent in old woods, hedgebanks and shady places, chiefly on heavy or moist soils.

C. ovalis Gooden. **Oval Sedge**

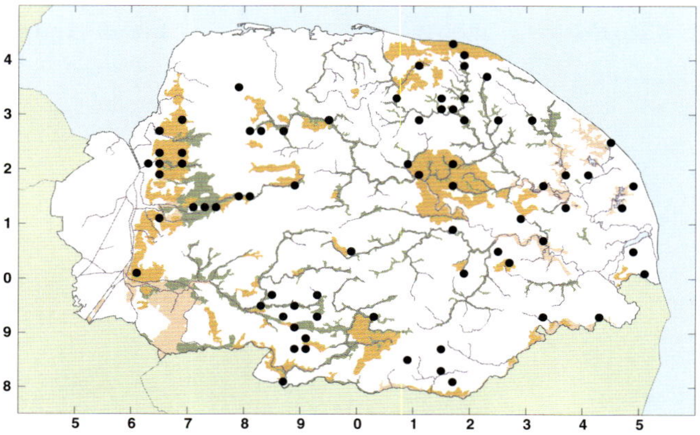

Scattered in damp, acid peaty ground.

C. echinata Murray **Star Sedge**

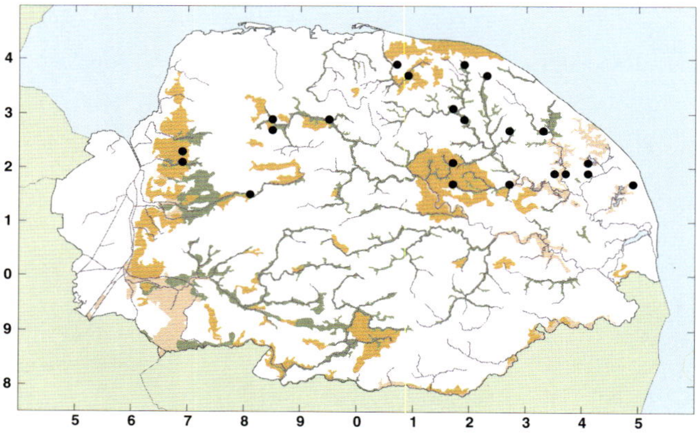

A scarce plant of bogs and acid marshes, much reduced in recent years.

C. dioica L. **Dioecious Sedge**

A rare plant of alkaline flushes in bogs and peaty soils. Found at Buxton Heath E12, comm. C.Penny; Beeston E14, FF; Foulden W79 in 1988; Roydon W62 in 1992, Flora Group; East Walton W71 in 1993; Badley Moor E01 in 1995, S.Smart and at Swanton Morley E01 in 1997, S.Henson. W3. E4.

C. curta Gooden. **White Sedge**

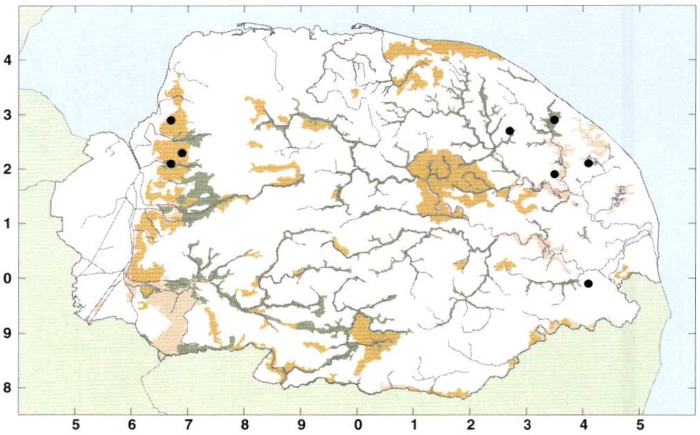

Although frequent in northern and western Britain, this sedge of acid, water-logged peat is very scarce in the east. It is usually found growing in pools and on dyke sides.

C. hirta L. **Hairy Sedge**

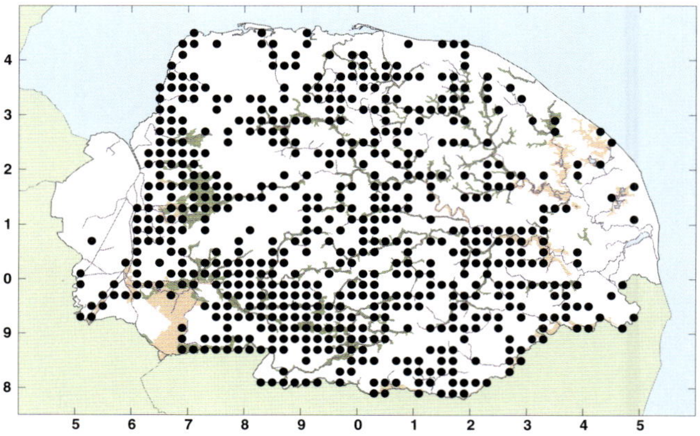

Common in neutral to base rich grassland in both damp and relatively dry sites. The commonest sedge in Norfolk

C. lasiocarpa Ehrh. **Slender Sedge**

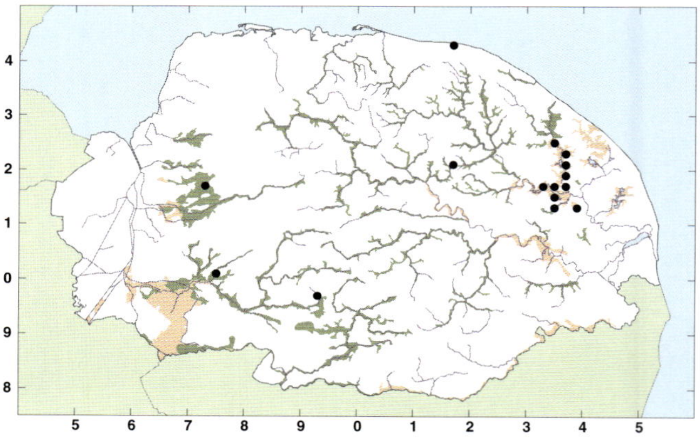

A rare plant in eastern and southern Britain, with us confined to peaty fens.

C. acutiformis Ehrh. **Lesser Pond Sedge**

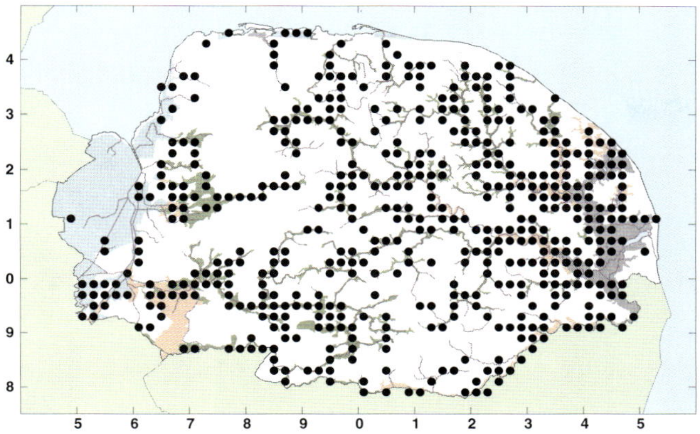

Frequent by still and flowing water, also in water-logged ground.

C. riparia Curtis **Greater Pond Sedge**

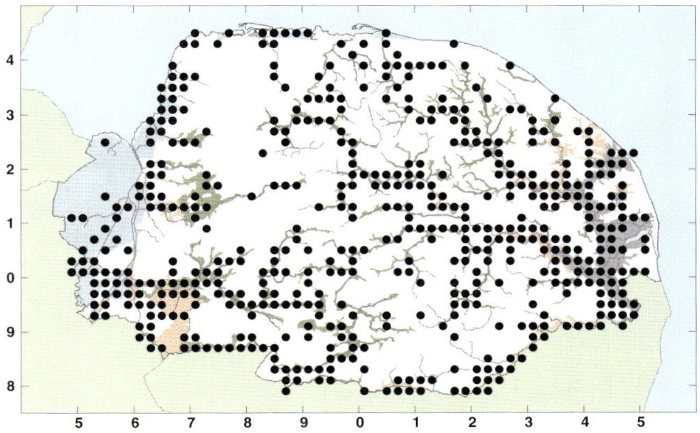

Frequent by still and flowing water and found in all but the most brackish dykes. Often grows with the preceding species.

C. pseudocyperus L. **Cypress Sedge**

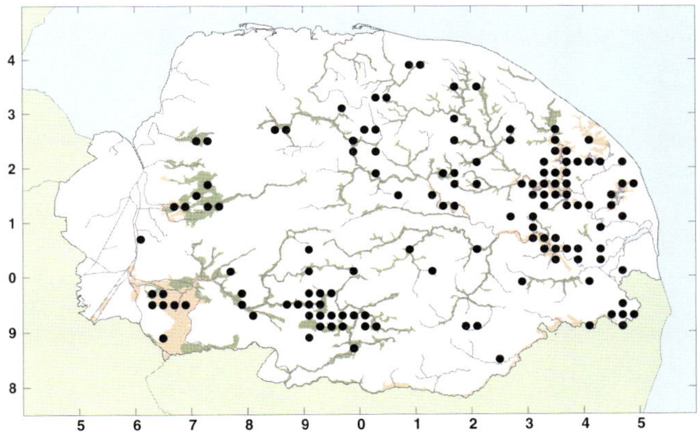

Widely scattered in ponds, marshes, fens and wet woods.

■ **C. x justi-schmidtii** Junge
(*C. pseudocyperus x C. rostrata*)
Recorded once at Gt.Hockham W99 in 1955, its only known British station.

C. rostrata Stokes **Bottle Sedge**

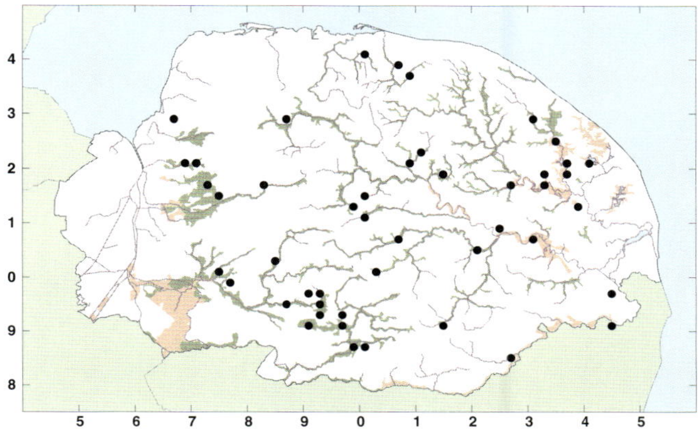

Occasional by ponds and in wet marshes, mostly on acid soils. Generally rare in eastern Britain.

Carex rostrata - Merton

C. vesicaria L. **Bladder Sedge**

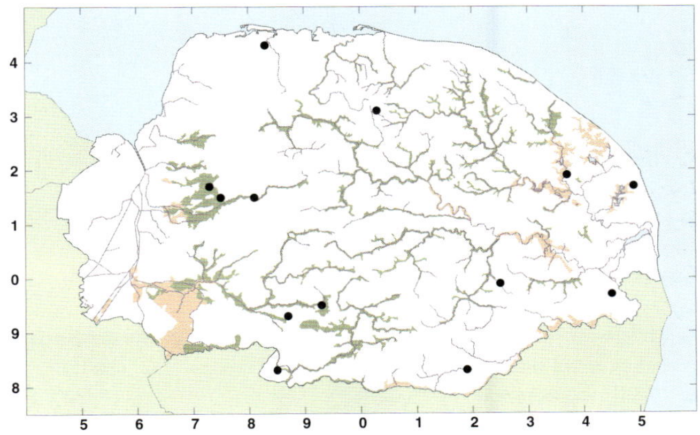

A scarce plant of marshes and wet meadows.

C. pendula Huds. **Pendulous Sedge**

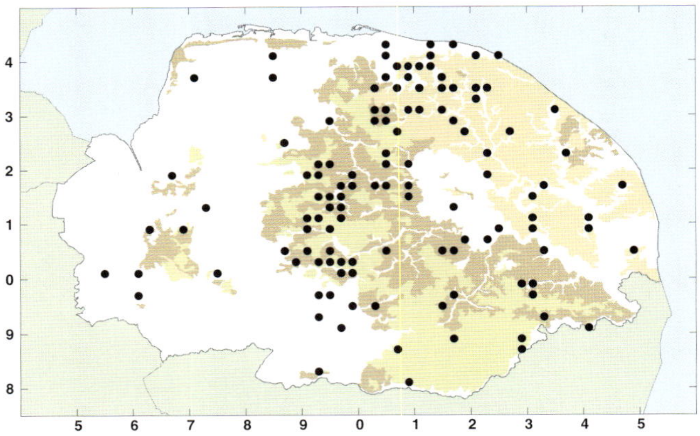

A mainly woodland species found on heavy, wet soils, also occasionally naturalised from gardens.

C. sylvatica Huds. **Wood Sedge**

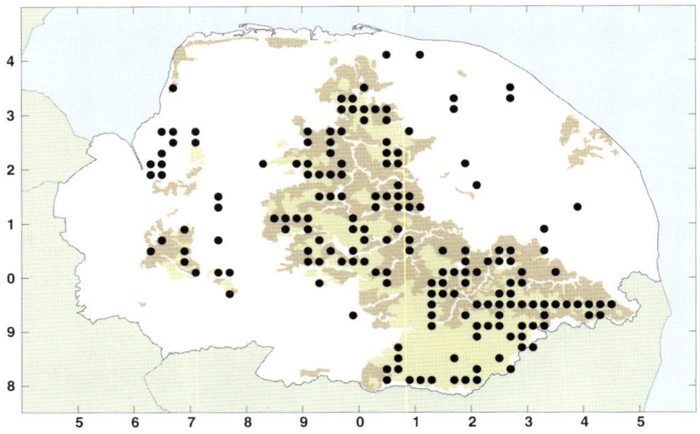

Confined to old woods and adjacent hedgebanks, a good indicator of past woodland distribution.

Carex sylvatica - Wolferton

C. strigosa Huds. **Thin-spiked Wood Sedge**
A rare plant of old woods, recorded only at Foxley E02, first by RPL in 1954, still present in 1996, ALB. WO. E1.

C. flacca Schreb. **Glaucous Sedge**

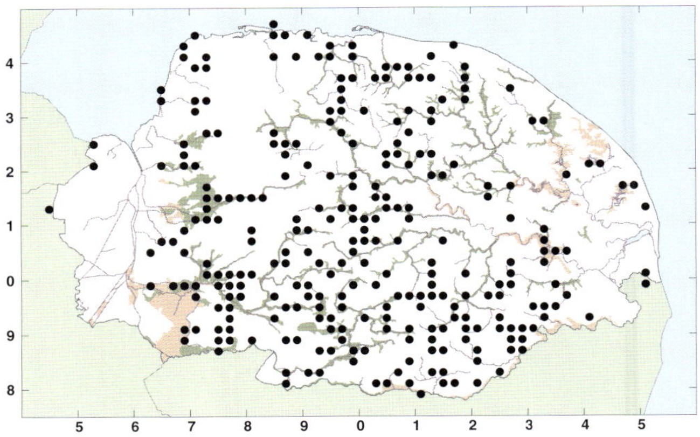

Frequent in grassland, chiefly on base rich soils.

C. panicea L. **Carnation Sedge**

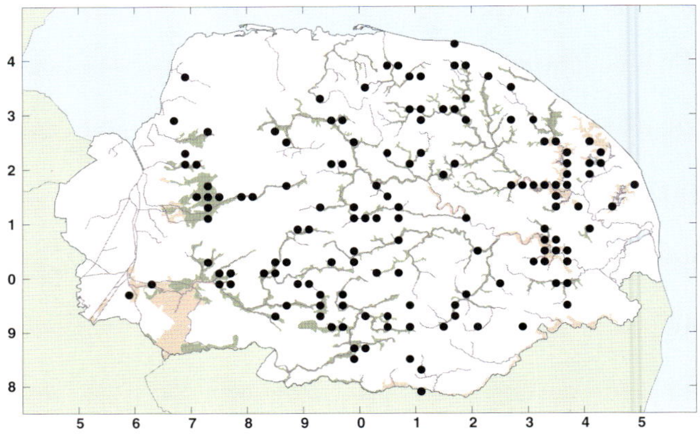

A plant of moist grassland, usually on neutral to acid soils, but sometimes growing with the preceding species.

C. binervis Sm. **Green-ribbed Sedge**

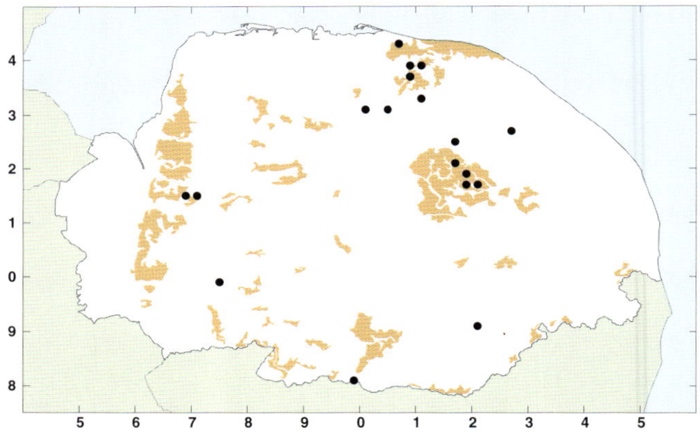

A common plant of northern Britain, scarce with us and confined to undisturbed moist, acid ground.

■ **C. trinervis**
Recorded at Ormesby St Michael E41 in 1869 by HG.Glasspoole. This record was doubted in Petch & Swann 1968, however, a plant descended from material collected at Ormesby has been seen by AC.Jermy who confirms that the identification is correct.

C. distans L. **Distant Sedge**

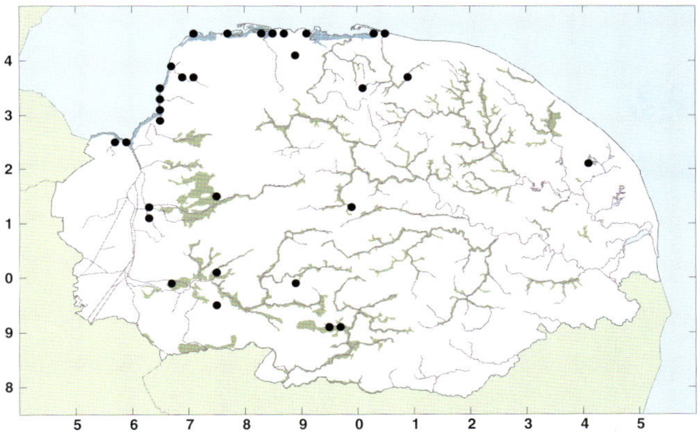

Scattered in maritime areas, also found inland in base rich fens.

C. extensa Gooden. **Long-bracted Sedge**
Confined to coastal marshes on the Wash and the North Coast from W62 as far east as Cley E04. W7. E2.

C. hostiana DC. **Tawny Sedge**

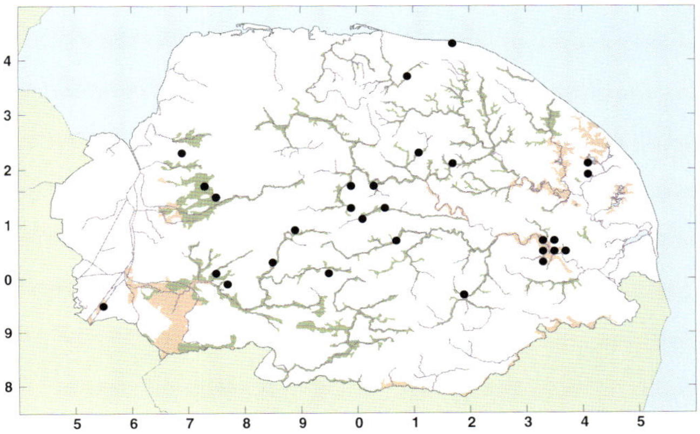

Scattered in base rich meadows and fens, but always scarce.

■ **C. x fulva** Gooden.
(*C. hostiana x C. viridula ssp. brachyrrhyncha*)
Considered by ELS in 1968 to be the most frequent sedge hybrid in the county, but both parents are now much less common and it was last recorded at East Walton W71 in 1978, ELS.

C. viridula Michx. ssp. **brachyrrhyncha** (Celak.) B. Schmid
Syn. *C. lepidocarpa* Tausch **Yellow Sedge**

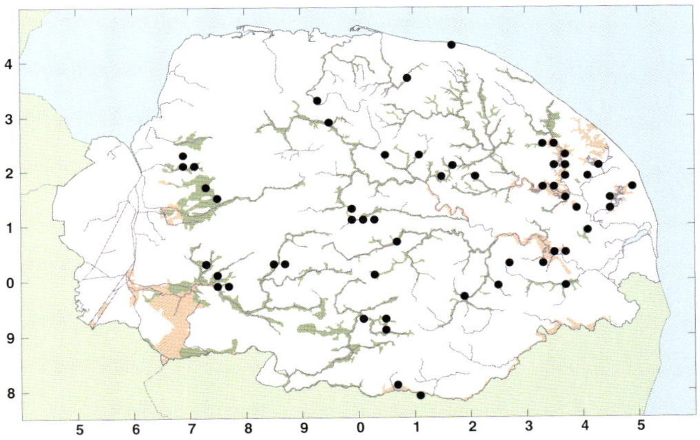

This sub-species is scattered in fens and marshes, particularly those which are base rich, but it can no longer be described as common as it was in 1968.

C. viridula Michx. ssp. **oedocarpa** (Andersson) B. Schmid
Syn. *C. demissa* Hornem.

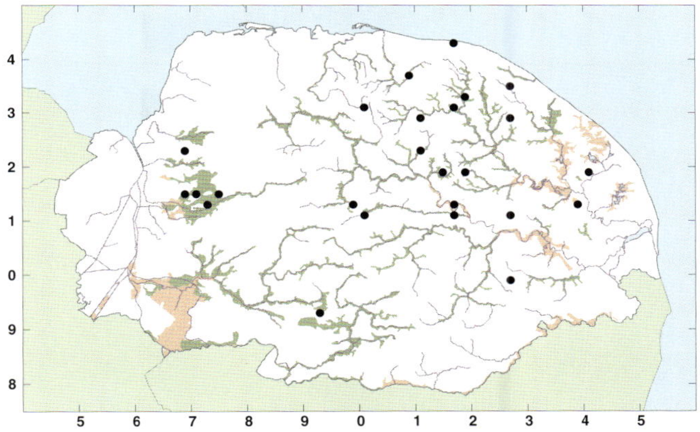

Found in similar habitats to the previous subspecies, but less confined to base rich soils.

C. viridula Michx. ssp. **viridula**
Syn. *C. serotina* Mérat
Always the scarcest of the three subspecies and found only on acid peat. It has been recorded during this survey only at East Walton W71. W1. E0.

C. pallescens L. **Pale Sedge**

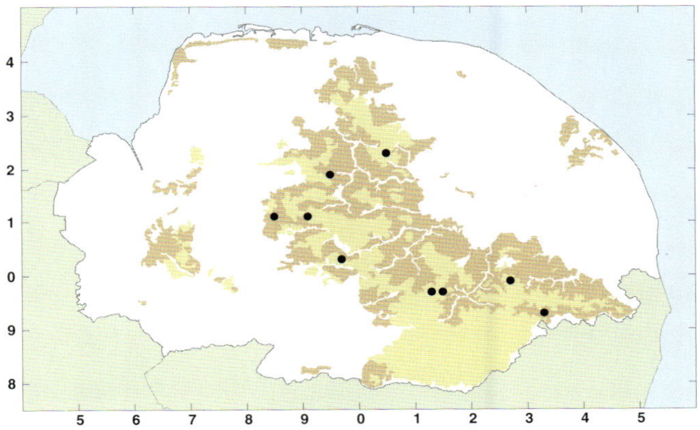

A rare plant of woodland rides on moist, heavy soils.

C. caryophyllea Latourr **Spring Sedge**

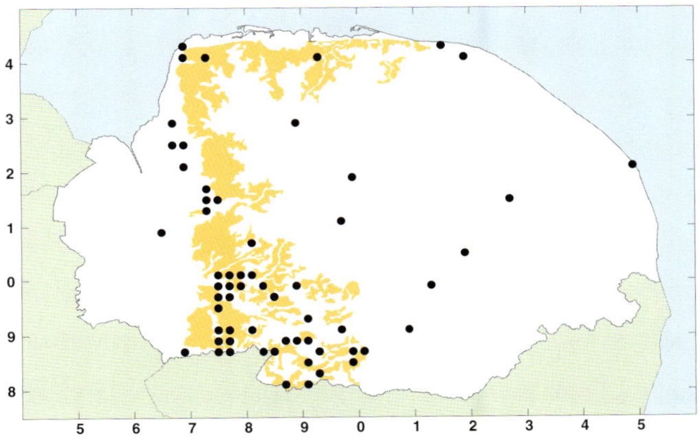

Scattered in dry, usually base rich grassland, apparently much decreased since Petch & Swann, 1968.

C. ericetorum Pollich **Rare Spring Sedge, Breckland Sedge**
In Norfolk confined to the Breckland where it grows on undisturbed, calcareous soils with sharp drainage, often on anthills. Found during this survey at Foulden W70 and W79, Weeting W78 and W89 and Stanford W89. A nationally scarce species. W7. E0.

Carex ericetorum - Weeting

C. pilulifera L. **Pill Sedge**

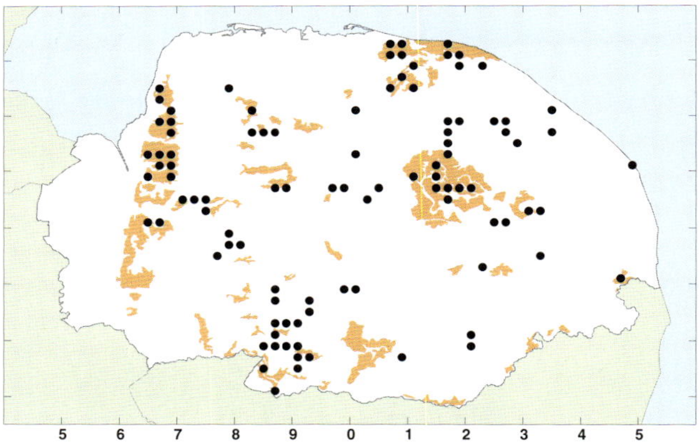

A plant of dry, acid soils, found on both heather and grass heaths.

■ **C. limosa** L. **Mud Sedge, Bog Sedge**
Recorded from four sites this century in acid bogs, last seen at Sutton E32 in 1963.

C. acuta L. **Slender Tufted Sedge**

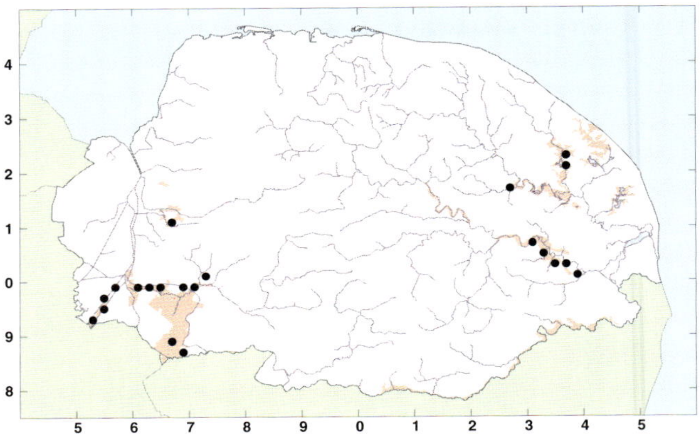

A scarce plant of wet, grassy places typical of washlands by rivers.

■ **C. x prolixa Fr.**
(*C. acuta x C. elata*)
Recorded in four sites before 1950.

C. nigra (L.) Reichard **Common Sedge**

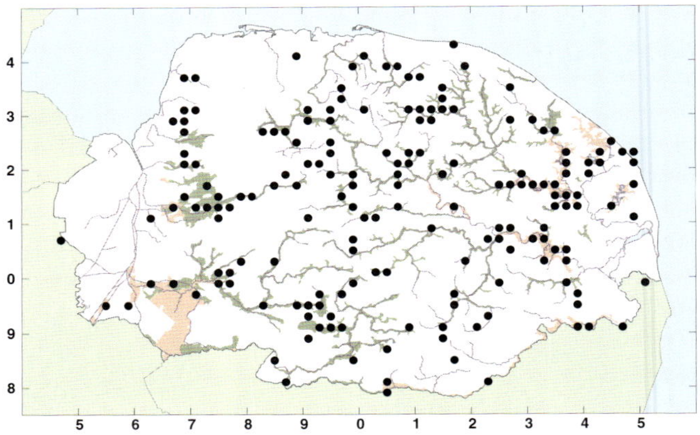

Widely scattered in marshy places, often, but not exclusively, on acid soils.

■ **C. x turfosa** Fr.
(*C. nigra x C. elata*)
Recorded four times before 1965.

C. elata All. **Tufted Sedge**

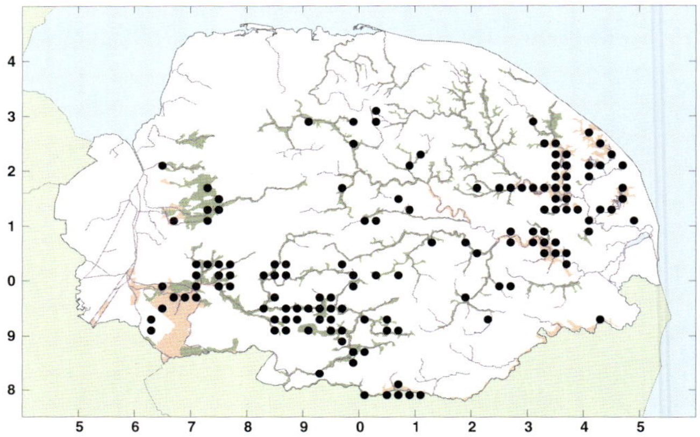

Frequent and constant in fens and reed swamps on base rich to mildly acid soils, commoner in East Anglia than in the rest of Britain.

C. pulicaris L. **Flea Sedge**

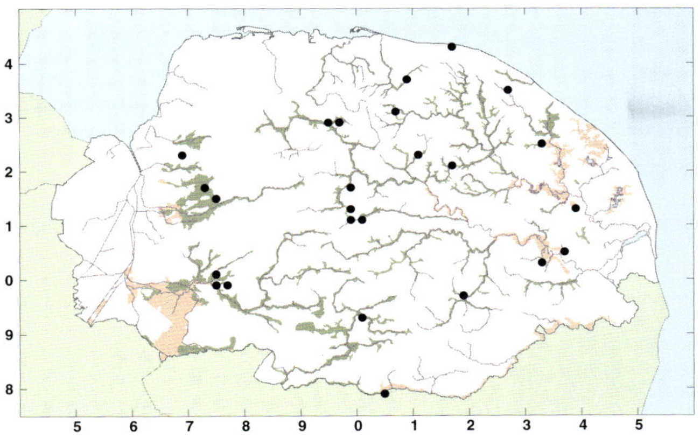

A scarce plant of base rich flushes in bogs and in fens.

POACEAE

Sasa palmata (Burb.) E.G.Camus **Broad-leaved Bamboo**
Native to Japan, grown in parks and gardens, occasionally naturalised in moist soils. W2. E4.

Pseudosasa japonica (Siebold & Zucc. ex Steud.) Makino ex Nakai **Arrow Bamboo**
Syn. *Arundinaria japonica* Siebold & Zucc. ex Steud.
Introduced to parks and gardens from Japan and Korea. Naturalised chiefly in damp sites. W2. E3.

Nardus stricta L. **Mat-grass**

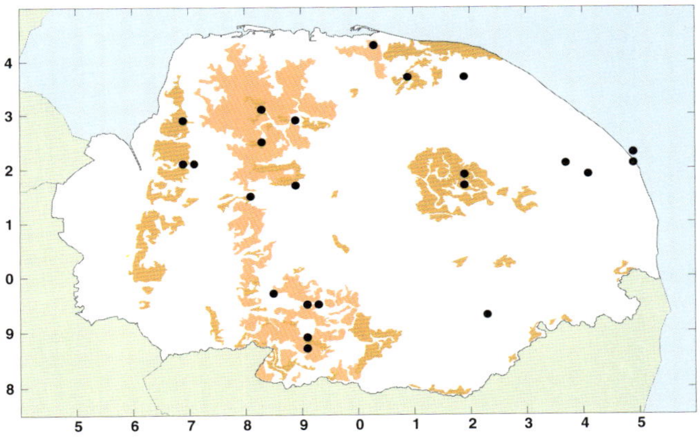

A perennial grass chiefly of acid grassland and heaths.

Carex elata - Foulden

Milium effusum L. **Wood Millet**

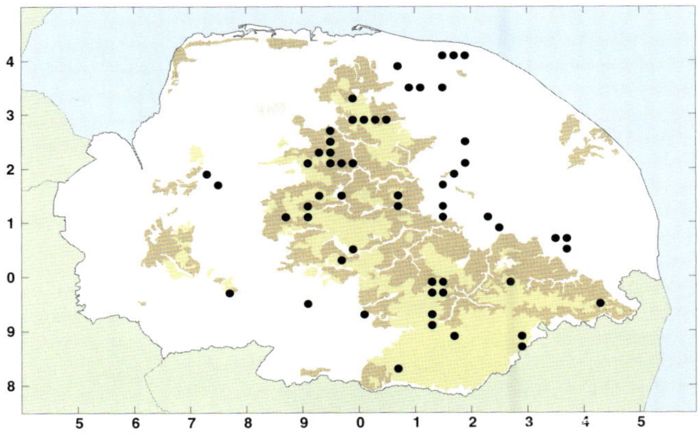

Perennial found in moist woods on heavy soils. A few sites may represent introductions from pheasant feed.

Festuca pratensis Huds. **Meadow Fescue**

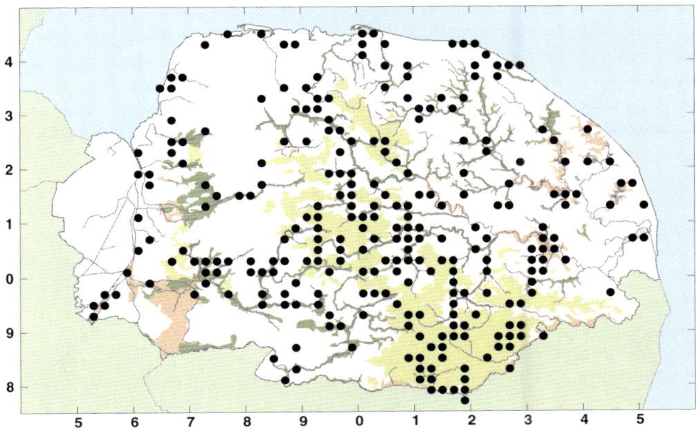

Perennial found in grassland chiefly on heavy, moist soils, not common.

■ **F. x schlickumii** Grantzow
(*F. pratensis x F. gigantea*)
Recorded twice, at Watton W90 in 1918 and at Appleton W72 in 1962.

F. arundinacea Schreb. **Tall Fescue**

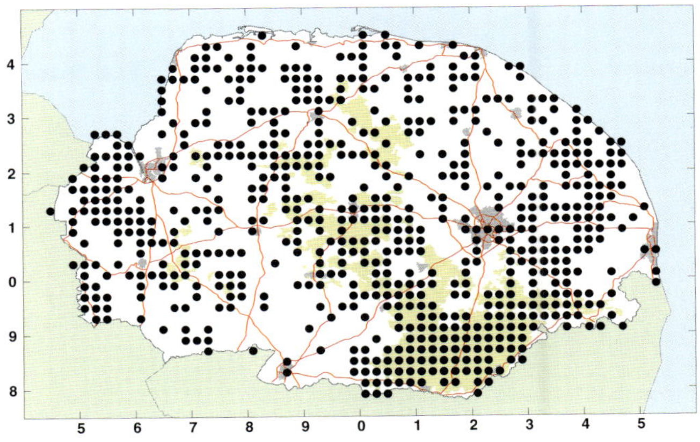

Perennial found in permanent grassland on moist soils where it is mostly native. Also widely used as a forage grass and the large patches found on roadside verges, even on dry ground, are likely to be from this source.

F. gigantea (L.) Vill. **Giant Fescue**

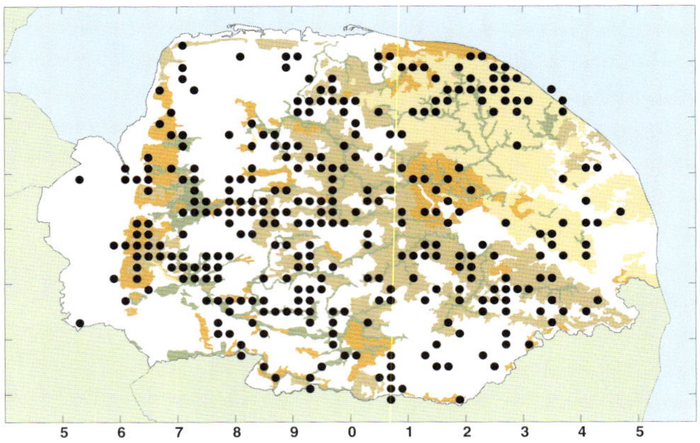

Perennial not infrequent by shady hedgerows and woods.

F. heterophylla Lam. **Various-leaved Fescue**
Doubtfully native to woods in parts of Britain, but with us a certain introduction. Recorded at Diss E08 in 1978, AC and still there 1990; Brundall E30 in 1995, SP and Wormegay W61 in 1997, SF.
W1. E2.

F. arenaria Osbeck **Rush-leaved Fescue**
Syn. *F. juncifolia* St-Amans

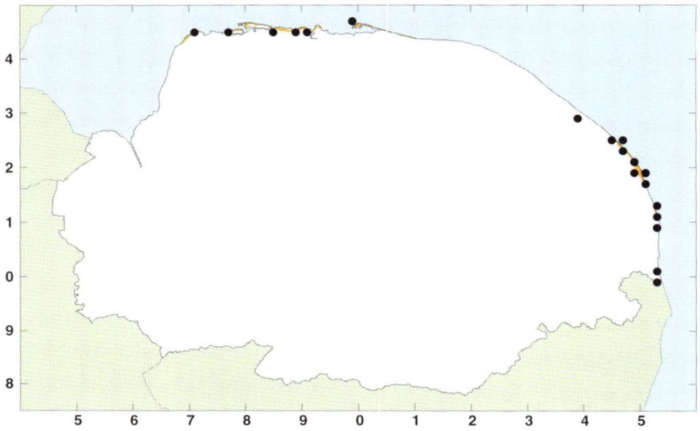

A scarce plant of sand dunes, usually growing with marram grass and found during this survey between Holme W64 and Blakeney E04 and from Waxham E33 to Yarmouth E50. A nationally scarce species.

F. rubra L. **Red Fescue**

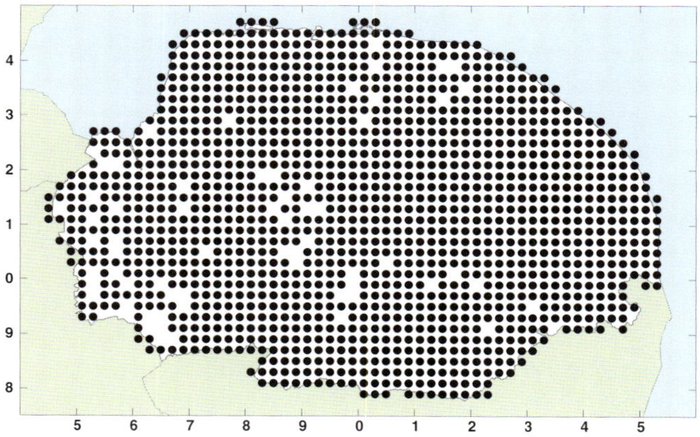

A common perennial grass on well-drained soils in a wide variety of habitats, very common on walls. Most of our plants are referable to ssp. **rubra**, but ssp. **juncea** (Hack.) K.Richt., was recorded at Wolferton W62 in 1944 and 1969. Ssp. **litoralis** (G.Mey.) Auquier, (**Cumberland Turf**), is frequent in salt marshes being recorded from Wolferton W62 around the coast to Blakeney, E04. Ssp. **megastachys** Gaudin, has been recorded in two sites, at Castle Rising golf-course in 1975 and 1977 by CEH and ELS and at Roydon E18 in 1998, AC. Ssp. **commutata** Gaudin (**Chewing's Fescue**) occurs on roadsides, especially on bypasses and newly sown verges.

F. ovina L. **Sheep's Fescue**

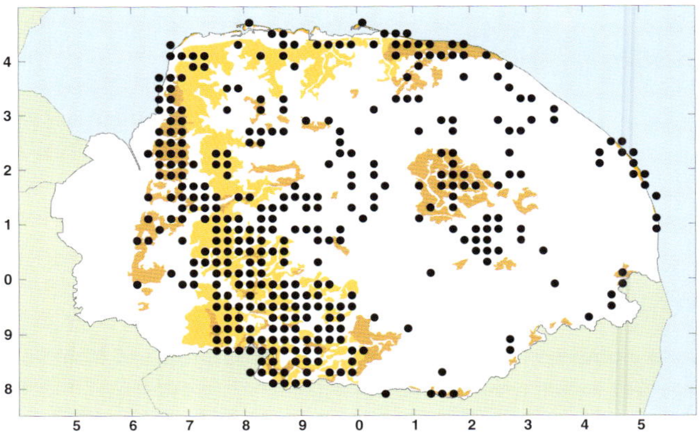

A perennial found on light, well-drained soils. The map shows the species in its broadest sense. Ssp. **ovina** is frequent on the chalk.

Festuca ovina - Stanton

F. filiformis Pourr. **Fine-leaved Sheep's Fescue**
Syn. *F. tenuifolia* Sibth.

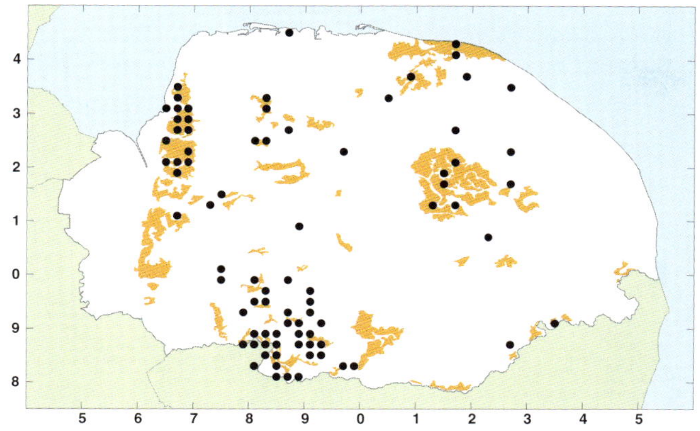

Scattered in dry, sandy places on acid soils, especially on the greensand and in Breckland.

F. brevipila R.Tracey **Hard Fescue**
Introduced from mainland Europe, this grass is now frequently sown beside new roads. It was noted on a grave at Banham E08 in 1958, AC and near the new A11 at Thetford W88 in 1996, N.Gibbons. W1. E0.

X Festulolium loliaceum (Huds.) P. Fourn. **Hybrid Fescue**
(*F. pratensis x Lolium perenne*)
The commonest of the series of intergeneric hybrids and scattered throughout the county. W10. E5.

■ **X F. braunii** (K.Richt.) A. Camus
(*F. pratensis x Lolium multiflorum*)
Recorded for west Norfolk in Stace, 1975.

X F. holmbergii (Dörfl.) P. Fourn.
(*F. arundinacea x Lolium perenne*)
Recorded at Harford tip, Norwich E20, in 1972, PGL & AC, and during this survey on a roadside bank at Wiggenhall St Mary, W51 in 1997, RMP. W1. E0.

Festuca arundinacea x Lolium multiflorum
Rare hybrid found once, in rough grassland on the banks of the New Bedford River at Denver W59 in 1998, RWE, JG & FS. det. TA.Cope. W1. E0.

■ **X Festulpia** Melderis ex Stace & R. Cotton
(*Festuca rubra x Vulpia myuros*)
Recorded once, on the shingle bank at Snettisham, W63 in 1974, RPL.

Lolium perenne L. **Perennial Rye-grass**
A ubiquitous grass of tracks, waste places, and grassland. Widely sown on roadside verges, usually with a commercial strain which is better able to withstand the harsh conditions than the native grass. Var. **longiglume** Grantzow was found at Longham, W91 in 1998, D&BM det. RMP. W707. E714.

■ **L. x boucheanum** Kunth
(*L. perenne x L. multiflorum*)
Found only once, at Beetley pit, W91 in 1973 by CEH, RPL and ELS. It is not easy to distinguish from its parents so may be more frequent than records suggest.

L. multiflorum Lam. **Italian Rye-grass**
Annual or biennial introduced from southern Europe, often sown on road verges. W131. E168.

■ **L. multiflorum x L. temulentum**
Recorded at Harford Tip, Norwich E20 in 1975 by GD.Watts det.CEH.

■ **L. temulentum** L. **Darnel**
Very scarce plant of waste ground and rubbish tips, recorded from six sites, the most recent at Saxthorpe E13 in 1980, MAB.

Vulpia fasciculata (Forssk.) Fritsch **Dune Fescue**
Syn. *V. membranacea* auct.non (L.) Dumort.
Annual, a rare plant of coastal sand dunes, seen by A.Watkinson at Holme W74 in 1991, where it has been known since 1950 and still survives, also on Yarmouth North denes E50 in 1991, ALB. A nationally scarce species. W1. E1.

V. bromoides (L.) Gray **Squirrel-tail Fescue**

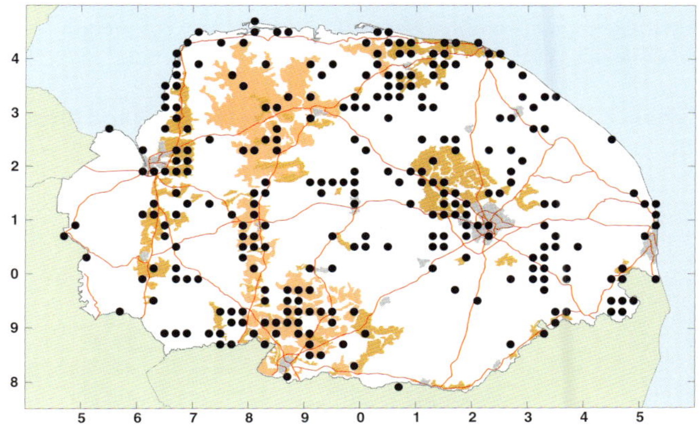

Annual, frequent on dry, sandy soils.

V. myuros (L.) C.C.Gmel. **Rat's-tail Fescue**
(Incl. *V. megalura* (Nutt.) Rydb.)

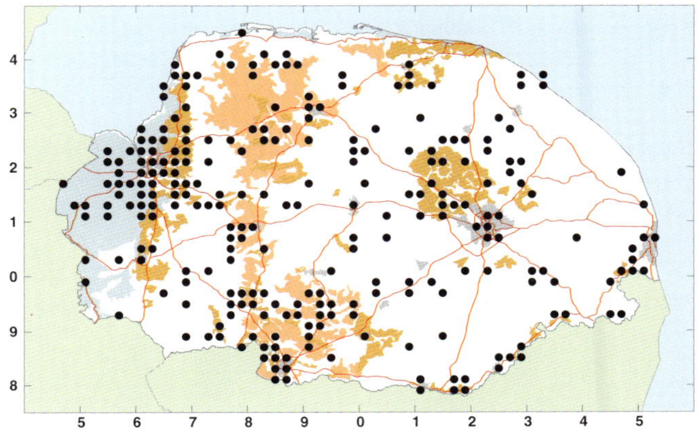

Annual, frequent on dry, sandy soils, shingle, walls and in gravel pits. These records do not differentiate the three forms of this taxon, forma *myuros*, forma *megalura* (Nutt.) Stace & R.Cotton and forma *hirsuta* (Hack.) Blom.

V. ciliata Dumort. **Bearded Fescue**

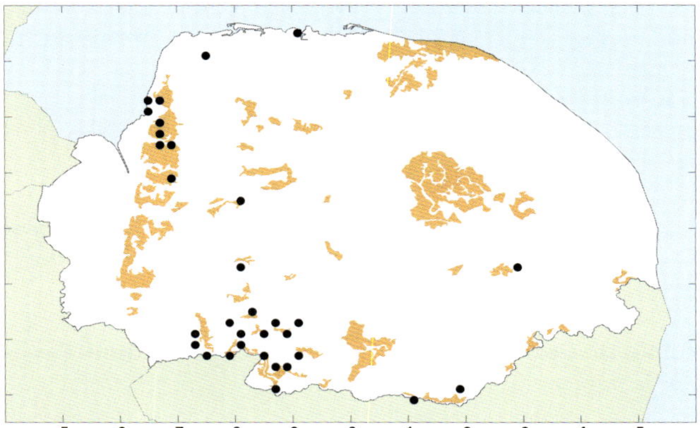

A scarce annual found on light, acid sands. Our plant is ssp. **ambigua** (Le Gall) Stace & Auquier. A nationally scarce species.

■ **V. unilateralis** (L.) Stace **Mat-grass Fescue**
Syn. *Nardurus maritimus* (L.) Murb.
Annual, seen in the county only at Sturston Warren W89 and at Riddlesworth W98, both in 1979.

Cynosurus cristatus L. **Crested Dog's-tail Grass**

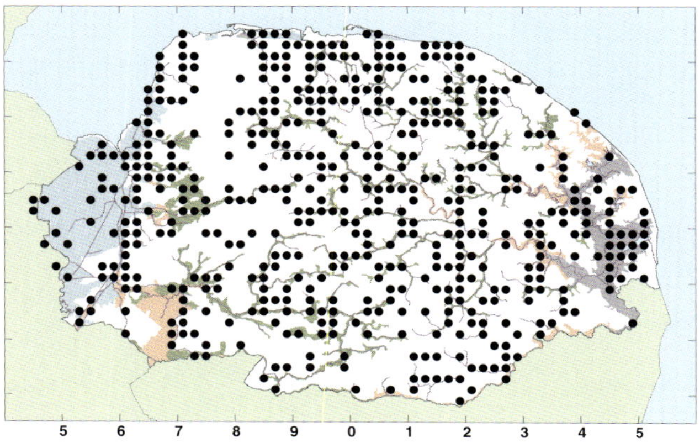

Evergreen perennial frequent in established grassland, but not on very dry or acid soils. A common constituent of grass seed mixes for rough grazing.

C. echinatus L. **Rough Dog's-tail Grass**
Spring germinating annual native to southern Europe. Recorded from King's Lynn W62 in 1990, CRS and Colney E10 in 1995, M.Gurney et al. W1. E1.

Puccinellia maritima (Huds.) Parl. **Common Saltmarsh-grass**

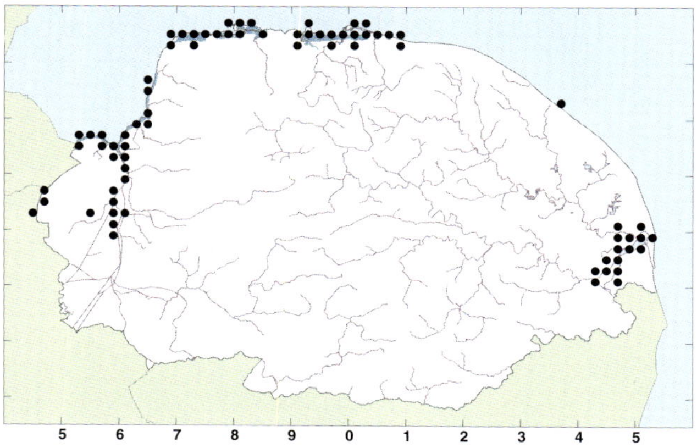

Herbaceous perennial scattered in muddy places by tidal rivers, estuaries and salt marshes, also occasionally inland on road verges.

■ **P. x hybrida** Holmb.
(*P. maritima x P. distans*)
Recorded once, at Burnham Overy W84 in 1967.

P. distans (Jacq.) Parl. **Reflexed Saltmarsh-grass**

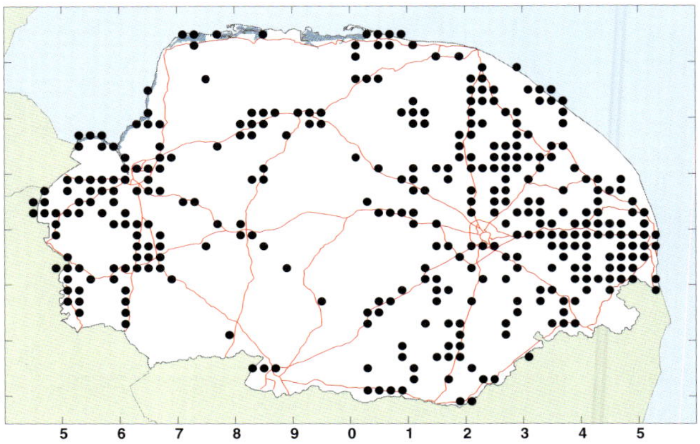

Herbaceous perennial found on coastal salt marshes, but now far commoner on inland roadside verges where it favours roundabouts, turnings and patches of firm, bare soil. First noted in Norfolk in this habitat by the new bypass around King's Lynn W62 in 1978, ELS. Recorded also as a weed of arable fields at West Caister E51 in 1998, ALB and on waste ground at Mundham E39 in 1998, Flora Group.

■ **P. x pannonica** (Hack.) Holmb.
(*P. distans x P. rupestris*)
Recorded at Norton Subcourse E40 in 1975, RPL.

P. fasciculata (Torr.) E.P.Bicknell **Borrer's Saltmarsh-grass**
Syn. *P. pseudodistans* (Crep.) Jansen & Wacht.
Rare on bare places in salt marshes and on sea walls, also rarely inland by roads. Found at Holme W74 in 1987, CPP; Salthouse E04 in 1993, ALB; Cley E04 in 1994, KKH; Breydon Wall E40 in 1997, RWE & ALB and Thornham W74 in 1998, JEG & FS. A nationally scarce species. W3. E3.

P. rupestris (With.) Fernald & Weath. **Stiff Saltmarsh-grass**
Rare on sea walls as at Reedham, E40 and the Halvergate Levels in 1990, L.Street; St.Olaves E49 in 1992 and at Terrington Bund W52 in 1996, SF, also occasionally on roadside verges as at Worstead E32 in 1995, KAB. A nationally scarce species. W1. E5.

Briza media L. **Quaking-grass**

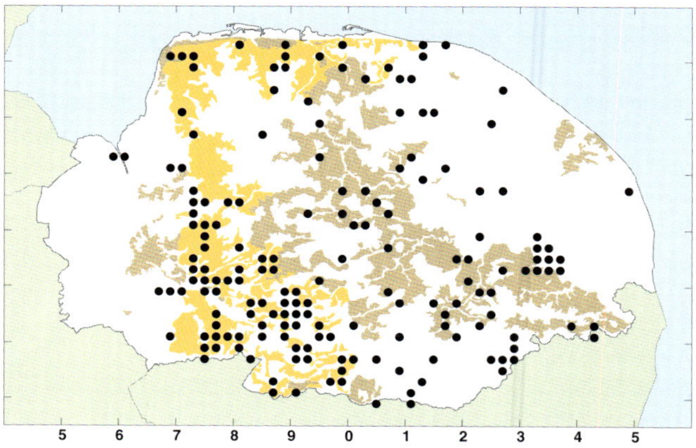

Semi-evergreen perennial scattered on base rich soils both in dry grassland and damp meadows.

B. maxima L. **Greater Quaking-grass**
Spring germinating annual introduced to gardens from the Mediterranean, occasionally found as an escape. Eight old records, four on tips, more recently at Wickmere E13 in 1997, KKH. W0. E1.

Poa annua L. **Annual Meadow-grass**
Annual, sometimes perennial, found abundantly in a wide range of habitats, flowering and seeding throughout the year W703. E716.

P. trivialis L. **Rough Meadow-grass**

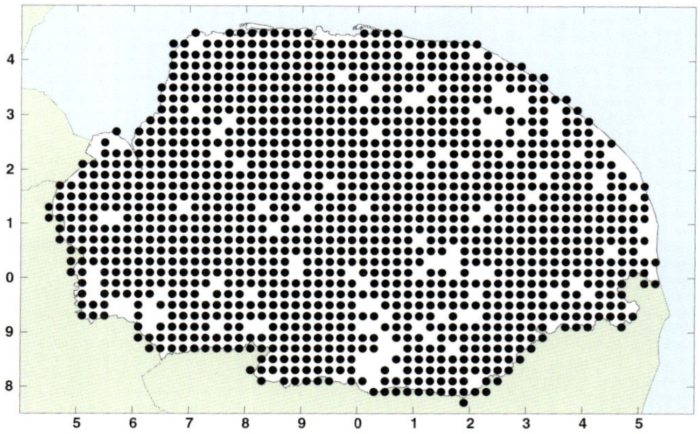

Winter green perennial common in cultivated ground and damp places, often in shade.

P. humilis Ehrh. ex. Hoffm. **Spreading Meadow-grass**
Syn. *P. subcaerulea* Sm.

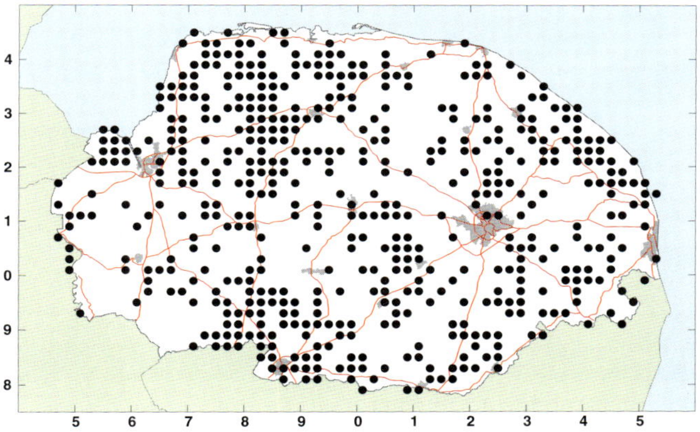

Frequent in grassland on sandy soils and on walls, also in moist meadows and on roadsides.

P. pratensis L. **Smooth Meadow-grass**

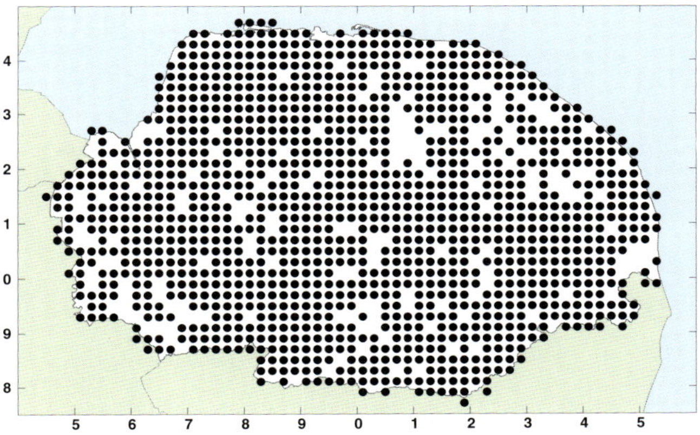

Perennial. In the broadest sense, to include both *P. humilis* and *P. angustifolia*, one of the most widespread grasses, but in the strict sense it appears to be most frequent in permanent grassland, chiefly on base rich soils.

P. angustifolia L. **Narrow-leaved Meadow-grass**

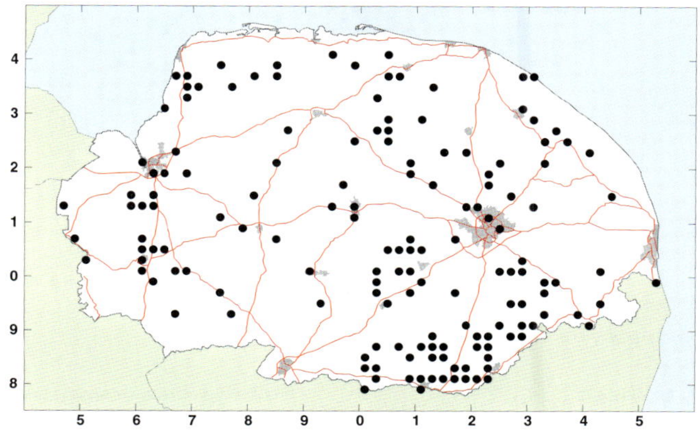

Perennial found in well-drained grassland, on banks and walls.

■ **P. chaixii** Vill. **Broad-leaved Meadow-grass**
Recorded twice, at Blickling E12 in 1960 and Guist W92 in 1972.

■ **P. chaixii x P. nemoralis**
Recorded once at Guist W92 in 1972 by RPL & ELS and confirmed by CEH as the first British record of this hybrid.

P. compressa L. **Flattened Meadow-grass**

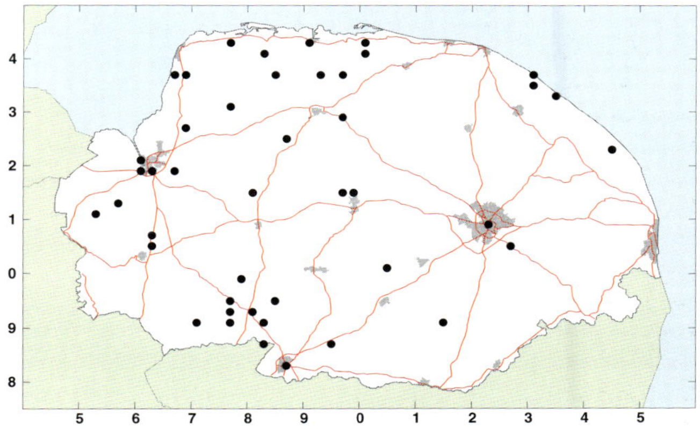

Perennial uncommon on walls and rare in dry, sandy grassland.

P. palustris L. **Swamp Meadow-grass**
Perennial recorded at Castle Acre W81 in 1945; Roydon Further Fen E08 in 1962 and at Harford tip, Norwich E20 in 1967. During this survey at Earlham E20 in 1990, ETD. W0. E1.

P. nemoralis L. **Wood Meadow-grass**

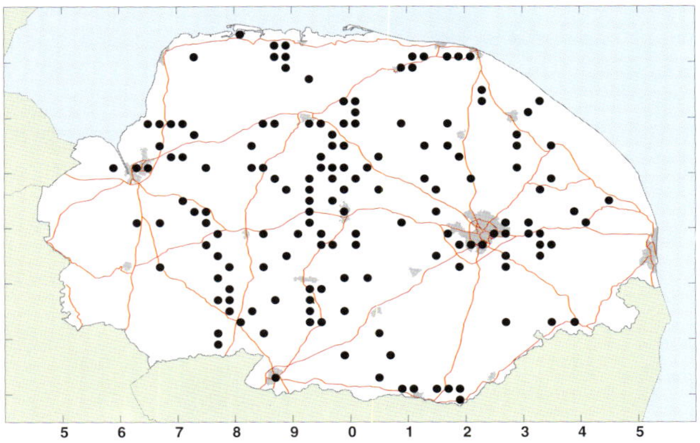

Perennial, not infrequent in woodland including dry plantations, occasionally on walls.

P. bulbosa L. **Bulbous Meadow-grass**

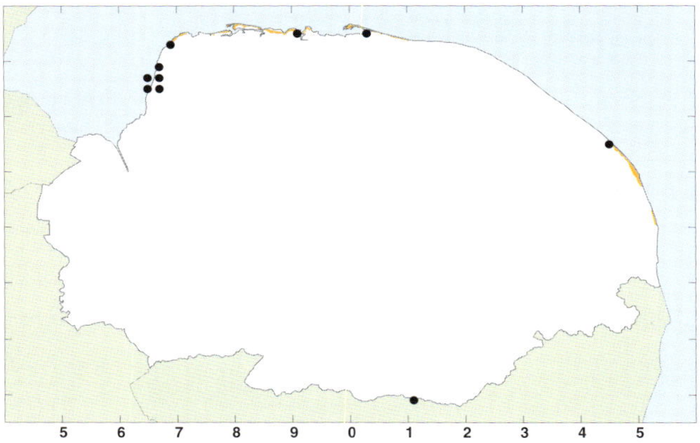

Bulbous-based perennial rare on coastal sands. It dies down in the autumn and the bulbs become dislodged and move with the sand. If spring storms come before they are sufficiently anchored, small plants can come loose, and with their young leaves acting as sails, bowl along the beach like tumbleweeds. obsv. PRC. Also found on sandy soil near Diss E18, AC.

Dactylis glomerata L. **Cock's-foot Grass**
Evergreen perennial grass, ubiquitous. W705. E716.

Catabrosa aquatica (L.) P. Beauv. **Water Whorl-grass**

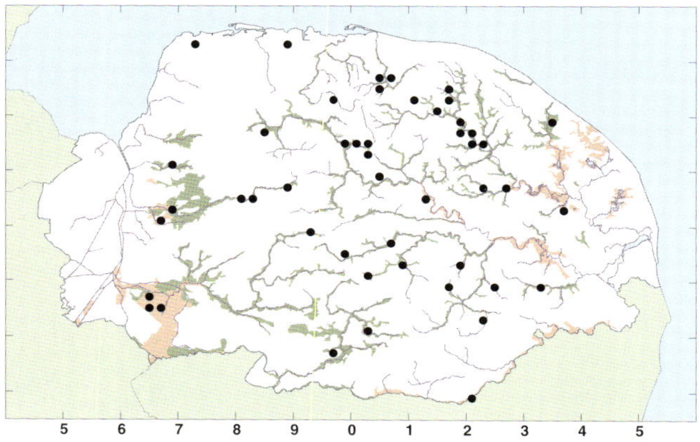

Perennial scattered in dykes and in pools in marshy ground.

Catapodium rigidum (L.) C.E. Hubb. **Fern-grass**
Syn. *Desmazeria rigida.* (L.) Tutin

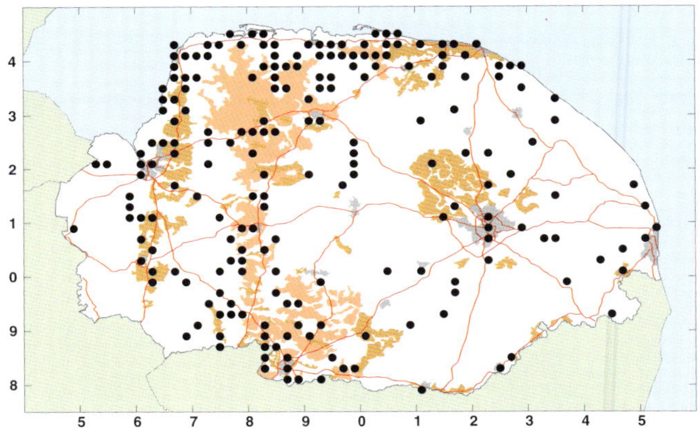

Annual, a frequent grass of walls, less often on dry, sandy soils.

Catapodium rigidum - Binham

C. marinum (L.) C.E. Hubb. **Sea Fern-grass**
Syn. *Desmazeria marina* (L.) Druce

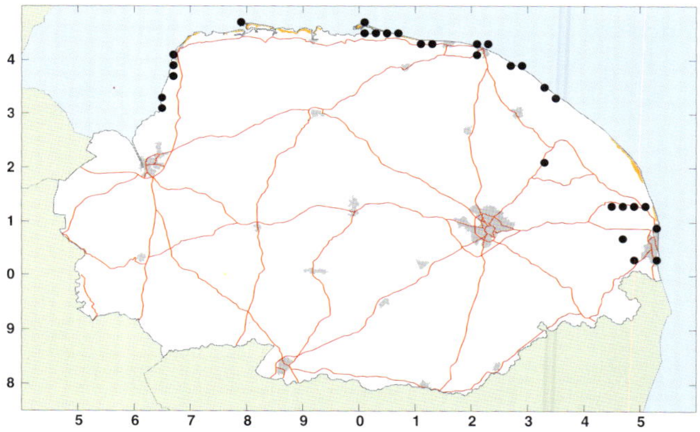

Annual, a scarce plant of shingle beaches and earthy places near the sea, also found on roadside verges at Beeston St. Lawrence E32 in 1994, D&BM and on the bypass at Caister E51 in 1998, ALB.

Parapholis strigosa (Dumort.) C.E. Hubb. **Hard-grass**

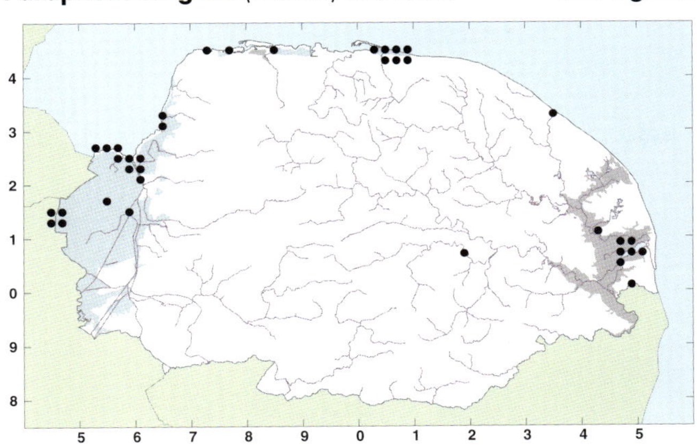

Annual, scarce in grassland adjacent to the sea and on the banks of estuaries.

P. incurva (L.) C.E. Hubb. **Curved Hard-grass**
Annual, a rare plant of coastal sands found at Scolt Head W84 in 1993, SW; Salthouse E04 also in 1993, ALB though perhaps lost after the beach was bulldozed to rebuild the shingle bank in 1996; Wolferton and Snettisham W63 in 1996, SF; Holme W74 in 1997, J.Masson and Yarmouth South Denes E50 in 1998, SP. A nationally scarce species. W6. E2.

■ **Hainardia cylindrica** (Willd.) Greuter
One-glumed Hard-grass
Syn. *Monerma cylindrica* (Willd.) Coss. & Durieu
Annual from southern Europe, a casual recorded once, on Harford tip Norwich E20 in 1972, PGL & AC.

Glyceria maxima (Hartm.) Holmb. **Reed Sweet-grass**

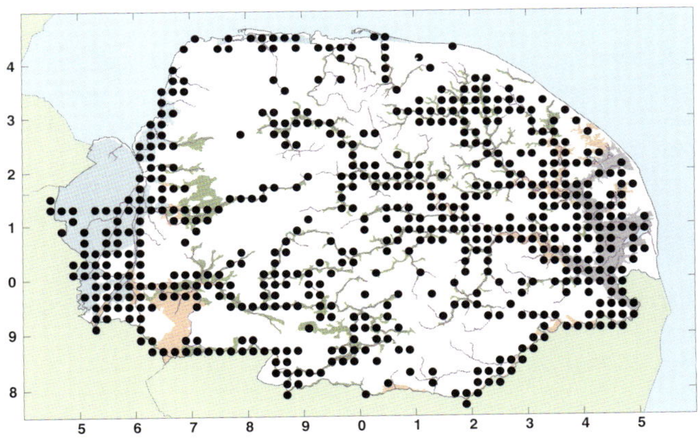

Deciduous perennial, an abundant grass by watercourses throughout the county, less common in brackish areas.

G. fluitans (L.) R. Br. **Floating Sweet-grass**

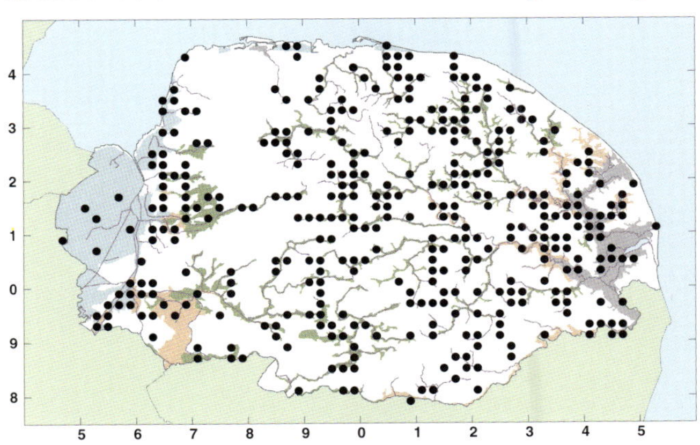

Winter green perennial frequent in slow-moving and still water in dykes, streams and pools in marshes.

■ **G. x pedicellata** F. Towns. **Hybrid Sweet-grass**
(*G. fluitans x G. notata*)
Winter green perennial recorded five times, the most recent at South Wootton W62 in 1972, RPL.

G. declinata Bréb. **Small Sweet-grass**

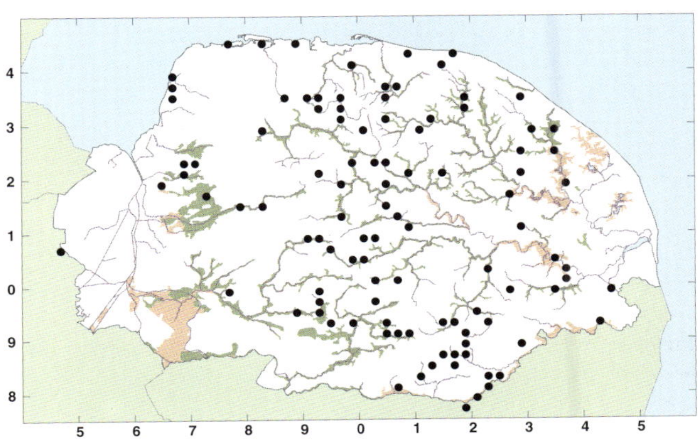

Winter green perennial scattered in damp places, particularly the muddy margins of ponds.

G. notata Chevall. **Plicate Sweet-grass**
Syn. *G. plicata* (Fr.) Fr.

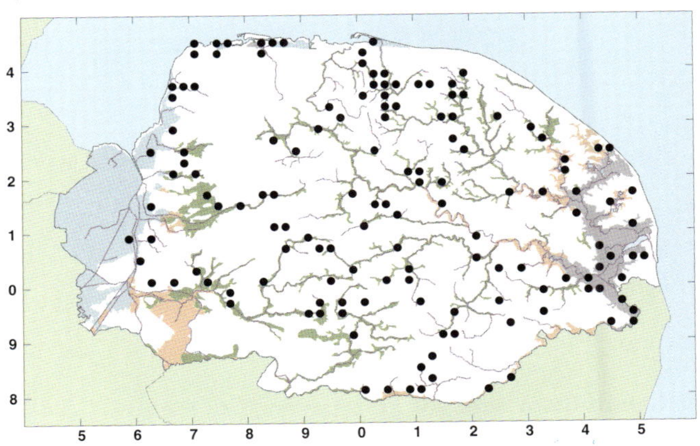

Winter green perennial scattered in slow-moving water and ditches.

Melica uniflora Retz. **Wood Melick**

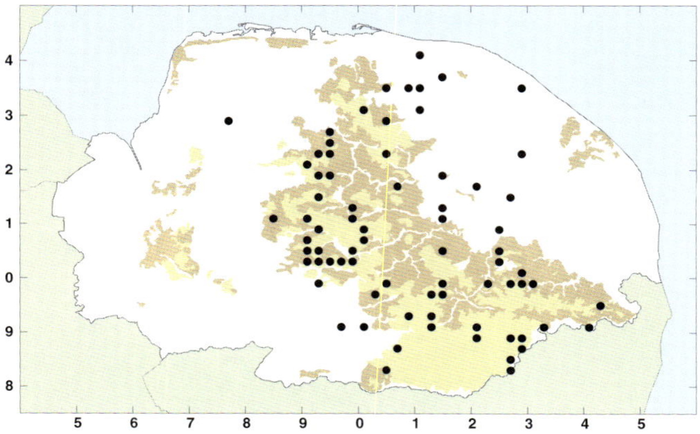

Winter green perennial more or less confined to long established woods, chiefly those on heavy soils.

Helictotrichon pubescens (Huds.) Pilg. **Downy Oat-grass**
Syn. *Avenula pubescens* (Huds.) Dumort.

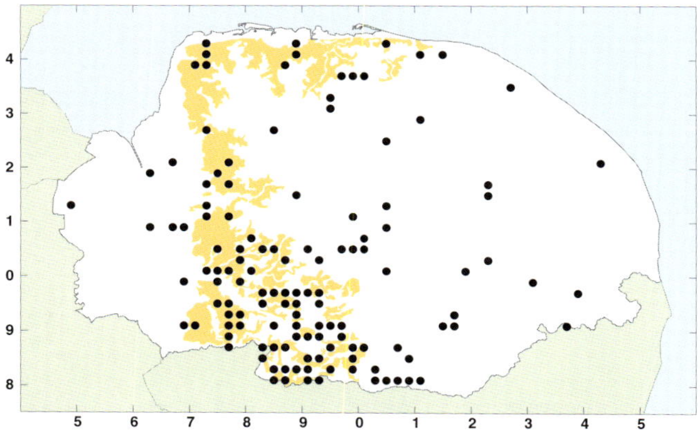

Perennial, widespread but scattered, most frequent on base rich soils, but not as critical in its requirements as the following species.

H. pratense (L.) Besser **Meadow Oat-grass**
Syn. *Avenula pratense* (L.) Dumort.

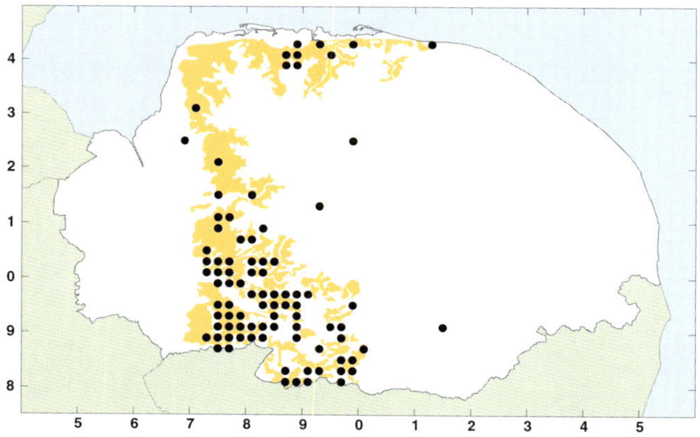

Perennial, found only on chalky soils where it can be locally frequent.

Arrhenatherum elatius (L.) P. Beauv. ex J.&C. Presl
False Oat-grass
Winter green perennial, ubiquitous. W706. E711.

Avena strigosa Schreb. **Bristle Oat**
Annual widely distributed in Europe, but of obscure origin, a casual weed in corn and other crops, found in a pea field at Sedgeford W73 in 1988, K&GB. W1. E0.

Koeleria macrantha - Cranwich

A. fatua L. **Wild-oat**
Annual, a widespread weed of arable land. W555. E611.

■ **A. sterilis** L. ssp. **ludoviciana** (Durieu) Gillet & Magne.
Syn. *A. ludoviciana* **Winter wild-oat**
Annual from southern Europe, a casual as a weed in grain, recorded once only, in King's Lynn docks W61 in 1971, RPL.

A. sativa L. **Oat**
Annual, a survivor from crops but does not persist. W15. E16.

■ **A. byzantina** K.Koch. **Algerian Oat**
Annual from the Mediterranean region, an occasional casual seen once, on Harford tip E20 in 1967.

Trisetum flavescens (L.) P. Beauv. **Yellow Oat-grass**

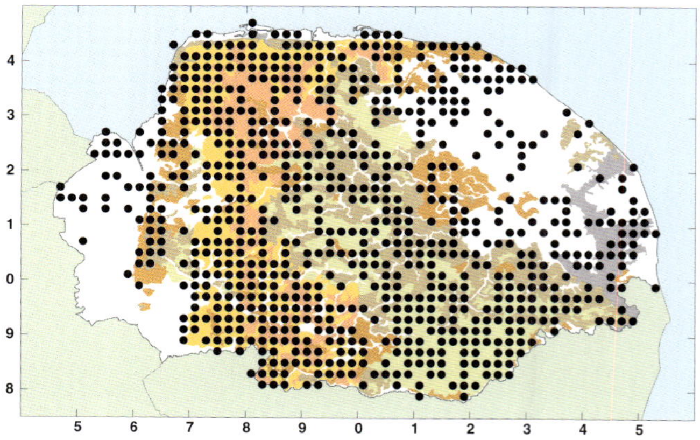

Perennial, frequent in grassland on well-drained soils, but avoiding very acid ones. Very conspicuous on some roadside verges in early summer.

Koeleria macrantha (Ledeb.) Schult. **Crested Hair-grass**

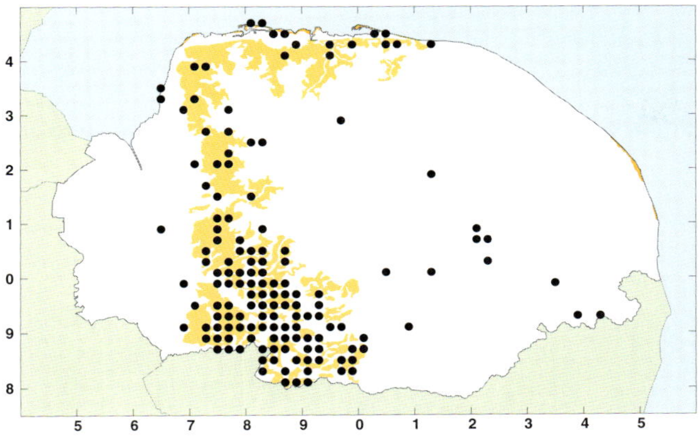

Perennial of well-drained grassland on light, sandy soils which are neutral to base rich.

Deschampsia cespitosa (L.) P. Beauv. **Tufted Hair-grass**

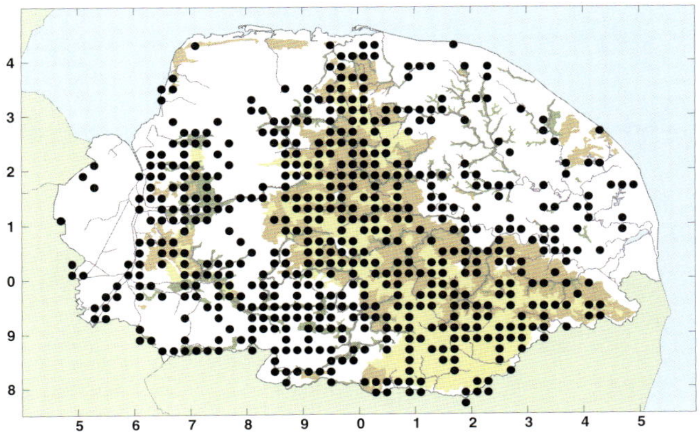

Winter green perennial found in a wide range of damp habitats. Ssp. **cespitosa** occurs throughout most of the county while ssp. **parviflora** (Thuill.) Dumort. is confined to ancient woodland.

■ **D. setacea** (Huds.) Hack. **Bog Hair-grass**
Perennial, a very rare grass of acid bogs, only certainly recorded twice for Norfolk, the first at Cawston Heath E12 in 1776 when it was described as growing with its leaves 'entirely immersed in water'. More recently it was found at East Winch W71 but has not been seen since 1974 and is presumed lost. A nationally scarce species.

D. flexuosa (L.) Trin. **Wavy Hair-grass**

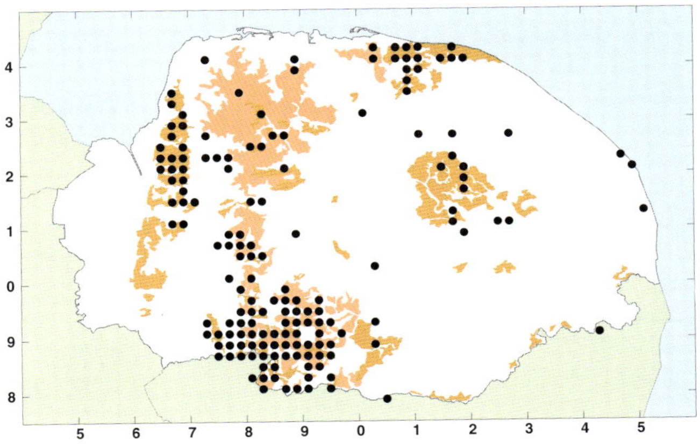

Deciduous perennial confined to dry, acid soils where it is sometimes abundant and invasive.

Holcus lanatus L. **Yorkshire Fog**
Winter green perennial abundant on all except the wettest soils.
W685. E702.

H. mollis L. **Creeping Soft-grass**

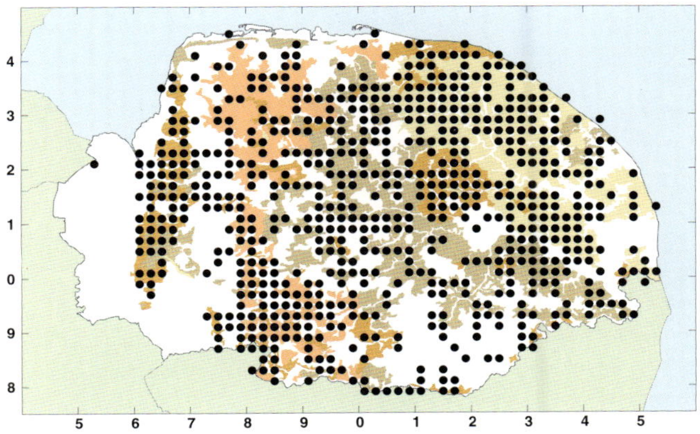

Winter green perennial frequent in shady sites, especially on acid soils.

Corynephorus canescens (L.) P. Beauv. **Grey Hair-grass**

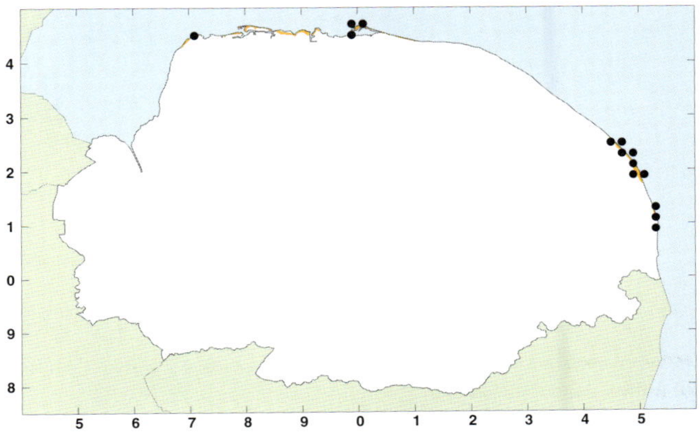

Perennial scattered around the coast but frequent only on the acid dunes of the north-east coast. It occurs in the Suffolk Breckland but has not been found inland in Norfolk. A nationally scarce species.

Deschampsia flexuosa - Syderstone

Aira caryophyllea L. **Silver Hair-grass**

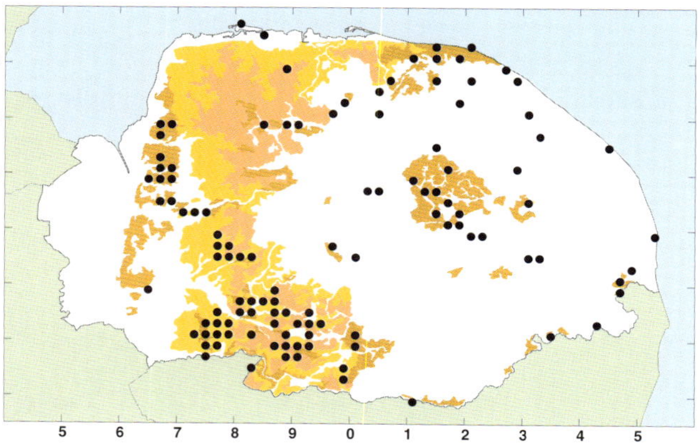

Spring germinating annual scattered on light, neutral to acid, sandy soils, often taking advantage of man-made sites such as sand-pits and disused railway tracks.

A. praecox L. **Early Hair-grass**

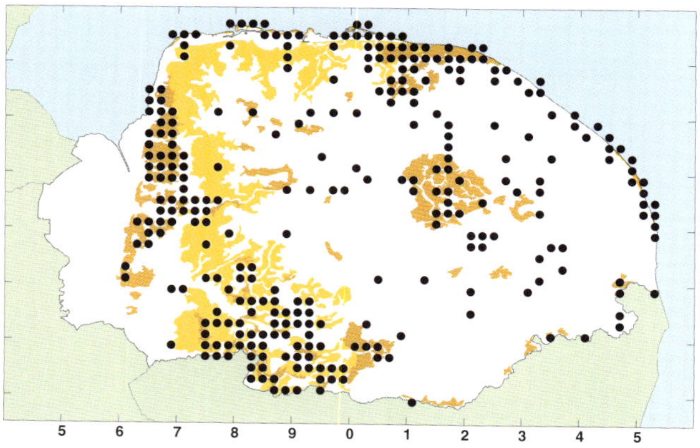

Spring germinating annual widely distributed on light, sandy soils, in gravel pits and on walls.

Anthoxanthum odoratum L. **Sweet Vernal-grass**

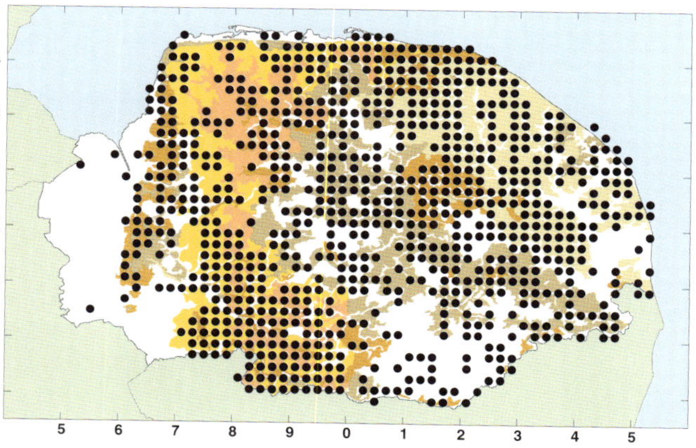

Winter green perennial frequent in established grassland on dry and moist but not wet soils.

■ **A. aristatum** Boiss. **Annual Vernal-grass**
Syn. *A. puelii* Lecoq. & Lamotte
Spring germinating annual from southern Europe recorded twice, Eaton E20 in 1935 and Quarles W83 in 1975, B.Taylor.

Phalaris arundinacea L. **Reed Canary-grass**

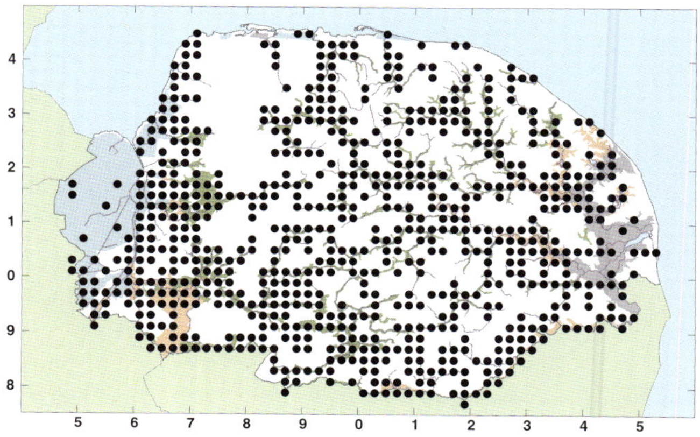

Deciduous perennial, the dead leaves staying on the stems through the winter. Abundant by water courses and in marshes.

P. aquatica L. **Bulbous Canary-grass**
Syn. *P. tuberosa* L.

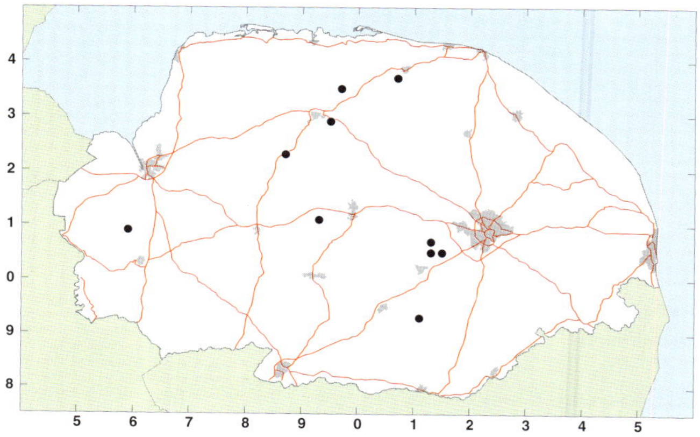

Perennial introduced from southern Europe, found in field edges and probably originally derived from pheasant-feed. First recorded in the county at Fakenham W92 in 1991, MK and becoming more frequent. A weed in a wheat field at Thursford E93.

P. canariensis L. **Canary-grass**
Annual from North Africa and the Canary Islands, usually found on waste ground and tips. W15. E25.

■ **P. minor** Retz. **Lesser Canary-grass**
Spring germinating annual from southern Europe, a casual found at Bawsey W61 in 1945 and Houghton W72 in 1950.

P. paradoxa L. **Awned Canary-grass**
Spring germinating annual from southern Europe, a casual found at King's Lynn W61 in 1945, at Fincham W60 in 1992, MK; in a separate site in the same parish in 1998, RMP and at Thersfield E08, also in 1998, RWE. An increasing arable weed in the south of England. W2. E1.

Agrostis capillaris L. **Common Bent**
Syn. *A. tenuis* Sibth.

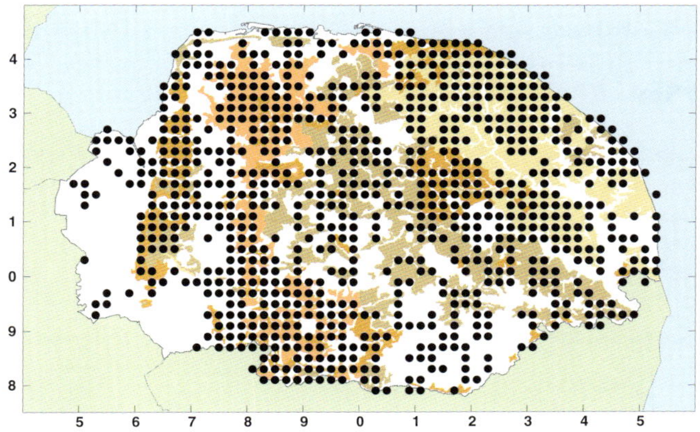

Winter green perennial widely distributed, commonest on dry acid soils.

■ **A. x murbeckii** Fouill. ex P. Fourn.
(*A. capillaris x A. stolonifera*)
Recorded at Hockham W99 in 1968 and at Derby Fen W62 in 1969, ELS.

A. gigantea Roth. **Black Bent**

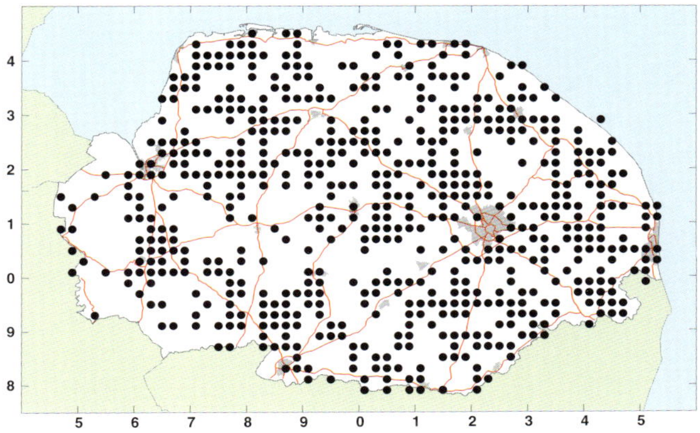

Perennial, a widespread and perhaps under-recorded weed of arable land.

A. castellana Boiss. & Reut. **Highland Bent**
Perennial chiefly from the Mediterranean region, recorded from five sites between 1972 and 1977. During this survey found at Roydon E08 and E18 in 1990, AC; King's Lynn W62 in 1993, RMP; at Ringstead W73 in 1996 in a field sown with 'native seed mix' K&GB and at Diss E17 in 1997 as a garden weed apparently introduced with young shrubs, AC. There are three named varieties of this species, var. **castellana,** var. **mixta** Hack to which the E08 & E18 records belong and var. **mutica** Hack, the variety found in E17.
W2. E3.

A. stolonifera L. **Creeping Bent**
Perennial, very common on all but the most acid soils. W667. E628.

A. canina L. **Velvet Bent**
Syn. *A. canina* ssp. *canina*

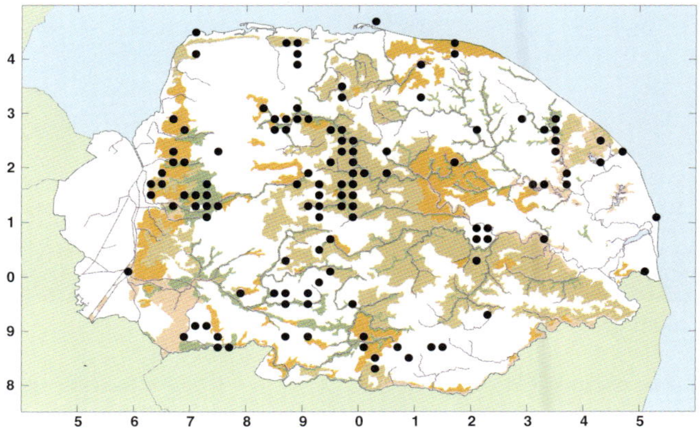

Perennial found in moist, usually acid grassland and less frequently in damp, shady woodland.

A. vinealis Schreb. **Brown Bent**
Syn. *A. canina* ssp. *montana* (Hartm.) Hartm.

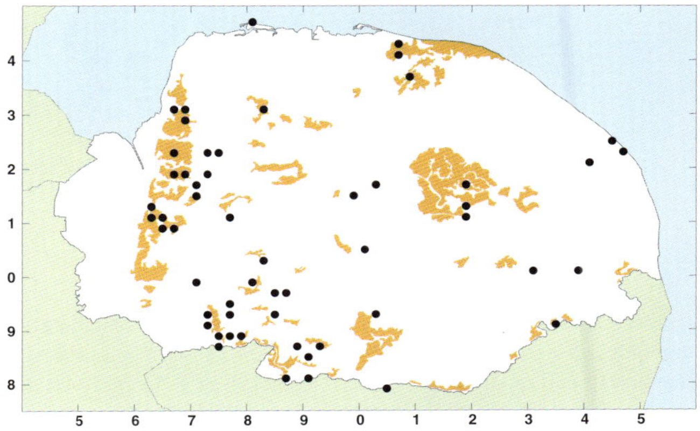

Perennial found on heaths and in sandy places, often with *A. capillaris.*

X Agropogon littoralis (Sm.) C.E.Hubb. **Perennial Beard-grass**
(*Agrostis stolonifera x Polypogon monspeliensis*)
An intergeneric hybrid found on two occasions, at Horsey E42 in 1958 and near Cantley Beet Factory E29 in 1995, A.Lewis conf. ALB.
W0. E1.

Calamagrostis epigejos (L.) Roth **Wood Small-reed**

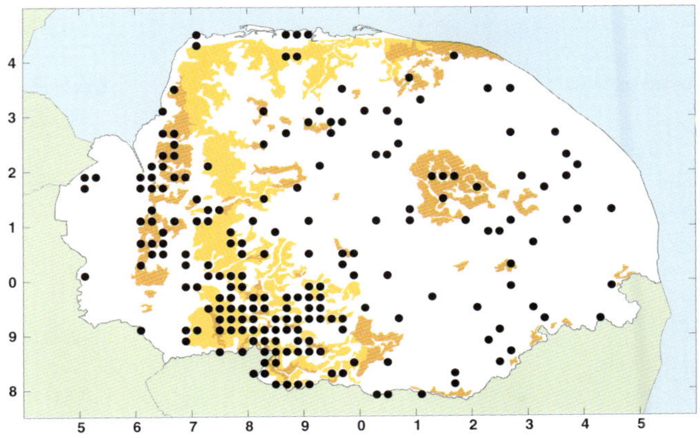

Deciduous perennial scattered in grassy places. It can form large, dense colonies in the coniferous forests of the Breckland, needing control to prevent it from smothering young trees.
Illustrated on the following page.

Calamagrostis epigejos - Cranwich page 245

C. canescens (F.H.Wigg.) Roth **Purple Small-reed**

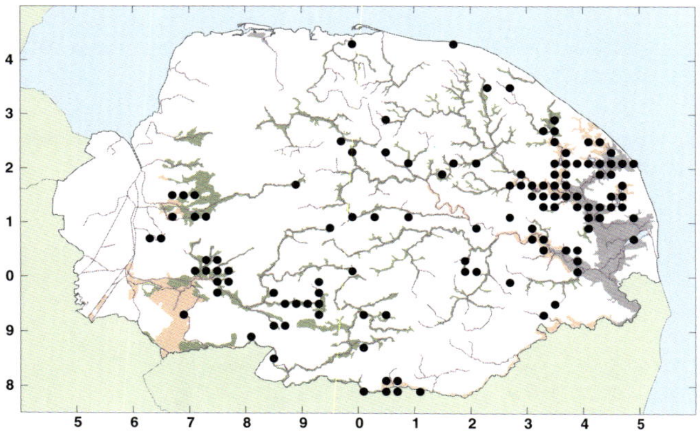

Perennial found in fens and marshes, often in shady sites.

C. stricta (Timm) Koeler **Narrow Small-reed**
Perennial confined to near waterlogged, peaty soils in three adjacent sites. First found at Cranberry Rough, Hockham in 1915 and still present, also at Thompson and Stow Bedon, all W99. Very sparse in all its sites, but showing some sign of recovery after scrub clearance. A nationally scarce species. W3. E0.

X Calammophila baltica (Flüggé ex Schrad.) Brand **Purple Marram**
Syn. *Ammocalamagrostis baltica* (Flüggé ex Schrad.) P.Fourn. (*Calamagrostis epigejos x Ammophila arenaria*)
First noted in 1805 between Winterton and Happisburgh, E42 where it occurred spontaneously. More recently introduced to several coastal sites because of its great value as a sand binder. Seen recently at Holme and Brancaster W64 and 74 on the north coast but frequent on the east coast between Happisburgh E33 and Winterton E51. W1. E15.

Ammophila arenaria (L.) Link **Marram**

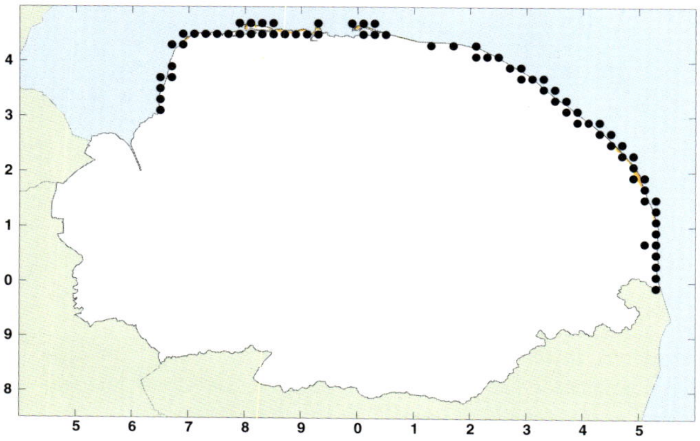

Winter green perennial common on sand dunes, the chief dune builder and flood protection of much of the coast.

■ **Gastridium ventricosum** (Gouan) Schinz & Thell. **Nit-grass**
Annual, a native of south-west England, recorded as a casual at Gillingham E49 and Cley E04 before 1945, not seen since.

Lagurus ovatus L. **Hare's-tail**
Annual introduced from southern Europe and grown for dried flower arranging, also occurring in bird-seed. Noted at King's Lynn W61 in 1993, RMP; Swaffham W80 in 1995, JEG and Thornham W74 in 1998, JEG&FS. W3. E0.

Apera spica-venti (L.) P. Beauv. **Loose Silky-bent**

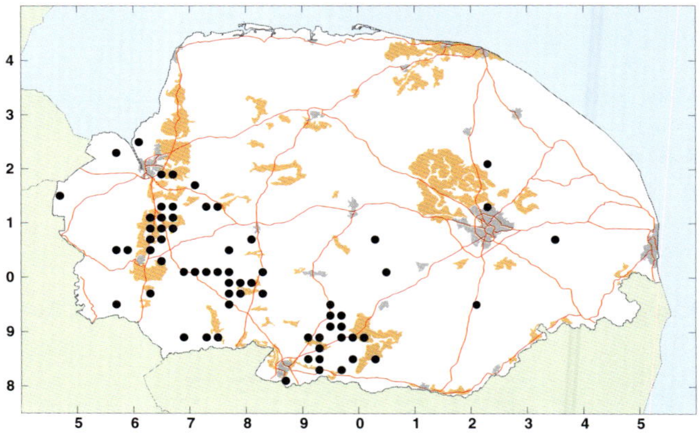

Annual. Rare in its natural habitat of sandy and gravelly places, but now also found on arable land and behaving as a weed, reaching pest proportion in a few wheat fields. A nationally scarce species.

A. interrupta (L.) P. Beauv. **Dense Silky-bent**

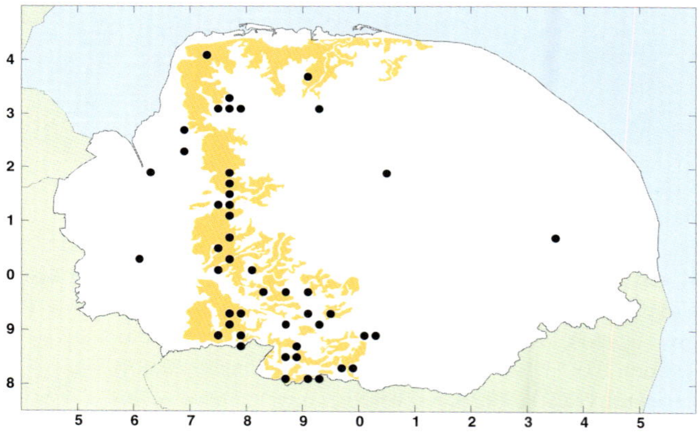

Annual, frequent only on dry, sandy soils, perhaps increasing. A nationally scarce species.

Polypogon monspeliensis (L.) Desf. **Annual Beard-grass**
Annual, scarce in the drier parts of salt marshes and adjacent damp grassland as at Cley E04 in 1990, KKH et al. and Sheringham E14 in 1995, Flora Group, also occasionally found inland by tidal rivers as by the Great Ouse near King's Lynn W62 in 1994, RMP. It is found as an urban weed at Strumpshaw Tip E30 in 1995, SP and Longwater Industrial Estate, Easton E11 in 1997, ALB&RWE, also around settling ponds at Cantley beet factory E30 in 1995, A.Lewis whence it was taken to Matlaske E13 with lime sludge, MAB. W1. E8.

P. viridis (Gouan) Breistr. **Water Bent**
Perennial from southern Europe, a rare casual but found on waste ground and walls at King's Lynn W61 in 1993, RMP and still surviving. W1. E0.

Alopecurus pratensis L. **Meadow Foxtail**

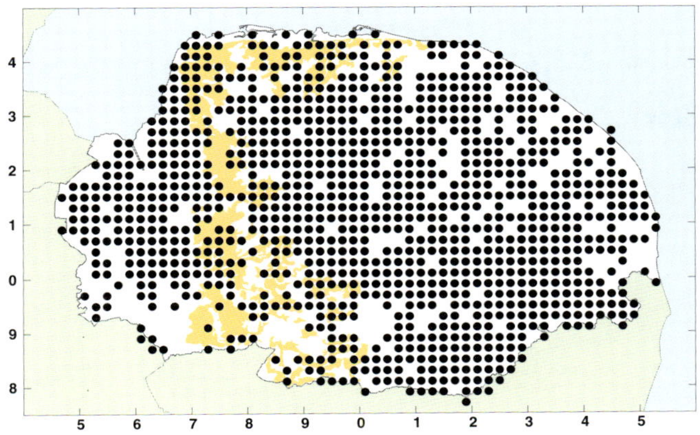

Perennial widespread on roadsides and in grassland on moist soils.

■ **A. x brachystylus** Peterm.
(*A. pratensis x A. geniculatus*)
Four records known, the last at Roydon, Diss E18 in 1960.

A. geniculatus L. **Marsh Foxtail**

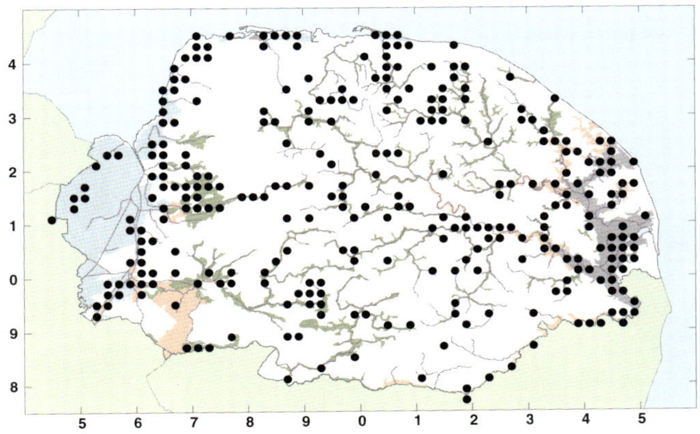

Perennial widespread in wet grassland and fens.

■ **A. x haussnechtianus** Asch. & Graebn.
(*A. geniculatus x A. aequalis*)
Noted as 'found in West Norfolk' by Hubbard, 1954.

A. bulbosus Gouan **Bulbous Foxtail**
Perennial, a rare plant of estuarine grassland, recorded near Great Yarmouth E50 in 1982, PJO.Trist, RPL & ELS and again in 1995, ALB & K.Rivett, also in the same area at E40 in 1995, A.Lewis. WO. E3.

A. aequalis Sobol. **Orange Foxtail**

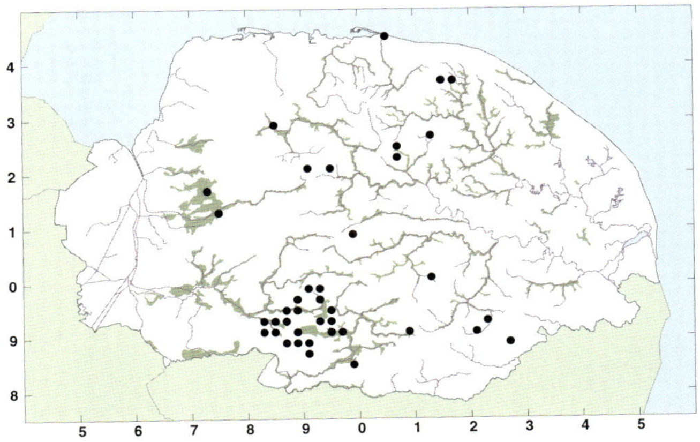

Annual or perennial, a scarce plant of pond margins and shallow pools, also a surprising weed of pots and pans in a water plant nursery.

Alopecurus aequalis - East Wretham Heath

A. myosuroides Huds. **Black-grass, Black-twitch**

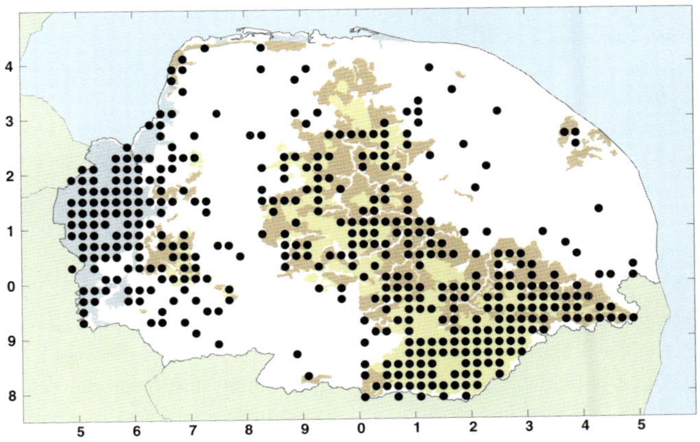

Annual, a serious weed of heavy soils.

■ **Beckmannia syzigachne** (Steud.) Fernand
American Slough-grass
Annual or short-lived perennial, a casual from North America recorded on Harford tip, Norwich E20 in 1972, AC and PGL.

Phleum pratense L. **Cat's-tail, Timothy**

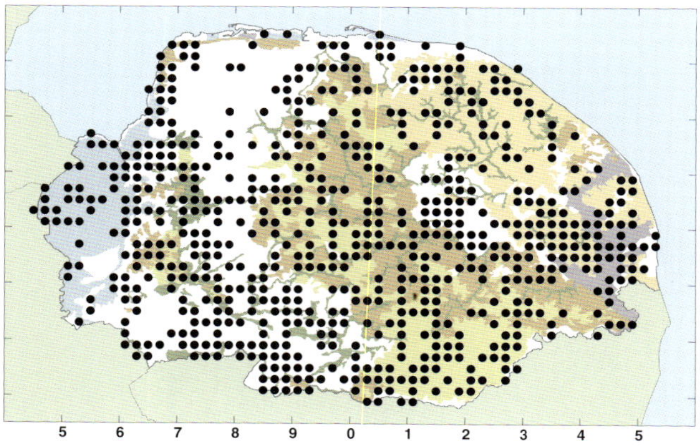

Perennial frequent on heavier and moist soils.

P. bertolonii DC. **Smaller Cat's-tail**
Syn. *P. pratensis* ssp. *bertolonii* (DC.) Bornm.

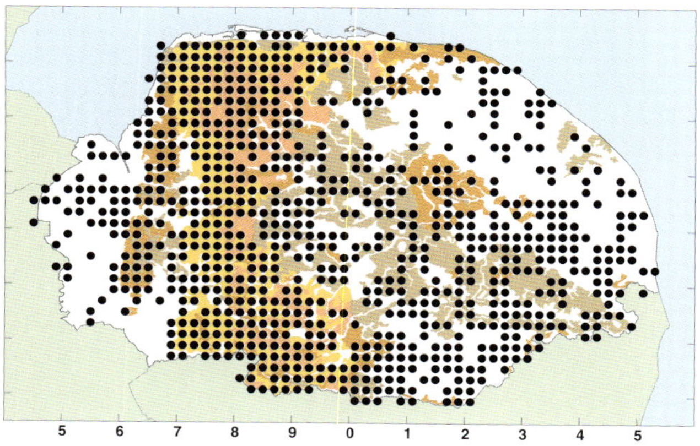

Perennial, frequent on lighter soils, but often growing with the previous species from which it is distinguished with difficulty.

P. phleoides (L.) H.Karst.
Purple-stemmed Cat's-tail, Breckland Cat's-tail (local)

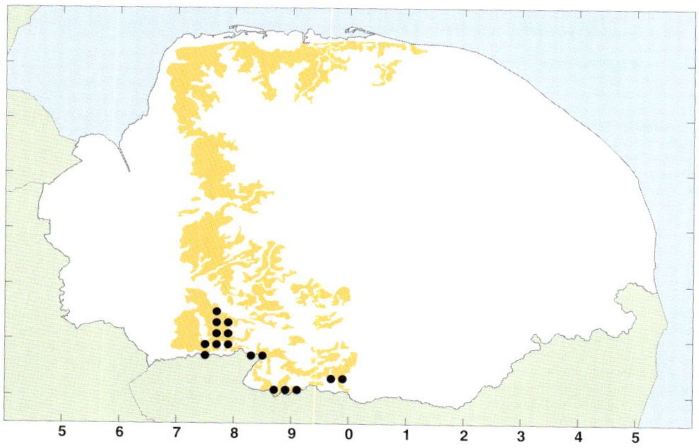

Perennial confined to permanent grassland on base rich soils, now only in the Breckland. Remarkably drought tolerant and the only green plant remaining on an area of grass heath at Brettenham W98 after the 1995 heatwave. Although abundant in some sites, it has been lost from a number of its more northerly stations since 1968. Red Data Book species.

P. arenarium L. **Sand Cat's-tail**

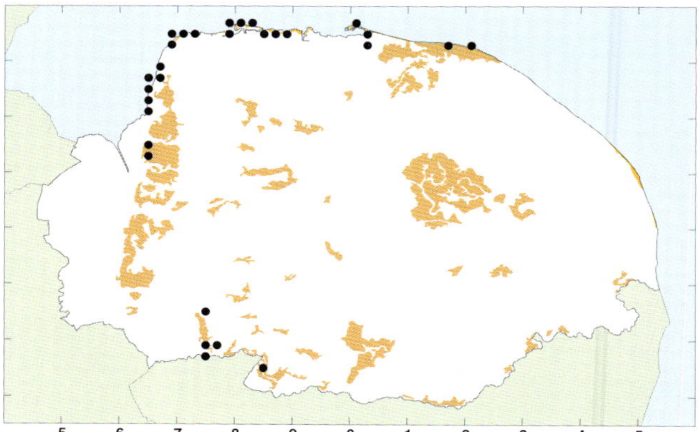

Annual, scarce on bare sand on coastal dunes, its usual habitat. It also grows on sandy heaths in the Breckland, its only inland stations in the country.

■ **Bromus arvensis** L. **Field Brome**
Annual from mainland Europe, a rare casual recorded three times before 1930.

B. commutatus Schrad. **Meadow Brome**

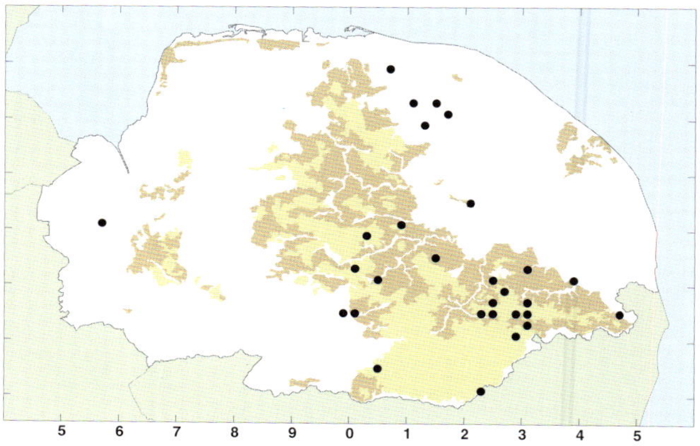

Annual, perhaps a native of damp meadows but many of its recent records refer to casual occurrences.

Phleum phleoides - Weeting

B. racemosus L. **Smooth Brome**

Annual, a casual recorded three times, most recently at King's Lynn W62 in 1993, RMP. W1. E0.

B. hordeaceus L. **Soft Brome**

Syn. *B. mollis* L.

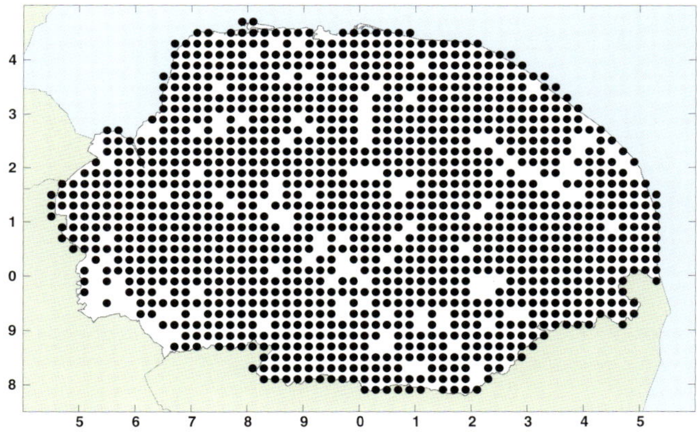

Annual. Ssp. **hordeaceus** is widespread on dry soils and is particularly associated with habitation. Ssp. **thominei** (Hardouin) Braun-Blanq. is a scarce plant of sandy ground and often misrecorded for the hybrid below. Verified records include Holme W74, 1973, Wolferton W62, 1977 and Overstrand E24, 1980.

B. x pseudothominei P.M. Sm. **Lesser Soft Brome**

(*B. hordeaceus x B. lepidus*)

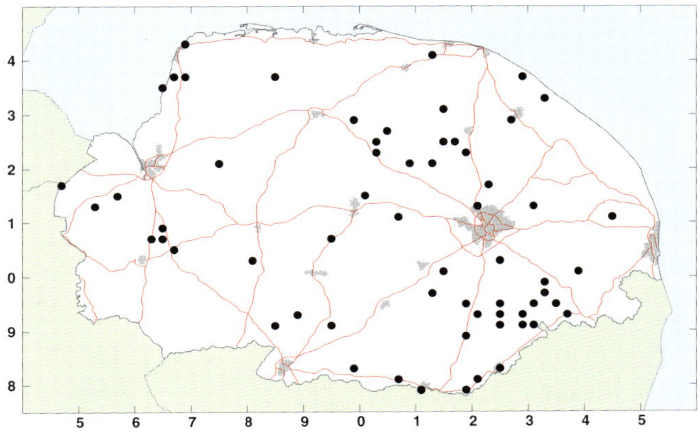

Widely distributed on well-drained soils, often far from any populations of *B. lepidus*.

B. lepidus Holmb. **Slender Soft Brome**

Annual, rare on roadsides and rough ground on light, sandy soils. Described in Petch & Swann, 1968 as locally frequent, but now quite scarce. W2. E5.

■ **B. interruptus** (Hack.) Druce **Interrupted Brome**

Annual, a rare plant formerly recorded from arable and waste ground, found at West Newton and Anmer, both W72 in 1946 by ELS. Not seen in Britain since 1972 and considered extinct.

B. secalinus L. **Rye Brome**

Annual from mainland Europe, recorded three times before 1968 as a casual weed. Seen at Hempnall E29 in 1992 and recently found to be locally abundant in wheat fields at Redenhall and Alburgh in E28 and Langley E30, all in 1998, ALB. It is spreading in similar habitats in Essex, J.J.Heath, pers.comm. W0. E16.

Bromopsis ramosa (Huds.) Holub **Hairy Brome**

Syn. *Bromus ramosus* Huds.

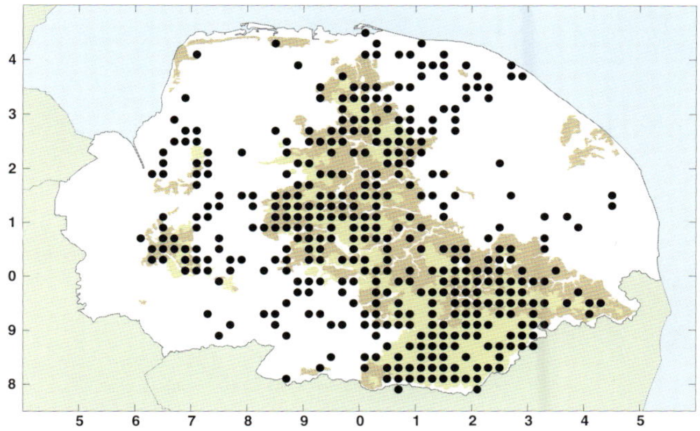

Perennial of woods and hedgerows on heavy, moist, neutral to acid soils.

B. erecta (Huds.) Fourr. **Upright Brome**

Syn. *Bromus erectus* Huds.

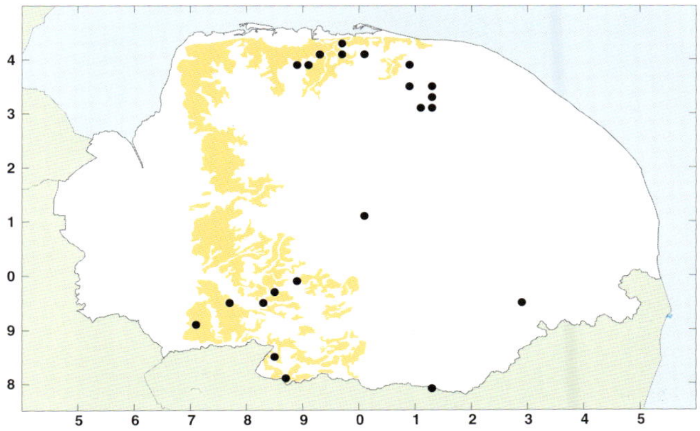

Perennial scattered but scarce in dry neutral to base rich grassland

B. inermis (Leyss.) Holub **Hungarian Brome**

Syn. *Bromus inermis* Leyss.

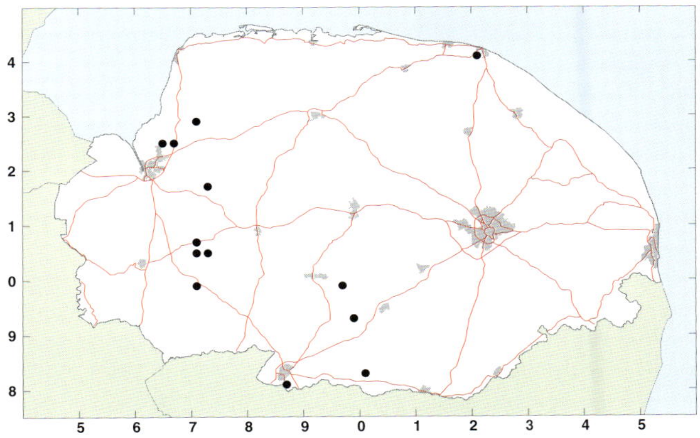

Evergreen perennial introduced for fodder from mainland Europe. Ssp. **inermis** was first noted as an escape at East Harling W98 in 1950. Now perhaps increasing on roadsides and rough grassy places where it can spread with its long rhizomes to form extensive and exclusive stands. Conspicuous around Barton Bendish W70.

Anisantha diandra (Roth) Tutin ex Tzvelev **Great Brome**
Syn. *Bromus diandrus* Roth.

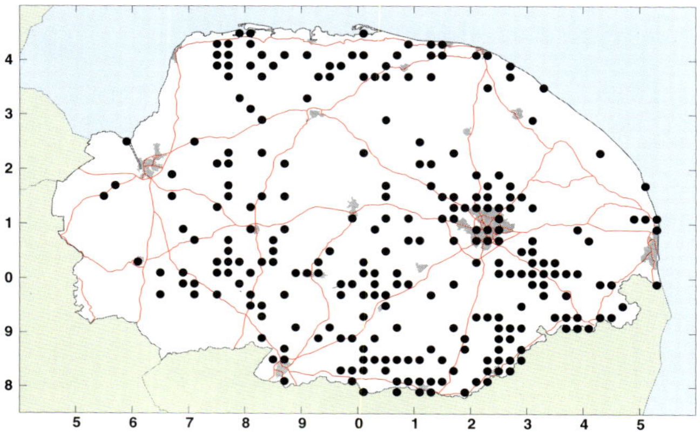

Annual, widely scattered and sometimes a serious weed of heavy arable land.

A. rigida (Roth) Hyl. **Ripgut Brome**
Syn. *Bromus rigidus* Roth.

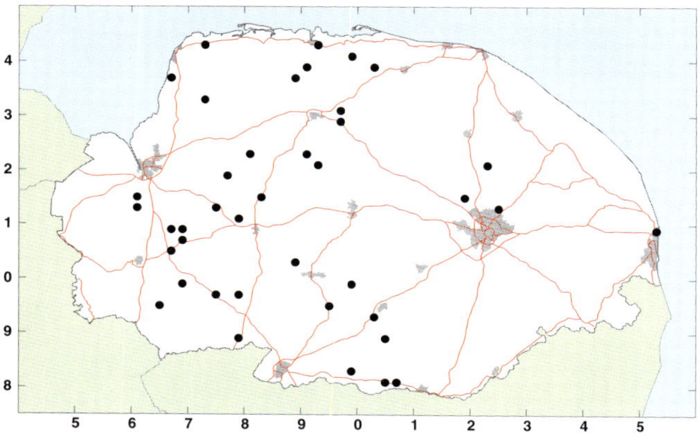

Annual, a scarce weed, by some authorities considered to be no more than a variety of *A. diandrus*, Sales 1993. The two can certainly be very difficult to separate and the same plants appear to differ according to the fertility of the soil and the time of the year.

A. sterilis (L.) Nevski **Barren Brome**
Syn. *Bromus sterilis* L.
Annual widespread in all habitats which provide a suitable seed bed, particularly common by roadsides, on walls and on waste ground. Sometimes a weed of arable fields. W695. E693.

A. tectorum (L.) Nevski **Drooping Brome**
Syn. *Bromus tectorum* L.
Annual from mainland Europe though considered by some to be a rare native of Breckland. Long known between Brandon W78 and Thetford W88 where it was formerly abundant but has not been seen since the early 1980s. Observed at Ringstead W74 in 1995, K&GB, in a field sown in a previous year with a so-called native seed-mix, so its status is doubtful. Red Data Book species. W1. E0.

Ceratochloa carinata (Hook. & Arn.) Tutin **Californian Brome**
Syn. *Bromus carinatus* Hook. & Arn..

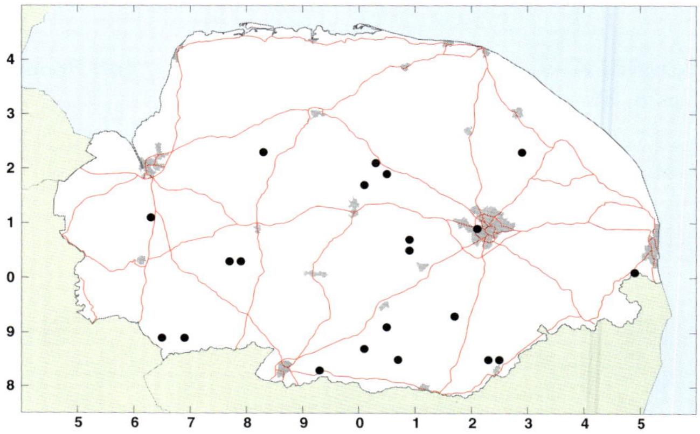

Perennial from South America, a casual in arable fields and waste ground, perhaps increasing.

C. cathartica (Vahl.) Herter **Rescue Grass, Rescue Brome**
Syn. *C. unioloides* (Willd.) Beauv., *Bromus unioloides* (Willd.) Kunth
Perennial from the Americas, a casual of roadsides and waste ground. Found at Roydon E17 in 1990, AC and Quidenham E08 in 1997, R.Chancellor. W0. E2.

Brachypodium pinnatum (L.) P. Beauv. **Tor-grass**

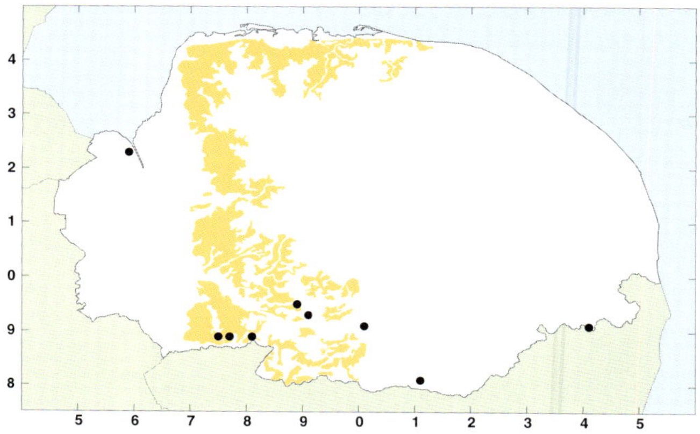

Perennial found in grassland on lime rich soils. In southern and parts of north-eastern England it can be very dominant but with us it seems confined to small, discrete patches. The colony at Terrington, W52 SF conf RMP, may have come in with material for the sea bank.

B. sylvaticum (Huds.) P. Beauv. **False Brome**

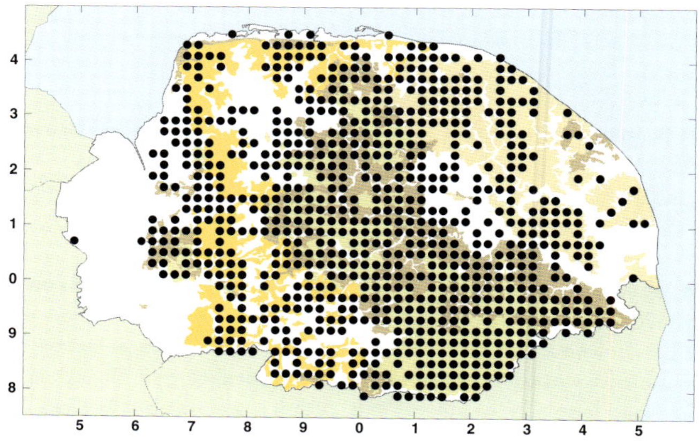

Deciduous perennial widespread in shady hedgebanks, ditches, woods and scrub.

Elymus caninus (L.) L. **Bearded Couch**
Syn. *Agropyron caninum* (L.) P. Beauv.

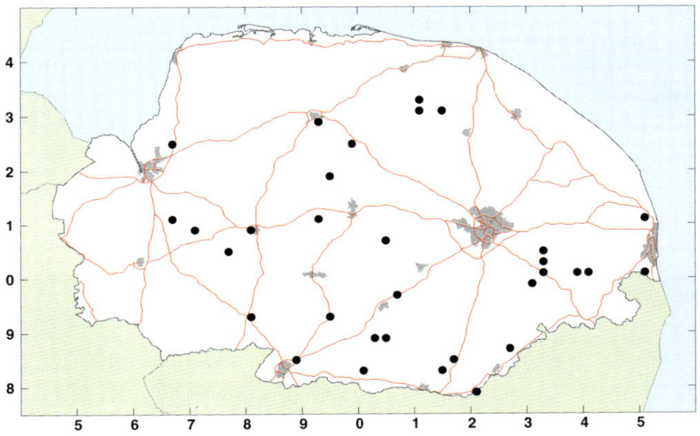

Deciduous perennial, scarce but scattered in shady places by woods and along hedgerows, usually on clay.

Elytrigia repens (L.) Desv. ex Nevski **Common Couch-grass**
Syn. *Elymus repens* (L.) Gould, *Agropyron repens* (L.) P. Beauv.
Ssp. **repens** is a herbacous perennial abundant in waste places, as an arable weed and on roadsides. Together with species of *Atriplex* it is one of the few species able to withstand the mineral rich soils adjacent to the tarmac of main roads and it is rapidly becoming dominant on the outer edge of many verges. Ssp. **arenosa** (Spenn.) A.Löve, was found at Holme W74 in 1995, PJO.Trist.
W683. E673.

■ **E. x oliveri** (Druce) Kerguélen ex Carreras Mart.
(*E. repens x E. atherica*)
A scarce hybrid occurring with the parents, found on five occasions, the last at Wolferton W62 in 1975, ELS.

■ **E. x laxa** (Fr.) Kerguélen
(*E. repens x E. juncea*)
A scarce hybrid found once, at Wolferton W62 in 1962.

E. atherica (Link) Kerguélen ex Carreras Mart. **Sea Couch**
Syn. *Agropyron pycnanthum* (Godr.) Godr., *A. pungens* auct. non (Pers.) Roem. & Schult.

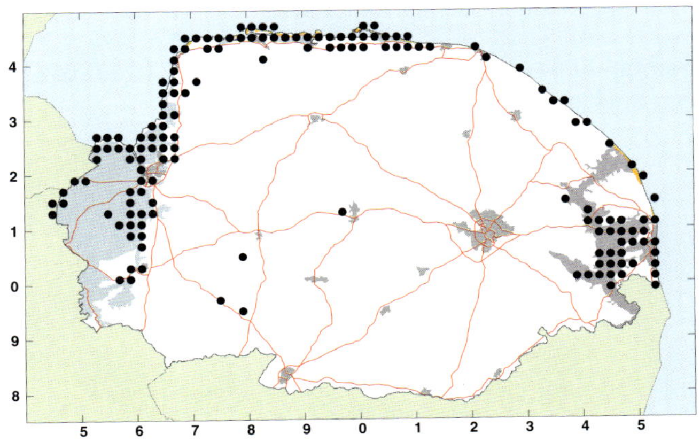

Perennial widespread on mud, sand and shingle around and above the high tide line, also on the banks of estuarine rivers extending beyond tidal influence. Occasionally found on road verges.

E. x obtusiuscula (Lange) Hyl.
(*E. atherica x E. juncea*)
Considered common by ELS, this hybrid occurs with the parents from Wolferton W62 around the coast to Blakeney E04. W3. E1.

E. juncea (L.) Nevski **Sand Couch**
Syn. *Agropyron junceum* (L.) P. Beauv., *Elymus farctus* (Viv.) Runemark ex Melderis

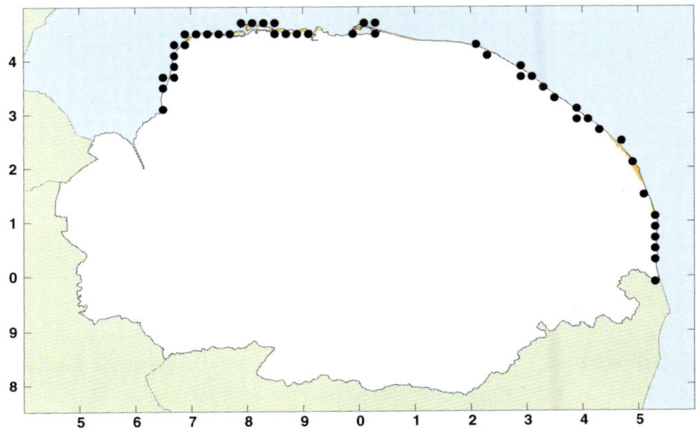

Semi-evergreen perennial of sand dunes, often growing with *Ammophila.*

■ **X Elytrordeum langei** (K.Richt.) Hyl.
(*Elytrigia repens x Hordeum secalinum*)
Scarce intergeneric hybrid recorded at Burnham Overy W84 in 1967 and inland on a roadside verge at Beeston-next-Mileham W91 in 1972, CEH.

Leymus arenarius (L.) Hochst. **Lyme-grass**
Syn. *Elymus arenarius* L.

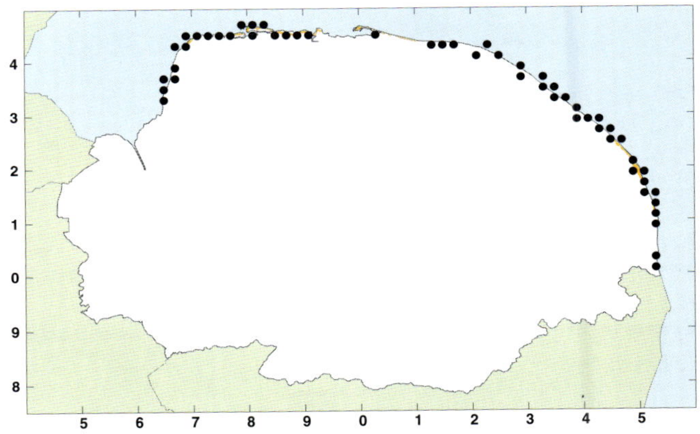

Perennial confined to sand dunes and shingle banks by the sea.

Hordeum vulgare L. **Six-rowed Barley**
Annual, formerly grown as a crop, now largely displaced by *H. distichon.* An occasional weed of roadsides and field edges.
W5. E3.

H. distichon L. **Two-rowed Barley**
Annual, the commonest cultivated Barley occurring on verges and as a volunteer in other crops. W50. E125.

H. murinum L. **Wall Barley**

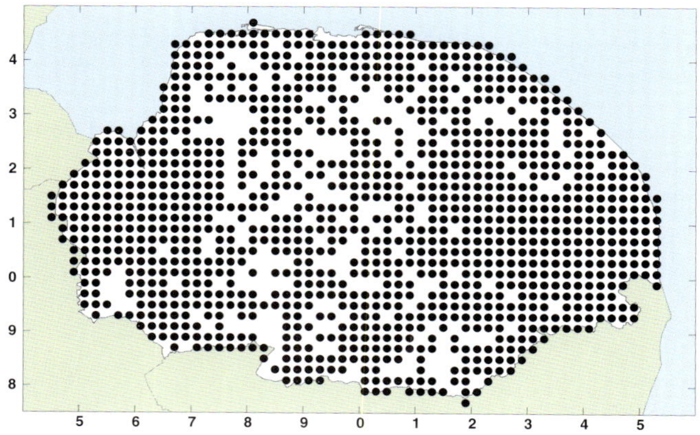

Annual, a frequent grass of waste ground, strongly associated with habitation. Ssp. **leporinum** (Link) Arcang, was recorded at Beetley Pit W91 in 1977, RPL.

H. jubatum L. **Foxtail Barley**

Perennial from the Americas, a casual often from birdseed, five records between 1967 and 1978. More recently recorded at King's Lynn W62 in 1996, SF; Nordelph W50 in 1997, RMP; the Longwater Estate E11 in 1997, R.Chancellor and on estuarine mud at Reedham behind Breydon Wall E40 in 1997, ALB & RWE. W2. E2.

H. secalinum Schreb. **Meadow Barley**

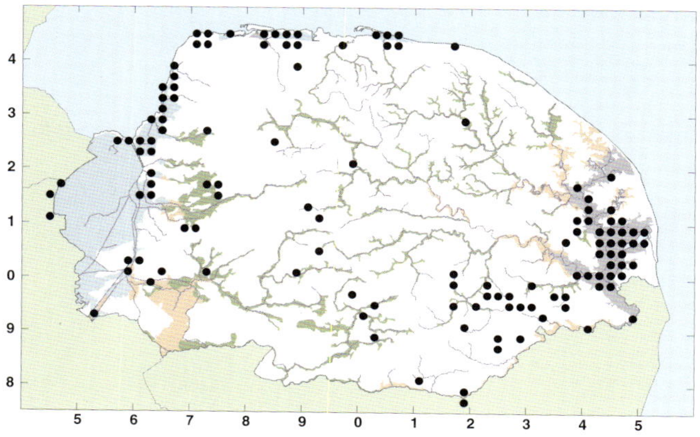

Annual found in base rich meadows, particularly those near the sea.

H. marinum Huds. **Sea Barley**

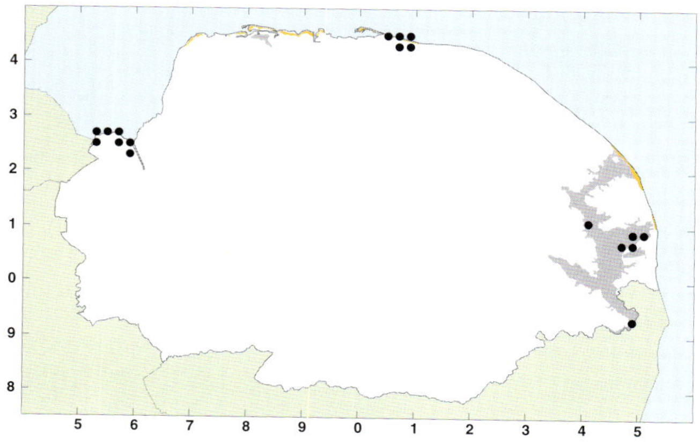

Annual, oddly scattered in its distribution, frequent on grassy sea walls along the southern shore of the Wash W52; at Cley and Salthouse E04 and around Breydon Water E40. Recorded before 1945 from the eastern shore of the Wash in W63 & W64 and along the north coast as far as Holkham, W74 & W84, but not seen in these areas since. A nationally scarce species.

Secale cereale L. **Rye**

Annual cultivated on a rather small scale, occurring as a casual from crops. W3. E1.

Triticum aestivum L. **Bread Wheat**

Annual, an escape on to roadside verges and as a volunteer in the following year's crops. Not long lasting. W83. E128.

■ **T. durum** Desf. **Pasta Wheat**

A casual recorded once, at Harford tip E20 in 1967.

Hordeum jubatum - Reedham Marshes

Danthonia decumbens (L.) DC. **Heath-grass**

Syn. *Sieglingia decumbens* (L.) Bernh.

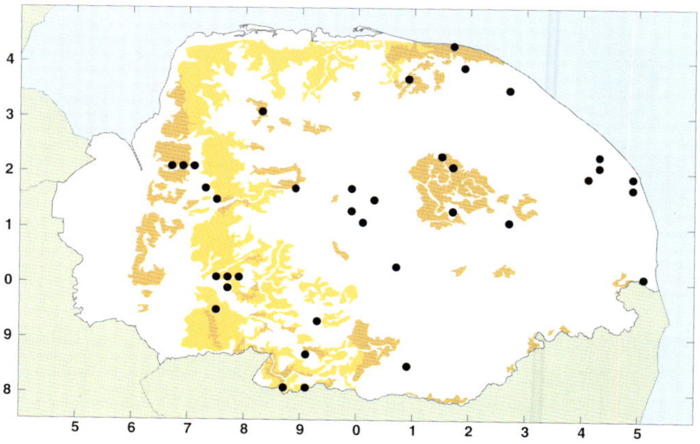

Perennial found in long established grassland on a wide range of soil types, but always scarce.

Cortaderia selloana (Schult. & Schult.f.) Asch. & Graebn. **Pampas-grass**
Evergreen perennial surviving where thrown out from gardens or purposefully planted as at Easton E11 in 1994, J.Humphris; Catfield E32 in 1996, ALB; Cawston E12 in 1997, ALB and Westwick E22 in 1977, Flora Group. W0. E4.

Molinia caerulea (L.) Moench **Purple Moor-grass**

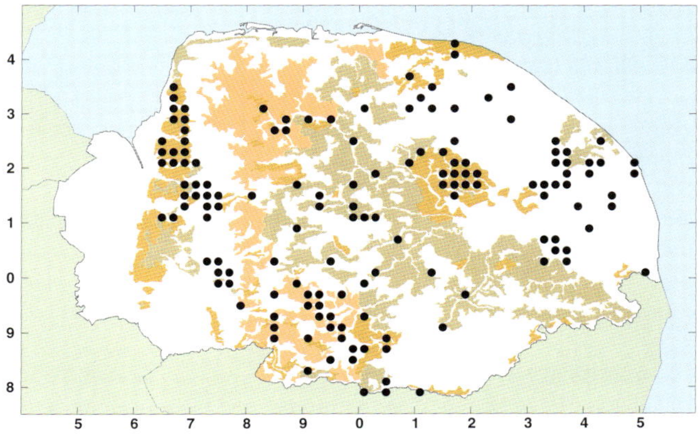

Deciduous perennial found on wet heaths and damp, acid grassland. Ssp. *arundinacea* Schrank (K.Richt.) was observed at Horsford E11 in 1998, A.Newton & ALB.

Phragmites australis (Cav.) Trin. ex Steud. **Common Reed**

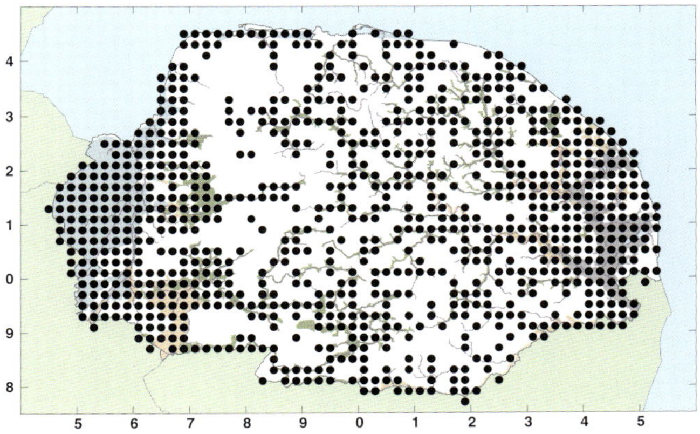

Deciduous perennial abundant in wet ground by rivers, streams, dykes and ponds, still managed for its use as the raw material of Norfolk thatch for roofs etc. A persistent weed on the edge of moist fields in the Fenland.

■ **Cynodon dactylon** (L.) Pers. **Bermuda-grass**
Perennial from warm areas of the world. Recorded by a road at Castle Rising W62 between 1944 and 1957.

■ **Spartina maritima** (Curtis) Fernald **Small Cord-grass**
Deciduous perennial, once not uncommon on muddy, tidal land, now replaced by *S. anglica* and not recorded in Norfolk since 1968.

■ **S. x townsendii** H.& J. Groves **Townsend's Cord-grass**
(*S. maritima x S. alternifolia*)
Old records do not differentiate between this, the original and male-sterile hybrid, and the following species which arose from it by chromosome doubling and is highly fertile. The only authenticated record for this first generation hybrid was made in 1962 when two colonies were found at Wolferton Creek by ELS&CEH.

S. anglica C.E. Hubb. **Common Cord-grass, Rice-grass**
(*S. maritima x S. alterniflora*)

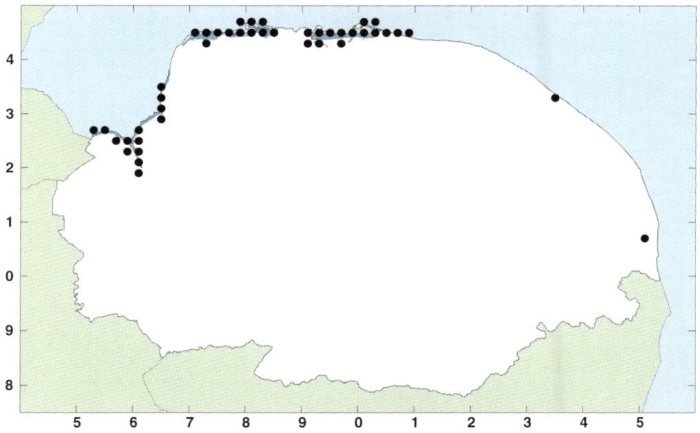

Deciduous perennial, the common hybrid which is scattered around our coast, but needs deeper mud to reach its potential than is present in Norfolk's largely shingle and sand-based salt marshes. Illustrated on the following page.

Phragmites australis - Wheatfen

■ **Panicum capillare** L. **Witch-grass**
Annual introduced from North America, a casual of waste places and formerly of carrot fields, recorded at Great Bircham W72 in 1948.

P. miliaceum L. **Common Millet**
Annual, introduced from Asia, an occasional casual from birdseed, mostly on waste tips and in pheasant feeding areas. W8. E8.

Echinochloa crus-galli (L.) P. Beauv. **Cockspur-grass**
Annual from warm-temperate to tropical regions, an occasional field weed somtimes sown for game feeding, more often on waste ground, probably originating from birdseed. W12. E5.

Spartina anglica - Cley page 253

■ **E. esculenta** (A.Braun) H.Scholz **Japanese Millet**
Syn. *E.utilis* Owhi & Yabuno
Annual of Asian origin, a constituent of birdseed. Six records were made between 1971 and 1984, all on refuse tips.

E. colona (L.) Link **Shama Millet**
Annual of tropical origin, three records between 1968 and 1981, all on refuse tips, more recently seen in Norwich E20 in 1997, ETD.
W0. E1.

■ **E. frumentacea** Link **White Millet**
Annual of Asian origin, recorded from waste tips three times in 1968.

Setaria pumila (Poir.) Schult. **Yellow Bristle-grass**
Syn. *S. lutescens* F.T.Hubb., *S. glauca* auct. non (L.) P. Beauv.
Annual from warm-temperate areas, a casual from birdseed, found on refuse tips, waste ground, field edges and in gardens. W10. E3.

■ **S. verticillata** (L.) P. Beauv. **Rough Bristle-grass**
Annual introduced from warm temperate areas of the old world, recorded three times between 1974 and 1982, all on refuse tips.

S. viridis (L.) P. Beauv. **Green Bristle-grass**
Annual, a casual from birdseed on refuse tips, waste ground, field edges and in gardens.
W5. E6.

■ **S. italica** (L.) P. Beauv. **Foxtail Bristle-grass**
Annual, a casual from birdseed on refuse tips and waste ground, last recorded in 1984.

■ **Digitaria ischaemum** (Schreb. ex Schweigg.) Muhl. **Smooth Finger-grass**
Annual from southern Europe, a casual from birdseed, seen on waste ground and refuse tips up to 1972.

D. sanguinalis (L.) Scop. **Hairy Finger-grass**
Annual from southern Europe , a casual of waste ground and refuse tips, probably originating from birdseed, seen at Heacham W63 in 1990, PRC.
W1. E0.

■ **D. ciliaris** (Retz.) Koeler **Tropical Finger-grass**
Syn. *D. adscendens* (Kunth) Henrard
Annual from the tropics of the old world, a casual of waste ground and refuse tips probably from birdseed, last seen in 1971.

Sorghum halepense (L.) Pers. **Johnson-grass**
Perennial introduced from North Africa, found as a casual in a potato field at Shouldham W60 in 1989, GM and on waste ground around the UEA 'Farm' E20 in 1996, M.Gurney.
W1. E1.

■ **S. bicolor** (L.) Moench **Great Millet**
Annual from Africa, casual of waste ground and refuse tips derived from birdseed, last seen in 1981.

SPARGANIACEAE

Sparganium erectum L. **Branched Bur-reed**

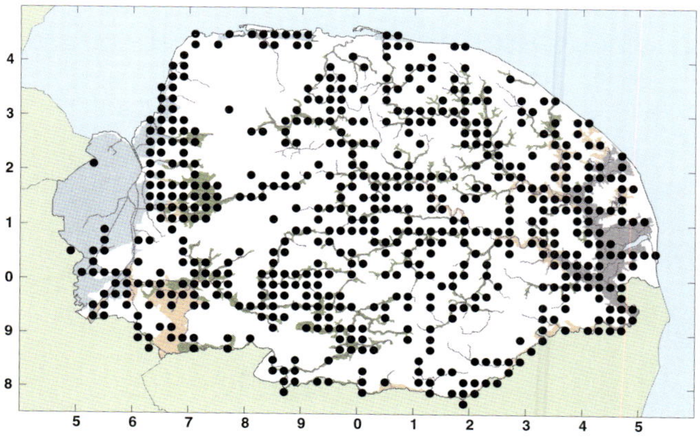

Herbaceous perennial, widespread on the banks of dykes, streams, ponds and slow-moving rivers. Both ssp. **erectum** and ssp. **neglectum** (Beeby) K.Richt. have been recorded, but not sufficiently to make any recognisable pattern.

S. emersum Rehmann **Unbranched Bur-reed**

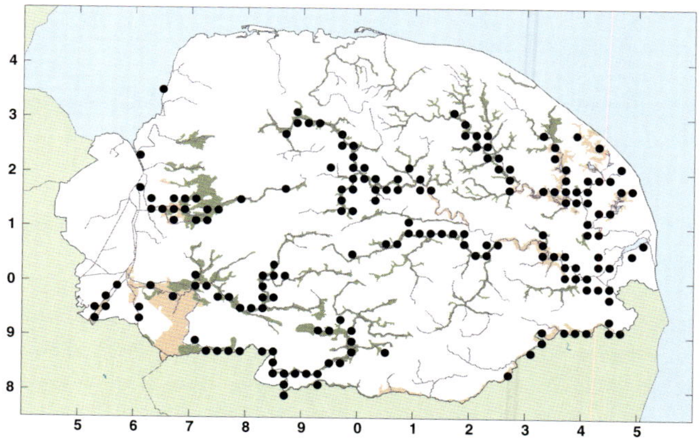

Herbaceous perennial, frequent along streams and dykes, usually growing in the water.

S. natans L. **Least Bur-reed**
Syn. *S. minimum* Wallr.
Herbaceous perennial, a rare plant of pools in base rich areas with seven sites known before 1968. During this survey recorded at Thompson W99 in 1987, CPP and Potter Heigham E41 in 1998, C.Doarks. W1. E1.

Typha latifolia L.
Reedmace, Bulrush, Cat's-tail, Pokers (local)

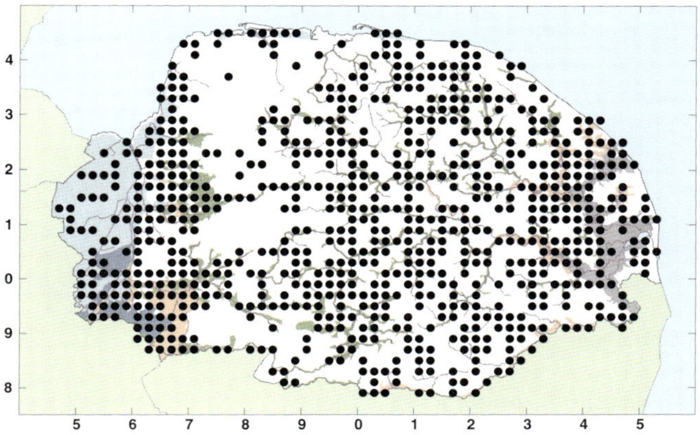

Deciduous perennial frequent in dykes, marshes, reed beds and along the banks of slow-moving streams and rivers, also in ponds where it can become dominant. Usually found on soils with a clay or inorganic base.

T. x glauca Godr. **Hybrid Reed-mace**
(*T. latifolia x T. angustifolia*)
Deciduous perennial. As both parents frequently grow together, it is surprisingly scarce with only two records, at Scoulton Mere W90 in 1918 and at Hilgay W69 in 1992, CD.Preston. W1. E0.

Spargnium emersum - Narborough

T. angustifolia L. **Lesser Reed-mace, Gladdon** (local)

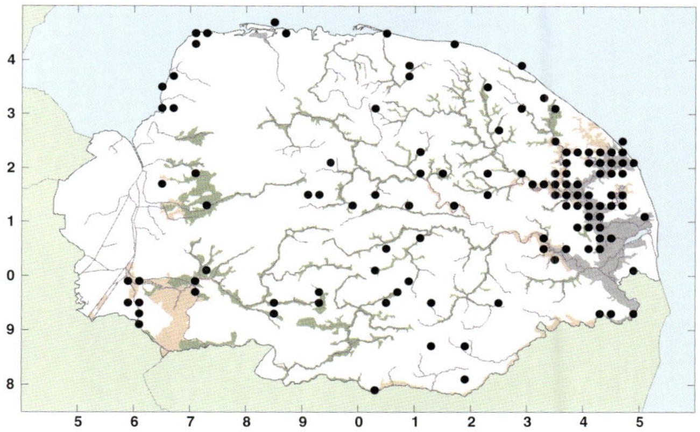

Deciduous perennial widely scattered in dykes, ponds and marshes, often with *T. latifolia* but usually on organic soils especially soft peat. Frequent enough in the Broads to gain its own local name.

PONTEDERIACEAE

Pontederia cordata L. **Pickerel-weed**
Herbaceous aquatic perennial introduced to gardens from North America, occasionally thrown into ponds and becoming established as at Marston Marshes E20 in 1990, R.Land. W0. E1.

LILIACEAE

Narthecium ossifragum (L.) Huds. **Bog Asphodel**
Deciduous perennial, a species with a largely northern distribution in Britain, found in bogs and on wet heaths. Formerly known from six sites, all on the greensand in W62, still present in the Roydon and Wolferton areas where it can sometimes be abundant. Illustrated on the following page. W4. E0.

Kniphofia uvaria (L.) Oken **Red-hot Poker**
Evergreen perennial introduced to gardens from South Africa, occasionally thrown out and established as on the shingle bank at Heacham W63, PRC, also planted on Blakeney Point W94 in 1912 and surviving. Elsewhere not usually long lasting. W5. E6.

K. praecox Baker. **Great Red-hot Poker**
Evergreen perennial introduced to gardens from South Africa, established on the shingle bank at Heacham W63 in 1986, PRC. W1. E0.

Colchicum autumnale L. **Meadow Saffron, Naked Ladies**
Deciduous perennial once native in south-east Norfolk in base rich, moist meadows. Recorded in the 19th century at Pulham E28, Starston E28, and Burston E18. Now found either in parkland as at Thelveton E18 or on roadside verges where colonies have usually started with bulbs dumped with garden rubbish. Some of these may prove to be garden hybrids which are now more widely grown than the true species. W1. E8.

Gagea lutea (L.) Ker Gawl. **Yellow Star of Bethlehem**
Summer dormant perennial of damp woods, known since at least 1941 in its one station at Wayland Wood, W99. W1. E0.

Erythronium dens-canis L. **Dog's-tooth-violet**
Late summer dormant perennial introduced from mainland Europe. Naturalised in parks and old gardens as at Reymerston E00 in 1998, ALB, known since 1958, comm. N.J.Stobbs. W0. E2.

Narthecium ossifragum - Roydon page 256

Tulipa sylvestris L. **Wild Tulip**
Summer dormant perennial from southern Europe, well naturalised on the clays of south Norfolk as at Reymerston E00 where it has been known for 40 years; at Lakenham E20 where a 19th century colony is now in the grounds of County Hall and also near Cromer where it has been known since 1831. W0. E4.

Fritillaria meleagris L. **Snakeshead Fritillary**
Summer dormant perennial, a scarce plant of moist meadows. In the past it was recorded as native in a number of sites on clay soils, the colony in a meadow in W91 being perhaps a survivor of these. Elsewhere an escape from cultivation. A nationally scarce species. W1. E8.

F. pyrenaica L. **Pyrenean Fritillary**
Summer dormant perennial introduced from southern Europe, planted on a roadside at Binham in about 1990 and still surviving. W1. E0.

Lilium martagon L. **Martagon Lily**
Bulbous perennial from mainland Europe, well naturalised in woods at Denton E28, ALB where it was first recorded towards the end of the 19th century; at Honingham E01, L.Laxton; Brakefield Green, Yaxham E00, ALB; Blofield E31, M.Helliwell and Framingham Earl E20, G.Florance. W0. E5.

L. pyrenaicum Gouan **Pyrenean Lily**
Bulbous perennial from the Pyrenees, well established in a wood in W98, S.Jeffreys. W1. E0.

Convallaria majalis L. **Lily-of-the-valley**

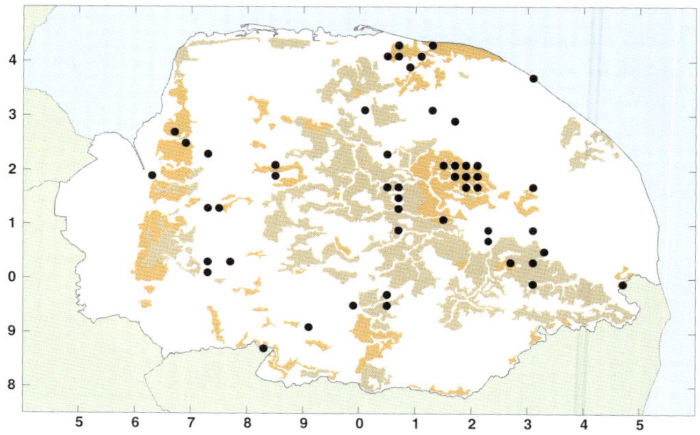

Herbaceous perennial, sometimes locally abundant in moist woodlands. Its soil requirements are sometimes described as base rich, but with us it is almost always on acid soils. Very persistent and surviving the felling of woods for some years. Found both under bracken and in woodland. Also as a garden escape.

Polygonatum multiflorum (L.) All. **Solomon's Seal**
Herbaceous perennial scattered in the county. Well naturalised in parkland, copses and churchyards though frequently confused both in and out of gardens with the hybrid which follows. W0. E8.

P. x hybridum Brügger **Garden Solomon's Seal**
(*P. multiflorum x P. odoratum*)
Herbaceous perennial, a hybrid establishing from rhizomes thrown out from gardens. By far the commonest Solomon's Seal in cultivation. W1. E11.

Maianthemum bifolium (L.) F.W. Schmidt **May Lily**
Herbaceous perennial of acid woodland, probably native in its one site in E03 where it grows with a similar assemblage of plants to those in its North Yorkshire stations. Introduced to the edge of woodland at Sheringham E14. Red Data Book species. W0. E2.

Maianthemum bifolium - Swanton Novers

Convallaria majalis - Swanton Novers

Paris quadrifolia L. **Herb Paris**

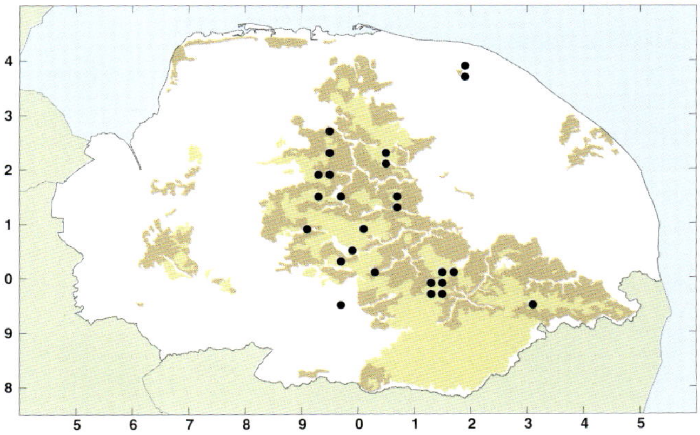

Herbaceous perennial, confined to ancient woods on heavy, wet soils where it can sometimes be frequent.

(The following group are all summer dormant bulbs)

Ornithogalum angustifolium Boreau **Star of Bethlehem**
Syn. *O. umbellatum* auct. non L.

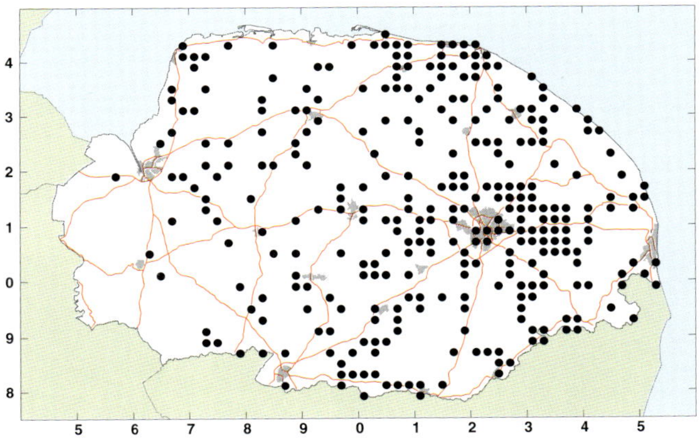

Widespread on grassy banks, perhaps native in the Breckland and on the light sandy soils of west Norfolk, elsewhere usually a garden escape. Very persistent.

O. nutans L. **Drooping Star of Bethlehem**
Introduced from southern Europe, widely scattered but often persistent as at Bodney W98 where it has been known since 1917.
W8. E16.

Scilla siberica Haw. **Siberian Squill**
Introduced to gardens from western Russia. Planted in semi-wild land and spreading, especially in churchyards. W0. E4.

Hyacinthoides non-scripta (L.) Chouard ex Rothm. **Bluebell**
Syn. *Endymion non-scriptus* (L.) Garcke.

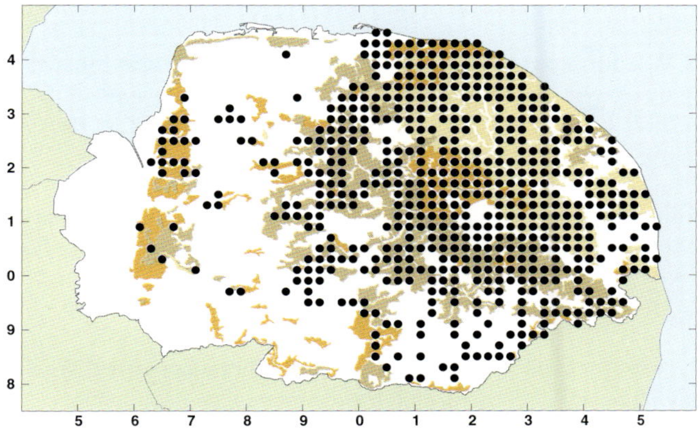

Native to ancient woods and meadows, most frequent on neutral to acid soils and often growing with bracken.

H. x variabilis P.D.Sell **Garden Bluebell**
(*H. non-scripta x H. hispanica*)
The common bluebell of gardens, naturalised from garden throw-outs, but also sometimes misguidedly planted in the wild in error for the true, native bluebell. W164. E152.

H. hispanica (Mill.) Rothm. **Spanish Bluebell**
A scarce garden plant from south-east Europe, confused with *H. x variabilis* both within gardens and without. Well established in the wild near Wicklewood E00 in 1998, Flora Group, but probably over recorded. W4. E74.

Puschkinia scilloides Adams
Introduced to gardens from south-west Asia, occasionally naturalised as in the churchyard at Framingham Earl E20, ALB. W0. E1.

Hyacinthus orientalis L. **Garden Hyacinth**
Introduced from south-west Asia, an occasional throw-out from gardens found on roadsides and waste places usually with other dumped garden rubbish, not long-lasting. W2. E1.

Chionodoxa forbesii Baker **Glory of the Snow**
Introduced from Turkey, an occasional garden throw-out on roadsides, also in some churchyards. W3. E9.

Muscari neglectum Guss. ex Ten. **Grape-hyacinth**
Syn. *M. atlanticum* Boiss. & Reut.

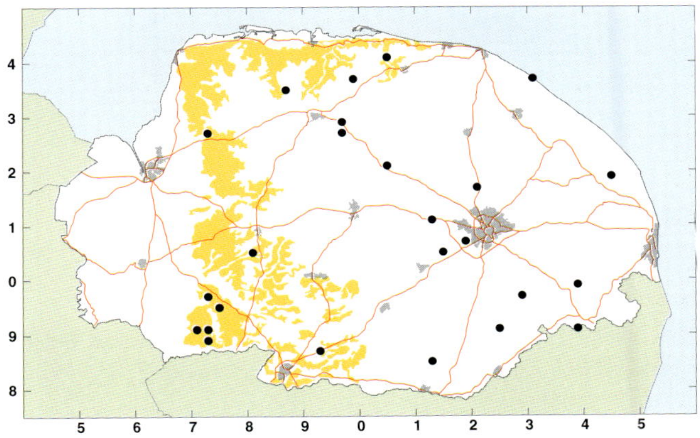

Native on light soils, especially in the Breckland, also in churchyards where it may be the remnants of a former local population. Self-sowing freely from local populations onto the central reservation of the A11 at Brettenham W98. Less commonly grown in modern gardens than the next species.

M. armeniacum Leichtlin ex Baker **Garden Grape-hyacinth**
Introduced from the Eastern Mediterranean, widely grown in gardens and often escaping on to road verges and hedgebanks. W28. E135.

■ **M. comosum** (L.) Mill. **Tassel Grape-hyacinth**
Introduced from mainland Europe and occasionally naturalised. Recorded three times, the most recent being at Mundford, W79 in 1980, M.Baxter.

Allium neapolitanum Cirillo **Neapolitan Garlic**
Introduced to gardens from southern Europe and long naturalised on a roadside bank in Newmarket Road, Norwich E20, ETD and at Roydon E17 & 18, AC. W0. E3.

A. triquetrum L. **Three-cornered Garlic**
Introduced to gardens from the Mediterranean, an occasional escape on grassy cliffs and in the village at Mundesley E33, also on a roadside bank at Alpington E20, Flora Group. W0. E2.

A. paradoxum (M. Bieb.) G. Don **Few-flowered Garlic**

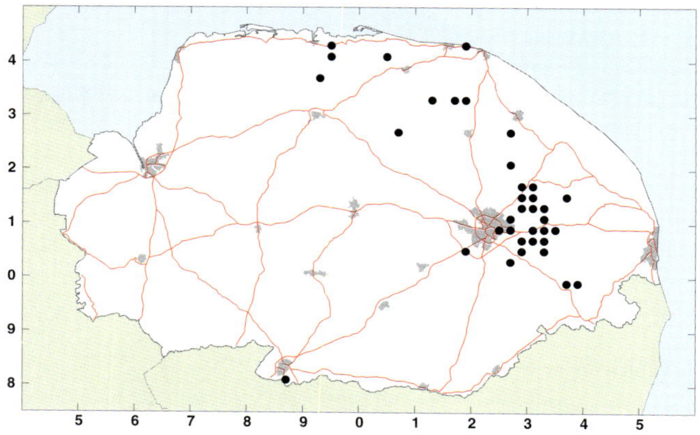

Introduced from the Caucasus, well established on roadsides and hedgebanks.

A. ursinum L. **Wild Garlic, Ramsons**

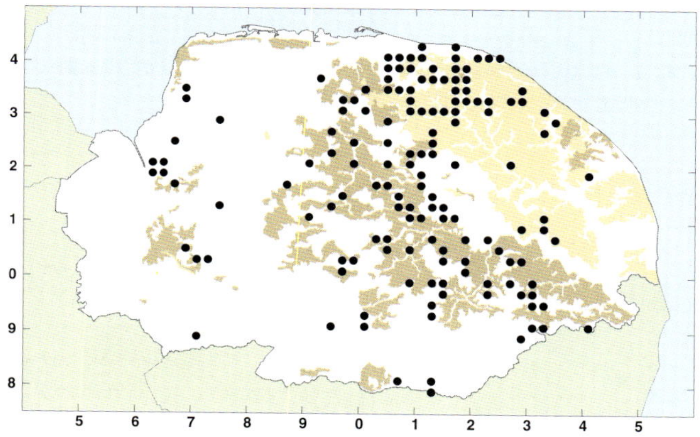

Native in wet woods and adjacent meadows on heavy soils, sometimes forming extensive patches. The origin of the name of Onion Corner at Castle Rising, W62.

■ **A. carinatum** L. **Keeled Garlic**
Introduced from Europe with only one record, Keswick E20 in 1955.

A. vineale L. **Crow Garlic**

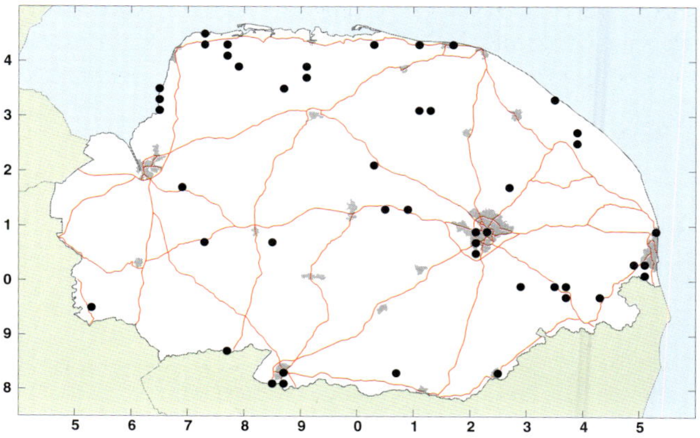

Widely but not commonly occurring on roadsides where it can form extensive patches.

Tristagma uniflorum (Lindl.) Traub **Spring Starflower, Ipheion**
Syn. *Ipheion uniflorum* (Graham) Raf.
Bulbous perennial introduced to gardens from South America, planted in semi-wild areas and becoming established, also an occasional garden throw-out as at Bradwell Industrial Estate E51, Flora Group. W0. E7.

Leucojum aestivum L. **Summer Snowflake, Loddon Lily**

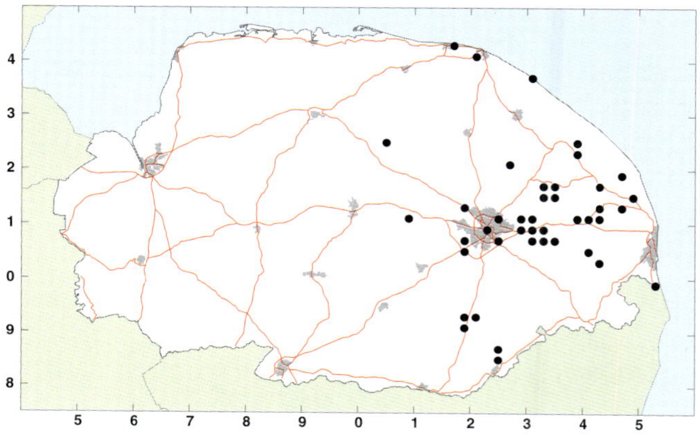

Native only in southern England (the name Loddon refers to the town in Berkshire), well naturalised on heavy soils, especially at Cringleford E00 together with the following species. Spreading downstream along the River Yare from this site.

Allium ursinum - Barton Bendish

L. vernum L. **Spring Snowflake**
Native in southern England, naturalised in a few damp sites, the most notable at Cringleford, E00 where it was first planted in 1873 (pers. comm. R.Harmer). W3. E8.

Galanthus nivalis L. **Snowdrop**

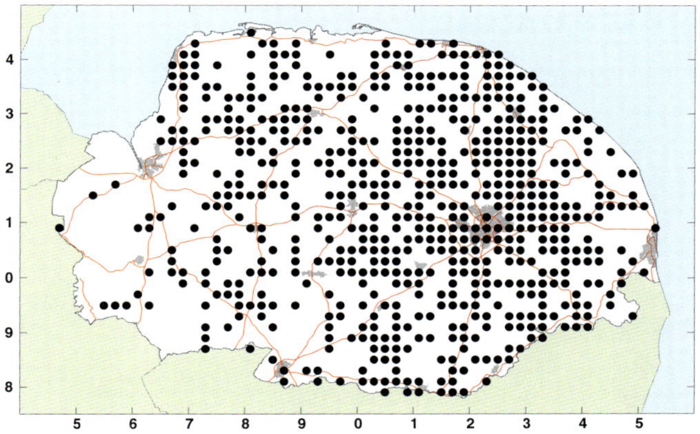

Abundant on roadsides, in parks, churchyards and woods, always introduced. Single and double forms, forma **pleniflorus** P.D.Sell, are equally common.

G. nivalis x G. plicatus
Naturalised in churchyards and found at Dunton W83; Colkirk W92 and East Rudham W82, in the latter without one of its parents, all KAB, also in five E. sites, all SP. W3. E5.

G. plicatus M. Bieb. **Pleated Snowdrop**
Introduced from the eastern Mediterranean observed to be thoroughly established at Dunton W83 and Colkirk W92, both KAB and at Framingham Earl E20, ALB & RWE. W2. E1.

G. elwesii Hook.f. **Greater Snowdrop**
Introduced from south-eastern Europe and Turkey, occasionally naturalised in churchyards and parks. Seen at Colkirk W92 MK; Framingham E20, ALB &. RWE; Chedgrave E39, ALB; Didlington W79, KAB and at Buckenham Tofts W89, a survivor of gardens derelict since 1942, ALB. W3. E2.

G. nivalis x G. elwesii
A garden hybrid found amongst its parents in parkland at Didlington W79, KAB. W1. E0.

G. ikariae Baker
Broad-leaved Snowdrop, Green Snowdrop
Introduced from southern Greece and well naturalised in parkland at Didlington W79, KAB and at Barton Turf E32, Flora Group. W1. E1.

Narcissus poeticus L. **Pheasant's-eye Narcissus**
Introduced from southern Europe and sometimes naturalised, especially in parks and churchyards. W3. E61.

N. pseudonarcissus L. ssp. **pseudonarcissus** **Wild Daffodil**
Rare as a true native, but also occurring as an introduction especially in churchyards where its double form, known in cultivation as 'Van Sion' occurs more widely than the type species. W4. E51.

N. cultivars **Garden Daffodils**
Many Daffodils, especially those belonging to Division 1 & 2 in the horticultural classification, are naturalised by roadsides, parks and churchyards. Some come from deliberate introduction, but many through the discard of unwanted bulbs, perhaps from one of the 'free with petrol' schemes which ran during several years of the 1990s. Some also come from crops grown commercially.

Asparagus officinalis L. **Garden Asparagus**

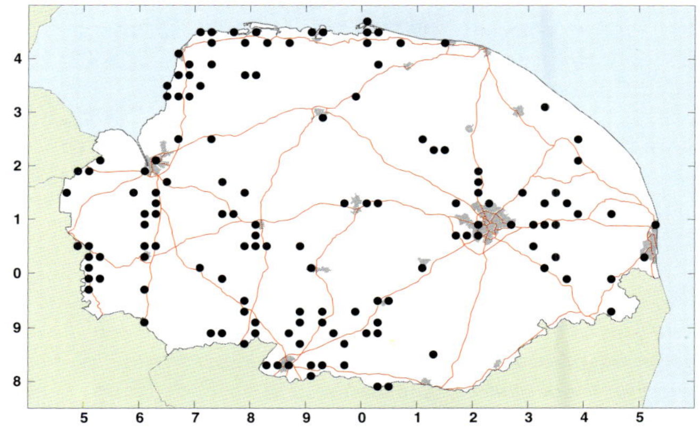

Herbaceous perennial introduced from Europe and grown as a crop on light soils where it is frequently self-sown on to field edges, road verges and rough grassland.

Ruscus aculeatus L. **Butchers' Broom**

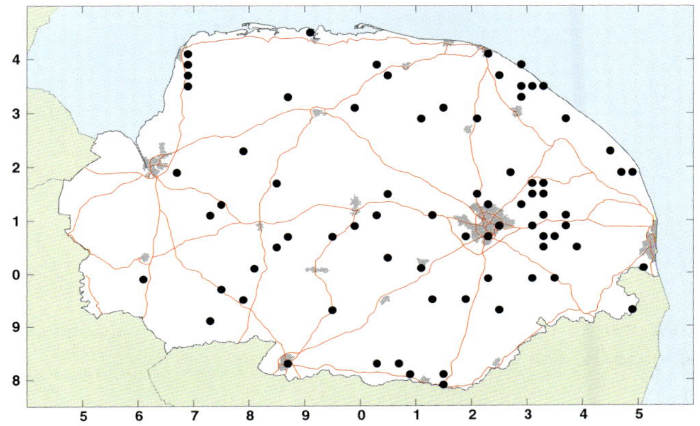

Evergreen shrub, perhaps native, but usually an escape or relic from 'wild' gardens, shrubberies and parks. Even large clumps often prove to be single-sexed clones derived from one original planting. These do not produce fruit.

IRIDACEAE

Sisyrinchium bermudiana L.
Introduced to gardens from western North America, an occasional escape found at Wolterton E13 in 1991, MAB. W0. E1.

S. californicum (Ker Gawl.) W.T.Aiton **Yellow-eyed-grass**
Introduced to gardens from western North America, an occasional escape, found on Beeston Common E14 in 1991, FF. W0. E1.

Iris germanica L. **Bearded Iris**
Introduced to gardens from the eastern Mediterranean, occasionally thrown out and surviving for some years. W4. E15.

■ **I. sibirica** L. **Siberian Iris**
Introduced from the eastern Mediterranean to Russia, an occasional throw-out recorded at Cranwich W79 in 1977, VM.Leather and at Strumpshaw tip, E30 in 1983, ETD.

I. pseudacorus L. **Yellow Flag, Yellow Iris**

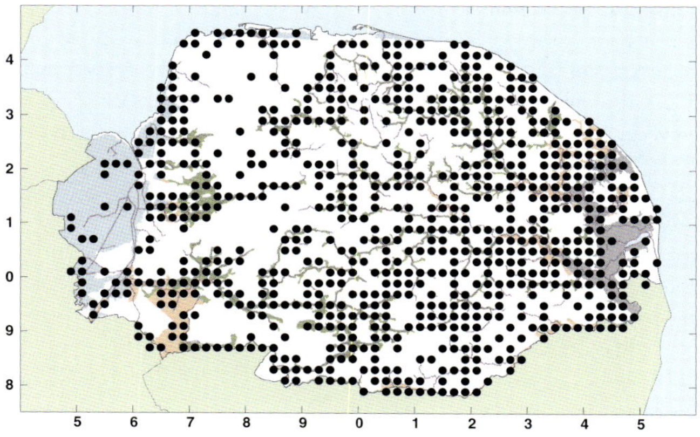

Herbaceous perennial widespread in wet places from pools in meadows to riversides. Introduced in some ponds and lakes.

I. foetidissima L. **Gladdon, Stinking Iris**

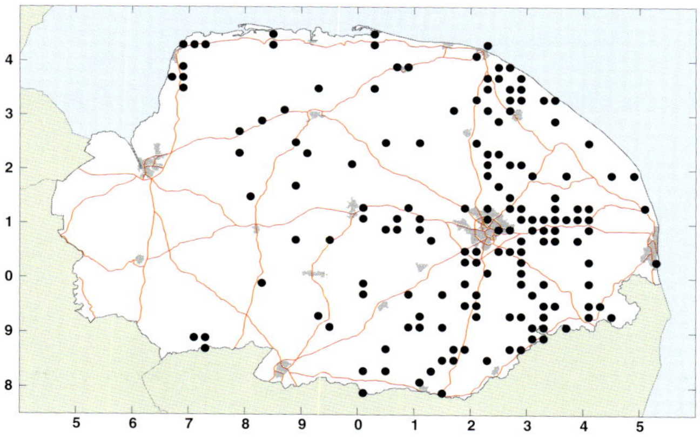

Evergreen perennial of woods, hedgebanks and churchyards, often on dry soils, also sometimes as an introduction.

I. latifolia (Mill.) Voss.
Introduced from the Pyrenees. A garden escape found naturalised in grassland at Seamere, Hingham E00 in 1996, ALB. W0. E1.

Crocus vernus (L.) Hill **Spring Crocus**
Syn. *C. purpureus* Weston
Introduced from southern Europe, established in churchyards and parks, also a casual on road verges. W4. E40.

C. chrysanthus (Herb.) Herb.
Summer dormant bulb from eastern Europe found naturalised on Abel Heath, Blickling E12 in 1995, ALB, det. KAB. W0. E1.

C. tommasinianus Herb. **Early Crocus**
Summer dormant bulb introduced from south-eastern Europe, occasional in churchyards and on roadside verges, particularly near habitation. W3. E15.

C. x stellaris Haw. **Yellow Crocus**
(*C. angustifolius x C. flavus*)
Summer dormant bulb of garden origin, its parents introduced from southern Europe, occasional in churchyards and on roadside verges, most near habitation, usually recorded as *C. flavus* Weston. W3. E0.

Crocosmia x crocosmiiflora (Lemoine) N.E.Br. **Montbretia**
(*C. aurea x C. pottsii*)
Herbaceous perennial, a purposeful hybrid made in the 1870s, now occasionally naturalised in damp places. W8. E46.

AGAVACEAE

Yucca recurvifolia Salisb. **Yucca, Spanish-dagger**
Evergreen perennial introduced to gardens from south-east USA. Planted on Blakeney Point E04 in 1912 and still surviving. W0. E1.

DIOSCORIACEAE

Tamus communis L. **Black Bryony**

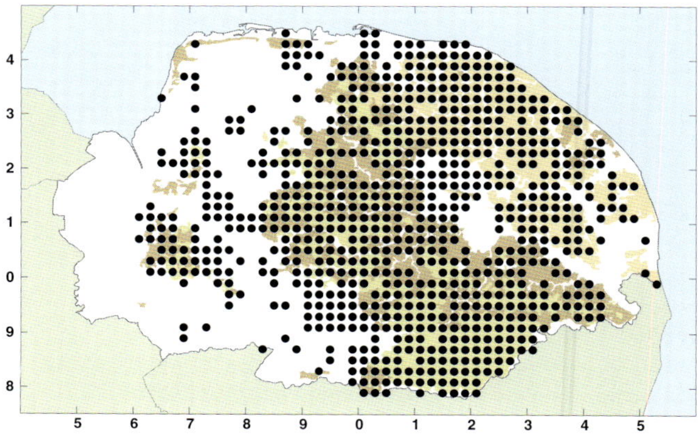

Herbaceous perennial frequent on heavy soils, absent from drier and lighter ones and also from more recently reclaimed land.

ORCHIDACEAE

Epipactis palustris (L.) Crantz **Marsh Helleborine**

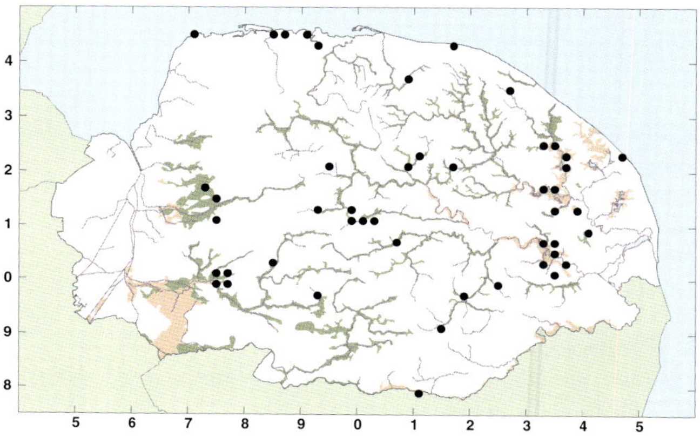

Perennial, widely distributed in marshes and fens but never common. Much reduced in number because of drainage and lost from ten of the sites known before 1960. Var. **ochroleuca** Barla with yellowish-white flowers was noted at Burnham Overy Staithe W84 in 1991, GB; Foulden W70 in 1993 H&JW and Beeston E04 in 1995, FF.

■ **E. purpurata** Sm. **Violet Helleborine**
This native species was formerly recorded on five occasions, the last at Swannington E11 in 1910.

E. helleborine (L.) Crantz **Broad-leaved Helleborine**

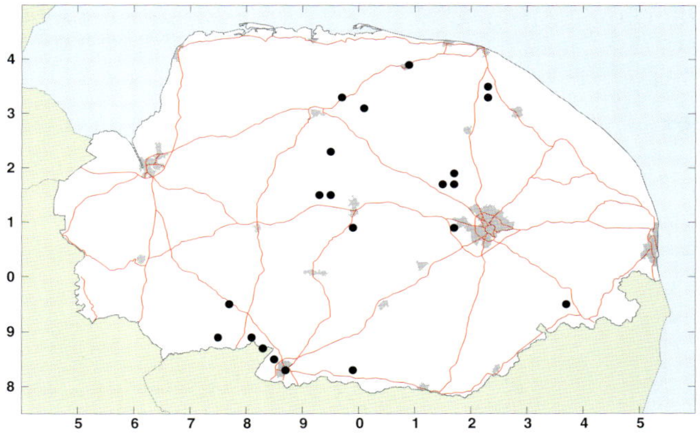

Perennial which can be found both in ancient woodland and in the coniferous woodlands of the Breckland where it grows either beneath or on the sites of long-established deciduous shelter belts. It is generally somewhat erratic in its appearance. A colony at Santon W88 had over 100 flowering spikes in 1997.

Epipactis palustris - Narborough

E. phyllanthes G.E.Sm. **Green-flowered Helleborine**
Very rare, discovered in Norfolk at Santon W88 in 1969 by A.J.Davison, growing in a scrubby area of largely deciduous trees within coniferous forest. Seen again at Santon in 1998, growing with *E. helleborine* in a mixed population under pine, a mile from the previous site, J.Squirrel. Found at Earlham E10 in 1985, still present in 1990, M.Gurney; Keswick E20 in 1990, R.Land and Costessey E11 in 1996, J.Humphris. A nationally scarce species. W1. E3.

Neottia nidus-avis (L.) Rich. **Bird's-nest Orchid**
Rare and very erratic in its appearance with many old records. Seen at Wayland Wood W99 in 1980, D.Holt and during this survey at Ashwellthorpe E19 in 1994, SR.Martin; Intwood E10 in 1997, S.Henson and Holt Country Park E03 in 1997 & 1998, S.Harrop & A.Starling. W0. E3.

Listera ovata (L.) R. Br. **Common Twayblade**

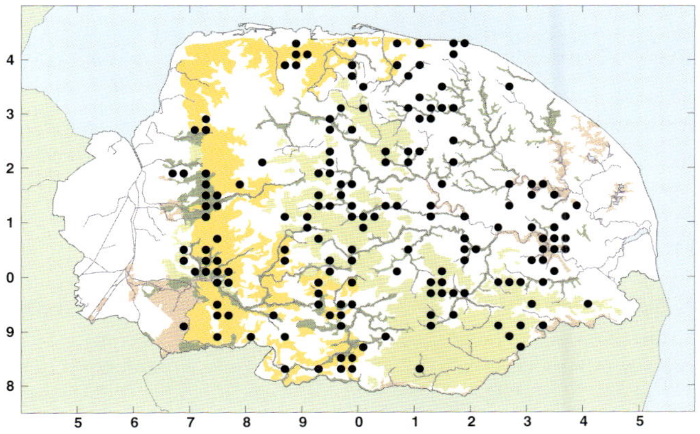

Frequent in a wide range of habitats from woods, hedgerows and fields to damp sand-pits.

Spiranthes spiralis (L.) Chevall. **Autumn Lady's-tresses**
A plant of short grassland with seven old records. It appears regularly on a tennis court at Gresham's School, Holt E03, A.Leech and certainly until 1992 on another lawn in Holt, S.Harrop & A.Starling. W0. E2.

Goodyera repens (L.) R. Br. **Creeping Lady's-tresses**
A perennial species of northern Britain and considered by some to have been introduced to plantations in Norfolk with pines brought from Scotland. It does, however, seem as likely that the orchids have derived from wind-blown seed finding a habitat to their liking in a new area. Eight different sites have been recorded in the past for this orchid, all in North Norfolk except for one colony in the Breckland, at Cranwich W79, which was last seen in 1977 by VM.Leather. During this survey recorded at Horsford Heath E11 and E21 J.Snape, still present in 1996, B.Martin and in Holt Country Park E03, A.Leech. Thriving under pines on the coastal dunes at Holkham W84 and Wells-next-the-Sea W94. A nationally scarce species. W2. E4.

Liparis loeselii (L.) Rich. **Fen Orchid**
A rare plant of pools on fen peat formerly known in E07 and W71, but last seen in both sites in 1965. Its loss is directly due to a falling water table. It is still extant at three sites in the Broads. A Red Data Book species. W0. E3.

Hammarbya paludosa (L.) Kuntze **Bog Orchid**
A rare plant of wet bogs thriving where it grows in shallow pools. It was formerly much more widespread but now survives at only one site in W62 where 117 flowering spikes were counted by WG.Clarke in 1910, but is now reduced to two or three. Again a plant which cannot survive in drying sites. A nationally scarce species. W1. E0.

■ **Herminium monorchis** (L.) R. Br. **Musk Orchid**
This rare orchid was recorded as a native from three stations by Nicholson, 1914, all on chalk.

Platanthera chlorantha (Custer) Rchb. **Greater Butterfly Orchid**

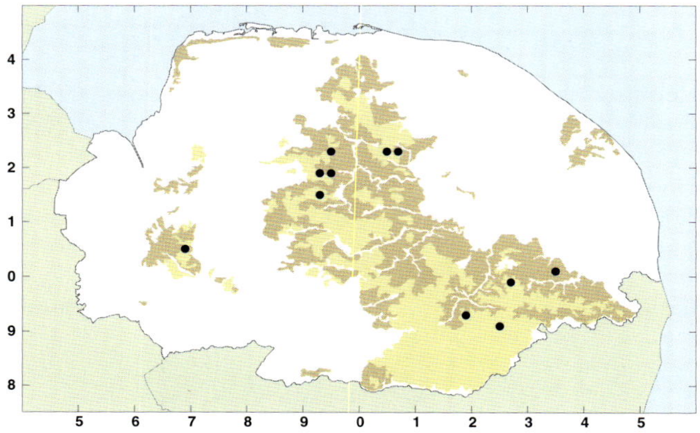

A scarce plant of ancient woodland, it has been recorded from 24 woods in the last hundred years, but a number of these sites have been coniferised and some clear-felled. It may, however, have been under-recorded during this survey as it does not flower every year and so is not always easy to find.

P. bifolia (L.) Rich. **Lesser Butterfly Orchid**
A rare plant of wet heaths and grassland, now much decreased in frequency and during this survey found only at Roydon W62, Beeston E14 and Holt E03. W1. E2.

Anacamptis pyramidalis (L.) Rich. **Pyramidal Orchid**

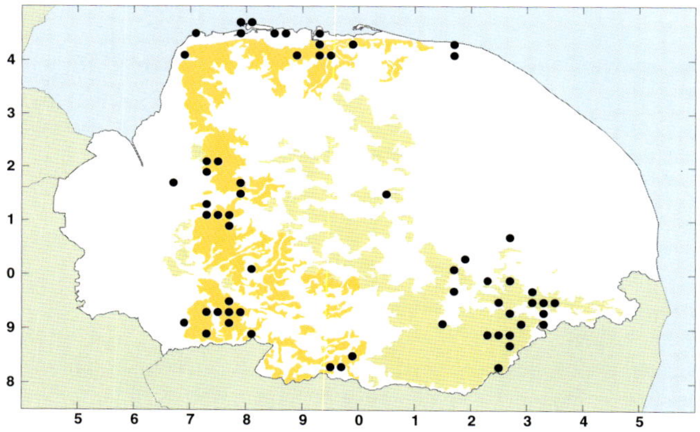

Winter green perennial of chalk grassland and calcareous dunes, still locally frequent.

Gymnadenia conopsea (L.) R. Br. **Fragrant Orchid**

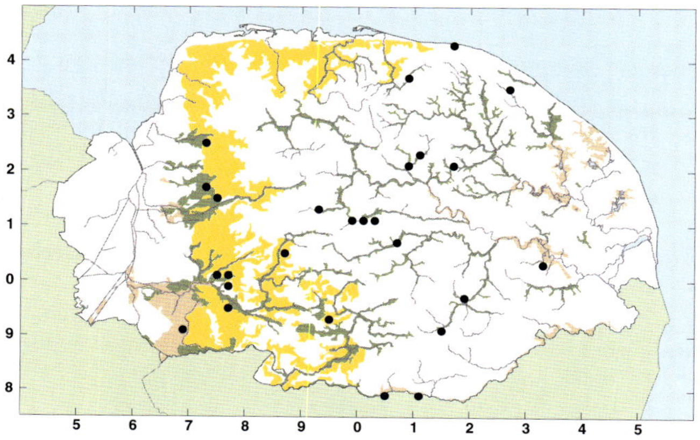

Perennial of chalk grassland. Ssp. **conopsea** is found on dry sites and the more frequent ssp. **densiflora** (Wahlenb.) E.G.Camus, Bergon & A.Camus, in moister grassland and fens. Both subspecies have decreased seriously during the last fifty years.

■ **Coeloglossum viride** (L.) Hartm. **Frog Orchid**
Perennial species of short grassland recorded four times during the 20th century, the most recent at Brisley W91 in 1954, J.Pankhurst.

Dactylorhiza fuchsii (Druce) Sóo **Common Spotted Orchid**

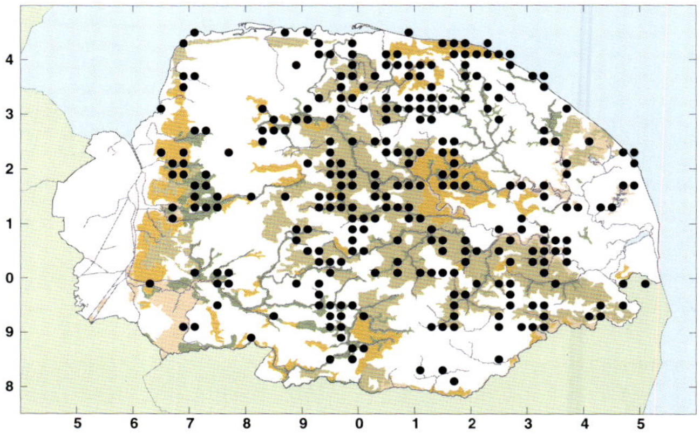

Herbaceous perennial, our most frequent orchid, occurring in woods, fens and damp grassland. Albinos with unspotted leaves as well as white flowers were noted at Carbrooke W90 in 1998, G.Nobes.

■ **D. x kerneriorum** (Sóo) Sóo
(*D. fuchsii x D. incarnata*)
This hybrid has been recorded on three occasions, the last at Reepham E12 in 1973, ETD, RPL & ELS.

D. x grandis (Druce) P. Hunt
Syn. *D. pardalina* (Pugsley) Nelson (*D. fuchsii x D. praetermissa*)
The commonest hybrid of this genus, found wherever the parents grow together. W4. E5.

D. maculata (L.) Sóo **Heath Spotted Orchid**

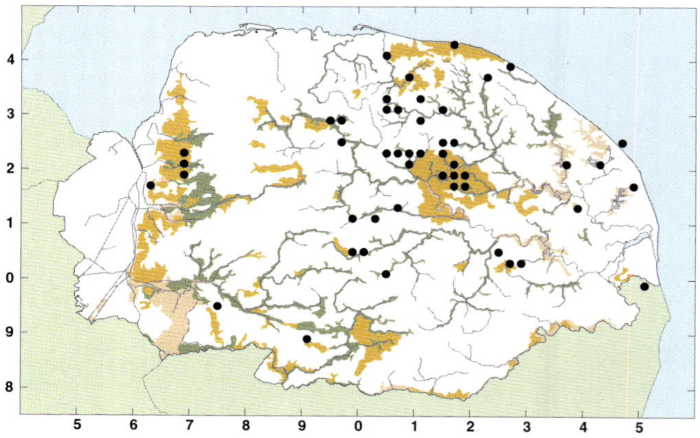

Perennial of moist, acid and heathy ground where it can be frequent. Our plant is ssp. **ericetorum** (E.F.Linton) P.Hunt & Summerh.

■ **D. x hallii** (Druce) Sóo
(*D. maculata x D. praetermissa*)
A rare hybrid as the parents require different habitats so seldom grow together. Recorded at Roydon W62 in 1946 and South Wootton W62 in 1959.

D. incarnata (L.) Sóo **Early Marsh Orchid**

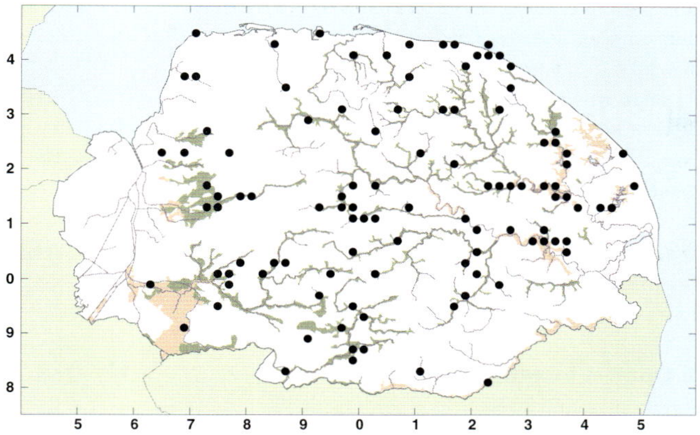

Perennial widely scattered in damp meadows and fens but never common. Ssp. **incarnata** is not infrequent where the ground is wet. Ssp. **coccinea** (Pugsley) Sóo, was recorded on slumped cliffs at Overstrand E24 in 1992, PWL and in dune slacks at Holme W74 where it was present in quantity in 1998, WAB. Ssp. **pulchella** (Druce) Sóo, was last recorded at Roydon W62 in 1973, J.O.Mountford and ssp. **ochroleuca** (Boll) P.F.Hunt & Summerh., first confirmed for Britain near Diss E07 in 1936, has five old records, but has not been seen since 1968.

Dactylorhiza incarnata coccinea - Holme Dunes

■ **D. x wintoni** (A. Camus) P.F. Hunt
(*D. incarnata x D. praetermissa*)
A scarce hybrid seen at Thompson W99 in 1979, ELS.

D. praetermissa (Druce) Sóo **Southern Marsh Orchid**

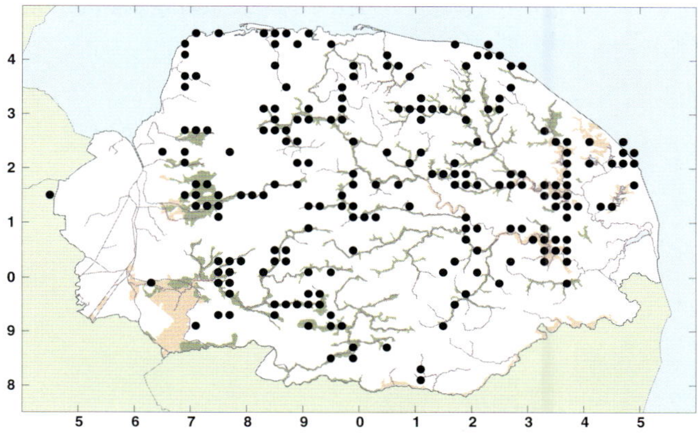

Perennial frequent in fens and wet grassland, often in quantity but variable in number according to the season. A fully peloric form with completely regular flowers lacking a differentiated lip was found at Hunstanton W64 in 1992, M.Fellowes, det. A.I.Denholm.

Dactylorhiza praetermissa (peloric from) **- Hunstanton**

D. traunsteineri (Saut. ex Rchb.) Sóo
Narrow-leaved Marsh Orchid

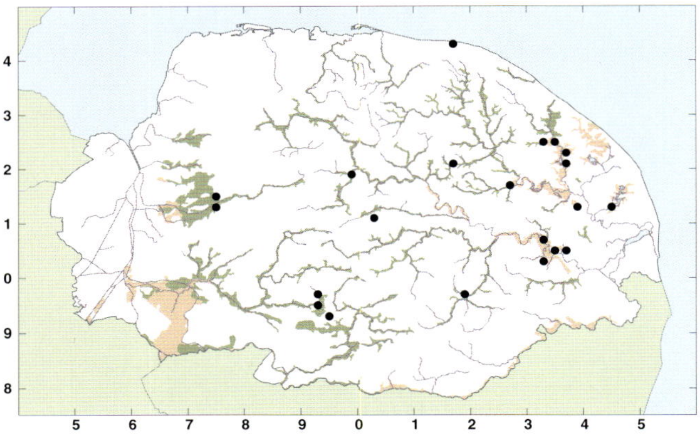

Perennial, a scarce plant of base rich fens and wet grassland, probably less frequent than formerly. A nationally scarce species.

Orchis mascula (L.) L. **Early-purple Orchid**

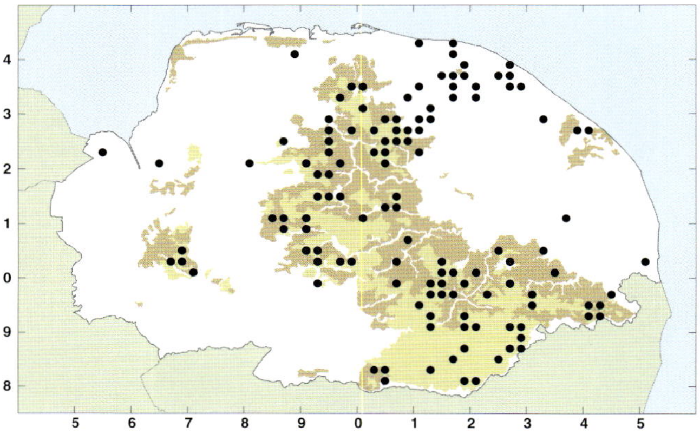

Perennial of long-established woodlands where it can sometimes be abundant, also on verges on clay soils. The flowering plant recorded on newly reclaimed land at Clenchwarton W52 in 1991, MK was presumed to be a chance seedling established in open grassland on a dyke bank. It did not survive long.

O. morio L. **Green-winged Orchid**

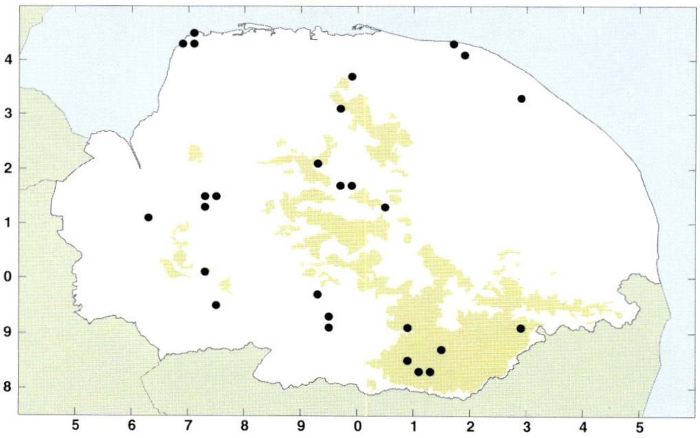

Perennial of damp, long established meadows, lost in many of its former sites through "improvement", either by drainage or the addition of fertilizer. Still abundant in a few sites.

■ **O. ustulata** L. **Burnt-tip Orchid**
Perennial of chalk grassland recorded at Shouldham W60 during the 19th century, not seen since and presumed extinct.

Aceras anthropophorum (L.) W.T. Aiton **Man Orchid**
Perennial recorded from two sites in dry grassland, Holt E03 in 1950 and at Holme W74,where it was first recorded in 1981 and is still present, WAB. A nationally scarce species. W1. E0.

■ **Himantoglossum hircinum** (L.) Spreng. **Lizard Orchid**
Perennial with several old records, the most recent at Long Stratton E19 in 1956.

■ **Ophrys insectifera** L. **Fly Orchid**
Perennial once known from three sites in the Harleston area E28, Galpin, 1888 and reported in the Weybourne to Overstrand area E14 & E24 between 1930 and 1965, but not confirmed.

Orchis morio - New Buckenham

O. apifera Huds. **Bee Orchid**

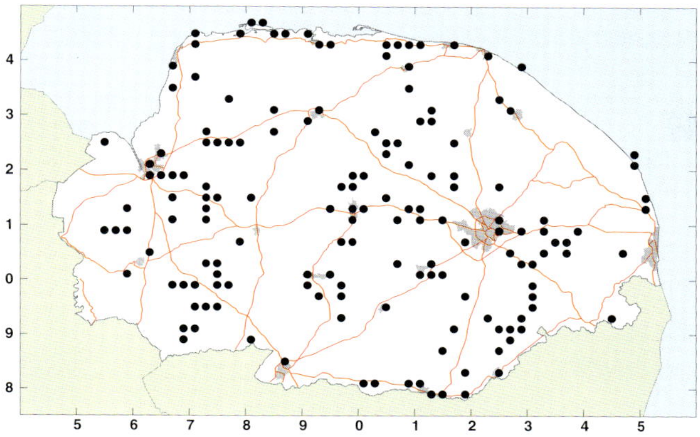

Perennial, widely scattered, chiefly on base rich soils. Found in dune slacks, forest rides, on dyke banks, factory sites and unmown lawns, free seeding and often colonising new ground. A white flowered form also occurs.

Ophrys apifera - Topcroft

The Liverworts and Hornworts

LEPIDOZIACEAE

Kurzia pauciflora (Dicks.) Grolle
Very rare in the county, though where present reasonably abundant. Grows mixed with *Sphagnum*, in bogs. W62; E03 [W61; W63; E11; E12; E22] Possibly under-recorded.

K. sylvatica (A. Evans) Grolle
Very rare. First recorded in 1995 from Roydon Common, King's Lynn, growing on acidic sandy soil W62. (1995, CRS).

Lepidozia reptans (L.) Dumort.

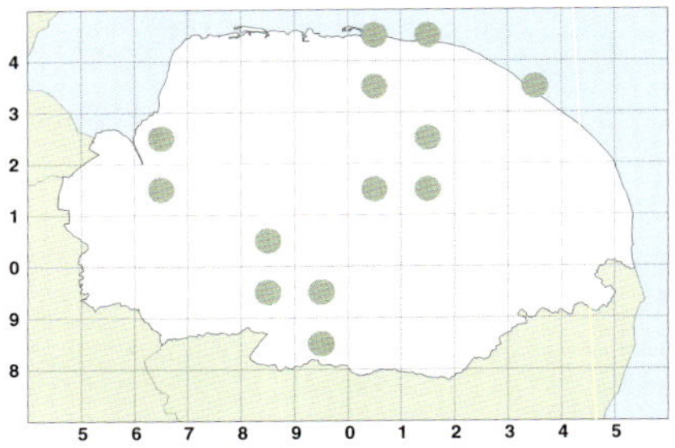

Reasonably common in suitable habitats, viz. rotting tree stumps, raw acid humus, and amongst other bryophytes in acid habitats.

CALYPOGEIACEAE

Calypogeia fissa (L.) Raddi

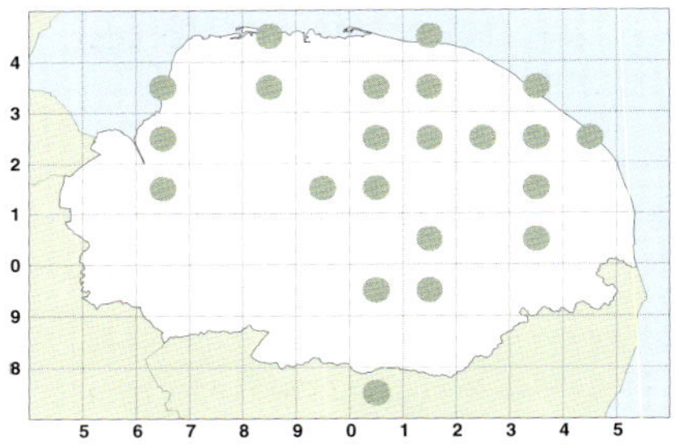

Fairly common in acidic habitats such as organic rich sands, on banks, and on decaying vegetation in heavily shaded habitats. Also amongst *Sphagnum*, in bogs.

C. muelleriana (Schiffn.) Müll. Frib.

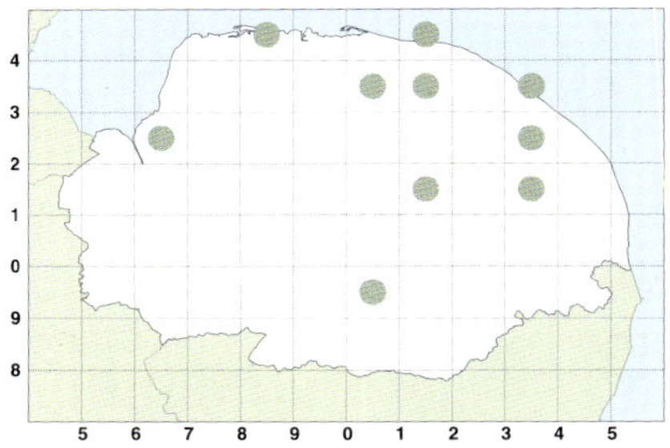

A plant of shaded acid habitats, being equally at home on woodland banks, rabbit burrows and mires.

C. sphagnicola (Arnell & J. Perss.) Müll. Frib.
Very rare. Known only from *Sphagnum* bogs and damp heaths in West Norfolk. Not infrequent in the right habitat. W62.

C. arguta Mont. & Nees
No recent records. [Rare by ditchsides & on sandy banks. W62 (1963); E13 (1977)]

CEPHALOZIACEAE

Cephalozia bicuspidata (L.) Dumort. (including var. **lammersiana** (Huebener) Breidl.)

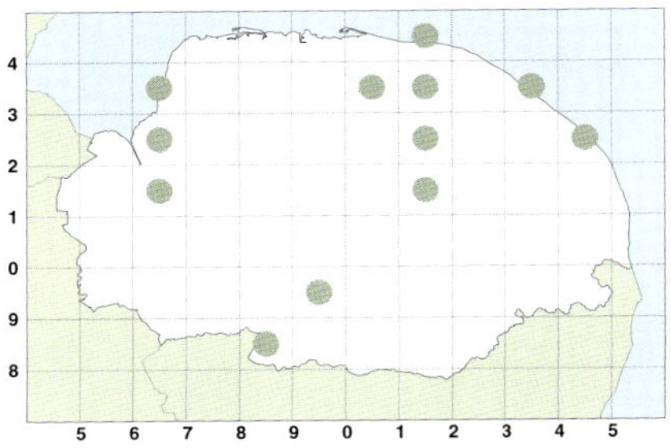

The commonest species of *Cephalozia* in the county. Often abundant on organic rich substrates, such as old stumps, in mires and fen carr. The var. in similar habitats to the type; less frequent.

C. macrostachya var. **macrostachya** Kaal.
Very rare. On damp heath. W62 [E12 (1953)].

C. pleniceps (Austin) Lindb.
No recent records. [Rare on damp heaths and on peat. E03; 12 (1967)].

C. connivens (Dicks.) Lindb.
Very rare. On acid sandy banks, rotting stumps, and *Sphagnum* in mires. Often abundant where it occurs. W62; E03; 12. [W61; 71; E11; 22]. Probably under-recorded.

Cladopodiella fluitans (Nees) H. Buch
Very rare. Locally abundant in bog pools and wet heaths in its few stations. The former habitat seems more important than the latter. W62; E03. [E12].

C. francisci (Hook.) H. Buch ex Jörg.
No recent records. [Wet heath. W61; 62 (1956); E12 (1962)]

Odontoschisma sphagni (Dicks.) Dumort.
Very rare. Amongst *Sphagnum* in mires. W62; E03; 12. [W61]

CEPHALOZIELLACEAE

Cephaloziella elachista (J.B.Jack ex Gottsche & Rabenh.) Schiffn.
No recent records. [Rare in bogs. E12 (1967)]

C. rubella (Nees) Warnst.
No recent records. [Rare amongst raw humus on damp heaths. W62 (1952)]

C. hampeana (Nees) Schiffn.
Rare. A plant of open acidic habitats, such as gravel paths and moist sand in pits and on heaths. W61; 62; 81; E13; 52. [W78; 98; E01; 12]. Probably under-recorded.

C. divaricata (Sm.) Schiffn.

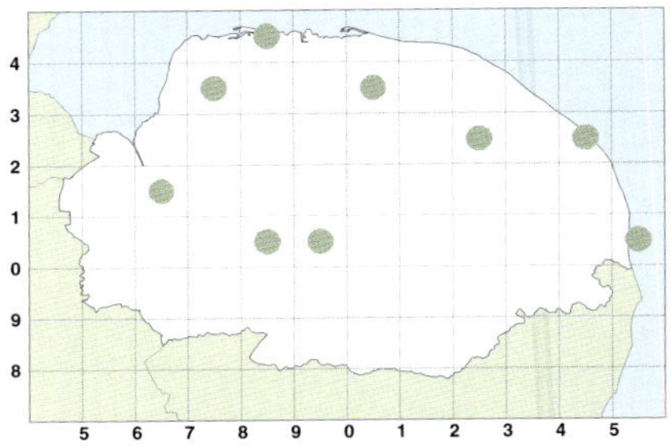

The commonest *Cephaloziella* in Norfolk. On acid, usually sandy soils; under heather, on dry heath; on trees in acid woodland; etc.

LOPHOZIACEAE

Barbilophozia floerkei (F. Weber & D. Mohr) Loeske
Very rare. First recorded in 1986 (RPL). It was found growing on an acid sandy bank in woodland, at South Wootton. Recent searches have failed to refind it. W62.

B. attenuata (Mart.) Loeske
Very rare. This is the only member of the genus which appears to be holding its own in the county, having been found several times in recent years. Its stronghold is Dersingham Bog NNR (W62), where it grows both on acid banks and tracks. Recently, however, it has been found growing, in small quantity, on the stump of a pine in Wootton Woods (W62) and in a similar habitat in Bacton Woods (E33). [W63].

B. hatcheri (A. Evans) Loeske
Very rare. Sand hill, Wolferton Warren W62 (RPL).

B. barbata (Schmidel ex Schreb.) Loeske
No recent records. [Ant hill on heath. W78 (1967)]

Lophozia ventricosa (Dicks.) Dumort.
Rare. A plant which Swann reported from a variety of habitats, including alder carrs and base rich soils in Breckland. Recent records are all from acid habitats such as heaths. W62; E03; 11; 12. [W89; 99; 08; E14]

L. longiflora (Nees) Schiffn.
No recent records. [In Petch & Swann (1968) as *L. porphyroleuca*. E03]

L. excisa var. **excisa** (Dicks.) Dumort.
Rare. In wet heaths, and on the shady banks & bottoms of old sand pits. Not found recently in 'depressions in chalk grassland', which Swann quotes as a habitat. Difficult to recognise, since the typical red gemmae are often absent. W61; 62; 78; E03; 12. [W08]

L. capitata (Hook.) Macoun
Very rare. A nationally rare species, which appears to be locally common in the King's Lynn area, in old sand and gravel workings. Some of the older records for **L. excisa** may belong to this species (1993, CRS). W61; 62; E01.

L. incisa (Schrad.) Dumort.
No recent records. [Very rare on wet heaths (1968) W62] *Not recorded for VC27, according to Corley & Hill (1981), yet two records are given in the Atlas.*

L. bicrenata (Schmidel ex Hoffm.) Dumort.
Very rare. In very short vegetation, on trampled path edge; damp hollow, Holkham W84 (1998, DLH) [W62; E03 (1969)]

Leiocolea rutheana (Limpr.) Müll. Frib.
Norfolk Flapwort
Very rare. A nationally rare (Red Data Book) species whose former centre of distribution lay in Norfolk. However, it has not been seen anywhere except Scarning Fen (W91) in recent years. Taxonomic revision, (Paton 1995), has recognised two varieties. Var. **rutheana** has been recorded from Buxton Heath [E12], whilst var. **laxa** (Schiffn. ex Burrell) Paton, in addition to occurring at Scarning and Buxton, has also been recorded from Swangey Fen, Shropham [E09 (1952)] and Holt Lowes [E03].

L. badensis (Gottsche.) Jörg.
Very rare. On damp calcareous soil on shaded north facing slope of old railway cutting, Swaffham Heath, W70 (1992, CRS).

L. turbinata (Raddi) H. Buch
Rare, In damp, shady, calcareous places. As pure patches on bare chalk, but also mixed with other calcicole species amongst long grass, etc. W70; 71; 82; 94; E11.

Gymnocolea inflata var. **inflata** (Huds.) Dumort.

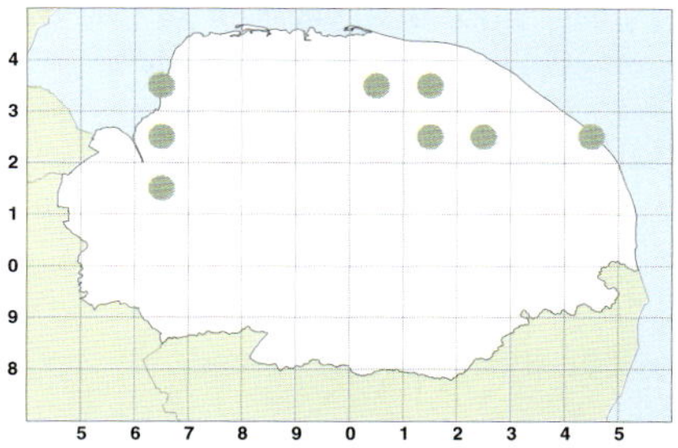

Colonises bare, wet peat hollows in heaths and mires, but also occurs, less frequently, as a pioneer species on bare sand, and on damp woodland banks. Capsules, which are very rare, found twice, at Winterton Dunes and at Bawsey.

Tritomaria exsectiformis (Breidl.) Loeske
No recent records. Recorded (1911), on moist heaths at Norwich, Felthorpe, Holt and Edgefield. Not seen since - presumably extinct. [E03; 11; 20].

JUNGERMANNIACEAE

Mylia anomala (Hook.) Gray
Very rare. In the very few localities where this plant occurs, it may be relatively frequent. A plant of wet heaths and acid mires. W62; E03; 12.

Jungermannia gracillima Sm.
Rare. In moist sandy hollows, banks & ditch sides in heaths and woods. W62; E03; 13; 14; 33. Probably under-recorded.

J. hyalina Lyell
Very rare. On woodland banks E13, and on sandstone base of the Monument, Holkham W84. [E01]

Nardia scalaris Gray
Very rare. Steep, damp, sandy banks & ditch sides. W61; 62. [W03; E22].

N. geoscyphus (De Not.) Lindb.
No recent records. First, and only, record: 1930. [Rare on peaty banks. W62].

SCAPANIACEAE

Diplophyllum albicans (L.) Dumort.

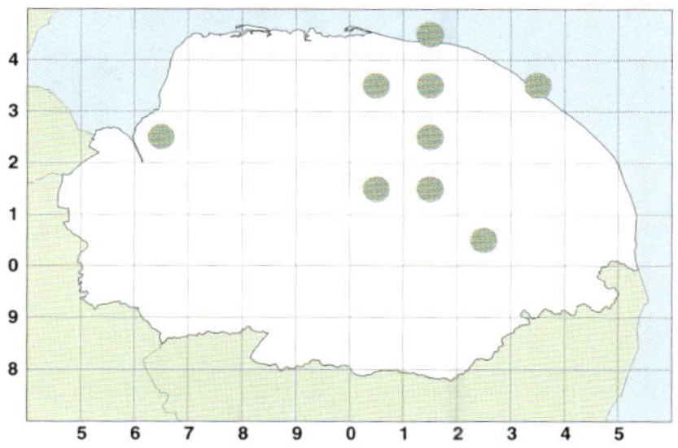

As the map indicates, this plant appears to be much commoner in East Norfolk. Calcifuge. Ditchsides, woodland banks, etc. [W82; 91; 99; E22; 29].

D. obtusifolium (Hook.) Dumort.
Very rare. Not seen since first found, in 1970, on a woodland bank on Weybourne Heath [E14].

Scapania compacta (Roth) Dumort.
Very rare. Short dry, grassy vegetation, by side of old railway line, Wolferton. W62.

S. nemorea (L.) Grolle
No recent records. [Rare on wet heaths. E03; 12 (1967); 22]

S. irrigua (Nees) Nees
Very rare. In damp acidic grassland, Sugar Fen, Leziate. W62. [E03; 12]

S. undulata (L.) Dumort.
No recent records. [Rare by streamsides E03 (1966)]

GEOCALYCACEAE

Lophocolea bidentata (L.) Dumort. (including var. **rivularis** (Raddi) Warnst.). Syn. *Lophocolea cuspidata* (Nees) Limpr., and *Lophocolea bidentata* auct. non (L.) Dumort.
Widespread and abundant throughout the county on both living and rotten tree trunks in woodland. The var. **rivularis** occurs more frequently on the ground, especially in fens, and dampish grasslands.

L. heterophylla (Schrad.) Dumort.
Mainly a plant of tree boles and rotting wood in acidic woodlands.

L. semiteres (Lehm.) Mitt.
Very rare. This alien species, hitherto known only from gardens in the Scillies and Argyll, has been found at Bawsey W61, growing on sand at the side of a track, and in adjacent heathland (1998, CRS).

Chiloscyphus polyanthos (L.) Dumort. and **C. pallescens** (Ehrh. ex Hoffm.) Dumort.

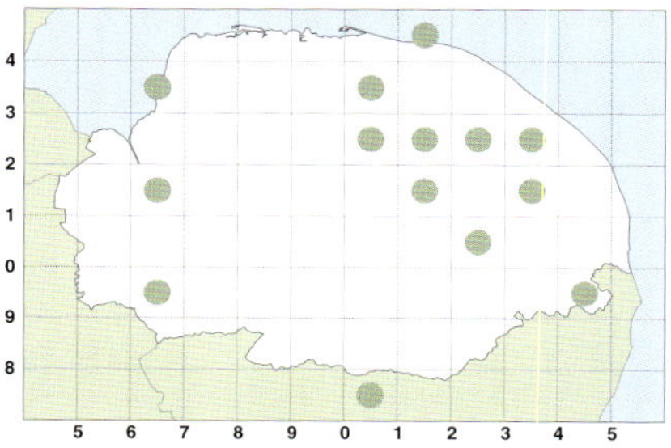

These two species have been, and are, much confused; accordingly, their records have been merged. **C. polyanthos** appears to be darker green, whilst **C. pallescens** is yellower and occurs in more basic habitats. Their habitats seem to be very similar: wet woodlands and fens, usually where there is at least some water movement.

PLAGIOCHILACEAE

Plagiochila asplenioides (L.) Dumort.

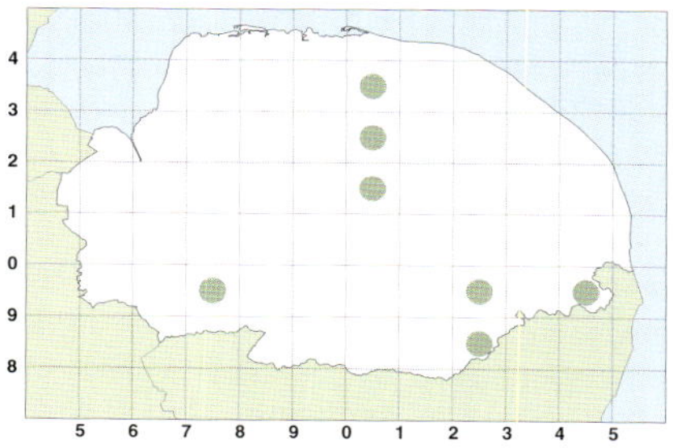

An uncommon plant of damp ditchsides and hollows in woodlands. Usually only occurs in small quantity, as isolated patches.

RADULACEAE

Radula complanata (L.) Dumort.

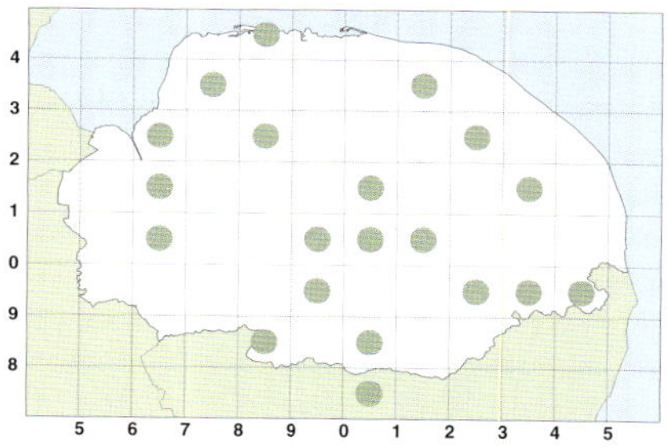

Widely scattered. Usually only occurs in small patches. Recorded on sallow, field maple, buckthorn, etc.

PTILIDIACEAE

Ptilidium ciliare (L.) Hampe
Rare. A plant which appears to have become less common, particularly in West Norfolk. Usually occurs as isolated stems, or small tufts, amongst heather, acidic grasses, or other bryophytes. W61; 62; 88; E01; 42. [W60; 78; 82; 89; 94; 98]

P. pulcherrimum (G. Weber) Vainio
Rare. A plant which is, allegedly, spreading nationally. If so, its progress into Norfolk is proving slow. Occurs as single patches on trees such as sallow and birch. W62; 88; 91; E07. [E03]

PORELLACEAE

Porella platyphylla (L.) Pfeiff.

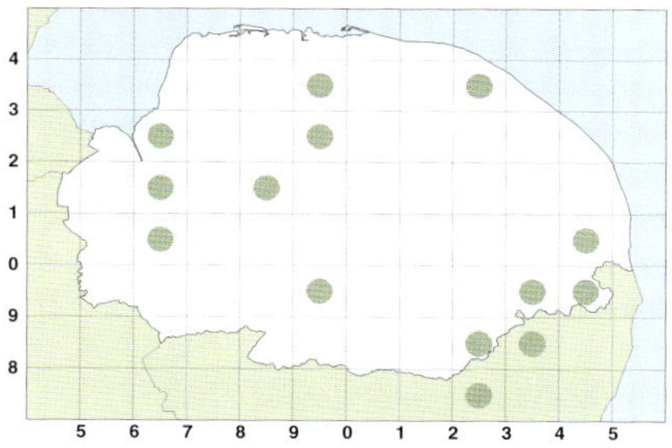

Most frequent on old mortar at the base of walls, usually those of churches, abbeys, etc. Occasionally on trees.

Porella platyphylla - Binham

FRULLANIACEAE

Frullania tamarisci (L.) Dumort.
Very rare. Recorded in the past from short calcareous grassland; however, the only recent record was as an epiphyte. W99. [W78].

F. dilatata (L.) Dumort.

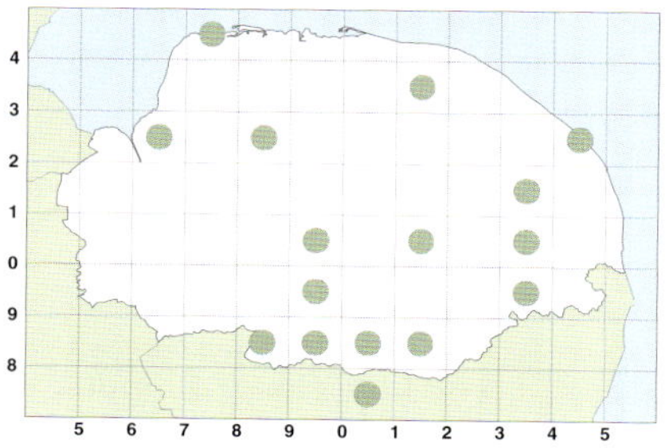

This is a species which appears to have declined in recent years; regarded as 'frequent' by Swann, it is, at best, occasional at present. It is rarely present in quantity. [W62; 69; 70; 73; 78; 83; E01; 02; 03; 09; 12; 22].

CODONIACEAE

Fossombronia foveolata Lindb.
Very rare. Moist, organic rich sand at the side of a ditch, Sugar Fen, Leziate W62. [E11]

F. pusilla var. **pusilla** (L.) Nees
Very rare. All recent records are from the east of the county. E04; 29; 31. [W62; 03]

F. wondraczekii (Corda) Dumort. ex Lindb.
Very rare. The only recent record was from damp clayey soil on the bottom of a dried out pool in an old quarry, Roydon Common W62. [E03]

Petalophyllum ralfsii (Wils.) Nees & Gottsche
Very rare. This Red Data Book species of damp hollows in sand dunes was found in 1957 by CJ Cadbury, at Holme. It was recently refound, in some quantity. W74. (1998, DLH)

PELLIACEAE

Pellia epiphylla (L.) Corda

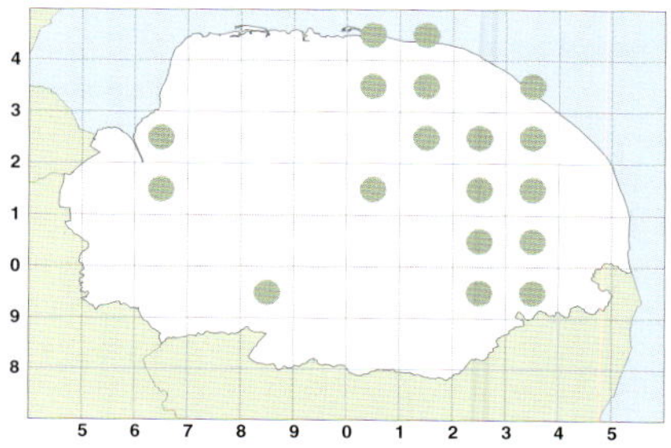

A calcifuge, of damp habitats, e.g. streamsides. Often abundant where present.

P. neesiana (Gottsche) Limpr.
Very rare? A single confirmed record. E03. (1987, MFVC). Almost certainly under-recorded. The *Atlas* (Hill, Preston & Smith 1991) shows it present in large areas in the NE of the county.

P. endiviifolia (Dicks.) Dumort.

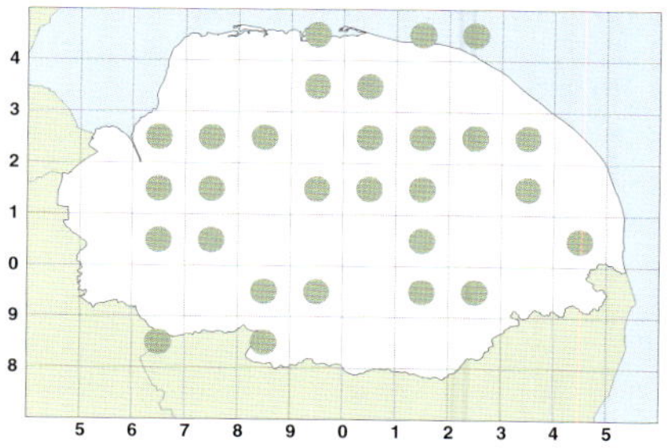

A common calcicole species of fens, and damp soils. Most conspicuous in autumn and winter when it bifurcates repeatedly around the edges of the thallus.

Moerckia hibernica (Hook.) Gottsche
Very rare. A calcicole of damp places. The only recent records are from Flordon (E19), Scarning Fen (W91), Southrepps Common (E23), and on the coast at Holkham (W84). Easily overlooked, so possibly still present at some old sites. [W62; 93; E07; 01; 09; 12; 31]

BLASIACEAE

Blasia pusilla L.
Very rare. An ephemeral pioneer species of moist sands whose appearance seems to depend on disturbance. W61; E13. [W62; E01; 41].

ANEURACEAE

Aneura pinguis (L.) Dumort.

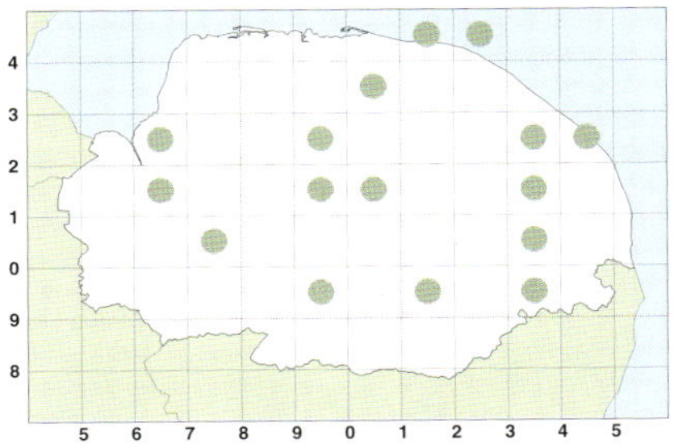

A pioneer species of both damp sands and clays, as well as occurring in fens. It is quite variable in appearance, plants from fens often being large, and of a rather yellowish green colour; other forms are darker green and may be much smaller.

Aneura pinguis - Roydon Common

Cryptothallus mirabilis Malmb.
No recent records. [Saprophytic under peat litter with *Sphagnum, Molinia, Betula & Polytrichum commune*. Rare. W62 (1967)]

Riccardia multifida (L.) Gray
Rare. In bogs and fens, usually in small quantity, mixed amongst other bryophytes. W62; 71; 99; E03; 31; 32.

R. chamedryfolia (With.) Grolle

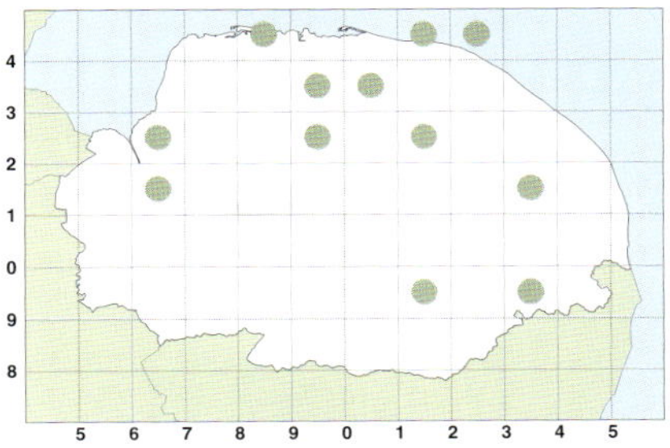

Occurs mainly in fens, on the sides of boulder clay dykes and similar damp habitats, such as the cliffs at Overstrand.

R. incurvata Lindb.
Very rare. On damp bottom of old sand pits, Blackborough End W61 (1993, CRS).

R. latifrons (Lindb.) Lindb.
Very rare. Uncommon in acid bogs E14. [W62; E12]

METZGERIACEAE

Metzgeria fruticulosa (Dicks.) A. Evans

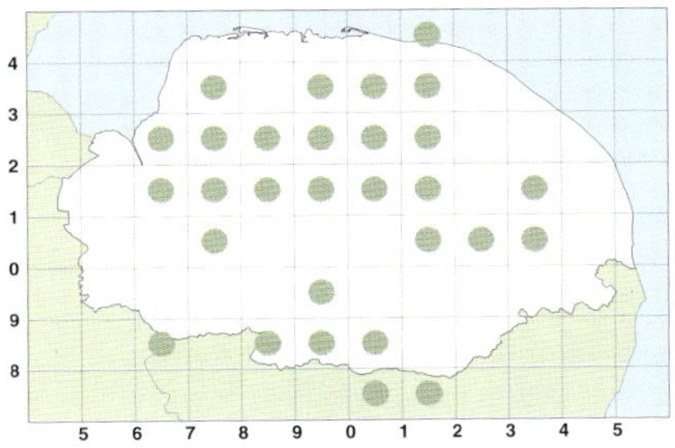

Widespread and often abundant in humid habitats, e.g. the branches of sallows in fen carr - often low down, close to water level, and on elders in deep pits, etc.

M. furcata (L.) Dumort.

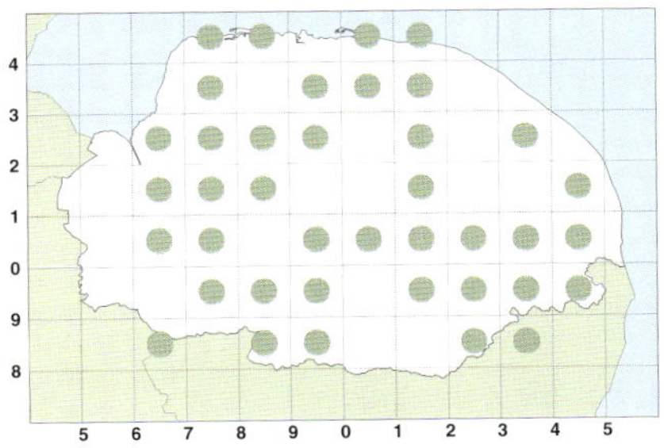

Widespread. An epiphyte, which usually occurs as small patches on a wide variety of tree species; also found on sand, at Holme Dunes.

SPHAEROCARPACEAE

Sphaerocarpos michelii Bellardi
Rare. Burrell regarded this plant as 'locally common on both clays and loams, in turnip and clover fields'. Swann regarded it as 'sporadic'. Occurs in stubble fields. E03; 04; 13; 23; 24. [W83; E01; 21]

S. texanus Austin
Rare. Red Data Book species. Occurs in stubble fields. E03; 04; 12; 13; 23; 31. [E01]

[Sphaerocarpos agg.**]**
These plants occurs mainly in the NE of the county. They are impossible to identify unless ripe spores are present; they are also difficult to 'grow on', so there are a number of records which cannot be assigned definitely to species level. E32; 34.

LUNULARIACEAE

Lunularia cruciata (L.) Dumort. ex Lindb.

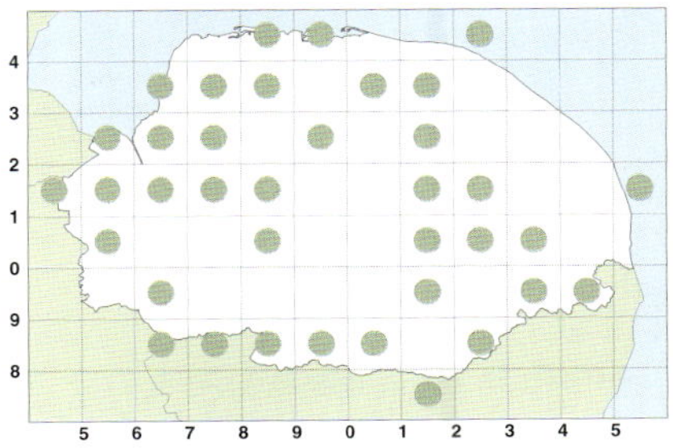

Widespread throughout the county, most often in damp, shady habitats close to man, though sometimes in more natural habitats such as streamsides in woodland.

CONOCEPHALACEAE

Conocephalum conicum (L.) Dumort.

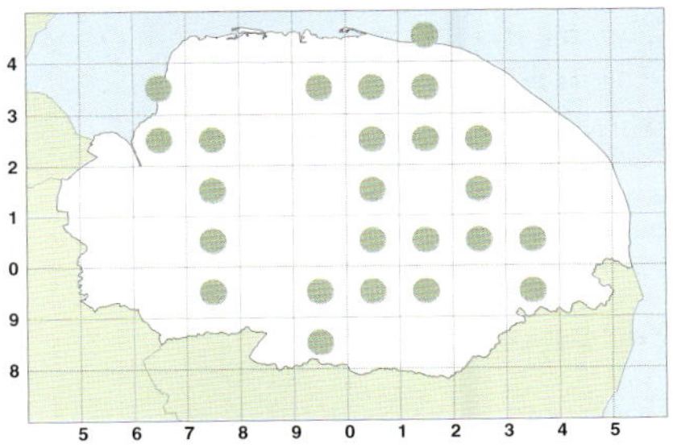

In similar habitats to the last species, though apparently less widespread. Often abundant where it does occur.

AYTONIACEAE

Reboulia hemisphaerica (L.) Raddi

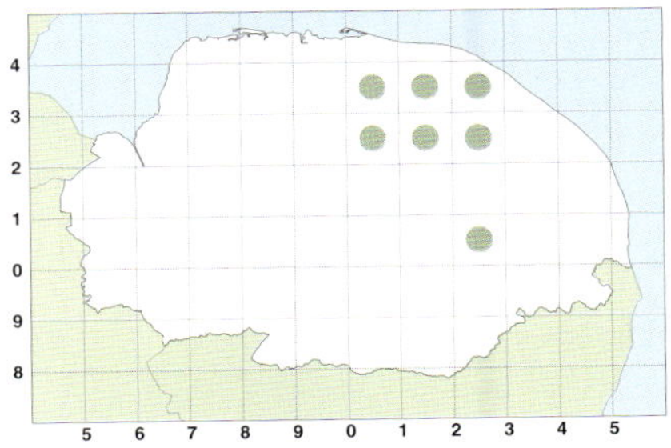

Uncommon. More-or-less limited to a fairly small area in the NE of the county. E02; 03; 12; 13; 20; 22; 23. (Nearly all records, MAB).

MARCHANTIACEAE

Preissia quadrata (Scop.) Nees
No recent records. [Moist hollows in base-rich fens. Rare. W91 (1956); 93; E12 (1951); 22].

Marchantia polymorpha agg.

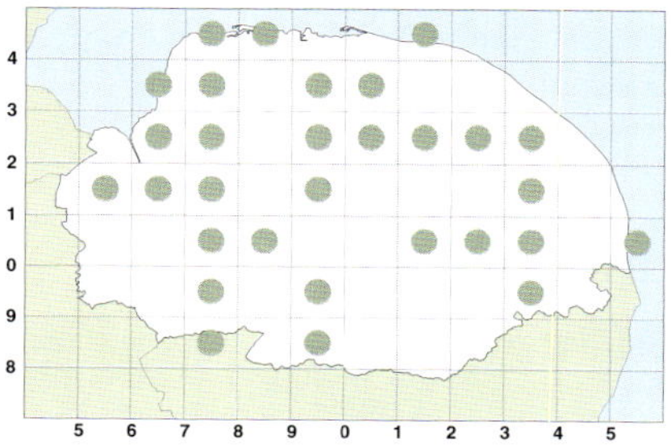

Almost certainly under-recorded. Recent taxonomic revision has recognised 3 subspecies, two of which have been recorded from Norfolk. **Marchantia polymorpha** (L.) ssp. **polymorpha** has been recorded from both VCs, ssp. **ruderalis** Bischl. & Boisselier from 27 alone.

Marchantia polymorpha - King's Lynn

RICCIACEAE

Ricciocarpus natans (L.) Corda
Rare. An aquatic, floating on the surface of ponds, lakes and ditches. Under-recorded. W62; 71; 88; 99; E32; 41. [W60; 89; 98].

Riccia cavernosa Hoffm.
Very rare. Found on mud, at the side of drying ponds and lakes in Breckland. When present it may occur in considerable quantity. W88; 98; 99.

R. crystallina L.
Old records for this species should be referred to *R. cavernosa*

R. fluitans L.

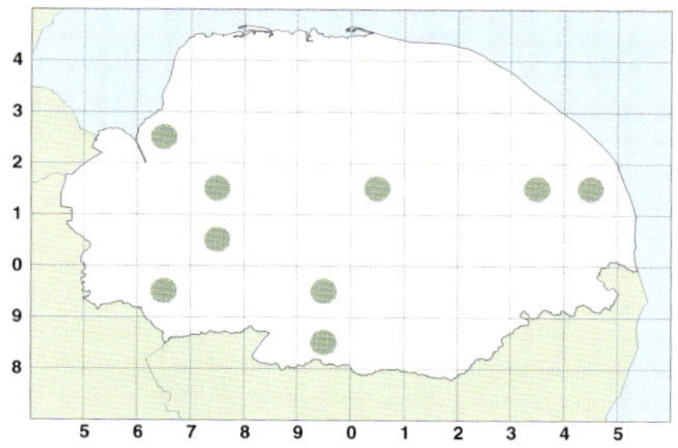

A widely scattered aquatic. Often present in quantity, floating on the surface of ponds or dykes, or growing on marginal muds. May also grow submerged, amongst other aquatic vegetation.

R. rhenana Lorb.
No recent records. [W82 (1967)].

R. sorocarpa Bisch.

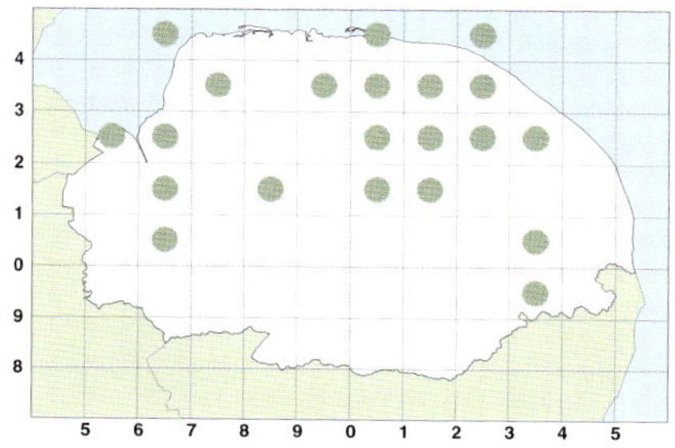

Probably the commonest of the *Riccia* species in the county. A dominantly stubble field plant, which can sometimes occur in quantity, though often absent from apparently suitable habitats.

R. glauca L.

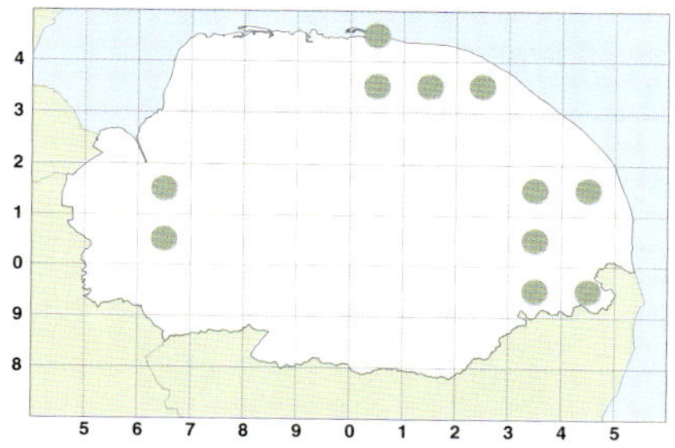

Another plant of stubble fields. Appears to be less common than **R. sorocarpa**.

R. subbifurca Croz.
Syn. *R. warnstorfii* auct. non Limpr. ex Warnst.
No recent records [Rare. Woodland and stubble fields. W03; E01 (1979)].

ANTHOCEROTACEAE

Anthoceros agrestis Paton
Very rare. This plant of arable habitats was regarded as rare by Burrell; it had not been seen since the beginning of the century, until recently refound in a stubble field near Alderfen Broad, Irstead, E31 (1998, JM).

The Sphagna or Bog Mosses

SPHAGNACEAE

Sphagnum papillosum Lindb.
Very rare. A plant of open bogs which appears to have declined. W62; E03; 12. [W61; E02; 11; 22]

S. palustre L.

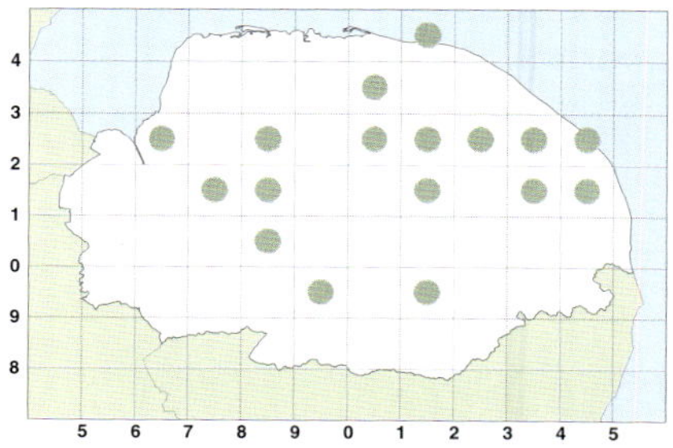

One of the commonest species of *Sphagnum* in the county. Where it occurs it is often very abundant; a plant of open habitats, as well as in carr.

Sphagnum palustre - Sugar Fen

S. magellanicum Brid.
Rare. This easily identifiable species forms dense dark-red cushions in wet bogs and heaths. W62; E03; 12; 32.

S. squarrosum Crome

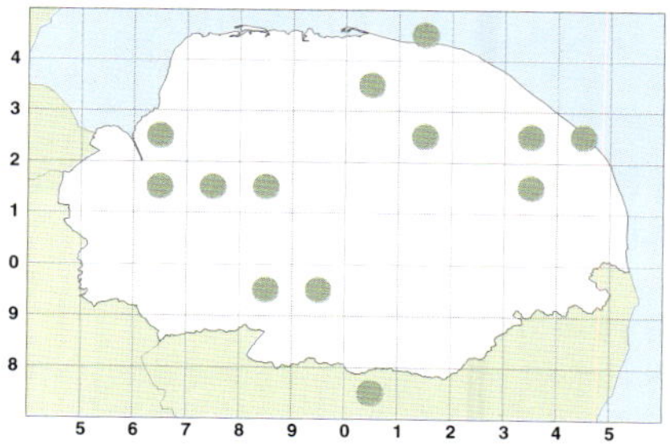

Widely distributed, and often abundant in fens, fen carr, & damp woods.

S. teres (Schimp.) Ångstr.
Very rare. Confined to East Norfolk. Found in less acid mires, especially where flushed. Possibly becoming more abundant, due to better management of its habitat. E03; 12; 31.

S. fimbriatum Wilson

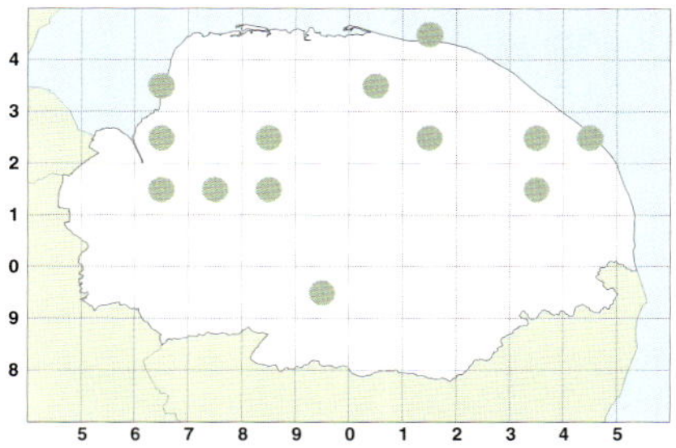

Often fairly abundant in fens and carrs; tolerant of a fairly wide range of pH.

S. russowii Warnst.
Very rare. Found in open boggy area on Alderfen Broad, Irstead. E31. (1998, DFS).

S. capillifolium (Ehrh.) Hedw. (including ssp. **capillifolium** & ssp. **rubellum** (Wilson) M.O. Hill)

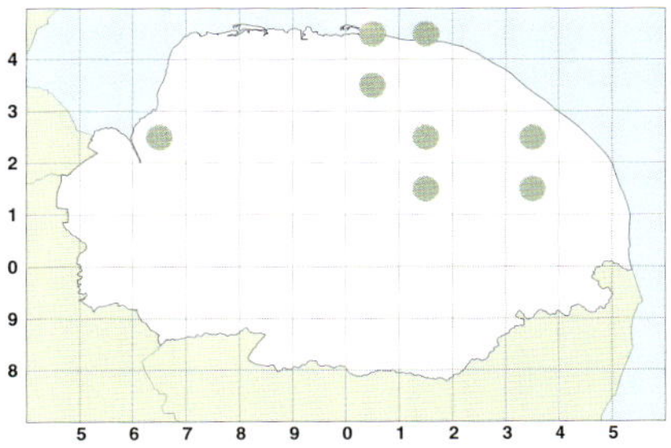

Records for these two ssp. have not been differentiated; plants of open bogs, & occasionally woodland. Less frequent than formerly?

S. subnitens Russow & Warnst.

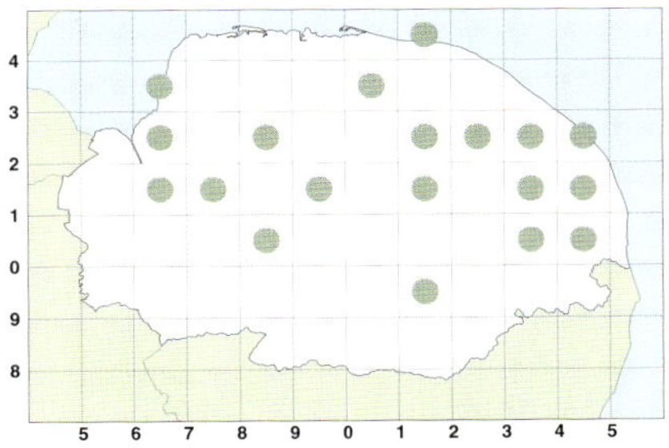

Often common in fens, wet heath & fen carr. Most frequent in open habitats.

S. molle Sull.
Very rare. A species of wet heaths, confined to a few localities in VC28. W62.

S. compactum Lam. & DC.

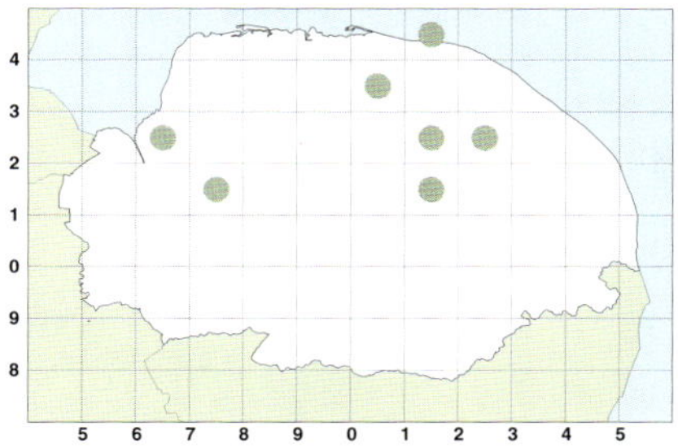

Usually abundant, where it occurs. A plant of wet heaths.

S. denticulatum Brid.
Syn. *S auriculatum* Schimp.

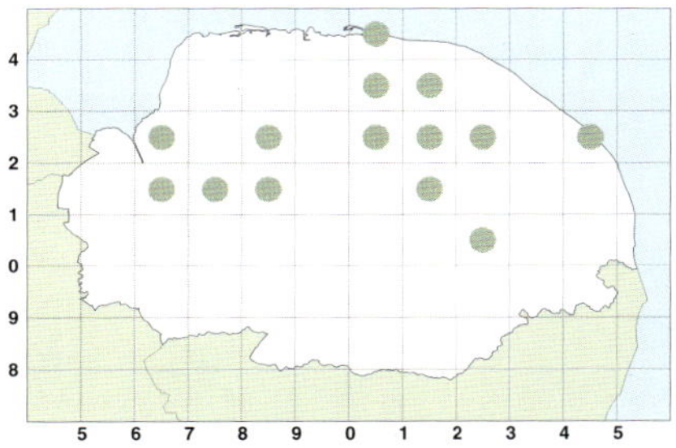

Occurs in a wide variety of boggy / flushed acid habitats, but usually only as small patches.

S. inundatum Russow
Syn. *S auriculatum* var. *inundatum* (Russow) M.O.Hill; *S. denticulatum* var. *inundatum* (Russow) Karttunen
Rare. Another species which seems to have become less frequent. W62; E02; 03; 12; 33. [W61; E11; 22].

S. subsecundum Nees.
Very rare. Usually in small quantity, in fenny areas. E12. [W61; 62; E03; 22; 32]

S. contortum Schultz
Very rare. In small quantity, in base-rich fen. W62 [E03]

S. cuspidatum Ehrh. ex Hoffm.
Rare. Another species which appears to be less common than formerly. Occurs in bog pools, either free-floating or submerged, where it may be locally abundant. W62; 71; E03; 12. [W61; E11; 22; 32].

Sphagnum cuspidatum - Roydon

S. tenellum (Brid.) Bory
Rare. A plant of wet heaths where it may be locally abundant. Normally associated with **S. compactum**. W62; 71; E12; 22; 31.

S. fallax (H. Klinggr.) H. Klinggr. (including ssp. **fallax**, & ssp. **brevifolium** (Flatberg) M.O. Hill)
Syn. *S. recurvum* P. Beauv. var. *mucronatum* (Russow) Warnst.

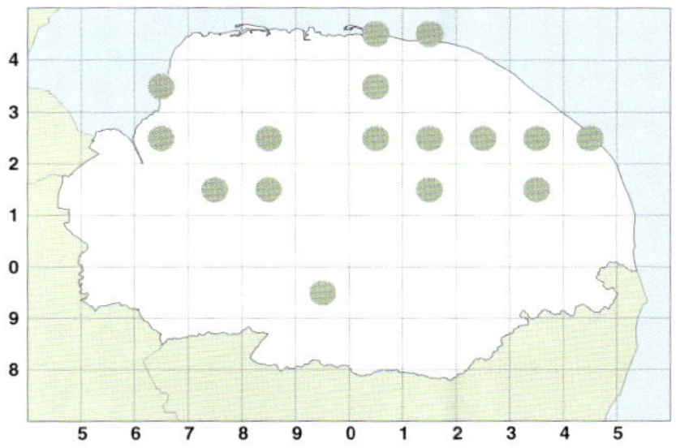

Widespread and usually abundant, forming carpets in bogs, ditches, damp woodland, etc. Often brownish in colour. (Ssp. **brevifolium**, only from Dersingham Bog. W62. (1993, MOH)).

S. flexuosum Dozy & Molk.
Syn. *S. recurvum* P. Beauv. var. *amblyphyllum* (Russow) Warnst..
Very rare, with only one recent record - E12. [E03] Possibly under-recorded.

The Mosses: Acrocarps

POLYTRICHACEAE

Pogonatum nanum (Hedw.) P. Beauv.
Very rare. A pioneer species of organic rich sandy soils. W61; E13 [E14; 33].

P. aloides (Hedw.) P. Beauv.

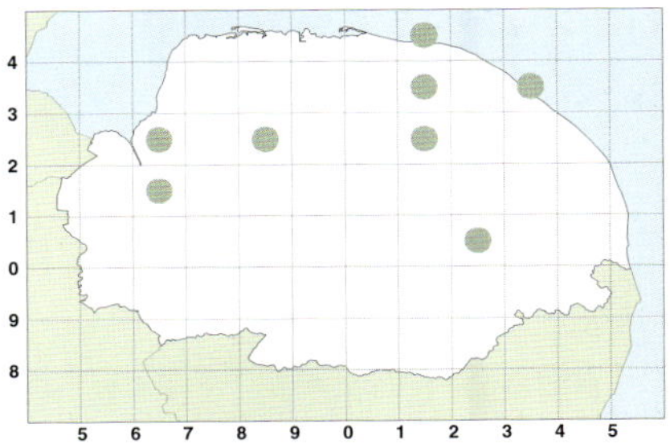

A pioneer species of acid sandy banks, pits, etc., which often occurs in considerable quantity. Sporophytes frequent.

P. urnigerum (Hedw.) P. Beauv.
Very rare. Recently found at Blackborough End in West Norfolk, W61. (1996, NGH), and, shortly afterwards in Bacton Woods, E33, (1996, RJF). *Old records exist for VC27.*

Polytrichum longisetum Sw. ex Brid.

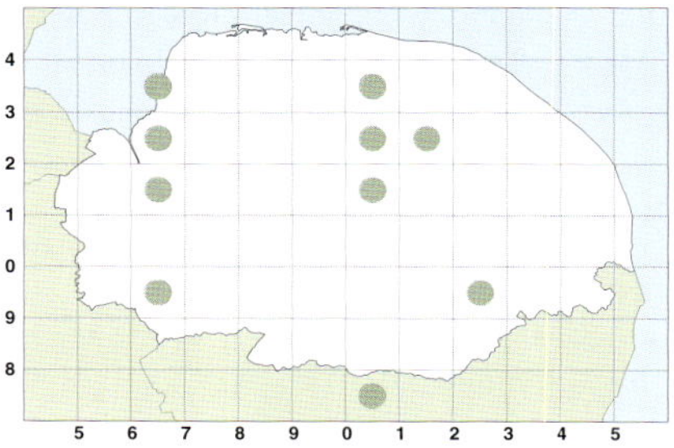

Not uncommon in wooded areas with acid soils. Usually forming short lawns.

P. formosum Hedw.

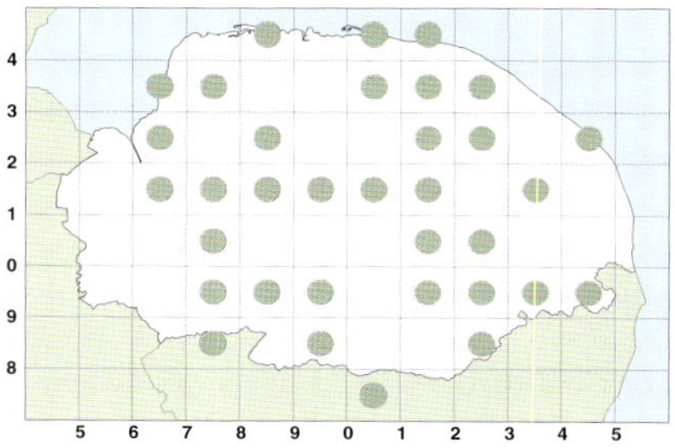

Widespread and abundant, mainly in woodland habitats.

P. commune var. **commune** Hedw.

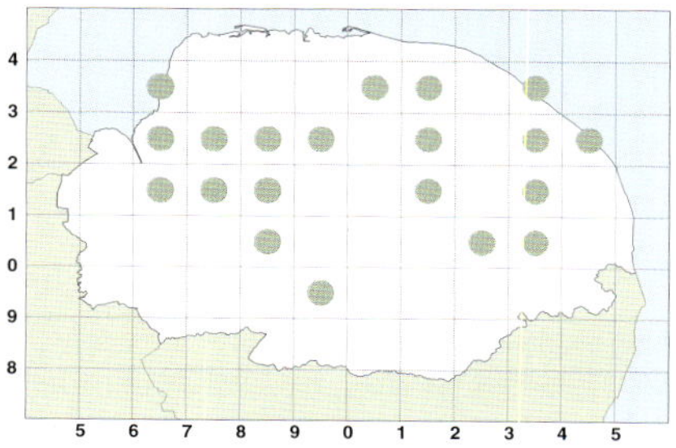

Often abundant in a variety of (usually) damp habitats on acid soil.

Polytrichum commune - Bawsey

P. piliferum Hedw.

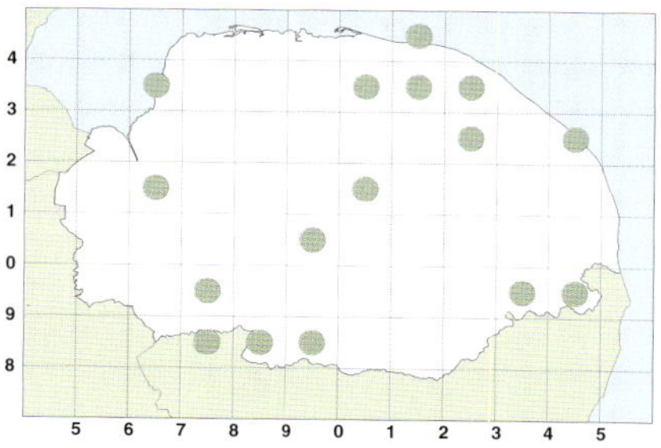

Much less common than the next species. This species is more of a pioneer colonist of open habitats, though the two do grow together sometimes.

P. juniperinum Hedw.

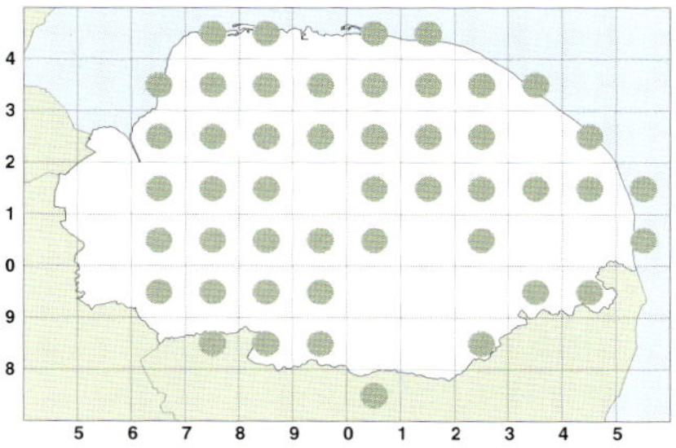

Usually abundant, usually as closed carpets on stabilised acidic open habitats. Clearly a better competitor than **P. piliferum**. Sporophytes common.

Polytrichum juniperinum - King's Lynn

P. strictum Brid.
Syn *P. alpestre* Hoppe
Very rare. A plant of damp acid habitats. W62; 71 [W72; 88; E12; 14; 23; 32]

Atrichum undulatum var. **undulatum** (Hedw.) P. Beauv
Very widespread. Common on shaded, mildly acid soils. Sporophytes common.

TETRAPHIDACEAE

Tetraphis pellucida Hedw.

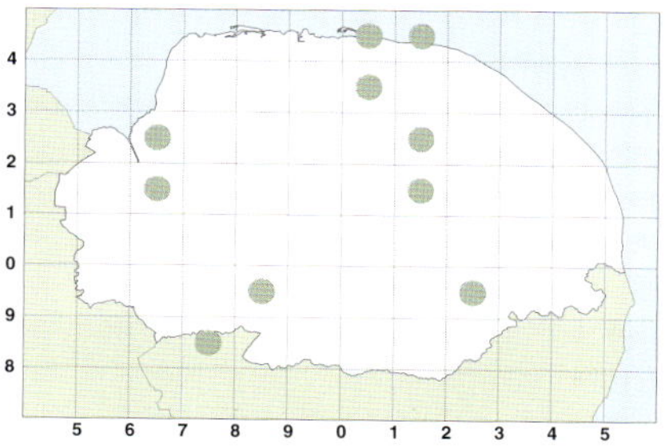

Usually occurring in small quantities on well rotted tree stumps and logs. Sporophytes rare.

BUXBAUMIACEAE

Buxbaumia aphylla Hedw.
No recent records. Last recorded, from Sprowston, [E21], in 1805.

DITRICHACEAE

Pleuridium acuminatum Lindb.

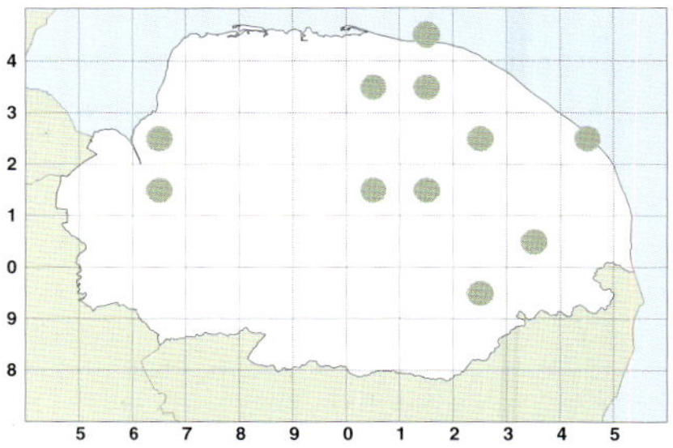

In rather open acidic habitats, including stubble fields. Commoner in VC27

P. subulatum (Hedw.) Rabenh.
Very rare. In similar habitats to the previous species. E13; 23. [W62; 90; E02]

Pseudephemerum nitidum (Hedw.) Loeske
Very rare. Bare clay in woodland rides, and damp ditch sides. W62; E03; 29. [W91]

Ditrichum cylindricum (Hedw.) Grout

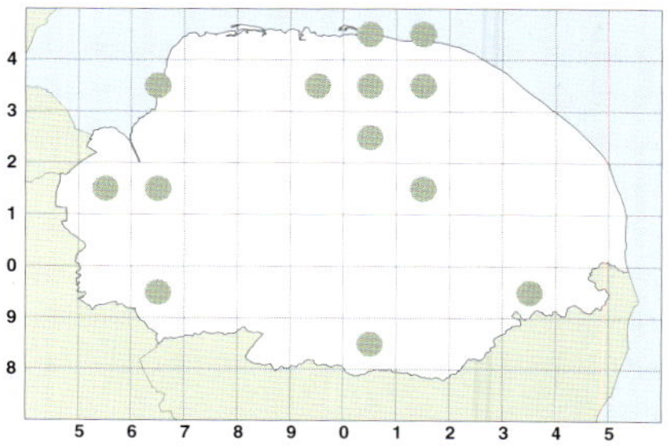

Usually occurs as scattered individuals, in stubble fields and waste ground. Only occasionally does it form larger patches.

D. heteromallum (Hedw.) E. Britton
No recent records. Recorded, in 1909, from Sheringham [E14]. *The Atlas (Hill, Preston & Smith 1992) shows one more recent record, from VC28.*

D. flexicaule (Schwägr.) Hampe
Rare. A plant of calcareous grassland, chalk pits and railway cuttings. W70; 78; 88; 89.

DICRANACEAE

Ceratodon purpureus (Hedw.) Brid.
Very widespread. A very common pioneer species, both in natural and urban habitats, found throughout the county. Sporophytes normally abundant.

C. conicus (Hampe ex Müll. Hal.) Lindb.
No recent records. This plant was recorded (as *C. purpureus* ssp *conicus*) in Warburg (1963). No further details are known.

Dicranella schreberiana (Hedw.) Dix.

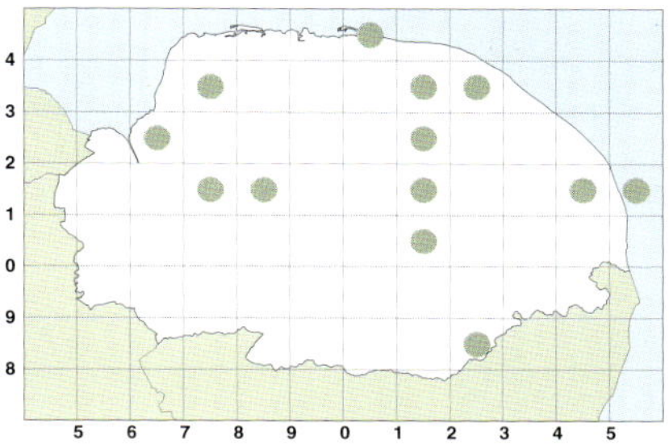

Exposed, usually rather damp places on both sandy and clay soils. Sporophytes rare.

D. varia (Hedw.) Schimp.

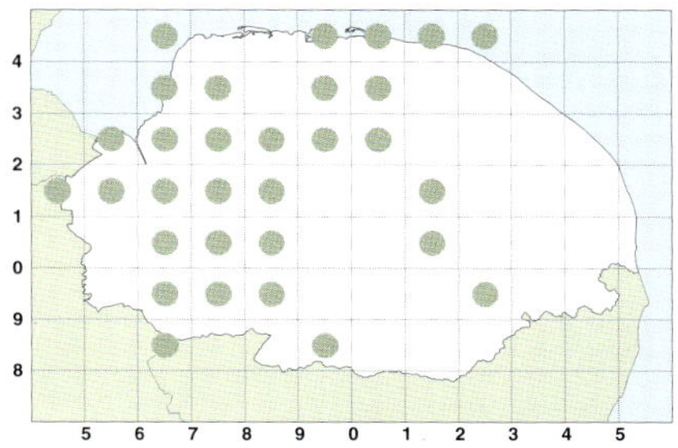

Common on calcareous substrates such as chalky till, as well as in chalk pits, etc. Sporophytes frequent.

D. staphylina H. Whitehouse

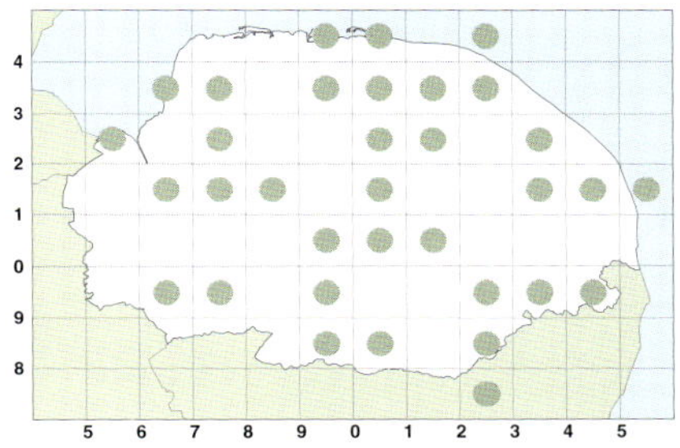

A common moss in arable fields, usually growing in extensive tufts.

D. rufescens (Dicks.) Schimp.
No recent records. [Rare. On detritus of disused chalk quarry / wet clay banks. W63 (1956); E14; 24]. *Record for VC28 not shown in Atlas (Hill, Preston & Smith 1992), nor given in Census Catalogue (Corley & Hill 1981).*

D. cerviculata (Hedw.) Schimp.

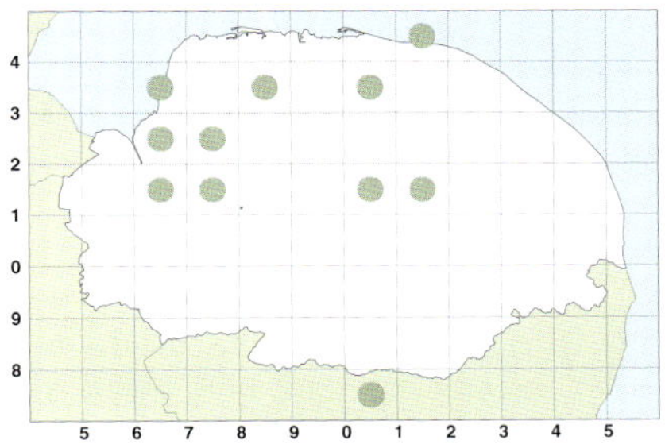

A major pioneer species on the damp margins of old sand workings in the King's Lynn area and elsewhere. Sporophytes frequent.

D. heteromalla (Hedw.) Schimp.
Very widespread; found almost everywhere except the Fens, wherever there are well drained acidic soils. Sporophytes frequent.

Dicranoweisia cirrata (Hedw.) Lindb. ex Milde
Very widespread. Common and usually abundant throughout the county. Occurs epiphytically, usually on acidic barks such as oak, and also on brick and stonework. Sporophytes normally abundant.

Dicranum polysetum Sw.
Very rare. A plant associated with conifer plantations. Possibly overlooked. W79; 62; E02.

D. bonjeanii De Not.

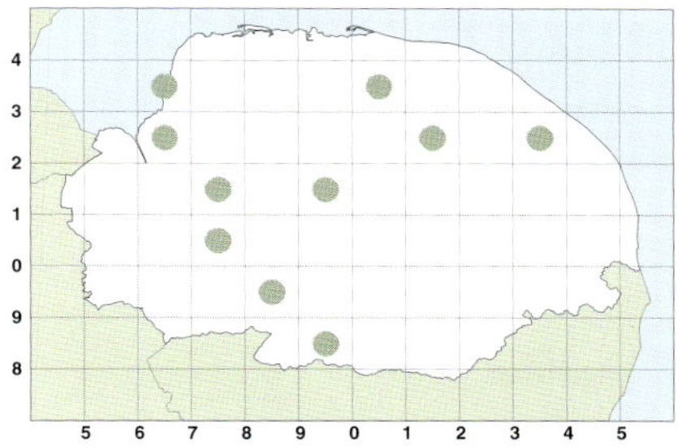

Widespread, but seldom common in appropriate habitats, i.e. bogs and poor fens. Possibly somewhat over-recorded in the past?

D. scoparium Hedw.
Very widespread. An extremely variable plant which can dominate in places. Commonest on the ground in woodlands, dunes & heaths, but it can also occur as an epiphyte in Fenland orchards, etc., thereby ensuring a very wide coverage. Sporophytes very rare.

D. majus Sm.
Rare. Occurs in acidic woodlands, where it may be locally abundant. W61; 62; E01; 03; 14; 33.

D. spurium Hedw.
No recent records. [Rare on heaths W61 (1958); 62; E03 (1961)].

D. fuscescens Sm.
Very rare. A plant which is probably on the edge of its range; found once, on Ingoldisthorpe Common W63 (1986, CRS), it has also been found at three separate localities in Suffolk.

D. tauricum Sapjegin
Very rare. Occurs as an epiphyte, usually on acidic barks, such as birch (1989, CRS). W82; 93; E14.

D. montanum Hedw.
Very rare. Another epiphytic species; found on oak, alder & willow (1988, RPL). E03; 07; 31.

Campylopus fragilis (Brid.) Bruch et al.
Very rare. Recorded in the past from VC 27. There were no recent records until it was found, on peaty acidic soil, where heath intergrades into fen, at Lopham 'Little' Fen, W07. (1996, DFS)

C. pyriformis (Schultz) Brid. (including var. **azoricus** (Mitt.) M.F.V. Corley)

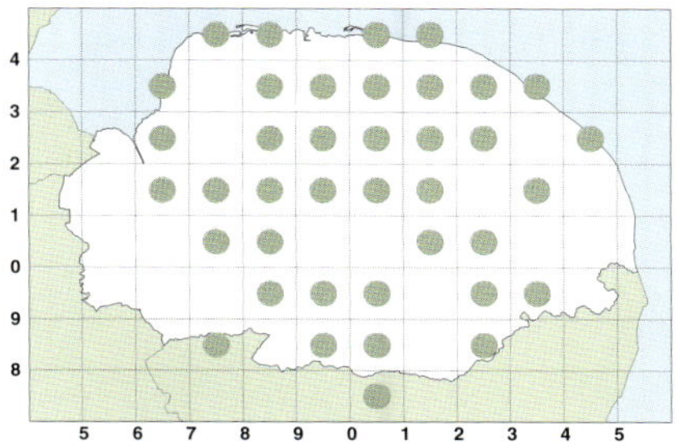

Often abundant in acidic habitats, particularly where rich in organic material, e.g. peat, decaying wood, etc. Tolerant of some trampling, so often found on path edges. (The var. **azoricus** has only been recorded from VC28).

C. flexuosus (Hedw.) Brid.

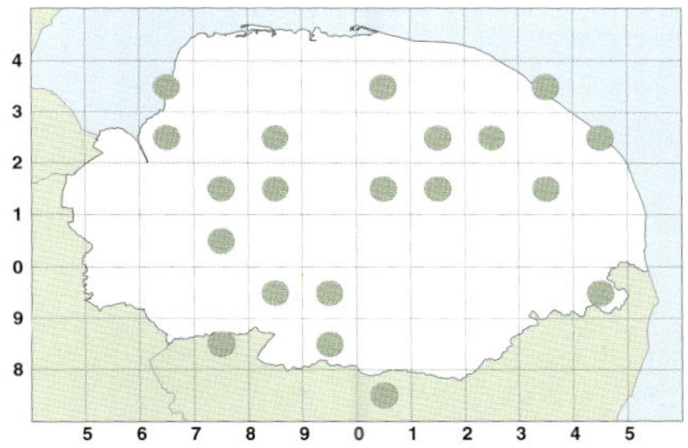

Occurs in similar habitats to the last species, but prefers more shaded and less disturbed sites, such as woods.

C. introflexus (Hedw.) Brid.
Very widespread; this southern African species is both very widespread and abundant in Norfolk. It occurs in a wide variety of acidic habitats, both wet and dry, such as heaths and woodlands, both as isolated tufts and as extensive totally dominant carpets. It often occurs as an epiphyte, though usually low down. It often produces sporophytes abundantly. It is quite variable in appearance, not all forms possessing obvious hair points

C. brevipilus Bruch et al.
No recent records. [A rare species of wet heaths but sometimes in dry places. W61; 62; 78; E03; 04; 12; 22]. This is a plant which seems to have gone missing. Unlike many of the other plants that have disappeared in recent years, it is a plant of acid ground. Some habitat loss has taken place, but it has either been overlooked, or may have succumbed to the aggressive newcomer **C. introflexus**.

LEUCOBRYACEAE

Leucobryum glaucum (Hedw.) Ångstr.

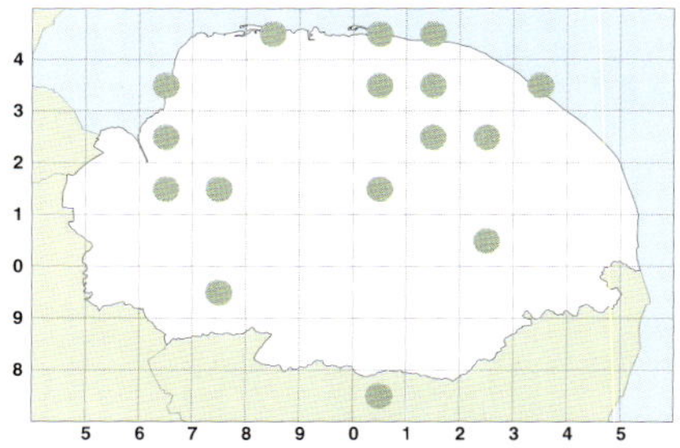

Often prominent on heaths and in acid woods. The well known detached balls of this moss, which are allegedly caused by scratching birds, are found most frequently in very young birch woods, where saplings are often close together. In the past very large mounds (up to 120cm width) have been recorded from bog habitats. Nothing of this size has been seen recently.

Leucobryum glaucum - Swanton Novers Great Wood

FISSIDENTACEAE

Fissidens viridulus (Sw.) Wahlenb.

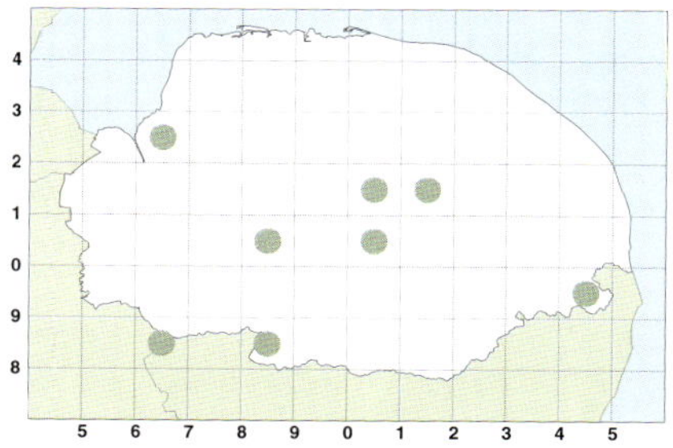

Found in a variety of habitats, e.g. shaded brickwork on churches, sides of ditches, and woodland floors; generally in small quantity.

F. pusillus (Wilson) Milde
Very rare. Shaded calcareous sandstone, near Lyng E01

F. gracilifolius Brugg.-Nan. & Nyholm
Syn. *F. pusillus* var. *tenuifolius* (Boul.) Podp.; *F. viridulus* var. *tenuifolius* (Boul.) A.J.E. Sm.
Very rare. On shaded brickwork at the base of a tomb. W71.

F. incurvus Starke ex Röhl.

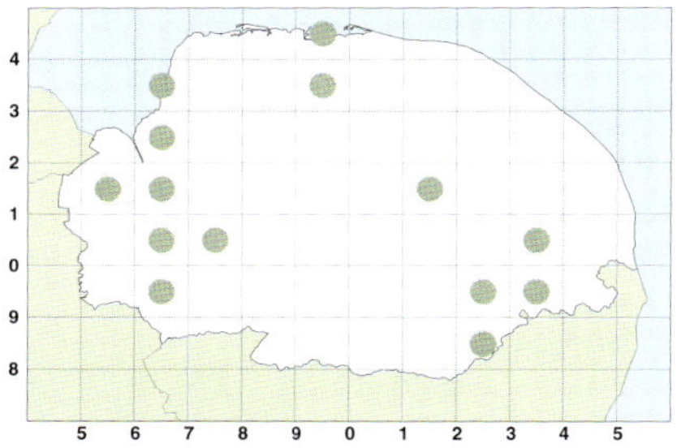

May occur in large quantity, as a dominant species, in rather open habitats; generally, however, it occurs as scattered individuals. Probably under-recorded. Sporophytes frequent.

F. bryoides Hedw.

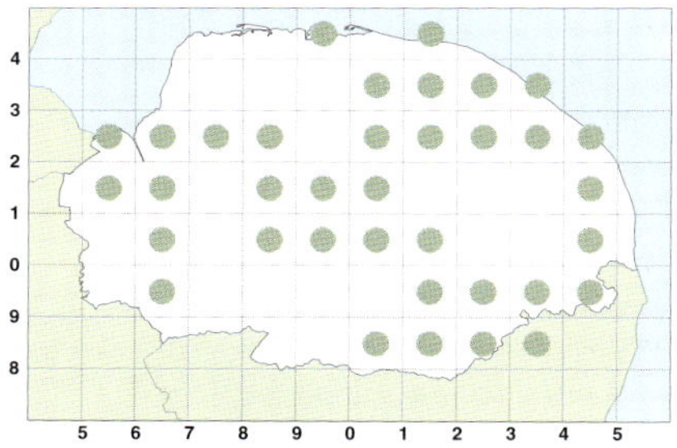

Occurs on both sandy and clay soils, usually in shaded positions. Often abundant where present. Sporophytes frequent.

F. crassipes Wilson ex Bruch et al.
Rare. This aquatic species occurs submerged on firm substrates including stonework and wooden embankments. Sporophytes not uncommon. E01; 17; 20; 28; 29; 38. [W80; 82; 89; 99]. Probably under-recorded, especially in the west of the county.

F. exilis Hedw.
Rare. A small and easily overlooked species of shaded sites on both clay and sandy soils. Most easily spotted when the conspicuous red peristomes are open; sporophytes are often abundant. (1985, CRS) W62; 72; 82; 92; 99; E02.

F. taxifolius Hedw.
Very widespread. Found in a wide variety of habitats, such as woodlands and grasslands, but usually most abundant on chalky boulder clay. Sporophytes frequent.

F. dubius P. Beauv.
Syn. *F. cristatus* Wilson ex Mitt.
Rare. This plant has been found growing in wet grasslands, on bare chalk, and in chalk grasslands, but always in small quantity. W62; 64; 72; 78; 88; E21.

F. adianthoides Hedw.

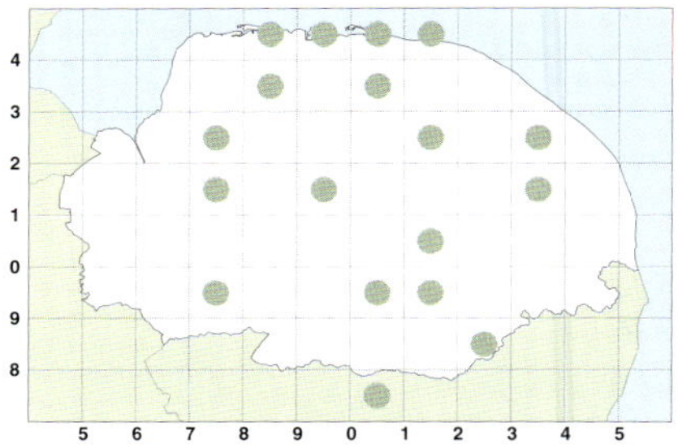

Often abundant in fens; less frequent in chalk grassland.

ENCALYPTACEAE

Encalypta streptocarpa Hedw.
Very rare. In chalk grassland. W78; 88. [W70; 80; 89].

E. vulgaris Hedw.
Very rare. A plant of chalk grasslands which seems to have become much rarer in recent years. Swann (1982) considered it 'frequent'. W70; 78; 89. [W64; 80; 88; 99].

POTTIACEAE

Eucladium verticillatum (Brid.) Bruch et al.
Very rare. In a well, and on damp brickwork near water. W79; 81. [W82; 89; E00].

Weissia controversa Hedw. (including var. **crispata** (Nees & Hornsch.) Nyholm)

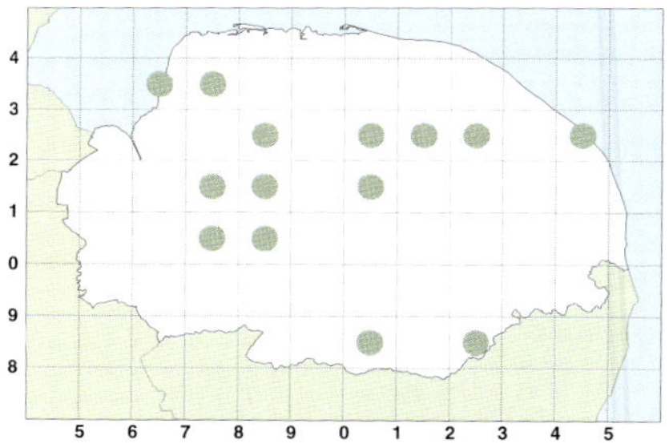

Most often found on light sandy soils, where it commonly occurs as dense colonies. Sporophytes frequent. No recent records for the var **crispata** [W79 (1965); E12 (1979)].

W. brachycarpa (Nees & Hornsch.) Jur.
Syn. *W. microstoma* var. *brachycarpa* (Nees & Hornsch.) C. Müll.
Rare. Chalk grassland. W64; 72; 94; E13.

W. longifolia Mitt. (including the var. **angustifolia** (Baumgartner) Crundw. & Nyholm)
Very rare. Swann (1982) records only var. **longifolia**, from [W74; 89; 90], where he notes it as a plant of chalk grassland. This is possibly an error, since the Census Catalogue (Corley & Hill 1981) records only the var. **angustifolia** for VC28. The only recent record, from bare soil on a grave, is from Hockering E01. No recent records for var. **angustifolia**.

Tortella inclinata (R. Hedw.) Limpr.
No recent records. Last recorded 1900. [W64].

T. flavovirens (Bruch.) Broth. (including var. **glareicola** (T.A. Chr.) Crundw. & Nyholm)
Very rare. A salt marsh plant, most frequently forming carpets under shrubby sea-blite. W64; 84; E04. *(The var.* ***glareicola*** *is given in the Census Catalogue (Corley & Hill 1981), for VC27, but is not mentioned by Swann at all).*

Trichostomum crispulum Bruch
Very rare. Open chalky ground on banks of Cut-Off Channel, Feltwell W68. (1997, CRS)

Pleurochaete squarrosa (Brid.) Lindb.
Very rare. Only two recent records - from chalk rubble on south facing slopes, Narborough Railway Line NWT Reserve W71, and Weeting Heath W78.

Pseudocrossidium hornschuchianum (Schultz) R.H. Zander
Syn. *Barbula hornschuchiana* Schultz

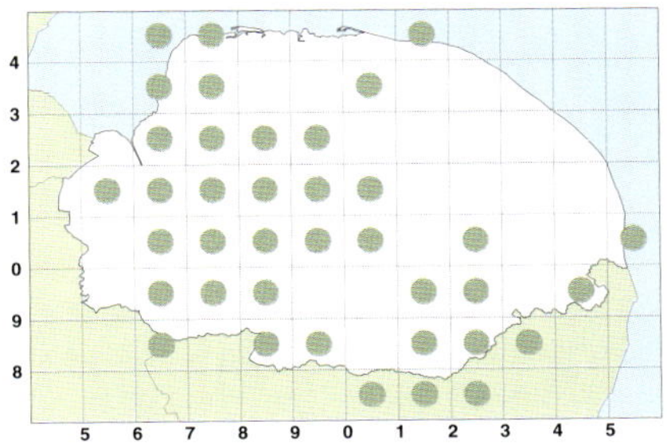

A weedy plant, which may occur as pure lawns, or mixed with other small acrocarps. Most frequently found on sandy soils, but also found on clays. Sporophytes rare.

P. revolutum (Brid.) R.H. Zander
Syn. *Barbula revoluta* Brid.

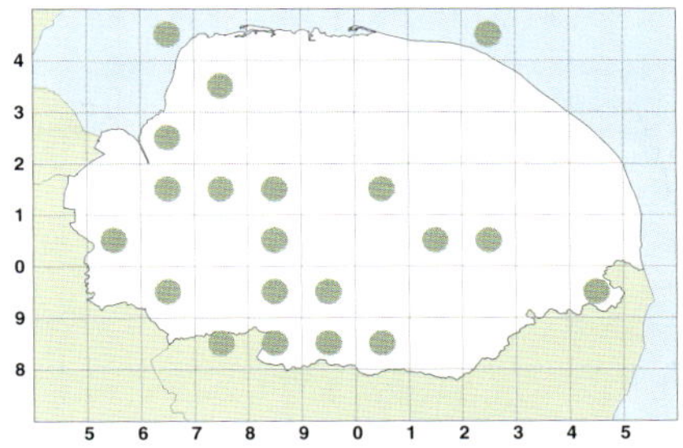

Usually occurs in fairly small quantity on oolitic stonework in churchyards.

Bryoerythrophyllum recurvirostrum (Hedw.) P.C. Chen
Syn. *Barbula recurvirostra* (Hedw.) Dixon

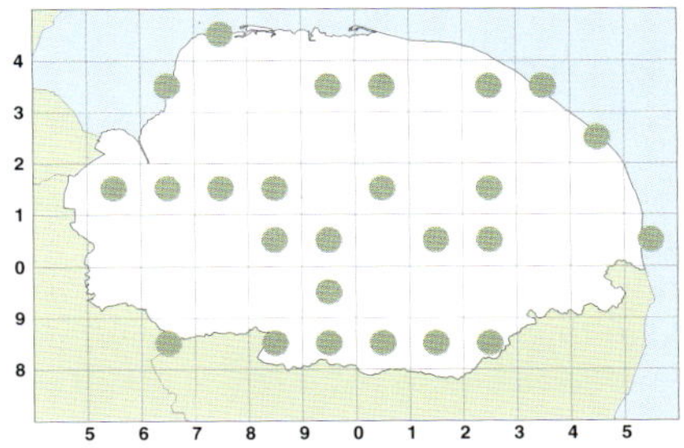

A plant which occurs in a range of habitats: tree roots in woods, brick walls, concrete. It normally occurs in fairly small patches. Sporophytes infrequent.

Leptodontium flexifolium (Dicks.) Hampe
No recent records. [Very rare. Heaths. W62 (1956)].

Gyroweisia tenuis (Hedw.) Schimp.
Rare. This tiny plant, which occurs in deeply shaded damp corners of buildings is, perhaps, rather overlooked. W62; 84; E01; 12. [W99; 82; E03].

Barbula convoluta Hedw.
Very widespread. Usually an abundant species in disturbed weedy habitats. Sporophytes infrequent.

B. unguiculata Hedw.
Very widespread. Occurs in very similar habitats to the last species, and in similarly large quantities. Tends to produce sporophytes more often than **B. convoluta**.

Didymodon rigidulus Hedw.
Syn. *Barbula rigidula* (Hedw.) Mitt.

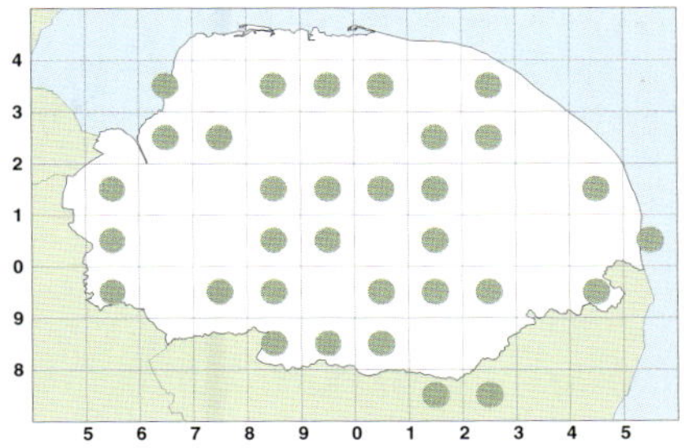

Normally found on calcareous stonework in churchyards and similar places. Sometimes very abundant, but usually only occurring in smallish patches.

D. umbrosus (Müll. Hal.) R.H. Zander
Syn. *Trichostomopsis umbrosus* (Müll. Hal.) H. Rob.
Very rare. Recorded from brickwork under a railway bridge. (1988, RCS). This is a species which is probably being overlooked. It occurs most frequently in urban habitats. W88.

D. vinealis (Brid.) R.H. Zander
Syn. *Barbula vinealis* Brid.
Very widespread. Usually one of the most abundant acrocarps on old brick and stonework in the county.

D. insulanus (De Not.) M.O. Hill
Syn. *Barbula cylindrica* (Taylor) Schimp.

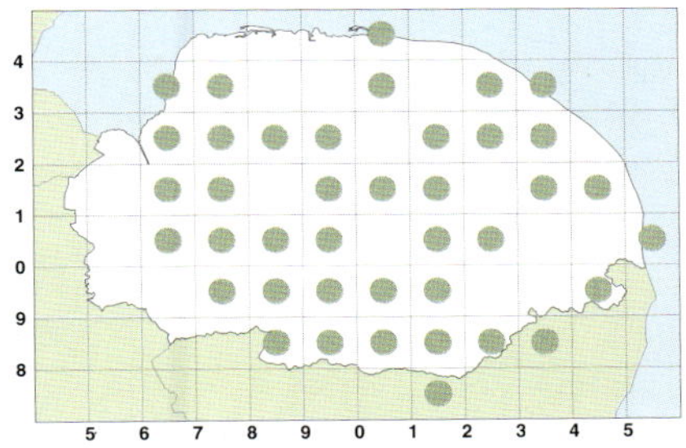

Found in a wide variety of habitats, e.g. brickwork, stonework, and soils - both sandy and limy; usually in fairly small quantity.

D. luridus Hornsch. ex Spreng
Syn. *Barbula trifaria* auct. non (Hedw.) Mitt.

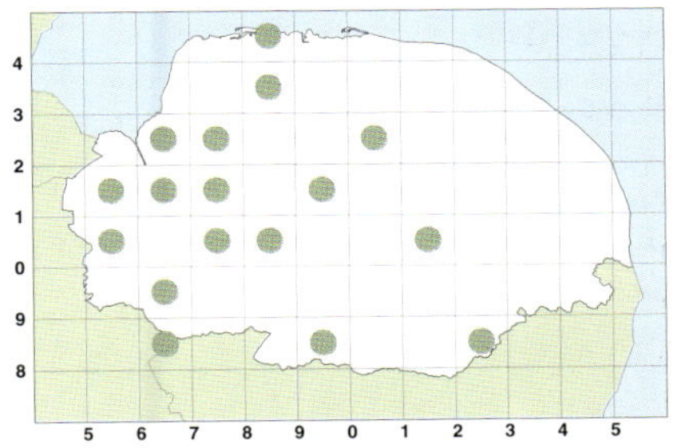

A plant most frequently found on limestone tombstones, though also found on flint. It usually occurs in small quantity and is therefore easily overlooked.

D sinuosus (Mitt.) Delogne
Syn. *Oxystegus sinuosus* (Mitt.) Hilpert

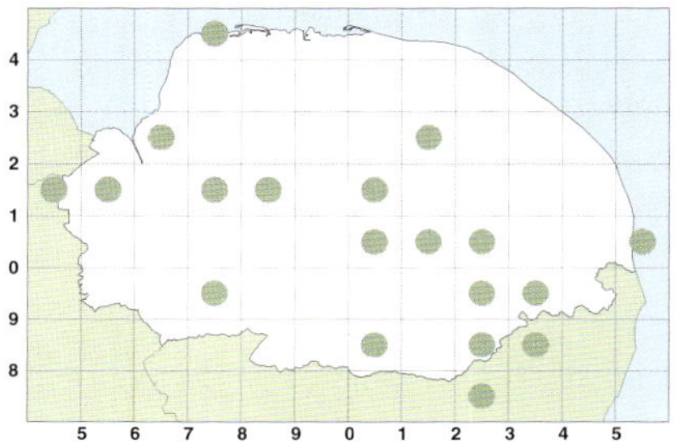

Occurs most frequently in somewhat dampish or shaded habitats, such as tree roots, sides of ditches, and the stonework of bridges. Rarely occurs in any great quantity.

D. tophaceus (Brid.) Lisa
Syn. *Barbula tophacea* (Brid.) Mitt.

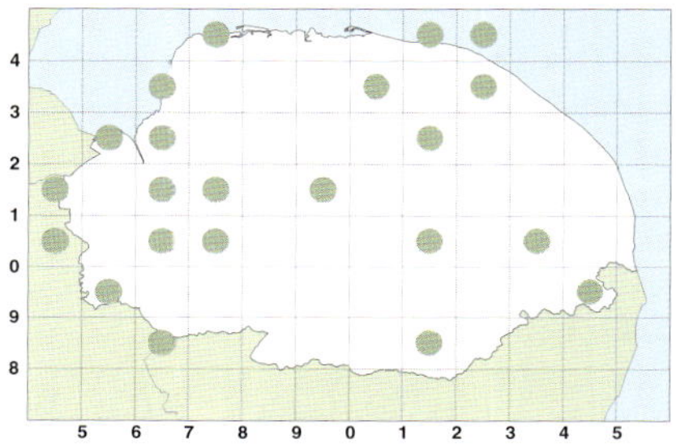

Another plant of damp habitats. This plant, however, usually occurs in some abundance, often forming large swelling tufts. The tips of the shoots are often discoloured by limy deposits left as a result of evaporation. Sporophytes frequent.

D. fallax (Hedw.) R.H. Zander
Syn. *Barbula fallax* Hedw.

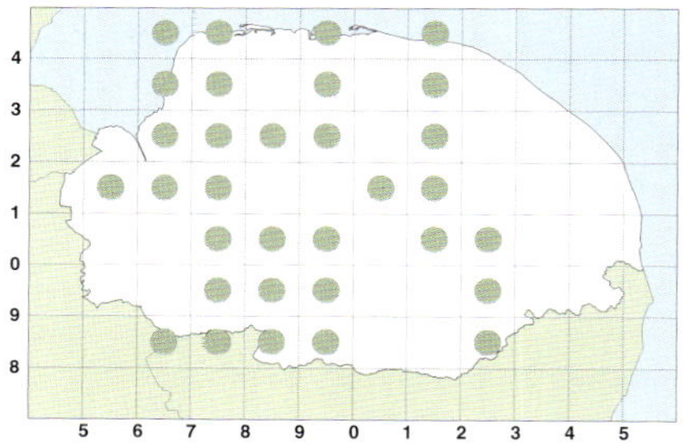

Often abundant on calcareous soils, in open sunny sites. More rarely it may occur on stonework.

D. ferrugineus (Schimp.) M.O. Hill
Syn. *Barbula reflexa* (Brid.) Brid.
No recent records. [Rare. Old chalk pits and dykesides on clay. W80; 92 (1979)] *Also in Census Catalogue (Corley & Hill 1981) for VC 27.*

Pterygoneuron ovatum (Hedw.) Dix.
Very rare. The only recent record was from chalky soil on the side of an old railway cutting - a site which has since been destroyed by landfill. W71. [W98].

P. lamellatum (Lindb.) Jur.
No recent records. [Very rare. Calcareous soil. W98]. Last seen in 1967, this plant may now be extinct in Britain (Blockeel, T.L., in Hill, Preston & Smith 1992)

Aloina rigida (Hedw.) Limpr.
Very rare. On chalk, Feltwell W68. [W78].

A. aloides (Schultz) Kindb.

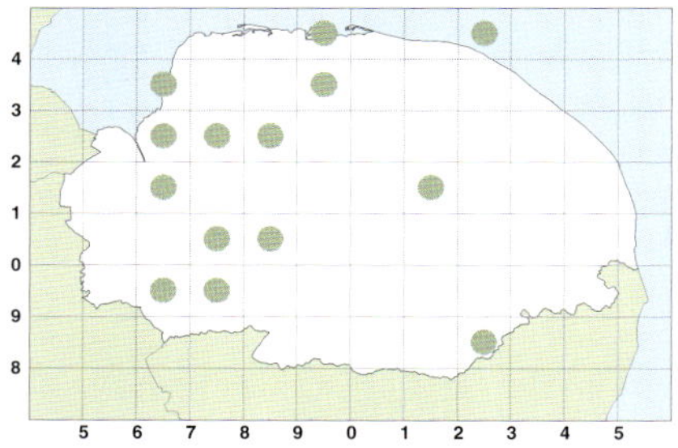

On calcareous clays and chalky soils, usually in fairly open habitats. Sporophytes common.

A. ambigua (Bruch & Schimp.) Limpr.
Syn. *Aloina aloides* var. *ambigua* (Bruch et al.) E.J. Craig
Very rare. In similar habitats to the last species. Possibly over-recorded in the past as the characters formerly used to separate the two species were not reliable. W68; 72. [W63; 70-73; 78; 80; 89; E14].

Leptobarbula berica (De Not.) Schimp.
Very rare. Growing on damp brickwork; site since lost due to renewal of guttering (1967, JGD); Corn Exchange, King's Lynn. W62.

Tortula subulata var. **subulata** Hedw. (including var. **angustata** (Schimp.) Limpr.)

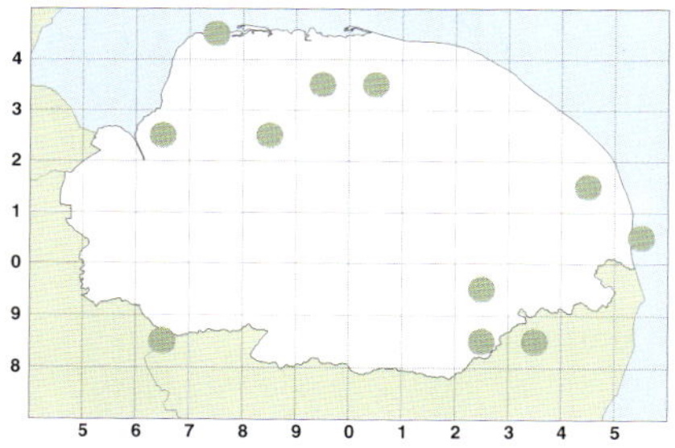

Found, normally in fairly small quantity, on sandy hedgebanks, sand dunes, tree roots and, rarely, as an epiphyte (on elder). Sporophytes common. No recent records for the variety [On acidic soil. Rare. E01 (1975)].

T. marginata (Bruch & Schimp.) Spruce
Rare. On stonework, etc. Probably overlooked. W72; E19; 20; 31. [E01]

T. muralis Hedw. (including var. **aestiva** (Hedw.))
Very widespread - present in every 10km square in the county. Very common on walls, old concrete, limestone tombstones, etc. More rarely as an epiphyte. The variety has not been recorded systematically; it is probably only a shade modification. Sporophytes both common and abundant.

T. lanceola R.H. Zander
Syn. *Pottia lanceolata* (Hedw.) Müll. Hal.
Rare. Chalk grassland and other calcareous soils. Usually occurs as isolated individuals. Sporophytes common. W68; 70; 78; 88; 92; 99.

T. wilsonii (Hook.) R.H. Zander
Syn. *Pottia wilsonii* (Hook.) Bruch & Schimp.
No recent records. [Very rare. On a sandy inland bank. E14 (1907)]

T. modica R.H. Zander
Syn. *Pottia intermedia* (Turner) Fürnr.

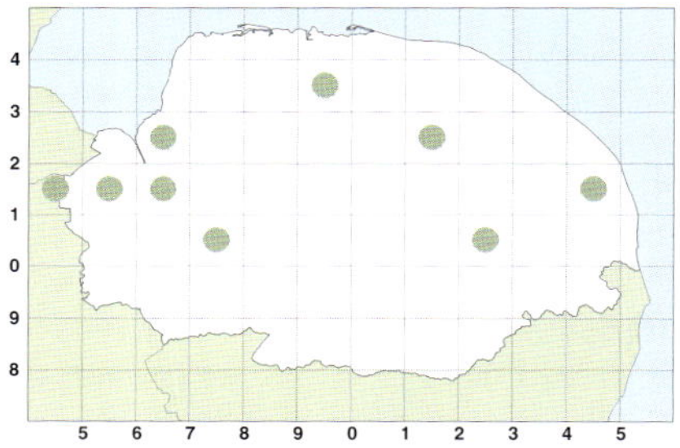

Found on a variety of soil types (sands / silts / clays), usually in open habitats, as small clusters of plants. Sporophytes frequent.

T. truncata (Hedw.) Mitt.
Syn. *Pottia truncata* (Hedw.) Fürnr.

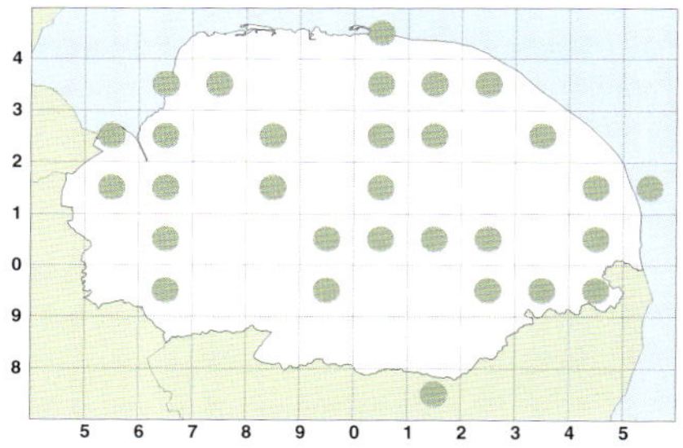

Occurs in stubble fields, by paths, and in other disturbed habitats. Often forms dense monocultures, though it may also occur as scattered individuals. Sporophytes normally abundant.

T. protobryoides R.H. Zander
Syn. *Pottia bryoides* (Dicks.) Mitt.

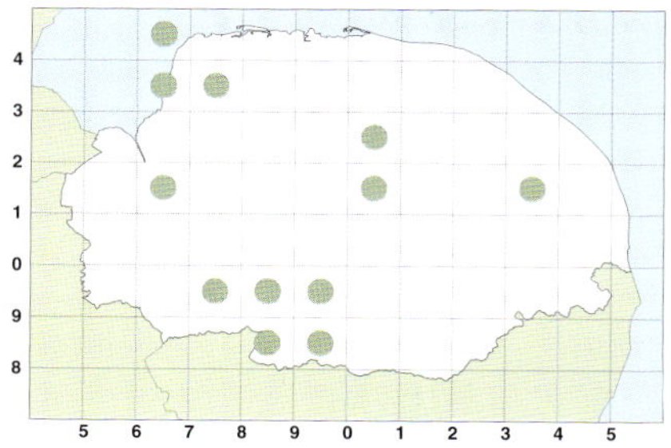

Found on disturbed soils, as in gravel pits,

on mole and ant hills, and track sides. Often forming small dense colonies with abundant fruit.

T. acaulon (L. ex With.) R.H. Zander (including var. **pilifera** (Hedw.) R.H. Zander, and var. **schreberiana** (Dicks.) R.H. Zander)
Syn. *Phascum cuspidatum* Hedw.
Very widespread and common weedy species, of arable fields and other disturbed ground. The varieties very rare: var. **schreberiana** E29 (1997, RJF); the var **pilifera** is a strictly coastal plant, and may well be overlooked. E04 (1997, DFS)

Microbryum starckeanum (Hedw.) R.H. Zander
Syn. *Pottia starckeana* var. *starckeana* (Hedw.) Müll. Hal.
Rare. On soil in stubble fields and other open places, as scattered isolated individuals. W51; 62; 79; E03.

M. davallianum (Sm.) R.H. Zander
Syn. *Pottia starckeana* subsp. *conica* (Schleich. ex Schwägr.) D.F. Chamb. and *Pottia starckeana* subsp. *minutula* (Schleich. ex Schwägr.) Bouvet.
Rare. On soil in stubble fields and other open places, as scattered individuals. W51; 61; 69; E18; 28.

M. rectum (With.) R.H. Zander
Syn. *Pottia recta* (With.) Mitt.
Rare. Occurs most frequently on chalky soils, but also on sands; usually occurs in small colonies. W79; 88; 92; 94; E03

M. curvicolle (Hedw.) R.H. Zander
Syn. *Phascum curvicolle* Hedw.
Very rare. On bare chalky soils. W70; 84; 88.

M. floerkeanum (F. Weber & D. Mohr.) Schimp.
Syn. *Phascum floerkeanum* F. Weber. & D. Mohr.
Very rare. Recorded in the past from VC27, the only recent records are all from the west. It occurs on open chalky soils and in stubble fields; probably under-recorded, as it is very small and difficult to spot. W68; 70; 79. [E14; 23].

Hennediella heimii (Hedw.) R.H. Zander
Syn. *Pottia heimii* (Hedw.) Fürnr.
Rare. A coastal species of salt marshes & muddy paths by the sea. W63; 74; 84; E01; 04; 40.

Acaulon muticum (Brid.) Müll. Hal.
No recent records. [Rare in bare heathy places. W78 (1967); 98; E23 (1960)]. *Not in Census Catalogue for VC28.*

Syntrichia ruralis (Hedw.) F. Weber & D. Mohr
Syn. *Tortula ruralis* (Hedw.) P. Gaertn., B.Mey & Scherb.
Very widespread. Often abundant in open sunny habitats. Very occasionally occurs epiphytically.

S. ruraliformis (Besch.) Cardot
Syn. *Tortula ruralis* ssp. *ruraliformis* (Besch.) Dixon.

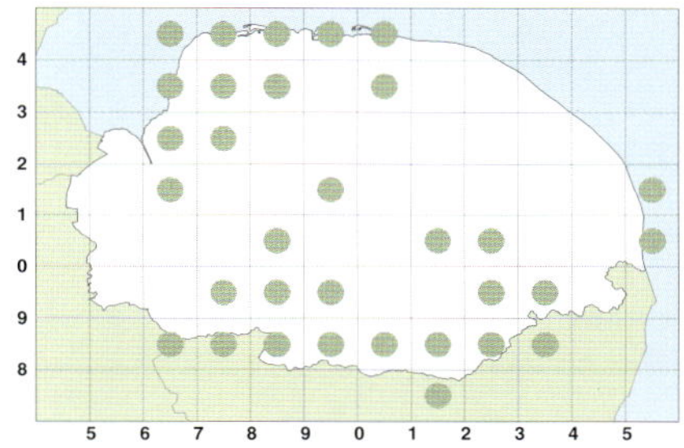

Normally regarded as coastal, this plant occurs throughout the county, usually, though not exclusively, associated with sandy soils.

S. intermedia Brid.
Syn. *Tortula intermedia* (Brid.) Berk.
Very widespread. A calcicole species which occurs most frequently on the steeply sloping, well drained, south facing oolitic buttresses of churches. Also on asbestos roofs, concrete, and occasionally as an epiphyte (on elder). Sporophytes produced frequently.

S. virescens (De Not.) Ochyra
Syn. *Tortula virescens* (De Not.) De Not.
Rare. Occurs, as small tufts, in a variety of habitats: epiphytically on willow and elder; on brick; on asbestos roofs, and on concrete by water. Recently found at Holkham with gemmae - for the first time in the UK. (1998, DLH). W62; 78; 84; E01; 07; 17.

S. laevipila var. **laevipila** Brid. (including var. **laevipilaeformis** (De Not.) Amman)
Syn. *Tortula laevipila* (Brid.) Schwägr.

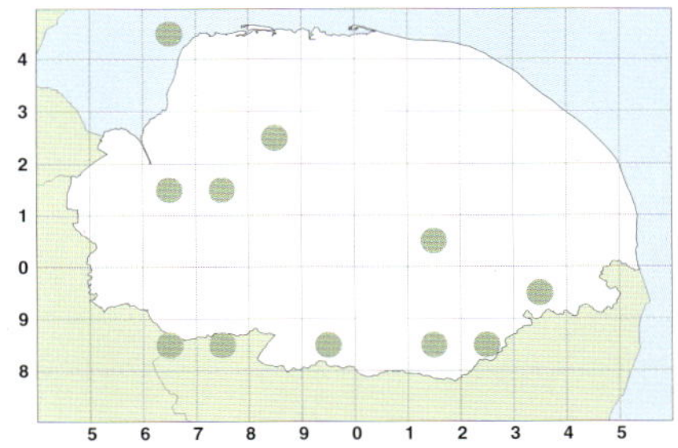

Occurs as an epiphyte on trees and hedgerow shrubs, including ash, elder, hawthorn and oak. The var. is rarer, but occurs in similar habitats, often with the species.

S. papillosa (Wilson) Jur.
Syn. *Tortula papillosa* Wilson
Rare. Swann (1982) noted a decline in the distribution of this dominantly epiphytic plant. It appears to have declined still further. Found on ash and elder, and on the stonework of churches. W68; E12; 18; 20; 31; 49. [W61; 62; 73; 84; 88; 99; E03].

S. latifolia (Bruch ex Hartm.) Huebener
Syn. *Tortula latifolia* Bruch ex Hartm.

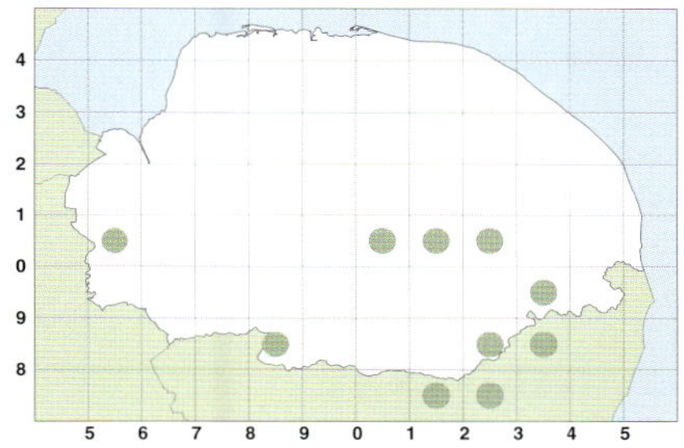

Found on stones, brickwork and tree roots in the flood zone of rivers. Absent from many suitable habitats.

GRIMMIACEAE

Schistidium apocarpum (Hedw.) Bruch et al.
Very widespread. Mainly on buildings where there is some lime present; also on grave markers. Occasionally on other substrates, such as tarmac. Sporophytes common.

The genus *Schistidium* has been subjected to a recent thorough revision (Blom 1996), and a considerable number of new segregates recognised. Preliminary work suggests that there are at least three species present in Norfolk, viz. **S. apocarpum** s.s., **S elegantulum** H.H. Blom, and **S. crassipilum** H.H. Blom.

Grimmia pulvinata (Hedw.) Sm.
Very widespread - found in every square. Usually abundant on brick and stonework; very occasionally occurring as an epiphyte. Sporophytes abundant.

G. trichophylla Grev.
Very rare. It is a plant of hard acid substrates, eg sandstones, which are rare in the county. Probably a casual, limited by lack of available habitat. W60; 84.

Racomitrium lanuginosum (Hedw.) Brid.
No recent records. This plant was pronounced extinct by Swann (1968). The following year it was refound at Bryants Heath, Felmingham, by Francis Rose. It has not been seen since [E22].

R. ericoides (Brid.) Brid.
Syn. *R. canescens* var. *ericoides* (Brid.) Hampe
No recent records. [In similar habitats to **R. canescens**. W62]. *The Atlas (Hill, Preston & Smith 1992) shows an old record for VC27.*

R. elongatum Frisvoll
Very rare. On a calcareous track, Weeting W78 (1987, HLKW & MOH)

R. canescens (Hedw.) Brid.
Very rare. Swann (1982) regarded this plant as 'locally abundant'; it is certainly not that nowadays. W61;62. [W60; 78; 84; 88; 89; 98; E21; 50].

SELIGERIACEAE

Seligeria calycina Mitt. ex Lindb.
Syn. *S. paucifolia* auct., non (Dicks.) Carruth.
Rare. Often abundant on blocks of chalk in pits, etc. W68; 70; 72; 80; 88.

S. calcarea (Hedw.) Bruch et al.
Rare. Often abundant on blocks of chalk in pits, etc. W68; 70; 72; 82; 94.

FUNARIACEAE

Funaria hygrometrica Hedw.
Very widespread. A common species on waste ground, bonfire sites, and, less commonly, in stubble fields.

Entosthodon fascicularis (Hedw.) Müll. Hal.
Syn. *Funaria fascicularis* (Hedw.) Lindb.
No recent records. [Rare. Bare earth. E14; 30]. Extinct?

E. obtusus (Hedw.) Lindb.
Syn. *Funaria obtusa* (Hedw.) Lindb.
No recent records. [Very rare. E04, Kelling Heath. 1909]. Extinct?

Physcomitrium pyriforme (Brid.) Brid.

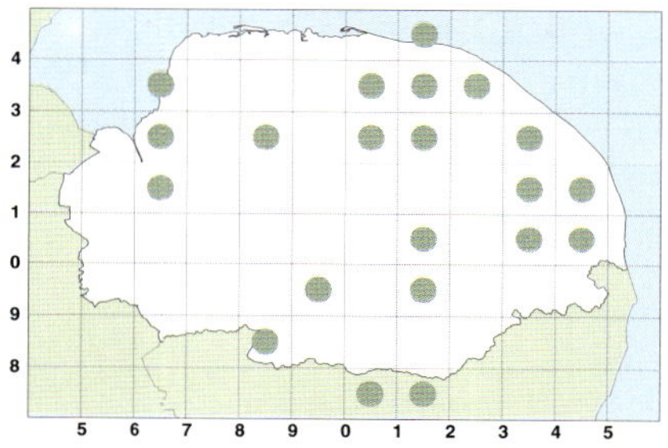

Normally found on clay or other soil dragged from ditches, drains and rivers. Sporophytes common.

P. eurystomum Sendtn.
Very rare. A nationally rare 'county speciality', associated with the dried out margins of Breckland meres, and therefore very sporadic in its appearance. W88; 98.

Aphanorhegma patens (Hedw.) Lindb.
Syn. *Physcomitrella patens* (Hedw.) Bruch et al.
No recent records. [Rare. Damp mud in a pond. W70 (1955)]. *Also in Census Catalogue (Corley & Hill 1981) for VC27.*

EPHEMERACEAE

Ephemerum recurvifolium (Dicks.) Boulay.
Very rare. This minute plant is found on rather bare chalky soils. W68; E29.

E. serratum var **minutissimum** (Lindb.) Grout
Very rare. Occurs on damp non-calcareous habitats. E31; 49. [W78; E03].

SPLACHNACEAE

Splachnum ampullaceum Hedw.
No recent records. [Rare. On cow dung in swamp (1973). The first record for about 100 years. Gooderstone Fen W70].

SCHISTOSTEGACEAE

Schistostega pennata (Hedw.) F. Weber & D. Mohr
Very rare. In sand martins' burrows, Wolferton W62.

BRYACEAE

Orthodontium lineare Schwägr.
Very widespread. Usually abundant in suitable habitats, viz. acid, earthy banks, rotting stumps, and the base of trees such as birch.

Leptobryum pyriforme (Hedw.) Wilson

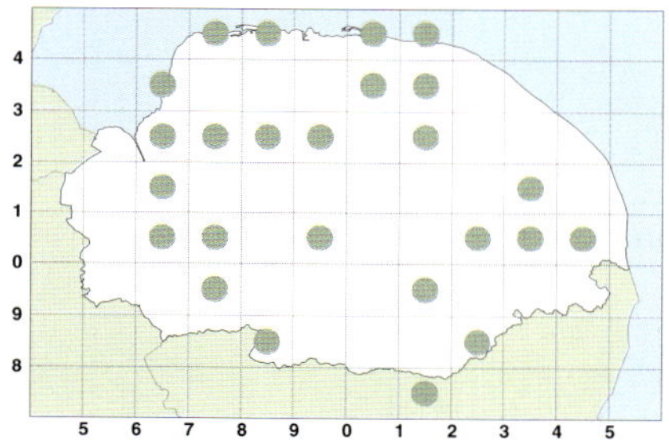

This plant is most frequently found in man made habitats, although it does occur elsewhere, mainly on light sands. Often found in greenhouses, it is probably under-recorded.

Pohlia nutans (Hedw.) Lindb.

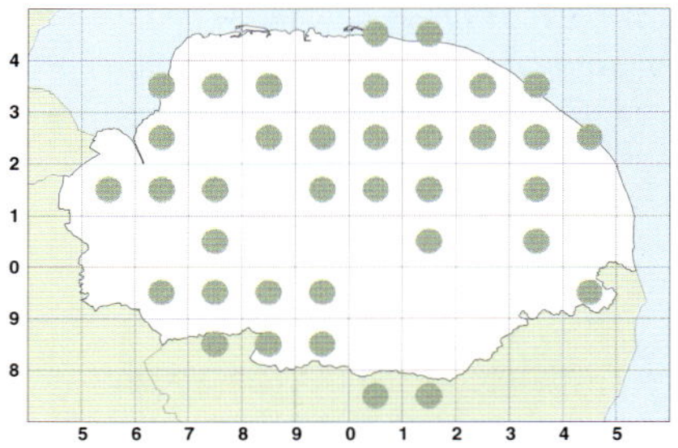

Common in acid habitats, colonising damp sands, on heaths, rotting thatch, etc.

P. bulbifera (Warnst.) Warnst.
Very rare. It occurs on muddy soils beside pools, Winterton Dunes. E42 (1995, DFS).

P. proligera (Lindb. ex Breidl.) Lindb. ex Arnell
Rare. A plant of open, bare, damp sandy habitats, it is often abundant in suitable spots. It may be under-recorded. W61; 62; 83; E33.

P. lutescens (Limpr.) H. Lindb.
Very rare. Stubble field margin E04. [E01].

P. melanodon (Brid.) A.J. Shaw
Syn. *P. delicatula* (Hedw.) Grout; *P. carnea* (Schimp.) Lindb.

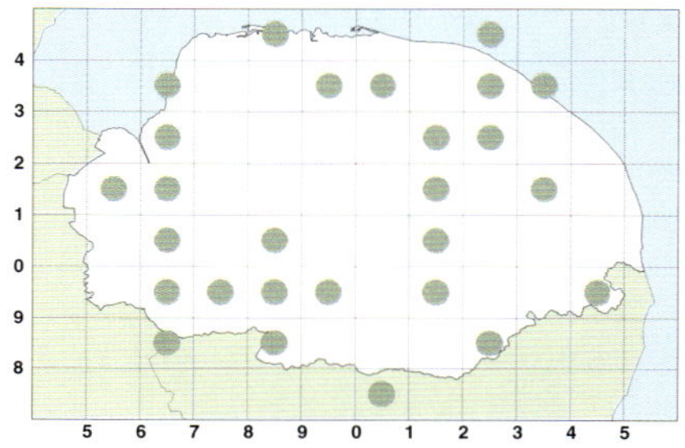

This plant normally occurs in fairly small quantity on damp, usually clay soils, often in quite shaded positions. Sporophytes occasional.

P. wahlenbergii (F. Weber. & D. Mohr) A.L. Andrews
Very rare. In damp habitats such as shaded forest tracks. W62; 99. [E01].

Bryum mamillatum Lindb.
No recent records. [Very rare. Hunstanton Golf Links, with *Juncus gerardi*, on damp sandy soil.1902. W64]. Extinct?

B. pallens Sw.
Rare. Occurs in damp places in the main, though also in sand dunes. W62; 64; 71; 89; 99; E19.

B. turbinatum (Hedw.) Turner
No recent records. [Very rare. Wet sandy places. Fakenham area, 1849. W93] Extinct?

B. algovicum Sendtn. ex Müll. Hal. var **rutheanum** (Warnst.) Crundw.
Very rare. A coastal species, of hollows in sand dunes. W64; 74; 84.

B. knowltonii Barnes
Very rare. Sandy hollow in infilled quarry, Bawsey. W62 (1985, RPL)

B. imbricatum (Schwägr.) Bruch & Schimp.
Syn. *B. inclinatum* (Brid.) Blandow
Very rare. Only one recent record, from Bawsey sand pits, W62. Mature capsules, which are essential for identification, ripen in summer when bryological recording activity tends to be low; probably under-recorded. [W63; 74; 84; 00; E03 (1955); 04; 11].

B. intermedium (Brid.) Blandow
No recent records. [Wet ground in old gravel pit. W88 (1959)]

B. capillare Hedw.
Very widespread - found in all squares. Usually abundant on stone and brick work, but also a frequent epiphyte. Sporophytes common.

B. subelegans Brid.
Syn. *B. flaccidum* Brid.

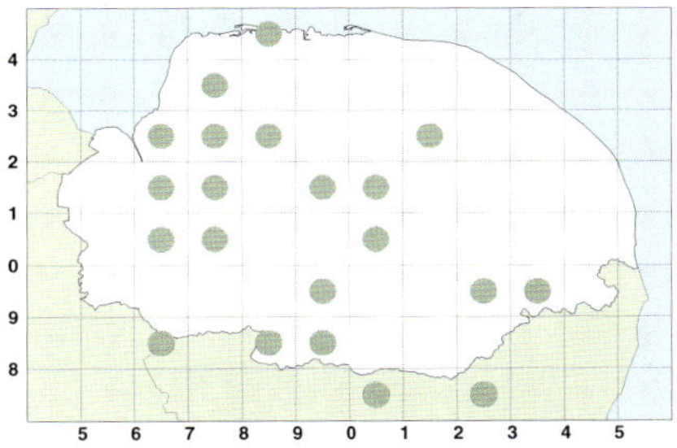

An epiphytic species most frequently found on elder. Often grows with **B. capillare**, with which it may be confused, resulting in possible under-recording.

B. torquescens Bruch ex De Not.
No recent records. *In the Census Catalogue (Corley & Hill 1981) for VC 27, and an old VC28 record is shown in the Atlas (Hill, Preston & Smith 1992), but the plant is not mentioned by Swann at all. No details known.*

B. pseudotriquetrum (Hedw.) P.Gaertn., B.Mey. & Scherb. (including var. **bimum** (Brid.) Lilj.)

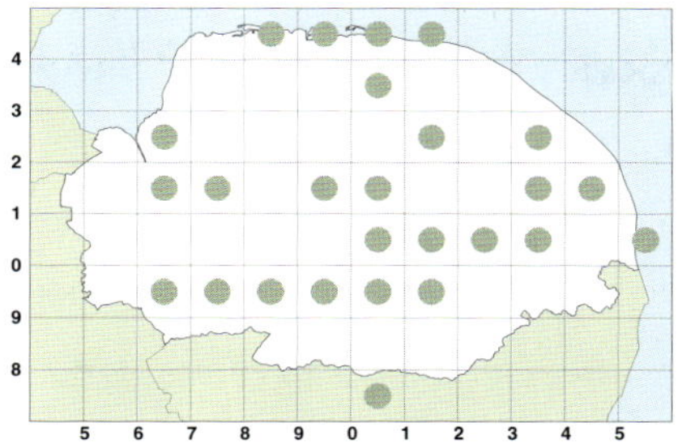

Occurs in fens and similar damp habitats. May occur as isolated strands, mixed with other mosses, or as dense tufts. The var. **bimum** is much rarer, and has only been recorded from W91 (1992, NGH) in recent years.

B. caespiticium Hedw.

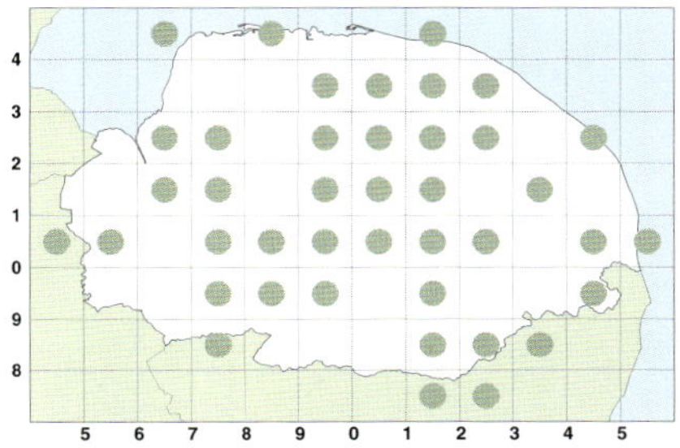

Usually occurs as densely packed carpets or tussocks, often in rather open dry habitats such as roadside banks.

B. argenteum Hedw.
Very widespread. An abundant weedy species occurring in the cracks between paving stones, in stubble fields, and other ruderal habitats.

B. gemmiferum R. Wilczek & Demaret

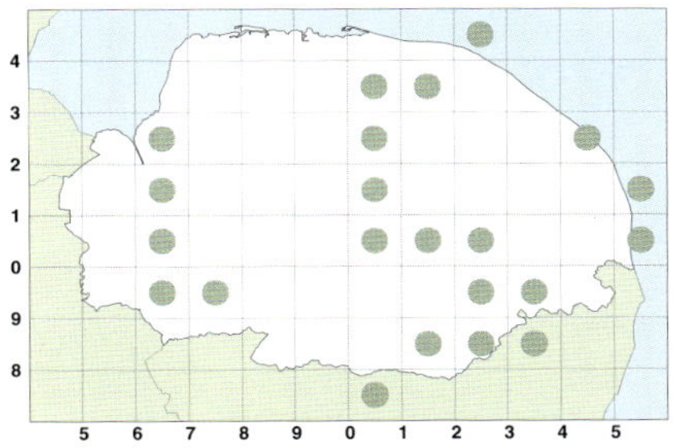

A pioneer species often found in bare, open, and often slightly damp habitats - on both sands & clays. Often occurs in abundance, but also as odd isolated plants.

B. gemmilucens R. Wilczek & Demaret
Very rare. Stubble field, on light sandy soil. W73. (1996, CRS).

B. bicolor Dicks.
Very widespread. A common weedy species both in man made and natural habitats; at its happiest when the site is at least slightly moist.

B. dunense A.J.E.Sm. & H. Whitehouse

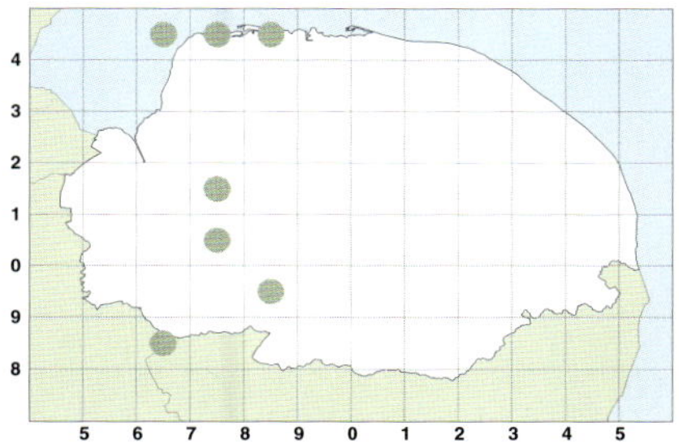

This plant usually occurs as scattered individuals, on sandy soils; found once, at Holme, with rhizoidal tubers. Many of the axillary bulbils in Norfolk material seem to be smaller than the dimensions given in the Flora (Smith 1978), although the plants agree in all other respects.

B. radiculosum Brid.

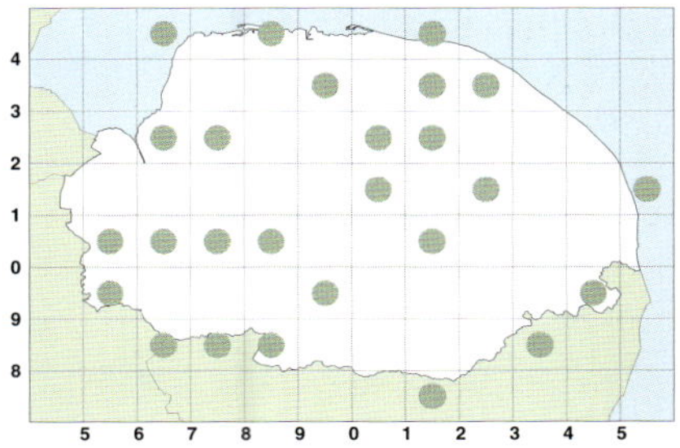

Occurs most frequently as small tight tufts on mortar and stonework; more rarely on calcareous soils.

B. ruderale Crundw. & Nyholm

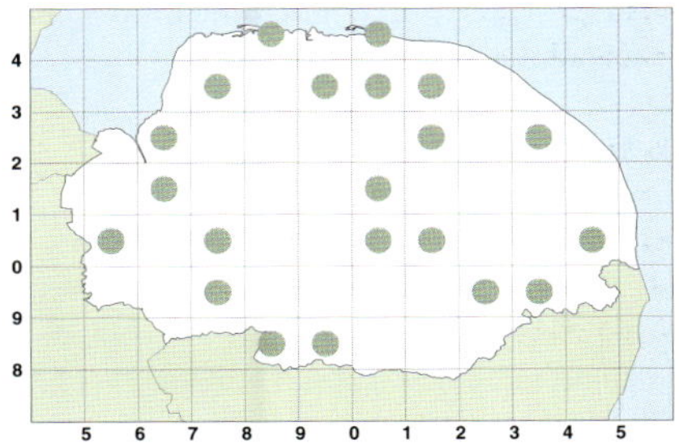

A plant of stubble fields and other disturbed habitats. Norfolk material rarely possesses violet rhizoids; they are usually reddish-brown.

B. violaceum Crundw. & Nyholm

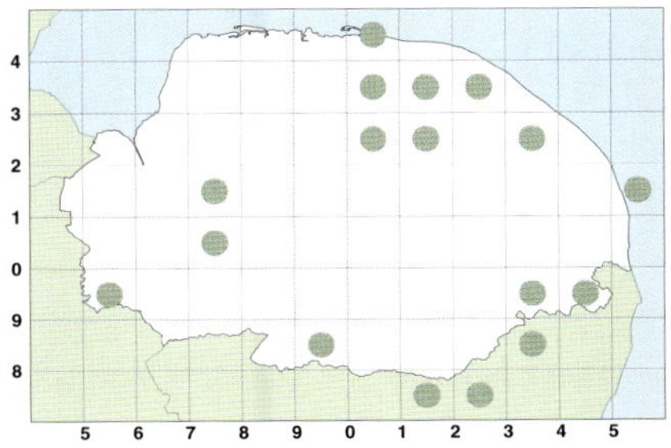

A stubble field moss which seems to occur on a variety of soil types. Usually in fairly small quantity, and therefore possibly overlooked. (1989, RJF)

B. klingraeffii Schimp.

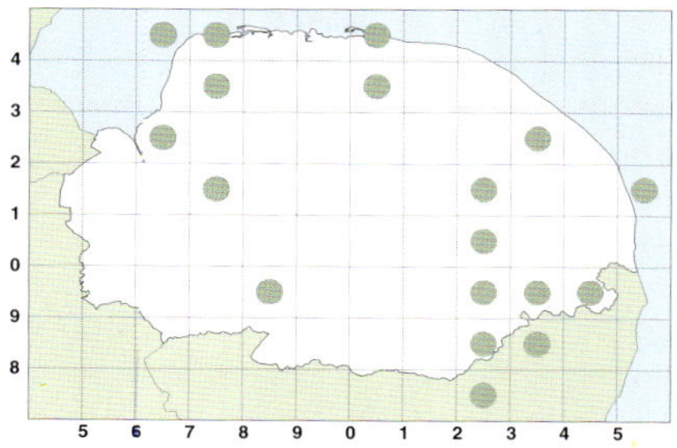

Another stubble field species, found mainly on sandy & silty soils.

B. sauteri Bruch at al.
Very rare. The rarity of this plant is strange since it occurs in very undistinguished habitats, and keeps very undistinguished company. E03; 11. [W72].

B. subapiculatum Hampe
Syn. *B. microërythrocarpum* Müll. Hal. & Kindb.

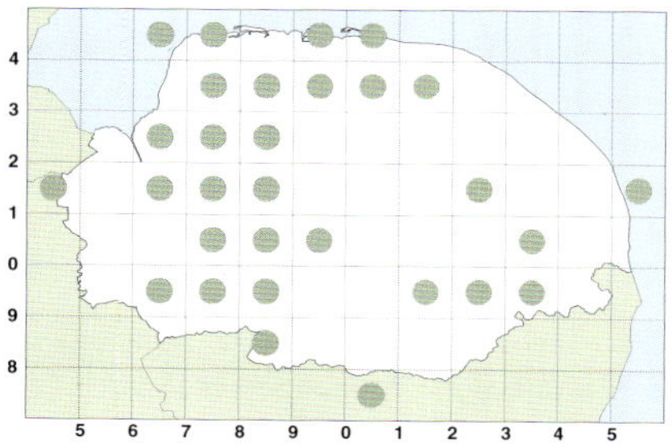

A stubble field species which also occurs in other disturbed sites where there are light, often calcareous, soils. Often quite common in such habitats.

B. bornholmense Wink. & R. Ruthe
Rare. Swann's statement, that this plant is 'frequent' must be revised; the published keys have proved ambiguous, resulting in many misidentifications. In fact it is rare, and confined to acid heathland habitats. W61; 62; 98; E11.

B. rubens Mitt.
Very widespread. The commonest member of the *Bryum erythrocarpum* complex, occurring in stubble fields, on ant and mole hills, and open patches in grasslands.

Rhodobryum roseum (Hedw.) Limpr.
Rare. Most recent records seem to have come from the sides of ant hills, where it may occasionally be abundant. However, it appears to be getting steadily scarcer. W62; 72; 78; 99. [W70; 71; 88; 08; E03; 10; 11; 14]. *The records from VC27 all date back to Burrell.*

MNIACEAE

Mnium hornum Hedw.
Very widespread; common in woodlands of all types, usually growing on slightly higher, or steeper, areas which remain clear of fallen leaves.

Mnium hornum (male) - Sheringham

Cinclidium stygium Sw.
Very rare. Grows more-or-less submerged in boggy pools. E32. [W09; E03; 12; 30].

Rhizomnium punctatum (Hedw.) T.J. Kop.

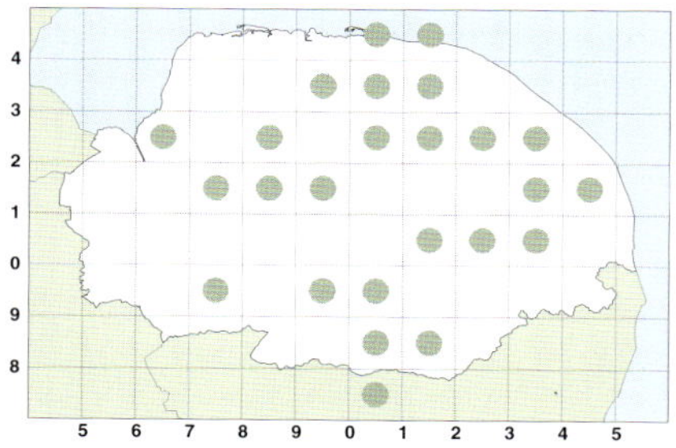

Often frequent in shaded damp habitats such as pond margins, stream sides and wet woodlands. It favours organic rich substrates.

R. pseudopunctatum (Bruch & Schimp.) T.J. Kop.

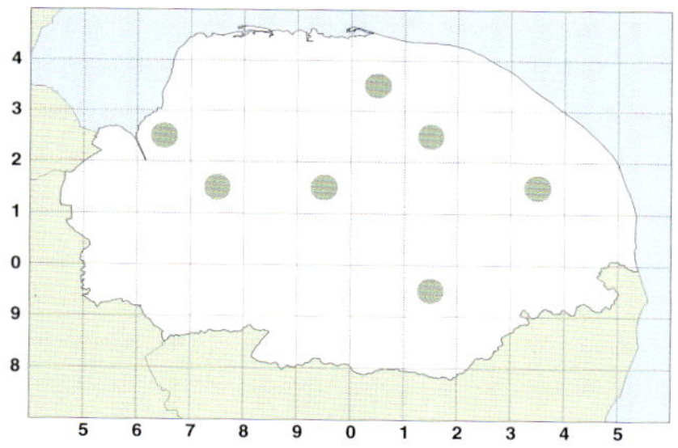

Occurs in much more open habitats than the last species, though the two can occur together.

Plagiomnium cuspidatum (Hedw.) T.J. Kop.
Very rare. Amongst grasses. W52; 82; E11.

P. affine (Blandow) T.J. Kop.

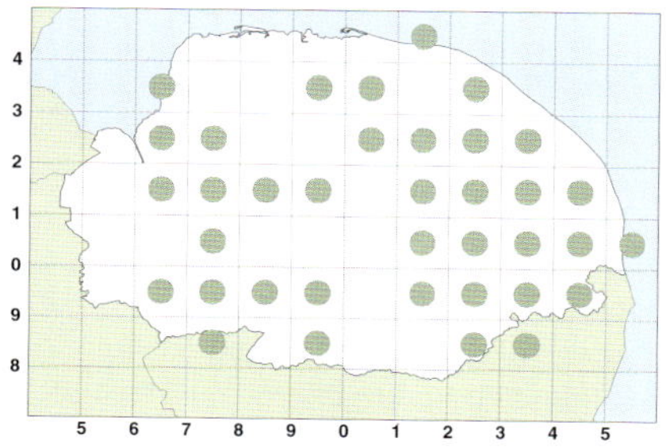

Commonly forms straggling mats on rides in woodland; in grassland, in reedbeds, and in fens. Often abundant over fairly extensive areas.

P. elatum (Bruch & Schimp.) T.J. Kop.

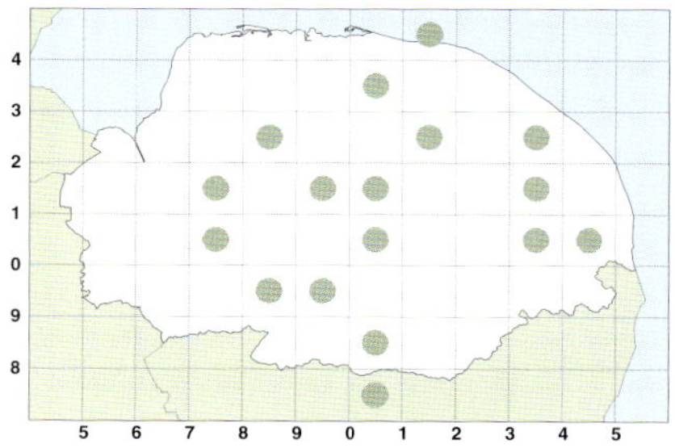

In rich fen communities, where it may locally form large patches.

P. ellipticum (Brid.) T.J. Kop.
Rare. Another plant of rich fen communities which again can occur in quite large patches. W98; 99; E11; 30; 31; 32.

P. undulatum (Hedw.) T.J. Kop.
Very widespread. The commonest species of *Plagiomnium* in the county. Often abundant in wet woodlands, and damp shaded grasslands - such as occur on the north side of churches.

P. rostratum (Schrad.) T.J. Kop.

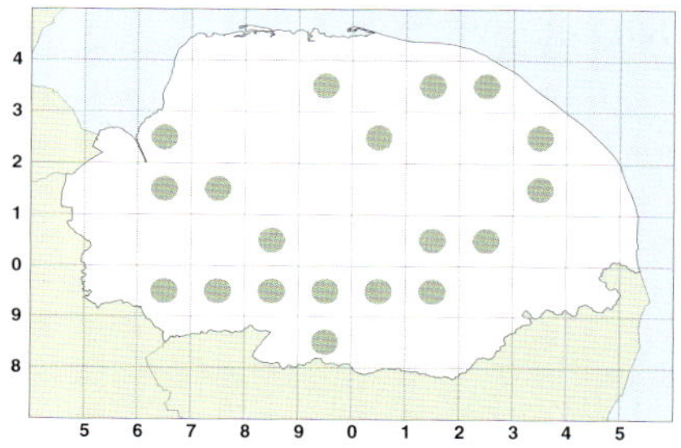

On soil in shaded habitats such as woodland.

AULACOMNIACEAE

Aulacomnium palustre (Hedw.) Schwägr.

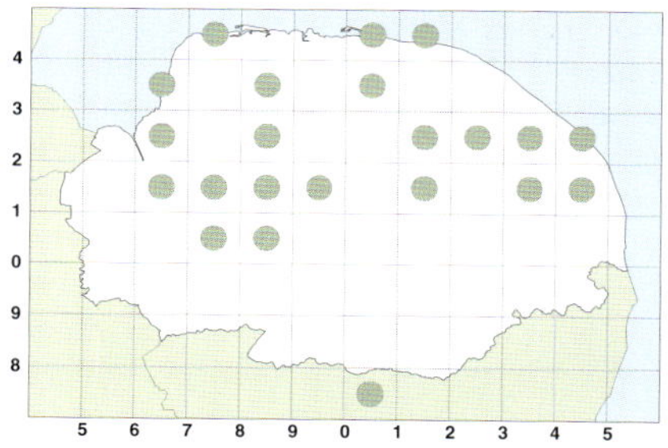

Often common in suitably damp habitats i.e. mires, damp heath and grassland, fen carr. More rarely as a colonist of old sand pits.

A. androgynum (Hedw.) Schwägr.
Very widespread. A common species in acid habitats, growing on soil, rotting wood, in deep shade at the base of dead grass tussocks, etc. Capsules found once, at Holkham Meals (W84).

MEESIACEAE

Amblyodon dealbatus (Hedw.) Bruch et al.
No recent records. Recorded from St. Faiths, near Norwich [E21], at the beginning of the last century, and not seen since. Extinct.

BARTAMIACEAE

Bartramia pomiformis Hedw.

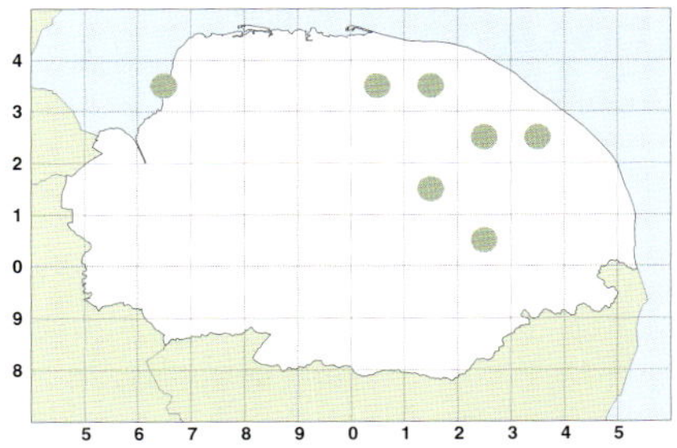

Widely scattered and locally abundant. Sporophytes produced fairly frequently.

Philonotis fontana (Hedw.) Brid.
Rare. A plant of fens & damp grasslands. Possibly more frequent than **P. calcarea**, although it tends to be much smaller, and therefore more easily overlooked. W61; 62; E04; 12. [E03].

P. calcarea (Bruch & Schimp.) Schimp.
Rare. This species, which is more typical of rich fen, appears to have shown a marked decline since Swann's days, both areally and in quantity. W70; 91; E04; 12; 14. [W61; 62; 79; 93; 99; 09; 01; E03; 22; 32; 40].

ORTHOTRICHACEAE

Zygodon viridissimus (Dicks.) R. Br. (including var. **stirtonii** (Schimp. ex Stirt.) I. Hagen)

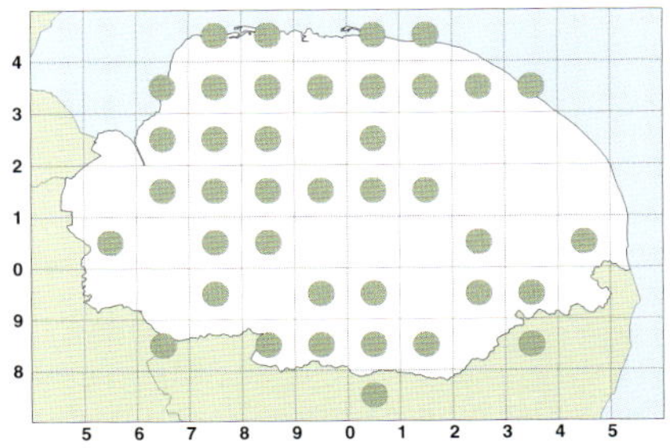

A species which, when present, is often abundant. It occurs both on stonework and as an epiphyte, most frequently on elder. There is only one recent record for the var. **stirtonii** (E03), which is odd, given the frequency with which the species occurs on stone.

Z. conoideus (Dicks.) Hook. & Taylor.
Rare. First recorded from Bacton Woods (1996, DFS), and since then recorded at several other sites. Probably widespread but overlooked as it is very similar to **Z. viridissimus** in the field. W62; 72; E10; 20; 33.

Orthotrichum lyellii Hook. & Taylor.
Rare. This epiphyte is possibly overlooked since the characteristic brown leaf gemmae are sometimes rather sparse. It appears to favour humid habitats. W62; 88; E07; 17; 22; 39.

O. striatum Hedw.
Very rare. Three records exist in Burrell, all from the east of the county [E13; 14; 50] but it was considered extinct, until found growing on a willow branch in Lopham Fen E07. (1994, DFS).

O. affine Brid.
Very widespread. The commonest purely epiphytic member of the genus. Occurs most frequently on elder, but also on a variety of other trees, such as apple, willow, hawthorn and ash. Sporophytes produced abundantly.

O. obtusifolium Brid.
Very rare. Found growing on elder in Blo' Norton Fen, W07 (1990, ACS). This was the first record for many decades from lowland Britain of this Red Data Book species.

O. anomalum Hedw.
Very widespread. Often abundant on calcareous stonework, most frequently in churchyards and cemeteries. However, it is sometimes absent from apparently ideal habitats. Sporophytes produced abundantly.

O. cupulatum Brid.

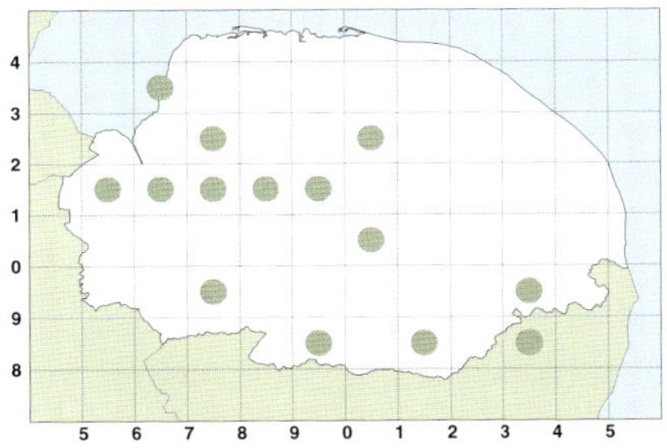

Occurs in similar habitats to **O. anomalum**, often immediately adjacent to it.

O. tenellum Bruch ex Brid.
Very rare. Burrell, who regarded this plant as 'Rare', cites three records: Swaffham [W80], Aylmerton [E13] and Holkham (W84), where it was recently refound, at the base of a poplar (1998, DLH).

O. pumilum Sw.
Syn. *O. schimperi* Hammar
No recent records. Burrell gave one record (1907), from Aylmerton [E13], where it occurred as 'minute tufts on Chestnut trees'. Preston, (in Hill, Preston & Smith 1994), notes than many of the older British records for this species have proved erroneous, and that the remainder require critical revision.

O. diaphanum Brid.
Very widespread. A pollution tolerant species which is common, and usually abundant, on basic stonework, asbestos roofs, and, as an epiphyte, on a wide variety of tree species.

O. pulchellum Brunt.

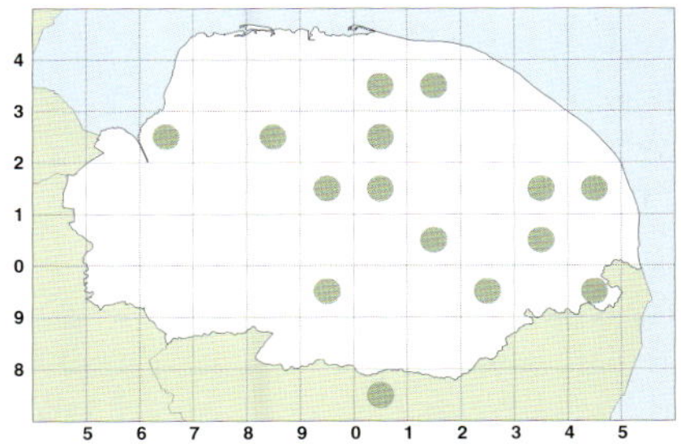

Another purely epiphytic species which may, in some places be very abundant. More frequently, however, it occurs in small quantity. Sporophytes produced abundantly. Mainly found on willow, in fen carr and similar humid habitats.

Ulota coarctata (P. Beauv.) Hammar
Syn. *U. ludwigii* (Brid.) Brid.
No recent records. Again, recorded by Burrell, from what appears to have been an exceptionally rich chestnut tree in Aylmerton [E13]. Extinct?

U. crispa (Hedw.) Brid.

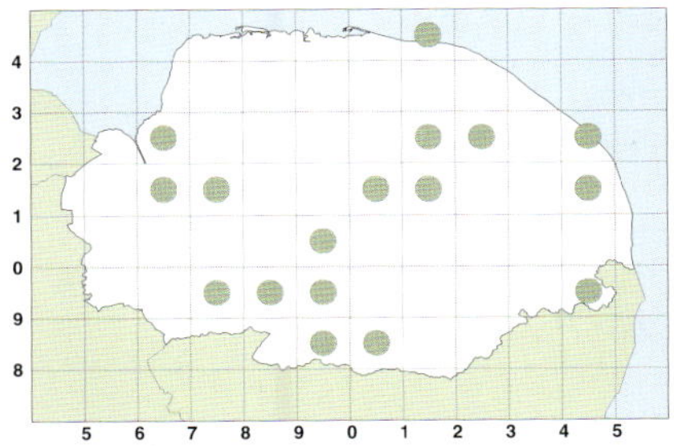

Swann (1982) regarded this species as 'rather rare & decreasing'. Since then its fortunes appear to have revived somewhat. It seems to occur more frequently, though rarely in great quantity. It is one of few epiphytes which occur on oak, though it is more frequent on other species, such as willow.

U. bruchii Hornsch. ex Brid.
Syn. *U. crispa* var. *norvegica* (Grönvall) A.J.E. Sm. & M.O. Hill
Swann (in Petch & Swann, 1968) gives one old record for this species. He does not mention it in his 1982 paper at all, yet it appears to be now tolerably common. Occurs in similar habitats to the last species, with which it may be confused, unless good ripe capsules are present.

U. phyllantha Brid.
Rare. This epiphytic species has been recorded from elm, willow and hornbeam. W72; E07; 14; 19.

HEDWIGIACEAE

Hedwigia stellata Hedenäs.
Very rare. Another example of a plant which is, inevitably, restricted by virtue of its habitat requirements, viz. hard acid rocks. This plant was found growing on syenite chippings covering a quarry ramp near King's Lynn (1984, CRS), a site which has since been destroyed. (It was first recorded as *Hedwigia ciliata* (Hedw.) P. Beauv.). W61.

The Pleurocarpous mosses

FONTINALACEAE

Fontinalis antipyretica Hedw. (including var. **gigantea** (Sull.) Sull.)

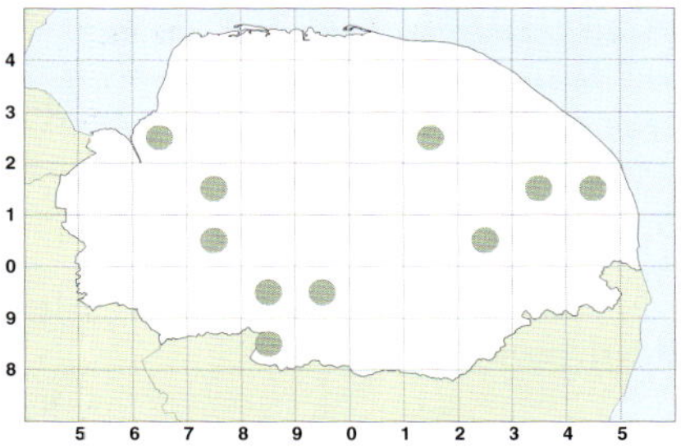

Usually abundant where present. Attached to stones in streams, lakes, etc but also to tree roots etc. Fruit rare. Almost certainly under-recorded. The var. is Very rare, though possibly overlooked. W60 (1987, CJL); E01.

CLIMACIACEAE

Climacium dendroides (Hedw.) F. Weber & D. Mohr

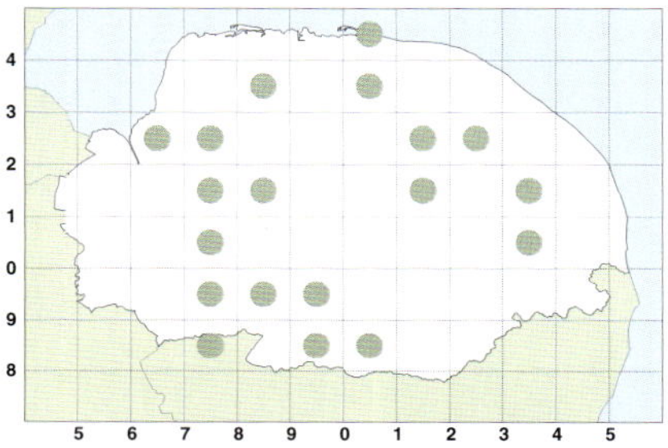

Normally found in at least periodically damp places; most abundant in fens, and permanently damp places.

CRYPHAEACEAE

Cryphaea heteromalla (Hedw.) D. Mohr

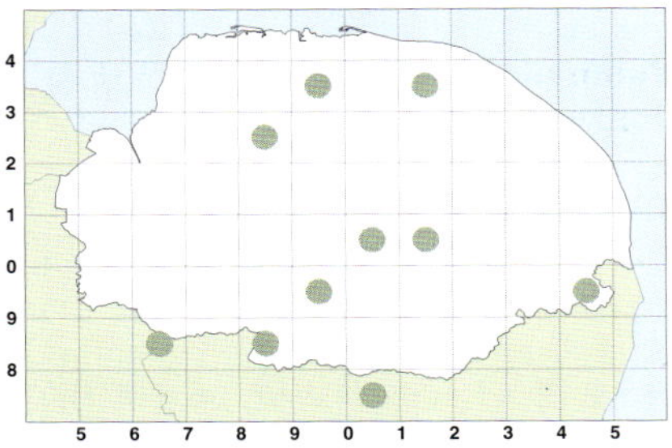

Usually found in small quantity. Normally fruiting. Most frequently on elder & ash.

LEUCODONTACEAE

Leucodon sciuroides (Hedw.) Schwägr.
Very rare. Regarded as common by Burrell, and rare by Swann, there are only three recent records: from stonework at Holkham Hall, W84; from Felbrigg Hall E13, and from the carstone wall surrounding Hillington Hall, from where it was recorded in 1961, by ELS. W72.

Antitrichia curtipendula (Hedw.) Brid.
No recent records. This plant has not been seen since the beginning of the century and was regarded as extinct by Swann (1982)

NECKERACEAE

Neckera pumila Hedw.
No recent records. Regarded as rare by Burrell, this plant is probably extinct. [E14; 20; 28].

N. complanata (Hedw.) Huebener.

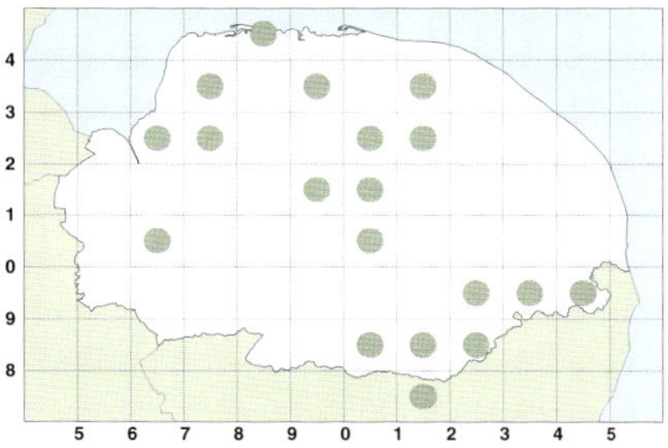

Burrell regarded this plant as very common; it has clearly become less so, as Swann (1982) noted. It seems to enjoy fairly exposed conditions, being most often found on tree roots at the edge of woodlands, more rarely in the wood itself. Also occurs on stonework in churchyards.

Homalia trichomanoides (Hedw.) Bruch et al.
Rare. Swann noted that this plant had become 'more frequent than formerly'. That trend appears to have been reversed. Usually occurs, in small quantity, on tree bases. W99; E03;19; 28. [W82; 91; 92].

HOOKERIACEAE

Hookeria lucens (Hedw.) Sm.
This plant was dedicated to W. J. Hooker, who first found the plant at Holt Lowes, E03. It still occurs there, on the sides of drainage channels feeding into the Glaven. Also recorded from Honing Common E32 (1991, ALB).

LESKEACEAE

Leskea polycarpa Hedw.

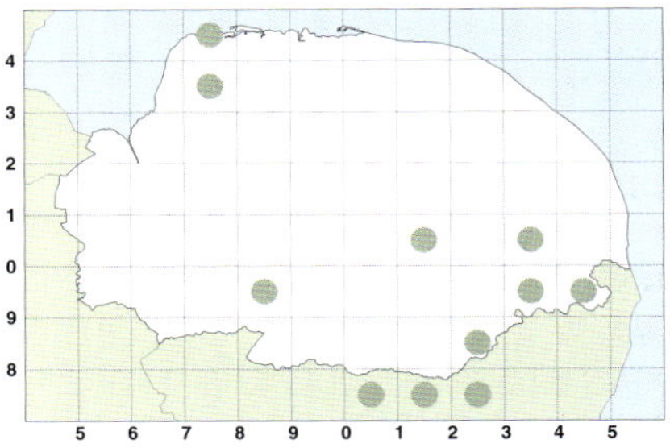

Although normally a species of tree roots prone to flooding, it does occur elsewhere, e.g. on elder, at Holme Dunes, this form being epapillose. Generally occurs in small quantity, though it may be abundant in suitable sites.

THAMNIACEAE

Thamnobryum alopecurum (Hedw.) Gangulee

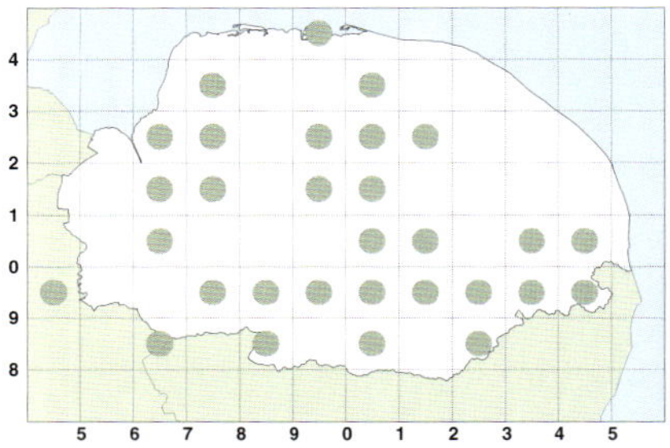

Widespread, and usually abundant. Burrell noted that it was used by gamekeepers as a packing for eggs. Sporophytes occasional.

THUIDIACEAE

Anomodon viticulosus (Hedw.) Hook. & Taylor
Rare. Burrell regarded this plant as common in clay districts, on banks, tree roots, etc. As noted by Swann (1982) it has become

much rarer. It normally occurs on tree bases in woods on clayey soil, but has also been found as an epiphyte, on elder. W71; 81; 91; E00; 27; 29.

Thuidium abietinum (Hedw.) Bruch et al.
Rare. A calcicole, it is sometimes present in quantity, but more usually as scattered stems. It seems to have become rarer. W70; 71; 78; 79; 88; 89. [W80; 03; 98; E08; 11]

T. tamariscinum (Hedw.) Bruch et al.

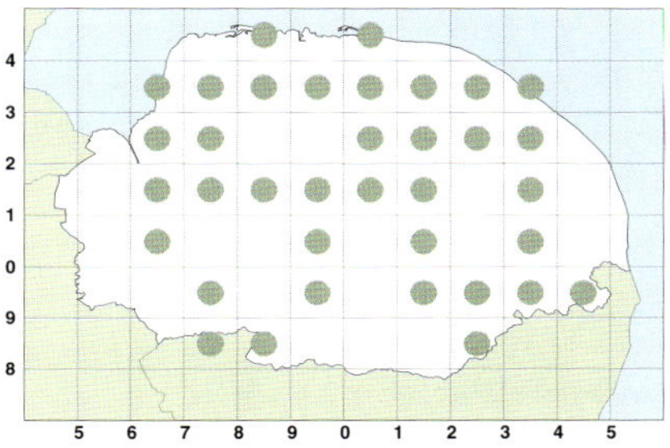

Sometimes abundant in woodland and in carrs, but may also occur, in smaller quantities, in damp places such as the north side of churchyards.

T. philiberti Limpr.
Very rare. Although both Burrell & Swann regarded this plant of chalk soils as rare it appears to have become even more so. W78; 89; E11.[W73; 78; 80; 89; E08].

AMBLYSTEGIACEAE

Cratoneuron filicinum (Sull.) Spruce

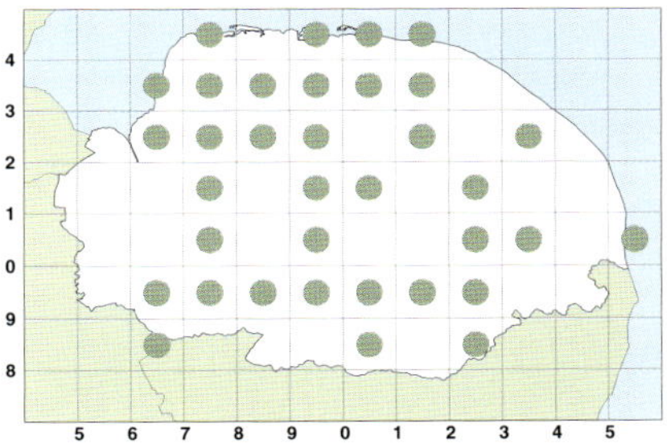

Common in a wide variety of habitats; most frequently where damp - as in fens, flushes, etc.

Palustriella commutata (Hedw.) Ochyra (including var. **falcata** (Brid.) Ochyra)
Syn. *Cratoneuron commutatum* (Hedw.) Roth

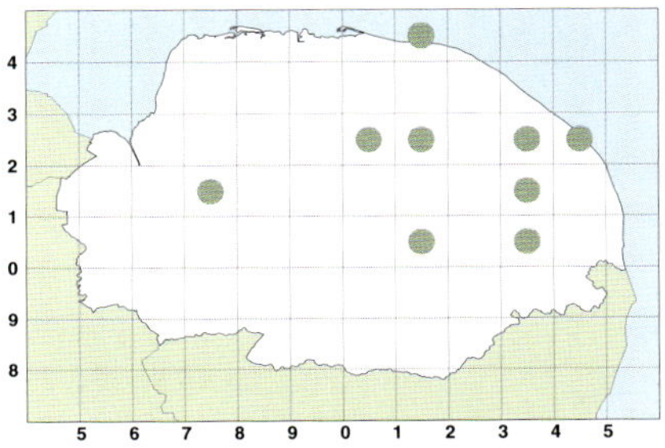

This plant of base-rich fens and calcareous flushes seems to have been fairly frequent in the 1950s. It has become much less common, though it is more frequent in the east of the county. The variety has not been recorded separately; it occurs in similar habitats to the type. [W70; 91; 93; 99; E03; 09; 22; 29].

Campylium stellatum var. **stellatum** (Hedw.) J. Lange & C.E.O.Jensen

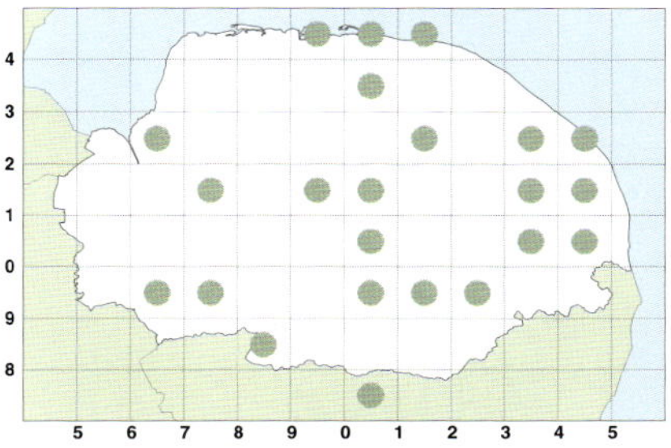

A common plant in rich fen habitats. Sporophytes occasional.

C. stellatum var. **protensum** (Brid.) Bryhn
Rare. In rather drier calcareous places than the type. W68; 71; 88; 94; E04; 17.

C. chrysophyllum (Brid.) J. Lange
Rare. On open chalky ground. W78; 70; 71; 72; 73; 88.

C. polygamum (Bruch et al.) J. Lange & C.E.O. Jensen.
Rare in fens. W94; E07; 31; 32; 42.

C. elodes (Lindb.) Kanda
Rare in rich fens. W71; E01; 31; 32.

Amblystegium serpens (Hedw.) Bruch et al.
Very widespread. This common, rather weedy species, occurs most abundantly as an epiphyte, particularly on and about the lower trunk of trees and bushes. It also occurs on brick and stonework, as well as on soil in hedgebanks. Sporophytes common.

A. varium (Hedw.) Lindb.

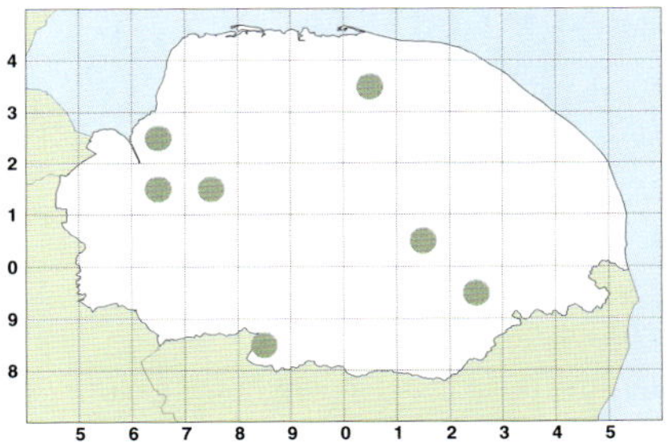

A plant of damp grassland, particularly where the habitat is fairly open. Probably overlooked.

A. humile (P. Beauv.) Crundw.
Very rare. Tree bases by streams; damp grassland; damp peat in calcareous fens. E07; 31.

Leptodictyum riparium (Hedw.) Warnst.
Syn. *Amblystegium riparium* (Hedw.) Bruch et al.

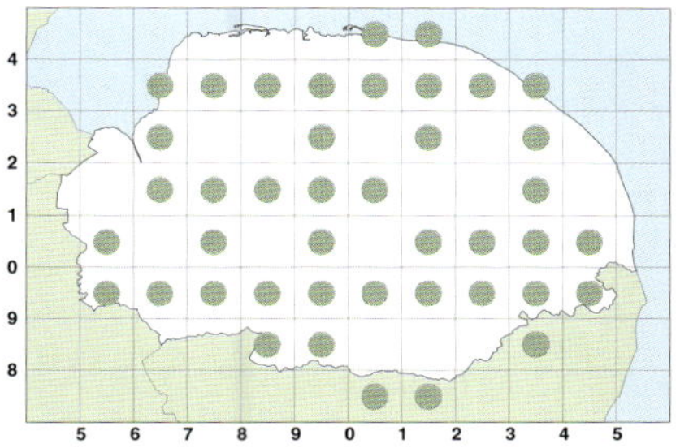

A widespread and abundant species. It normally occurs close to slow moving or static water, where it occurs on roots, prostrate branches and stonework. Occasionally it may occur as an epiphyte in odd, drier, places - causing much confusion. Sporophytes abundant.

Warnstorfia fluitans (Hedw.) Loeske
Syn. *Drepanocladus fluitans* (Hedw.) Warnst.
This species is locally abundant in a few places, e.g. ditches and pools on acid ground & colonising pools in sandpits, but it does not seem to be as common as formerly. It often lacks the denticulate margin specified by most descriptions, except at the very tip of young leaves. W61; 62; 63; E31; 41; 42. [W79; 82; 98; E 02; 03; 11; 12; 32].

W. exannulata (Bruch et al.) Loeske
Syn. *Drepanocladus exannulatus* (Bruch et al.) Warnst.
No recent records. [Rare. In similar habitats to the previous species, Holt Lowes. E03 (1956)].

Drepanocladus aduncus (Hedw.) Warnst.

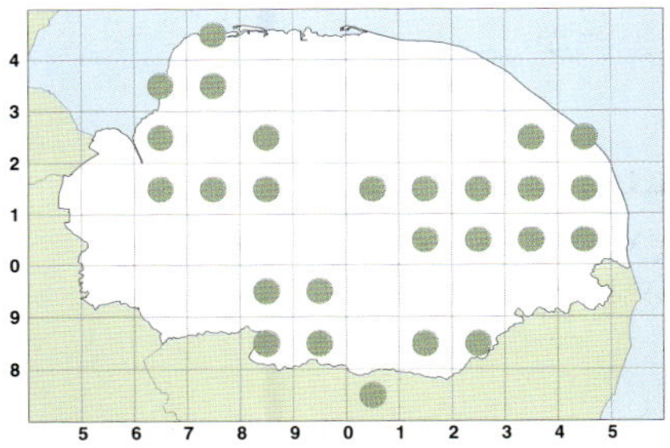

A common plant of damp eutrophic habitats; it may occur submerged, as well as on land.

D. sendtneri (Schimp. ex H. Müll.) Warnst.
No recent records. [Rare. Submerged in a pool, Foulden Common. W79 (1956)].

D. lycopodioides (Brid.) Warnst.
Very rare. Submerged in pools in rich fen, Scarning Fen, W91. [W70].

D. cossonii (Schimp.) Loeske
Syn *Drepanocladus revolvens* var. *intermedius* (Lindb.) P.W.Richards & E.C.Wallace (including **D. revolvens** (Sw.) Warnst.)

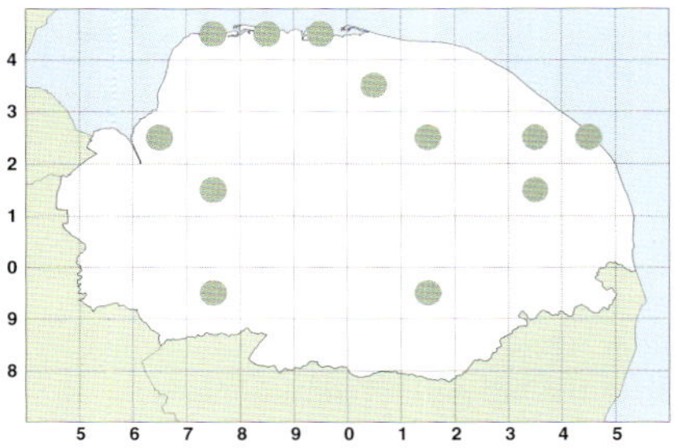

Swann 'lumped' these two species. However, they have been split recently, the var. *intermedius* having been segregated off as **D. cossonii**. Because of the unreliability of some field records, the two are mapped together. **D. revolvens** ss is a plant of more acid conditions than **D. cossonii**, which is probably the commoner of the two in Norfolk.

Hamatocaulis vernicosus (Mitt.) Hedenäs
Syn. *Drepanocladus vernicosus* (Mitt.) Warnst.
No recent records. This is a European Red Data book species; in the past it has been (erroneously) recorded from many sites in the county. Only three of these records were correct viz. Buxton Heath [E12]; Great Witchingham Common [E01], and Brumstead Common [E32]. The last two records are from 1910.

Tomenthypnum nitens (Hedw.) Loeske
Syn. *Homalothecium nitens* (Hedw.) H. Rob.
No recent records. This is a species which appears to have become extinct in the county. All recent searches at its former known localities have been without success. [W62; 09; E03; 12; 40].

Sanionia uncinata (Hedw.) Loeske
Syn. *Drepanocladus uncinatus* (Hedw.) Warnst.
No recent records. Formerly recorded from round Yarmouth, in VC27, this plant may be justifiably regarded as extinct [E50].

Hygrohypnum luridum (Hedw.) Jenn.
Rare. This plant occurs on damp, usually calcareous stonework, usually in small quantity. W50; 69; 80; E00; 01.

Scorpidium scorpioides (Hedw.) Limpr.
Rare. In pools in fens; less frequent than formerly, due to loss of habitat. W62; 71; E14; 19; 31; 32. [W79; 08; 09; E18; 22; 32].

Calliergon stramineum (Brid.) Kindb.
Rare. In acidic mires, often associated with *Sphagnum* spp. Usually in small quantity. W62; 81; E03; 12; 31; 41.

C. cordifolium (Hedw.) Kindb.

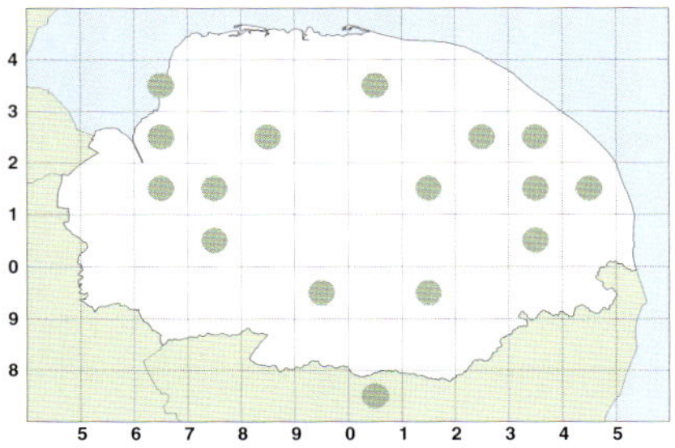

Damp habitats in carrs and marshes.

C. giganteum (Schimp.) Kindb.

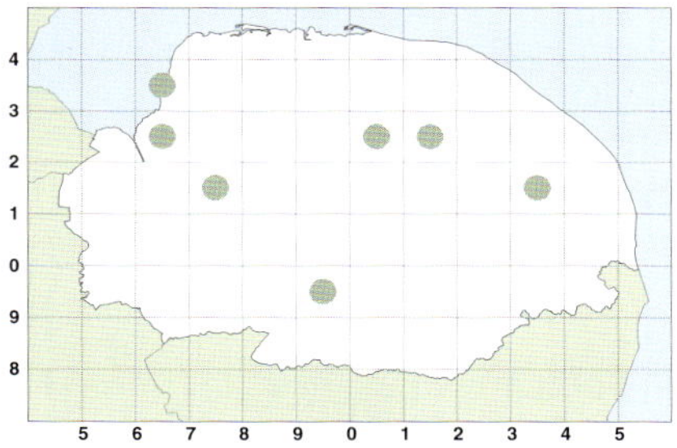

This plant of marshes and fens was recorded from 16 10km squares by Swann (1982), however, there are only 7 recent records. W62; 63; 71; 99; E02; 12; 31. [W70; 72; 91; 93; 01; 08; 09; E03; 09; 22; 30].

Calliergonella cuspidata (Hedw.) Loeske.
Syn. *Calliergon cuspidatum* (Hedw.) Kindb.
Very widespread. Commonest in wet / damp places, such as fens. However, it may also occur abundantly in fairly dry places such as chalk grassland, though usually on north facing slopes.

BRACHYTHECIACEAE

Isothecium myosuroides Brid.

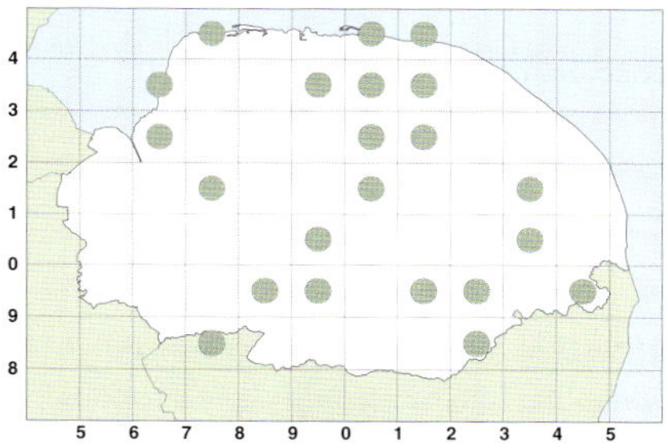

Epiphytic on tree bases; usually only present in fairly small quantity.

I. alopecuroides (Dubois) Isov.
Syn. *Isothecium myurum* Brid.

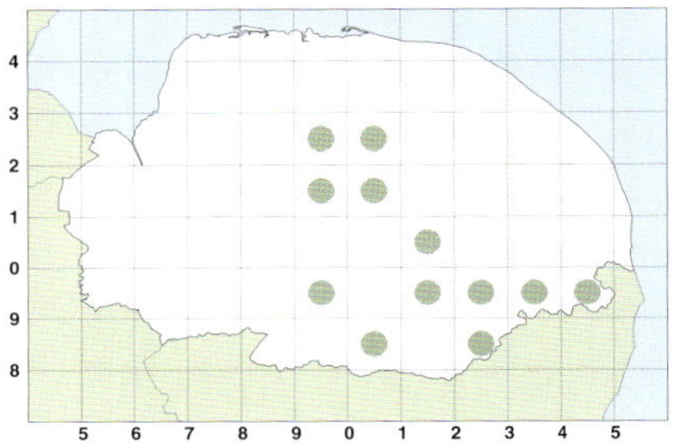

In similar habitats to the last species, though less common.

Homalothecium sericeum (Hedw.) Bruch et al.
Very widespread; it appears to be commoner on walls & stonework than it is on trees. Sporophytes occasional.

H. lutescens (Hedw.) H. Rob.

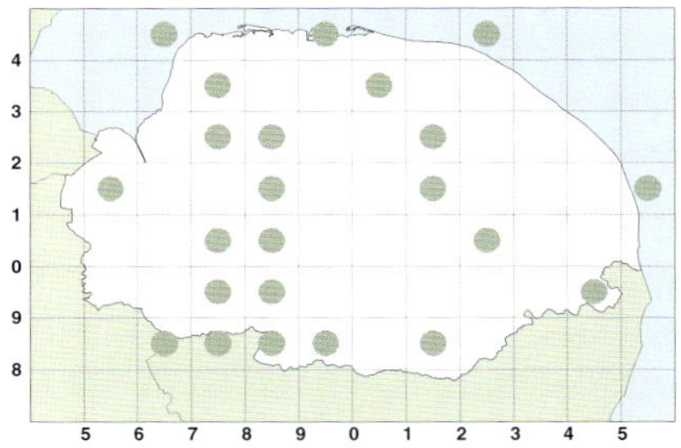

A robust calcicole species, found most frequently in open chalky habitats, as well as in chalk grassland.

Brachythecium albicans (Hedw.) Bruch et al.
Very widespread; usually common in acidic grassland habitats where the vegetation is open or low.

B. glareosum (Spruce) Bruch et al
No recent records. [Uncommon. Marl and Chalk pits but usually in small amounts W71; 72; 78 (1957); 80; 89]

B. salebrosum (F. Weber & D. Mohr) Bruch et al.

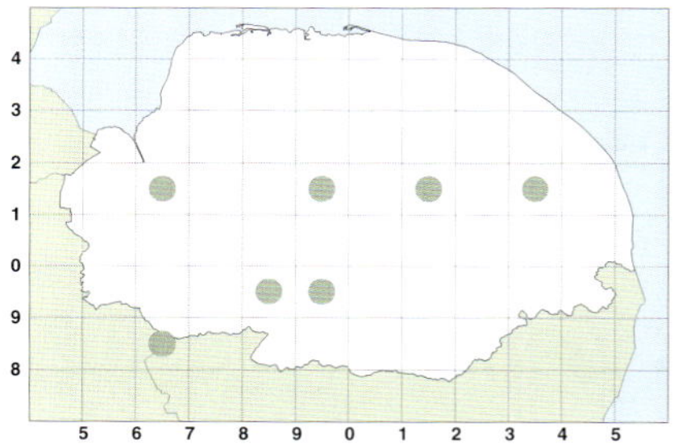

In wet woodlands, on rotting wood, but also in fens and on elder.

B. mildeanum (Schimp.) ex Milde

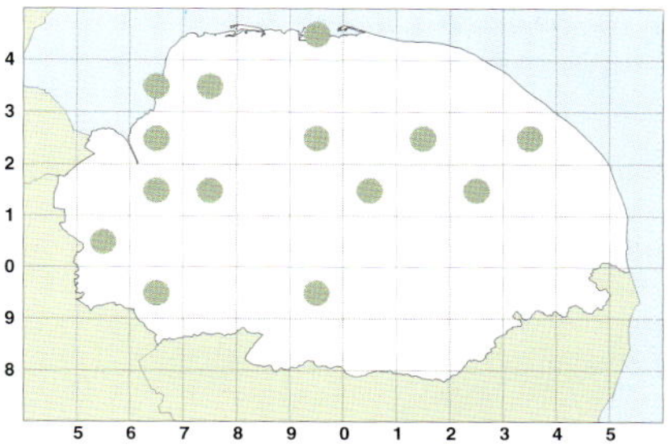

Not recorded in the last Flora, this moss has proved to be of widespread occurrence throughout the county. It is usually found in grassy areas that are at least reasonably moist.

B. rutabulum (Hedw.) Bruch et al.
Very widespread. Common throughout the county. It occurs in a wide variety of habitats - as an epiphyte, as well as on the ground. It is usually most obvious in winter, often surviving concealed under dense vascular cover during the summer months.

B. rivulare Bruch et al.

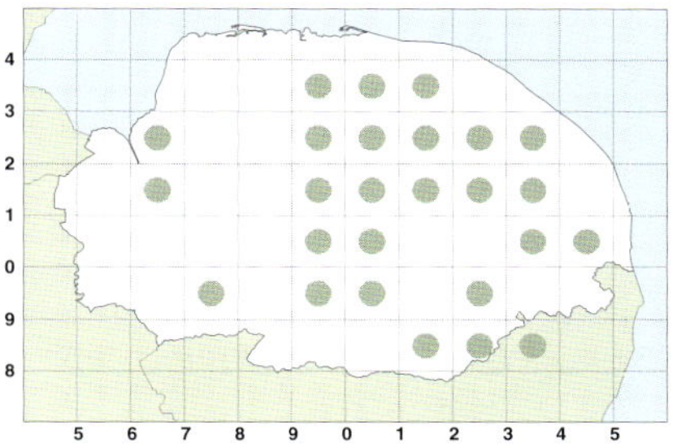

A common species in suitable damp habitats. It seems to favour at least some degree of water movement.

B. velutinum (Hedw.) Bruch et al.

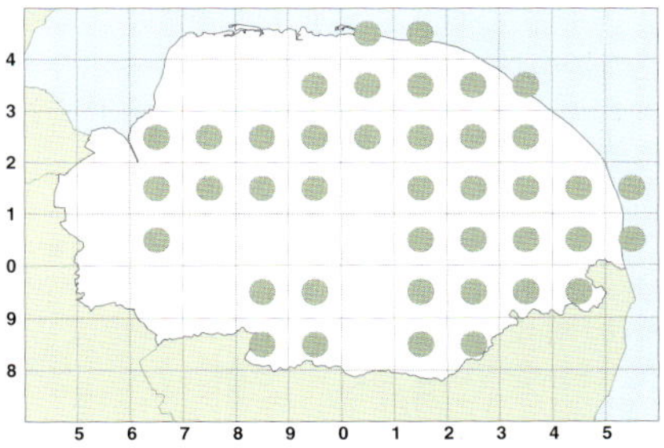

Occurs in a variety of habitats: most typically epiphytic on trees, usually round the bases, but also on railway ballast, and soil.

B. populeum (Hedw.) Bruch et al.
Very rare. This plant was regarded as rather common in East Norfolk by Burrell [E10; 11; 14; 22; 39], but it had not been seen since, until it was found very recently at Stowbridge W50, growing on a grave edging stone (1998, CRS).

Scleropodium purum (Hedw.) Limpr.
Syn. *Pseudoscleropodium purum* (Hedw.) M. Fleisch.
Very widespread and common throughout the county, mainly in grasslands, both wet and dry, calcareous and acid.

S. tourettii (Brid.) L. F. Koch
No recent records [Very rare. Stony ground on heath. W08 (1956)].

Cirriphyllum piliferum (Hedw.) Grout

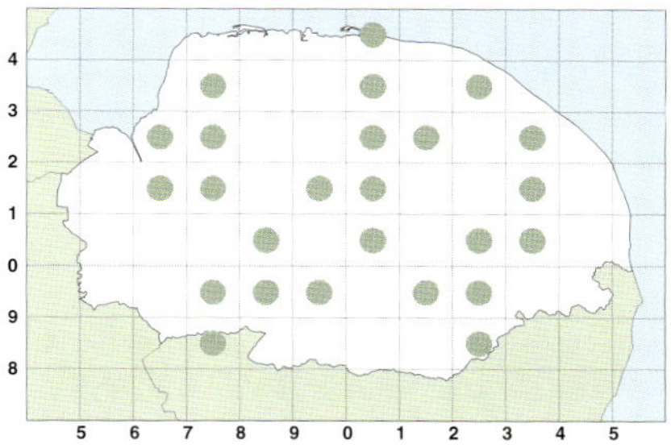

Normally regarded as a woodland plant, it also occurs quite often in fens. It seldom occurs in great quantity.

Rhynchostegium riparioides (Hedw.) Cardot

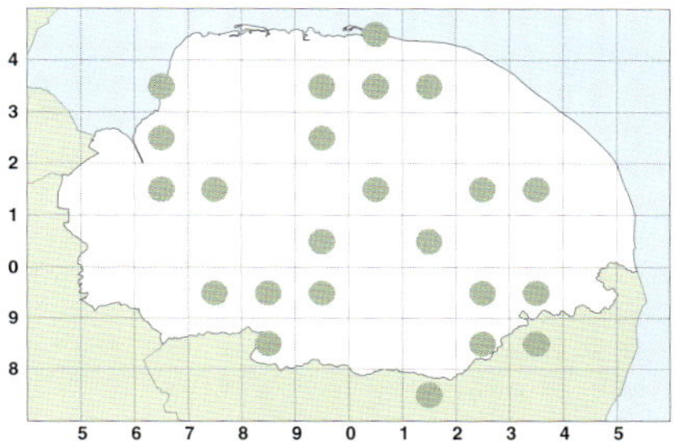

A plant of running water, and therefore restricted in habitat. Often missing from apparently suitable habitats.

R. murale (Hedw.) Bruch et al.

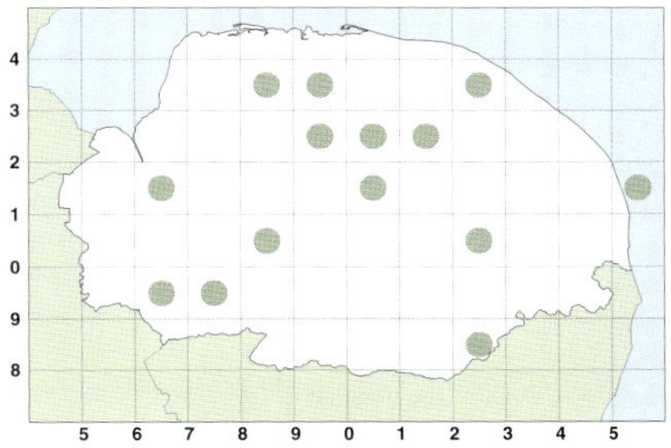

A relatively uncommon species of hard calcareous substrates, ranging from concrete to oolite tombstones.

R. confertum (Dicks.) Bruch et al.
Very widespread, and usually common, this is a rather weedy species occurring in a wide variety of habitats though it is commonest on stonework, and as an epiphyte.

R. megapolitanum (F.Weber & D.Mohr) Bruch et al.
Very rare; though probably much overlooked as, unless it is in fruit, it is very undistinguished. Grows amongst long rank grasses. W64; 98. [W74; 78; E01].

Eurhynchium striatum (Hedw.) Schimp.

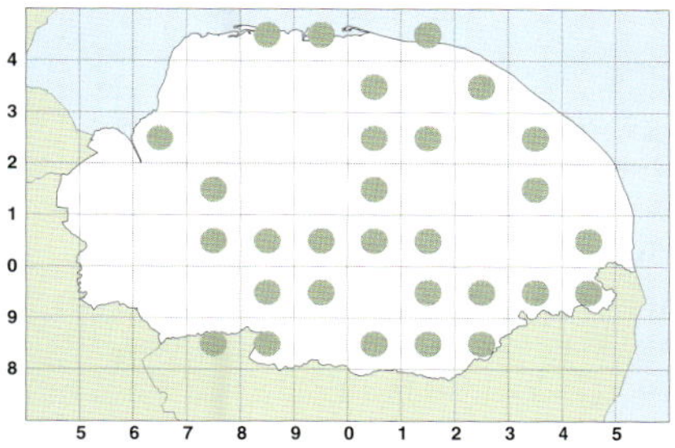

A plant of woodlands, on a variety of soils, as well as, less frequently, calcareous grassland.

E. pumilum (Wilson) Schimp.

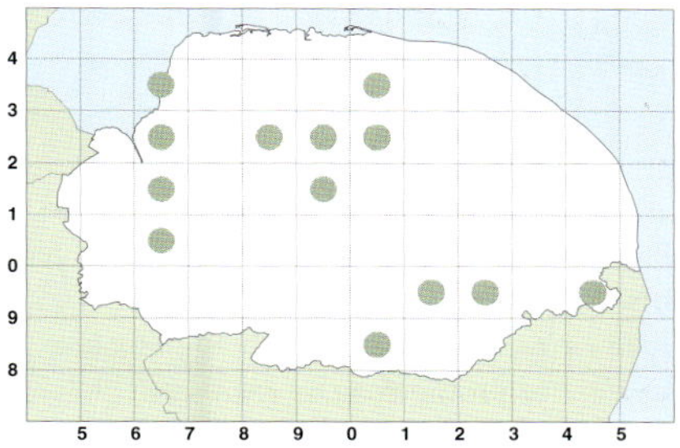

Possibly overlooked because of its small size, this plant is often abundant in suitable habitats, such as woodbanks and the edges of woods. Occurs on both sandy and clay soils.

E. praelongum (Hedw.) Bruch et al. (including var. **stokesii** (Turn.) Dix.)
Very widespread. One of the commonest species in the county. It is very variable in appearance; particularly luxuriant forms occur on damp woodland floors, but it is equally happy as an epiphyte, or amongst grasses. Sporophytes reasonably common. Var. **stokesii** infrequent.

E. hians (Hedw.) Sande Lac. (including var. **rigidum** (Boul.))
Syn. *E. swartzii* (Turner) Curn.
Very widespread. Usually a plant of open well lit habitats, providing they are not too acid. No recent records for the var. **rigidum**. [Rare. Chalk grassland. W78 (1956)].

E. schleicheri (R. Hedw.) Jur.
No recent records. [Apparently very rare. E33].

E. speciosum (Brid.) Jur.

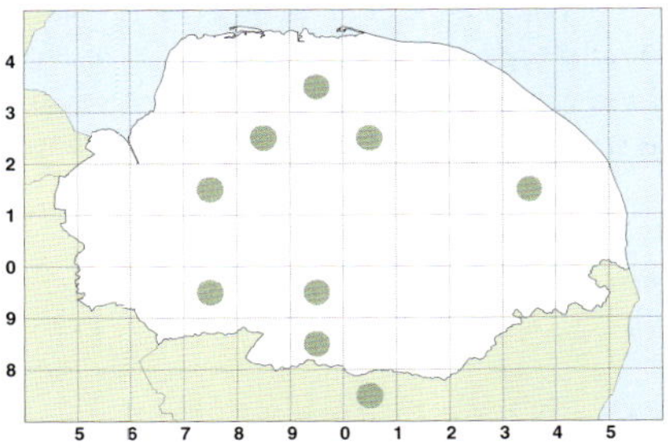

This plant favours very damp conditions in fens, wet woodlands, etc. Floras stress its dark green colour, however, in open habitats it can be a much yellower green, and then closely resembles **E. hians**.

Rhynchostegiella tenella (Dicks.) Limpr.

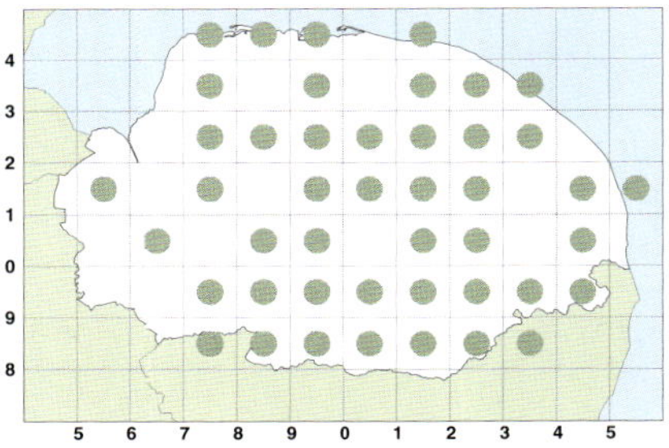

Widespread throughout the county. Usually found adhering to oolite gravestones, low down, amongst grasses shading the base; only occasionally in more exposed positions. Sporophytes common.

PLAGIOTHECIACEAE

Plagiothecium latebricola Bruch et al.
Very rare. On the base of an alder in fen carr, and on decaying stumps. E31 [E01].

P. denticulatum (Hedw.) Bruch et al.

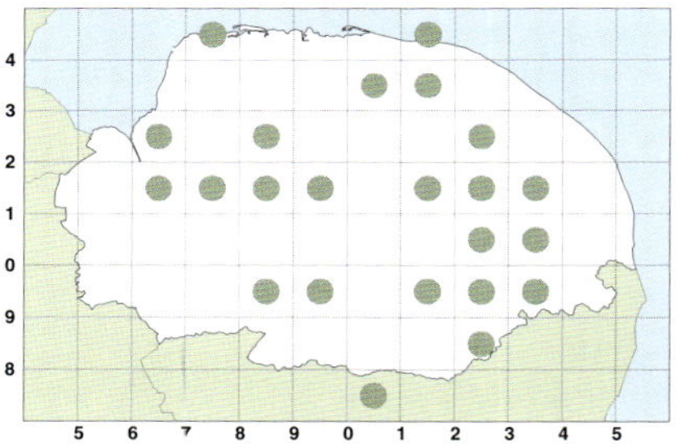

Occurs on organic rich substrates, in relatively acid habitats, such as rotting stumps, banksides, etc.

P. ruthei Limpr
No recent records. [Occasional in carrs and damp woods on leaf litter. W62; 81; 90; 91; 99; 09; E03;11;12;14]. A critical taxon. The available keys involve the use of some finely shaded criteria, which may account for the lack of records. Hodgetts (in Hill, Preston & Smith 1994) expressed doubts as to the value of this taxon.

P. curvifolium Schlieph. ex Limpr.

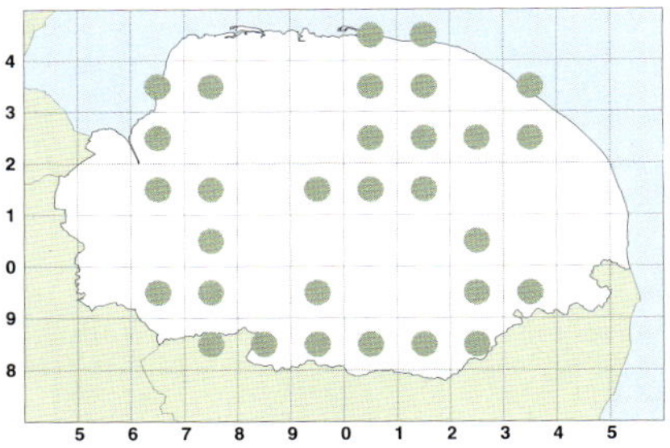

Usually common in suitable habitats, i.e. banks, and rotting stumps in acidic woods and conifer plantations. Very tolerant of deep shade, it is one of the commonest members of the genus in Norfolk. Sporophytes common and abundant.

P. cavifolium (Brid.) Z. Iwats.
Very rare. A remarkable recent addition to the Norfolk flora. Elsewhere in Britain it is usually a plant of moist, basic ledges in the uplands. In Norfolk it occurs on a roadside bank, on organic rich sands. Fellbrigg E14. (1998, CRS & DFS)

P. succulentum (Wilson) Lindb.

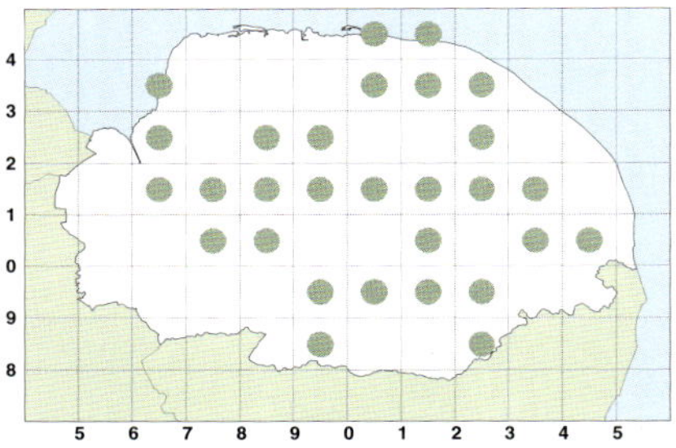

Normally abundant where it occurs, i.e. on damp sandy banks in woodlands.

P. nemorale (Mitt.) A.Jaeger

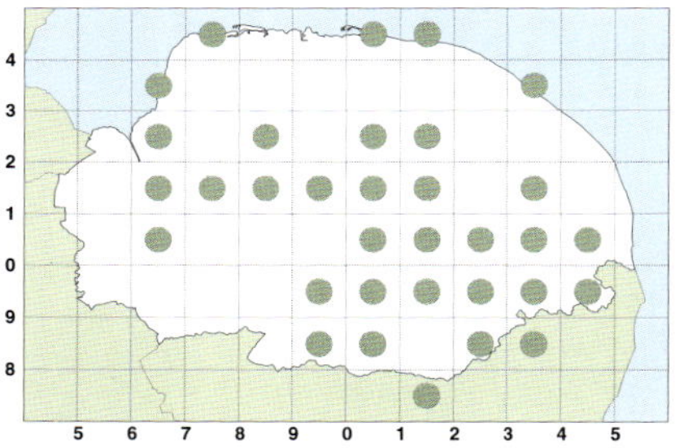

In similar habitats, and similar quantities, to the last species.

P. undulatum (Hedw.) Bruch et al.

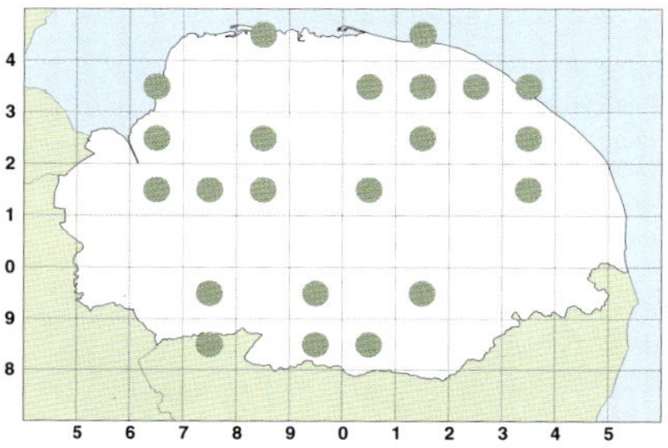

This conspicuous plant usually occurs on peaty acid soils and litter in woods and pine plantations. Often represented by a single colony in a wood.

Plagiothecium undulatum - Sandringham

Herzogiella seligeri (Brid.) Z. Iwats.
Very rare. First recorded in the 1940s, it was not seen again until fairly recently. It has been found on rotting conifer and birch. Sporophytes common. W62; 07; E14. [E03].

Pseudotaxiphyllum elegans (Brid.) Z. Iwats.
Syn. *Isopterygium elegans* (Brid.) Lindb.

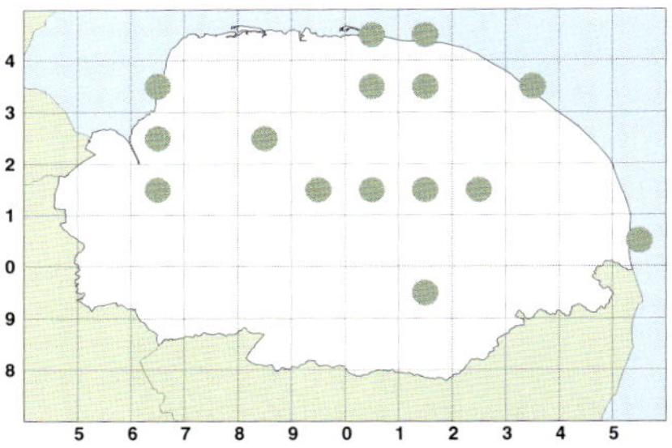

This rather undistinguished plant may well be somewhat under-recorded. Where present, i.e. on sandy acid banks, and tree stumps and bases in woodland, it is often abundant.

HYPNACEAE

Platygyrium repens (Brid.) Bruch et al.
Rare. This epiphyte appears to be spreading. Found on ash, willow and hornbeam. W62; 73; 91; E29.

The ***Hypnum cupressiforme*** *complex has been subject to recent revision (Ando 1992), as a result of which most of the former varieties have been elevated to specific status.*

Hypnum cupressiforme Hedw.
Very widespread throughout the county, in a variety of habitats: chiefly as an epiphyte, but also on stonework, and, particularly near the coast, on the ground. Sporophytes common.

H. lacunosum (Brid.) Hoffm. ex Brid.
Syn. *H. cupressiforme* var. *lacunosum* Brid.

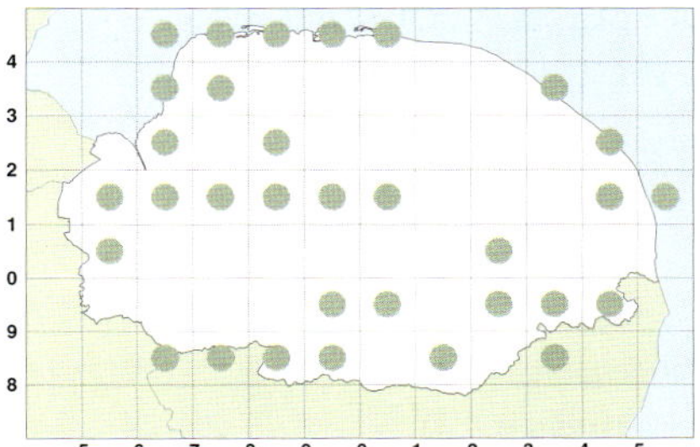

This robust coppery coloured plant occurs most frequently on the ground, where a closed bryophyte community has established itself on a sandy substrate.

H. resupinatum Taylor
Syn. *H. cupressiforme* var. *resupinatum* (Tayl.) Schimp.
Very widespread. One of the commonest epiphytes in the county as it seems able to survive on substrates such as hawthorn which will support few other species. It also occurs on damp stonework, etc. Sporophytes occasional.

H. andoi A.J.E. Sm.
Syn. *H. mammilatum* (Brid.) Loeske
Very rare. Small, parallel sided pendent branched plants of *H. 'mammilatum'* occur occasionally; in the absence of capsules, however, it is difficult to confirm that they are specifically this plant. E13; 14. (1987, JA)

H. jutlandicum Holmen & E.Warncke

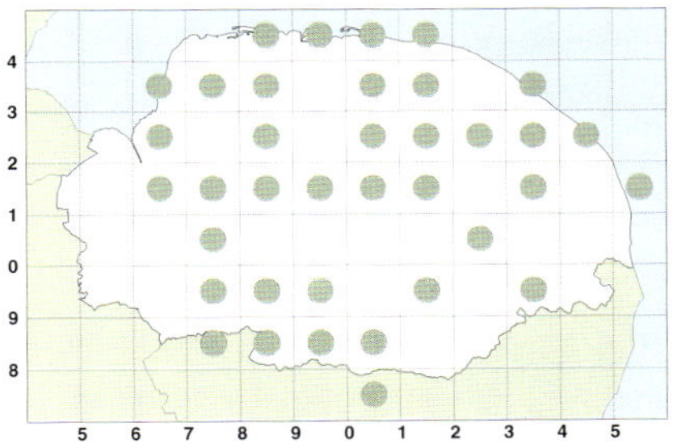

Not listed by Swann (1982), presumably in error. Usually both common and abundant in acid habitats, particularly heaths and light woodland. Sporophytes occasional.

H. imponens Hedw.
Very rare. Wet Heath. East Winch Common. W71.

Ptilium crista-castrensis (Hedw.) De Not.
No recent records. A plant recorded in the Census Catalogue, but not mentioned by Swann. A plant of northern / upland woodlands it was, presumably, imported with Forestry Commission stock. It has been searched for on several occasions, but not found. It may confidently be assumed to have died out [W79].

Ctenidium molluscum (Hedw.) Mitt.

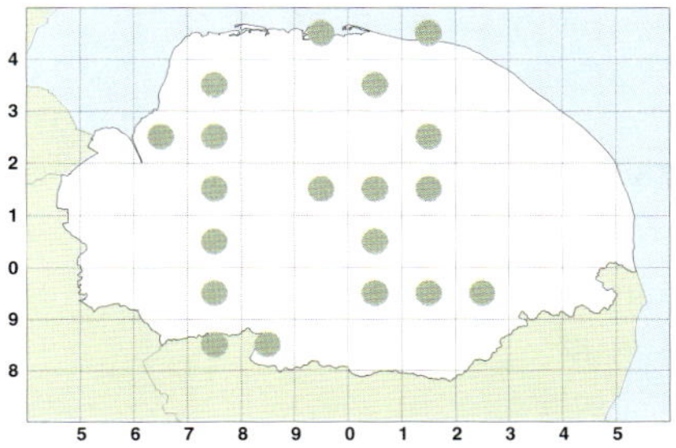

Sometimes abundant and luxuriant in fens; where it occurs in chalk grassland it is usually both less abundant and less luxuriant.

Rhytidium rugosum (Hedw.) Kindb.
Very rare. This plant appears to have become much rarer. Swann states it was occasional in most of the chalk grassland of the south-west Breckland. Only recorded in W78 & 88. [W79; 89]

Rhytidium rugosum - Grimes Graves

Rhytidiadelphus triquetrus (Hedw.) Warnst.

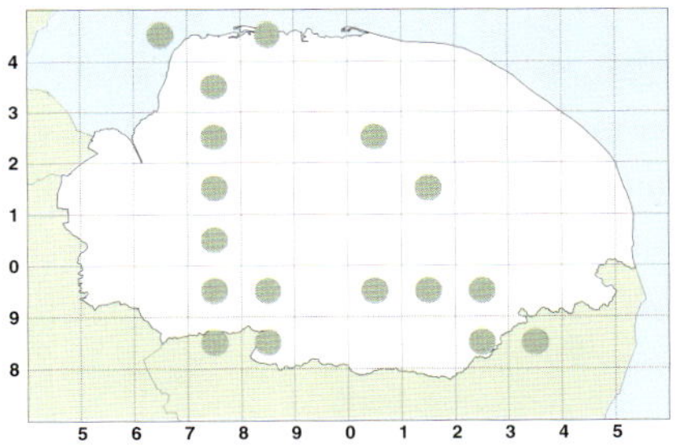

Woodland rides, mixed scrub, chalk grassland, pits in Breckland and occasionally in coastal sand dunes. Sometimes occurs in quantity, but more frequently as fairly small patches.

Rhytidiadelphus triquetrus - Hilborough

R. squarrosus (Hedw.) Warnst.
A common plant of old grasslands, usually on acidic ground, it appears to be intolerant of too much trampling. Sporophytes found on two occasions, both in very wet habitats.

R. loreus (Hedw.) Warnst.
No recent records. [Very rare. Acid humus of woodland. E11 (1962); 12].

Pleurozium schreberi (Brid.) Mitt.

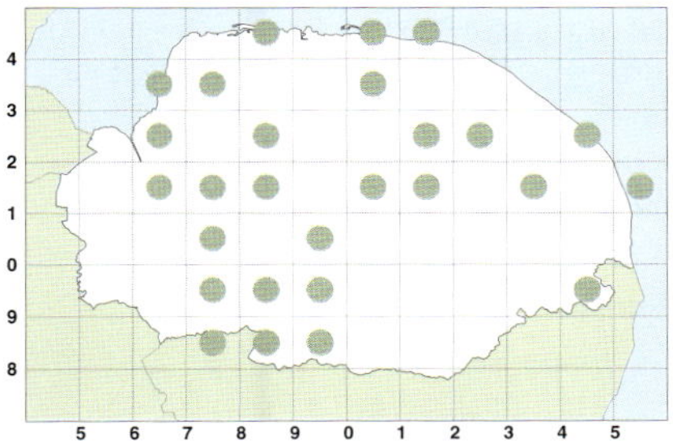

Often abundant amongst heather on heaths, in acid grassland, and in forestry plantations.

Hylocomium splendens (Hedw.) Bruch et al.

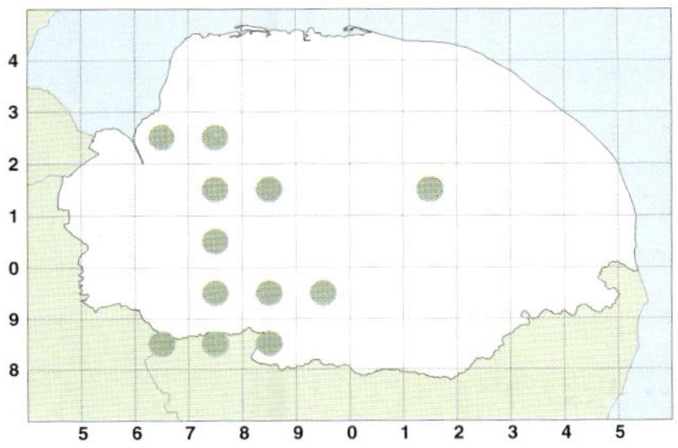

Usually occurs in fairly small quantities in grassland, heaths, etc.

A

- Abele 93
- **Abies**
  - **cephalonica** 57
  - *concolor* 56
  - **grandis** 57
  - *nobilis* 56
  - *procera* 56
- Abraham, Isaac and Jacob 169
- **Abutilon theophrasti** 90
- **Acaena**
  - *anserinifolia* 128
  - **novae-zelandiae** 128
- **Acanthus**
  - **mollis** 188
  - **spinosus** 188
- **Acaulon muticum** *44*, 277
- **Acer**
  - **campestre** 152
  - **platanoides** 151
  - **pseudoplatanus** *24*, 152
- **Aceras anthropophorum** *29*, 264
- **Achillea**
  - **filipendulina** 208
  - **millefolium** 209
  - **ptarmica** 209
- *Acinos arvensis* 175
- **Aconitum**
  - **napellus** 59
  - *x cammarum* 59
- **Acorus calamus** 220
- **Actinidia**
  - *chinensis* 87
  - **deliciosa** 87
- Adder's Tongue 52
- **Adiantum capillus-veneris** 52
- **Adonis**
  - **aestivalis** 63
  - **annua** *43*, 63
- **Adoxa moschatellina** 194
- **Aegopodium podagraria** *23*, 158
- **Aesculus**
  - **hippocastanum** 151
  - **x carnea** 151
- **Aethusa cynapium** 159
- **Agrimonia**
  - **eupatoria** 127
  - *odorata* 128
  - **procera** 128
- Agrimony 127
- Agrimony, Fragrant 128
- **X Agropogon littoralis** 245
- *Agropyron caninum* 251
  - *junceum* 251
  - *pungens* 251
  - *pycnanthum* 251
  - *repens* 251
- **Agrostemma githago** *23*, 81
- **Agrostis**
  - **canina** 245
  - **capillaris** *26*, 245
  - **castellana** 245
  - **gigantea** 245
  - **stolonifera** 245
  - *tenuis* 245
  - **vinealis** 245
  - **x murbeckii** 245
- **Ailanthus altissima** 152
- **Aira**
  - **caryophyllea** 244
  - **praecox** *30*, 244
- **Ajuga**
  - **chamaepitys** 174
  - **reptans** 174
- **Alcea rosea** 90
- **Alchemilla**
  - **filicaulis** 128
  - **glabra** *43*, 128
  - **mollis** 128
  - *vestita* 128
- Alder, Common *25*, *39*, 69
  - European 69
  - Grey 69
- Ale-hoof 174
- Alexanders *23*, **157**, 157
- Alfalfa 139
- **Alisma**
  - **gramineum** 215
  - **lanceolatum** 215
  - **plantago-aquatica** 215
  - *ranunculoides* see Baldellia *37*, 215
- Alison, Golden 101
  - Hoary 101
  - Small 101
  - Sweet 101
- Alkanet, Green 169
  - Yellow 169
- **Alliaria petiolata** 97
- **Allium**
  - **carinatum** 258
  - **neapolitanum** 258
  - **paradoxum** 258
  - **triquetrum** 258
  - **ursinum** *25*, **258**, 258
  - **vineale** 258
- Allseed 151
- **Alnus**
  - **cordata** 69
  - **glutinosa** *26*, 69
  - **incana** 69
- **Aloina**
  - **aloides** 276
  - **ambigua** 276
  - **rigida** 276
- **Alopecurus**
  - **aequalis** *29*, **247**, 247
  - **bulbosus** 247
  - **geniculatus** *29*, 247
  - **myosuroides** 247
  - **pratensis** 247
  - **x brachystylus** 247
  - **x haussknechtianus** 247
- **Althaea**
  - **hirsuta** 90
  - **officinalis** *22*, **90**, 90
  - *rosea* 90
- **Alyssum**
  - **alyssoides** *43*, 101
  - **saxatile** 101
- Amaranth, Bouchon's **75**, *75*
  - Common *74*
  - Green *75*
  - Mucronate *75*
- **Amaranthus**
  - **albus** *75*
  - **blitoides** *75*
  - **bouchonii** **75**, *75*
  - **caudatus** *75*
  - **cruentus** *75*
  - **hybridus** *75*
  - *hypochondriacus* *75*
  - **quitensis** *75*
  - **retroflexus** 74
- **Amblyodon dealbatus** *44*, 281
- **Amblystegium**
  - **humile** 283
  - *riparium* 283
  - **serpens** 283
  - **varium** 283
- **Ambrosia**
  - **artemisiifolia** 212
  - **trifida** 212
- **Amelanchier lamarckii** 132
- **American Slough-grass** 247
- American Weed 216
- **Ammi majus** 161
- *X Ammocalamagrostis baltica* 246
- **Ammophila arenaria** *29*, *30*, 246
- **Amsinckia**
  - **lycopsoides** 169
  - **micrantha** 169
- **Anacamptis pyramidalis** *22*, *26*, 262

Anagallis
- **arvensis** 110
- **minima** 110
- **tenella** 110

Anchusa
- **arvensis** 169
- **ochroleuca** 169

Andromeda polifolia *43*, 107

Anemone
- **apennina** 59
- **blanda** 59
- **nemorosa** *25*, **59**, 59
- *pulsatilla* 59
- **ranunculoides** 59

Anemone, Apennine 59
- Balkan 59
- Wood **59**, 59
- Yellow Wood 59

Aneura pinguis *29*, **269**, 269

Angelica
- **archangelica** 161
- **sylvestris** *28*, 161

Angelica, Garden 161
- Wild 161

Anisantha
- **diandra** 250
- **rigida** 250
- **sterilis** 250
- **tectorum** 250

Anomodon viticulosus 282

Antennaria dioica *39*, *43*, 204

Anthemis
- **arvensis** 209
- **cotula** 209
- **tinctoria** 209

Anthoceros agrestis 270

Anthoxanthum
- **aristatum** 244
- **odoratum** 244
- *puelii* 244

Anthriscus
- **caucalis** 156
- **cerefolium** 156
- **sylvestris** 156

Anthyllis vulneraria *30*, **134**, 134

Antirrhinum
- *asarina* 182
- **majus** 182
- *orontium* 182

Antitrichia curtipendula *44*, 282

Apera
- **interrupta** 246
- **spica-venti** 246

Aphanes
- **arvensis** 128
- **australis** 128
- *inexspectata* 128
- *microcarpa* 128

Aphanorhegma patens *44*, 278

Apium
- **graveolens** 160
- **inundatum** *22*, 160
- **nodiflorum** 160

Aponogeton distachyos 216

Apple 131
- Crab 131

Apple of Peru 164

Aquilegia vulgaris 63

Arabidopsis thaliana **97**, 97

Arabis
- **caucasica** 100
- **glabra** **101**, 100
- **hirsuta** 100

Arabis, Garden 100

Araucaria araucana 56

Archangel, Variegated 172
- Yellow **172**, 172

Arctium
- **lappa** 196
- **minus** 196
- *nemorosum* 196

Arenaria
- **balearica** 76
- **serpyllifolia** 76

Aristolochia clematitis 58

Armeria maritima *30*, 87

Armoracia rusticana 99

Arnoseris minima *43*, 199

Arrhenatherum elatius *30*, 242

Arrowgrass, Marsh 216
- Sea 216

Arrowhead 214

Artemisia
- **absinthium** 208
- **biennis** 208
- **campestris** *19*, *26*, **208**, 208
- *maritima* 208
- **verlotiorum** 208
- **vulgaris** 208

Arum
- **italicum** 220
- **maculatum** **220**, 220

Arundinaria japonica 235

Asarina procumbens 182

Ash *16*, *25*, *39*, 179

Asparagus officinalis 259

Asparagus, Garden 259

Aspen 93

Asperula
- **cynanchica** *26*, 191
- **taurina** 191

Asplenium
- **adiantum-nigrum** 53
- **ruta-muraria** 54
- *scolopendrium* 53
- **trichomanes** 53

Aster
- **lanceolatus** 206
- **novae-angliae** 206
- **novi-belgii** 206
- **tripolium** *30*, **206**, ***207***, 206
- **x salignus** 206
- **x versicolor** 206

Aster, Sea **206**, **207**, 206

Astragalus
- **danicus** *26*, **133**, 133
- **glycyphyllos** 133

Astrantia major 156

Athyrium filix-femina 54

Atrichum undulatum 272

Atriplex
- **glabriuscula** 72
- **hortensis** 71
- **laciniata** *29*, 72
- **littoralis** *24*, 72
- **longipes** 72
- **patula** 72
- **pedunculata** *43*, 73
- **portulacoides** *30*, **72**, 72
- **prostrata** *24*, *30*, 71
- **x gustafssoniana** 72
- **x hulmeana** 72

Atropa belladonna *33*, 164

Aubretia 100

Aubrieta deltoidea 100

Aucuba japonica 148

Aulacomnium
- **androgynum** 281
- **palustre** 280

Aurinia saxatile 101

Avena
- **byzantina** 242
- **fatua** 242
- *ludoviciana* 242
- **sativa** 242
- **sterilis** ssp. **ludoviciana** 242
- **strigosa** 242

Avens, Hybrid 127
- Water 127
- Wood 127

Avenula pratense *26*, 242
- *pubescens* *26*, 242

Azolla
caroliniana 56
filiculoides 56

B

Baldellia ranunculoides 215
Ball Mustard 102
Ballota nigra 171
Balm of Gilead 94
Balm, Bee 174
Lemon 174
Balsam, Himalayan *23,* 155
Indian *23,* 155
Orange *23,* 155
Small 155
Bamboo, Arrow 235
Broad-leaved 235
Barbarea
intermedia 98
stricta *43,* 98
verna 98
vulgaris 98
Barberry 64
Box-leaved 64
Chinese 64
Darwin's 64
Hybrid 64
Mrs Wilson's 64
Thunberg's 64
Barbilophozia
attenuata 266
barbata *44,* 266
floerkei 266
hatcheri 266
Barbula
convoluta 275
cylindrica 275
fallax 276
hornschuchiana 275
recurvirostra 275
reflexa 276
revoluta 275
rigidula 275
tophacea 276
trifaria 275
unguiculata 275
vinealis 275
Barley, Foxtail **252**, 252
Meadow 252
Sea 252
Six-rowed 251
Two-rowed 251
Wall 252
Barren Strawberry 127
Bartramia pomiformis 281
Bartsia, Red 187
Yellow **187**, 187
Basil Thyme 175
Bassia scoparia 71
Bastard Cabbage 105
Bastard Toadflax 148
Batchelors' Buttons 115
Bay 57
Beak-sedge, White 227
Bean-weed 217
Beard-grass, Annual 246
Perennial 245
Bear's-breech 188
Spiny 188
Beckmannia syzigachne 247
Bedstraw, Fen 191
Heath 192
Hedge 192
Lady's **192**, 191
Marsh 191
Wall 192
Beech *25, 30,* 68
Beet, Sea 73
Sugar 73
Bell Heather *19, 20,* 107
Bellflower, Adriatic 189
Clustered **189**, 189
Creeping 190
Giant *16,* 190
Nettle-leaved *16,* 190
Peach-leaved 189
Trailing 189
Bellis perennis 207
Bent, Black 245
Brown 245
Common 245
Creeping 245
Highland 245
Velvet 245
Berberis
buxifolia 64
darwinii 64
julianae 64
thunbergii 64
vulgaris 64
wilsoniae 64
x lologensis 64
Bergenia
cordifolia 113
crassifolia 113
Bermuda-grass 253
Berry Alder **150**, 150
Berry Catchfly 82
Berteroa incana 101
Berula erecta 158
Beta vulgaris 73
Betonica officinalis 171
Betony 171
Betula
pendula 69
pubescens *26,* 69
x aurata 69
Bidens
cernua **213**, 213
pilosa 214
tripartita 213
Bidi-Bidi 128
Bilberry 107
Bilderdyckia aubertii 85
Bindweed, Field 165
Hairy 166
Hedge 166
Large 166
Sea 166
Birch 39
Downy 69
Hybrid 69
Silver 69
Warty 69
Bird Cherry *25,* **131**, 131
Bird-in-a-bush 65
Bird's foot 134
Bird's-foot Clover 140
Bird'sfoot Trefoil, Common *24,* 134
Greater 134
Narrow-leaved 134
Bird'snest, Yellow **108**, 108
Birthwort 58
Bistort, Amphibious 83
Common 83
Red 83
Bitter-cress, Hairy 100
Large 100
Wavy 100
Bittersweet 165
Black Bindweed 85
Black Bog-rush 227
Black Bryony 260
Black-eyed Susan 212
Black-grass 247
Black-jack 214
Blackstonia perfoliata 163
Blackthorn *39,* 130

Black-twitch 247
Bladder Ketmia 90
Bladder-senna 133
Bladderwort 189
  Greater *22,* 189
  Intermediate 189
  Lesser 189
Blasia pusilla 269
Blechnum spicant 56
Blinks *75*
Blood-drop-emlets 182
Blue Eryngo 156
Blue Sow-thistle 201
Bluebell, *16,* 257
  Garden 257
  Hybrid 257
  Spanish 257
Blysmus, Broad 227
Blysmus compressus 227
Bog Asphodel *20,* **256**, 255
Bog Myrtle 68
Bog Rosemary 107
Bog-rush, Black 227
Bog-bean **167**, 167
Bolboschoenus maritimus *30,* 226
Bolder 226
Borage 169
Borago officinalis 169
Boston Ivy 151
Botrychium lunaria *29,* 52
Box *24, 39,* 149
Brachyglottis 'Sunshine' 212
Brachypodium
  pinnatum 250
  sylvaticum 250
Brachythecium *37*
  albicans 284
  glareosum *44,* 284
  mildeanum 285
  populeum 285
  rivulare 285
  rutabulum 285
  salebrosum 284
  velutinum 285
Bracken *18,* 53
Bramble *39,* **118**, **120**, 115
Brandy-bottle **58**, 58
Brassica
  juncea 104
  napus 104
  nigra 105
  oleracea 104
  rapa 104
Bridewort 114
  Confused 114
  intermediate 114
Bristle-grass *24*
  Foxtail 254
  Green 254
  Rough 254
  Yellow 254
Bristly Coneflower 213
Brittle Bladder Fern 54
Briza
  maxima 239
  media *26, 28,* 238
Broad Arrow 214
Broad Blysmus 227
Brome, Barren 250
  Californian 250
  Drooping 250
  False 250
  Field 248
  Great 250
  Hairy 249
  Hungarian *24,* 249
  Interrupted 249
  Lesser Soft 249
  Meadow 248
  Rescue 250
  Ripgut 250
  Rye 249
  Slender Soft 249
  Smooth 249
  Soft 249
  Upright 249
Bromopsis
  erecta 249
  inermis *24,* 249
  ramosa 249
Bromus
  arvensis 248
  carinatus 250
  commutatus 248
  diandrus 250
  erectus 249
  hordeaceus *26,* 249
  inermis 249
  interruptus *43,* 249
  lepidus 249
  mollis 249
  racemosus 249
  ramosus 249
  rigidus 250
  secalinus *24,* 249
  sterilis 250
  tectorum 250
  unioloides 250
  x pseudothominei 249
Brooklime 184
Brookweed **110**, 110
Broom 143
  Hairy-podded 143
  Montpellier 143
  Spanish 143
  Woolly-podded 143
Broomrape, Common 188
  Greater 188
  Hemp 188
  Purple **188**, 188
  Tall 188
Brunnera macrophylla 169
Bryoerythrophyllum recurvirostrum 275
Bryonia dioica 93
Bryum
  algovicum 279
  argenteum 279
  bicolor 279
  bornholmense 280
  caespiticium 279
  capillare 279
  dunense 279
  flaccidum 279
  gemmiferum 279
  gemmilucens 279
  imbricatum 279
  inclinatum 279
  intermedium *43,* 279
  klingraeffii 280
  knowltonii 279
  mamillatum *44,* 278
  microërythrocarpum 280
  pallens 278
  pseudotriquetrum 279
  radiculosum 279
  rubens 280
  ruderale 279
  sauteri 280
  subapiculatum 280
  subelegans 279
  torquescens *44,* 279
  turbinatum *44,* 278
  violaceum 279
Buckler Fern, Broad 56
  Crested **55**, 55
  Narrow 56
Buckthorn, Alder **150**, 150
  Purging 150
Buckwheat 84
  Green 84
Buddleja davidii 179
Buffalo Weed 212
Buffalo-bur 165
  red 165
Bugle 174

Bugloss 169
Buglossoides arvensis 168
Bullace 130
Bullwort 161
Bulrush 255
Grey 226
True 226
Bunias orientalis 97
Bupleurum
lancifolium 160
rotundifolium 160
subovatum 160
tenuissimum *22,* 160
Burdock, Greater 196
Lesser 196
Bur-marigold, Nodding **213**, 213
Trifid 213
Burnet, Fodder 128
Great 128
Salad 128
Burnet-saxifrage 157
Greater 157
Burning Bush, 71
Bur-parsley, Greater 155
Small 155
Bur-reed, Branched 254
Least *18,* 255
Unbranched **255**, 254
Bushgrass **246**, 245
Butchers' Broom 259
Butomus umbellatus **214**, 214
Butter and Eggs 134, 183
Butterbur 212
Buttercup, Bulbous 60
Bur 60
Celery-leaved 61
Corn *23,* **60**, 60
Creeping 60
Hairy *22,* 60
Meadow 60
Small-flowered 60
Butterfly Bush *16, 22,* 179
Butterfly Orchids 262
Butterwort, Common *20,* 189
Buttonhole Flower 75
Buttonweed 210
Buxbaumia aphylla *37, 44,* 272
Buxus sempervirens *24,* 149

Cabbage 104
Cakile maritima *29,* 105
Calamagrostis
canescens *28,* 246
epigejos **246**, 245
stricta 246
Calamint, Common 175
Lesser 175
Calamintha ascendens 175
nepeta 175
X Calammophila baltica 246
Calendula officinalis 212
Californian Poppy 65
Calliergon
cordifolium 284
cuspidatum *28,* 284
giganteum *28,* 284
stramineum 284
Calliergonella cuspidata *28,* 284
Callistephus chinensis 207
Callitriche
hamulata 178
intermedia 178
obtusangula 178
platycarpa 178
stagnalis 178
Calluna vulgaris *19, 22, 27, 30,* **107**, 107
Caltha palustris *59*
Calvary Clover 140
Calypogeia
arguta *44,* 266
fissa 266
muelleriana 266
sphagnicola 266
Calystegia
dahurica 166
pulchra 166
sepium 166
silvatica 166
soldanella 166
Camelina sativa 102
Campanula
glomerata *26,* **189**, 189
latifolia *16, 25,* 190
medium 189
persicifolia 189
portenschlagiana 189
poscharskyana 189
rapunculoides 190
rotundifolia 190
trachelium *16, 25,* 190

Campion, Bladder 81
Hybrid 81
Red 82
Rose 80
Sea *22,* 81
White 81
Campylium
chrysophyllum 283
elodes 283
polygamum 283
stellatum *29,* 283
Campylopus
brevipilus *44,* 274
flexuosus 273
fragilis 273
introflexus 273
pyriformis 273
Canadian Pondweed *23,* 216
Canary-grass 244
Awned 244
Bulbous *24,* 244
Lesser 244
Reed *23,* 244
Candytuft, Garden 103
Perennial 103
wild 103
Cankerweed 210
Cannabis sativa *67*
Canterbury Bells 189
Cape Gooseberry 164
Cape Pondweed 216
Capsella bursa-pastoris 102
Caraway 161
Corn *23,* 160
Cardamine
amara 100
corymbosa 100
flexuosa 100
hirsuta 100
pratensis 100
uniflora 100
Cardaria draba 104
Carduus
acanthoides 197
crispus 197
nutans 197
tenuiflorus 197
x dubius 197
Carex
acuta 234
acutiformis 231
appropinquata *28,* **228**, 228
arenaria *29, 30,* 229
binervis *27,* 232
caryophyllea 233
contigua 228

curta 230
demissa 233
diandra *28*, 228
dioica *28*, 230
distans 233
disticha 229
divisa 229
divulsa 229
echinata 230
elata *28*, **235**, 234
ericetorum *26*, **234**, 234
extensa 233
flacca *26*, 232
hirta *37*, 230
hostiana 233
lasiocarpa *28*, 230
lepidocarpa *29*, 233
limosa *37*, *43*, 234
muricata 229
nigra *38*, 234
otrubae 228
ovalis 230
pairei 229
pallescens 233
panicea 232
paniculata *26*, 228
pendula 232
pilulifera *27*, 234
polyphylla 229
pseudocyperus *28*, 231
pulicaris 235
remota 230
riparia 231
rostrata *28*, **231**, 231
serotina 233
spicata 228
strigosa 232
sylvatica *16*, *25*, **232**, 232
trinervis *38*, *43*, 232
vesicaria *37*, 231
viridula ssp. **brachyrrhyncha** *28*, *29*, 233
viridula ssp. **oedocarpa** 233
viridula ssp. **viridula** 233
x fulva 233
x justi-schmidtii 231
x prolixa 234
x pseudoaxillaris 228
x rotae 228
x turfosa 234
Carlina vulgaris 196
Carpinus betulus 69
Carrot, Wild 162
Carthamus tinctorius 199
Carum carvi 161
Castanea sativa *23*, 68
Catabrosa aquatica 240
Catapodium
marinum 240
rigidum **240**, 240
Catchfly, Night-flowering 81
Sand 82
Small-flowered 82
Spanish *19*, *20*, **81**, 81
Caterpillar Plant 135
Catmint 174
Garden 174
Cat's-ear 199
Smooth *20*, *21*, **199**, 199
Cat's-tail 255
Cat's-tail, (grass) 248
Breckland **248**, 248
Purple-stemmed *20*, **248**, 248
Sand 248
Smaller 248
Caucalis
lappula 155
latifolia 155
platycarpos 155
Caucasian Wingnut 68
Cedrus atlantica 56
deodara 56
libani 56
Celandine, Greater 65
Lesser 61
Celery, Wild 160
Centaurea
aspera 198
calcitrapa 198
cyanus 198
dealbata *23*, 198
diluta 198
melitensis 198
montana 198
nemoralis 198
nigra *26*, 198
scabiosa 198
solstitialis 198
Centaurium
erythraea 163
littorale 163
minus 163
pulchellum 163
Centaury, Common 163
Lesser 163
Yellow 163
Centranthus ruber 195
Centunculus minimus 110
Cephalozia
bicuspidata 266
connivens 266
macrostachya 266
pleniceps *44*, 266
Cephaloziella
divaricata 266
elachista *44*, 266
hampeana 266
rubella *44*, 266
Cerastium
arvense *77*
arvense x C. tomentosum 78
atrovirens 78
diffusum 78
fontanum 78
glomeratum 78
holosteoides 78
semidecandrum 78
tetrandrum 78
tomentosum 78
x mauereri 78
Ceratocapnos claviculata *17*, 66
Ceratochloa
carinata *24*, 250
cathartica 250
unioloides 250
Ceratodon
conicus *44*, 272
purpureus *30*, 272
Ceratophyllum
demersum *29*, 58
submersum 58
Cerinthe major 168
Ceterach officinarum **54**, 54
Chaenomeles speciosa 131
Chaenorhinum minus 182
Chaerophyllum temulum 156
Chaffweed 110
Chamaecyparis
lawsoniana *57*
nootkatensis 56
Chamaemelum nobile *43*, 209
Chamaenerion angustifolium 147
Chamerion angustifolium 147
Chamomile 209
Corn 209
Stinking 209
Yellow 209
Chara intermedia *29*
Charlock 105
Cheiranthus cheiri *23*, 98
Chelidonium majus 65

Chenopodium
- **album** 71
- **bonus-henricus** 70
- *botryoides* 70
- **chenopodioides** *43,* 70
- **ficifolium** 71
- **glaucum** *43,* 70
- **hircinum** 71
- **hybridum** 71
- **murale** 71
- **polyspermum** 70
- **probstii** 71
- **rubrum** 70
- **suecicum** 71
- **urbicum** *43,* 71
- **vulvaria** *43,* 71

Cherry Pie 212
Cherry *39*
- Bird *25,* **131**, 131
- Dwarf 131
- Rum 131
- Wild 130

Cherry-plum 130
Chervil, Bur 156
- Garden 156
- Rough 156

Chestnut, Horse *39,* 151
- Sweet *23, 39,* 68

Chewing's Fescue 236
Chick Pea 135
Chickweed *76*
- Greater *77*
- Jagged *77*
- Lesser *76*
- Upright *78*
- Water *78*

Chicory *35,* 199
Chilean Glory Flower 188
Chiloscyphus
- **pallescens** 268
- **polyanthos** 268

Chinese Aster 207
Chinese Gooseberry 87
Chinese Lanterns 164
Chionodoxa forbesii 257
Chrysanthemum
- **coronarium** 209
- *leucanthemum* 210
- *maximum* 210
- *parthenium* 207
- **segetum** **209**. 209
- *uliginosum* 209

Chrysanthemum, Annual 209

Chrysosplenium
- **alternifolium** 114
- **oppositifolium** 114

Cicendia filiformis *43,* 163
Cicer arietinum 135
Cicerbita bourgaei 201
Cichorium intybus 199
Cicuta virosa *28,* **161**, 161
Cinclidium stygium 280
Cinquefoil, Creeping 127
- Hoary **126**, 126
- Hybrid 126
- Marsh 126
- Russian 126
- Spring 126
- Sulphur 126
- Ternate-leaved 126
- Trailing 126

Circaea lutetiana 148
Cirriphyllum piliferum 285
Cirsium
- **acaule** *26,* **197**, 197
- **arvense** 198
- **dissectum** *28, 29,* 197
- **eriophorum** *43,* 197
- **palustre** *28, 29,* 198
- **vulgare** 197
- **x forsteri** 197
- **x woodwardii** 197

Cladium mariscus *22, 28, 29,* **227**, 227
Cladonia sp. *26,* 30
Cladopodiella
- **fluitans** 266
- **francisci** *44,* 266

Clarkia 147
Clarkia
- **amoena** 147
- **unguiculata** 147

Clary, Annual 177
- Meadow 177
- Whorled 177
- Wild **177**, 177

Claytonia
- **perfoliata** *75*
- **sibirica** *75*

Cleavers 192
- Corn 192

Clematis
- *orientalis* 60
- **tangutica** 60
- **vitalba** 60

Clematis, Orange-peel 60
Climacium dendroides 282

Clinopodium
- **acinos** 175
- **ascendens** 175
- **calamintha** 175
- **vulgare** 175

Clover, Alsike 140
- Bird's-foot 140
- Burrowing 142
- Calvary 140
- Clustered 141
- Crimson 142
- Hare'sfoot 142
- Knotted 142
- Narrow 142
- Red 141
- Reversed 141
- Rough 142
- Sea 142
- Strawberry *22,* 141
- Subterranean 142
- Suffocated 141
- Sulphur *17,* **142**, 142
- White 140
- Zig-zag 141

Clubmoss, Common *20, 51*
- Fir *20, 51*
- Krauss's *51*
- Marsh *20, 51*

Club-rush, Bristle **226**, 226
- Common 226
- Floating 226
- Grey 226
- Sea 226
- Slender 226
- Wood 225

Cochlearia
- **anglica** 101
- **danica** *24, 30,* **102**, 102
- **officinalis** 102

Cock's-foot Grass 240
Cockspur-grass *24,* 253
Cockspur-thorn, Broad-leaved 132
Codlins and Cream 145
Coelocaulon aculeatum *26, 29*
Coeloglossum viride *43,* 262
Coincya monensis 105
Colchicum autumnale 255
Coltsfoot 212
Columbine 63
Colutea arborescens 133
Comarum palustre *37,* 126
Comfrey, Common 168
- Creeping 169
- Hidcote 169
- Rough 168
- Russian 168

Tuberous 168
Turkish 169
White 169
Coneflower, Bristly 213
Conium maculatum 160
Conocephalum conicum 269
Conopodium majus 157
Conringia orientalis 104
Consolida
ajacis 59
orientalis 59
regalis 59
Convallaria majalis *17, 25,* **257**, 256
Convolvulus
arvensis 165
tricolor 166
Convolvulus, Trailing 166
Conyza
bonariensis 207
canadensis *23,* 207
sumatrensis 207
Cord-grass, Common **254**, 253
Small 253
Townsend's 253
Coriander 157
Coriandrum sativum 157
Corn Cockle *23,* 81
Corn Marigold **209**, 209
Cornelian Cherry 148
Cornflower *23,* 198
Perennial 198
Cornsalad, Broad-fruited 194
Common **194**, 194
Keeled-fruited 194
Narrow-fruited 194
Cornus
alba 148
mas 148
sanguinea 148
sericea 148
stolonifera 148
Coronilla
scorpioides 134
varia 135
Coronopus
didymus 104
squamatus 104
Corrigiola littoralis 79
Cortaderia selloana 253
Corydalis
bulbosa 65
cava 65
claviculata *17,* 66
lutea 65
ochroleuca 65
solida 65
Corydalis, Climbing *17,* 65
Pale 65
Yellow 65
Corylus
avellana 70
maxima 70
Corynephorus canescens *29,* 243
Cosmea 214
Cosmos 214
Cosmos bipinnatus 214
Cotoneaster
bullatus 132
dielsianus 132
divaricatus 132
franchetii 132
frigidus 132
hjelmqvistii 132
horizontalis 132
integrifolius 132
lacteus 132
microphyllus 132
salicifolius 132
simonsii 132
sternianus 132
x watereri 132
Cotoneaster, Diel's 132
Entire-leaved 132
Franchet's 132
Himalayan 132
Hjelmqvist's 132
Hollyberry 132
Late 132
Spreading 132
Stern's 132
Tree 132
Wall 132
Waterer's 132
Willow-leaved 132
Cotton Thistle 198
Cotton-grass, Broad-leaved **225**, 225
Common *20,* 224
Hare's-tail 225
Slender 225
Cotula coronopifolia 210
Couch, Bearded 251
Common *24,* 251
Sand 251
Sea 251
Couch-grass, Common 251
Cow Mumble 162
Cow Parsley 156
Cow Parsnip 162
Cowbane **161**, 161
Cowberry 107
Cowherb 82
Cowslip 108
Japanese 108
Cow-wheat, Common 186
Crested *17,* 186
Field 186
Crambe maritima *22, 39,* **105**, **106**, 105
Cranberry *20,* 107
Cranesbill, Bloody 153
Caucasian 153
Cut-leaved 153
Dove's-foot 154
Druce's 153
Dusky 154
French 153
Hedgerow 154
Knotted 153
Long-stalked **153**, 153
Meadow 153
Pencilled 153
Purple 153
Rock 154
Round-leaved 153
Shining 154
Small-flowered 154
Crassula
helmsii *23,* 112
tillaea *21,* **111**, 111
Crataegus *26*
laevigata 133
monogyna 132
oxyacanthoides 133
persimilis 132
prunifolia 132
x macrocarpa 133
x media 133
Cratoneuron
commutatum 283
filicinum 283
Creeping Jenny 109
Crepis
biennis 202
capillaris *29, 30,* 202
foetida 203
nicaeensis 202
setosa 203
taraxacifolia 203
tectorum 202
vesicaria 203
Cress, Garden 103
Hoary 104
Crested Hair-grass **243**, 243
Crithmum maritimum 158
Crocosmia x crocosmiiflora 260

Crocus
- **chrysanthus** 260
- *flavus* 260
- *purpureus* 260
- **tommasinianus** 260
- **vernus** 260
- **x stellaris** 260

Crocus, Early 260
- Spring 260
- Yellow 260

Crosswort 192
Crowberry 106
Crowfoot, Corn 60
- Meadow 60

Crowfoot, see also Water Crowfoot 61
Crown Vetch 135
Cruciata
- *chersonensis* 192
- **laevipes** 192

Cryphaea heteromalla 282
Cryptomeria japonica 56
Cryptothallus mirabilis *44,* 269
Ctenidium molluscum 287
Cuckoo Flower 100
Cuckoo-pint **220**, 220
Cucubalus baccifer 82
Cucumber 93
Cucumis
- **melo** 93
- **sativa** 93

Cucurbita pepo 93
Cudweed, Broad-leaved 204
- Common 204
- Heath 205
- Jersey **205**, 205
- Marsh 205
- Red-tipped 204
- Small 204

Cumberland Turf 236
Cupressus macrocarpa 56
Curly Waterweed 216
Currant, Black 111
- Mountain 111
- Red 111

Cuscuta
- **campestris** 166
- **epilinum** *43,* 166
- **epithymum** 166
- **europaea** *43,* 166
- **trifolii** *43,* 166

Cyclamen 109
Cyclamen
- **coum** 109
- **hederifolium** 109
- *neapolitanum* 109

Cydonia oblonga 131
Cymbalaria muralis *23,* 182
Cynara cardunculus 198
Cynodon dactylon 253
Cynoglossum officinale 170
Cynosurus
- **cristatus** *26,* 238
- **echinatus** 238

Cyperus
- **eragrostis** 227
- **longus** 227

Cystopteris fragilis 54
Cytisus
- **scoparius** 143
- **striatus** 143

D

Dactylis glomerata 240
Dactylorhiza
- **fuchsii** *22,* 262
- **incarnata** *22, 29, 30,* **263**, 263
- **maculata** 262
- *pardalina* 262
- **praetermissa** **263**, 263
- **traunsteineri** *28,* 264
- **x grandis** 262
- **x hallii** 262
- **x kerneriorum** 262
- **x wintoni** 263

Daffodil, Wild 259
- Garden 259

Daisy, Common 207
Daisy-bush 207
Dame's Rocket 98
Dame's Violet *23,* 98
Damson 130
Dandelion 202
Danewort 193
Danthonia decumbens 252
Daphne
- **laureola** 145
- **mezereum** 145

Darnel 237
Datura
- **stramonium** 165
- *tatula* 165

Daucus carota 162
Deadly Nightshade *23,* 164
Dead-nettle, Cut-leaved 172
- Red 172
- Spotted 172
- White 172

Deer-grass 225
Delphinium ambiguum 59
- *orientale* 59

Deptford Pink 83
Deschampsia
- **cespitosa** 243
- **flexuosa** *27,* **243**, 243
- **setacea** *43,* 243

Descurainia sophia 97
Desmazeria marina 240
- *rigida* 240

Devil's Cherry 164
Devil's Guts 168
Devil's-bit Scabious 195
Dewberry 125
Dianthus
- **armeria** 83
- **barbatus** 83
- **deltoides** 83
- *plumarius* 83
- **superbus** 83

Dicranella
- **cerviculata** 273
- **heteromalla** 273
- **rufescens** *44,* 273
- **schreberiana** 273
- **staphylina** 273
- **varia** 273

Dicranoweisia cirrata 273
Dicranum
- **bonjeanii** 273
- **fuscescens** 273
- **majus** 273
- **montanum** 273
- **polysetum** 273
- **scoparium** *30,* 273
- **spurium** *44,* 273
- **tauricum** 273

Didymodon
- **fallax** 276
- **ferrugineus** *44,* 276
- **insulanus** 275
- **luridus** 275
- **rigidulus** 275
- **sinuosus** 276
- **tophaceus** 276
- **umbrosus** 275
- **vinealis** 275

Digitalis purpurea 183

Digitaria
- *adscendens* 254
- **ciliaris** 254
- **ischaemum** 254
- **sanguinalis** 254

Diplophyllum
- **albicans** 267
- **obtusifolium** 267

Diplotaxis
- **erucoides** 104
- **muralis** 104
- **tenuifolia** 104

Dipsacus
- **fullonum** 195
- **pilosus** **195**, 195
- *sylvestris* 195

Ditrichum
- **cylindricum** 272
- **flexicaule** 272
- **heteromallum** *44*, 272

Dittander 104
Dock, Broad-leaved 86
- Clustered 85
- Curled 85
- Fiddle 86
- Golden 86
- Great Water 85
- Marsh 86
- Shore 85
- Wood 86

Dodder, Clover 166
- Common **166**, 166
- Flax 166
- Great 166
- Yellow 166

Dog's-tail Grass, Crested 238
- Rough 238

Dog's-tooth-violet 255
Dog-violet, Common 91
- Early 92
- Heath 92

Dogwood 148
- Red Osier 148
- White 148

Doronicum
- **columnae** 212
- **pardalianches** 212

Douglas Fir *39*, 57

Drepanocladus
- **aduncus** 283
- **cossonii** 284
- *exannulatus* 283
- *fluitans* 283
- **lycopodioides** 284
- *revolvens* 284
- **sendtneri** *44*, 284
- *uncinatus* 284
- *vernicosus* 284

Dropwort 114

Drosera *20*
- **anglica** *29*, 91
- **intermedia** 91
- *longifolia* 91
- **rotundifolia** *22, 26-29*, 90
- **x beleziana** 91
- **x obovata** 91

Dryopteris
- **affinis** 55
- *austriaca* 56
- *borreri* 55
- **carthusiana** *26*, 56
- **cristata** *26, 29*, **55**, 55
- **dilatata** 26, 56
- **filix-mas** 55
- *lanceolata-cristata* 56
- *pseudomas* 55
- *spinulosa* 56
- **x deweveri** 56
- **x uliginosa** 56

Duchesnea indica 127
Duckweed, Common 221
- Fat 221
- Greater 220
- Ivy-leaved 221
- Least *23*, 221

Duke of Argyll's Tea-plant 164
Dutch Rush 51
Dyer's Greenweed *17*, **143**, 143
Dyer's Rocket 106

E

Earthnut 157
Eccremocarpus scaber 188

Echinochloa
- **colona** *24*, 254
- **crus-galli** *24*, 253
- **esculenta** 254
- **frumentacea** 254
- *utilis* 254

Echinops
- *commutatus* 196
- **exaltatus** 196
- **sphaerocephalus** 196

Echium vulgare **168**, 168

Eel-grass 219
- Dwarf 219
- Narrow-leaved 219

Elder *39*, 192
- Dwarf 193
- Red-berried 192

Elecampane 205

Eleocharis
- **multicaulis** 225
- **palustris** 225
- **quinqueflora** 225
- **uniglumis** 225

Eleogiton fluitans 226
Elephants' Ears 113
Elm *39*
- East Anglian **67**, 67
- English 67
- Huntingdon 67
- Wych 66

Elodea
- **canadensis** *23*, 216
- **nuttallii** *23*, 216

Elymus
- *arenarius* 251
- **caninus** 251
- *farctus* 251
- *pycnanthus* *29, 30*, 251
- *repens* 251

Elytrigia
- **atherica** *29, 30*, 251
- **juncea** *29*, 251
- **repens** *24*, 251
- **x laxa** 251
- **x obtusiuscula** 251
- **x oliveri** 251

X Elytrordeum langei 251
Empetrum nigrum 106

Encalypta
- **streptocarpa** 274
- **vulgaris** 274

Enchanter's Nightshade 148

Endymion
- *hispanicus* 257
- *non-scriptus* 257

Entosthodon
- **fascicularis** *44*, 278
- **obtusus** *44*, 278

Ephemerum
- **recurvifolium** 278
- **serratum** 278

Epilobium
- *adenocaulon* 147
- *adnatum* 146
- *angustifolium* 147
- **ciliatum** *23*, 147
- **hirsutum** 145

(Epilobium)
- *lamyi* 146
- lanceolatum 146
- montanum 146
- montanum x E. ciliatum 146
- obscurum 146
- obscurum x E. ciliatum 146
- palustre 147
- parviflorum 146
- parviflorum x E. ciliatum 146
- roseum 146
- tetragonum 146
- x dacicum 146
- x limosum 146
- x novae-civitatis 146
- x palatinum 146
- x rivulare 146
- x schmidtianum 146
- x semiobscurum 146
- x subhirsutum 145

Epipactis
- helleborine 261
- palustris *22, 28, 29,* **261**, 260
- phyllanthes 261
- purpurata *43,* 260

Equisetum
- arvense 51
- fluviatile 51
- hyemale *43,* 51
- palustre *30,* 51
- sylvaticum 51
- telmateia 51
- variegatum 51

Eranthis hyemalis 59

Erica
- cinerea *19, 20, 27,* 107
- tetralix *22, 27,* 107

Erigeron
- acer 207
- *canadensis* 207
- glaucus 207
- karvinskianus 207
- *mucronatus* 207

Erinus alpinus 183

Eriophorum
- angustifolium *20, 29,* 224
- gracile *40, 43,* 225
- latifolium **225**, 225
- vaginatum *20,* 225

Erodium
- cicutarium *26,* **155**, 154
- *glutinosum* 154
- *lebelii* 154
- moschatum 154

Erophila
- glabrescens 101
- majuscula 101
- *praecox* *43,* 101
- verna *29,* 101

Erucastrum gallicum 105

Eryngium
- maritimum *29,* **156**, 156
- planum 156

Eryngo 156

Erysimum
- cheiranthoides 98
- cheiri *23,* 98
- x marshalii **36**, 98

Erythraea pulchella -
- see Centaurium pulchellum 35

Erythronium dens-canis 255

Escallonia 111

Escallonia macrantha 111

Eschscholzia californica 65

Eucladium verticillatum 274

Euonymus
- europaeus 148
- japonicus 148

Eupatorium cannabinum *28,* 214

Euphorbia
- amygdaloides *17,* 150
- characias 150
- cyparissias 150
- esula 150
- exigua 150
- helioscopia 149
- lathyris 149
- paralias 150
- peplus 150
- platyphyllos 149
- *robbiae* 150
- x pseudovirgata 150

Euphrasia
- anglica 186
- confusa 187
- micrantha 187
- nemorosa 186
- officinalis 186
- pseudokerneri 186
- rostkoviana 187

Eurhynchium
- hians 285
- praelongum 285
- pumilum 285
- schleicheri *44,* 285
- speciosum 286
- striatum 285
- *swartzii* 285

Evening Primrose, Common 147
- Fragrant 147
- Intermediate 147
- Large-flowered 147
- Small-flowered **147**, 147

Everlasting Flower 205

Eyebright 186

F

Factory Weed *75*

Fagopyrum
- esculentum 84
- tataricum 84

Fagus sylvatica 68

Fairy Flax 151

Fairy Foxglove 183

Falcaria vulgaris 161

Fallopia
- *aubertii* 85
- baldschuanica 85
- convolvulus 85
- japonica *23,* 84
- sachalinensis *23,* 85

False Acacia 133

False Grass-poly 145

False Oxlip 108

Fat Hen 71

Felicia petiolata 206

Fennel *23,* 159

Fen Sedge, Great *22, 28, 29,* **227**, 227

Fenugreek 139

Fern-grass **240**, 240
- Sea 240

Ferret's-tail 58

Fescue, Bearded 238
- Dune 237
- Fine-leaved Sheep's 237
- Giant 236
- Hard 237
- Hybrid 237
- Mat-grass 238
- Meadow 235
- Rat's-tail 237
- Red *24,* 236
- Rush-leaved 236
- Sheep's **236**, 236
- Squirrel-tail 237
- Tall 235
- Various-leaved 236

Festuca
- **arenaria** 236
- **arundinacea** 235
- **arundinacea x Lolium multiflorum** 237
- **brevipila** *24,* 237
- **filiformis** 237
- **gigantea** 236
- **heterophylla** 236
- *juncifolia* 236
- **ovina** *26, 30,* **236**, 236
- **pratensis** 235
- **rubra** *24, 29, 30,* 236
- **rubra x Vulpia myuros** 237
- *tenuifolia* 237
- **x schlickumii** 235

X Festulolium
- **braunii** 237
- **holmbergii** 237
- **loliaceum** 237

X Festulpia 237

Feverfew *23,* 207

Fiddleneck, Common 169
- Scarce 169

Field Bean 136

Field Madder 190

Figwort, Balm-leaved 181
- Common 181
- Green **181**, 181
- Water 181
- Yellow 181

Filago
- *apiculata* 204
- *germanica* 204
- **lutescens** 204
- **minima** 204
- **pyramidata** 43, 204
- *spathulata* 204
- **vulgaris** 204

Filbert 70

Filipendula
- **ulmaria** *28,* 115
- **vulgaris** *26,* 114

Finger-grass, Hairy 254
- Smooth 254
- Tropical 254

Fir, Grand 57
- Greek 57
- Silver *39*

Firethorn 132

Fissidens
- **adianthoides** 274
- **bryoides** 274
- **crassipes** 274
- *cristatus* 274
- **dubius** 274
- **exilis** 274
- **gracilifolius** 274
- **incurvus** 274
- **pusillus** 274
- **taxifolius** 274
- **viridulus** 274

Flag, Yellow *23,* 260

Flat Sedge 227

Flax 151
- Fairy 151
- Pale 151
- Perennial 151

Fleabane, Argentinian 207
- Blue 207
- Canadian *23,* 207
- Common 205
- Guernsey 207
- Mexican 207
- Small 205

Flixweed 97

Flower of an Hour 90

Flowering Currant 111

Flowering Rush **214**, 214

Fluellen, Round-leaved 182
- Sharp-leaved 182

Foeniculum vulgare *23,* 159

Fontinalis antipyretica 282

Fool's Parsley 159

Forget-me-not, Changing 170
- Creeping 170
- Early 170
- Field 170
- Giant 169
- Tufted 170
- Water 169
- Wood 170

Forsythia 179

Forsythia x intermedia 179

Fossombronia
- **foveolata** 268
- **pusilla** 268
- **wondraczekii** 268

Fox and Cubs 203

Foxglove 183

Fox-tail 178

Foxtail Bulbous 247
- Marsh 247
- Meadow 247
- Orange **247**, 247

Fragaria
- **moschata** 127
- *muricata* 127
- **vesca** 127
- **x ananassa** 127

Frangula alnus **150**, 150

Frankenia laevis *22, 30,* **93**, 93

Fraxinus excelsior *26,* 179

French Marigold 214

Fringed Waterlily *22,* 167

Fritillaria
- **meleagris** 256
- **pyrenaica** 256

Fritillary, Pyrenean 256
- Snakeshead 256

Frogbit **215**, 215

Frullania
- **dilatata** 268
- **tamarisci** 268

Fuchsia 147

Fuchsia magellanica 147

Fumaria
- **bastardii** 66
- **capreolata** 66
- **densiflora** 66
- *micrantha* 66
- **muralis** 66
- **officinalis** 66
- **parviflora** 66
- **vaillantii** 66

Fumitory, Climbing *17,* 65, 66
- Common 66
- Common Ramping 66
- Dense-flowered 66
- Few-flowered 66
- Fine-leaved 66
- Tall Ramping 66
- White Ramping 66

Funaria
- *fascicularis* 278
- **hygrometrica** 278
- *obtusa* 278

Furze 144

G

Gagea lutea *25,* 255

Gaillardia x grandiflora 214

Galanthus
- **elwesii** 259
- **ikariae** 259
- **nivalis** 259
- **nivalis x G. elwesii** 259
- **nivalis x G. plicatus** 259
- **plicatus** 259

Galega officinalis 133

Galeobdolon argentatum 172
- *luteum* 172

Galeopsis
angustifolia 172
bifida 173
speciosa **173**, 172
tetrahit 173
Galingale 227
Fale 227
Galinsoga
ciliata 213
parviflora 213
quadriradiata 213
Galium
album 192
aparine 192
cruciata 192
mollugo 192
odoratum **191**, 191
palustre *28*, 191
parisiense 192
saxatile *27*, 192
tricornutum 192
uliginosum 191
verum *26*, *29*, **192**, 191
x pomeranicum 192
Gallant Soldier 213
Gantweed 210
Gardeners' Garters *23*, 244
Garlic Mustard 97
Garlic, Crow 258
Few-flowered 258
Keeled 258
Neapolitan 258
Wild **258**, 258
Three-cornered 258
Gastridium ventricosum 246
Genista
anglica 143
monspessulana 143
tinctoria *17*, **143**, 143
Gentian, Autumn *22*, 163
Field 163
Marsh *23*, **163**, 163
Gentiana pneumonanthe
20, *27*, *37*, **163**, 163
Gentianella
amarella *22*, *26*, 163
campestris 163
Geranium
columbinum **153**, 153
dissectum 153
endressii 153
ibericum 153
lucidum 154
macrorrhizum 154
molle 154
nodosum 153
phaeum 154
pratense 153
pusillum 154
pyrenaicum 154
robertianum 154
rotundifolium 153
sanguineum 153
versicolor 153
x magnificum 153
x oxonianum 153
Germander, Wall 174
Water 174
Geum
rivale 127
urbanum 127
x intermedium 127
Giant Rhubarb 144
Gladdon 255, 260
Glasswort, Common 73
Glaucous 73
Long-spiked 74
One-flowered 73
Perennial 73
Purple 73
Shining 74
Yellow 74
Glaucium
corniculatum 65
flavum *22*, *30*, **64**, 65
Glaux maritima *30*, 110
Glechoma hederacea 174
Globe Artichoke 198
Globe thistle, Glandular 196
Tall 196
Glory of the Snow 257
Glyceria
declinata 241
fluitans 241
maxima 241
notata 241
plicata 241
x pedicellata 241
Gnaphalium
luteoalbum *29*, **205**, 205
sylvaticum 205
uliginosum 205
Goat's Rue 133
Goat's-beard 200
Godetia 147
Gold of Pleasure 102
Golden Glow 212
Golden Saxifrage, Alternate-leaved 114
Opposite-leaved 114
Goldenrod 206
Canadian 206
Early 206
Goldilocks 61
Golds 209
Good King Henry 70
Goodyera repens *29*, 261
Gooseberry 111
Goosefoot, Fig-leaved 71
Many-seeded 70
Many-spiked 70
Maple-leaved 71
Nettle-leaved 71
Oak-leaved 70
Probst's 71
Red 70
Stinking 71
Swedish 71
Upright 71
Goosegrass 192
Gorse, Common *19*, 144
Dwarf 144
Western *19*, *20*, 144
Gourd 93
Goutweed 158
Grape Vine 151
Grape-hyacinth 257
Garden 258
Tassel 258
Grass of Parnassus 114
Grass-weed 219
Great Fen-sedge 227
Greater Celandine 65
Greater Honeywort 168
Grey Hair-grass 243
Grimmia
pulvinata 277
trichophylla 277
Groenlandia densa 219
Gromwell, Common 167
Field 168
Ground Elder *23*, 158
Ground Ivy 174
Ground Pine 174
Groundsel 211
Heath 211
Sticky 211
Guelder Rose 193
Guizotia abyssinica 212
Gunnera tinctoria 144
Gymnadenia conopsea *28*, 262
Gymnocarpium
dryopteris 54
robertianum 54
Gymnocolea inflata 267
Gypsywort 176
Gyroweisia tenuis 275

H

- **Hainardia cylindrica** 241
- Hair-grass, Bog 243
  - Crested **242**, 243
  - Early *30*, 244
  - Grey 243
  - Silver 244
  - Tufted 243
  - Wavy **243**, 243
- Hairy Rocket 105
- *Halimione pedunculata* 73
  - *portulacoides* 72
- **Hamatocaulis vernicosus** *44*, 284
- **Hammarbya paludosa** *29*, 261
- Hard Fern 56
- Hard-grass 241
  - Curved 241
  - One-glumed 241
- Hardhack 114
- Hardheads 198
- Harebell 190
- Hare's-ear Mustard 104
  - Slender *22*, 160
- Hare's-tail 246
- Hart's-tongue Fern 53
- Hawkbit, Autumn 199
  - Lesser 199
  - Rough 199
- Hawk's-beard, Beaked 203
  - Bristly 203
  - French 202
  - Narrow-leaved 202
  - Rough 202
  - Smooth 202
  - Stinking 203
- Hawkweed 203
  - Mouse-ear 203
  - Shrubby 203
- Hawthorn 132
  - Midland 133
- Hazel *16, 25, 39*, 70
- Heath, Cross-leaved *22*, 107
- Heather *19, 22*, **107**, 107
- Heath-grass 252
- **Hebe**
  - **brachysiphon** 186
  - **parviflora** 186
  - **parviflora x H. salicifolia** 186
  - **salicifolia** 186
  - Hooker's 186
- **Hedera helix** 155
- Hedge Mustard 97
- Hedge Parsley, Knotted 162
  - Spreading 162
  - Upright 162
- **Hedwigia stellata** 282
- **Helianthemum**
  - *chamaecistus* 91
  - **nummularium** *19, 26*, **91**, 91
- **Helianthus**
  - **annuus** 213
  - **decapetalus** 213
  - **tuberosus** 213
  - **x laetiflorus** 213
- **Helichrysum bracteatum** 205
- **Helictotrichon**
  - **pratense** *26*, 242
  - **pubescens** *26*, 242
- Hellebore, Green 59
  - Stinking 59
- Helleborine, Broad-leaved 261
  - Green-flowered 261
  - Marsh *22*, **261**, 260
  - Violet 260
- **Helleborus**
  - **foetidus** *38*, 59
  - **viridis** 59
- *Helxine soleirolii* 68
- **Hemizonia pungens** 212
- Hemlock 160
- Hemp 67
- Hemp Agrimony 214
- Hemp-nettle, Bifid 173
  - Common 173
  - Greater **173**, 172
  - Red 172
- Henbane **164**, 164
- Henbit 172
- **Hennediella heimii** 277
- **Heracleum**
  - **mantegazzianum** *23*, 162
  - **sphondylium** 162
  - **villosum** 162
- Herb Bennet 127
- Herb Paris *16*, 257
- Herb Robert 154
- **Herminium monorchis** *43*, 262
- **Herniaria**
  - **glabra** *20*, **79**, 79
  - **hirsuta** 80
- **Herzogiella seligeri** 286
- **Hesperis matronalis** *23*, 98
- **Hibiscus trionum** 90
- **Hieracium**
  - **acuminatum** 204
  - *anglorum* 204
  - *aurantiacum* 203
  - *brunnocroceum* 203
  - **cheriense** 204
  - **diaphanum** 204
  - **eboracense** 204
  - **exotericum** 204
  - *grandidens* 204
  - *lachenalii* 204
  - **lanatum** 204
  - **maculatum** 204
  - *perpropinquum* 203
  - *pilosella* *26*, 203
  - **sabaudum** 203
  - **scotistichum** 204
  - **speluncarum** 204
  - *strumosum* 204
  - **trichocaulon** 204
  - *tridentatum* 204
  - *tunbridgense* 204
  - **umbellatum** 204
  - **vagum** **203**, 203
  - **vulgatum** 204
- Himalayan Giant 120
- **Himantoglossum hircinum** *43*, 264
- **Hippocrepis comosa** *26*, **135**, 135
- **Hippophae rhamnoides** *29*, 144
- **Hippuris vulgaris** *29*, **178**, 178
- **Hirschfeldia incana** 105
- Hoary Mustard 105
- Hogweed 162
  - Giant *23*, 162
- **Holcus**
  - **lanatus** *29*, 243
  - **mollis** 243
- Hollow-root 65
- Holly *39*, 149
  - Highclere 149
- Hollyhock 90
- Holly-leaved Naiad 219
- **Holosteum umbellatum** *35, 43, 77*
- **Homalia trichomanoides** 282
- **Homalothecium**
  - **lutescens** 284
  - *nitens* 284
  - **sericeum** 284
- **Honckenya peploides** *29, 30*, 76
- Honesty *23*, 100
- Honeysuckle 193
  - Fly 193
  - Garden 193
  - Himalayan 193
  - Japanese 193
  - Perfoliate 193
  - Wilson's 193
- **Hookeria lucens** 282
- Hop 67

Hordeum
- **distichon** 251
- **jubatum** **252**, 252
- **marinum** 252
- **murinum** 252
- **secalinum** 252
- **vulgare** 251

Horehound, Black 171
Hornbeam *16, 25, 39,* 69
Horned-poppy, Red 65
- Yellow *22,* **64**, 65

Hornwort, Rigid 58
- soft 58

Horse Bean 136
Horse Chestnut *39,* 151
- Red 151

Horse-radish 99
Horseshoe Vetch *26,* **135**, 135
Horsetail, Field 51
- Great 51
- Marsh 51
- Rough 51
- Variegated 51
- Water 51
- Wood 51

Hottonia palustris **109**, 109
Hound's Tongue 170
Houseleek 112
- Cobweb 112

Hulver 149
Humulus lupulus 67
Huperzia selago *43,* 51
Hyacinth, Garden 257
Hyacinthoides
- **hispanica** 257
- **non-scripta** *25,* 257
- **x variabilis** 257

Hyacinthus orientalis 257
Hydnum auriscalpum *35*
- imbricatum *36*

Hydrocharis morsus-ranae *29,* **215**, 215
Hydrocotyle vulgaris *28,* 156
Hygrohypnum luridum 284
Hylocomium splendens 287
Hyoscyamus niger **164**, 164
Hypericum
- **androsaemum** 87
- **calycinum** 87
- **elodes** *28,* 89
- **hircinum** 87
- **hirsutum** 89
- **humifusum** **88**, 88
- **maculatum** 88
- **montanum** 89
- **perforatum** 88
- **pulchrum** 89
- **tetrapterum** 88
- **x desetangsii** 88
- **x inodorum** 88

Hypnum
- **andoi** 287
- **cupressiforme** *29,* 286
- **imponens** 287
- **jutlandicum** 287
- **lacunosum** 287
- *mammilatum* 287
- **resupinatum** 287

Hypochaeris
- **glabra** *20, 21,* **199**, 199
- **maculata** *43,* 199
- **radicata** 199
- **x intermedia** 199

Hyssop 175
Hyssopus officinalis 175

I

Iberis
- **amara** *43,* 103
- **sempervirens** 103
- **umbellata** 103

Ice Plant 112
Ilex
- **aquifolium** *25,* 149
- **x altaclerensis** 149

Impatiens
- **capensis** *23,* 155
- **glandulifera** *23,* 155
- **parviflora** 155

Indian Poke *70*
Inula
- *conyza* 205
- **conyzae** 205
- **helenium** 205

Ipheion 258
Ipheion uniflorum 258
Ipomoea hederacea 166
Iris
- **foetidissima** 260
- **germanica** 259
- **latifolia** 260
- **pseudacorus** *23,* 260
- **sibirica** 259

Iris, Bearded 259
- Siberian 259
- Stinking 260
- Yellow *23,* 260

Isatis tinctoria 97
Isolepis
- **cernua** 226
- **setacea** **226**, 226

Isopterygium elegans 286
Isothecium
- **alopecuroides** 284
- **myosuroides** 284
- *myurum* 284

Iva xanthiifolia 212
Ivy 155
Ivy-leaved Toadflax *23,* 182

J

Jack by the Hedge 97
Jacob's Ladder 167
Japanese Lantern 164
Jasione montana 190
Jasmine, Winter 179
Jasminum nudiflorum 179
Jerusalem Artichoke 213
Jewel-weed 155
Johnson-grass 254
Juglans regia 68
Juncus *30*
- **acutiflorus** 223
- **acutus** 223
- **ambiguus** 222
- **articulatus** 222
- **bufonius** 222
- **bulbosus** 223
- **compressus** *22,* 222
- **compressus x J. gerardii** 222
- **conglomeratus** 224
- *dudleyi* 221
- **effusus** 223
- **gerardii** *22,* 222
- **inflexus** 223
- *kochii* 223
- **maritimus** *30,* 223
- **squarrosus** 221
- **subnodulosus** *28, 29,* **222**, 222
- **tenuis** 221
- **x diffusus** 223
- **x surrejanus** 223

Juneberry 132
Jungermannia
- **gracillima** 267
- **hyalina** 267

Juniper *57*

Juniperus communis *57*

K

Kenilworth Ivy 182
Kerria 115
Kerria japonica 115
Kickxia
  elatine 182
  spuria 182
Kidney Vetch **134**, 134
Kingcup 59
Kiwi Fruit 87
Knapweed, Common 198
  Greater 198
Knautia arvensis 195
Knawel, Annual *79*
  Perennial *79*
Kniphofia
  praecox 255
  uvaria *39*, 255
Knotgrass 84
  Cornfield 84
  Equal-leaved 84
  Ray's 84
Knotweed, Bell-flowered 83
  Giant *23*, 85
  Himalayan 83
  Japanese *23*, 84
Kochia densiflora 71
  scoparia 71
Koeleria macrantha **242**, 243
Kohlrauschia prolifera 82
Kurzia
  pauciflora 266
  sylvatica 266

L

Laburnum 143
Laburnum anagyroides 143
Lace-weed 219
Lactuca
  saligna 201
  sativa 201
  serriola *23*, 201
  virosa **201**, 201
Ladies' Fingers 134
Ladies' Smock 100
Lady Fern 54
Lady's Mantle 128
  Garden 128
Lady's Slipper 134
Lady's-tresses, Autumn 261
  Creeping 261
Lagarosiphon major 216
Lagurus ovatus 246
Lamb's-ear 171
Lamiastrum galeobdolon *25*, **172**, 172
Lamium
  album 172
  amplexicaule 172
  hybridum 172
  maculatum 172
  purpureum 172
Lamb's Succory 199
Lapsana communis 199
Larch *39*
  European *57*
  Hybrid *57*
  Japanese *57*
Larix
  decidua *57*
  kaempferi *57*
  x eurolepis *57*
  x marschlinsii *57*
Larkspur *59*
Larkspur-forking *59*
Lathraea clandestina 188
Lathyrus
  aphaca 138
  grandiflorus 137
  heterophyllus 137
  hirsutus 138
  japonicus *39*, 137
  latifolius 137
  nissolia *22*, *26*, 138
  odoratus 138
  palustris *22*, **137**, 137
  pratensis 137
  sativus 138
  sylvestris 137
  tuberosus 137
Laurel, Cherry *39*, 131
  Portugal *39*, 131
Laurestinus 193
Laurus nobilis *57*
Lavandula angustifolia 177
Lavatera
  arborea 90
  olbia 90
  thuringiaca 90
Lavender 177
Lawson's Cypress *57*
Legousia hybrida 190
Leiocolea
  badensis 267
  rutheana 267
  turbinata 267
Lemna
  gibba 221
  minor 221
  minuscula 221
  minuta *23*, 221
  polyrhiza 220
  trisulca 221
Lemon-scented Fern 53
Lens culinaris 136
Lentil 136
Leontodon
  autumnalis 199
  hispidus 199
  saxatilis 199
  taraxacoides 199
Leonurus cardiaca 171
Leopard's-bane 212
  Eastern 212
Lepidium
  campestre 103
  draba 104
  heterophyllum 103
  latifolium 104
  ruderale 103
  sativum 103
  smithii 103
Lepidozia reptans 266
Leptobarbula berica 276
Leptobryum pyriforme 278
Leptodictyum riparium 283
Leptodontium flexifolium *44*, 275
Leskea polycarpa 282
Lettuce, Garden 201
  Greater Prickly **201**, 201
  Least 201
  Prickly *23*, 201
Leucanthemella serotina 209
Leucanthemum
  vulgare *26*, 210
  x superbum 210
Leucobryum glaucum *25*, **274**, 274
Leucodon sciuroides 282
Leucojum
  aestivum 258
  vernum 259

Leycesteria formosa 193
Leymus arenarius *29,* 251
Ligustrum
  ovalifolium 179
  vulgare *29,* 179
Lilac 179
Lilium
  martagon 256
  pyrenaicum 256
Lily, Martagon 256
  Pyrenean 256
Lily-of-the-valley *17,* **257**, 256
Lime *39*
  Common 89
  Hybrid 89
  Large-leaved 89
  Small-leaved *16, 17, 25,* 89
Limestone Fern 54
Limnanthes douglasii 155
Limonium
  bellidifolium *22, 30,* **87**, 87
  binervosum *30,* 87
  humile 86
  vulgare *30,* 86
Limosella aquatica 182
Linaria
  maroccana 183
  purpurea 183
  repens 183
  vulgaris 183
Linden 89
Ling *19,* 107
Linseed 151
Linum
  anglicum 151
  bienne 151
  catharticum *26,* 151
  perenne *43,* 151
  radiola see Radiola linoides *37,* 151
  usitatissimum 151
Liparis loeselii *18, 28, 36,* **36**, 261
Listera ovata 261
Lithospermum
  arvense 168
  officinale 167
Littorella uniflora *29,* 179
Live-long 112
Lobelia
  erinus 190
  siphilitica 190
Lobelia, Garden 190
Lobularia maritima 101
Loddon Lily 258
Lolium
  multiflorum 237
  multiflorum x temulentum 237
  perenne 237
  temulentum 237
  x boucheanum 237
London Plane 66
London Pride 113
London Rocket 96
  False 96
Longleaf 161
Lonicera
  caprifolium 193
  japonica 193
  nitida 193
  periclymenum 193
  x americana 193
  x italica 193
  xylosteum 193
Loosestrife, Dotted 109
  Purple **145**, 145
  Yellow 109
Lophocolea
  bidentata 267
  cuspidata 267
  heterophylla 267
  semiteres 267
  bicrenata 267
Lophozia
  capitata 267
  excisa 267
  incisa *44,* 267
  longiflora *44,* 266
  porphyroleuca 266
  ventricosa 266
Lords and Ladies **220**, 220
  Italian 220
Lotus
  corniculatus *24,* 134
  glaber 134
  pedunculatus 134
  tenuis 134
  uliginosus 134
Lousewort 187
  Marsh 187
Love Lies Bleeding 75
Love-in-a-mist 59
Lucerne 139
  Sand 140
Lunaria
  annua *23,* 100
  rediviva 101
Lungwort 168
Lunularia cruciata 269
Lupin, Annual Yellow 143
  Narrow-leaved 143
  Russell 143
  Tree 143
Lupinus
  angustifolius 143
  arboreus 143
  luteus 143
  polyphyllus 143
  x regalis 143
Luronium natans 215
Luzula
  campestris 224
  multiflora 224
  pilosa *16,* 224
  sylvatica *17, 25,* 224
Lychnis
  coronaria 80
  flos-cuculi 80
Lycium
  barbarum 164
  chinense 164
  halimifolium 164
Lycoperdon coliforme *37*
Lycopersicum esculentum 164
Lycopodiella inundata *20, 27, 43,* 51
Lycopodium
  clavatum *20, 43,* 51
  inundatum *20,* 51
  selago *20,* 51
Lycopsis arvensis 169
Lycopus europaeus 176
Lyme-grass 251
Lysimachia
  nemorum *26,* 109
  nummularia 109
  punctata 109
  vulgaris *26,* 109
Lythrum
  junceum 145
  portula 145
  salicaria 145

M

Madia sativa 212
Mahonia
  aquifolium *24,* 64
  x decumbens 64
Maianthemum bifolium *25,* **256**, 256
Maidenhair Fern 52
Malcolmia maritima 98
Male Fern 55
  Scaly 55

Mallow, Common 89
Dwarf 90
Least 90
Marsh *22*, **90**, 90
Musk 89
Rose 90
Small 90
Malus
domestica 131
sylvestris 131
Malva
moschata 89
neglecta 90
parviflora 90
pusilla 90
sylvestris 89
Maple, Field *16*, *25*, *39*, 152
Norway 151
Marchantia polymorpha **270**, 270
Mare's-tail **178**, 178
Marigold, Pot 212
Marjoram *19*, 175
Marram 246
Purple 246
Marrubium vulgare 173
Marsh Fern 53
Marsh Fleabane 211
Marsh Mallow *22*, **90**, 90
Hispid 90
Rough 90
Marsh Marigold 59
Marsh Orchid, Early *22*, **263**, 263
Narrow-leaved 264
Southern **263**, 263
Marsh-elder 212
Marshwort, Lesser *22*, 160
Marvel of Peru 70
Masterwort 156
Mat-grass 235
Matricaria
discoidea 210
matricarioides 210
recutita 210
Matteuccia struthiopteris 54
Matthiola
bicornis 98
incana 98
longipetala 98
May Lily **256**, 256
Mayweed, Scented 210
Scentless 210
Sea 210
Meadow Saffron 255
Meadow-foam 155
Meadow-grass, Annual 239
Broad-leaved 239
Bulbous 240
Flattened 239
Narrow-leaved 239
Rough 239
Smooth 239
Spreading 239
Swamp 239
Wood 240
Meadow-rue, Greater 63
Lesser 63
Meadowsweet 115
Meconopsis cambrica 65
Medicago
arabica 140
falcata 139
hispida 140
lupulina 139
minima **140**, 140
polymorpha 140
sativa ssp. **sativa** 139
sativa ssp. **falcata** **139**, 139
sativa ssp. **varia** 140
x varia 140
Medick, Black 139
Bur **140**, 140
Sickle **139**, 139
Spotted 140
Toothed 140
Melampyrum
arvense *43*, 186
cristatum *17*, *25*, 186
pratense 186
Melandrium rubrum 82
Melica uniflora *16*, *25*, 242
Melick, Wood 242
Melilot, Ribbed *16*, *25*, 139
Small 139
Tall 139
White 139
Melilotus
albus 139
altissimus 139
indicus 139
officinalis 139
Melissa officinalis 174
Melon 93
Mentha
aquatica *28*, 176
arvensis 176
dumetorum 176
gentilis 176
pulegium *43*, 177
rotundifolia 177
spicata 176
suaveolens 177
x citrata 176
x cordifolia 177
x gracilis 176
x niliaca 177
x piperita 176
x rotundifolia 177
x smithiana 176
x verticillata 176
x villosa *23*, 177
x villosonervata 176
Menyanthes trifoliata *28*, *29*, *37*, **167**, 167
Mercurialis
annua 149
perennis *18*, *25*, **149**, 149
Mercury, Annual 149
Dog's *18*, **149**, 149
Mertensia maritima *43*, 169
Metasequoia glyptostroboides 56
Metzgeria
fruticulosa 269
furcata 269
Mezereon 145
Michaelmas Daisy, Common 206
Confused 206
Hairy 206
Late 206
Narrow-leaved 206
Willow-leaved 206
Microbryum
curvicolle 277
davallianum 277
floerkeanum 277
rectum 277
starckeanum 277
Mignonette, Garden 106
Upright 106
White 106
Wild 106
Milium effusum *25*, 235
Milk Parsley *22*, 161
Milk Thistle 198
Milkmaids 100
Milk-vetch, Purple **133**, 133
Milkwort, Common 151
Heath 151
Millet 24
Common 253
Great 254
Japanese 254
Shama 254
White 254
Wood *25*, 235

- **Mimulus**
  - **guttatus** *23*, 181
  - **luteus** 182
  - **x robertsii** 182
- Mind-your-own-business 68
- Miners' Lettuce *75*
- Mint, Apple- *23*, *177*
  - Bowles' 177
  - Bushy 176
  - Corn 176
  - Lamb 177
  - Round-leaved 177
  - Sharp-toothed 176
  - Spear 176
  - Tall 176
  - Water 176
  - Whorled 176
- Mintweed 177
- **Minuartia hybrida** *76*
- **Mirabilis jalapa** *70*
- **Misopates**
  - **calycinum** 182
  - **orontium** 182
- Mistletoe 148
- **Mnium hornum** **280**, 280
- Mock Orange 111
- **Moehringia trinervia** *76*
- **Moenchia erecta** 78
- **Moerckia hibernica** 269
- **Molinia caerulea** *27-29* 253
- *Monerma cylindrica* 241
- Monkeyflower *23*, 181
  - Hybrid 182
- Monk's-hood 59
- **Monotropa hypopitys** *29*, **108**, 108
- Montbretia 260
- **Montia**
  - **fontana** *75*
  - *perfoliata* *75*
  - *sibirica* *75*
- Moonwort 52
- Morning-glory, Ivy-leaved 166
- Moschatel 194
- Mossy Stonecrop *21*, **111**, 111
- Mother Die 156
- Motherwort 171
- Mountain Everlasting 204
- Mountain Fern 53
- Mouse-ear, Common 78
  - Field 77
  - Little 78
  - Sea 78
  - Sticky 78
- Mousetail *23*, **63**, 63
- Mudwort 182
- Mugwort 208
  - Breckland 208
  - Chinese 208
- Mullein, Caucasian 180
  - Dark 180
  - Dense-flowered 180
  - Great 180
  - Hoary **180**, 180
  - Hungarian 180
  - Moth 180
  - Orange 180
  - Twiggy 180
  - White 181
- Mung Bean 133
- **Muscari**
  - **armeniacum** 258
  - *atlanticum* 257
  - **comosum** 258
  - **neglectum** *43*, 257
- Mustard, Black 105
  - Chinese 104
  - White 105
- **Mycelis muralis** 201
- **Mylia anomala** 267
- **Myosotis**
  - **arvensis** 170
  - *caespitosa* 170
  - **discolor** 170
  - *hispida* 170
  - **laxa** 170
  - **ramosissima** *29*, 170
  - **scorpioides** 169
  - **secunda** 170
  - **sylvatica** 170
- **Myosoton aquaticum** *29*, *78*
- **Myosurus minimus** *23*, **63**, 63
- **Myrica gale** *26*, 68
- **Myriophyllum**
  - **alterniflorum** 144
  - **aquaticum** 144
  - *brasiliense* 144
  - **spicatum** *29*, 144
  - **verticillatum** 144
- Myrobalan 130
- **Myrrhis odorata** 157

N

- Naiad, Holly-leaved 157
- **Najas marina** *29*, *39*, 219
- Naked Ladies 255
- **Narcissus**
  - **cultivars** 259
  - **poeticus** 259
  - **pseudonarcissus** 259
- Narcissus, Pheasant's-eye 259
- **Nardia**
  - **geoscyphus** *44*, 267
  - **scalaris** 267
- *Nardurus maritimus* 238
- **Nardus stricta** *27*, 235
- Narrow-leaved Pepperwort 103
- **Narthecium ossifragum** *20*, *29*, **256**, 255
- Nasturtium 155
- *Nasturtium microphyllum* 99
  - *officinale* 98
- Navelwort 112
- **Neckera**
  - **complanata** 282
  - **pumila** *44*, 282
- Needle Furze 143
- **Neottia nidus-avis** 261
- **Nepeta**
  - **cataria** 174
  - **x faassenii** 174
- **Neslia paniculata** 102
- Net-rein 229
- Nettle, Roman *67*
  - Small *67*
  - stinging *67*
- New Year's Gift 59
- New Zealand Bur 128
- New Zealand Holly 207
- New Zealand Pigmyweed *23*, 112
- New Zealand Spinach *70*
- **Nicandra physalodes** 164
- **Nicotiana**
  - **alata** 165
  - **x sanderae** 165
- **Nigella damascena** 59
- Niger 212
- Night-scented Stock 98
- Nightshade, Black 165
  - Deadly *23*, 100
  - Green 165
  - Leafy-fruited 165
  - Small 165
  - Woody 165
- Nipplewort 199
- Nit-grass 246
- Norfolk Flapwort 267
- Norway Spruce *57*
- **Nuphar**
  - **advena** 58
  - **lutea** *29*, **58**, 58
- Nuttall's Waterweed 216

Nymphaea
- **alba** *23, 29,* 58
- *x laydeckeri* 58
- *x marliacea* 58

Nymphoides peltata *22,* 167

O

Oak Fern 54
Oak *25, 39*
- Evergreen 68
- Hybrid 69
- Lucombe 68
- Pedunculate *16,* 69
- Red 69
- Sessile *11,* **68**, 68
- Turkey 68

Oat 242
- Algerian 242
- Bristle 242
- Wild 242
- Winter, Wild 242

Oat-grass, Downy 242
- False 242
- Meadow 242
- Yellow 242

Odontites vernus 187
Odontoschisma sphagni 266
Oenanthe
- **aquatica** **159**, 159
- **crocata** 158
- **fistulosa** 158
- **fluviatilis** 159
- **lachenalii** 158
- **silaifolia** 158

Oenothera
- **biennis** 147
- **cambrica** **147**, 147
- *erythrosepala* 147
- **fallax** 147
- **glazioviana** 147
- **perangusta** 147
- **stricta** 147

Old Man's Beard 60
Olearia
- **macrodonta** 207
- **x haastii** 207

Onobrychis
- **caput-gallii** 133
- **viciifolia** 133

Ononis
- **repens** 138
- **spinosa** **138**, 138

Onopordum acanthium 198
Ophioglossum vulgatum *26,* 52
Ophrys
- **apifera** **265**, 265
- **insectifera** *43,* 264
- *paludosa* - see Hammarbya paludosa *35,* 261

Orache, Babington's 72
- Common 72
- Frosted 72
- Garden 71
- Grass-leaved *24,* 72
- Halberd-leaved *24,* 71
- Kattegat 72
- Long-stalked 72
- Mealy 72
- Spear-leaved *24,* 71

Orchid, Bee **265**, 265
- Bird's-nest 261
- Bog 261
- Burnt-tip 264
- Common Spotted, *22,* 262
- Early-purple *16,* 264
- Fen *18,* 261
- Fly 264
- Fragrant 262
- Frog 262
- Greater Butterfly *16,* 262
- Green-winged *17,* **264**, 264
- Heath-spotted 262
- Lesser Butterfly *22,* 262
- Lizard 264
- Man 264
- Musk 262
- Pyramidal *22,* 262

Orchis
- *conopsea* - see Gymnadenia conopsea *37,* 262
- **mascula** *16, 25,* 264
- **morio** *17, 26,* **264**, 264
- **ustulata** *43,* 264

Oregano 175
Oregon Grape *24,* 64
- Hybrid 64

Oreopteris limbosperma 53
Origanum vulgare *19,* 175
Ornithogalum
- **angustifolium** 257
- **nutans** 257
- *umbellatum* 257

Ornithopus perpusillus 134
Orobanche
- **elatior** *37,* 188
- **minor** 188
- **purpurea** *30,* **188**, 188
- **ramosa** *43,* 188
- **rapum-genistae** *43,* 188

Orpine 112
Orthodontium lineare 278
Orthotrichum
- **affine** 281
- **anomalum** 281
- **cupulatum** 281
- **diaphanum** 281
- **lyellii** 281
- **obtusifolium** 281
- **pulchellum** 281
- **pumilum** *44,* 281
- *schimperi* 281
- **striatum** 281
- **tenellum** 281

Osier 95
- Eared 95
- Silky-leaved 95

Osmunda regalis *22, 26, 39,* 52
- *spicant* see Blechnum spicant *37,* 56

Ostrich Fern 54
Oxalis
- **acetosella** *17,* **152**, 152
- **articulata** 152
- **corniculata** 152
- *corymbosa* 152
- **debilis** 152
- *europaea* 152
- **exilis** 152
- *floribunda* 152
- **incarnata** 152
- *rosea* 152
- **stricta** 152

Oxeye Daisy 210
- Autumn 209

Oxlip 108
- False 108

Ox-tongue, Bristly 200
- Hawkweed 200

Oxycoccus palustris 107
Oxystegus sinuosus 276
Oysterplant 169

P

Palustriella commutata 283

Pampas-grass 253
Panicum *24*
  capillare 253
  miliaceum 253
Pansy, Field 93
  Garden 93
  Horned 91
  Sand *20,* **92**, 92
  Wild 92
Papaver
  argemone 65
  atlanticum 64
  bracteatum 64
  dubium 64
  hybridum 65
  lateritum 64
  orientale 64
  pseudoorientale 64
  rhoeas 64
  somniferum *23,* 64
Parapholis
  incurva 241
  strigosa 241
Parentucellia viscosa **187**, 187
Parietaria
  diffusa 68
  judaica 68
Paris quadrifolia *16, 25, 257*
Parnassia palustris *28,* 114
Parrot's Feather 144
Parsley Piert 128
  Slender 128
Parsley, Corn 160
  Garden 160
Parsnip 162
Parthenocissus
  inserta 151
  quinquefolia 151
  tricuspidata 151
Pasque Flower 59
Pastinaca sativa 162
Pea, Broad-leaved Everlasting 137
  Fyfield 137
  Garden 138
  Indian 138
  Marsh *22,* **137**, 137
  Narrow-leaved Everlasting 137
  Norfolk Everlasting 137
  Sea 137
  Sweet 138
  Tuberous 137
  Two-flowered Everlasting 137
Pear, Domestic 131
  Wild 131
Pearlwort, Annual 79
  Heath 79
  Knotted 78
  Mossy 79
  Sea 79
Pedicularis
  palustris 187
  sylvatica 187
Pellia
  endiviifolia 268
  epiphylla 268
  neesiana 268
Pellitory-of-the-wall 68
Penny-cress, Field 103
  Garlic 103
Pennyroyal 177
Pennywort 156
Pentaglottis sempervirens 169
Peplis portula *37,* 145
Pepper Saxifrage 159
Peppermint 176
  Black 176
Pepperwort, Field 103
  Smith's 103
Periwinkle, Greater 164
  Lesser 164
Persicaria
  amphibia 83
  amplexicaulis 83
  bistorta 83
  campanulata 83
  capitata 84
  hydropiper 84
  lapathifolia 83
  laxiflora 84
  maculosa 83
  minor 84
  mitis 84
  patulum 83
  pensylvanica 83
  senegalense 83
  wallichii 83
Persicaria, Pale 83
Petalophyllum ralfsii *29,* 268
Petasites
  fragrans 212
  hybridus 212
Petrorhagia prolifera *19,* **82**, 82
Petroselinum
  crispum 160
  segetum *23,* 160
Petty Whin 143
Petunia x hybrida 165
Peucedanum palustre *22, 28,* 161
Phacelia 167
Phacelia tanacetifolia 167
Phalaris
  aquatica *24* ,244
  arundinacea *23,* 244
  canariensis 244
  minor 244
  paradoxa 244
  tuberosa 244
Phascum curvicolle 277
  cuspidatum 277
  floerkeanum 277
Phaseolus aureus 133
Pheasant's-eye 63
Phegopteris robertiana 54
Philadelphus coronarius 111
Philonotis
  calcarea 281
  fontana 281
Phleum
  arenarium *30,* 248
  bertolonii 248
  phleoides *20,* **248**, 248
  pratense 248
Phragmites australis *26, 28-30,* **253**, 253
Phyllitis scolopendrium 53
Physalis
  alkekengi 164
  peruviana 164
  philadelphica 164
Physcomitrella patens 278
Physcomitrium
  eurystomum *29,* 278
  pyriforme 278
Phytolacca acinosa 70
Picea
  abies 57
  sitchensis 56
Pickerel Weed 215, 255
Picris
  echioides 200
  hieracioides 200
Pignut 157
Pigweed, Prostrate 75
  White 75
Pilewort 61
Pillwort 52
Pilosella
  aurantiaca 203
  officinarum *26, 29,* 203
Pilularia globulifera 52
Pimpernel, Blue 110
  Bog 110
  Scarlet 110
  Yellow 109
Pimpinella
  major 157
  saxifraga 157

Pine, Austrian *39, 57*
Corsican 57
Maritime 57
Scots *16*, **57**, 57
Weymouth *39, 57*
Pineapple Weed 210
Pinguicula vulgaris *20*, 189
Pink 83
Maiden 83
Proliferous *19*, **82**, 82
proliferous 82
Superb 83
Pinus
contorta 56
nigra *29, 57*
pinaster *57*
pinea 56
radiata 56
strobus *57*
sylvestris **57**, *57*
wallichiana 56
Pipperidge 64
Pirri-Pirri 128
Pisum sativum 138
Plagiochila asplenioides 268
Plagiomnium
affine 280
cuspidatum 280
elatum 280
ellipticum 280
rostratum 280
undulatum 280
Plagiothecium
cavifolium 286
curvifolium 286
denticulatum 286
latebricola 286
nemorale 286
ruthei **44**, 286
succulentum 286
undulatum **286**, 286
Plane 66
Plantago
arenaria 179
coronopus *30*, 178
indica 179
lanceolata *30*, 179
major 179
maritima *30*, 179
media 179
Plantain, Buck's-horn 178
Greater 179
Hoary 179
Rat's-tail 179
Ribwort 179
Sea 179
Platanthera
bifolia *22*, 262
chlorantha *16, 25*, 262
Platanus x hispanica 66
Platygyrium repens 286
Pleuridium
acuminatum 272
subulatum 272
Pleurochaete squarrosa 275
Pleurozium schreberi 287
Ploughman's Spikenard 205
Plum 130
Poa
angustifolia 239
annua 239
bulbosa 240
chaixii 239
chaixii x nemoralis 239
compressa 239
humilis 239
nemoralis 240
palustris 239
pratensis 239
subcaerulea 239
trivialis 239
Pogonatum
aloides 271
nanum 271
urnigerum 271
Pohlia
bulbifera 278
carnea 278
delicatula 278
lutescens 278
melanodon 278
nutans 278
proligera 278
wahlenbergii 278
Pokers 255
Pokeweed 70
Polemonium caeruleum 167
Policeman's Helmet 155
Polygala
serpyllifolia 151
vulgaris 151
Polygonatum
multiflorum 256
x hybridum 256
Polygonum
amphibium 83
amplexicaule 83
arenastrum 84
aubertii 85
aviculare 84
baldschuanicum 85
bistorta 83
campanulatum 83
capitatum 84
convolvulus 85
cuspidatum 84
hydropiper 84
lapathifolium 83
laxiflorum 84
minus 84
nodosum 83
oxyspermum 84
persicaria 83
polystachyum 83
raii 84
rurivagum 84
sachalinense 85
Polypodium
filix-mas -
see Dryopteris filix-mas *37*, 55
interjectum 52
oreopteris -
see Oreopteris limbosperma *37*, 53
vulgare *29*, 52
x mantoniae 52
Polypody, Common *29*, 52
Hybrid 52
Polypogon
monspeliensis 246
viridis 246
Polystichum
aculeatum 55
setiferum 54
x bicknellii 55
Polytrichum
alpestre 272
commune **272**, 272
formosum 272
juniperinum **272**, 272
longisetum 272
piliferum 272
strictum 272
Pondweed, Blunt-leaved 218
Bog 217
Broad-leaved **217**, 217
Curled 219
Fen *18*, 217
Fennel 219
Flat-stalked 218
Grass-wrack 218
Hair-like 218
Horned 219
Lesser 218
Long-leaved 217
Long-stalked 217
Opposite-leaved 219
Perfoliate 218
Red 217

(Pondweed)
Sharp-leaved 218
Shining 217
Small 218
Various-leaved 217
Willow-leaved 217
Pontederia cordata 255
Poplar, Black *39*, **94**, 94
Grey 93
Hybrid Black 94
Western Balsam 94
White *39*, 93
Poppy Welsh 65
Atlas 64
Common 64
Long-headed 64
Opium *23*, 64
Oriental 64
Prickly 65
Rough 65
Yellow-juiced 65
Populus
alba 93
'Balsam Spire' 94
candicans 94
nigra **94**, 94
tremula 93
trichocarpa 94
x canadensis 94
x canescens 93
x euramericana 94
x gileadensis 94
x jackii 94
Porella platyphylla **268**, 268
Potamogeton
acutifolius *29*, 218
alpinus 217
berchtoldii 218
coloratus *18*, 217
compressus 218
crispus *29*, 219
friesii *29*, 218
gramineus 217
lucens 217
natans *29*, **217**, 217
obtusifolius 218
pectinatus *29*, 219
perfoliatus *29*, 218
polygonifolius *29*, 217
praelongus *43*, 217
pusillus *29*, 218
trichoides 218
x fluitans 217
x grovesii 218
x pseudofriesii 218
x salicifolius 217
x zizii 217
Potato 165
Potentilla
anglica 126
anserina 126
argentea **126**, 126
erecta *27*, 126
intermedia 126
neumanniana *43*, 126
norvegica 126
palustris *28*, 126
recta 126
reptans 127
sterilis 127
tabernaemontani 126
verna 126
x italica 126
x mixta 126
Poterium officinalis 128
polygamum 128
sanguisorba 128
Pottia bryoides 276
heimii 277
intermedia 276
lanceolata 276
recta 277
starckeana 277
truncata 276
wilsonii 276
Preissia quadrata *44*, 270
Primrose 108
Primula
acaulis 108
elatior *43*, 108
japonica 108
pulverulenta 108
veris *26*, 108
vulgaris *25*, 108
x polyantha 108
x tommasinii 108
x variabilis 108
Privet *39*
Common 179
Garden 179
Prunella vulgaris *29*, 174
Prunus
avium 130
cerasifera 130
cerasus 131
domestica 130
insititia 130
laurocerasus 131
lusitanica 131
padus *25*, **131**, 131
serotina 131
spinosa 130
Pry 89
Pseudephemerum nitidum 272
Pseudocrossidium
hornschuchianum 275
revolutum 275
Pseudofumaria
alba 65
lutea 65
Pseudosasa japonica 235
Pseudoscleropodium purum 285
Pseudotaxiphyllum elegans 286
Pseudotsuga menziesii *57*
Psoralea
americana 133
dentata 133
Pteridium aquilinum *25*, 53
Pteris cretica 52
Pterocarya fraxinifolia 68
Pterygoneuron
lamellatum *44*, 276
ovatum 276
Ptilidium
ciliare 268
pulcherrimum 268
Ptilium crista-castrensis *44*, 287
Puccinellia *30*
distans *24*, 238
fasciculata 238
maritima *30*, 238
pseudodistans 238
rupestris 238
x hybrida 238
x pannonica 238
Pulicaria
dysenterica 205
vulgaris *43*, 205
Pulmonaria officinalis 168
Pulsatilla vulgaris *43*, 59
Pumpkin 93
Purple Moor-grass 253
Purslane, Pedunculate Sea 73
Pink 75
Sea **72**, 72
Puschkinia scilloides 257
Pyracantha coccinea 132
Pyrola
minor 107
rotundifolia *26*, *29*, 107
Pyrus
communis 131
pyraster 131

Q

- Quaking-grass 238
  - Greater 239
- Queen Anne's Lace 156
- **Quercus** *17*
  - **cerris** 68
  - **ilex** *29*, 68
  - **petraea** *25*, **68**, 68
  - **robur** *25*, 69
  - **rubra** 69
  - *x hispanica* 68
  - **x pseudosuber** 68
  - **x rosacea** 69
- Quince 131
  - Chinese 131

R

- **Racomitrium**
  - **canescens** 278
  - **elongatum** 278
  - **ericoides** *44*, 278
  - **lanuginosum** *44*, 277
- **Radiola linoides** *37*, 151
- Radish, Cultivated 106
  - Wild 106
- **Radula complanata** 268
- Ragged Robin 80
- Ragweed 212
  - Great 212
- Ragwort 210
  - Broad-leaved 210
  - Great Fen 210
  - Hoary 211
  - Marsh 210
  - Oxford **211**, 211
  - Silver 210
- Ramsons **258**, 258
- **Ranunculus**
  - **acris** 60
  - **aquatilis** 62
  - **arvensis** *23*, **60**, 60
  - **auricomus** *25*, 61
  - **baudotii** 62
  - **bulbosus** 60
  - **circinatus** 63
  - **ficaria** 61
  - **flammula** 61
  - **fluitans** 62
  - **hederaceus** 61
  - *heterophyllus* 62
  - *lenormandii* 62
  - **lingua** *22*, *28*, 61
  - **omiophyllus** *43*, 62
  - **parviflorus** *43*, 60
  - **peltatus** 62
  - **penicillatus** *29*, 62
  - **repens** 60
  - **sardous** *22*, 60
  - **sceleratus** 61
  - **trichophyllus** 62
- Rape, Oil-seed 104
- **Raphanus**
  - **raphanistrum** 106
  - **sativus** 106
- **Rapistrum**
  - *orientale* 105
  - **perenne** 105
  - **rugosum** 105
- Raspberry 115
- Rattle Baskets 187
- **Reboulia hemisphaerica** 269
- Red-hot Poker 255
  - Great 255
- Redshank 83
- Reed, Common **253**, 253
- Reedmace 255
  - Hybrid 255
  - Lesser 255
- Rescue Grass 250
- **Reseda**
  - **alba** 106
  - **inodora** 106
  - **lutea** 106
  - **luteola** 106
  - **odorata** 106
- Restharrow, Common 138
  - Spiny **138**, 138
- *Reynoutria japonica* 84
  - *sachalinensis* 85
- **Rhamnus cathartica** 150
- **Rheum**
  - *rhabarbarum* 85
  - *x cultorum* 85
  - **x hybridum** 85
- **Rhinanthus minor** 187
- **Rhizomnium pseudopunctatum** 280
  - **punctatum** 280
- **Rhodobryum roseum** 280
- Rhododendron *24*, *39*, 106
- **Rhododendron ponticum** *24*, 106
- Rhubarb 85
- **Rhus**
  - **hirta** 152
  - *typhina* 152
- *Rhynchosinapis cheiranthos* 105
- **Rhynchospora alba** *29*, 227
- **Rhynchostegiella tenella** 286
- **Rhynchostegium**
  - **confertum** 285
  - **megapolitanum** 285
  - **murale** 285
  - **riparioides** 285
- **Rhytidiadelphus**
  - **loreus** *44*, 287
  - **squarrosus** 287
  - **triquetrus** **287**, 287
- **Rhytidium rugosum** **287**, 287
- Ribbon Fern 52
- **Ribes**
  - **alpinum** 111
  - **nigrum** 111
  - **rubrum** 111
  - **sanguineum** 111
  - **uva-crispa** 111
- **Riccardia**
  - **chamedryfolia** 269
  - **incurvata** 269
  - **latifrons** 269
  - **multifida** 269
- **Riccia**
  - **cavernosa** 270
  - **crystallina** 270
  - **fluitans** 270
  - **glauca** 270
  - **rhenana** *44*, 270
  - **sorocarpa** 270
  - **subbifurca** *44*, 270
  - *warnstorfii* 270
- **Ricciocarpus natans** 270
- Rice-grass **254**, 253
- **Robinia pseudoacacia** *39*, 133
- Rock Rose *19*, **91**, 91
- Rock Samphire 158
- Rock-cress, Hairy 100
- Rocket, Eastern 97
  - Tall 97
- **Roemeria hybrida** *43*, 64
- **Rorippa**
  - **amphibia** **99**, 99
  - **islandica** 99
  - **microphylla** 99
  - **nasturtium-aquaticum** 98
  - **palustris** 99
  - **sylvestris** 99
  - **x anceps** 99
  - **x sterilis** 98

Rosa
- agrestis 130
- arvensis **129**, 129
- caesia 129
- canina 129
- 'Dorothy Perkins' 128
- ferruginea 129
- *glauca* 129
- *lucida* 129
- micrantha 130
- mollis 130
- multiflora 128
- obtusifolia 129
- pimpinellifolia *20*, 129
- rubiginosa 130
- *rubrifolia* 129
- rugosa 129
- sherardii 130
- stylosa 129
- tomentosa 129
- *villosa* 130
- virginiana 129
- x andegavensis 129
- x belnensis 129
- x bigeneris 130
- x dumalis 129
- x dumetorum 129
- x nitidula 129
- x praegeri 129
- x verticillacantha 129

Rose of Heaven 82

Rose of Sharon 87

Rose, Burnet *20*, 129
- Dog 129
- Field **129**, 129
- Glaucous Dog- 129
- Harsh Downy- 129
- Japanese 129
- Many-flowered 128
- Red-leaved 129
- Round-leaved Dog- 129
- Sherard's Downy- 130
- Short-styled Field- 129
- Soft Downy- 130
- Virginian 129

Rose-bay Willowherb 147

Rosemary 177

Rosmarinus officinalis 177

Rowan 132

Royal Fern *22*, 52

Rubus *25*
- adamsii 122
- adscitus 120
- adspersus 116
- amplificatus 118
- anglocandicans 119
- anisacanthos 122
- armeniacus 120
- armipotens 119
- arrheniiformis 115
- atrebatum 124
- bertramii 115
- boraeanus 120
- boudiccae **118**, 118
- britannicus 124
- caesius 125
- cantianus 122
- cardiophyllus 118
- cissburiensis 118
- conjungens 124
- criniger **120**, 120
- cromerensis 122
- crudelis 116
- dasyphyllus 123
- eboracensis 125
- echinatoides 122
- egregius 121
- elegantispinosus 118
- fissus 116
- flexuosus 122
- garrianensis 116
- glareosus 121
- gratus 117
- hindii 125
- hylocharis 123
- *hypoleucos* 120
- iceniensis 124
- idaeus 115
- incurvatiformis 118
- incurvatus 118
- infestus 122
- insectifolius 123
- intensior 125
- laciniatus 117
- *lentiginosus* 119
- leptothyrsos 117
- leucostachys 120
- leyanus 122
- lindleianus 117
- lintonii 121
- londinensis 118
- macrophyllus 117
- malvernicus 123
- marshallii 124
- *menkei* 124
- moylei 121
- mucronulatus 121
- nemoralis 119
- nemorosus 125
- nessensis 116
- nitidiformis 116
- norvicensis 122
- pallidus 123
- pergratus 116
- phaeocarpus 124
- *pistoris* 119
- platyacanthus 117
- plicatus 116
- poliodes 117
- polyanthemus 119
- *procerus* 120
- pruinosus 125
- pyramidalis 118
- radula 123
- raduloides 122
- rhombifolius 119
- rudis 123
- rufescens 123
- scaber 124
- scrabrosus 125
- septentrionalis 119
- sprengelii 119
- subinermoides 119
- subtercanens 123
- sulcatus 116
- 'Superhystrix' 124
- surrejanus 121
- *tereticaulis* 124
- trichodes 122
- tuberculatus 125
- ulmifolius 120
- vestitus 121
- vigorosus 116
- villosior 121
- watsonii 124
- winteri 120

Rudbeckia
- *bicolor* 213
- hirta 213
- laciniata 212

Rumex
- acetosa 85
- acetosella *26*, *30*, **85**, 85
- conglomeratus 85
- crispus *30*, 85
- hydrolapathum 85
- maritimus *29*, 86
- obtusifolius 86
- palustris 86
- pulcher 86
- sanguineus 86
- *x callianthemus* 86
- *x dufftii* 86
- *x knafii* 86
- *x lingulatus* 86
- *x muretii* 86
- *x ogulinensis* 86
- *x pratensis* 86

x sagorskii 86
x schulzei 86
x wirtgenii 86
Ruppia
cirrhosa 219
maritima 219
spiralis 219
Rupturewort, Hairy 80
Smooth *20*, **79**, *79*
Ruscus aculeatus 259
Rush, Blunt-flowered **222**, 222
Bulbous 223
Compact 224
Great Sharp 223
Hard 223
Heath 221
Jointed 222
Round-fruited *22*, 222
Saltmarsh *22*, 222
Sea 223
Sharp-flowered 223
Slender 221
Soft 223
Toad 222
Russian Mustard 96
Russian Vine 85
Rusty-back **54**, 54
Rye 252
Rye-grass, Italian 237
Perennial 237

S

Safflower 199
Sage, Garden 177
Sagina
apetala 79
ciliata 79
maritima 79
nodosa 78
procumbens 79
subulata 79
Sagittaria sagittifolia *29*, 214
Sainfoin 133
Salad Burnet *19*, 128
Salicornia *30*
disarticulata 73
dolichostachya 74
europaea **73**, 73
fragilis 74
nitens 74
obscura 73
perennis 73
pusilla 73
ramosissima 73
stricta 73
Salix *36*
alba 95
aurita 96
caprea 96
cinerea *26*, 96
eleagnos 95
fragilis 94
pentandra 94
purpurea 95
repens *29*, 96
triandra 95
viminalis 95
x ambigua 96
x basfordiana 94
x calodendron 95
x capreola 96
x ehrhartiana 94
x meyeriana 94
x mollissima 95
x multinervis 96
x pontederiana 95
x reichardtii 96
x rubens 94
x rubra 95
x sericans 95
x smithiana *36*, 95
x sordida 95
x stipularis 95
x subsericea 96
Salsify 200
Slender 200
Salsola
kali *29*, 74
pestifer 74
Salt-marsh-grass, Borrer's 238
Common 238
Reflexed *24*, 238
Stiff 238
Saltwort 74
Salvia
horminoides 177
horminum 177
officinalis 177
pratensis 177
reflexa 177
verbenaca **177**, 177
verticillata 177
viridis 177
x sylvestris 177
Sambucus
ebulus 193
nigra 192
racemosa 192
Samolus valerandi **110**, 110
Samphire **73**, 73
Sandwort, Fine-leaved 76
Mossy 76
Sea 76
Three-nerved 76
Thyme-leaved 76
Sanguisorba
minor *19*, *26*, 128
officinalis 128
Sanicula europaea 156
Sanionia uncinata 44, 284
Saponaria
ocymoides 82
officinalis *23*, 82
Sarcocornia perennis 73
Sarothamnus scoparius 143
Sasa palmata 235
Satureja montana 174
Satyrium viridis -
see Coeloglossum viride *37*, 262
Savory, Winter 174
Saw Sedge *22*, **227**, 227
Saw-wort 198
Saxifraga
cymbalaria 113
granulata 113
tridactylites 113
x urbium 113
Saxifrage, Celandine 113
Meadow 113
Rue-leaved 113
Scabiosa columbaria *26*, **196**, 196
Scabious, Devil's-bit 195
Field 195
Small **196**, 196
Scandix
pecten-veneris *23*, 157
stellata 157
Scapania
compacta 267
irrigua 267
nemorea *44*, 267
undulata *44*, 267
Schistidium
apocarpum 277
crassipilum 277
elegantulum 277
Schistostega pennata 278
Schoenoplectus
lacustris 226
tabernaemontani 226

Schoenus nigricans *28, 29,* 227
Scilla siberica 257
Scirpus
caespitosus 225
cernuus 226
fluitans 226
lacustris 226
maritimus 226
setaceus 226
sylvaticus *43,* 225
tabernaemontani 226
Scirpus, Bristle 226
Floating 226
Nodding 226
Scleranthus
annuus *79*
perennis *43, 79*
Scleropodium
purum 285
tourettii *44,* 285
Scorpidium scorpioides 284
Scorpion-grass 169
Scorpion-vetch, Annual 134
Scorpiurus muricatus 135
Scotch Thistle 198
Scrophularia
aquatica 181
auriculata 181
nodosa 181
scorodonia 181
umbrosa **181**, 181
vernalis 181
Scurfy Pea 133
Scurvy-grass, Common 102
Danish **102**, 102
Early *24,* **102**, 102
English 101
Scutellaria
galericulata *28,* 173
hastifolia 173
minor 173
Sea Aster **206**, **207**, 206
Sea Beet *73*
Sea Buckthorn 144
Sea Heath *22,* **93**, 93
Sea Holly **156**, 156
Sea Kale *22,* **105**, **106**, 105
Sea Lavender, Common 86
Lax-flowered 86
Matted *22,* **87**, 87
Rock 87
Sea Milkwort 110
Sea Pea 137
Sea Rocket 105
Sea spurrey, Greater 80
Lesser 80
Rock 80
Seablite, Annual *74*
Shrubby *22,* **74**, *74*
Seaside Daisy 207
Secale cereale 252
Securigera varia 135
Sedge 227
Bladder 231
Bog 234
Bottle **231**, 231
Breckland **234**, 234
Brown 229
Carnation 232
Common 234
Cypress 231
Dioecious 230
Distant 233
Divided 229
False Fox 228
Fibrous Tussock **228**, 228
Flea 235
Glaucous 232
Greater Pond 231
Greater Tussock 228
Green-ribbed 232
Grey 229
Hairy 230
Lesser Pond 231
Lesser Tussock 228
Long-bracted 233
Mud 234
Oval 230
Pale 233
Pendulous 232
Pill 234
Prickly 229
Rare Spring **234**, 234
Remote 230
Sand 229
Slender 230
Slender Tufted 234
Spiked 228
Spring 233
Star 230
Tawny 233
Thin-spiked wood 232
Tufted **235**, 234
White 230
Wood **232**, 232
Yellow 233
Sedum
acre *30,* 112
album 113
anglicum 113
dasyphyllum 113
forsterianum 112
reflexum 112
rupestre 112
rupestre in part 112
spectabile 112
spurium 112
telephium 112
Selaginella kraussiana 51
Self-heal 174
Seligeria
calcarea 278
calycina 278
paucifolia 278
Sempervivum
arachnoideum 112
tectorum 112
Senecio
aquaticus 210
bicolor 210
cineraria 210
erucifolius 211
fluviatilis 210
greyi 212
jacobaea *30,* 210
maritimus 210
paludosus *43,* 210
palustris 211
squalidus **211**, 211
sylvaticus 211
viscosus 211
vulgaris 211
x albescens 210
x baxteri 211
x ostenfeldii 210
x subnebrodensis 211
Sequioadendron giganteum 56
Sequoia sempervirens 56
Serapias palustris -
see Epipactis palustris *37,* 260
Seriphidium maritimum **208**, 208
Serpent's Tongue 214
Serratula tinctoria 198
Service-tree, Wild *17,* 132
Setaria *24*
glauca 254
italica 254
lutescens 254
pumila 254
verticillata 254
viridis 254
Shaggy Soldier 213
Shasta Daisy 210
Sheep's Bit 190
Sheep's Parsley 156
Sheep's Samphire *73*

Shepherd's Cress **103**, 102
Shepherd's Needle *23, 157*
Shepherd's Purse 102
Sherardia arvensis 190
Shield fern, Hard 55
Soft 54
Shoreweed 179
Shrubby Ragwort 212
Sieglingia decumbens 252
Silaum silaus *26,* 159
Silene
alba 81
coeli-rosea 82
conica 82
dioica 82
gallica 82
latifolia 81
maritima 81
noctiflora 81
otites *19, 20,* **81**, 81
uniflora *22, 30,* 81
vulgaris 81
x hampeana 81
Silky-bent, Dense 246
Loose 246
Silverweed 126
Silybum marianum 198
Sinapis
alba 105
arvensis 105
Sison amomum *17,* 161
Sisymbrium
altissimum 97
irio 96
loeselii 96
officinale 97
orientale 97
polyceratum 97
volgense 96
bermudiana 259
californicum 259
Sium latifolium *22, 28,* **158**, 158
Skullcap, Common 173
Lesser 173
Norfolk 173
Sloe 130
Small Toadflax 182
Small-leaved Lime 89
Small-reed, Narrow 246
Purple 246
Wood **246**, 245
Smyrnium olusatrum *23,* **157**, 157
Snapdragon 182
Sneezewort 209
Snowberry *39,* 193
Pink 193
Snowdrop 259
Broad-leaved 259
Greater 259
Green 259
Pleated 259
Snowflake, Spring 259
Summer 258
Snow-in-summer 78
Soapwort *23,* 82
Alpine 82
Soft-grass, Creeping 243
Solanum
cornutum 165
dulcamara 165
nigrum 165
nitidibaccatum 165
physalifolium 165
rostratum 165
sarachoides 165
sisymbriifolium 165
triflorum 165
tuberosum 165
Soleirolia soleirolii 68
Solidago
canadensis 206
gigantea 206
virgaurea 206
Solomon's Seal 256
Garden 256
Sonchus
arvensis 201
asper 201
oleraceus 201
palustris **200**, 200
Sorbaria 114
Sorbaria sorbifolia 114
Sorbus
aria 132
aucuparia *25,* 132
intermedia 132
torminalis *17, 25,* 132
Sorghum
bicolor 254
halepense 254
Sorrel, Common 85
Large-flowered Pink 152
Least Yellow 152
Pale Pink 152
Pink 152
Procumbent Yellow 152
Sheep's **85**, 85
Upright Yellow 152
Wood *17,* **152**, 152
Southernwood, Field *19,* 208
Slender 208
Sowbane 71
Sow-thistle, Corn 201
Marsh **200**, 200
Perennial 201
Prickly 201
Smooth 201
Spanish-dagger 260
Sparganium
emersum **255**, 254
erectum 254
minimum *18,* 255
natans *18,* 255
Spartina *30*
anglica **254**, 253
maritima *43,* 253
x townsendii 253
Spartium junceum 143
Spatterdock 58
Spearmint 176
Spearwort, Greater *22,* 61
Lesser 61
Speedwell, Bird's-eye 183
Blue Water- 184
Breckland **184**, 184
Common Field 185
Fingered 184
Garden 185
Germander 183
Green Field 185
Grey Field 185
Heath 183
Hybrid Water- 184
Ivy-leaved 185
Marsh 184
Pink Water- 184
Slender *23,* 185
Spiked *20, 38,* **38**, **185**, 185
Spring 185
Thyme-leaved 183
Wall 184
Wood 183
Spergula
arvensis 80
rupicola 80
Spergularia
marina 80
media 80
rubra 80
Sphaerocarpos michelii **269**, 269
Sphagnum
auriculatum *29,* 271
capillifolium 271
compactum *27,* 271
contortum 271
cuspidatum *29,* **271**, 271
denticulatum *29,* 271
fallax 271

(Sphagnum)
fimbriatum *26*, 271
flexuosum *26*, *29*, 271
inundatum 271
magellanicum 270
molle 271
palustre *26*, **270**, 270
papillosum *29*, 270
recurvum *26*, *29*, 271
russowii 271
squarrosum *28*, 270
subnitens *29*, 271
subsecundum 271
tenellum 271
teres 270
Spikerush, Common 225
Few-flowered 225
Many-stalked 225
Slender 225
Spikeweed 212
Spinacea oleracea 71
Spinach, Garden 71
Spindle 148
Evergreen 148
Spiny Cocklebur 212
Spiraea 114
Spiraea
douglasii 114
salicifolia 114
tomentosa 114
x billardii 114
x pseudosalicifolia 114
x rosalba 114
x vanhouttei 114
Spiraea, Van Houtte's 114
Spiranthes spiralis 261
Spirodela polyrhiza *29*, 220
Splachnum ampullaceum *44*, 278
Spleenwort, Black 53
Maidenhair 53
Spotted Cat's-ear 199
Spotted Laurel 148
Spotted Orchid, Common 262
Spotted Orchid, Heath 262
Spring Beauty 75
Spring Star-flower 258
Spruce *39*, *57*
Spurge Laurel 145
Spurge, Broad-leaved 149
Caper 149
Cypress 150
Dwarf 150
Leafy 150
Mediterranean 150
Petty 150
Sea 150
Sun 149
Twiggy 150
Warted 149
Wood *17*, 150
Spurrey, Corn 80
Sand 80
Squash 93
Squill, Siberian 257
Squinancywort 191
St John's-wort, Common 88
Hairy 89
Imperforate 88
Marsh 89
Pale 89
Perforate 88
Slender 89
Square-stemmed 88
Trailing **88**, 88
Stachys
annua 171
arvensis 171
byzantina 171
lanata 171
officinalis 171
palustris 171
sylvatica 171
Star of Bethlehem 257
Drooping 257
Yellow 255
Star-thistle, Lesser 198
Maltese 198
Red 198
Rough 198
Yellow 198
Steeple Bush 114
Stellaria
alsine *77*
graminea *77*
holostea *77*
media *76*
neglecta *77*
nemorum *76*
pallida *76*
palustris *77*
uliginosa *77*
Steppe Cabbage 105
Stitchwort, Bog *77*
Greater *77*
Lesser *77*
Marsh *77*
Wood *76*
Stock, Hoary 98
Stone Parsley *17*, 161
Stonecrop, Biting 112
Butterfly 112
Caucasian 112
English 113
Mossy *21*, **111**, 111
Reflexed 112
Rock 112
Thick-leaved 113
White 113
Stork's-bill, Common **155**, 154
Musk 154
Strapwort *79*
Stratiotes aloides *22*, *29*, **216**, 215
Strawberry, Garden 127
Hautbois 127
Wild 127
Yellow-flowered 127
Suaeda fruticosa *74*
Suaeda
maritima *30*, *74*
vera *22*, *30*, **73**, *74*
Succisa pratensis 195
Succory *35*, 199
Sugar Beet *73*
Sumach, Stag'shorn 152
Summer Cypress *71*
Sundew 20
Great 91
Oblong-leaved 91
Round-leaved *22*, 90
Sunflower 213
Sunflower, Perennial 213
Sweet Briar 130
Small-flowered 130
Small-leaved 130
Sweet Chestnut *23*, *39*, 68
Sweet Cicely 157
Sweet Flag 220
Sweet Gale 68
Sweet Vernal-grass 244
Sweet William 83
Sweet-grass, Floating 241
Hybrid 241
Plicate 241
Reed 241
Small 241
Swine Cress 104
Lesser 104
Sycamore *24*, *39*, 152
Symphoricarpos
albus 193
asperum 168
grandiflorum 169
'Hidcote Blue' 169
officinale 168
orientale 169

rivularis 193
tuberosum 168
x chenaultii 193
x uplandicum 168
Syntrichia
intermedia 277
laevipila 277
latifolia 277
papillosa 277
ruraliformis 277
ruralis 277
virescens 277
Syringa 111
Syringa vulgaris 179

T

Tagetes patula 214
Tamarisk 93
Tamarix
anglica 93
gallica *39,* 93
Tamus communis 260
Tanacetum
parthenium *23,* 207
vulgare 208
Tansy 208
Taraxacum spp. 202
Tare, Hairy 135
Smooth 136
Tasselweed, Beaked 219
Spiral 219
Taxodium distichum 56
Taxus baccata *24, 57*
Teaplant *39*
Chinese 164
Teasel, Small **195**, 195
Wild 195
Tea-tree, False 164
Teesdalia nudicaulis *26,* **103**, 102
Telekia speciosa 205
Tellima grandiflora 113
Tenchweed 217
Tephroseris palustris *43,* 211
Tetragonia tetragonioides 70
Tetraphis pellucida 272
Teucrium
chamaedrys 174
scordium *43,* 174
scorodonia 174
Thale-cress **97**, 97
Thalictrum
flavum 63
minus 63
Thamnobryum alopecurum 282
Thanet Cress 104
Thelycrania sanguinea 148
Thelypteris
dryopteris 54
limbosperma 53
oreopteris 53
palustris *26, 28,* 53
thelypteroides 53
Thesium humifusum *43,* 148
Thistle, Carline 196
Creeping 198
Dwarf **197**, 197
Marsh 198
Meadow 197
Musk 197
Slender 197
Spear 197
Stemless **197**, 197
Welted 197
Woolly 197
Thlaspi
alliaceum 103
arvense 103
Thorn-apple 165
Thorow-wax 160
False 160
Thrift 87
Thuidium
abietinum 283
philiberti 283
tamariscinum 283
Thuja plicata 57
Thyme, Breckland **176**, 175
Garden 176
Greater Wild *19,* 175
Wild 175
Thymus
drucei 175
polytrichus 175
praecox 175
pulegioides *19, 26,* 175
serpyllum **176**, 175
vulgaris 176
Tilia
cordata *17,* 89
platyphyllos 89
x europaea 89
x vulgaris 89
Tillaea muscosa *35, 37,* 111
Timothy 248
Toad Pipes 51
Toad Rush 222
Toadflax, Annual 183
Common 183
Ivy-leaved *23,* 182
Pale 183
Purple 183
Small 182
Tobacco, Sweet 165
Tomato 164
Tomenthypnum nitens *44,* 284
Toothwort, Purple 188
Tor-grass 250
Torilis
arvensis 162
japonica 162
nodosa 162
Tormentil 126
Tortella
flavovirens 275
inclinata *44,* 275
Tortula
acaulon 277
intermedia 277
laevipila 277
lanceola 276
latifolia 277
marginata 276
modica 276
muralis 276
papillosa 277
protobryoides 276
ruralis 277
subulata 276
truncata 276
virescens 277
wilsonii *44,* 276
Tower Mustard **101**, 100
Town Hall Clock 194
Trachystemon orientalis 169
Tragopogon
hybridus 200
porrifolius 200
pratensis 200
x mirabilis 200
Trailing Snapdragon 182
Travellers' Joy 60
Treacle Mustard 98
Tree Mallow 90
Tree of Heaven 152
Trefoil, Hop 141
Lesser 141
Slender 141
Trichophorum cespitosum 225
Trichostomopsis umbrosus 275
Trichostomum crispulum 275

Trifolium
angustifolium 142
arvense 142
campestre 141
dubium 141
fragiferum *22*, 141
glomeratum 141
hybridum 140
incarnatum 142
medium 141
micranthum 141
ochroleucon *17*, **142**, 142
ornithopodioides 140
pratense 141
repens 140
resupinatum 141
scabrum 142
squamosum *43*, 142
striatum 142
subterraneum 142
suffocatum 141
Triglochin
maritimum 216
palustre *28*, 216
Trigonella foenum-graecum 139
Tripleurospermum
inodorum 210
maritimum 210
Trisetum flavescens *26*, 242
Tristagma uniflorum 258
Triticum
aestivum 252
durum 252
Tritomaria exsectiformis *44*, 267
Tropaeolum majus 155
Trumpets 211
Tsuga
canadensis 56
heterophylla 57
Tulip, Wild 256
Tulipa sylvestris 256
Tumbling Mustard 97
Turgenia latifolia 155
Turnip 104
Turritis glabra 100
Tussilago farfara *30*, 212
Tutsan 87
Stinking 87
Twayblade, Common 261
Typha
angustifolia *28*, 255
latifolia 255
x glauca 255

U

Ulex
europaeus *19*, 144
europaeus x U. gallii 144
gallii *19*, *20*, *27*, 144
minor 144
Ulmus
carpinifolia 67
glabra 66
minor **67**, 67
procera 67
x hollandica 67
x vegeta 67
Ulota
bruchii 282
coarctata *44*, 281
crispa 281
ludwigii 281
phyllantha 282
Umbilicus rupestris 112
Urtica
dioica *26*, *28*, 67
galeopsifolia 67
pilulifera 67
urens 67
Utricularia
australis 189
intermedia 189
minor 189
neglecta 189
vulgaris *22*, 189

V

Vaccaria
hispanica 82
pyramidata 82
Vaccinium
myrtillus 107
oxycoccos *20*, *29*, 107
vitis-idaea 107
Valerian, Common 194
Marsh 194
Red 195
Valeriana
dioica 194
officinalis 194
Valerianella
carinata 194
dentata 194
locusta **194**, 194
rimosa *43*, 194
Velvet Leaf 90
Venus's Looking Glass 190
Verbascum
blattaria 180
densiflorum 180
longifolium 180
lychnitis 181
nigrum 180
olympicum 180
phlomoides 180
pulverulentum **180**, 180
pyramidatum 180
sinuatum 179
speciosum 180
thapsus 180
virgatum 180
x duernsteinense 180
x godronii 180
x mixtum 180
x semialbum 180
x wirtgenii 180
Verbena, Wild **171**, 170
Verbena officinalis **171**, 170
Vernal-grass, Annual 244
Sweet 244
Veronica
agrestis 185
anagallis-aquatica 184
arvensis 184
beccabunga 184
catenata 184
chamaedrys 183
filiformis *22*, 185
hederifolia 185
longifolia 185
montana 183
officinalis 183
persica 185
polita 185
praecox **184**, 184
scutellata 184
serpyllifolia 183
spicata *20*, *38*, **38**, **185**, 185
sublobata 185
triphyllos 184
verna *43*, 185
x lackschewitzii 184
Vervain **171**, 170

Vetch, Bithynian 136
Bush 136
Common 136
Cultivated 136
Fine-leaved 135
Fodder 135
Horse-shoe *26*, **135**, 135
Narbonne 136
Purple 135
Slender 136
Spring 136
Tufted 135
Yellow 136
Vetchling, Grass-leaved *22*, 138
Hairy 138
Meadow 137
Yellow 138
Viburnum
lantana *25*, *43*, 193
opulus 193
rhytidophyllum 193
tinus 193
Vicia
angustifolia 136
benghalensis 135
bithynica 136
cracca *28*, 135
faba 136
hirsuta 135
lathyroides 136
lutea 136
narbonensis 136
parviflora 135
sativa 136
sativa ssp. **nigra** 136
sativa ssp. **segetalis** 136
sepium 136
tenuifolia 135
tenuissima 135
tetrasperma 136
villosa 135
Vigna
radiata 133
Vinca
major 164
minor 164
Viola
arvensis 93
canina 92
cornuta 91
hirta 91
odorata 91
palustris 92
persicifolia *43*, 92
reichenbachiana 92
riviniana 91
stagnalis 92
tricolor 92
tricolor ssp. **curtisii** *22*, **92**, 92
x contempta 92
x wittrockiana 93
Violet, Fen 92
Hairy 91
Marsh 92
Sweet 91
Viper's Bugloss **168**, 168
Virginia Creeper 151
False 151
Virginian Stock 98
Viscum album 148
Vitis vinifera 151
Vulpia
bromoides 237
ciliata 238
fasciculata 237
megalura 237
membranacea 237
myuros 237
unilateralis *43*, 238

W

Wall Lettuce 201
Wall Rocket, Annual 104
Perennial 104
White 104
Wall Rue 54
Wallflower *23*, 98
Siberian 98
Wallflower Cabbage 105
Walnut 68
Warnstorfia
exannulata *44*, 283
fluitans 283
Warty Cabbage 97
Water Bent 246
Water Betony 181
Water-cress 98
Fool's 160
Hybrid 98
Narrow-fruited 99
Water Crowfoot, Brackish 62
Common 62
Fan-leaved 63
Ivy-leaved 61
Pond 62
River 62
Round-leaved 62
Stream 62
Thread-leaved 62
Water-dropwort, Fine-leaved **159**, 159
Hemlock 158
Narrow-leaved 158
Parsley 158
River 159
Tubular 158
Water Fennel 62
Water Fern 56
Water Hyacinth 216
Waterlily, Fringed *22*, 167
White *23*, 58
Yellow **58**, 58
Water-milfoil, Alternate 144
Spiked 144
Whorled 144
Water Parsnip, Greater *22*, **158**, 158
Lesser 158
Water-pepper, Small 84
Tasteless 84
Water Plantain 215
Floating 215
Lesser 215
Narrow-leaved 215
Ribbon-leaved 215
Water Purslane 145
Water Soldier *22*, **216**, 215
Water-starwort, Blunt-fruited 178
Common 178
Intermediate 178
Various-leaved 178
Water Violet **109**, 109
Water Whorl-grass 240
Waterweed *23*, 216
Canadian *23*, 216
Curly 216
Nuttall's *23*, 216
Wayfaring Tree 193
Weasel's-snout 182
Pale 182
Weissia
brachycarpa 274
controversa 274
longifolia 275
microstoma 274
Weld 106
Western Hemlock 57
Western Red Cedar 57
Wheat, Bread 252
Pasta 252
Whin 144
White Beak-sedge 227
White Bryony 93
White Horehound 173

- Whitebeam 132
  - Swedish 132
- Whitethorn *39*, 132
- Whitlow-grass Common 101
  - Glabrous 101
  - Hairy 101
- Whitten Tree 193
- Whorl-grass, Water 240
- Wigeon-grass 219
- Wild Basil 175
- Wild Liquorice 133
- Wild Maidenhair 63
- Wild Marguerite 210
- Wild Rhubarb 212
- Wild Service Tree 132
- Wild-oat 242
  - Winter 242
- Willow, Almond 95
  - Bay 94
  - Bedford 94
  - Broad-leaved 95
  - Crack 94
  - Creeping 96
  - Cricket bat 95
  - Eared 96
  - Ehrhardt's 94
  - Goat 96
  - Green-leaved 95
  - Grey 96
  - Holme 95
  - Hybrid Crack 94
  - Olive 95
  - Purple 95
  - Sharp-stipuled 95
  - Shining 94
  - White 95
  - White-varnished 94
- Willowherb, American *23*, 147
  - Broad-leaved 146
  - Great 145
  - Hoary 146
  - Marsh 147
  - Pale 146
  - Short-fruited 146
  - Spear-leaved 146
  - Square-stemmed 146
- Winter Aconite 59
- Winter Heliotrope 212
- Winter-cress, American 98
  - Common 98
  - Medium-flowered 98
  - Small-flowered 98
- Wintergreen, Common 107
  - Round-leaved 107
- Witch-grass 253
- Woad *35*, *97*
- Wolfbane 59
- Wood Club-rush 225
- Wood Melick *16*, 242
- Wood Millet 235
- Wood Sage 174
- Wood Sanicle 156
- Woodbine 193
- Woodruff, Pink 191
  - Sweet **191**, 191
- Woodrush, Field 224
  - Great *17*, 224
  - Hairy *16*, 224
  - Heath 224
- Wormwood 208
  - Field **208**, 208
  - Sea **208**, 208
- Woundwort, Annual Yellow 171
  - Field 171
  - Hedge 171
  - Marsh 171
- Wych Elm *16*, *66*

X

- **Xanthium spinosum** 212

Y

- Yarrow 209
- Yellow Bartsia 187
- Yellow Bird'snest **108**, 108
- Yellow Flag 260
- Yellow Oat-grass 242
- Yellow Ox-eye 205
- Yellow Pimpernel 109
- Yellow-cress, Creeping 99
  - Great **99**, 99
  - Marsh 99
  - Northern 99
- Yellow-eyed-grass 259
- Yellow-rattle 187
- Yellow-wort 163
- Yew *16*, *24*, *39*, *57*
- Yorkshire Blackberry 125
- Yorkshire Fog 243
- Yucca *39*, 260
- **Yucca recurvifolia** 260

Z

- **Zannichellia palustris** *29*, 219
- **Zostera**
  - **angustifolia** 219
  - **marina** 219
  - **noltii** 219
- **Zygodon**
  - **conoideus** 281
  - **viridissimus** 281